Algorithms for Intelligent Systems

This book series publishes research on the analysis and development of algorithms for intelligent systems with their applications to various real world problems. It covers research related to autonomous agents, multi-agent systems, behavioral modeling, reinforcement learning, game theory, mechanism design, machine learning, meta-heuristic search, optimization, planning and scheduling, artificial neural networks, evolutionary computation, swarm intelligence and other algorithms for intelligent systems.

The book series includes recent advancements, modification and applications of the artificial neural networks, evolutionary computation, swarm intelligence, artificial immune systems, fuzzy system, autonomous and multi agent systems, machine learning and other intelligent systems related areas. The material will be beneficial for the graduate students, post-graduate students as well as the researchers who want a broader view of advances in algorithms for intelligent systems. The contents will also be useful to the researchers from other fields who have no knowledge of the power of intelligent systems, e.g. the researchers in the field of bioinformatics, biochemists, mechanical and chemical engineers, economists, musicians and medical practitioners.

The series publishes monographs, edited volumes, advanced textbooks and selected proceedings.

All books published in the series are submitted for consideration in Web of Science.

More information about this series at https://link.springer.com/bookseries/16171

Kedar Nath Das · Debasish Das ·
Anjan Kumar Ray ·
Ponnuthurai Nagaratnam Suganthan
Editors

Proceedings of the International Conference on Computational Intelligence and Sustainable Technologies

ICoCIST 2021

Springer

Editors
Kedar Nath Das
Department of Mathematics
National Institute of Technology Silchar
Silchar, Assam, India

Debasish Das
Department of Civil Engineering
JIS College of Engineering
Nadia, West Bengal, India

Anjan Kumar Ray
Department of Electrical and Electronics Engineering
National Institute of Technology Sikkim
Ravangla, Sikkim, India

Ponnuthurai Nagaratnam Suganthan
School of Electrical and Electronics Engineering
Nanyang Technological University
Singapore, Singapore

ISSN 2524-7565 ISSN 2524-7573 (electronic)
Algorithms for Intelligent Systems
ISBN 978-981-16-6892-0 ISBN 978-981-16-6893-7 (eBook)
https://doi.org/10.1007/978-981-16-6893-7

This Springer imprint is published by the registered company Springer Nature Singapore Pte Ltd.
The registered company address is: 152 Beach Road, #21-01/04 Gateway East, Singapore 189721, Singapore

Organization

Conference Organizing Committee

Chief Patrons

Prof. Sivaji Bandyopadhyay, Director, NIT Silchar, Assam
Prof. Mahesh Chandra Govil, Director, NIT Sikkim, Assam

Patrons

Prof. A. K. Barbhuya, NIT Silchar, Assam
Prof. M. Ali Ahmed, NIT Silchar, Assam

Honorary Chair

Prof. Kusum Deep, IIT Roorkee

General Chairs and Steering Committee

Dr. Kedar Nath Das, NIT Silchar
Dr. Sadhan Gope, Mizoram University, Aizawl
Dr. Subhojit Dawn, VRSEC, Vijayawada
Dr. Debasish Das, JIS College of Engineering, West Bengal

Conference Chairs

Dr. Mausumi Sen, HOD, Mathematics, NIT Silchar
Dr. Sourav Mallick, HOD, EEE, NIT Sikkim

Organizing Secretaries

Dr. Kedar Nath Das, NIT Silchar
Dr. Aurobinda Panda, NIT Sikkim

Organizing Committees

Dr. S. Roy, NIT Silchar, Assam
Dr. P. K. De, NIT Silchar, Assam
Mr. B. Nath, NIT Silchar, Assam
Dr. G. Ramesh, NIT Silchar, Assam
Dr. P. K. Gupta, NIT Silchar, Assam
Dr. S. Bera, NIT Silchar, Assam
Dr. Maqbul, NIT Silchar, Assam
Dr. B. H. S. Raju, NIT Silchar, Assam
Dr. P. Biswas, NIT Silchar, Assam
Dr. Juthika Mahanta, NIT Silchar
Dr. Santan Kumar, NIT Silchar, Assam
Dr. Susanta Kumar Tripathy, NIT Silchar
Dr. Prabina Pattanayak, NIT Silchar, Assam
Mr. S. K. Pattanayak, NIT Silchar, Assam
Dr. Trupti Ranjan Lenka, NIT Silchar
Dr. Sudarsan Sahoo, NIT Silchar
Dr. Anjan Kumar Ray, NIT Sikkim
Dr. Pratyay Kuila, NIT Sikkim
Mr. B. Balaji Naik, NIT Sikkim
Mr. Jogi Paul, NIT Sikkim
Dr. Amit Kumar Yadav, NIT Sikkim
Dr. Pradeep Kumar, NIT Sikkim
Dr. Abhishek Rajan, NIT Sikkim
Dr. Anulekha Saha, NIT Sikkim
Dr. Prasenjit Dey, NIT Sikkim
Mr. Tarun Biswas, NIT Sikkim
Mr. Arup Das, SIT Siliguri
Dr. Biplab Chaudhuri, G. C. College, Silchar

Dr. Debashish Sharma, G. C. College, Silchar
Dr. Rajashree Mishra, KIIT University
Dr. Raghav Parouha, IGNTU, Madhya Pradesh

Advisory Committee (International)

Dr. Sanjeevikumar Padmanaban, Professor, Department of Energy Technology, Aalborg University's Esbjerg Campus, Denmark
Dr. Vincenzo Piuri, Professor, Universita' degli Studi di Milano, Dipartimento di Informatica, Via Celoria, 18—20133 Milano—Italy
Dr. Liu Haiwen, Professor, Xi'an Jiaotong University, China
Dr. Akhtar Kalam, Professor, Victoria University, Australia
Dr. Chi-Hou Chan, Chair Professor, Electronic Engineering, City University of Hong Kong, Hong Kong
Dr. Atulya Nagar, Dean of Faculty of Science, Liverpool Hope University, UK
Dr. Frede Blaabjerg, Professor, Aalborg University, Denmark
Dr. Md. Atiqur Rahman Ahad, Specially Appointed Associate Professor, Osaka University, Japan
Dr. Jens Bo Holm-Nielsen, Associate Professor, Aalborg University, Denmark
Dr. Suman Nag, Independent Biophysics/Biochemistry Lead Scientist, South San Francisco, California, USA
Dr. John K. Pedersen, Professor, Aalborg University, Denmark
Dr. Wang Lipo, Associate Professor, Nanyang Technological University, Singapore
Dr. Atma Sahu, Professor, Coppin State University, USA
Dr. Pierluigi Siano, Professor, University of Salerno, Italy
Dr. Stig Munk-Nielsen, Professor, Aalborg University, Denmark
Dr. Haitham Abu-Rub, Professor, Texas A&M University, Qatar
Dr. Gabriele Grandi, Professor, University of Bologna, Italy
Dr. Atif Iqbal, Associate Professor, Qatar University, Qatar
Dr. David Infield, Research Professor, University of Strathclyde, Glasgow, UK
Dr. Joseph Olorunfemi Ojo, Professor, Tennessee Technological University, USA
Dr. Innocent Kamwa, Research Scientist and Registered Professional Engineer, Hydro-Quebec Research Institute, USA
Dr. Christian Rehtanz, Professor, TU Dortmund, Germany
Dr. Luigi Martirano, Professor, Università di Roma Sapienza, Italy
Dr. T. Sudhakar Babu, Professor, Institute of Power Engineering, Universiti Tenaga Nasional, Malaysia
Dr. Ivana Bhddhinska, Director, SAS, Slovak Academy of Science, Slovak
Dr. Mohammad Shorif Uddin, Professor, Jahangirnagar University, Dhaka, Bangladesh
Prof. Montaz Ali, Professor, University of the Witwatersrand

Advisory Committee (National)

Dr. Sivaji Chakravorti, Director, NIT Calicut, India
Dr. Nidul Sinha, Professor, NIT Silchar, India
Dr. Binoy Krishna Roy, Professor, NIT Silchar, India
Dr. Pramod Kumar Singh, Associate Professor, Indian Institute of Information Technology and Management Gwalior, India
Dr. Sandip Banerjee, Associate Professor, IIT Roorkee, India
Dr. Subindu Kumar, Associate Professor, IIT (Indian School of Mines), Dhanbad, India
Dr. Radhakant Padhi, Professor, IISc Bangalore, India
Dr. Nalin Behari Dev Choudhury, Professor, NIT Silchar, India
Dr. Pradip Kumar Sadhu, Professor, IIT (Indian School of Mines), Dhanbad, India
Dr. Praveen Kumar, Associate Professor, IIT Guwahati, India
Dr. Himani Saini, Space Scientist, ISRO, Bangalore
Dr. R. Balasubramanian, Professor, IIT Roorkee, India

Technical Committee

Dr. Sujit K. Biswas, Vice Chair, Technical Activities, IEEE India Council, Former Professor & Head, Jadavpur University, West Bengal, India
Dr. Debi Prassana Acharjya, Professor, VIT University, Vellore, India
Dr. Sourav Mukhopadhyay, Associate Professor, IIT Kharagpur, West Bengal, India
Dr. Arnab Sarkar, Associate Professor, IIT BHU, Uttar Pradesh, India
Dr. Chandan Kumar Chanda, Professor, IIEST Shibpur, India
Dr. Atanu Banerjee, Associate Professor, NIT Meghalaya, India
Dr. Sudipta Chattopadhyay, Associate Professor, Mizoram University, Mizoram, India
Dr. Sukanta Roga, Assistant Professor, Visvesvaraya NIT Nagpur, India
Dr. Mantosh Biswas, Assistant Professor, NIT Kurukshetra, India
Dr. L. Lolit Kumar Singh, Professor, Mizoram University, Mizoram, India
Dr. Manish Kumar Thukral, Associate Professor, Manipal University Jaipur, India
Dr. Jayendra Kumar, Associate Professor, VIT Andhra Pradesh, India
Dr. Prodipto Das, Assistant Professor, Assam University, India
Dr. Harish Sharma, Associate Professor, RTU, Kota, India
Dr. Sunil Dutt Purohit, Associate Professor, RTU, Kota, India
Dr. Sourav Bose, Assistant Professor, NIT Uttarakhand, India
Dr. G. S. Mahapatra, Associate Professor, NIT Puducherry, India
Dr. Anupam Yadav, Assistant Professor, NIT Jalandhar, India
Dr. Neha Yadav, Assistant Professor, NIT Hamirpur, India
Dr. Barnali Dey, Assistant Professor, Sikkim Manipal Institute of Technology, India
Dr. Ashim Saha, Assistant Professor, NIT Agartala, India

Dr. Asadur Rahman, Assistant Professor, NIT Srinagar, India
Dr. Jagdish Chand Bansal, Assistant Professor, South Asian University, Delhi, India
Dr. Somen Debnath, Assistant Professor, Mizoram University, Mizoram, India
Dr. Abhijit Sinha, Assistant Professor, NIT Mizoram, India
Dr. Ranjita Das, Assistant Professor, NIT Mizoram, India
Dr. Molay Roy, Assistant Professor, NIT Sikkim, India
Dr. Prashant Kumar Tiwari, Assistant Professor, NIT Allahabad, India
Dr. Dipankar Debnath, Assistant Professor, IIT Kharagpur, India
Dr. Tanmoy Roy Choudhury, Assistant Professor, KIIT Bhubaneswar, India
Dr. Kuntal Mandal, Assistant Professor, NIT Sikkim, India
Dr. Sangram Ray, Assistant Professor, NIT Sikkim, India
Dr. Manashita Borah, Assistant Professor, Tezpur University, Assam, India
Dr. Alak Majumder, Assistant Professor, NIT Arunachal Pradesh, India
Dr. Arnab Ghosh, Assistant Professor, NIT Rourkela, India
Dr. Nitai Pal, Associate Professor, IIT Dhanbad, India
Dr. Saheli Ray, Assistant Professor, NIT Silchar, India
Dr. Debashish Bhowmik, Assistant Professor, Tripura Institute of Technology Agartala, India
Dr. Chaitali Koley, Assistant Professor, NIT Mizoram, India
Dr. Aparesh Datta, Assistant Professor, NIT Durgapur, India
Dr. Badal Soni, Assistant Professor, NIT Silchar, India
Dr. Rakesh Roy, Assistant Professor, NIT Meghalaya, India
Dr. Arup Goswami, Associate Professor, NIT Silchar, India
Dr. Arup Ratan Bhowmik, Assistant Professor, NIT Agartala, India
Dr. Shuma Adhikari, Assistant Professor, NIT Manipur, India
Dr. Dipti Singh, Assistant Professor, Gautam Buddha University, India
Dr. Champa Nandi, Assistant Professor, Tripura University, Agartala, India
Dr. Rituparna Mitra, Assistant Professor, Adamas University, Kolkata, India
Dr. Akshya Ojha, Associate Professor, IIT Bhubaneswar, India
Dr. Manoj Thakur, Associate Professor, IIT Mandi
Dr. Prem Nath, Associate Professor, Garhwal University, Srinagar, Uttarakhand
Dr. Galiveeti Hemakumar Reddy, Associate Professor, MVJ College of Engineering Bangalore, Bangalore, India
Dr. Pabitra Kumar Biswas, Assistant Professor, NIT Mizoram, India
Dr Krishna Pratap Singh, Associate Professor, IIIT Allahabad

Preface

ICoCIST stands for "International Conference on Computational Intelligence and Sustainable Technologies." Computational intelligence (CI) (usually refers to "soft computing") is a new concept for advanced information processing, in order to realize a new approach to analyze and create flexible information processing of humans such as sensing, understanding, learning, recognizing and thinking. CI plays a vital role in improving the quality technology that can be sustained for a wider range of time than usual. Therefore, ICoCIST aims to provide an opportunity to the global participants on an amalgamation platform on both sustainable technologies and computational intelligence. The key attempt of ICoCIST is to bring together the global participants for exploring and sharing an intact idea not only to bridge the research gaps, but also to establish international research collaborations in their career path. ICoCIST focuses on quality research and targets to various institutions of repute to organize its following series, in the years to come.

We are delighted that the 1st "ICoCIST-2021" is conducted in online/virtual mode, organized jointly by NIT Silchar, Assam, and NIT Sikkim, India, during October 28–30, 2021. This proceedings as an outcome of the 1st meeting of ICoCIST community includes a collection of selected high-quality articles on various topics related to the applications of computational intelligence, smart computer network and security and sustainable technologies in societal development. The articles are placed in one concise volume on the themes of artificial intelligence in machine learning, big data analysis, soft computing techniques, pattern recognitions, sustainable infrastructural development, sustainable grid computing and innovative technology for societal development, renewable energy and innovations in Internet of things (IoT). Further, we expect for its mainstream impact to intensify our efforts to reach out non-academic beneficiaries and users in the real-life scenario. This edited volume must be helpful as a comprehensive reference source for researchers, practitioners, students and professionals.

The editors would like to express their sincere gratitude to its patrons, plenary speakers, invited speakers, authors, reviewers, Program Committee members, National and International Advisory Committee, and Local Organizing Committee, without whose support the quality and standards of the conference could not be maintained. Special thanks to Springer and its team for this valuable publication.

Over and above, we would like to express our deepest sense of gratitude to our own institutions NIT Silchar and NIT Sikkim to host this conference successfully.

Silchar, Assam
October 2021

Kedar Nath Das
Debasish Das
Anjan Kumar Ray
Ponnuthurai Nagaratnam Suganthan

Contents

About the Editors

Dr. Kedar Nath Das is currently working as Assistant Professor in the Department of Mathematics, NIT Silchar, Assam. He qualified the GATE examination and was awarded his Ph.D. degree from IIT Roorkee in the year 2008. His area of research interest includes operations research, evolutionary optimization techniques, networking optimization and multi-objective optimization. As of now, he has over 50 research papers to his credit including papers in international journals of repute, which has around 400 Google scholar citations till date. There is 1 books and 3 book chapters to his credit. He has already guided 5 Ph.D. scholars, and recently 4 are working under him. Also, he has successfully guided about 6 M.Sc. projects and 2 undergraduate internships. He has visited around 11 foreign countries (including Las Vegas, USA; London, UK; Amsterdam, Netherlands; Bangkok, Thailand; Singapore; Dubai; Kuching, Malaysia; Bratislava; Budapest, Vienna, Prague, Liverpool, UK; Wales, UK) and delivered about 35 expert lectures at different institutions of repute, in India. As coordinator, he has organized 12 events including STTP, international conference, workshops and self-financed courses. He has already reviewed more than 100 research papers. He is the life member of Operational Research Society of India (ORSI), The Indian Society for Technical Education (ISTE), Odisha Mathematical Society (OMS) and member of IEEE. He has been recently awarded 'Slovak National Scholarship' in the year 2019. He also acts as Investigator to a project, Government of India.

Dr. Debasish Das is presently working as Head of the Department in the Department of Civil Engineering, JIS College of Engineering (JISCE), Kalyani, West Bengal. Prior to joining JISCE, he worked as Assistant Professor and Head of the Department in the Department of Civil Engineering, Siliguri Institute of Technology, Siliguri. He has completed his Ph.D. in Transportation Engineering from National Institute of Technology Silchar, India; Master of Technology (M-Tech) in Transportation Engineering from National Institute of Technology, Silchar, India, and Bachelor of Technology (B.Tech.) in Civil Engineering from National Institute of Technology, Durgapur. He has published more than 20 research papers in reputed international journals/conferences in the field of traffic engineering and transportation planning.

His current research includes traffic engineering, transportation demand modelling, effect of COVID-19 on mode choice, effect of e-rickshaw in environmental pollution, plastic waste management, etc. He is a continuous reviewer of many reputed national/international journals. He also entitled with a project from WB-DST&BT, Government of West Bengal. He was participating in many international conferences as organizing member, session chair and member in technical programme committee.

Dr. Anjan Kumar Ray is now working as Assistant Professor in the Electrical and Electronics Engineering Department at NIT Sikkim, India. He has obtained his Ph.D. degree from IIT Kanpur. He was observed as a research associate at University of Ulster, UK. He has been posted as senior project engineer at IIT Kanpur and has been a visiting research scholar at University of Texas at San Antonio, USA. He is now the Dean of Research and Consultancy at NIT Sikkim. Along with, he has shouldered many administrative responsibility in his parent institution, like Senate Member, FIC Control Systems Laboratory, Chairperson, Innovation Cell, Start-up Cell Coordinator, FIC Library, Warden of Hostel (Admin). He has been working as HoD In-charge, EEE Department and FIC ICTI. He has completed one project under TEQIP-III, and one more MeitY project is ongoing. To his credit, he has awarded one Ph.D. scholar and more three are ongoing. He published 14 journal papers and 4 book chapters. He has been invited at the Rashtrapati Bhavan (President Office), India, for Festival of Innovation and Entrepreneurship 2018 to represent NIT Sikkim for project display and interaction with exhibitor groups with innovation clubs in March 2018. In addition, he acted as a reviewer and technical programme committee member of a number of journals and conferences.

Dr. Ponnuthurai Nagaratnam Suganthan is now working as an associate professor in the School of Electrical and Electronic Engineering, NTU, Singapore. He finished schooling at Union College (Tellippalai, Jaffna) and subsequently received the B.A. degree, Postgraduate Certificate and M.A. degree in Electrical and Information Engineering from the University of Cambridge, UK in 1990, 1992 and 1994, respectively. He received an honorary doctorate (i.e. Doctor Honoris Causa) in 2020 from University of Maribor, Slovenia. After completing his Ph.D. research in 1995, he served as Pre-doctoral Research Assistant in the Department of Electrical Engineering, University of Sydney, in 1995–96 and a lecturer in the Department of Computer Science and Electrical Engineering, University of Queensland in 1996–99. He moved to Singapore in 1999. He was Editorial Board Member of the *Evolutionary Computation Journal*, MIT Press (2013–2018) and Associate Editor of the *IEEE Trans. on Cybernetics* (2012–2018). He is Associate Editor of *Applied Soft Computing* (Elsevier, 2018–), *Neurocomputing* (Elsevier, 2018–), *IEEE Trans. on Evolutionary Computation* (2005–), *Information Sciences* (Elsevier, 2009–), *Pattern Recognition* (Elsevier, 2001–) and *IEEE Trans. on SMC: Systems* (2020–). He is a founding co-editor-in-chief of *Swarm and Evolutionary Computation* (2010–), an SCI Indexed Elsevier Journal. His co-authored SaDE paper (published in April 2009) won the "IEEE Trans. on Evolutionary Computation outstanding paper award" in 2012. His former Ph.D. student, Dr. Jane Jing Liang, won the IEEE CIS Outstanding Ph.D. dissertation award, in 2014. IEEE CIS Singapore Chapter won the best chapter award in Singapore in 2014 for its achievements in 2013 under his leadership. His research interests include swarm and evolutionary algorithms, pattern recognition, forecasting, randomized neural networks, deep learning and applications of swarm, evolutionary and machine learning algorithms. His publications have been well cited (Googlescholar Citations: ~45k). He was selected as one of the highly cited researchers by Thomson Reuters every year from 2015 to 2020 in computer science. He is ranked worldwide 300–400 among all computer science and electronics researchers (also include some control and communication engineering researchers) with public Google Scholar profiles. He served as General Chair of the IEEE SSCI 2013. He is an IEEE CIS distinguished lecturer (DLP) in 2018–2021. He has been a member of the IEEE (S'91, M'92, SM'00, Fellow'15) since 1991 and an elected AdCom member of the IEEE Computational Intelligence Society (CIS) in 2014–2016.

Multi-choice Programming with Benefits Using Kriging Interpolation Method

S. Dutta and Ramya V. Nair

Abstract Most of the real-life problems involves multi-choice or more than one option, and in such a situation out of many choices, one choice needs to be selected for making decision. In this paper, we propose a novel problem involving multiple choice parameter with benefit and to handle such multiple-choice parameter, Kriging interpolation method is used, as other existing mathematical method cannot be used. An illustration is provided to explain the proposed method.

Keywords Multi-objective · Multi-choice · Kriging interpolation

2000 Mathematics Subject Classification. MSC 90B50 · MSC 90C90 · MSC 65D05

1 Introduction

Multi-choice plays a significant role in everyone's life, as most of the decision-making process involves multi-choice and selecting a proper choice leads us to a better way in life. Decision-making process plays a vital role for its growth, improvement, and future stability. Healy [5] introduced the concept of multi-choice in 1964 and thereafter [4] used this concept in decision making and proposed multi-choice goal programming problem. In recent years, many researchers [2, 3, 6, 9, 12, 14, 15] studied and worked in multi-choice as parameters in mathematical programming problem under different environments and with different approaches. Mohanty et al. [11] developed a solution procedure to handle multi-choice using genetic algorithm without converting into its deterministic equivalent.

S. Dutta (✉)
Department of Mathematics, Assam Don Bosco University, Tapesia Campus, Assam, India

R. V. Nair
St. Philomena's Public School, Bangalore, Karnataka, India

K. N. Das et al. (eds.), *Proceedings of the International Conference on Computational Intelligence and Sustainable Technologies*, Algorithms for Intelligent Systems,
https://doi.org/10.1007/978-981-16-6893-7_1

The mathematical model of multi-choice programming problem is given below:

$$\max/min : z^k = \sum_{m=1}^{q} (c_m)^k x_m, k = 1, 2, \ldots, K \tag{1}$$

Subject to

$$\sum_{m=1}^{q} a_{lm} x_m \leq g_l, l = 1, 2, \ldots, p \tag{2}$$

$$x_m \geq 0, m = 1, 2, \ldots, q \tag{3}$$

where a_{lm}, h_l and $(c_m)^k$ are real numbers.

The multi-choice parameters may be present either in objective function, or/ and in constraints.
D G Krige developed the novel an interpolation technique known as Kriging. The basic idea about the Kriging interpolation method are presented in [7, 13, 18]. Kriging method has been applied in many fields like geo-statistics, computer modeling, design and analysis of computer experiment, simulation, sequential meta-modeling, etc., [1, 8, 10, 16, 17]. In this research paper, we propose a novel multi-choice mathematical programming model with benefit and also the solution procedure to solve such mathematical model. To evaluate the multi-choice in the objective function, Kriging interpolation method is used for the first time.
The research paper is assembled as follows: Beginning with the introduction, the proposed novel mathematical model formulation is furnished in Sect. 2. Solution procedure is delineated in Sect. 3. A numerical example is provided in support of the novel model and method in Sect. 4. At the end, the conclusions are drawn in Sect. 5.

2 Mathematical Models for Multi-choice Programming Problem with Benefits

$$\max/min\ Z = \sum_{l=1}^{p} \sum_{m=1}^{q} \begin{cases} c_{lm}^{1}, c_{lm}^{2}, c_{lm}^{3}, \ldots, c_{lm}^{klm} \\ c_{lm}^{1'}, c_{lm}^{2'}, c_{lm}^{3'}, \ldots, c_{lm}^{klm'} \end{cases} x_{lm}, \; k = 1, 2, \ldots, K \tag{4}$$

Subject to

$$\sum_{m=1}^{q} x_{lm} \le g_l, l = 1, 2, \ldots, p \quad (5)$$

$$\sum_{l=1}^{p} x_{lm} \ge h_m, m = 1, 2, \ldots, q \quad (6)$$

$$x_{lm} \ge 0, \forall l, m \quad (7)$$

where $g_l(l = 1, 2, \ldots, p)$ and $h_m(m = 1, 2, \ldots, q)$ are real numbers and $c_{lm}^{k} = c_{lm}^{1}, c_{lm}^{2}, \ldots, c_{lm}^{km}$ as a multi- choice parameter with $c_{lm}^{k'} = c_{lm}^{1'}, c_{lm}^{2'}, \ldots, c_{lm}^{km'}$ as benefits, ∀ l, m, k.

3 Solution Methodology

The mathematical models are difficult to solve directly and with existing methods, so an equivalent deterministic model are derived using Kriging method to obtain the compromise solution for MCPPP problem.

3.1 Kriging

Basic Kriging: It is an interpolation method to predict the undetermined values of an irregular field, process, or function $W(x) : x \in M$, where M is a fixed subset of $\mathbb{R}$. These predictions are weighted linear combinations of the observed values. The main idea of Kriging is based on the assumption that if the predictive points are closer to the sample data point, then the prediction errors are more positively correlated. The weights are based on covariance among the sample points and between the sample and predictive point.

1. Sample Covariances:

$$Cov_{lm} = Cov(z_l, z_m) = Cov(Z_l, Z_m) \forall l, m$$

$$K_1 = \begin{bmatrix} K_{11} & K_{12} & \ldots & K_{1p} \\ K_{12} & K_{22} & \ldots & K_{2p} \\ . & . & \ldots & . \\ . & . & \ldots & . \\ . & . & \ldots & . \\ K_{1p} & K_{2p} & \ldots & K_{pp} \end{bmatrix}$$

The above matrix is known as covariance matrix of the p points.

2. To find the covariance column matrix between the sample points and the prediction point:

$$K_{l0} = Cov(Z(z_l), Z(z_0)) = K(z_l, z_0) = Cov(Z_l, Z_0) \forall l$$

$$K_0 = \begin{bmatrix} K_{10} \\ K_{20} \\ . \\ . \\ . \\ K_{n0} \end{bmatrix}$$

Column vector of covariance between the sample points and the prediction point.

Next, we need to find the estimator $\widehat{Z_0} = w'Z$ such that:

(i) $\widehat{E(Z_0)} = E(Z_0)$. This relation will hold if the mean is constant and the sum of the weights is 1, i.e., $\sum w_i = 1$.
(ii) The prediction variance should be minimized, i.e., $\sigma_\epsilon^2 = E\left[Z_0 - \widehat{Z_0}\right]^2 = Var(Z_0 - \widehat{Z_0})$ is minimized.
(iii) In order to minimize the variance error or the prediction error, the Kriging estimator is the best estimator.
(iv) Ordinary Kriging can be taken as a predictor if the mean of the entire region is unknown, which can be assumed to be constant.
(v) If the value of the mean is constant, then simple Kriging method can be used for optimal predictions.
(vi) If any estimator meeting condition (i) and (ii) above, then it is purported to be the best linear unbiased predictor(BLUP).

The mean squared prediction error (MSPE) can be written as:

$$\begin{aligned} \sigma_\epsilon^2 = Var(Z_0, \widehat{Z_0}) &= Var(Z_0) + Var(\widehat{Z_0}) - 2Cov(\widehat{Z_0}, Z_0) \\ &= \sigma^2 + Var\left(\sum w_l Z_l\right) - 2Cov\left(\sum w_l Z_l, Z_0\right) \\ &= \sigma^2 + \sum_{l=1}^{p}\sum_{m=1}^{p} w_l w_m Cov(Z_l, Z_m) - 2\sum_{l=1}^{p} w_l Cov(Z_l, Z_0) \end{aligned} \tag{8}$$

$$= \sigma^2 + \sum_{l=1}^{p}\sum_{m=1}^{p} w_l w_m K_{lm} - 2\sum_{l=1}^{p} w_l K_{l0} \tag{9}$$

To minimize $Var(Z_0, \widehat{Z_0})$ with $\sum_{l=1}^{p} w_l = 1$, we use variogram.

Variogram plays a crucial role in Kriging which is the variance of the difference between the two points. The second-order stationary covariance process implies that the variogram is a function of the distance between two points (s). To fit the variogram, curve analyst applies different model such as exponential and spherical. In this paper, we are using exponential model, and the exponential model is given by:

$$\Gamma(s) = \begin{cases} k_0 + k_1(1 - e^{||s||/a}), & ifs \neq 0 \\ 0, & ifs = 0 \end{cases}$$

The variogram $2\Gamma(s)$ approaches a constant K(0), for large values of $||s||$ which is called as the sill; and the differences $W(x+s) - W(x)$ are invariant w.r.t 'h'. Covariogram can be defined as:

$$\begin{aligned} K(s) &= Cov(W(x), W(x+s)) \\ \Gamma(s) &= K(0) - K(s) \end{aligned}$$

The following steps are used to find the weights associated to different points in order to find the unobserved value.

1. Form the Lagrangian:

$$L = Var(Z_0, \widehat{Z_0}) + 2\lambda \left(\sum w_l - 1\right)$$

where the final term in this entirety ensures that the unbiasedness imperative $\sum w_l = 1$ is met through the Lagrange multiplier λ.

2. Differentiating L partial w.r.t the w_l's and λ, and setting it equal to zero, and then simplifying and solving it to obtain:

$$\begin{bmatrix} 2\sum_{m=1}^{p} w_m K_{1m} - 2K_{10} + 2\lambda = 0 \\ \vdots \\ 2\sum_{m=1}^{p} w_m K_{pm} - 2K_{p0} + 2\lambda = 0 \\ 2\left(\sum_{m=1}^{p} w_m - 1\right) = 0 \end{bmatrix}$$

$$\Leftrightarrow \begin{bmatrix} \sum_{m=1}^{p} w_m K_{1m} + \lambda = K_{10} \\ \cdot \\ \cdot \\ \cdot \\ \sum_{m=1}^{p} w_m K_{pm} + \lambda = K_{p0} \\ \sum_{m=1}^{p} w_m.1 + 0 = 1 \end{bmatrix}$$

$$\Leftrightarrow \begin{bmatrix} K_{11} & \dots & K_{1p} & 1 \\ \cdot & \cdots & \cdot & \cdot \\ \cdot & \cdots & \cdot & \cdot \\ \cdot & \cdots & \cdot & \cdot \\ K_{1p} & \dots & K_{pp} & 1 \\ 1 & \dots & 1 & 0 \end{bmatrix} \begin{bmatrix} w_1 \\ \cdot \\ \cdot \\ \cdot \\ w_n \\ \lambda \end{bmatrix} = \begin{bmatrix} K_{10} \\ \cdot \\ \cdot \\ \cdot \\ K_{p0} \\ 1 \end{bmatrix}$$

$$\Leftrightarrow \begin{bmatrix} K_1 & 1 \\ 1' & 0 \end{bmatrix} \begin{bmatrix} w_1 \\ \lambda \end{bmatrix} = \begin{bmatrix} k_0 \\ 1 \end{bmatrix}$$

Hence, the following system of linear equation can be solved to determine the weights for ordinary Kriging. It is also known as Kriging equation:

$$= \begin{bmatrix} K_{11} & \dots & K_{1p} & 1 \\ \cdot & \cdots & \cdot & \cdot \\ \cdot & \cdots & \cdot & \cdot \\ \cdot & \cdots & \cdot & \cdot \\ K_{1p} & \dots & K_{pp} & 1 \\ 1 & \dots & 1 & 0 \end{bmatrix}^{-1} \begin{bmatrix} K_{10} \\ \cdot \\ \cdot \\ \cdot \\ K_{p0} \\ 1 \end{bmatrix} = \begin{bmatrix} K_1 & 1 \\ 1' & 0 \end{bmatrix}^{-1} \begin{bmatrix} K_0 \\ 1 \end{bmatrix}$$

The values of the weights can be obtained as:

$$\begin{bmatrix} w_1 \\ \cdot \\ \cdot \\ \cdot \\ w_p \\ \lambda \end{bmatrix} = \begin{bmatrix} K_1 & 1 \\ 1' & 0 \end{bmatrix}^{-1} \begin{bmatrix} K_0 \\ 1 \end{bmatrix}$$

So, w_1 can be calculated as

$$w_1 = K_1^{-1} k_0 - \frac{K_1^{-1} 1 1' K_1^{-1} k_0}{1' K_1^{-1} 1} + \frac{K_1^{-1} 1}{1' K_1^{-1} 1}$$

Similarly, we can calculate the others weight for l = 2, …,p

4 Numerical Example

$$\max z = 6x_1 + x_2 + 2x_3 \quad (10)$$

Subject to

$$2x_1 + 2x_2 + 0.5x_3 \leq \begin{bmatrix} 4 & 3 & 5 \\ 1 & 2 & 2 \end{bmatrix} \quad (11)$$

$$-4x_1 - 2x_2 - 1.5x_3 \leq \begin{bmatrix} 3 & 4 & 5 \\ 1 & 2 & 3 \end{bmatrix} \quad (12)$$

$$x_1 + 2x_2 + 0.5x_3 \leq 1 \quad (13)$$

$$x_1, x_2, x_3 \geq 0 \quad (14)$$

where $\{4, 3, 5\}$ and $\{3, 4, 5\}$ are the multi-choices and $\{1, 2, 2\}$ and $\{1, 2, 3\}$ are the benefits for the first and second constraint, respectively. Consider the first constraint, where s is the difference between choices and benefit.

$$\begin{bmatrix} x & y & s \\ 4 & 1 & 3 \\ 3 & 2 & 1 \\ 5 & 2 & 3 \end{bmatrix}$$

Then, computing the covariance function and the covariance matrix among the observed values

$$K(s) = 100e^{0.3s_o}$$
$$K_{10} = 100e^{0.3*3} = 40.656$$
$$K_{20} = 100e^{0.3*1} = 74.082$$
$$K_{30} = 100e^{0.3*3} = 40.656$$

$$K_0 = \begin{bmatrix} K_{10} \\ K_{20} \\ K_{30} \end{bmatrix} = \begin{bmatrix} 40.656 \\ 74.082 \\ 40.656 \end{bmatrix}$$

$$K(s) = \begin{bmatrix} 100 & 74.082 & 100 \\ 74.082 & 100 & 100 \\ 74.082 & 74.082 & 100 \end{bmatrix}$$

$$K^{-1}(s) = \begin{bmatrix} 0.03858 & 0 & -0.038529 \\ 0 & 0.03852 & -0.03858 \\ -0.038529 & -0.03858 & 0.067165 \end{bmatrix}$$

$$w_1 = k_1^{-1}k_0 - \frac{k_1^{-1}11'k_1^{-1}k_0}{1'k_1^{-1}1} + \frac{k_1^{-1}}{1'k_1^{-1}1}$$

$$= \begin{bmatrix} 0.002073450 \\ 1.28756266 \\ -0.5486646 \end{bmatrix} - \begin{bmatrix} 0.002073456 \\ 1.28756266 \\ -0.5486646 \end{bmatrix} + \begin{bmatrix} 0.0050 \\ 0 \\ +0.99500 \end{bmatrix}$$

$$= \begin{bmatrix} 0.0050 \\ 0 \\ +0.99500 \end{bmatrix}$$

Similarly, for the second constraint, the weights are

$$w_2 = \begin{bmatrix} -0.0000002002 \\ -0.00000033203 \\ 1.0 \end{bmatrix}$$

Accordingly, the most weighted parameter is chosen from the given choices or if we add the choice after multiplying with respective weights then also we get same result. The final formulation of the problem using the above weight is

$$\max z = 6x_1 + x_2 + 2x_3 \tag{15}$$

Subject to

$$2x_1 + 2x_2 + 0.5x_3 \leq 5 \tag{16}$$

$$-4x_1 - 2x_2 - 1.5x_3 \leq 5 \tag{17}$$

$$x_1 + 2x_2 + 0.5x_3 \leq 1 \tag{18}$$

$$x_1, x_2, x_3 \geq 0 \tag{19}$$

Solving the above problem using Lingo 18, we get, the objective function value Z = 6 and the decision variables are $x_1 = 1; x_2 = x_3 = 0$

5 Conclusion

In this research article, a novel solution procedure is proposed to solve the multi-choice programming problems with benefit. As the proposed mathematical model is the extension of the existing mathematical model of multi-choice programming where the decision makers not only make a choice from the given choices but also has to analyze the benefit associated with the given choices. So, Kriging method is used to solve the mathematical model of multi-choice programming problem with benefit. In the numerical example, the multi-choice parameters with benefit are considered in the constraints. We considered only two choice parameters with benefits in this paper, and this method can be extended to handle a greater number of choices and in other existing problems like solid transportation, vehicle routing, game theory, etc. This research paper will open new door for the computational methods or developing new algorithm for optimization as well.

Acknowledgements The authors are thankful to the reviewers for the suggestion in improving the quality of the paper.
Conflict of interest The authors declare that they have no conflict of interest.

References

1. Abdullah, A.S., Matoha, S., Lubis, D.A., Falah, A.N., Jaya, I., Hermawan, E., Ruchjana, B.N.: Implementation of generalized space time autoregressive (gstar)-kriging model for predicting rainfall data at unobserved locations in west java. Appl. Math. Information Sci. **12**(3), 607–615 (2018)
2. Belay, B., Acharya, S.: Multi-choice multi-objective fractional probabilistic programming problem. In: Proceedings of the Fifth International Conference on Mathematics and Computing, Springer, Heidelberg, pp. 169–180 (2021)
3. Benoist, T., Bourreau, E., Caseau, Y., Rottembourg, B.: Towards stochastic constraint programming: a study of onine multi-choice knapsack with deadlines. In: International Conference on Principles and Practice of Constraint Programming. Springer, Heidelberg, pp. 61–76 (2001)
4. Chang, C.T.: Multi-choice goal programming. Omega **35**(4), 389–396 (2007)
5. Healy, W., Jr.: Multiple choice programming (a procedure for linear programming with zero-one variables). Oper. Res. **12**(1), 122–138 (1964)
6. Hwang, Y.A., Liao, Y.H.: Equivalence theorem, consistency and axiomatizations of a multi-choice value. J. Global Optimization **45**(4), 597–613 (2009)
7. Kleijnen, J.P.: Kriging metamodeling in simulation: a review. Eur. J. Oper. Res. **192**(3), 707–716 (2009)
8. Kleijnen, J.P., Sargent, R.G.: A methodology for fitting and validating metamodels in simulation. Eur. J. Oper. Res. **120**(1), 14–29 (2000)
9. Maiti, S.K., Roy, S.K.: Analysing interval and multi-choice bi-level programming for stackelberg game using intuitionistic fuzzy programming. Int. J. Math. Oper. Res. **16**(3), 354–375 (2020)
10. Martin, J., Simpson, T.: A Monte Carlo simulation of the kriging model. In: 10th AIAA/ISSMO Multidisciplinary Analysis and Optimization Conference, p. 4483 (2004)
11. Mohanty, D., Jana, R., Biswal, M.: Multi-choice chance-constrained programming problems using genetic algorithm. In: Mathematical Modeling and Computational Tools: ICACM 2018, Kharagpur, India, November 23–25, vol. 320, p. 375 (2020)

12. Nasseri, S., Bavandi, S.: Multi-choice linear programming in fuzzy random hybrid uncertainty environment and their application in multi-commodity transportation problem. Fuzzy Information Eng. **12**(1), 109–122 (2020)
13. Oliver, M.A., Webster, R.: Basic Steps in Geostatistics: The Variogram and Kriging. Springer, Heidelberg (2015)
14. Pradhan, A., Biswal, M.: A bi-level multi-choice programming problem. Int. J. Math. Oper. Res. **7**(1), 1–18 (2015)
15. Rout, P.K., Nanda, S., Acharya, S.: Computation of multi-choice multi-objective fuzzy probabilistic two stage programming problem. Int. J. Comput. Sci. Math. **11**(2), 168–191 (2020)
16. Santner, T.J., Williams, B.J., Notz, W.I., Williams, B.J.: The Design and Analysis of Computer Experiments, vol. 1. Springer, Heidelberg (2003)
17. Simpson, T., Maurey, T., Korte, J., Mistree, F.: Kriging meta-models for global approximation in simulation-based design. Eng. Comput. **17**(2), 129–150 (2001)
18. Van Beers, W.C., Kleijnen, J.P.: Kriging for interpolation in random simulation. J. Oper. Res. Soc. **54**(3), 255–262 (2003)

Prediction of Alzheimer's Disease Using Machine Learning Algorithm

Rahul Sharma, Tripti Goel, and R. Murugan

Abstract Accurate analysis of Alzheimer's Disease (AD) from healthy patients is a significant concern. The knowledge about the condition's seriousness and formative dangers due to AD leads to take prudent steps before the permanent brain damage takes place. Lately, there has been incredible benefits of computer-aided diagnosis (CAD) which uses various machine learning (ML) techniques for assisting the medical staff for AD diagnosis. In this paper, we proposed a model based on support vector machine (SVM) and made a comparison with three other classification algorithms for predicting normal control (NC) patients from AD patients. Lastly, the performances of the different algorithms are compared, and results of the model clearly indicate that gives quality performance for NC and AD classification.

Keywords Support vector machine · Decision tree · K-nearest neighbour · Regression · Machine learning

1 Introduction

In current scenario, dementia is one of the most serious health issues. Dementia is not a single disease but a group of symptoms which is directly related to brain degradation. AD is the most conventional cause of dementia [4]. It has been seen that the AD patients are mostly old age persons with age of 60 years and above. The disease causes brain degeneration which leads to memory dysfunction, thereby loss in short time memory, forgetting social behaviour, decline in thinking capability and

R. Sharma · T. Goel (✉) · R. Murugan
Department of ECE, National Institute of Technology Silchar, Silchar, Assam 788010, India
e-mail: triptigoel@ece.nits.ac.in

R. Sharma
e-mail: rahul_rs@ece.nits.ac.in

R. Murugan
e-mail: murugan.rmn@ece.nits.ac.in

K. N. Das et al. (eds.), *Proceedings of the International Conference on Computational Intelligence and Sustainable Technologies*, Algorithms for Intelligent Systems,
https://doi.org/10.1007/978-981-16-6893-7_2

in the worse case may even leads to death [10]. Across the globe, an estimated more than 50 millions people are enduring dementia, out of which 70

Various work has already been done in this area but the handling data plays a vital role. Zhang et al. [14] proposed a machine learning method in which several brain parameter were extracted from 3D NiFTi images and multi-kernel support vector machine (SVM) was used as a classifier. The method gives highest accuracy using polynomial kernel in SVM. Another paper of Beheshti et al. [2] proposed SVM based classifier with genetic algorithm optimization used for feature shortlisting.

The general approach for AD diagnosis is to conduct a magnetic resonance imaging (MRI) and conduct some genetic or pathological test which is basically not economical for majority of patients. Instead, our proposed method uses cognitive approach as many experts take the benefits of this approach for upgrading there results [13]. Also, study-related data along with brain data helps in better understanding of AD for early diagnosis. In this paper, section two consists of data set used in the model. Section three gives the details about the methodology implemented and idea about the proposed as well as the compared algorithms. Section four provides us the results and discussion made over the results. Section five concludes the paper and gives the idea about the future directions to the proposed model.

2 Data set

The data set was acquired from Open Access Series of Imaging Studies (OASIS) project [7, 8]. OASIS is originated by Dr. Randy Buckner which provides free availability of the brain MRI data sets. OASIS project consist of cross-sectional and longitudinal MRI data. The longitudinal MRI data set used in our method contains 371 subjects aged 60 to 96 which is used as training set. The cross-sectional data set contains collection of 216 subjects aged 18 to 96 which acts as testing test for our model. All the patients data used are right-handed. The data set consists of general, demographics, clinical and derived anatomical information related to the patients.

3 Methodology

For preparing the model and evaluating the results with other model, we have taken k-nearest neighbour (k-NN), regression model, decision tree and SVM model and applied it to the data set, and the comparisons were made over several parameters. The model consists of input data set acquired from OASIS project which is fed into multi-kernel SVM, and classification of disease are made into two category. The first class stands for CN resembling the normal patient and the second class, i.e. AD, resembling AD patient. The basic block diagram for the model is represented in Fig. 1. SVM with four different kernels is used in our proposed model, and the results are compared with three different classification techniques.

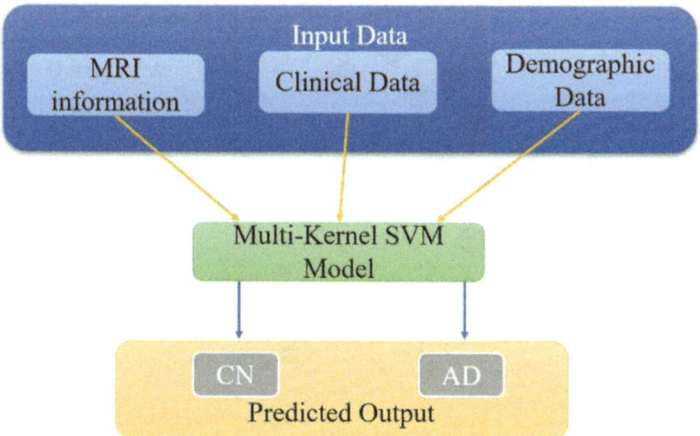

Fig. 1 Proposed model

3.1 Classification Techniques

The four classification techniques used are compared and explained.

3.1.1 SVM

SVM [1] is the most stable and commonly used technique for classification. SVM implements maximum margin principle for classification using its decision-making function which is given by the equation

$$g(x) = \mathrm{sgn}(w^T \Delta x + b) \quad (1)$$

where w stands for weights and b stands for bias. SVM algorithm calculations yields an ideal hyperplane which is a line partitioning a plane in sections, wherein individual class settles on either side and helps in sorting models as shown in Fig. 2. The higher the value of the margin, the minimum will be the generalization error of SVM.

SVM main highlights is handling the non-linear data which requires kernels. In proposed methods, four types of kernel were used, i.e. linear, Gaussian, radial basic function (RBF) and polynomial.

Each kernels has its own applications and merits. In linear kernel, there is complexity of parameter tuning because the function is always a form of line. Equation 2 gives the idea about the function for linear kernel.

$$l(p, q) = 1 + p.q + p.q.\min(p, q) - \frac{p+q}{2}.\min(p, q)^2 + \frac{1}{3}.\min(p, q)^3 \quad (2)$$

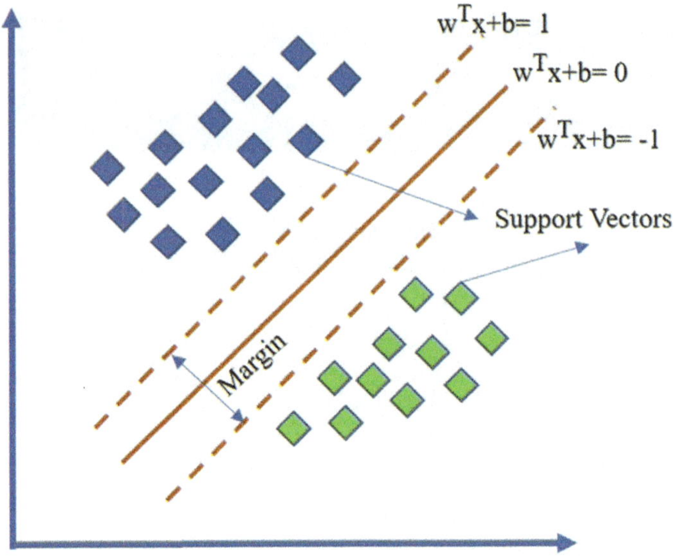

Fig. 2 SVM classification

For high dimensionality data, RBF kernel improves the performance. RBF kernel is used when the boundaries are tentatively curved shape. Equation 3 represents the function for the RBF kernel.

$$r(p_i, q_j) = \exp(-\gamma||p_i - q_j||^2) \tag{3}$$

The Gaussian kernel is an exponentially decreasing function which is used with relevant regularization parameter to predict global optimum and, hence, led to reduction in estimation and approximation errors. Gaussian is a special case of RBF function. The expression for the kernel is given by Eq. 4.

$$k(x, y) = \exp(-\frac{||x - y||^2}{2\sigma^2}) \tag{4}$$

Polynomial Kernels allow much efficient understanding of the non-linear model. This kernel tries to find the similarity between the given features. Equation 5 is the equation for the polynomial kernel.

$$k(p_i, q_j) = (p_i.q_j + a)^d \tag{5}$$

All the above four kernels are applied in our proposed model, and the classification is done on the basis.

3.1.2 K-Nearest Neighbour

k-NN [12] is the most simplest and well-known classifier used. As the name resembles, the algorithm deduce that common things exist in vicinity. Simply we can say that analogous things are near to each other. k-NN match the similarity between training and testing data. k-NN uses the idea of similarity and calculate the distances. The most nearest distance is used for classification on testing set.

3.1.3 Types of Regression

Regression [6] is a type of supervised machine learning classification algorithms and is basically of two types, linear and logistic. In comparison, we have used logistic regression(LR). LR is used to predict the categorical dependent variable using a set of independent variables. The relationship between dependent variable and independent variable should be non-linear for logistic regression. It has been found that logistic regression surpass the linear model due to capability of non-linear classification.

3.1.4 Decision Tree Classifier

Decision tree classifier (DTC) [3] is basically a representation for classifying examples. It is also a type of supervised learning algorithms where the data is continuously split according to a certain parameter. DTC is an iterative process based on the outcome generated by comparing the input under certain boundary. The main disadvantage of this algorithms is that it easily gets overfit but is still used because it is extremely fast in classifying.

In our proposed model, we have taken SVM as our classification algorithms, and other models are compared with the same data set. The model validation was done with a fivefold cross-validation. The data set is split into 70% and 30% as training and validation, respectively.

4 Results and Discussion

From Table 1, it is very much clear that the proposed model attains the highest accuracy of 96.40%. Also, in terms of precision, recall and F-1 score, the output by model has shown the highest values when compared to all other three algorithms, respectively. The lowest performance was given by decision tree algorithm. SVM with all the four kernels outperforms the compared classification techniques. Also, linear kernel outperforms in all the aspects. Although, accuracy of Linear and RBF kernel was same but precision was much high. Hence, SVM outperforms with several other machine learning over different performance metrics.

Table 1 Performance comparison of the various classifiers

Classifier	Performance metrics			
	Accuracy (%)	Precision	Recall	F-1 score
Decision tree	90.99	0.9429	0.9167	0.9296
k-NN	93.46	0.92	0.9388	0.9074
Logistic regression	94.59	0.9778	0.9464	0.9167
SVM with linear kernel	96.40	1.00	0.925	0.9505
SVM with Gaussian kernel	95.5	0.9804	0.9259	0.9524
SVM with RBF kernel	96.40	0.9792	0.9259	0.9615
SVM with polynomial kernel	95.33	0.9787	0.92	0.9485

5 Conclusion

AD is an utmost global neurological degenerative problem and prognosis or diagnosis is still a challenge for research even tough lot of work has already been done. This paper concludes that despite of using invasive or non-invasive imaging techniques, we can still use clinical test and demographics data for early prediction in a most economical way. Also from algorithm point, SVM always outperforms and hence can be used as a tool by physician for much more accurate treatment.

Future work requires to enhance the feature set as in our model, we have taken 7 features which is less, but still effective. Secondly, we can directly try to access the MRI images along with clinical and demographic data which will leads to increase in classification and early detection of AD.

References

1. Alam, S., Kwon, G.R., Initiative, A.D.N.: Alzheimer disease classification using kpca, lda, and multi-kernel learning svm. Int. J. Imaging Syst. Technol. **27**(2), 133–143 (2017)
2. Beheshti, I., Demirel, H., Matsuda, H., Initiative, A.D.N., et al.: Classification of alzheimer's disease and prediction of mild cognitive impairment-to-alzheimer's conversion from structural magnetic resource imaging using feature ranking and a genetic algorithm. Comput. Biol. Med. **83**, 109–119 (2017)
3. Dana, A.D., Alashqur, A.: Using decision tree classification to assist in the prediction of alzheimer's disease. In: 2014 6th International Conference on Computer Science and Information Technology (CSIT), pp. 122–126. IEEE (2014)
4. Dening, T., Sandilyan, M.B.: Dementia: definitions and types. Nursing Standard (2014+) **29**(37), 37 (2015)

5. Jenkins, R., Fox, N.C., Rossor, A.M., Harvey, R.J., Rossor, M.N.: Intracranial volume and alzheimer disease: evidence against the cerebral reserve hypothesis. Arch. Neurol. **57**(2), 220–224 (2000)
6. Johnson, P., Vandewater, L., Wilson, W., Maruff, P., Savage, G., Graham, P., Macaulay, L.S., Ellis, K.A., Szoeke, C., Martins, R.N., et al.: Genetic algorithm with logistic regression for prediction of progression to alzheimer's disease. BMC Bioinf. **15**(S16), S11 (2014)
7. Marcus, D.S., Fotenos, A.F., Csernansky, J.G., Morris, J.C., Buckner, R.L.: Open access series of imaging studies: longitudinal mri data in nondemented and demented older adults. J. Cogn. Neurosc. **22**(12), 2677–2684 (2010)
8. Marcus, D.S., Wang, T.H., Parker, J., Csernansky, J.G., Morris, J.C., Buckner, R.L.: Open access series of imaging studies (oasis): cross-sectional mri data in young, middle aged, nondemented, and demented older adults. J. Cognit. Neurosc. **19**(9), 1498–1507 (2007)
9. Patterson, C.: World Alzheimer Report 2018: The State of the Art of Dementia Research: New Frontiers. Alzheimer's Disease International (ADI): London, UK (2018)
10. Reitz, C., Brayne, C., Mayeux, R.: Epidemiology of alzheimer disease. Nat. Rev. Neurol. **7**(3), 137–152 (2011)
11. Sluimer, J., Vrenken, H., Blankenstein, M., Fox, N., Scheltens, P., Barkhof, F., Van Der Flier, W.: Whole-brain atrophy rate in alzheimer disease: identifying fast progressors. Neurology **70**(19 Part 2), 1836–1841 (2008)
12. Tufail, A.B., Abidi, A., Siddiqui, A.M., Younis, M.S.: Automatic classification of initial categories of alzheimer's disease from structural mri phase images: a comparison of psvm, knn and ann methods. Age **2012**, 1731 (2012)
13. Wang, S.H., Du, S., Zhang, Y., Phillips, P., Wu, L.N., Chen, X.Q., Zhang, Y.D.: Alzheimer's disease detection by pseudo zernike moment and linear regression classification. CNS & Neurological Disorders-Drug Targets (Formerly Current Drug Targets-CNS & Neurological Disorders) **16**(1), 11–15 (2017)
14. Zhang, Y., Dong, Z., Phillips, P., Wang, S., Ji, G., Yang, J., Yuan, T.F.: Detection of subjects and brain regions related to alzheimer's disease using 3d mri scans based on eigenbrain and machine learning. Front. Comput. Neurosc. **9**, 66 (2015)

A Study of the Caputo-Fabrizio Fractional Model for Atherosclerosis Disease

Praveen Kumar Gupta and Asish Adak

Abstract In this article, we developed a fractional-order atherosclerosis disease model and discussed the inflammatory and control criteria during inflation of the disease. Here, we also discuss the existence of the solution for a proposed model using the fixed-point theorem and carried out the qualitative analysis for the model that shows it also has a unique solution. We have done all these analyses for the kinetic system. This fractional-order mathematical model will be more effective to find the exact parameter values than the standard differential equation model. The authors believe that the newly established fractional-order mathematical model for a disease will help to detect atherosclerosis inflammation. The analytical results shown in this article will generate a new idea for the future experiments and tests.

Keywords Atherosclerosis disease · Mathematical modeling · Reaction–diffusion equation · Fractional calculus

AMS Subject Classification 35K57 · 92C45

1 Introduction

In the current scenario, cardiovascular disease is one of the fastest-growing diseases on our planet, and this is merely due to our unhealthy lifestyle. In general, if the heart or blood vessels are not functioning well, the disease is called cardiovascular disease (CVD). It is usually linked with the formation of fatty deposits, such as fat, cholesterol, calcium, within the arteries (is called atherosclerosis) which increases the risk of blood clots. In short, atherosclerosis is a state where the arteries become contracted and toughened due to the development of plaque in some areas of the artery wall. It may also be related to arterial injuries in major organs such as the brain, eyes, kidneys, and heart too. The CVD is mainly four types: TIAs and strokes,

P. K. Gupta (✉) · A. Adak
Department of Mathematics, National Institute of Technology Silchar, Assam, India
e-mail: praveen@math.nits.ac.in

K. N. Das et al. (eds.), *Proceedings of the International Conference on Computational Intelligence and Sustainable Technologies*, Algorithms for Intelligent Systems,
https://doi.org/10.1007/978-981-16-6893-7_3

peripheral arterial disease, aortic disease, and the most dangerous coronary heart disease.

As we know, a mathematical modeling idea provides an improvement to existing diagnostic tools and supports the development of drugs and other therapies. Concerning the mathematical formulations in cardiovascular disease, recently, Wu and Yang [1] developed mathematical model of inflammation for atherosclerosis subject to no-flux boundary conditions and used the dynamical theory, spectrum analysis of operators for elaborating the steady-state bifurcations. Hidalgo and Tello [2] provided a porous medium-type atherosclerosis mathematical model for tracing the early stages of atherosclerosis development, including analytical and numerical findings. Karimi et al. [3] numerically investigated the relationship between crack initiation and propagation in atherosclerosis and spontaneous coronary artery and estimated fracture parameters of healthy and atherosclerosis-infected arteries in the human coronary path [4]. Recently, Cusimano et al. [5] studied the feature of cardiac tissue microstructure in order to determine complex spatio-temporal excitation dynamics. They described the possible projecting power of modeling tools and revealed the stoutness of physiological models.

The application of fractional calculus has been one of the most popular research areas in the last two decades, reflecting the enormous interest in mathematical physics, biology, medicine, economics, finance, and computer science, etc. (see [6–8]). Wharmby and Bagley [9] demonstrate the capability to predict the dielectric behavior of objects to terahertz radiation using a viscoelastic model. In this sequence, Machado and Mata [10] employ fractional calculus theory to deduce the likely causes of the western world economic crisis, concluding that by 2020, the global ranking will be less unusual, and the US will advance even farther. In epidemiological problem, Gupta [11] presented a dynamical behavior of the HIV/AIDS dynamics model with Caputo sense fractional derivative. He et al. [12] developed a fractional-order coronary artery model with the inclusion of two different disturbances such as rich dynamics and complex environment of the human body and exhibited the bifurcation diagram and phase portrait through dynamical analysis. Recently, Singh et al. [13] discuss a q-homotopy analysis Elzaki transform method (q-HAETM) and solve fractional multidimensional diffusion equations.

In most of the studies of atherosclerosis disease, the models are standard derivative ordinary differential equation models, which may not show the behavior of the variables when the derivative is fractional. So many models are described in the form of ordinary differential equation which is actually failed to provide the real behavior of this disease in the initial stage. The ability of the model parameters is not very much related to the medical scenario. Therefore, the fractional-order model gives new insights to these studies. With these results and motivation, we have selected the latest and most accepted fractional-order derivative of Caputo-Fabrizio and proposed a novel Caputo-Fabrizio fractional atherosclerosis intimation model with consideration of few terminology from Panfilov et al. [14] and Khatib et al. [15], which reflects many mysterious features that are not in previous existing models.

2 Preliminaries

Before giving the description of the model, we have defined some basic definition of Caputo-Fabrizio fractional operators.

Definition 1 Let $f \in H^1(0, b), b > 0$ and $0 < \alpha < 1$, then the fractional Caputo-Fabrizio fractional differential operator is defined as [16]

$$ {}_0^{CF}D_t^\alpha f(t) = \frac{(2-\alpha)F(\alpha)}{2(1-\alpha)} \int_0^t \exp\left[-\frac{\alpha(t-\tau)}{1-\alpha}\right] f\prime(\tau)d\tau, t \geq 0 $$

with a normalization function $F(\alpha)$ which is depending on α such that function $F(0) = F(1) = 1$.

Definition 2 Let $0 < \alpha < 1$ and $f \in H^1(0, b), b > 0$, then fractional integral of order α of the function f is defined by [16]

$$ {}_0^{CF}I_t^\alpha (f(t)) = \frac{2(1-\alpha)}{(2-\alpha)F(\alpha)} f(t) + \frac{2\alpha}{(2-\alpha)F(\alpha)} \int_0^t f(x)dx, t \geq 0 $$

where $F(\alpha)$ is normalization function.

3 Mathematical Model

In this paper, we consider a Caputo-Fabrizio fractional-order reaction–diffusion model on finite interval, representing the arterial intima

$$ {}_0^{\mathrm{CF}}D_t^\alpha M = D_M \frac{\partial^2 M}{\partial x^2} + f_M(A) - \lambda_M M, \tag{1} $$

$$ {}_0^{\mathrm{CF}}D_t^\alpha A = D_A \frac{\partial^2 A}{\partial x^2} + f_A(A)M - \lambda_A A. \tag{2} $$

Here, the density of immune cells, which includes monocytes, macrophages, is represented as M, and the density of the cytokines secreted by the immune cells is represented as A. The function $f_M(A)$ defines rate of the immune cells secreted from the bloodstream and endorsed by the inflammatory cytokines

$$ f_M(A) = \frac{a_1 + b_1 A}{1 + \frac{A}{\xi_1}} \tag{3} $$

where $a_1 = f_M(0)$ is the number of monocytes at starting of inflammation due to the occurrence of oxidized LDL, which is encouraged by chemo-attractants, endothelial connected molecules, and standard development factors. The second term b_1 in numerator stands for the auto-amplifications rate of monocytes because of inflammatory cytokines (which is secreted by existing monocytes). The denominator factor $1 + \frac{A}{\xi_1}$ represents the dispersion of the recruitment of M, with ξ_1 being the development time of the fibrous cap. Even though, this fact is relatively complex, which increases the proliferation and relocation of smooth muscle cells [15].

In the second equation, the $f_A(A)\,M$ reflects the cytokines fabrication rate, and the function is

$$f_A(A) = \frac{a_2 A}{1 + \frac{A}{\xi_2}} \tag{4}$$

where, $a_2 A$ is the self-secretion part of pro-inflammatory cytokines and, in the denominator, $1 + \frac{A}{\xi_2}$ corresponds to the inhibition of the pro-type of inflammatory cytokines, which is secreted by the anti-type inflammatory cytokines, with ξ_2 being the required time for the inhibition to perform. The factors $-\lambda_M M$ and $-\lambda_A A$ represent the transience rate of the immune cells M and the cytokines A, respectively. And, $D_M \frac{\partial^2 M}{\partial x^2}$ and $D_A \frac{\partial^2 A}{\partial x^2}$ express their cell displacement (diffusion term) in the intima. The parameter values of the model are taken as positive or zero. Also, f_M to be an increasing function of A, and $\xi_1 > \frac{a_1}{b_1}$.

3.1 Kinetic System

Now, we establish the conditions to sustain the inflammatory reaction. So, we first concentrate on the reaction term of the Caputo-Fabrizio fractional-order reaction–diffusion system (1)–(2); that is, we look at the kinetic system of equations

$${}_{0}^{CF}D_t^{\alpha} M = \left(\frac{a_1 + b_1 A}{1 + \frac{A}{\xi_1}}\right) - \lambda_M M \tag{5}$$

$${}_{0}^{CF}D_t^{\alpha} A = \left(\frac{a_2 A}{1 + \frac{A}{\xi_2}}\right) M - \lambda_A A \tag{6}$$

with the initial conditions,

$$M(0) = M_0 \text{ and } A(0) = A_0. \tag{7}$$

In this complete study, the total density of immune cells and cytokines has taken to be constant $I_{\max} = M + A$. Assume B is the set of continuous-valued function from the Banach space, $\Re \to \Re$, on the finite interval Γ with the specified norm $\|(M, A)\| = \|M\| + \|A\|$, where $\|M\| = \sup\{\,|S(t)| : t \in \Gamma\,\} \le \upsilon$ and $\|A\| = \sup\{\,|A(t)| : t \in \Gamma\,\} \le \omega$. In particular, $B = V(\Gamma) \times V(\Gamma)$, where $V(\Gamma)$ is the set of continuously $\Re$ valued functions in the Banach space, is defined at the domain Γ and fulfilling the supremum norm property.

4 Analysis of the Model

In this section, we shall discuss the existence and uniqueness of the solution of the Caputo-Fabrizio fractional cardiovascular system using fixed-point theory [17].

4.1 Existence of the Solution

In this sequence, first we employ the Caputo-Fabrizio integral definitions (see [18–20]), we get

$$M(t) - M(0) = {}_{0}^{\mathrm{CF}}I_t^{\alpha}\left[\left(\frac{a_1 + b_1 A}{1 + \frac{A}{\xi_1}}\right) - \lambda_M M\right], \tag{8}$$

$$A(t) - A(0) = {}_{0}^{\mathrm{CF}}I_t^{\alpha}\left[\left(\frac{a_2 A}{1 + \frac{A}{\xi_2}}\right) M - \lambda_A A\right]. \tag{9}$$

Now, we apply the Caputo-Fabrizio fractional integral definition, and we get

$$\begin{aligned} M(t) - M(0) =& \frac{2(1-\alpha)}{(2-\alpha)F(\alpha)}\left[\frac{a_1 + b_1 A}{1 + \frac{A}{\xi_1}} - \lambda_M M\right] \\ &+ \frac{2\alpha}{(2-\alpha)F(\alpha)}\int_0^t \left[\frac{a_1 + b_1 A(s)}{1 + \frac{A(s)}{\xi_1}} - \lambda_M M(s)\right]\mathrm{ds}, \end{aligned} \tag{10}$$

$$\begin{aligned} A(t) - A(0) =& \frac{2(1-\alpha)}{(2-\alpha)F(\alpha)}\left[\frac{a_2 A M}{1 + \frac{A}{\xi_2}} - \lambda_A A\right] \\ &+ \frac{2\alpha}{(2-\alpha)F(\alpha)}\int_0^t \left[\frac{a_2 A(s) M(s)}{1 + \frac{A(s)}{\xi_2}} - \lambda_A A(s)\right]\mathrm{ds}. \end{aligned} \tag{11}$$

For simplicity and clarity, we assume the two kernel values

$$K_1(t, M) = \frac{a_1 + b_1 A(t)}{1 + \frac{A(t)}{\xi_1}} - \lambda_M M(t), \tag{12}$$

$$K_2(t, A) = \frac{a_2 A(t) M(t)}{1 + \frac{A(t)}{\xi_2}} - \lambda_A A(t). \tag{13}$$

Theorem 1 *The above two kernels* $K_1(t, M)$ *and* $K_2(t, A)$ *satisfy the Lipchitz condition and contraction if the inequality conditions* $0 \leq \lambda_M < 1$ *and* $0 \leq a_2 \upsilon + \lambda_A < 1$ *holds.*

Proof Now first, we check the Lipchitz condition for the kernel K_1. Assume M and M_1 are two functions, and find out the following norm for corresponding these functions

$$\begin{aligned}
&\|K_1(t, M) - K_1(t, M_1)\| \\
&= \left\| \left(\frac{a_1 + b_1 A(t)}{1 + \frac{A(t)}{\xi_1}} - \lambda_M M(t) \right) - \left(\frac{a_1 + b_1 A(t)}{1 + \frac{A(t)}{\xi_1}} - \lambda_M M_1(t) \right) \right\|, \\
&\|K_1(t, M) - K_1(t, M_1)\| \leq \gamma_1 \|M(t) - M_1(t)\|,
\end{aligned} \tag{14}$$

where $\lambda_M \leq \gamma_1$. Hence, the Lipschitz condition is satisfied for K_1, other than $0 \leq \lambda_M < 1$.

Similarly, we verify the Lipschitz condition for the kernel K_2. Let A and A_1 be two different functions, and evaluate the similar norm

$$\begin{aligned}
&\|K_2(t, A) - K_2(t, A_1)\| \\
&= \left\| \left(\frac{a_2 A(t) M(t)}{1 + \frac{A(t)}{\xi_2}} - \lambda_A A(t) \right) - \left(\frac{a_2 A_1(t) M(t)}{1 + \frac{A_1(t)}{\xi_2}} - \lambda_A A_1(t) \right) \right\|, \\
&\leq \left(a_2 \|M(t)\| \left\| \frac{1}{\left(1 + \frac{A(t)}{\xi_2}\right)\left(1 + \frac{A_1(t)}{\xi_2}\right)} \right\| + \lambda_A \right) \|A(t) - A_1(t)\| \\
&\leq \gamma_2 \|A(t) - A_1(t)\|,
\end{aligned} \tag{15}$$

where $\left(1 + \frac{A(t)}{\xi_2}\right) \geq 1$, $\left(1 + \frac{A_1(t)}{\xi_2}\right) \geq 1$, and $a_2 \upsilon + \lambda_A \leq \gamma_2$.

On the evidence of the abovementioned kernels, Eqs. (10) and (11) become

$$M(t) - M(0) = \frac{2(1-\alpha)}{(2-\alpha)F(\alpha)} K_1(t, M) + \frac{2\alpha}{(2-\alpha)F(\alpha)} \int_0^t K_1(s, M) \mathrm{ds} \tag{16}$$

$$A(t) - A(0) = \frac{2(1-\alpha)}{(2-\alpha)F(\alpha)} K_2(t, A) + \frac{2\alpha}{(2-\alpha)F(\alpha)} \int_0^t K_2(s, A) \mathrm{ds}. \tag{17}$$

Therefore, we can define the following recursive formula with the help of (16) and (17)

$$M_n(t) = \frac{2(1-\alpha)}{(2-\alpha)F(\alpha)}K_1(t, M_{n-1}) + \frac{2\alpha}{(2-\alpha)F(\alpha)}\int_0^t K_1(s, M_{n-1})\mathrm{ds}, \quad (18)$$

$$A_n(t) = \frac{2(1-\alpha)}{(2-\alpha)F(\alpha)}K_2(t, A_{n-1}) + \frac{2\alpha}{(2-\alpha)F(\alpha)}\int_0^t K_2(s, A_{n-1})\mathrm{ds}. \quad (19)$$

The first value of these iterations, i.e., initial conditions are given by

$$M_0(t) = M(0) \text{ and } A_0(t) = A(0) \quad (20)$$

The difference between the consecutive terms is shown as follows

$$\begin{aligned} X_n(t) =& M_n(t) - M_{n-1}(t) = \frac{2(1-\alpha)}{(2-\alpha)F(\alpha)}(K_1(t, M_{n-1}) - K_1(t, M_{n-2})) \\ &+ \frac{2\alpha}{(2-\alpha)F(\alpha)}\int_0^t (K_1(s, M_{n-1}) - K_1(s, M_{n-2}))\mathrm{ds} \end{aligned} \quad (21)$$

$$\begin{aligned} Y_n(t) =& A_n(t) - A_{n-1}(t) = \frac{2(1-\alpha)}{(2-\alpha)F(\alpha)}(K_2(t, A_{n-1}) - K_2(t, A_{n-2})) \\ &+ \frac{2\alpha}{(2-\alpha)F(\alpha)}\int_0^t (K_2(s, A_{n-1}) - K_2(s, A_{n-2}))\mathrm{ds} \end{aligned} \quad (22)$$

Eqs. (14) and (15) show that the kernel satisfies the Lipschitz condition; therefore, we can write these equations as follows

$$\begin{aligned} \|M_n(t) - M_{n-1}(t)\| \leq & \frac{2(1-\alpha)}{(2-\alpha)F(\alpha)}\gamma_1\|M_{n-1} - M_{n-2}\| \\ &+ \frac{2\alpha}{(2-\alpha)F(\alpha)}\gamma_1\int_0^t \|M_{n-1} - M_{n-2}\|\mathrm{ds} \end{aligned} \quad (23)$$

Then, from (21), we can write

$$\|X_n(t)\| \leq \frac{2(1-\alpha)}{(2-\alpha)F(\alpha)}\gamma_1\|X_{n-1}(t)\| + \frac{2\alpha}{(2-\alpha)F(\alpha)}\gamma_1\int_0^t \|X_{n-1}(s)\|\mathrm{ds}. \quad (24)$$

Similarly, we get the following result

$$\|Y_n(t)\| \leq \frac{2(1-\alpha)}{(2-\alpha)F(\alpha)}\gamma_2\|Y_{n-1}(t)\| + \frac{2\alpha}{(2-\alpha)F(\alpha)}\gamma_2\int_0^t \|Y_{n-1}(s)\|\mathrm{ds}. \quad (25)$$

After deliberating the above results, we state the next theorem.

Theorem 2 *The analytical solution exists for the Caputo-Fabrizio fractional mathematical model of cardiovascular disease under the following conditions that we can find t_0 such that.*

$$\frac{2(1-\alpha)}{(2-\alpha)F(\alpha)}\gamma_1 + \frac{2\alpha}{(2-\alpha)F(\alpha)}\gamma_1 t_0 < 1 \text{ and}$$
$$\frac{2(1-\alpha)}{(2-\alpha)F(\alpha)}\gamma_2 + \frac{2\alpha}{(2-\alpha)F(\alpha)}\gamma_2 t_0 < 1 \tag{26}$$

Proof Since, the functions $M(t)$ and $A(t)$ are enclosed in some finite region. Moreover, in Eqs. (14) and (15), we already proved that these kernels satisfy the Lipschitz condition. Consequently, using the results of the last two Eqs. (24) and (25) and applying the recursive method, we write the following relation:

$$\|X_n(t)\| \leq \|M_n(0)\|\left[\left(\frac{2(1-\alpha)}{(2-\alpha)F(\alpha)}\gamma_1\right) + \left(\frac{2\alpha}{(2-\alpha)F(\alpha)}\gamma_1 t\right)\right]^n, \tag{27}$$

$$\|Y_n(t)\| \leq \|A_n(0)\|\left[\left(\frac{2(1-\alpha)}{(2-\alpha)F(\alpha)}\gamma_2\right) + \left(\frac{2\alpha}{(2-\alpha)F(\alpha)}\gamma_2 t\right)\right]^n. \tag{28}$$

Therefore, the abovementioned solutions will be exist and continuous. On the other side, to prove that the above functions are the solution of the proposed model, we assume

$$M(t) - M(0) = M_n(t) - R_n(t),$$

$$A(t) - A(0) = A_n(t) - S_n(t).$$

Therefore, we have

$$\|R_n(t)\| = \left\| \frac{2(1-\alpha)}{(2-\alpha)F(\alpha)}(K_1(t, M) - K_1(t, M_{n-1})) + \frac{2\alpha}{(2-\alpha)F(\alpha)}\int_0^t (K_1(s, M) - K_1(s, M_{n-1}))\mathrm{ds} \right\|$$

$$\|R_n(t)\| = \left\| \frac{2(1-\alpha)}{(2-\alpha)F(\alpha)}(K_1(t, M) - K_1(t, M_{n-1})) + \frac{2\alpha}{(2-\alpha)F(\alpha)}\int_0^t (K_1(s, M) - K_1(s, M_{n-1}))\mathrm{ds} \right\| \tag{29}$$

After using this iteration recursively, it gives up

$$\|R_n(t)\| \leq \left(\frac{2\,(1-\alpha)}{(2-\alpha)\,F(\alpha)} + \frac{2\,\alpha}{(2-\alpha)\,F(\alpha)}\,t\right)^{n+1} \gamma_1^{n+1} \upsilon$$

Then at t_0, we have

$$\|R_n(t)\| \le \left(\frac{2(1-\alpha)}{(2-\alpha)F(\alpha)} + \frac{2\alpha}{(2-\alpha)F(\alpha)}t_0\right)^{n+1} \gamma_1^{n+1}\upsilon. \tag{30}$$

If n is the extremely large number, i.e., if it tends to infinity, then we get

$$\|R_n(t)\| \to 0. \tag{31}$$

In the same way, we can derive

$$\|S_n(t)\| \to 0. \tag{32}$$

Hence, this evidence shows that these solutions exist.

4.2 *Uniqueness of the Solution*

Now, we focus on proving the solution is unique or not. So first, we speculate that another solution to the proposed model exists, and it is $M_1(t)$ and $A_1(t)$, then

$$\begin{aligned} M(t) - M_1(t) =& \frac{2(1-\alpha)}{(2-\alpha)F(\alpha)}(K_1(t, M) - K_1(t, M_1)) \\ &+ \frac{2\alpha}{(2-\alpha)F(\alpha)}\int_0^t (K_1(s, M) - K_1(s, M_1))\text{ds} \end{aligned}$$

Applying norm on the above equation and simplifying, we get

$$\|M - M_1\|\left(1 - \frac{2(1-\alpha)}{(2-\alpha)F(\alpha)}\gamma_1 + \frac{2\alpha}{(2-\alpha)F(\alpha)}\gamma_1 t\right) \le 0. \tag{33}$$

Theorem 3 *For the following condition, the model has a single solution.*

$$\left(1 - \frac{2(1-\alpha)}{(2-\alpha)F(\alpha)}\gamma_1 + \frac{2\alpha}{(2-\alpha)F(\alpha)}\gamma_1 t\right) > 0. \tag{34}$$

Proof Let us assume the above condition (34) is true, and from (33),

$$\|M - M_1\|\left(1 - \frac{2(1-\alpha)}{(2-\alpha)F(\alpha)}\gamma_1 + \frac{2\alpha}{(2-\alpha)F(\alpha)}\gamma_1 t\right) \le 0,$$

Hence, it shows that $\|M(t) - M_1(t)\| = 0$.

Therefore, we easily say that

$$M(t) = M_1(t). \tag{35}$$

Now, we apply the same procedure for another function, so we get

$$A(t) = A_1(t). \tag{36}$$

Hence, this proof shows that the proposed model also has unique solution.

5 Conclusion

In this manuscript, we proposed the Caputo-Fabrizio fractional-order mathematical model for atherosclerosis disease and examined the kinetic behavior of this model with the help of fractional calculus theory. Using the fixed-point theorem, we demonstrated that the proposed model has at least one solution, i.e., whether the disease will remain or not. The Lipschitz condition reflects that the solution is unique for some initial density of immune cells and cytokines secreted by the immune cells. We reveal that under some conditions and for a certain value of inflammatory cytokines, the inflammation of atherosclerosis is fixed. The authors confidently believed that this fractional model could be tremendously valuable for academics working on cardiovascular modeling challenges.

References

1. Wu, D., Yang, W.: Analysis on bifurcation solutions of an atherosclerosis model. Nonlinear Anal. Real World Appl. **39**, 396–410 (2018)
2. Hidalgo, A., Tello, L.: Numerical simulation of a porous medium type atherosclerosis initiation model. Comput. Fluids **169**, 380–387 (2018)
3. Karimi, A., Razaghi, R., Koyama, M.: A patient specific numerical modeling of the spontaneous coronary artery dissection in relation to atherosclerosis. Comput. Meth. Prog. Biomed. **182**, 105060 (2019)
4. Yao, S.W., Faridi, W.A., Asjad, M.I., Jhangeer, A., Inc, M.: A mathematical modelling of a Atherosclerosis intimation with Atangana-Baleanu fractional derivative in terms of memory function. Results Phys. 104425 (2021)
5. Cusimano, N., Gizzi, A., Fenton, F.H., Filippi, S., Gerardo-Giorda, V.: Key aspects for effective mathematical modelling of fractional-diffusion in cardiac electrophysiology: A quantitative study. Commun. Nonlinear Sci. Numer. Simul. **84**, 105512 (2020)
6. Singh, J., Kumar, D., Hammouch, Z., Atangana, A.: A fractional epidemiological model for computer viruses pertaining to a new fractional derivative. Appl. Math. Comput. **316**, 504–515 (2018)
7. Agarwal, R., Purohit, S.D.: Mathematical model pertaining to the effect of buffer over cytosolic calcium concentration distribution. Chaos Solitons Fractals **143**, 110610 (2021)
8. Habenom, H., Suthar, D.L., Baleanu, D., Purohit, S.D.: A numerical simulation on the effect of vaccination and treatments for the fractional hepatitis b model. J. Comput. Nonlinear Dyn. **16**(1), 011004 (2021)
9. Wharmby, A.W., Bagley, R.L.: The application of the fractional calculus model for dispersion and absorption in dielectrics I. Terahertz waves. Int. J. Eng. Sci. **93**, 1–12 (2015)

10. Machado, J.A.T., Mata, M.E.: Pseudo phase plane and fractional calculus modelling of western global economic downturn. Commun. Nonlinear Sci. Numer. Simul. **22**, 396–406 (2015)
11. Gupta, P.K.: Local and global stability of fractional order HIV/AIDS dynamics model. In: Ghosh, D., Giri, D., Mohapatra, R., Savas, E., Sakurai, K., Singh, L. (eds.) Mathematics and Computing. ICMC 2018, Communications in Computer and Information Science, vol. 834. Springer, Singapore (2018)
12. He, S., Fataf, N.A.A., Banerjee, S., Sun K.: Complexity in the muscular blood vessel model with variable fractional derivative and external disturbance. Physica A **526**, 120904 (2019)
13. Singh, J., Kumar, D., Purohit, S.D., Mishra, A.M., Bohra, M.: An efficient numerical approach for fractional multidimensional diffusion equations with exponential memory. Numer. Meth. Partial Diff. Eqn. **37**(2), 1631–1651 (2021)
14. Panfilov, A.V., Dierckx, H., Volpert, V.: Reaction diffusion waves in cardiovascular diseases. Physica D **399**, 1–34 (2019)
15. Khatib, N., Genieys, S., Volpert, V.: Atherosclerosis initiation modeled as an inflammatory process. Math. Model. Nat. Phenom. **2**, 126–141 (2007)
16. Dave, S., Khan, A.M., Purohit, S.D., Suthar, D.L.: Application of green synthesized metal nanoparticles in the photocatalytic degradation of dyes and its mathematical modelling using the caputo-fabrizio fractional derivative without the singular kernel. J. Math. **8**, 9948422 (2021)
17. Agarwal, R., Yadav, M.P., Baleanu, D., Purohit, S.D.: Existence and uniqueness of miscible flow equation through porous media with a nonsingular fractional derivative. AIMS Math. **5**(2), 1062–1073 (2020)
18. Baleanu, D., Mousalou, A., Rezapour, S.: A new method for investigating approximate solutions of some fractional integro-differential equations involving the Caputo-Fabrizio derivative. Adv. Difference Equ. **51**, 1–12 (2017)
19. Caputo, M., Fabrizio, M.: A new definition of fractional derivative without singular kernel. Prog. Fract. Diff. Appl. **1**, 73–85 (2015)
20. Atanacković, T.M., Pilipović, S., Zorica, D.: Properties of the Caputo-Fabrizio fractional derivative and its distributional settings. Fract. Calc. Appl. Anal. **21**(1), 29–44 (2018)

Fourth-Order Computations of Forced Convection Heat Transfer Past an Isothermal/isoflux Cylinder in Cylindrical Geometry with Pseudo-Time Iteration Technique

Dipjyoti Nath and B. Hema Sundar Raju

Abstract Numerical simulation is employed to investigate the forced convection heat transfer from an isoflux and isothermal circular cylinder for two Newtonian fluids (air and water). The cylindrical geometry fourth-order compact scheme (CGFOCS) is applied to solve the Navier–Stokes (N–S) and energy equations in the pseudo-time iteration technique. The numerical computations are elucidated with the distribution of heat transfer as local Nusselt number (Nu) along cylinder surface together with average Nusselt number ($\overline{\text{Nu}}$) in the ranges of following parameters: Reynolds number (Re) from 1 to 40, for specific Pr (Prandtl number) of air (Pr = 0.72) and water (Pr = 7). It is found that the local Nusselt number is slightly greater at front stagnation point for isothermal case while compare to isoflux case, whereas opposite effect is seen at rear stagnation point. Further, as step size reduces, the present results approach to fourth-order accuracy. $\overline{\text{Nu}}$ values are more in isoflux case while compared to isothermal case.

1 Introduction

Forced convection from a cylinder involves many industrial and engineering applications like contact cylinder driers is used for paper making and plastics industries, cylindrical cooling device is used in glass industries, etc. [1]. The applications of computational fluid dynamics such as aerodynamic models, turbulence models, some specific flows, heat and mass transfer are used in the development of computational intelligence/sustainable technology [2]. Dennis and Chang [3], Fornberg [4], Dennis et al. [5], Baranyi [6], Raju and Sekhar [7], Sekhar and Raju [8], Erturk and Gökçöl [9–11], Kumar and Kalita [12] studied numerically the flow and forced convection heat transfer from an isothermal cylinder with the presence of low and high parameters.

D. Nath · B. H. S. Raju (✉)
Department of Mathematics, National Institute of Technology, Silchar, Assam 788010, India
e-mail: raju@math.nits.ac.in

K. N. Das et al. (eds.), *Proceedings of the International Conference on Computational Intelligence and Sustainable Technologies*, Algorithms for Intelligent Systems,
https://doi.org/10.1007/978-981-16-6893-7_4

The forced convection heat transfer in air (Pr = 0.7) from an isoflux circular cylinder has been studied by Ahmad [13] by implementing the finite difference methods and hybrid scheme with the presence of the parameters $100 \leq \text{Re} \leq 500$. A boundary layer analysis (integral approach) was utilized by Khan et al. [14] to investigate the forced convection from an isoflux and isothermal cylinder. The similar problem has been considered by Bharti et al. [15] using semi-implicit finite volume method (FVM) and QUICK scheme in the ranges of parameters $10 \leq \text{Re} \leq 45$ and $0.7 \leq \text{Pr} \leq 400$. Dhole et al. [16] investigated the forced convection past a isoflux and isothermal sphere using FVM for different Prandtl number values. The forced convection past an isoflux/isothermal circular cylinder was analyzed by Paramane et al. [17] using FVM-based transient heat-line method. Sufyan et al. [1] investigated the forced convection heat transfer from an isothermal/isoflux circular cylinder by employing FVM. The forced convection past a isothermal and isoflux sphere with linear slip velocity was investigated by Ramteke and kishore [18] using FVM. Mohamed and Wood [19] studied the influences of isothermal and isoflux boundary cases on forced convection past a circular cylinder applying FVM.

In the above literature, the numerical schemes applied by the researchers are mainly second-order accurate. To improve the scheme more accurate, Raju and Sekhar [7], Sekhar et al. [8, 20, 21], Raju et al. [22], Nath et al. [23], Nath and Raju [24], Raju et al. [25, 26] considered Navier–Stokes and energy equations in cylindrical and spherical coordinates using fourth-order compact (FOC) scheme, which is fourth-order accurate. Raju [27] investigated the problem of forced convective heat transfer from a flat plate for different Newtonian fluids using elliptical cylindrical geometry fourth-order compact scheme in the pseudo-time iteration technique (PTIT). The aim of the current work is to analyze the forced convective heat transfer past an isoflux and isothermal circular cylinder for two Newtonian fluids (air and water) by applying CGFOCS in pseudo-time iteration technique.

2 Problem Formulation

A steady, laminar, and incompressible motion of fluid past a circular cylinder of radius $d/2$ with uniform stream velocity (U_∞) and temperature (T_∞) has been considered. The non-dimensional vorticity, stream, and energy equations together with velocity components after applying the transformation $r = e^{\pi\xi}$ and $\theta = \pi\lambda$ [4, 28] are listed below.

Velocity components:

$$v_r = \frac{e^{-\pi\xi}}{\pi}\frac{\partial\psi}{\partial\lambda}, \quad v_\theta = -\frac{e^{-\pi\xi}}{\pi}\frac{\partial\psi}{\partial\xi} \tag{1}$$

Stream function equation:

$$\frac{\partial^2\psi}{\partial\xi^2} + \frac{\partial^2\psi}{\partial\lambda^2} = -G\omega \tag{2}$$

Vorticity equation:

$$\frac{\partial^2 \omega}{\partial \xi^2} + \frac{\partial^2 \omega}{\partial \lambda^2} = R_1 \, J(\omega, \psi) \tag{3}$$

Energy equation:

$$\frac{\partial^2 \Theta}{\partial \xi^2} + \frac{\partial^2 \Theta}{\partial \lambda^2} = P_1 \, J(\Theta, \psi) \tag{4}$$

where $G = \pi^2 e^{2\pi\xi}$, $R_1 = \frac{\text{Re}}{2}$, $P_1 = \frac{\text{Pe}}{2}$, $\text{Pe} = \text{Re}Pr$, $J(\phi_1, \phi_2) = \left(\frac{\partial \phi_1}{\partial \xi}\frac{\partial \phi_2}{\partial \lambda} - \frac{\partial \phi_2}{\partial \xi}\frac{\partial \phi_1}{\partial \lambda}\right)$

Equations (2)–(4) are solved under the following boundary conditions:

$\psi = \frac{\partial \psi}{\partial \xi}=0$, $\omega = -\frac{1}{\pi^2}\frac{\partial^2 \psi}{\partial \xi^2}$, $\frac{\partial \Theta}{\partial \xi} = -1$ (isoflux) and $\Theta = 1$ (isothermal) at $\xi = 0$

$\psi = 0, \omega = 0, \frac{\partial \Theta}{\partial \lambda} = 0$ at $\lambda = 0, \ \lambda = 1$

$\psi = e^{\pi\xi} \sin \pi\lambda, \omega = \Theta = 0$ as $\xi \to \infty$

The drag coefficient is determined from the field of vorticity as follows

$$C_D = -\frac{4\pi}{\text{Re}} \int_0^1 \omega(0, \lambda) \sin(\pi\lambda) d\lambda + \frac{4}{\text{Re}} \int_0^1 \left(\frac{\partial \omega}{\partial \xi}\right)_{\xi=0} \sin(\pi\lambda) d\lambda \tag{5}$$

The heat transfer (local and average Nusselt number) on the cylinder surface is determined for two thermal boundary conditions as follows

$$\text{Nu}(\lambda) = \begin{cases} \left(\frac{2}{\Theta}\right)_{\xi=0}, & \text{for isoflux case} \\ -\frac{2}{\pi}\left(\frac{\partial \Theta}{\partial \xi}\right)_{\xi=0}, & \text{for isothermal case} \end{cases}$$

and

$$\overline{\text{Nu}} = \int_0^1 Nu(\lambda) d\lambda. \tag{6}$$

3 Numerical Methodology

The discretizations of the partial derivatives in fourth order are as follows:

$$\frac{\partial Z}{\partial \xi} = Z_\xi - 2 f_1 \frac{\partial^3 Z}{\partial \xi^3} + O(h^4) \tag{7}$$

$$\frac{\partial^2 Z}{\partial \xi^2} = Z_{\xi\xi} - f_1 \frac{\partial^4 Z}{\partial \xi^4} + O(h^4) \tag{8}$$

$$\frac{\partial Z}{\partial \lambda} = Z_{\lambda} - 2g_1 \frac{\partial^3 Z}{\partial \lambda^3} + O(k^4) \tag{9}$$

$$\frac{\partial^2 Z}{\partial \lambda^2} = Z_{\lambda\lambda} - g_1 \frac{\partial^4 Z}{\partial \lambda^4} + O(k^4), \tag{10}$$

where Z_{ξ}, Z_{λ}, $Z_{\xi\xi}$ and $Z_{\lambda\lambda}$ are the second-order central differences (SOCD) and $f_1 = \frac{h^2}{12}$ and $g_1 = \frac{k^2}{12}$ with the step sizes h and k are in the directions of ξ and λ, respectively. Substitute equations (8) and (10) in equation (2), we obtain

$$\psi_{\xi\xi} + \psi_{\lambda\lambda} = f_1 \frac{\partial^4 \psi}{\partial \xi^4} + g_1 \frac{\partial^4 \psi}{\partial \lambda^4} - G\omega + O(h^4, k^4) \tag{11}$$

By taking the partial derivatives with respect to ξ and then λ on both sides of stream function equation (2),

$$\frac{\partial^3 \psi}{\partial \xi^3} = -\left(\omega G_{\xi} + G\omega_{\xi} + \psi_{\xi\lambda\lambda}\right) \tag{12}$$

$$\frac{\partial^4 \psi}{\partial \xi^4} = -\left(\omega G_{\xi\lambda} + 2G_{\xi}\omega_{\xi} + G\omega_{\xi\xi} + \psi_{\xi\xi\lambda\lambda}\right) \tag{13}$$

$$\frac{\partial^3 \psi}{\partial \lambda^3} = -\left(\omega G_{\lambda} + G\omega_{\lambda} + \psi_{\xi\xi\lambda}\right) \tag{14}$$

$$\frac{\partial^4 \psi}{\partial \lambda^4} = -\left(G\omega_{\lambda\lambda} + 2G_{\lambda}\omega_{\lambda} + \omega G_{\lambda\lambda} + \psi_{\xi\xi\lambda\lambda}\right) \tag{15}$$

Substitute Eqs. (13) and (15) in Eq. (11), we get

$$\psi_{\xi\xi} + \psi_{\lambda\lambda} = -G\omega - \lambda_1 + O(h^4, k^4), \tag{16}$$

where

$$\lambda_1 = f_1 \left(\omega G_{\xi\lambda} + 2G_{\xi}\omega_{\xi} + G\omega_{\xi\lambda} + \psi_{\xi\xi\lambda\lambda}\right) + g_1 \left(\omega G_{\lambda\lambda} + 2G_{\lambda}\omega_{\lambda} + G\omega_{\lambda\lambda} + \psi_{\xi\xi\lambda\lambda}\right)$$

Equation (16) is FOC discretization form of stream function equation (2).

By the same strategy, FOC discretization form of vorticity and energy equations is listed below:

$$\omega_{\xi\xi}(1+\beta_1) + \omega_{\lambda\lambda}(1+\gamma_1) = \omega_{\xi}(R_1\psi_{\lambda} + \delta_1) + \omega_{\lambda}\left(-R_1\psi_{\xi} + \epsilon_1\right) + \zeta + O(h^4, k^4) \tag{17}$$

$$\Theta_{\xi\xi}(1+\beta\beta_1)+\Theta_{\lambda\lambda}(1+\gamma\gamma_1)=\Theta_\xi(P_1\psi_\lambda+\delta\delta_1)+\Theta_\lambda(-P_1\psi_\xi+\epsilon\epsilon_1)+\zeta\zeta+O(h^4,k^4) \tag{18}$$

where

$$\begin{aligned}
\beta_1 &= f_1{R_1}^2\psi_\lambda\psi_\lambda-2f_1\mathrm{Re}\psi_{\xi\lambda}\\
\gamma_1 &= g_1{R_1}^2\psi_\xi\psi_\xi+2g_1R_1\psi_{\xi\lambda}\\
\delta_1 &= (f_1+g_1)R_1\psi_{\xi\xi\lambda}-f_1{R_1}^2\psi_\lambda\psi_{\xi\lambda}+g_1R_1G_\lambda\omega+g_1{R_1}^2\psi_\xi\psi_{\lambda\lambda}\\
\epsilon_1 &= -f_1R_1G_\xi\omega-(f_1+g_1)R_1\psi_{\xi\lambda\lambda}+f_1{R_1}^2\psi_\lambda\psi_{\xi\xi}-g_1{R_1}^2\psi_\xi\psi_{\xi\lambda}\\
\zeta &= -(f_1-g_1)R_1G\omega_\xi\omega_\lambda-2f_1R_1\psi_{\xi\lambda}\omega_{\xi\lambda}-(f_1+g_1)R_1\psi_\xi\omega_{\xi\xi\lambda}\\
&\quad+(f_1+g_1)R_1\psi_\lambda\omega_{\xi\lambda\lambda}+(f_1+g_1){R_1}^2\psi_\xi\psi_\lambda\omega_{\xi\lambda}\\
&\quad-(f_1+g_1)\omega_{\xi\xi\lambda\lambda}+2g_1R_1\psi_{\lambda\lambda}\omega_{\xi\lambda}
\end{aligned}$$

$$\begin{aligned}
\beta\beta_1 &= f_1{P_1}^2\psi_\lambda\psi_\lambda-2f_1P_1\psi_{\xi\lambda}\\
\gamma\gamma_1 &= g_1{P_1}^2\psi_\xi\psi_\xi+2g_1P_1\psi_{\xi\lambda}\\
\delta\delta_1 &= (f_1+g_1)P_1\psi_{\xi\xi\lambda}-f_1{P_1}^2\psi_\lambda\psi_{\xi\lambda}+g_1P_1G_\lambda\omega+g_1{P_1}^2\psi_\xi\psi_{\lambda\lambda}\\
\epsilon\epsilon_1 &= -f_1P_1G_\xi\omega-(f_1+g_1)P_1\psi_{\xi\lambda\lambda}+f_1{P_1}^2\psi_\lambda\psi_{\xi\xi}-g_1{P_1}^2\psi_\xi\psi_{\xi\lambda}\\
\zeta\zeta &= -f_1P_1G\omega_\xi\Theta_\lambda+g_1P_1G\omega_\lambda\Theta_\xi-2fP_1\psi_{\xi\lambda}\Theta_{\xi\lambda}\\
&\quad-(f_1+g_1)P_1\psi_\xi\Theta_{\xi\xi\lambda}+(f_1+g_1)P_1\psi_\lambda\Theta_{\xi\lambda\lambda}+(f_1+g_1)P_1^2\psi_\xi\psi_\lambda\Theta_{\xi\lambda}\\
&\quad-(f_1+g_1)\Theta_{\xi\xi\lambda\lambda}+2gP_1\psi_{\lambda\lambda}\Theta_{\xi\lambda}.
\end{aligned}$$

4 Pseudo-Time Iteration Technique

The fourth-order discretized governing equations (16)–(18) are solved using PTIT [29]. Apply this technique to the vorticity equation (3),

$$\begin{aligned}
&\omega_t=\beta\omega_{\xi\xi}+\gamma\omega_{\lambda\lambda}-\delta\omega_\xi-\epsilon\omega_\lambda-\zeta\\
\Rightarrow\ &\frac{\partial\omega}{\partial t}=\beta\frac{\partial^2\omega}{\partial\xi^2}+\gamma\frac{\partial^2\omega}{\partial\lambda^2}-\delta\frac{\partial\omega}{\partial\xi}-\epsilon\frac{\partial\omega}{\partial\lambda}-\zeta\\
\Rightarrow\ &\frac{\omega^{n+1}-\omega^n}{\Delta t}=\beta^n\frac{\partial^2\omega^{n+1}}{\partial\xi^2}+\gamma^n\frac{\partial^2\omega^{n+1}}{\partial\lambda^2}-\delta^n\frac{\partial\omega^{n+1}}{\partial\xi}-\epsilon^n\frac{\partial\omega^{n+1}}{\partial\lambda}-\zeta^n\\
\Rightarrow\ &\left(1-\Delta t\beta^n\frac{\partial^2}{\partial\xi^2}-\Delta t\gamma^n\frac{\partial^2}{\partial\lambda^2}+\Delta t\gamma^n\frac{\partial}{\partial\xi}+\Delta t\epsilon^n\frac{\partial}{\partial\lambda}\right)\omega^{n+1}=\omega^n-\Delta t\zeta^n
\end{aligned} \tag{19}$$

Repeat the above technique to stream Eq. (2) and energy Eq. (4), we get

$$\left(1-\Delta t\frac{\partial^2}{\partial\xi^2}-\Delta t\frac{\partial^2}{\partial\lambda^2}\right)\psi^{n+1}=\psi^n+G\,\Delta t\omega^n+\Delta t\alpha^n, \tag{20}$$

and

$$\left(1 - \Delta t\beta\beta^n \frac{\partial^2}{\partial \xi^2} - \Delta t\gamma\gamma^n \frac{\partial^2}{\partial \lambda^2} + \Delta t\delta\delta^n \frac{\partial}{\partial \xi} + \Delta t\epsilon\epsilon^n \frac{\partial}{\partial \lambda}\right) \Theta^{n+1} = \Theta^n - \Delta t\zeta\zeta^n \tag{21}$$

Due to the presence of larger band matrix, Eqs. (19)–(21) are not computationally stable and so splits the LHS operator of Eq. (19)–(21) as follows.

$$\begin{aligned}&\left(1 - \Delta t \frac{\partial^2}{\partial \xi^2}\right)\left(1 - \Delta t \frac{\partial^2}{\partial \lambda^2}\right)\psi^{n+1} \\ &= \psi^n + G \,\Delta\, t\omega^n + \Delta t\alpha^n + \left(\Delta t \frac{\partial^2}{\partial \xi^2}\right)\left(\Delta t \frac{\partial^2}{\partial \lambda^2}\right)\psi^n,\end{aligned} \tag{22}$$

$$\begin{aligned}&\left(1 - \Delta t\beta^n \frac{\partial^2}{\partial \xi^2} + \Delta t\delta^n \frac{\partial}{\partial \xi}\right)\left(1 - \Delta t\gamma^n \frac{\partial^2}{\partial \lambda^2} + \Delta t\epsilon^n \frac{\partial}{\partial \lambda}\right)\omega^{n+1} \\ &= \omega^n - \Delta t\zeta^n + \left(\Delta t\beta^n \frac{\partial^2}{\partial \xi^2} - \Delta t\delta^n \frac{\partial}{\partial \xi}\right)\left(\Delta t\gamma^n \frac{\partial^2}{\partial \lambda^2} - \Delta t\epsilon^n \frac{\partial}{\partial \lambda}\right)\omega^n\end{aligned} \tag{23}$$

and

$$\begin{aligned}&\left(1 - \Delta t\beta\beta^n \frac{\partial^2}{\partial \xi^2} + \Delta t\delta\delta^n \frac{\partial}{\partial \xi}\right)\left(1 - \Delta t\gamma\gamma^n \frac{\partial^2}{\partial \lambda^2} + \Delta t\epsilon\epsilon^n \frac{\partial}{\partial \lambda}\right)\Theta^{n+1} \\ &= \Theta^n - \Delta t\zeta\zeta^n + \left(\Delta t\beta\beta^n \frac{\partial^2}{\partial \xi^2} - \Delta t\delta\delta^n \frac{\partial}{\partial \xi}\right)\left(\Delta t\gamma\gamma^n \frac{\partial^2}{\partial \lambda^2} - \Delta t\epsilon\epsilon^n \frac{\partial}{\partial \lambda}\right)\Theta^n\end{aligned} \tag{24}$$

By introducing new variables f_ψ, g_ω and h_Θ, rearrange Eqs. (22)–(24) in two steps (ξ and λ sweeps) such that
Stream function (ξ and λ sweeps):

$$\left(1 - \Delta t \frac{\partial^2}{\partial \lambda^2}\right)\psi^{n+1} = f_\psi \tag{25}$$

$$\left(1 - \Delta t \frac{\partial^2}{\partial \xi^2}\right) f_\psi = \psi^n + G \,\Delta\, t\omega^n + \Delta t\alpha^n + \left(\Delta t \frac{\partial^2}{\partial \xi^2}\right)\left(\Delta t \frac{\partial^2}{\partial \lambda^2}\right)\psi^n \tag{26}$$

Vorticity equation (ξ and λ sweeps):

$$\left(1 - \Delta t\gamma^n \frac{\partial^2}{\partial \lambda^2} + \Delta t\epsilon^n \frac{\partial}{\partial \lambda}\right)\omega^{n+1} = g_\omega \tag{27}$$

$$\left(1 - \Delta t\beta^n \frac{\partial^2}{\partial \xi^2} + \Delta t\delta^n \frac{\partial}{\partial \xi}\right) g_\omega$$
$$= \omega^n - \Delta t\zeta^n + \left(\Delta t\beta^n \frac{\partial^2}{\partial \xi^2} - \Delta t\delta^n \frac{\partial}{\partial \xi}\right)\left(\Delta t\gamma^n \frac{\partial^2}{\partial \lambda^2} - \Delta t\epsilon^n \frac{\partial}{\partial \lambda}\right)\omega^n \quad (28)$$

Energy equation (ξ and λ sweeps):

$$\left(1 - \Delta t\gamma\gamma^n \frac{\partial^2}{\partial \lambda^2} + \Delta t\epsilon\epsilon^n \frac{\partial}{\partial \lambda}\right)\Theta^{n+1} = h_\Theta \quad (29)$$

$$\left(1 - \Delta t\beta\beta^n \frac{\partial^2}{\partial \xi^2} + \Delta t\delta\delta^n \frac{\partial}{\partial \xi}\right) h_\Theta$$
$$= \Theta^n - \Delta t\zeta\zeta^n + \left(\Delta t\beta\beta^n \frac{\partial^2}{\partial \xi^2} - \Delta t\delta\delta^n \frac{\partial}{\partial \xi}\right)\left(\Delta t\gamma\gamma^n \frac{\partial^2}{\partial \lambda^2} - \Delta t\epsilon\epsilon^n \frac{\partial}{\partial \lambda}\right)\Theta^n \quad (30)$$

where $\beta = 1 + \beta_1,\ \gamma = 1 + \gamma_1,\ \delta = R_1\psi_\lambda + \delta_1,\ \epsilon = -R_1\psi_\xi + \epsilon_1$. and $\beta\beta = 1 + \beta\beta_1,\ \gamma\gamma = 1 + \gamma\gamma_1,\ \delta\delta = P_1\psi_\lambda + \delta\delta_1,\ \epsilon\epsilon = -P_1\psi_\xi + \epsilon\epsilon_1$. Eqs. (25)–(30) yield a tridiagonal system of equations and is numerically stable. By employing TDMA technique, we first solve the equation (26) for f_ψ and then solve the equation (25) for stream function (ψ). In similar manner, we solve the vorticity and energy by using Eqs. (27)–(30).

5 Discretisation of Boundary Conditions

Briley's formula [30] is utilized to discretize the vorticity boundary condition on the cylinder surface as

$$\omega_{0,j} = \frac{1}{18\,h^2\pi^2}\left(-108\psi_{1,j} + 27\psi_{2,j} - 4\psi_{3,j}\right).$$

On the cylinder surface, usual fourth order forward (FOF) formula is used to approximate the isoflux boundary condition. The usual FOF as well as backward difference formula has been used to approximate the temperature gradient boundary condition along axis of symmetry.

6 Results and Discussions

To verify the grid sensitivity analysis of CGFOCS, numerical experiments are conducted in distinct grids, namely 41 × 41, 61 × 61, 81 × 81, 101 × 101, 121 × 121, and 141 × 141 at Re = 30 and 40. The C_D values attained from the CGFOCS in the above referred grids are furnished in Table 1. The values of C_D referred in Table 1

Table 1 C_D values acquired from CGFOCS for distinct grid systems at Re = 30 and 40

	41 × 41 (0.038471)	61 × 61 (0.023247)	81 × 81 (0.017435)	101 × 101 (0.013984)	121 × 121 (0.011623)	141 × 141 (0.099631)
Re = 30	1.670305	1.700623	1.706137	1.707531	1.707984	1.708163
Re = 40	1.451854	1.503488	1.514227	1.517133	1.518108	1.518494

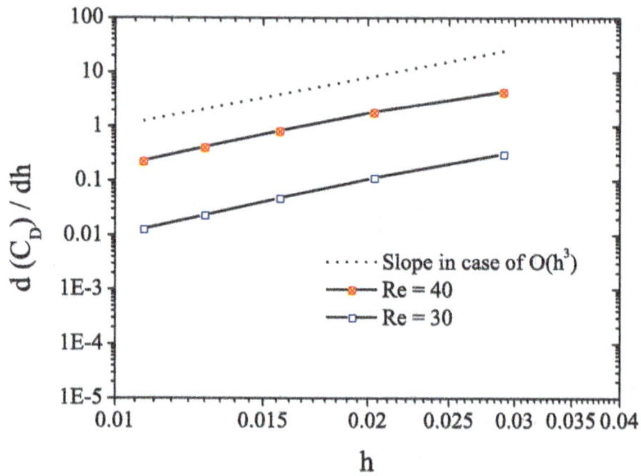

Fig. 1 Variation of $d(C_D)/dh$ with h (step sizes) at Re = 30, 40

are almost identical for two grids, namely 121 × 121 and 141 × 141. Thus, in the current work, 141 × 141 grid has been taken as a resultant grid size for all numerical solutions.

To verify the order of accuracy of the CGFOCS, the first-order divided differences formula for C_D values with regard to step size (h) are arranged together with fourth-order reference line as can be seen in Fig. 1. The slopes of the current results are identical to the reference line, which is fourth-order accurate. It shows that as $h \to 0$, the present results approach to fourth-order accuracy.

The $\overline{\text{Nu}}$ values calculated from the CGFOCS is validated with Bharti et al. [15] with isoflux boundary condition in Table 2 for Pr = 0.7 and Re = 10, 20, 40 together with difference (in percentage). Table 2 shows that the current results vary with a maximum difference of 0.9879%, and it proves that current results are consistent with the results of Bharti et al. [15].

The angular variation of Nu at Re = 20, 40 for Pr = 0.72 and 7 in different thermal boundary conditions is exhibit in Fig. 2. Figure 2 noted that the pattern of Nu on the cylinder surface in both isoflux and isothermal cases is qualitatively equivalent. Figure 2 shows that at $\theta = 0$, Nu values are larger in isoflux boundary condition in compared to isothermal case, but the opposite effect is seen at $\theta = \pi$. Moreover, the difference between local Nusselt number values of isoflux and isothermal cases increases with an increase of Pr.

Table 2 Comparison of present results (isoflux case) of $\overline{Nu}$ with [15] for Pr = 0.7 and distinct Re

Re	Bharti et al. [15]	Present scheme	%Diff
10	2.0400	2.0241	0.7794
20	2.7788	2.7548	0.8637
40	3.7755	3.7382	0.9879

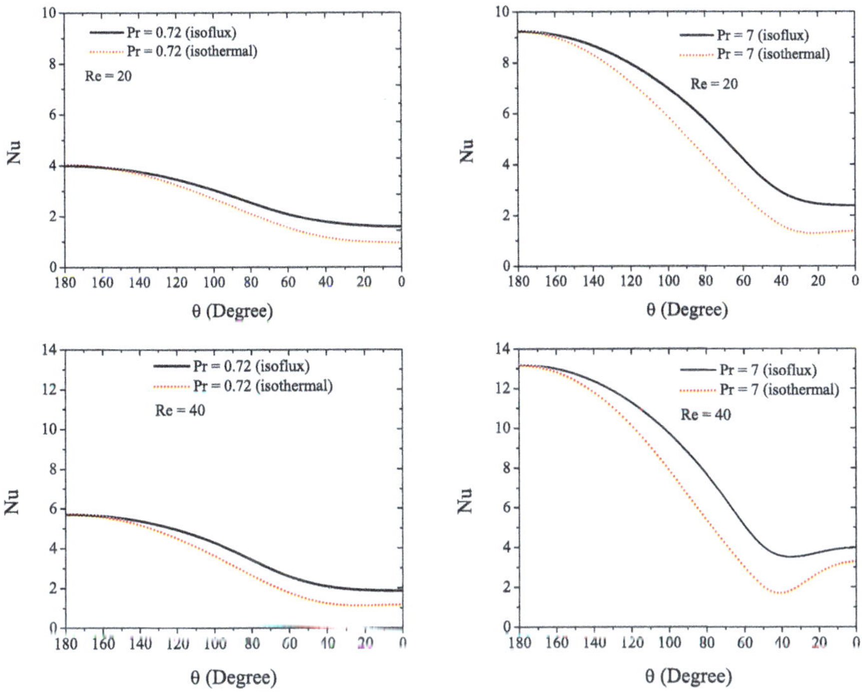

Fig. 2 Angular variation of of Nu in isothermal and isoflux case for Re = 20, 40 and Pr = 072, 7

Figure 3 delineates the variation of $\overline{Nu}$ with various Re and Pr = 0.72 and 7 for two thermal boundary conditions. For Pr = 0.72, 7 and low Re, there is a slight difference between $\overline{Nu}$ values for isothermal and isoflux cases. Again, as Re and Pr increase, the difference among $\overline{Nu}$ values of isothermal and isoflux cases are increasing, because with the increase of Re and Pr, the flow separation phenomena is increased.

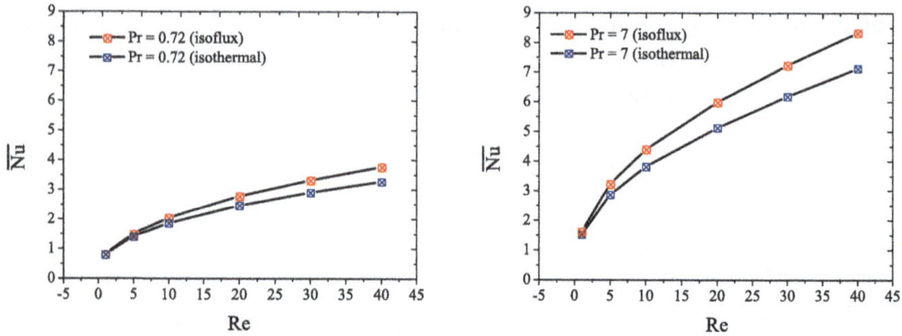

Fig. 3 $\overline{\text{Nu}}$ values evaluated from CGFOCS in isothermal and isoflux case for distinct values Re and Pr = 072, 7

7 Conclusions

The forced convective heat transfer from a circular cylinder depends on the surrounding fluids and varies with the different thermal conditions on the object surface. Therefore in the current work, the forced convection heat transfer past an isoflux and isothermal cylinder has been studied by implementing cylindrical geometry fourth-order compact scheme (CGFOCS) in pseudo-time iteration technique. The important findings from the current numerical investigations are highlighted as follows:

1. As grid improves, the present computations approach to fourth-order accuracy.
2. The local Nusselt number is slightly greater at front stagnation point for isothermal case while compare to isoflux case, whereas opposite effect is seen at rear stagnation point.
3. For Prandtl number = 0.72, 7 and larger values of Re, the difference between average Nusselt number values of isoflux and isothermal cases are increasing.
4. The heat transfer is more for isoflux case while compared to isothermal case.

References

1. Muhammad, S., Manzoor, S., Nadeem, A.S.: Free stream flow and forced convection heat transfer across rotating circular cylinder in steady regime: effects of rotation, Prandtl number and thermal boundary condition. J. Mech. Sci. Tech. **29**(4), 1781–1797 (2015)
2. Wang, B., Wang, J.: Application of artificial intelligence in computational fluid dynamics. Ind. Eng. Chem. Res. **60**(7), 2772–2790 (2021)
3. Dennis, S.C.R., Chang, G.Z.: Numerical solution for study flow past a circular cylinder at Reynolds numbers up to 100. J Fluid Mech. **42**, 471–489 (1970)
4. Fornberg, B.A.: Numerical study of steady viscous flow past a circular cylinder. J. Fluid Mech. **98**, 819–855 (1980)
5. Dennis, S.C.R., Hudson, J.D., Smith, N.: Steady laminar forced convection from a circular cylinder at low Reynolds numbers. Phys. Fluid. **11**, 933–940 (1968)

6. Baranyi, L.: Computation of unsteady momentum and heat transfer from a fixed circular cylinder in laminar flow. J. Comp. Appl. Mech. **4**, 13–25 (2003)
7. Raju, B.H.S., Sekhar, T.V.S.: Higher order compact scheme combined with multigrid method for momentum, pressure Poisson and energy equations in cylindrical geometry. The Open Num. Meth. J. **2**, 46–58 (2012)
8. Sekhar, T.V.S., Raju, B.H.S.: Fourth order accurate free convective heat transfer solutions from a circular cylinder. Int. J. Math. Comp. Phy. Elc. Comp. Engg. **6**(8), 1437–1441 (2012)
9. Erturk, E., Gökçöl, C.: Fourth order compact scheme for stream function-vorticity formulation of the Navier-Stokes equations in polar coordinates. Int. J. Mech. Engg. Tech. **9**(13), 1327–1336 (2018)
10. Erturk, E., Gökçöl, C.: Fourth order compact formulation of steady Navier-Stokes equations on non-uniform grids. Int. J. Mech. Engg. Tech. **9**(10), 1379–1389 (2018)
11. Erturk, E., Gökçöl, C.: Numerical solutions of steady incompressible flow around a circular cylinder up to Reynolds number 500. Int. J. Mech. Engg. Tech. **9**(10), 1368–1378 (2018)
12. Kumar, P., Kalita, J.: A comprehensive study of secondary and tertiary vortex phenomena of flow past a circular cylinder: a Cartesian grid approach. Phys. Fluids **33**, 053608 (2021)
13. Ahmad, R.A.: Steady-state numerical solution of the Navier-Stokes and energy equations around a horizontal cylinder at moderate Reynolds numbers from 100 to 500. Heat Trans. Engg. **17**(1), 31–81 (1996)
14. Khan, W.A., Culham, J.R., Yovanovich, M.M.: Fluid flow around and heat transfer from an infinite circular cylinder. J. Heat Trans. **127**(7), 785–795 (2005)
15. Bharti, R.P., Chhabra, R.P., Eswaran, V.: A numerical study of the steady forced convection heat transfer from an unconfined circular cylinder. Heat Mass Trans. **43**(7), 639–648 (2007)
16. Dhole, S.D., Chhabra, R.P., Eswaran, V.: A numerical study on the forced convection heat transfer from an isothermal and isoflux sphere in the steady symmetric flow regime. Int. J. Heat Mass Trans. **49**(5–6), 984–994 (2006)
17. Paramane, S.B., Sharma, A.: Heat and fluid flow across a rotating cylinder dissipating uniform heat flux in 2D laminar flow regime. Int. J. Heat Mass Trans. **53**(21–22), 4672–4683 (2010)
18. Ramteke, R.R., Kishore, N.: Effects of uniform heat flux and velocity-slip conditions at interface on heat transfer phenomena of smooth spheres in Newtonian fluids. J. Heat Trans. (2017). https://doi.org/10.1115/1.4036598
19. Mohamed, A., Wood, D.: Effect of thermal boundary condition on forced convection from circular cylinders. Num. Heat Transfer, Part A **76**(6), 420–437 (2019)
20. Sekhar, T.V.S., Raju, B.H.S.: An efficient higher order compact scheme to capture heat transfer solutions in spherical geometry. Comp. Phys. Commun. **183**, 2337–2345 (2012)
21. Sekhar, T.V.S., Raju, B.H.S., Murthy, P.V.S.N.: Higher order compact scheme for laminar natural convective heat transfer from a sphere. Appl. Math. Modelling **40**(3), 2039–2055 (2016)
22. Raju, B.H.S., Nath, D., Pati, S.: Effect of Prandtl number on thermo-fluidic transport characteristics for mixed convection past a sphere. Int. Commun. Heat Mass Trans. **98**, 191–199 (2018)
23. Nath, D., Pati, S., Raju, B.H.S.: Analysis of mixed convection past a heated sphere. J. Proc. Mech. Engg. **233**(3), 601–616 (2019)
24. Nath, D., Raju, B.H.S.: Effect of isoflux thermal boundary condition on mixed convective heat transfer from a sphere for liquid metals. Int. J. Amb. Energy (2019). https://doi.org/10.1080/01430750.2019.1636881
25. Raju, B.H.S., Nath, D., Pati, S.: Analysis of mixed convective heat transfer past an isoflux/isothermal sphere: influence of Prandtl number. Phy. Scr. **95**(8), 085211 (2020)
26. Raju, B.H.S., Nath, D., Pati, S., Baranyi, L.: Analysis of mixed convective heat transfer from a sphere with an aligned magnetic field. Int. J. Heat Mass Trans. **162**, 120342 (2020)
27. Raju, B.H.S.: Analysis of forced convection from a horizontal flat plate for liquid metals. Int. J. Amb. Energy (2019). https://doi.org/10.1080/01430750.2019.1614986
28. Ahmad, R.A., Qureshi, Z.H.: Laminar mixed convection from a uniform heat flux horizontal cylinder in a crossflow. J. thermophysics heat transfer **6**(2), 277–287 (1992)

29. Erturk, E., Gökçöl, C.: Fourth-order compact formulation of Navier-Stokes equations and driven cavity flow at high Reynolds numbers. Int. J. Num. Meth. Fluids **50**(4), 421–436 (2006)
30. Briley, W.R.: A numerical study of laminar separation bubbles using the Navier-Stokes equations. J. Fluid Mech. **47**(4), 713–736 (1971)

Implementation of Arithmetic Logic Unit Using Area-Efficient Adder

Abhishek Zade, Sumit Kumar Singha, Jagritee Talukdar, Addanki Prathima, and Kavicharan Mummaneni

Abstract Central processing unit (CPU) is the brain of the computer, and arithmetic and logic unit (ALU) serves as the major component of the CPU. In ALU, for the addition of multiple bits, ripple carry adder (RCA) with carry select adder (CSA) configuration is used in this paper. These configurations require various blocks of full adders in cascaded form leading to enhanced area of the circuit. Therefore, this paper mainly focuses on reducing the area of such adders, so that implementation of ALU can be done in minimum space. Here, the area of 4-bit and 8-bit adders with CSA is designed and optimized in such a way that adders can improve the area consumption. The designing of the proposed adder is implemented in Verilog hardware description language (HDL) and simulated using ISIM simulator. It is synthesized and implemented using Xilinx ISE 14.7 by selecting Spartan 3E family and XC3S100E as the device. The proposed 4-bit adder and 8-bit adder reduces the area by 16.8 and 7.38%, respectively.

Keywords Arithmetic and logic unit · Carry propagation delay · Ripple carry adder · Carry select adder

1 Introduction

ALU is an important part of the CPU, and it is responsible for carrying out various arithmetic and logical operations [1]. Various arithmetic operations [2, 3] carried by an ALU such as addition, subtraction, increment, decrement, and various logical operations, viz AND, OR, XOR, and complement. Out of these, any operation can be selected based on the given opcode by the user. In this work, we have opted for 3-bit opcode and assigned opcode from [000] to [011] to arithmetic operator

A. Zade (✉) · S. K. Singha · J. Talukdar · A. Prathima · K. Mummaneni
Department of Electronics and Communication Engineering, National Institute of Technology, Silchar, India

K. Mummaneni
e-mail: kavicharan@ece.nits.ac.in

K. N. Das et al. (eds.), *Proceedings of the International Conference on Computational Intelligence and Sustainable Technologies*, Algorithms for Intelligent Systems,
https://doi.org/10.1007/978-981-16-6893-7_5

and from [100] to [111] to logical operator. For carrying out all these arithmetic operations, there is an extensive need for adders. For the addition of more than two one-bit numbers, full adders are cascaded resulting in ripple carry adder (RCA) [4, 5]. However, this RCA creates the carry propagation delay (CPD) problem, which means that the next stage has to wait for the carry of the previous stage until then the next stage has to wait for sum and carry outputs. Therefore, CPD problem reduces the speed of the system, which is undesired by the user. So to eliminate the CPD problem, carry select adder (CSA) is used, by using the RCA-RCA configuration in which two RCA blocks of full adder one with $C_{in} = 0$ and the other with $C_{in} = 1$ at LSB is used and correct sum output is selected by 2 × 1 Multiplexer [6]. It reduces the CPD, but due to the use of two RCA blocks where a single RCA block uses 'n' full adders, increases the area.

To overcome the area problem, we have proposed a new idea where a half adder has been used in place of the RCA block with $C_{in} = 0$. These half adders are used in place of full adders in RCA block with $C_{in} = 0$, and it has been connected in a parallel way with full adder, so that the input to each adder can be provided simultaneously and to select correct output sum and carry bit from half adder and full adder 2 × 1 mux will be used which selects the correct carry and sum based on the previously selected carry bit.

The rest of the paper work is directed as below: Part-II explains the RCA-RCA configuration for 4-bit and 8-bit adder with carry select adder. Part-III explains the working of the proposed adder configuration for 4-bit and 8-bit and determines the number of gates being used by it. Part-IV shows the simulation results obtained through Xilinx software. Part-V concludes the paper.

2 Review of RCA-RCA with CSA

To overcome the problem of carry propagation delay (CPD), carry select adder (CSA) has been used. In RCA, the next stage has to wait till the arrival of carry from the previous stage which is the major cause of delay. As carry bit can be either 0 or 1, so in this work, the sum and carry output for both the carry, i.e., 0 and 1, has been calculated. For this, two ripple carry adder (RCA) blocks are used [7, 8]. For '*n*' bit input, '*n*' full adders are needed which are cascaded. So for RCA-RCA CSA configuration, one RCA block with pre-assumed carry $C_{in} = 0$ and the other RCA block with pre-assumed carry $C_{in} = 1$ are provided at the least significant bit (LSB) of the blocks as shown in Fig. 1. By this pre-assumption of carry, the output sum for both carries (0 and 1) is computed, and to the get correct sum, the output of both the adder is given as the input to 2 × 1 multiplexer, and the correct sum is decided by 2 × 1 multiplexer based on the previous stage carry which is given as the select line. The carry output of each adder is passed to the adder of the next stage as shown in Fig. 1. In Fig. 2 for the first two LSB bits, full adders are connected in a cascaded way, and the carry of the second stage is passed to the third stage through AND gate and fourth stage carry is selected by 2 × 1 multiplexer based on the previous stage carry and

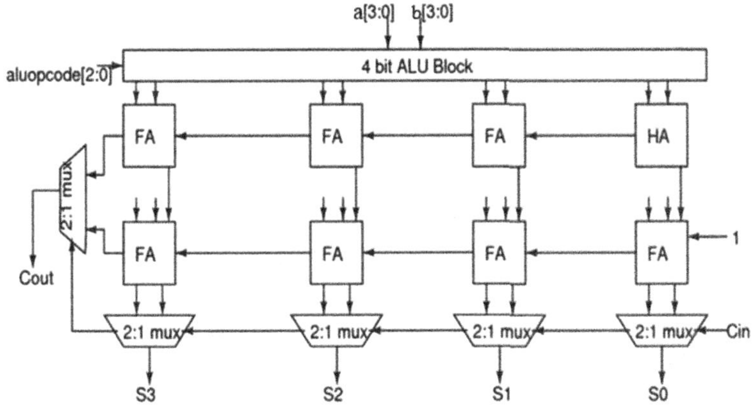

Fig. 1. 4-bit RCA-RCA with CSA configuration

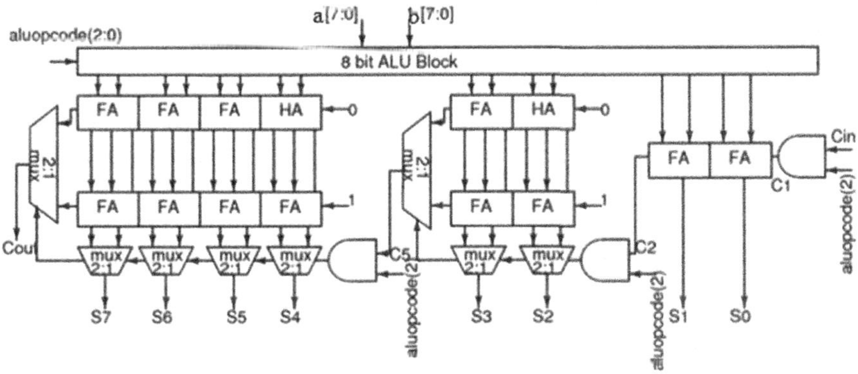

Fig. 2. 8-bit RCA-RCA with CSA configuration

then again passed to fifth stage. Finally, by pre-assuming carry, carry propagation delay is reduced, but the area utilized by the configuration increased significantly [9].

3 Design of Various Adders

3.1 Design and Working Principle of Proposed 4-bit Adder with ALU

This section presents the design and working operation of 4-bit adder with arithmetic and logic unit. To overcome the problem of area constraints, the block of RCA

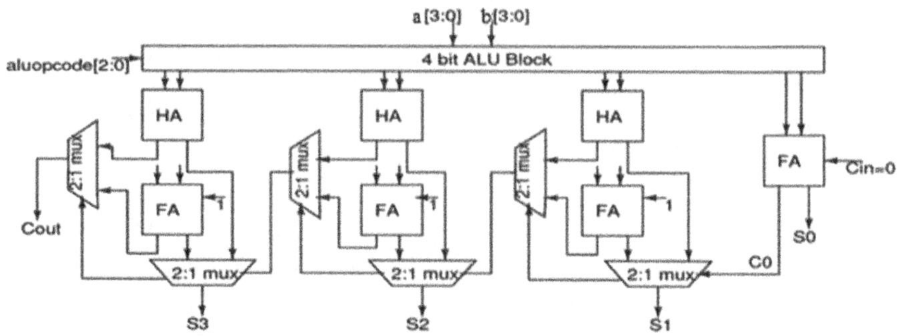

Fig. 3 Proposed 4-bit adder along with ALU

with $C_{in} = 0$ at its LSB is replaced by half adders. However, instead of cascaded connection, both half adder and full adder are connected in parallel form, so that simultaneously inputs can be given to each adder as shown in Fig. 3. The carry $C_{in} = 1$ is provided to each full adder except at LSB position. Now, half adders calculate the sum of input bits same as done by the RCA block with $C_{in} = 0$, and full adders with carry = 1 calculate the sum of input bits same as done by the RCA block with $C_{in} = 1$ in the above-discussed RCA-RCA configuration. The sum output and carries output of half adder and full adder are given as input to 2 × 1 Multiplexer as shown in Fig. 3, whose select line is the output of carry multiplexer of the previous stage. So according to the previous stage carry, the correct sum and carry of the present stage is selected, and now the present stage selected carry will act as a select line for the next stage multiplexer. In this way, the process repeats in the proposed configuration.

3.1.1 Area Comparison Between RCA-RCA with CSA and the Proposed Adder Without ALU Block

This part of the work carries a comparison of the proposed adder with RCA-RCA with CSA configuration in terms of area efficiency. But before that, we will tabulate the number of basic gates used by various blocks [10] as given in Table1.

Figure 1 shows that the stage 1 (with S0 as output) of RCA-RCA with CSA configuration has one full adder, one half adder, and one 2 × 1 multiplexer; hence, stage1 utilizes a total of 23 basic gates. For stage 2 (with S1 as output), it has two full adders and one 2 × 1 multiplexer; therefore, 30 basic gates are utilized, similarly for stage 3 (with S2 as output) and stage 4 (with S3 as output). Hence, it can be observed

Table 1 Number of basic gates with blocks

Blocks	Number of gates
2 × 1 Mux	4
Half adder	6
Full adder	13

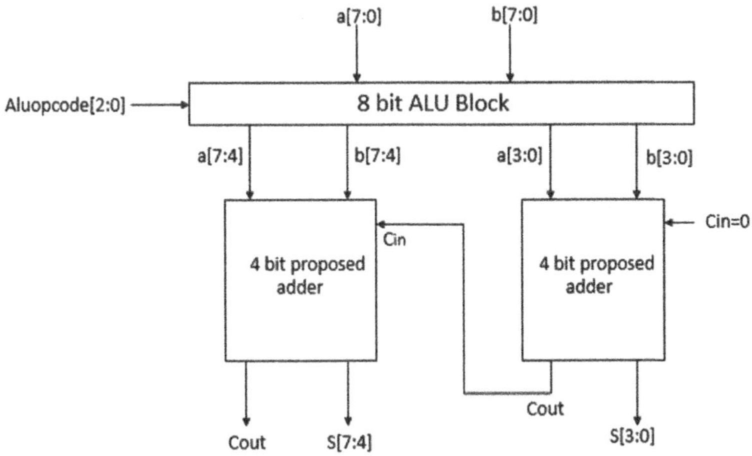

Fig. 4 Proposed 8-bit ALU using two 4-bit proposed adders

that complete 4-bit adder (without ALU), utilizes 113 basic gates. However, in case of the proposed adder as shown in Fig. 3 for stage 1(with S0 as output), it uses 1 full adder, so stage 1 utilizes 13 basic gates. Stage 2 (with S1 as output) uses one half adder, one full adder, and two multiplexers, so it utilizes 27 basic gates and similarly for stage 3 (with S2 as output) and stage 4 (with S3 as output) also utilizes 27 basic gates, respectively. Hence, it can be observed that the proposed 4-bit adder utilizes total 94 basic gates instead of 113 basic gates in existing design. Therefore, it is evident that by using the proposed 4-bit adder configuration, the number of basic gates can be reduced up to 19 as compared to RCA-RCA with CSA configuration. Hence, it can be portrayed that the proposed adder will consume less space than the RCA-RCA with CSA configuration.

3.2 *Design and Working Principle of Proposed 8-bit Adder with ALU*

The 8-bit adder is implemented by cascading two proposed 4-bit adders as shown in Fig. 4. The C_{out} of the first 4-bit adder block will become C_{in} of the second 4-bit adder block. The first adder block provides the lower nibble of the 8-bit output sum, while the second nibble will give the upper nibble of the output sum.

3.2.1 Area Calculation for 8-bit Adder with ALU Block

As evident from Fig. 2, 8-bit RCA-RCA with CSA configuration uses 2 half adders, 12 full adders, 8 multiplexers, and 3 AND gates. The existing 8-bit adder requires

203 basic gates, while the proposed 8-bit adder uses 188 (94 × 2) basic gates. Hence, the proposed 8-bit adder requires 15 basic gates lesser and is more area efficient as compared to the 8-bit RCA-RCA with CSA configuration.

4 Simulation Results

The 4-bit and 8-bit adder with ALU is designed by using Verilog HDL, and simulation is done by using ISIM simulator 14.7 by selecting Spartan 3E as family XC3S100E as the device. The area occupancy is estimated by Xilinx software and measured in terms of four inputs LUT.

Figures 5 and 6 show the simulation results of ALU with 4-bit RCA-RCA carry select adder and an ALU with 4-bit proposed adder, respectively. In both cases, the simulation results are same (a and b are 4-bit input which is fed to the ALU). In Fig. 5, carryha and sumha are the outputs of the half adder which is placed at LSB of RCA block 1. The carry0 [3:1] and sum0 [3:1] are the carry and sum of the full adders placed in RCA block 1(with $C_{\text{in}} = 0$), while carry1 [3:0] and sum1 [3:0] are the output carry and output sum of the full adders placed in RCA block 2 (with $C_{\text{in}} = 1$) as depicted in Fig. 1. The sum s[3:0] is provided by the multiplexer after selecting

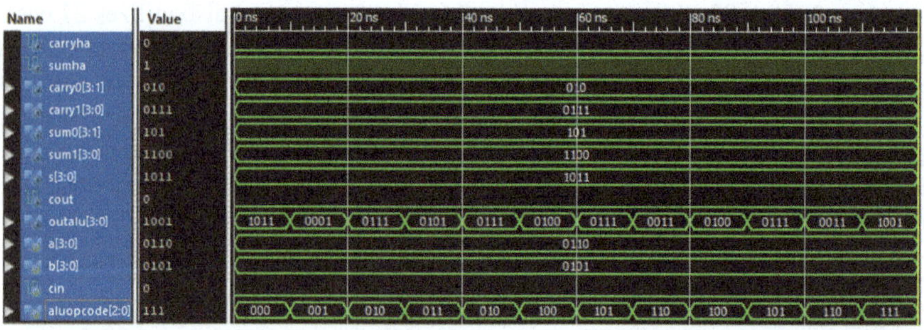

Fig. 5 Simulation results of 4-bit RCA-RCA with CSA ALU configuration

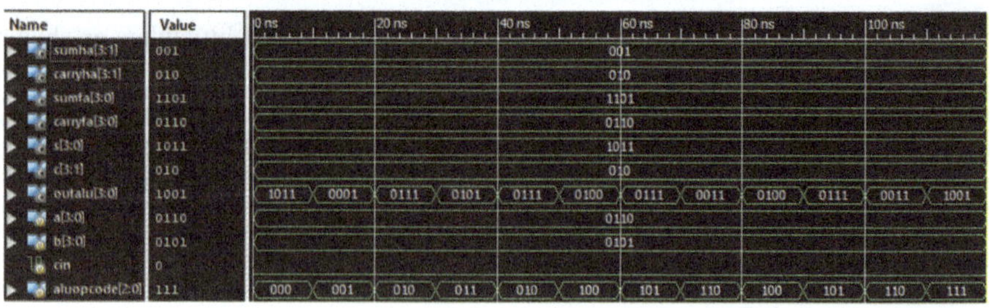

Fig. 6 Simulation results of proposed 4-bit Adder with ALU

the correct sum bit, and outalu [3:0] is the output of ALU for different opcode ranging from 000 to 111.

In Fig. 6, sumha [3:1] and carryha [3:1] are the sum and carry of the half adders. Sumfa [3:0] and carryfa [3:0] are the output sum and output carry of the full adder with $C_{in} = 1$. The output sum s[3:0] is obtained by the multiplexers, accepting output sum from half adders and full adders as input. The carry c [3:1] is provided by multiplexers which takes the output carry from half adder and full adder as input as depicted in Fig. 3.

As evident from the simulation results, the output of ALU outalu [3:0] and the output of 4-bit adder: s[3:0] for 4-bit ALU with RCA-RCA with CSA configuration (in Fig. 5) are exactly as proposed work of 4-bit adder with ALU (in Fig. 6). The proposed configuration results perfectly match with existing configuration, but the existing design consumes more number of gates. Hence, the proposed adder is more area efficient than the existing.

Fig. 7 presents the simulation results of 8-bit RCA-RCA CSA configuration with ALU, where sumha [1:0] and carryha [1:0] are the sum and carry outputs of half adders which is placed at third and fifth stage of RCA block with $C_{in} = 0$, sumha [11:0] and carryha [11:0] are the sum and carry of full adders which are placed at different stages. The sum output s[7:0] is fed by the eight multiplexers and y[2:0] is the output of the AND gate which is placed to ensure that, the carry propagation is 0 for logical operations.

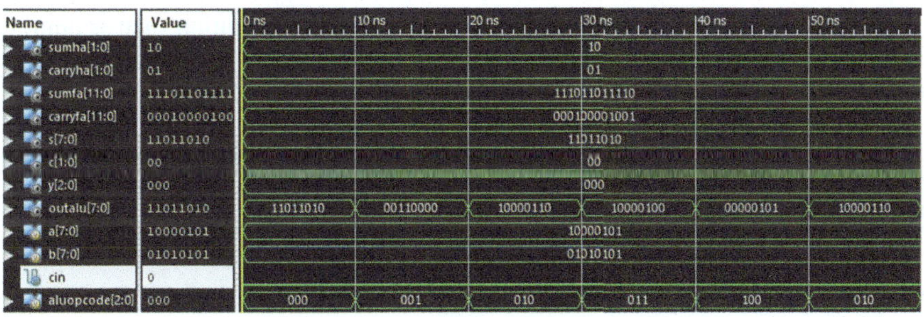

Fig. 7 Simulation results of 8-bit RCA-RCA CSA configuration with ALU

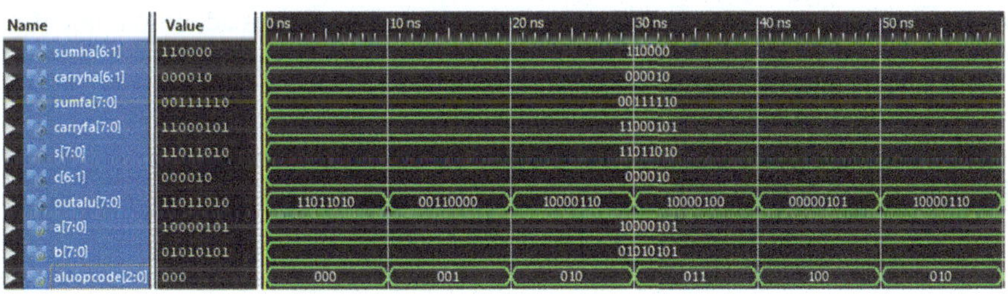

Fig. 8 Simulation results of proposed work for 8-bit adder with ALU

Table 2 Comparison of proposed adders and existing adders in terms of number of slices, number of four input LUTs and delay

Title of adder	Number of slices	Number of 4 inputs LUTs	Delay (ns)
Proposed work 4-bit adder with ALU	25	45	11.779
4-bit RCA-RCA CSA with ALU [6]	27	49	11.977
Proposed work 8-bit adder with ALU	54	98	12.832
8-bit RCA-RCA CSA with ALU [6]	56	101	12.419

Fig. 8 shows the simulation results of the proposed 8-bit Adder with ALU, where sumha [6:1] and carryha [6:1] are the sum and carry of the half adders while sumfa [7:0] and carryfa [7:0] are the sum and carry signal of full adders. The multiplexer input is the sum output of full adder and half adder. The carry output c[6:1] is the output of the multiplexer whose input is the carry signal from half adder and full adder as shown in Fig. 4.

As evident from Figs. 7 and 8, the ALU output (aluout[7:0]) and the sum output (s[7:0]) for 8-bit RCA-RCA CSA adder with ALU are exactly same as proposed work of 8-bit adder with ALU. In addition, as compared to the existing configuration, the proposed work stands better as it uses fewer numbers of gates. From Table 1, it can be observed that the performance metrics of proposed configuration such as number of slices and area is better as compared to the existing RCA-RCA CSA configuration in case of both 4-bit and 8-bit configuration (Table 2).

5 Conclusion

The simulation of the proposed work are compared with existing RCA-RCA with carry select adder (CSA) configuration and found that the proposed 4-bit and 8-bit adder occupies less area without much change in delay. An ALU based area-efficient adder is implemented in lesser space than existing works. Further, as a future work, we would like to extend it for 16-, 32-, and 64-bit area-efficient adder to implement ALU in lesser space.

References

1. Babaie, S., Sadoghifar, A., Bahar, A.N.: Design of an efficient multilayer arithmetic logic unit in quantum-dot cellular automata (QCA). IEEE Trans. Circuits Syst. II Express Briefs **66**(6), 963–967 (2019). https://doi.org/10.1109/TCSII.2018.2873797

2. Kirichenko, A.F., et al.: ERSFQ 8-bit parallel arithmetic logic unit. IEEE Trans. Appl. Supercond. **29**(5), 1–7 (2019). https://doi.org/10.1109/TASC.2019.2904484
3. Barla, P., Joshi, V.K., Bhat, S.: A novel low power and reduced transistor count magnetic arithmetic logic unit using hybrid STT-MTJ/CMOS circuit. IEEE Access **8**, 6876–6889 (2020). https://doi.org/10.1109/ACCESS.2019.2963727
4. Stephen, B., Zvonko, V.: Fundamental of digital logic with VLSI design, 3rd edition, McGraw Hills companies. Inc (2009)
5. Balali, M., Rezai, A.: Design of low-complexity and high-speed coplanar four-bit ripple carry adder in QCA technology. Int. J. Theor. Phys. **57**, 1948–1960 (2018). https://doi.org/doi.org/10.1007/s10773-018-3720-8
6. Nautiyal, P., Madduri, P., Negi, S.: Implementation of an ALU using modified carry select adder for low power and area-efficient applications. In: 2015 International Conference on Computer and Computational Sciences (ICCCS), pp. 22–25 (2015). https://doi.org/10.1109/ICCACS.2015.7361316
7. Bedrji, O.J.: Carry-Select adder. IRE Transactions on Electronic Computers. EC-11(1962) 340–346
8. Kandula, B.S., Kalluru, P.V., Inty, S.P.: Design of area efficient VLSI architecture for carry select adder using logic optimization technique. Comput. Intell. 1– 11. doi.org/https://doi.org/10.1111/coin.12347
9. Gopinath, B., Sangeetha, S., Nancy, J., Umarani, T.: Design and implementation of high speed carry select adder. IJERT **4**, 419–422 (2015)
10. Mano, M., Morris, Ciletti, M.D.: Digital Design with an introduction of Verilog HDL. 5th edition, Pearson edu. Inc, (2011).

Optimized Design of ALU Using Reversible Gates

Pragya Pandey, Kajal Kumari, Malvika, Addanki Prathima, and Kavicharan Mummaneni

Abstract This paper presents a new approach to enhance the speed of the Arithmetic logical unit using different basic reversible gates through instruction management. Generally, Arithmetic logical unit is divided into three fundamental components, which include the Arithmetic, Logical, and Control unit. The proposed architecture of Arithmetic and logical unit yields efficient results compared with traditional design in terms of various performance parameters such as Ancilla input, Garbage output, and Quantum cost. The Ancilla input, Garbage output, and Quantum cost of the proposed design are much better as compared to the existing design in the order of 64, 37 and 35%, respectively.

Keywords Arithmetic logical unit · Reversible gate · Garbage output · Quantum cost · Ancilla input

1 Introduction

Technological advancement has been increased manifolds in the last few decades. The number of transistors fabricated on-chip is continuously increasing according to Moore's principle [1]. As billions of transistors are fabricated on single chip, the overall delay and power dissipation is continuously increasing. There is a plethora of techniques dealing with the same and aims to reduce both overall delay and

P. Pandey (✉) · K. Kumari · Malvika · A. Prathima · K. Mummaneni
Department of Electronics and Communication Engineering, National Institute of Technology Silchar, Silchar, Assam, India
e-mail: pragya_pg@ece.nits.ac.in

K. Kumari
e-mail: kajal_pg@ece.nits.ac.in

A. Prathima
e-mail: addanki_rs@ece.nits.ac.in

K. Mummaneni
e-mail: kavicharan@ece.nits.ac.in

K. N. Das et al. (eds.), *Proceedings of the International Conference on Computational Intelligence and Sustainable Technologies*, Algorithms for Intelligent Systems,
https://doi.org/10.1007/978-981-16-6893-7_6

power dissipation. Reversible gates are the remedy for the problems mentioned above. Nowadays, nearly all digital circuits are achieved by applying Reversible logic approach due to minimized power dissipation [2], approximately energy-free computation and capability of higher circuit densities. Reversible means input can be mainly determined from the output of the circuit. Every individual output has unique input. Classical computing is irreversible, which is done by conventional logic gates [3] like AND, Ex-OR, OR, etc. In 1961, a theorem was given by Landauer [4] related to irreversibility operation. It states that, if technological factors and properties of the material used in the manufacturing of computers are not considered, there will be a lot of energy utilization in the computer, which is primarily due to logical irreversibility operation.

In traditional gates for each bit of information loss a fixed amount of heat generated, which is expressed in terms of $K \times T \times \ln 2$ joules. Here, K and T is referred as Boltzmann constant and absolute temperature, respectively. Further, Bennett [5] suggested a theorem, which states that usage of reversible gates for computational purposes can lead to no information loss, reduced power dissipation, and hence zero heat dissipation. In literature, there are various reversible logic gates [6–12] available for performance enhancement of digital circuits.

In this paper, we have used CNOT [8], PERES [10, 11], and COG reversible gate [12] for the development of optimized ALU because of having the advantage of their less quantum cost. A truly low power circuit can be designed using reversible gates, which enhances the performance of processors. The Arithmetic logic unit is an essential component of processor or controller which performs arithmetical operations like addition, subtraction, etc. and logical operations such as AND, OR, NOT, etc. hence, there is a great demand of developing high speed ALU.

2 Review of main attributes of Reversible logic gate

The Reversible gate is $n \times n$ input/output circuit and is uniquely mapped. In other words, a Reversible gate acquires the similar count of inputs and outputs and one to one mapping among input and output [6, 7]. It is a memory less logic gate with the further advantage of lossless information transfer.

2.1 Ancilla Inputs

The Reversible gate's inputs are locked to one of two logic '0' or logic '1' to obtain the required Boolean equation. It is a type of constant input called Ancilla inputs [13].

2.2 Garbage Output

Outputs that are needed to retain a gate's reversibility and are of no logical use are called garbage output. It is an undesired output [14].

2.3 Quantum Cost

It is mainly the sum of 1×1 or 2×2 primitive gates used in the circuit. It is determined by counting the sum of primitive gates namely NOT gate, V quantum gate, CNOT, and V + quantum gate [15].

3 Proposed Design

An ALU is a circuit which is able to perform multiple functions depending on selection input. The selection inputs help ALU to distinguish between arithmetic and logic operations. In this paper, Arithmetic Unit and Logical Unit of ALU are designed separately.

3.1 Arithmetic Unit

Arithmetic unit (AU) is used to perform arithmetic operations such as addition, subtraction, multiplication. In the proposed design, arithmetic unit is implemented with reversible logic gates, which causes negligible information loss. A, B are inputs, X, Y, and Cin are selection bits and S, Cout are the outputs. The final arithmetic results are denoted by S and Cout. The expressions for S and Cout are given below.

$$C_{\text{out}} = ((\overline{A}\&X) \wedge (A\&Y))\,\&((B \wedge C_{\text{in}}) \wedge (B\&C_{\text{in}})) \tag{1}$$

$$S = ((A\&X) \wedge (A\&Y)) \wedge B \wedge C_{\text{in}} \tag{2}$$

The symbol ∧ represents Ex-or logical operation and symbol and represents AND logical operation. Figure 1 illustrates the proposed architecture of AU which contains two types of reversible gates, namely, PERES and COG gate. In this design, the count of garbage outputs is 4, Ancilla input is 1, and its quantum cost is 12. In the indicated diagram, the quantum cost is reduced compared to the existing design quantum cost of 13 [16].

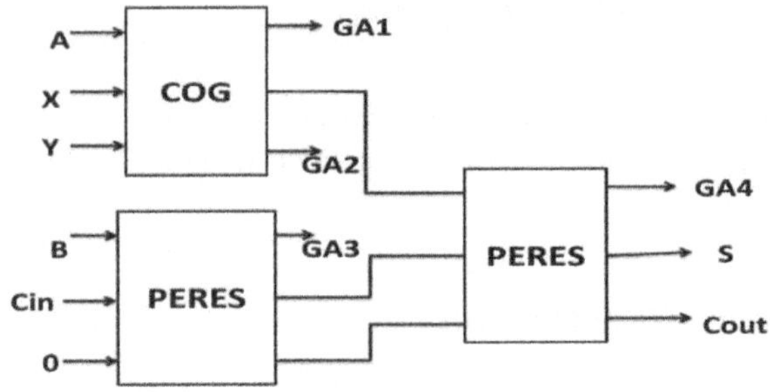

Fig. 1 Proposed design of arithmetic unit

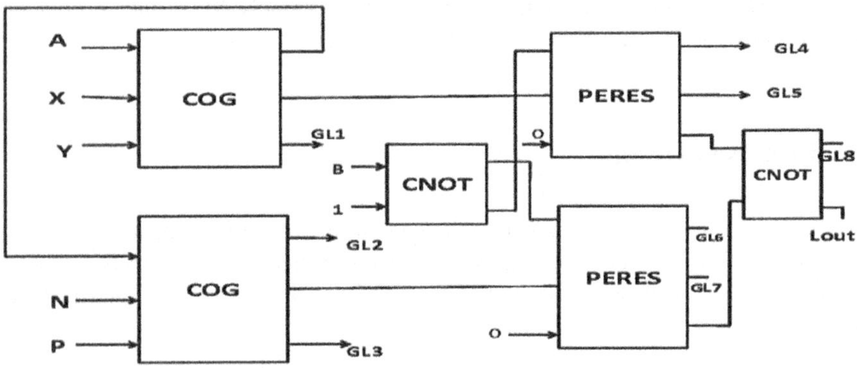

Fig. 2 Proposed design of logical unit (LU)

3.2 Logical Unit

Logical unit (LU) performs operations such as AND, OR, Ex-OR. Figure 2 denotes proposed design of logical unit where *A* and *B* are inputs, *P*, *N*, *Y*, *X* are selection lines and *L*out is the logical unit output. Equation for final output is given by *L*out.

$$L_{\text{out}} = (\overline{A}\&\overline{B}\&X) \wedge (A\&\overline{B}\&Y) \wedge (\overline{A}\&B\&N) \wedge (A\&B\&P) \quad (3)$$

The proposed logical unit is realized using various reversible gates such as COG, PERES, and CNOT. The design has 8 garbage outputs, 3 Ancilla inputs, and 18 quantum cost. It is observed that the quantum cost is reduced as compared to the existing design quantum cost of 38 [16].

3.3 Control Unit

Figure 3 shows the control unit (CU) of ALU. The SEL pin decides whether Arithmetic or Logical function is to be executed. The control unit includes one COG gate

Fig. 3 Block diagram of control unit

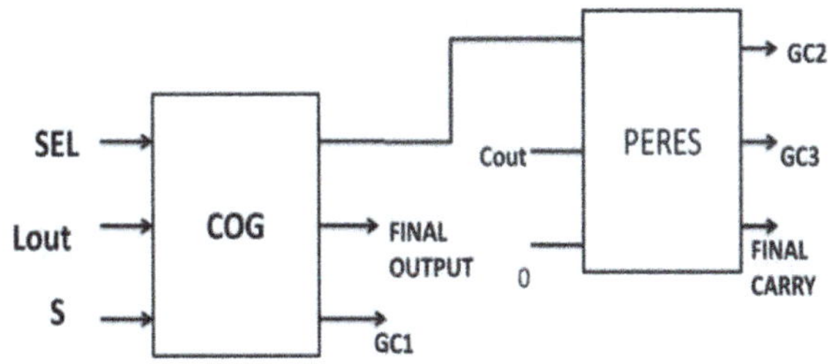

and one PERES gate. The COG gate consists of three inputs and three outputs and the output of COG gate is given as input to PERES gate. The final output of PERES is GC2, GC3, and final carry.

3.4 ALU Design

The combination of CU and the proposed designs of AU and LU, leads to final design of ALU which is shown in Fig. 4. The output of ALU relies on the present state of inputs, not on the previous state. The described design has 12 inputs and 2 outputs. In the diagram, GAn, GLn, and GCn are the nth Garbage outputs of the arithmetic, logic, and control unit, respectively. The number of garbage outputs are diminished to 15 as compared to the existing design [16] garbage outputs of 24. Finally, as compared to the existing design [16] and design 1 of paper [17], the proposed design quantum cost is reduced from 59 and 41 to 38, respectively which indicates the overall system performance.

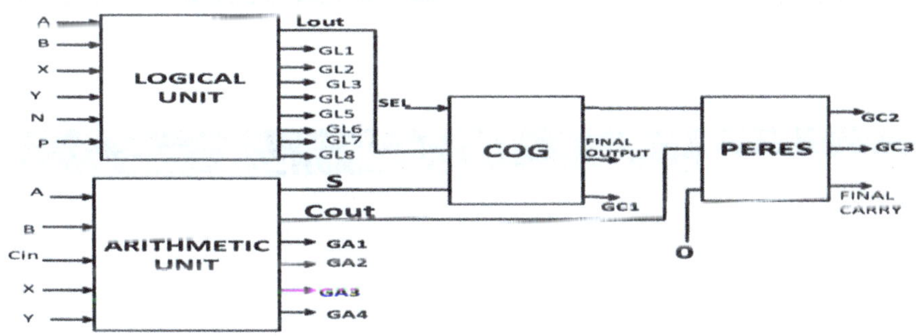

Fig. 4 Block diagram of complete ALU

4 Results and Analysis of ALU

The proposed logical and arithmetic units are implemented in Xilinx ISE 14.7 using Verilog HDL code. Figures 5 and 6 denote the simulation results of AU and LU. In AU and LU, based on selection bits X, Y, N, P, and Cin various functions are performed on inputs A and B using Eqs. (1)–(3). The quantum cost, Ancilla inputs and garbage outputs are reduced by more than half of the value in the proposed design of ALU and LU which are much lesser than the existing design. Figure 7 illustrates the simulation results of the proposed ALU. The output of arithmetic and logical operations is observed on the output bits Fcarry and Fout using SEL pin.

Table 1 summarizes the different performance matrices of the proposed designs. Tables 2, 3, and 4 present the comparisons between the proposed design and the existing design of AU, LU, and ALU in terms of various performance matrices. From Tables 2 and 3 it can be observed that the percentage improvement of various

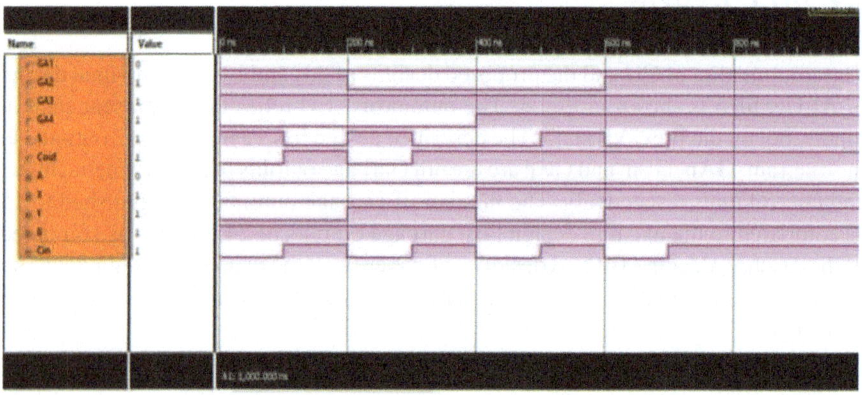

Fig. 5 Simulation results of arithmetic unit

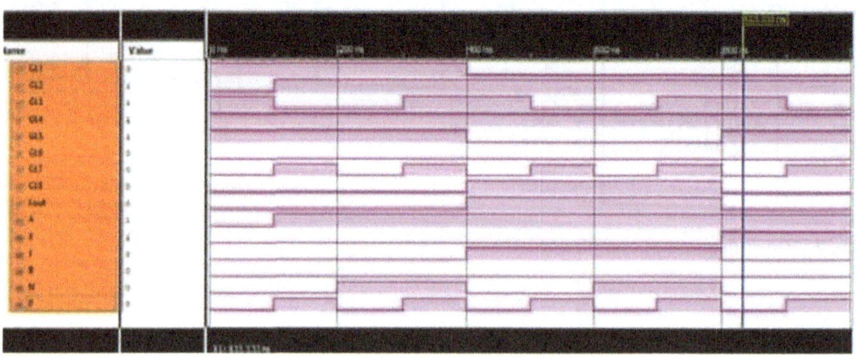

Fig. 6 Simulation results of logical unit

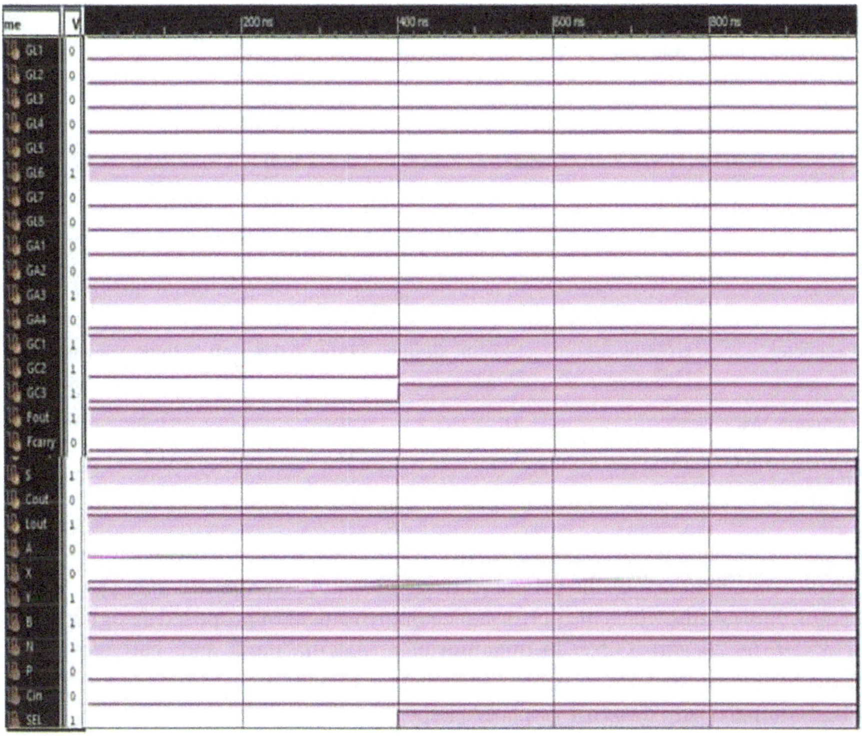

Fig. 7 Simulation results of arithmetic and logical unit

Table 1 Performance matrices of various blocks of ALU

Design	Quantum cost	Garbage output	Ancilla input
Arithmetic unit	12	4	1
Logical unit	18	8	3
Control unit	8	3	1
ALU	38	15	5

Table 2 Comparison of various parameters of arithmetic unit with existing design

Parameter	Existing design [16]	Proposed design	Improvement (%)
Quantum cost	13	12	7.69
Garbage output	5	4	20
Ancilla input	2	1	50

Table 3 Comparison of various parameters of logical unit with existing design

Parameter	Existing design [16]	Proposed design	Improvement (%)
Quantum cost	38	18	52.63
Garbage output	16	8	50
Ancilla input	11	3	72.72

Table 4 Comparison of various parameters of ALU with existing designs

Parameter	Existing design [16]	Existing design (1) [17]	Proposed design	Improvement (%)	
				[16]	[17]
Quantum cost	59	41	38	35.59	5.08
Garbage output	24	6	15	37.5	–
Ancilla input	14	1	5	64.29	–

performance matrices of the proposed design is more appropriate than the existing design. Table 4 illustrates the performance improvement of proposed design in terms of 35.59, 64.29 and 37.5% in quantum cost, Ancilla input and, garbage output, respectively, as compare to paper [16] whereas quantum cost improved by 5.08% when compare with design 1 of paper [17].

5 Conclusion

A new ALU implementation using reversible gates is presented, which outperforms the existing models and best suited for arithmetic and logical operations. The proposed design is compared with the extant structure, which is based on the Toffoli gate. The quantum cost, garbage output, and Ancilla input of ALU are 38, 15, and 5, respectively, which are much lesser than the existing ALU designs. The proposed design requires fewer reversible gates and reduces circuit complexity. The proposed ALU approach minimizes various quantum parameters like Quantum Cost, Ancilla input, and Garbage output over various prevailing designs, making it a valuable candidate over its counterparts.

References

1. Moore, G.E.: Cramming more components onto integrated circuits. Electron. Mag. **38**(8), 114 (1965)
2. Anamika, R.B.: Reversible logic gates and its performances. In: Second International Conference on Inventive Systems and Control (ICISC 2018)

3. Vasudevan, D.P., Lala, P.K., Parkerson, J.P.: Online testable reversible logic circuit design using NAND blocks. In: IEEE International Symposium on Defect and Fault Tolerance in VLSI Systems, Cannes, France, October 10–13, pp. 324–331 (2004)
4. Landauer, R.: Irreversibility and heat generation in the computing process. IBM J. Res. Dev. **3**, 183–191 (1961)
5. Bennett, C.H.: Logical reversibility of computation. IBM J. Res. Dev. **17**, 525–532 (1973)
6. Wille, R., Drechsler, R.: Towards a design flow for reversible logic, Springer (2010)
7. Thapliyal, H., Ranganathan, N.: Design of reversible sequential circuits optimizing quantum cost, delay and garbage outputs. ACM J. Emerg. Technol. Comput. Syst. **6**(4–14), 14:1-14:35 (2010)
8. Feynman, R.: Quantum mechanical computers. Optic News **11**(2), 11–20 (1985)
9. Toffoli, T.: Reversible computing. Tech memo MIT/LCS/TM151, MIT Lab for Comp. Sci. (1980)
10. Peres, A.: Reversible logic and quantum computers. Phys. Rev. A **32**(6), 3266–3276 (1985)
11. Shor, P.: Algorithms for quantum computation-discrete log and factoring. In: SFCS'94 Proceedings of 35th Annual Symposium On Foundations of Computer Science, IEEE Computer Society, pp. 124–134. Los Alamitos (1994)
12. Deeptha, A., Muthanna, D., Dhrithi, M., Pratiksha, M., Kariyappa, B.S.: Design and optimization of 8-bit ALU using reversible logic. In: IEEE International Conference on Recent Trends in Electronics, Information & Communication Technology, pp. 1632–1636 (2016)
13. Kanth, B.R., Krishna, B.M., Sridhar, M., Swaroop, V.G.S.: A distinguish between reversible and conventional logic GATES. Int. J. Eng. Res. Appl. (IJERA) **2**(2), 148–151 (2012)
14. Smoline, J., Vincenzo, D.P.: Five two-qubit gates are sufficient to implement the quantum fredkin gate. Physics Review A **53**(4), 2855–2856 (1996)
15. Thapliyal, H., Ranganathan, N.: A new reversible design of BCD Adder. In: IEEE International Conference on Design, Automation and Test, pp. 1180–1183, 14–18 March, 2011
16. Khatter, P., Pandey, N., Gupta, K.: An arithmetic and logical unit using reversible gates. In: 2018 International Conference on Computing, Power and Communication technologies (GUCON), pp. 476–480. IEEE (2018)
17. Syamala, Y., Tilak, A.: Reversible arithmetic logic unit. 101109/ICECTECH20115941987. Kanyakumari: IEEE, pp. 207–211 (2011)

On-Street Parking in Residential Areas: A Review

Ashwani Bokadia, Mokaddes Ali Ahmed, and Debasish Das

Abstract In India, parking is one of the biggest problems for all the developing or developed urban city. The developing populace of India has made numerous issues, one of the difficult ones being car parking which we face consistently. Other than the issue of space for vehicles moving onward the street, more is the issue of space for a parked vehicle thinking about that private vehicles remain parked for the vast majority of their time. Vehicle parking is a major issue in metropolitan cities and developing nations. Due to various reasons among which comfort is a major issue, people start using their own vehicle, and it leads to increase of vehicles on road. When towns were planned, such abnormal growth of vehicle is not considered, and this leads to parking problem in developing or developed urban cities. This irregularity is mainly due to ineffectual land use during planning and miscounts of area requirements during the first phases of planning. Shortage of parking area, high parking charges, and traffic congestion because of vehicle users in search for a parking area are few examples of consistent parking issues. The paper reviews new arranging patterns and imaginative innovative arrangements that can help alleviate the strain of the issue. Because vehicle parking arrangements are not an end in itself, but instead a method for achieving larger local area objectives to improve metropolitan transportation and make urban communities more reasonable and effective. This paper also analyzes parking literature with a focus on economic problems. Parking is not only one of the primary interim resources in the economy; it is also immense land use. Many theoretical and scientific articles examine the number and price of parking, focusing on some aspects of the problem. The issues discussed here are on-street parking, spatial competitiveness, (minimum and maximum) parking criteria, intelligent car parking, regression model, car park pricing and path pricing, and time–space pricing. Different parking forms are investigated before identification of subjects which were under-studied should be on the study agenda, including residential parking, shared parking, shopping car parks, and workplace sponsored parking.

A. Bokadia (✉) · M. A. Ahmed
Department of Civil Engineering, NIT, Silchar, India

D. Das
Department of Civil Engineering, JIS College of Engineering, Kalyani, India

K. N. Das et al. (eds.), *Proceedings of the International Conference on Computational Intelligence and Sustainable Technologies*, Algorithms for Intelligent Systems,
https://doi.org/10.1007/978-981-16-6893-7_7

Keywords On-street parking · Economy · Night parking · SPSS · Residential area · Parking modeling

1 Introduction

If the social economy [1] grows and people's quality of life rises, residents in the suburban area essentially meet the living standard of one car or even two vehicles per household [2, 3]. However, the initial car park services are unable to accommodate current parking requirements any longer, which means there are convenient parking loads and inadequate parking capacity [4] in both the daytime and nighttime car parking areas [5]. And demand for parking [6, 7] is fewer at night on the commercial square, on the office and at other locations close to residential areas, as a vehicle requires some a parking spot for them at the origin and destination of a journey. In recent decades, the need for parking [8] in urban areas has risen as traffic in cars has risen rapidly. While in most metropolitans, the number of dedicated parking lots has risen, and car parking areas also struggle regularly to meet peak hours [9]. As a result, more car parks (and more in the future) will park their cars on street even though they are illegal [10]. In developed nations, these parking-related issues are more troubling and could become terrifying if the parking policies do not pass [10, 11]. As car spends most of their time as parked, so parking is still an under-investigated area of transport [12] every day on an average in three different parking places [13]. In the lack of enough requirements and test findings, it is impossible for decision makers to explore alternate parking solutions. Pricing policies are a series of and management policies intended to maximize available capital usage. Parking charges [14] may be attributable to complete policy tool for directing decreasing externalities, and the parking market generated the general traffic and in particular, parking pricing procedures are given—a supply control mechanism, directly available and relevant, which can be used to adjust side and levels here, to maximize the pricing [15].

Use of available parking space is in great demand locations. In the case of a lower-priced parking lot [16], increased noise, mobility and environmental effects [17] as well as effect on the primary modal option also have negative impacts. The implementation of parking policies is faster, cheaper, and politically appropriate, thus popular, in contrast to congestion pricing [15].

2 Literature Review

John G., George Y., Michel H., 2002 "Off-street parking Choice Sensitivity". Transportation Planning and Technology [18] in this study, determinant of decision among on-and-off-street parking. Right now, questionnaire-based review was directed, and the expressed inclination technique was utilized to build up an explanatory model.

The model evaluation indicated that parking cost has, true to form, the most significant effect on the decision of parking options. Louise Eriksson, Annika M. Nordlund, Jörgen Garvill 2010 expected car use reduction in response to structural travel demand management measures paper [19] in this investigation of Swedish drivers, push (higher fuel tax) and pull measures (better travel service) are insufficient all alone (Woo Lee, 2014. "Establishment of On-Street Parking Demand Model According to Total Building Floor Area") [20]. By comparing and analyzing the on-site examination of present on-street parking conditions and total building floor areas by use, this study produced an on-street parking demand estimation model. The developed model will be utilized in forecasting on-street parking demand and suggesting alternative measures for the estimation and implementation of parking policies, district unit planning, and others. Helen Micheaux Shaheen et al. 2015 "The Evolution of One-Way Carsharing and Operator Perspectives from 2015," Americas Paper [21], in this study, find out the car sharing in the USA saves users between 154 and 435 euros a month after they become members. (Debases Das, Dr. M Ali Ahmed, Saptarshi Sen [10], 2016) Parking on the street demands estimation for four wheeler in urban CBD model: [22] Kolkata as a case study by this study, two CBDs (Gariahat and Dalhousie) area of Kolkata is selected for study. This paper shows parking demand can be determined by different factors like average number of four-wheeler owned average duration parking (in hour) and probability choosing automobile a mode. This paper is very helpful to separately from land use and also studies about the parking demands analysis and primarily focus on the CBD area also the different lad utilize considered parking generation rate from those distinctive land use are considered, Elizabeth Jean Taylor. Who's been parking on my street in 2018? Legislative difficulties and a disproportionate use of residential parking spaces [23]. This examination takes a gander at residential parking usage in Melbourne, Australia, and was investigated using two sources: a current travel investigation of 8994 nuclear families and a custom online survey of 307 people. Notwithstanding the way that they have limits, the two exam inations give new information into where and by whom the utilization of on-road parking spaces around the house is more normal and additionally discovers the larger part share. Separated lodging occupants account for 77–83% of on-street residential parking. The majority of on-street parking customers have enough off-street parking, and half of them utilize carports for temporary storage or accommodation. (Gonzalo Antolín,. Matthias N. Sweet, Imprint R. Ferguson, 2019) Parking demands the executives in a generally uncongested college setting [24] in this paper. This examination portrays and gauges how much available TDM gadget can truly incite auto travel diminishes by investigating information from a 2015 online travel audit of workforce, understudy, and staff from McMaster School in Hamilton, Ontario (Canada). Additionally, this examination assembles essential study data and usage on discrete choice models to explore improving future parking demand circumstances. Results from circumstances made using mode decision models recommend that improving non-auto travel administration can be significant. C.J. Gabbea, Gregory Pierceb, Gordon Clowersc, 2020 "parking policy" The impacts of residential least parking necessities in Seattle [25]. This examination analyzes Seattle, Washington (USA), and the site of 2012 plan changes that diminished or discarded parking prerequisites for the most part focal and travel arranged areas. Dataset involves 868 enhancements with 60,361 lodging units acknowledged somewhere in the range of 2012 and 2017. They use particular measurements, develop a counterfactual circumstance, and demonstrate indicated least squares (OLS) regression models to recognize associations between parking necessities and veritable parking plan.

3 Parking-Related Study

3.1 Characteristics of Parking

It is critical to have data on the availability of parking spaces, the quantity that is being used, the amount of time spent in parking, and the assessment of parking demand during the critical period of study in order to take any useful changes to improve parking circumstances. Different outlines are used to determine various parking attributes, which are referred to as parking qualities or metrics. After that, parking attributes are employed [26]. The total number of automobiles left at a given time interval is known as parking accumulation [27]. The accumulation bend/profile reference diagram is usually used to look after it. It depicts the variation in parking collection for a specific parking office over a set time period or overview period and made a methodology for the parking collecting profile for the entire day, utilizing an overview at parking regions and group examination, which takes less time than a full-day insight review. The total number of vehicles left throughout a certain time period or overview period is referred to as parking volume [28]. It is important that it does not take into account a similar vehicle's reiteration. In this line, it simply estimated the number of cars entered throughout the research period. The maximum number of parking spaces/bays available at a parking lot is known as the limit [29]. The total number of left automobiles at a predetermined time is the occupancy factor or parking index for a certain parking facility. It can also be obtained by dividing the parking load by the limit over a certain time period. It is a measure of a parking lot's effectiveness in terms of how well it is being used. Under the accumulation curve, the parking load tends to fill the entire zone. It is obtained by increasing the number of vehicles occupying parking space for a given time interval [30].

3.2 Residential Parking

Residential normally has given curb side to address overflow parking around shopping centers, places of business, stadiums, and so forth, to help inhabitant's residents living close by. In many nations, residents need to pay for parking rights (or parking permits) in residential areas [31]. Residential parking also includes on-site garage parking available in residential homes, the availability of which is usually governed by parking zoning restrictions. Residential parking may be an l, so incorporate drive way parking. The solitary existing hypothetical to date the work on residential parking has been done by [32]. They look at the trade-off between the ease of living near the city's facilities and the difficulty of finding a parking spot in the city's core. From both the resident's and the social planner's perspectives, they decide when it is desirable to all to find resident curbside parking's point of view. Utilizing a household study in New York City Guo (2013a, 2013b) [33] discovers that residential parking availability has a significant impact on automobile ownership. Typically, the cost of residential parking is included in the cost of lodging. Van Omme renting takes advantage of property pricing in Amsterdam and a year ago [34] evaluates resident's day-by-day ability to pay for parking permits to be about 10 euros. They argue that although this sum is larger than what residents pay for parking permits, it is lower than what non-residents pay for parking there. Along these lines, permit of parking is a wasteful utilization of control side parking space. In a similar vein, van Ommerenetal (2014) [35] fixation is the connection between residential parking permits and parking supply. Parking permits are estimated to cause a welfare loss of roughly 275 euros per permit [36].

3.3 Parking Demand Models

Past investigations have shown that parking price is by all accounts perhaps the most affecting factors for demand of parking in the case of a certain parking lot and modular decision. Lord and Hensher (2001) [35] evaluated the aspect of parking price in age of interest of parking at explicit parking area. They utilized an expressed inclination information of private vehicle clients and public travel clients to get tests of individual conduct shifting with parking cost and long of activity at different areas around there. McGuiness and McNeil (1991) [36] additionally viewed as the arrangement of cost of parking in creating measurable parking spot occupancy model [37] and talked about accepting consistent interest. As the cost of parking rises, so does the elasticity curve build as well as the occupancy additionally increments. End is by all accounts wrong from the outset eye. However, when calculating the whole cost of parking, [38] expenses, availability includes features such as access time, security, and safety, could be conceivable that parking with high charges is sensible to utilize and clarified the past segments, the blended uses land improvement, and the metropolitan region requests enormous parking places which are available. Regular if there is a separate parking lot spots accommodated extraordinary however connecting land utilizes, the parking space is hardly used [39].

3.4 Policy Recommendations and Enhancements

Parking policies are expected to play a vital role in the development of a well-functioning traffic system structure for a location to improve road users' driving conditions [40]. They also help to improve the utilization of available parking resources and are beneficial from the standpoint of car maintenance [41]. All around organized parking approaches contribute to the advancement of effective transportation organization, lower outflow, higher densities, and better metropolitan arrangement in a variety of ways (Foundation of Parkways and Transportation, 2005 [42]; Shoup, 2005 [43]; Stubbs, 2002) [44]. Parking policies should be developed in conjunction with local and territorial spatial and transportation planning measures (Marsden and May, 2006) [Page 79]. As previously stated, past regions essentially worry when there is a great demand for parking expanding development pace of automobile possession wide utilization of personal automobiles. Shoup (1999, 2005) [45] considered how much parking strategy had been brought down as a result of the continued usage of private and commercial land, and the thickness of private and commercial land use has increased of never-ending suburbia which empowered the further vehicle reliance. Demand of parking [46] is profoundly defenseless in the face of parking estimates, which can be backed by various people from the aforementioned groups looked into writing on chip-based request models (e.g., Hensher and Ruler, 2001).

4 Conclusion

This paper discusses the earlier research, which highlighted both aggregate and disaggregated parking features and trends in parking preference patterns. The paper also addresses numerous approaches to estimate parking space demand. Models for parking demand provisioning have been developed using a variety of methodologies, including linear regression, low-square regression, logistic models, and key factor analysis, smart parking analysis, and gray correlation analysis. In developed countries, parking concerns are more serious, and little research in their context has until now been discussed. Consequently, urban transit designers need to properly consider the relationship between parking criteria during the planning and construction processes. A variety of factors are critical in shaping parking preference behavior and demand which can narrowly be categorized as psychological and socioeconomic features for drivers, parking characteristics and guidance systems, alternate mode features as well as parking policy impacts. Finally, the major consequences of implementing car sharing, which will offer significant economic, social, and environmental benefits, have been addressed in short. Both these findings together provide readers with valuable knowledge on the parking scheme as a whole.

References

1. Button, K.: The political economy of parking charges in '"first"' and '"second-best"' worlds". Trans. Policy **13**, 470–478 (2006)
2. Guo, Z.: Home parking convenience, household car usage, and implications to residential parking policies. Trans. Policy **29**, 97–106 (2013)
3. Guo, Z.: Does residential parking supply affect household car ownership? The case of New York City. J. Trans. Geogr. **26**, 18–28 (2013)
4. Arnott, R., Inci, E., Rowse, J.: Downtown curbside parking capacity. J. Urban Econ. **86**, 83–97 (2015)
5. Benenson, I., Martens, K., Birfir, S.: PARKAGENT: an agent-based model of parking in the city. Comput. Environ. Urban Syst. **32**, 431–439 (2008)
6. Azari, K.A., Arintono, S., Hamid, H., Rahmat, R.A.O.K.: Modelling demand under parking and cordon pricing policy. Trans. Policy **25**, 1–9 (2013)
7. Parmar, J., Das, P., Dave, S.M.: Study on demand and characteristics of parking system in urban areas: a review. J. Traffic Transp. Eng. (Engl. Ed.) **7**(1):111e124 (2020)
8. Barter, P.A.: Off-street parking policy without parking requirements: a need for market fostering and regulation. Taylor & Francis 0144-1647 print/1464-5327 online/10/050571-18.
9. Ottosson, D.B., Chen, C., Wang, T., Lin, H.: The sensitivity of on-street parking demand in response to price changes: a case study in Seattle WA. Trans. Policy **25**, 222–232 (2013)
10. Spiliopouloua, C., Antonioub, C.: Analysis of illegal parking behavior in Greece. Procedia Soc. Behav. Sci. **48**, 1622–1631 (2012)
11. Marsden, G.: The evidence base for parking policies—a review. Trans. Policy **13**, 447–457 (2006)
12. Shiftan, Y., Burd-Eden, R.: Modeling response to parking policy. Trans. Res. Rec. 1765-27 Paper No. 01–2316
13. Li, C., Huang, S.W.X., Li, X., Yu, R., Zhao, F.: Parked vehicular computing for energy-efficient internet of vehicles: a contract theoretic
14. Caliskan, M., Graupner, D., Mauve, M.: Decentralized discovery of free parking places. Los Angeles, California, USA. Copyright 2006 ACM 1-59593-540-1/06/0009
15. Davis, A.Y., Pijanowski, B.C., Robinson, K., Engel, B.: The environmental and economic costs of sprawling parking lots in the United States. Land Use Policy **27**, 255–261 (2010)
16. Simićević, J., Vukanović, S., Milosavljević, N.: The effect of parking charges and time limit to car usage and parking behavior. Trans. Policy **30**, 125–131 (2013)
17. Aoun, A., Abou-Zeid, M., Kaysi, I., Myntti, C.: Reducing parking demand and traffic congestion at the American University of Beirut. Trans. Policy **25**, 52–60 (2013)
18. Ibeas, Á., Cordera, R., Dell'olio L., Moura, J.L.: Modeling demand in restricted parking zones. Transp. Res. Part A **45**, 485–498
19. Tiexin, C., Miaomiao, T., Ze, M.: The model of parking demand forecast for the urban CCD. In: 2012 International Conference on Future Energy, Environment, and Materials, Energy Procedia, vol. 16, pp. 1393–1400 (2012)
20. Lee, Y.W.: Establishment of on street parking demand model according to total building floor area. Adv. Sci. Technol. Lett. (Archit. Civ. Eng.) **69**, 26–29 (2014)
21. Shaheen, S.A., Chan, N.D., Micheaux, H.: One-way carsharing's evolution and operator perspectives from the Americas. Springer Sci. Bus. Media New York. **42**, 519–536 (2015)
22. Das, D., Ahmed, M.A., Sen, S.: On-street parking demand estimation model: a case study of Kolkata. (2016) researchgate.net/publication/312621346
23. Taylor, E.J.: Who's been parking on my street? The politics and uneven use of residential parking space (2018)
24. Sweet, M.N., Ferguson, M.R.: Parking demand management in a relatively uncongested university setting. In: World Conference on Transport Research Society. Published by Elsevier Ltd. pp. 453–462 (2019)
25. Gabbea, C.J., Pierceb, G, Clowersc, G.: Parking policy: the effects of residential minimum parking requirements in Seattle. Land Use Policy **91** (2020)

26. Chen, Q., Wang, Y., Pan, S.: Characteristics of parking in central Shanghai. China. J. Urban Plann. Dev. **142**(3), 05015012 (2016)
27. Tong, C.O., Wong, S.C., Leung, B.S.Y.: Estimation of parking accumulation profiles from survey data. Kluwer Academic Publishers Transportation **31**, 183–202 (2004)
28. Setiawati, D.N., Intari, D.E., Wulandari, R.D.: Analysis of characteristics and parking needs in Sudimara station South Tangerang. Mater. Sci. Eng. **673**, 012021 (2019)
29. Cats, O., Zhang, C., Nissan, A.: Survey methodology for measuring parking occupancy: Impacts of an on-street parking pricing scheme in an urban center. Trans. Policy **47**, 55–63 (2016)
30. Huayan, S., Wenji, L., Haijun, H.: Empirical study of parking problem on university **7**(2) (2007)
31. Wei, Z., Xiaokunb, W., Dapeng, Z.: Truck crash severity in New York city: An investigation of the spatial and the time of day effects. Accident Anal. Preven. **99**, 249–261 (2017)
32. Litman, T.: Parking management strategies, evaluation and planning. Victoria Transport Policy Institute (2006)
33. Inci, E., van Ommeren, J., Kobus, M.: The external cruising costs of parking. J. Econ. Geogr. **17**, 1301–1323 (2017)
34. Van Ommeren, J., Wentink, D., Dekkers, J.: The real price of parking policy. J. Urban Econ. **70**, 25–31 (2011)
35. Hensher, D.A., King, J.: Parking demand and responsiveness to supply, pricing and location in the Sydney central business district. Transp. Re. Part A **35**, 177–196 (2001)
36. Guiness, E.M., Neil, S.M.: statistical models to predict commercial- and parking-space occupancy. J. Urban Plan. Dev. **117**(4), 29–139 (1991)
37. Sen, S., Ahmed, M.A.: A case study on on-street parking demand estimation for 4-wheelers in urban CBD. J. Basic Appl. Eng. Res. (2016)
38. Lima, H., Williamsa, G.T., Abdelqadera, D., Amagliania, J., Linga, Z., Priestera, D.W., Cherry, C.R.: Alternative approach for forecasting parking volumes. Transp. Re. Procedia **25**, 4175–4188 (2017)
39. Eric Laurier: A parking space. iera.ed.ac.uk (2005)
40. Jingheng, Z., Jun, C., Kai, X., Xingchen, Y., Le, C.: Parking demand forecasting for (2013)
41. Johnsen, S., Fitzpatrick, S.: The impact of enforcement ion street users in England. In: The impact of enforcement ion street users in England. Policy Press, Bristol (2007)
42. Greene, D.L.: Sustainable transport PII: SO966–6923(1997)00013
43. Manville, M., Shoup, D.: Parking, people, and cities. J. Urban Plan. Dev. **131**(4), 233–245 (2005)
44. Stubs, M.: Car parking and residential development: sustainability, design and planning policy, and public perceptions of parking provision. J. Urban Des. **7**(2), 213–237 (2002)
45. Marsden, G.: The evidence base for parking policies—a review. Transp. Policy **13**, 447–457 (2006)
46. Willson, R.W.: Estimating the travel and parking demand effects of employer- paid parking. Reg. Sci. Urban Econ. **22**, 133–145 (1992)

On-Street Parking Management in Urban CBD: A Critical Review

Prasenjit Das, Mokaddes Ali Ahmed, and Debasish Das

Abstract Now a days, parking management is one of the most important factors for transport planner because of shortage of parking space and ill-legal on-street parking. This paper incorporates all these viewpoints and presents a cutting-edge audit of models and studies on the stopping framework. Issues were identified with and because of the stopping, benefits, and negative marks of stopping different stopping attributes and their applications, improvement of interest models thinking about different factors, and audit of stopping approaches as a basic piece of the metropolitan vehicle framework are talked about in detail. While immature, creators found the writing propose that more prominent consideration have to be given to measurements like the simplicity of contact, walk time, stopping controls, stopping direction, and data framework, the board, and so on, at all phases of preparation and strategy definition. Taken together, referenced investigations show helpful data concerning the whole stopping framework.

Keywords Parking · Parking attributes · Effects of parking · Parking demand · Parking management

1 Introduction

Due to increase the population, the demand of parking is higher than supply. The parking is an under-research area [1, 2]. The economy of the people has become higher thus with the gradual increase in the population in developing countries like India, especially in metropolitan cities where living standard is higher compared to the rural areas. Travel demand is also higher in urban area and subsequently the

P. Das (✉)
NIT Silchar, Assam, India

M. A. Ahmed
Department of Civil Engneering, NIT Silchar, Assam, India

D. Das
Department of Civil Engneering, JIS College of Engineering, Kalyani, WB, India

K. N. Das et al. (eds.), *Proceedings of the International Conference on Computational Intelligence and Sustainable Technologies*, Algorithms for Intelligent Systems,
https://doi.org/10.1007/978-981-16-6893-7_8

traffic volume is high compare to the rural areas. But, the places for parking the vehicles in metropolitan cities especially in the CBD area where parking places are limited. But the overflowing of the vehicle in the CBD area and for less parking space, peoples tend to park their vehicle on on-street. And for that lots of problems comes in a real scene like traffic congestion increase; the width of carriageway decrease; air pollution and noise pollution increase; fuel consumption increases, and many more. To overcome the problem related to parking need to make good parking policies and appropriate parking management strategy which will not be a benefit for parking only, it will affect the mobility of the traffic and the socioeconomic of the city. The parking policy like car parks fee, it must be designed so that it will not affect the people need to analyze inappropriate strategy-wise. To find the chances of managing the parking lots properly, circulating the details to the commuters and different ITS technologies can be used in the city [3]. A proper parking management system can be reduced the parking problems and can be profitable for the social, economy, and environment. Need to increase the curb on the road to increase the parking space.

With the gradual increase in private and public vehicle, one of the utmost issues is confronted by the large number of traffic is the drastic scarcity of parking space. The urging for parking has gradually expand in frightening proportion specially in CBD area and the other workable center of the city. As a result, in the lack of proper parking space organization, the main road is occupied by the parked vehicle. Consumers park their vehicle on road, due to that the width of the road reduced, and thus, vehicle is forced to move through that short width road and simultaneously increasing the on-street traffic congestion and for that the accident rate increased and the convenience of the people reduced, and this issue must be require to solve by managing the parking specially on-street parking at important places of the city. As per researcher by 2040, the number of highway vehicles in India would be more than 300 million. And vehicle is gradually increasing. According to MoRTH, Table 1 shows the vehicle growth from 2003 to 2017. Though vehicle increases, the number of accidents also increases.

2 Parking

Parking is nothing but an act of stopping and remove of vehicle and leave the occupied and convert the place in unoccupied. Generally parking is two types:

(a) On-street parking
(b) Off-street parking

2.1 On-Street Parking

Parallel parking: When the consumers parked their vehicle along the length of the street, then it is called parallel parking. Hence, there is no backside movement when

Table 1 Registered vehicles in India: 2003–2017

Year	All automobile	Two wheelers	Cars, jeeps, taxis	Busses	Goods vehicles	Others
2003	67,007	47,519	8599	721	3492	6676
2004	72,718	51,922	9451	768	3749	6828
2005	81,499	58,799	10,320	892	4031	7457
2006	89,618	64,743	11,526	992	4436	7921
2007	96,707	69,129	12,649	1350	5119	8460
2008	105,353	75,336	13,950	1427	5601	9039
2009	114,951	82,402	15,313	1486	6041	9710
2010	127,746	91,598	17,109	1527	6432	1108 0
2011	141,866	101,865	19,231	1604	7064	12,102
2012	159,491	115,419	21,568	1677	7658	13,169
2013	176,044	127,830	24,056	1814	8307	14,037
2014	190,704	139,410	25,998	1887	8698	14,712
2015	210,023	154,298	28,611	1971	9344	15,799
2016	230,031	168,975	30,242	1757	10,516	18,541
2017	253,311	187,091	33,688	1864	12,256	18,411
CAGR (2007to 2017	10.11	10.47	10.29	3.28	9.12	8.09

the vehicle is parked and unparked. As a result, it is the safest stopping location in terms of mishaps.

30° parking: The automobiles are parked at a 30° angle as per the roadway arrangement in 30° type perking. When compared to equal leaving, many vehicles can be left in this case. Similarly, mobility has improved. The traffic detour generated by this type of halt is also the smallest.

45° parking: As the result of leaving builds, a more prominent quantity of automobiles can be left. Henceforth, contrasted with equal leaving and 30° leaving, a more prominent quantity of cars can be obliged in this kind of leaving.

60° parking: The automobiles are parked with their backs to the street at a 60-degree angle. In this form of departure, a larger number of cars can be required.

Right angle parking: Automobiles are left in the opposite direction of the street in right-point or 90° departure. Despite the fact that it burns-through, the needed kerb length is practically none.

2.2 *Off-Street Parking*

It includes (a) Single level parking (b) Multi-level parking (c) Roof parking (d) Automated parking garages (e) Underground garages.

3 Parking Statistics

Parking accumulation: It is defined as the number of vehicles on the road at any particular time. Typically, this is communicated by collection bend. Amassing bend is the diagram gotten by plotting the quantity of coves busy as for time. According to Tong el al. [4], stopping amassing profile is time reliant just as tedious perception overview, and group investigation is applied for assessment stopping gathering profile (Fig. 1).

The number of parking spaces available in a parking complex determines its capacity. The parking accumulation divided by the capacity equals the occupancy of a parking facility [4].

Parking volume: The total number of cars left at a given point in time is referred to as the leaving volume. This ignores the importance of automobiles. The number of cars entering the space that can be verified is noted.

Parking load: The region under the collection bend receives a stopping load. It can also be obtained by simply increasing the number of cars involved in the leaving territory at each time interval with the time stretch.

Average parking duration: It is the ratio of absolute vehicle hours to the number of vehicles still on the road.

$$\text{Parking duration} = \frac{\text{Parking load}}{\text{parking volume}}$$

Peak parking saturation: It is the ratio of the number of vehicles left during peak times to the parking space limit in terms of the number of bayous [5].

Peak parking ratio: It is the ratio of the number of vehicles remaining at peak times to the total number of vehicles on the road at any one time [5].

Parking turnover: It is the ratio of the number of vehicles left in a span to the number of cars leaving narrows open. The increased value of parking turnover means that the revenue from the parking lot will increase. The estimation of parking turnover

Fig. 1 The typical parking accumulation curve. Source: google

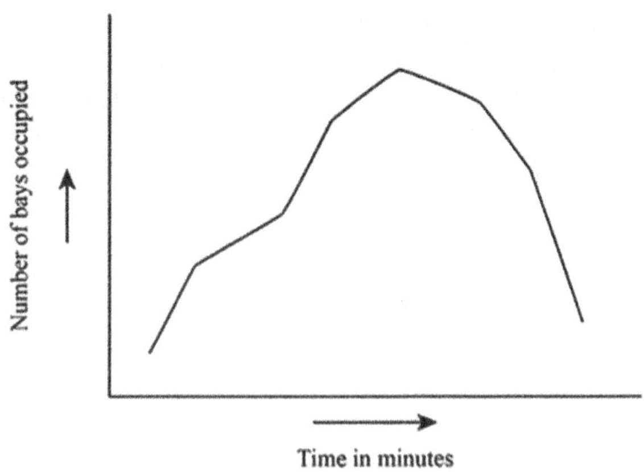

is through the license plate surveys, in–out survey, and fixed time sampling methods.

$$\text{Parking turnover} = \frac{\text{Parking volume}}{\text{no. of bays available}}$$

Parking index: Stopping file is likewise termed inhabitancy or effectiveness. It is characterized as the proportion of number of sounds involved in a period length to the all-out space accessible. It gives a total proportion of how successfully the parking spot is used. Stopping list can be discovered as follows:

$$\text{Parking Index} = \frac{\text{Parking load}}{\text{Parking capacity}} \times 100$$

Above all, the parking characteristics are useful for developing a validated parking demand model. One study has done by Parmar et al. [6] in Delhi city of India, and they have found the typical parking conditions, and study evaluated that the parking demand and supply is not equal; it is in spill over condition, and the reason behind it is nominal charge in parking.

4 Merits and Demerits of On-Street Parking

4.1 Merits

4.1.1 Impact On-Street Client's Security

Many researchers have discussed about the on-street parking that it does not create any convenient to the road users and it creates only high traffic congestion and accident. And for comfort mobility and road user's safety the on-street parking should be prohibited in urban area. But some of researcher believe that it helps in improving the attitudinal and cultural behaviors of population also [7, 8]. Some investigation says that streets with on-road stopping show an altogether decreased accident rate for the most extreme kinds of accidents [8]. Notwithstanding traffic quieting, on-road stopping goes about as a buffer between moving vehicles and people on foot strolling along the street; causes them to feel more got and loose what's more, in general, establishes a protected climate for street clients [8]. A volume of work, accordingly, has pointed that the on-road stopping holds some intrinsic commitments toward street clients' wellbeing despite the fact that stopping itself can't improve the wellbeing without any assistance

4.1.2 Effect on Monetary Improvement of Business Region

On-street parking helps to develop a city or area. Specially, it helps to the street vendors and road side shop owner because it helps to them to communicate with the commuters.

4.2 Demerits

Congestion: Because stopping takes up a lot of road space, the street limit has been lowered. As a result, speed will be reduced, but venture time and postponement will also be increased. The vehicle's operating costs rise, causing enormous financial hardship for the community.

Accidents: Stopping and unparking with haste causes mishaps, which are referred to as stopped mishaps. The most common types of leaving mishaps occur when driving a vehicle out of the leaving area, opening the doors of left automobiles in an indiscreet manner, and remembering to get the automobiles into the parking garage for departing.

Environmental pollution: Parking also one reason to pollute the environment because the movement of vehicles during the car parks and unparking creates sound and smokes and on parking are reason for air pollution also as per Cold starts are the main constitution of all emissions (specially Carbon oxides and VOC) and pPAH emitted from parking places and if the vehicle is 10 to 12 years old than it gives more pollutant.

The vehicle emission and air pollution during the parking and developed some model (BRAHE and simpark) that kerb parking can reduced 40% emission rather than garage parking. Parking contributes to the increase of CO_2 and NOx emissions, as well as influencing the esthetic quality of structures, as vehicles parked in every available spot give the impression that the structure is rising from a plinth of vehicles on environmental effect on parking infrastructure and analyze different emission of pollutants in US by four categories, i.e., scenario A to D. The stopping framework assessed in Scenarios A through C costs the US somewhere in the range of $4 and $20 billion every year (Table 2).

Table 2 Parking spaces in the US (in millions)

	On-street	Off-street		Total	Parking area as % of US land area
		surface	Structure		
Scenario A	92	520	110	722	0.64
Scenario B	180	520	110	810	0.66
Scenario C	150	610	84	844	0.68
Scenario D	1100	790	120	2010	0.90

Create problem to firefighting operations: Parked vehicles on the road side may create problem for the firefighting vehicles during its movement. Due to that population is suffering in the point of safety issue.

5 On-Street Parking Demand and Management

Lots of parking studies have been done yet which are aggregate and disaggregate. Developed an on-street parking demand now a days one of the most important factor for urban planner as well as administrative authority also, besides that how to manage that demand also important. According to this concerned, several authors developed some models. Table 3 shows the different model in different conditions and different places.

Priyanka kolhar [3], examines the problems with current stopping rehearses are concentrated with the stopping gathering and supply study in Dharwad. Willing to pay (WTP), survey was conducted to develop the parking demand model by the using of SPSS software. Pal and Bhattacharya [9], this paper manages observing and displaying of the unsettling influences delivered because of vehicular street traffic discontinuous by traffic stream conditions on close to home work execution around there in Agartala city. In this paper, MATLAB and statistical analysis are used. Arjun

Table 3 The different model in different conditions and different places

Author	Year	Model/analysis type
McGuiness-s and McNeil	1991	Least square regression
Wong et al	2000	LRM
Hollander et al	2006	Game theory (logit model)
Chakrabarti and Mazumdar	2010	LRM
Hilven et al	2012	Discrete choice
Cheng et al	2012	LRM
Aderamo and Salau	2013	LRM
Tong et al	2013	Unit graph technique
Lee	2014	Multiple regression model
Sen et al	2016	LRM
Chuz.lan et al	2016	Regression analysis
Das et al	2016	Multiple regression and AHP
Ramli et al	2017	Correlation and regression analysis
Lim et al	2017	Four-stage modeling approach
Fiez et al	2018	Gaussian mixture model
Ajeng and Cim	2018	LRM
LRM	Linear regression model	

and Nagakumar [10] investigate the state of stopping and level of service (LOS) delay in signalized crossing point and additionally gather the conduct reaction from the workers. Volume check, stopping length, collection, request review, and willingness to pay overview were all done as part of the important research. (WTP). Relapse examination was utilized to investigate the stopping information.

Rajat Rastogi [11] found out the parking duration of driver in Kota city, India in CBD, social visit and housing area by the conventional method like intention survey, questioners survey. The analysis involves the parking characteristics, parking accumulation, person accumulation, and parking duration. Two types of analysis done short term and long-term analysis. Result shows that full term analysis is better than short-term analysis. Tembhurkar and Khobragade [12] developed a parking demand forecasting model using some factors like trip characteristics, parking characteristics, and level of parking service. Parking demand forecasting model.

$$Q = \frac{\alpha \times R}{\rho \times \gamma} \times \eta \times \delta$$

Das and Ahmed [13] show that the problem of the parking can be manage by utilization of an effective parking structure. This study estimated the LOS for four CBDs of Kolkata metropolitan area based on the three characteristics, i.e., parking, safety, and design characteristics. The data are analyzed by SPSS.

Das and Ahmed [13], in this paper, two parking demand model estimate, i.e., the fee index and the cost factor for the city of Kolkata, based on the regression analysis by using SPSS software. FI model shows that the request will gratify the prevailing supply, and CF shows that transit fare need to increase for both the CBD, i.e., Dalhousie and Gariahat area 52 and 26%, respectively. This paper recommended to the consumers need to be moved from private vehicle (PV) to public transit (PT).

(1) Fee Index model
$Y = \left[\sum_{i=1}^{n} \frac{a1}{\mu 1} \times r_i\right] \times \mathrm{FI} \times \frac{\mathrm{AIp}}{\mathrm{AIt}} \times \beta \times \alpha$

(2) Cost Factor model

$$Y = \left[\sum_{i=1}^{n} \frac{a1}{\mu 1} \times r_i\right] \times \mathrm{CF} \times \frac{\mathrm{AIPL}}{\mathrm{AIt}} \times \beta \times \gamma$$

Fadeyev [14] investigates the focused on financial assessment of stopping the executive's instruments, and examination of clients' conduct with a look to applied administration devices. The examination permitted acquiring near assessments of effectiveness of paid stopping and time imperative by the case of Irkutsk. Plus, the examination decided the ideal rate and the way how expenses of looking for a parking space impact the productivity of the board apparatuses.

$$N_x t_x t_Y^{\mathrm{Max}} W = N_X \int_0^{tx} (\alpha - \beta_q)\mathrm{dq} + (N - N_x) \int_0^{ty} (\alpha - \beta_q - C)\mathrm{dq}$$

Wang et. al. [15], this investigation referenced that stopping estimating is an amazing asset for stopping the board and diminished gridlock in CBD regions. In this paper, information was gathered in four stages, i.e., when execution of the new stopping estimating strategy in Nanning in China to examine the impact of on-road stopping attributes. Results show that stopping term diminishes as stopping cost increments. Here, SPSS software is used to determine the K-S test to find out difference between parking characteristics before and after the implementation of parking price.

6 Conclusion

This paper audits the previous investigations that stressed the stopping qualities and improvements in models for stopping decision conduct, both at total and disaggregate levels. The paper additionally examines various techniques for assessing interest for parking spot. The investigations examining the arrangements, which might be helpful for viable use of accessible assets and having a positive effect on the manageable transport idea, have been looked into. Astoundingly, stopping strategy ought to be considered as a fundamental piece of transport arranging and the board. All in all, to decrease the private vehicle traffic flow in metropolitan regions, vehicle controlled leaving strategy also, enhancements out in the open travel have key jobs. Private vehicle traffic limitation can be accomplished through the leaving limitations, which are all the more ecologically ideal since they create less commotion, air contamination, and driver pressure. Improvement in stopping execution and regard for supportability will help transport organizers to build up an economical stopping environment. At last, the significant effect of the presentation of carsharing has been talked about in short, which can possibly give huge financial, social, and ecological advantages. Taken together, every one of these investigations present helpful data to per users for the whole stopping framework.

7 Future Scope of the Study

- To estimate the parking statistics, as well as further to estimate the parking demand model.
- To find parking fee to manage on-street parking during peak hour.
- The possessions of on-street car parks price-changing strategy on parking characteristics.
- To study, the standing position of pollution stages in the urban area by copy the pollution concentration at different location.

References

1. Marsden, G.: The evidence base for parking policies—a review. Transp. Policy **13**, 447–457 (2006)
2. Manville, M., Shoup, D., Bacon, F.: Parking, people, and cities. J. Urban. Plan. Dev. ASCE **131**, 233–245 (2005)
3. Kolhar, P.: On street parking management plan and cost-benefit analysis for Dharwad City Karnataka, India. IJERA **2**(3), 1654–1665 (2012)
4. Tong, C.O., Mice, F.M., Wong, S.C., Masce, F.F.M.M., W W T Lau BEng MHKIE R.P.E. MICE PE MITSHK: A demand-supply equilibrium model for parking services in Hong Kong. HKIE Trans. **11**(1), 48-53 (2004)
5. Chen, Q., Wang, Y., Pan, S.: Characteristics of parking in central Shanghai, China. J. Urban Plan. Dev. **142**(1), 05015012 (2015)
6. Parmar, J., Das, P., Dave, S.M.: Study on demand and characteristics of parking system in urban areas: a review. J.Traffic Transp. Eng. (Engl. Ed.) **7**(1), 111–124 (2019)
7. Naseri, A.: A method for estimating average delay time due on-street parking. Aust. J. Basic Appl. Sci. **7**(8), 767–774 (2013)
8. Marshall, W.E., Garrick, N.W., Hansen, G.: Reassessing on-street parking. transportation research record: journal of the transportation research board, No. 2046, Transportation Research Board of the National Academies, Washington, D.C., 45–52 (2008)
9. Pal, D., Bhattacharya, D.: Effect of road traffic noise pollution on human work efficiency in government offices, private organizations, and commercial business centres in Agartala City using fuzzy expert system: a case study. Adv. Fuzzy Syst. **828593** (2012)
10. Arjun, C.A., Nagakumar, M.S.: Studies on On-street parking using license plate method in Basavangudi Bangalore. In: ICRTIET-2014 Conference Proceeding, 30th-31st August (2014)
11. Rastogi, R.: Validating stated parking duration of drivers in Kota City, India (2014)
12. Tembhurkar, A., Khobragade, P.: Parking demand forecast model for institutional campus, ICTCE'15 London, pp. 39–42 (2015)
13. Das, D., Ahmed, M.A.: Level of service for on-street parking. KSCE J. Civ. Eng. **22**(1), 330–340 (2017)
14. Fadeyev, D.: Method for evaluating economic efficiency of parking management tools. Trans. Res. Procedia **20**(2017), 193–199 (2017)
15. Wang, H., Li, R., Wang, X., Shang, P.: Effect of on-street parking pricing policies on parking characteristics: a case study of Nanning. Transp. Res. Part A **137**(2020), 65–78 (2020)
16. Das, D., Ahmed, M.A.: On-street parking demand estimation in urban CBD using FI and CF model: a case study—Kolkata, India. Indian J. Sci. Technol. **10**(12) (2017)
17. Ministry of Road Transport and Highways (MoRTH), India (2019–20)

Optimal Location of IPFC on LFC Studies Considering PI-TIDN Controller and RT-Lab

Naladi Ram Babu, Lalit Chandra Saikia, Sanjeev Kumar Bhagat, and Biswanath Dekaraja

Abstract This article demonstrates the effect of interline power flow controller (IPFC) and high voltage direct current (HVDC) on two-area LFC thermal-DSTS system. The investigations are carried on the real-time simulation laboratory (RT-Lab). For the first time, an attempt was made to design a cascade PI-TIDN as a secondary controller with crow-search algorithm (CA) in the AGC field. Performance comparisons of PI-TIDN controller are found to be better over other conventional controllers like PIDN and TIDN. Moreover, the comparison of PI-TIDN controller with CA converges and settles faster than FA and PSO. Further, system dynamics with integration of HVDC and IPFC shows better dynamics. Furthermore, the optimal location of IPFC is found in the disturbance or less capacity area.

Keywords AC-HVDC · CA technique · Firefly algorithm · LFC · PSO · PI-TIDN controller and IPFC

Nomenclature

TF	Transfer function
PS	Power system
Δ	Deviation
F/Ptie	Frequency/Tie-power
RT-Lab	Real-time simulation laboratory
PSO	Particle swarm optimization
LFC	Load frequency control

N. R. Babu (✉) · L. C. Saikia · S. K. Bhagat · B. Dekaraja
EED, NIT Silchar, Assam, India

K. N. Das et al. (eds.), *Proceedings of the International Conference on Computational Intelligence and Sustainable Technologies*, Algorithms for Intelligent Systems,
https://doi.org/10.1007/978-981-16-6893-7_9

1 Introduction

In modern PS, LFC plays a prominent role in supplying consumers with quality, adequate, and reliable power. It is a group of individual or utility components that have been synchronized to form a large network, which is controlled by control areas with interconnected tie-lines. The PS frequency and power has to stay within acceptable limits for smooth execution. LFC maintains the power balance by automatically changing the governor speed settings [1]. Authors in Nanda et al. [2] and Javad et al. [3] integrated hydro, gas with thermal considering single reheat turbine and GRC. Fossil fuel decrement and excess carbon emissions tend to the penetration of renewable. Authors in [4–9] integrated thermal with STPP, wind, and DSTS in two-area system but considered in one area only. Hence, integration of DSTS in both the areas needs further investigation.

Continuously increased load demand leads to technology development in the PS. The HVDC helps bulk power inter-areas transfer capability which lead to adoption in PS. Authors in [10, 11] integrated HVDC in LFC studies. Sharma et al. [12] proposed parallel AC-HVDC in LFC. Further, system dynamics get enhanced with the integration of IPFC [13, 14]. The LFC studies don't comprise AC-HVDC and IPFC in two-area DSTS-thermal LFC system which require further study.

Present scenario, the design of the controller is the central area for the researcher in LFC studies. Various control aspects like cascade [4, 6], integer [5], tilt [13, 15], fuzzy [16], fractional [17, 18] controller are explored in LFC. A new control aspect named by PI-TIDN controller is proposed, which has advantages over integer and fractional controllers. This gives an opportunity in LFC studies.

Optimization of secondary controllers provides better results. Optimization techniques like whale optimization [4], gray wolf [6], sin-cosine [7], firefly algorithm (FA) [19], PSO [20], coyote [21], hybrid pattern search [22], grass hoper [23], Harris's hawks [24], and artificial flora [25] are accessible. A newly optimization technique called crow-search algorithm (CA) [26] is available. The properties of utilizing tools, food hiding, and stealing when the owner birds leaves make crow a clever bird. It work based on the crow's food-seeking behavior. Surprisingly, its application in LFC studies is to be carried.

The mentioned stated LFC studies are simulated in Matlab. However, the LFC studies can also be conducted in RT-Lab. This provides scope for further investigations.

The objectives are as follows:

(a) Develop a thermal two-area system integrated with DSTS, HVDC, and IPFC in a RT-Lab software.
(b) Utilization of CA for optimization of PIDN, TIDN, and PI-TIDN.
(c) Application of AC-HVDC and IPFC system.
(d) Optimal location of IPFC in two-area system.

2 System Investigated

The two-area thermal-DSTS system of 1:2 is taken for the investigation and is shown in Fig. 1. GRC for thermal system is considered for the practical aspect. The HVDC unit is associated in parallel to AC system. The change in power through AC-HVDC system is given by (1)–(3).

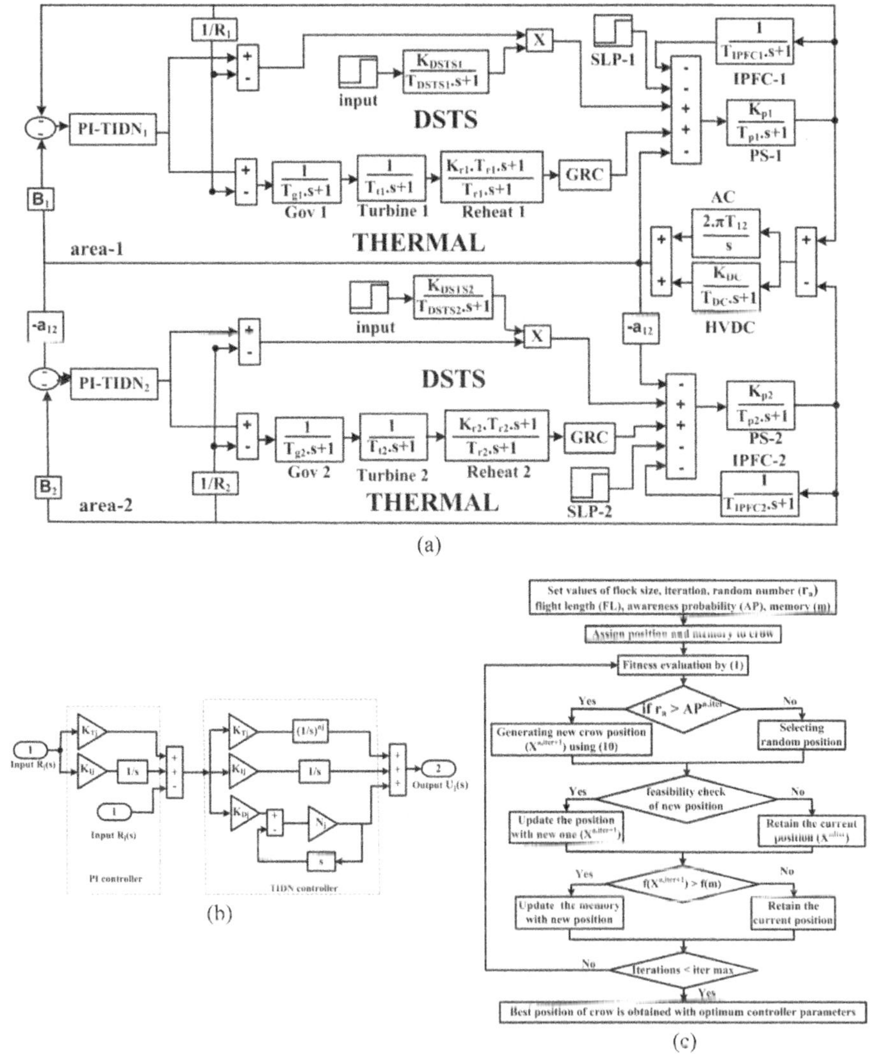

Fig. 1 **a** The investigated system. **b** The proposed PI-TIDN controller. **c** Flowchart of CA optimization technique

$$\Delta P_{\text{tie 12 AC}} = \frac{2\Pi T_{12}}{s}(\Delta F_1 - \Delta F_2) \tag{1}$$

$$\Delta P_{\text{tie 12 DC}} = \frac{K_{\text{DC}}}{1 + sT_{\text{DC}}}(\Delta F_1 - \Delta F_2) \tag{2}$$

$$\Delta P_{\text{tie 12}} = \Delta P_{\text{tie 12 AC}} + \Delta P_{\text{tie 12 DC}} \tag{3}$$

The system in Fig. 1a is investigated with PI-TIDN controller, and its parameters are optimized by CA. The RT-Lab software is used for the investigation subjected to 1% SLP in area-1 considering ISE in (4).

$$\eta_{\text{ISE}} = \int_0^T \{(\Delta F_1)^2 + (\Delta F_2)^2 + (\Delta P_{\text{tie 1-2}})^2\}.\text{d}t \tag{4}$$

3 Proposed Cascade PI-TIDN Controller

The design of TID is analogous to PID, and it has tilt characteristic which provides better degree of stability than the other integer type controller. The TF model of cascaded PI and TID is in Fig. 1b, and the TF equation of Fig. 1b is given by (5) and (6)

$$\text{TF}_{\text{PI}} = K_{\text{Pj}} + K_{\text{Ij}}/s \tag{5}$$

$$\text{TF}_{\text{TIDN}} = K_{\text{Tj}}(1/s)^{n_j} + K_{\text{Ij}}/s + K_{\text{Dj}}.s\left(\frac{N_j}{s + N_j}\right) \tag{6}$$

The controller parameters are augmented by CA with constraints in (7)

$$\begin{gathered} 0 \le K_{\text{Tj}}/K_{Pj}(\text{Tilt/Proportional}) \le 1,\ 0 \le n_j(\text{real number}) \le 7 \\ 0 \le K_{\text{Ij}},\ K_{\text{I1j}}(\text{Integral}) \le 1, 0 \le K_{\text{Dj}}(Derivative) \le 1, \text{and } 0 \le N_j(Filter) \le 100 \end{gathered} \tag{7}$$

4 Crow-Search Algorithm (CA)

Crows are considered as one of the clever entities. It follows other birds for food hiding and stealing it when the owner bird leaves. A. Askarzadeh proposed a revolutionary swarm-based population-based algorithm named CA with crow as a hunt agent [19]. The expression for evaluation in all ways by crow-search agent is in (8)

$$[X^{y,\text{iter}} = X_1^{y,\text{iter}},\ X_2^{y,\text{iter}},\ \ldots.,\ X_d^{y,\text{iter}}] \tag{8}$$

Each crow's food hiding place is remembered and traced by the hunt agents. In iteration, if crow-z is willing to follow crow-y, then the two prospects occur:

Prospect-1: Crow-y willing to follow crow-z, but they don't know how to approach to crow-z; the approaches for optimal place of crow-y is in (9).

$$X^{y,\text{iter}+1}{=}X^{y,\text{iter}} + r_y \times \text{FL}^{y,\text{iter}} \times (m^{z,\text{iter}} - X^{z,\text{iter}})r_b \geq \text{AP}^{z,\text{iter}} \tag{9}$$

where random (r_y), flight length (FL) of crow-y in the iteration toward the local as well as global optima. Awareness probability (AP) with small step size will obtain best solution.

Prospect-2: Crow-z recognizes the approach of crow-y and flews to a random position and is given by (10).

$$X^{a,\text{iter}+1} = \text{random position} \tag{10}$$

In iteration, the best fitness of crow is stowed and is updated with the best fitness found in the next iteration. The flowchart of CA is in Fig. 1c.

5 Results and Analysis

LFC system comprising of DSTS-thermal system integrated with HVDC, IPFC and is provided by PIDN, TIDN, and PI-TIDN. Investigations are carried with PI-TIDN controller with ode4, Considered system with controller optimum values is linked with RT Lab. The execution of LFC model in RT-Lab is compiled and is shown in Fig. 2.

5.1 System Dynamics with TIDN, PIDN, and PI-TIDN Controllers.

The considered system in Fig. 1a is provided with PIDN, TIDN, and PI-TIDN controller and are optimized by CA. Values are listed in Table 1. The dynamics of system corresponding to optimum value are shown in Fig. 3. Critical observations disclose that responses with PI-TIDN controller show better responses over other controllers.

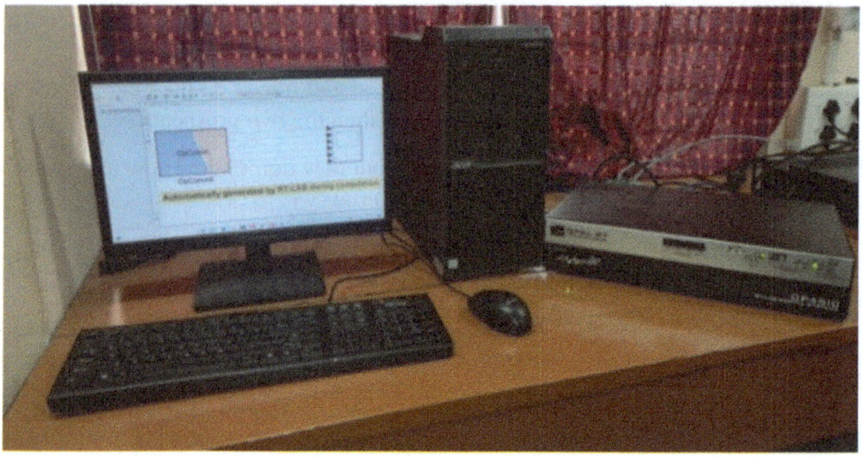

Fig. 2 The validated system with RT-Lab software

5.2 *System Dynamics with Various Meta-Heuristic Algorithms*

5.2.1 Dynamic Performance Comparison of PI-TIDN Controller with FA, PSO, and CA Algorithm

The best controller obtained in Sect. 5.1 is considered for investigation, and its parameter is simultaneously optimized by FA, PSO, and the proposed CA. The obtained optimum values of controller is listed in Table 2. Its dynamics are revealed in Fig. 4a. It is evident that the response with PI-TDN controller and CA algorithm performs better over other algorithms.

5.2.2 Performance Comparison of PI-TIDN Controller with Various Algorithms like FA, PSO and CA

The two-area thermal-DSTS system with PI-TIDN is considered. It is subjected to ISE as performance index, and PI-TIDN gains are optimized with FA, PSO, and the proposed CA. The respective convergence value is listed in Table 2. Its curves are plotted in Fig. 4b and are observed that the CA converges faster over FA and PSO.

5.3 *System with HVDC Integration*

The system in Sect. 5.2 is integrated with HVDC in both areas. The best controller and algorithm from Sect. 5.1 and 5.2 are utilized for the investigation. The controller

Table 1 CA augmented controller parameters of thermal-DSTS system

PIDN controller							
Area-1	$K_{P1} = 0.5069$	$K_{I1} = 0.6304$	$K_{D1} = 0.7870$	$N_1 = 67.8901$			
Area-2	$K_{P2} = 0.3300$	$K_{I2} = 0.0940$	$K_{D2} = 0.1100$	$N_2 = 84.5670$			
TIDN controller							
Area-1	$K_{T1} = 0.7014$	$n_1 = 3.5900$	$K_{I1} = 0.6000$	$K_{D1} = 0.9200$	$N_1 = 47.8026$		
Area-2	$K_{T2} = 0.3570$	$n_2 = 3.5700$	$K_{I2} = 0.3640$	$K_{D2} = 0.9404$	$N_2 = 98.0120$		
PI-TIDN controller							
Area-1	$K_{P1} = 0.5161$	$K_{II1} = 0.6833$	$K_{T1} = 0.1101$	$n_1 = 6.3051$	$K_{I1} = 0.3021$	$K_{D1} = 0.3301$	$N_1 = 88.0124$
Area-2	$K_{P2} = 0.6514$	$K_{II2} = 0.5182$	$K_{T2} = 0.7601$	$n_2 = 6.0014$	$K_{I2} = 0.5450$	$K_{D2} = 0.4503$	$N_2 = 75.0462$

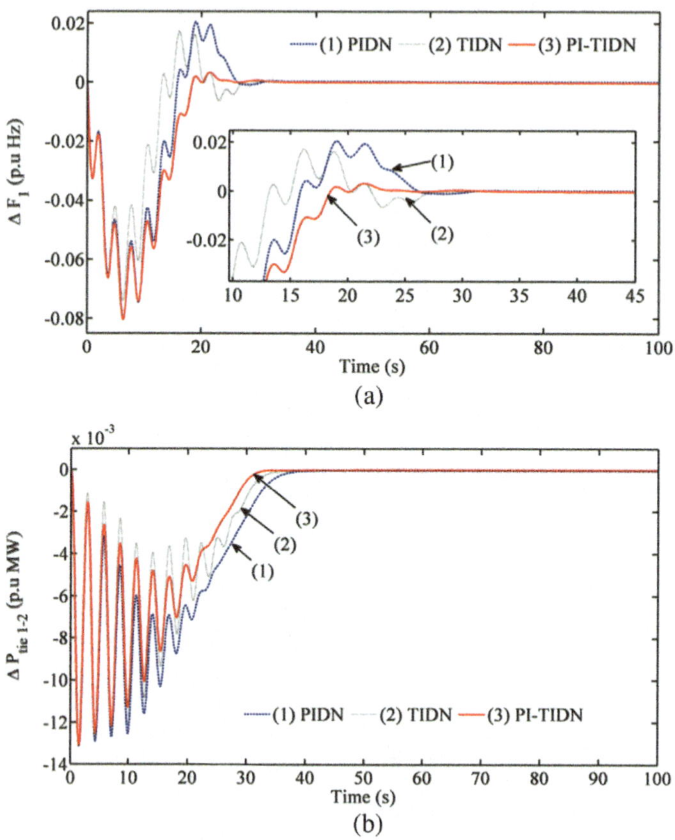

Fig. 3 Response comparison with PIDN, TIDN, and PI-TIDN controllers. **a** ΔF_1 verses time. **b** $\Delta P_{\text{tie 1-2}}$ verses time

optimum values are listed in Table 4, and its responses are associated with AC tie-line and are displayed in Fig. 5. It is observed that dynamics with AC-HVDC enhances over AC alone (Table 3).

5.4 *Effect of Facts Devices Integration*

5.4.1 Effect of IPFC Integration

The considered system in Sect. 5.3 is integrated with IPFC in both areas. The PI-TIDN and CA are utilized for the investigation. The optimum gains are recorded in Table 4, and its corresponding dynamics are compared with the system having

Table 2 Optimized PI-TIDN controller values of two thermal area system using FA, PSO, and CA

PI-TIDN controller with FA							
Area-1	$K_{P1} =$ 0.5161	$K_{I11} =$ 0.6833	$K_{T1} =$ 0.1101	$n_1 =$ 6.3051	$K_{I1} =$ 0.3021	$K_{D1} =$ 0.3301	$N_1 =$ 88.0124
Area-2	$K_{P2} =$ 0.6514	$K_{I12} =$ 0.5182	$K_{T2} =$ 0.7601	$n_2 =$ 6.0014	$K_{I2} =$ 0.5450	$K_{D2} =$ 0.4503	$N_2 =$ 75.0462
$ISE_{FA} = 0.008$							
PI-TIDN controller with PSO							
Area-1	$K_{P1} =$ 0.1906	$K_{I11} =$ 0.5842	$K_{T1} =$ 0.1009	$n_1 =$ 4.2054	$K_{I1} =$ 0.2056	$K_{D1} =$ 0.2991	$N_1 =$ 77.0250
Area-2	$K_{P2} =$ 0.1774	$K_{I12} =$ 0.5746	$K_{T2} =$ 0.5900	$n_2 =$ 5.1144	$K_{I2} =$ 0.4990	$K_{D2} =$ 0.3987	$N_2 =$ 35.1023
$ISE_{PSO} = 0.005$							
PI-TIDN controller with CA							
Area-1	$K_{P1} =$ 0.4723	$K_{I11} =$ 0.5431	$K_{T1} =$ 0.1101	$n_1 =$ 6.3051	$K_{I1} =$ 0.3021	$K_{D1} =$ 0.3301	$N_1 =$ 88.0124
Area-2	$K_{P2} =$ 0.2510	$K_{I12} =$ 0.8987	$K_{T2} =$ 0.7601	$n_2 =$ 6.0014	$K_{I2} =$ 0.5450	$K_{D2} =$ 0.4503	$N_2 =$ 75.0462
$ISE_{CA} = 0.0038$							

AC-HVDC in Fig. 6 It is evident that the system with IPFC-AC-HVDC shows better responses.

5.4.2 Optimal Location of IPFC

To find the IPFC optimal location, the configuration is as follows:

i. IPFC connection across area-1,
ii. IPFC connection across area-2, and
iii. IPFC connection across both areas.

In all the above configurations, system in Sect. 5.4 is investigated with PI-TIDN controller and CA. The obtained dynamics are compared with Fig. 6a and b and plotted in Fig. 6c and d. From Fig. 6c and d, critical observations reveal that, when IPFC is connected across low capacity area shows better responses and is more or less same considering IPFC in both areas.

6 Conclusions

An attempt was made to find the ideal location of IPFC in LFC studies. Studies are investigated with a new controller, namely cascade PI-TIDN controller with RT-Lab

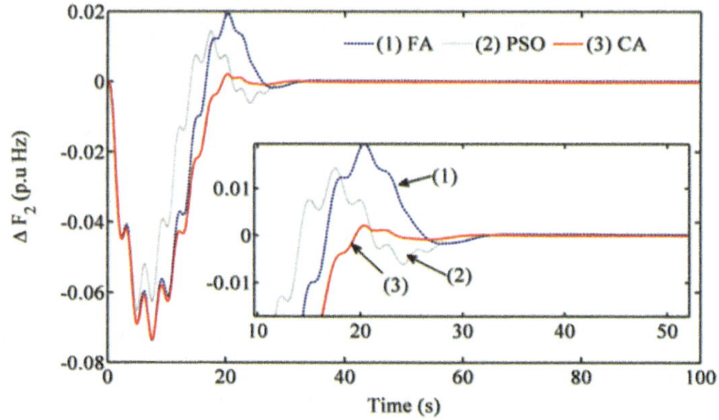

(a)

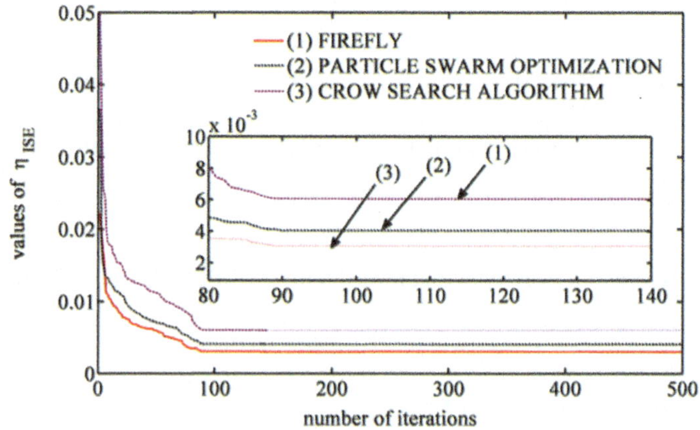

(b)

Fig. 4 Comparison of various algorithms like FA, PSO, and CA. **a** ΔF_1. **b** convergence curve

software in LFC. CA augments the controller values, and its outcomes are linked with FA and PSO. Investigations with PI-TIDN controller outperform over PIDN and TIDN. Moreover, it is also observed that integration of thermal-DSTS with HVDC and IPFC shows better dynamics over thermal-DSTS with AC tie-line alone, AC-HVDC tie-line, and IPFC. Further, it is observed that CA with PI-TIDN controller shows better convergence than PSO and FA algorithms. Furthermore, the optimal location of IPFC is found in the less capacity region.

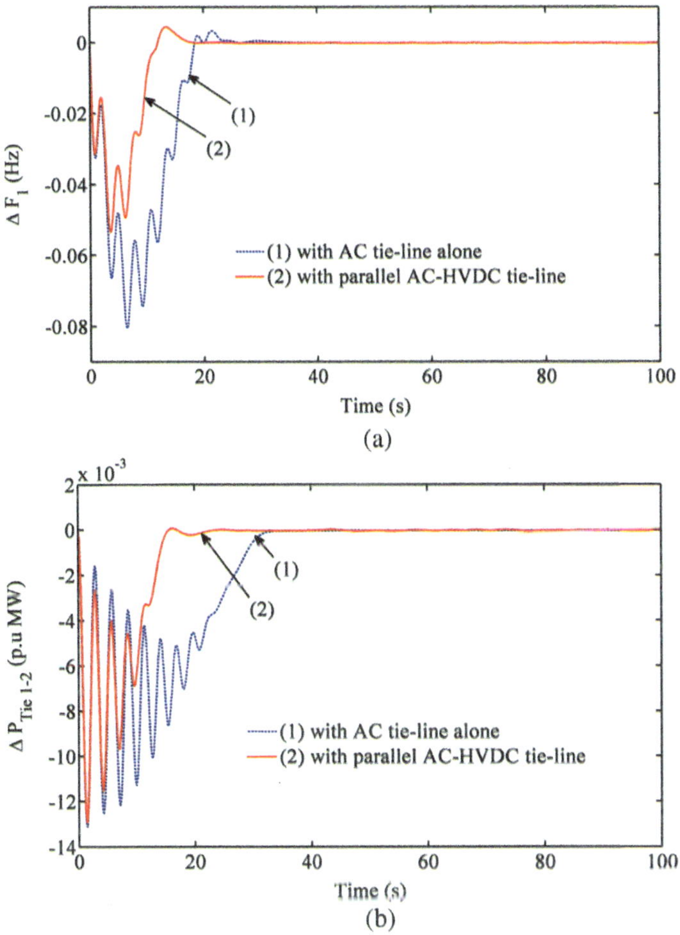

Fig. 5 Responses with AC-HVDC system. **a** ΔF_1. **b** $\Delta P_{tie\ 1-2}$

Table 3 CA optimized PI-TIDN values with HVDC integration

Area-1	$K_{P1} =$ 0.6211	$K_{I11} =$ 0.5015	$K_{T1} =$ 0.2690	$n_1 =$ 3.001	$K_{I1} =$ 0.7031	$K_{D1} =$ 0.5920	$N_1 =$ 64.0250
Area-2	$K_{P2} =$ 0.7478	$K_{I12} =$ 0.4578	$K_{T2} =$ 0.2130	$n_2 =$ 4.4301	$K_{I2} =$ 0.8903	$K_{D2} =$ 0.4013	$N_2 =$ 58.4578

Table 4 CA optimized PI-TIDN values with integration of HVDC and IPFC

Area-1	$K_{P1} =$ 0.8801	$K_{I11} =$ 0.6254	$K_{T1} =$ 0.5001	$n_1 =$ 3.4005	$K_{I1} =$ 0.3501	$K_{D1} =$ 0.3301	$N_1 =$ 88.0124
Area-2	$K_{P2} =$ 0.5178	$K_{I12} =$ 0.7845	$K_{T2} =$ 0.3201	$n_2 =$ 6.8950	$K_{I2} =$ 0.7603	$K_{D2} =$ 0.5013	$N_2 =$ 75.0462

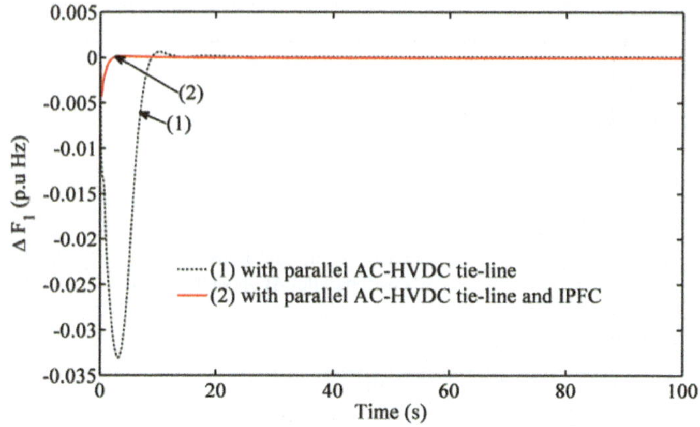

(a)

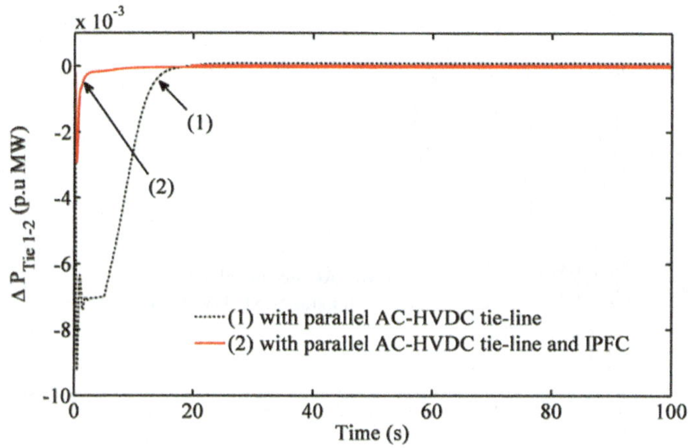

(b)

Fig. 6 Dynamic comparison with IPFC integration. **a** ΔF_1 considering with and without IPFC. **b** $\Delta P_{tie\ 1-2}$ considering with and without IPFC. **c** ΔF_1 with various IPFC configurations. **d** $\Delta P_{tie\ 1-2}$ with various IPFC configurations

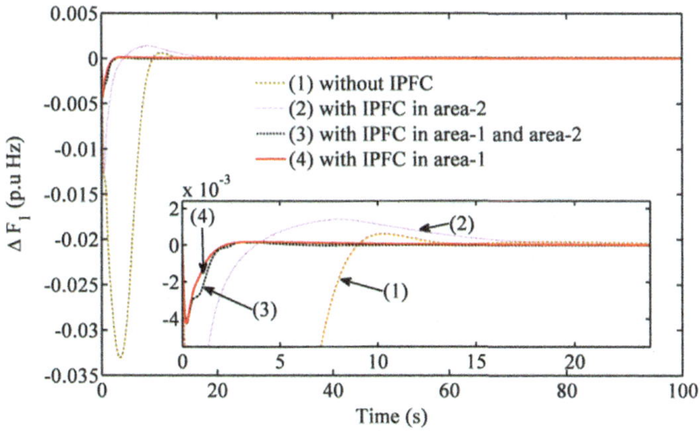

(c)

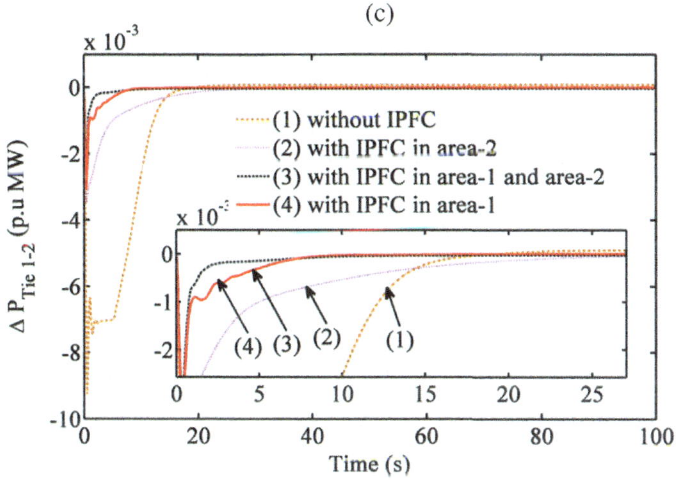

(d)

Fig. 6 (continued)

References

1. Elgerd, O.I., Fosha, C.E.: Optimum megawatt-frequency control of multiarea electric energy systems. IEEE Trans. Power Appar. Syst. **89**, 556–563 (1970)
2. Nanda, J., Mangla, A., Suri, S.: Some new findings on automatic generation control of an interconnected hydrothermal system with conventional controllers. IEEE Trans. Energy Convers. **21**, 187–194 (2006)
3. Javad, M., Kazem, Z., Mehrdad, T.H.: Applying fractional order PID to design TCSC-based damping controller in coordination with automatic generation control of interconnected multi-source power system. Eng. Sci. Technol. Int. J. **20**(1), 1–17 (2017)
4. Saha, A., Saikia, L.C.: Utilization of Ultra-capacitor in load frequency control under restructured STPP-thermal power systems using WOA optimized PIDN-FOPD controller. IET Gener. Transm. Distrib. **11**, 3318–3331 (2017)

5. Sharma, Y., Saikia, L.C.: Automatic generation control of a multi-area ST–thermal power system using grey wolf optimizer algorithm based classical controllers. Int. J. Electr. Power Energy Syst. **73**, 853–862 (2015)
6. Tasnin, W., Saikia, L.C.: Maiden application of an sine–cosine algorithm optimised FO cascade controller in automatic generation control of multi-area thermal system incorporating dish-Stirling solar and geothermal power plants. IET Renew. Power Gener. **12**, 585–597 (2018)
7. Saikia, L.C., Chowdhury, A., Shakya, N., Shukla S., Soni, P.K.: AGC of a multi area gas-thermal system using firefly optimized IDF controller. In: Annual IEEE India Conference (INDICON), pp. 1–6. Mumbai (2013)
8. Ram Babu, N., Saikia, L.C.: AGC of a multiarea system incorporating accurate HVDC and precise wind turbine systems. Int. Trans. Electric. Energy Syst. **30**(4), 1–27 (2019)
9. Ram Babu, N., Saikia, L.C.: Load frequency control of a multi-area system incorporating realistic high-voltage direct current and dish-Stirling solar thermal system models under deregulated scenario. IET Renew. Power Gener. **15**, 1116–1132 (2021)
10. Adeuyi, O.D., et al.: Frequency support from modular multilevel converter based multi-terminal HVDC schemes. In: IEEE Power & Energy Society General Meeting, pp. 1–5. Denver, CO (2015)
11. Rakhshani, E., Rodriguez, P.: Inertia emulation in AC/DC interconnected power systems using derivative technique considering frequency measurement effects. IEEE Trans. Power Syst. **32**, 3338–3351 (2017)
12. Bhagat, S.K., Saikia, L.C., Babu, N.R., Saha, D.: Impact of PLL and virtual inertia on deregulated agc system integrated with parallel AC/HVDC. IETE J. Res. (2021). https://doi.org/10.1080/03772063.2021.1894249
13. Boutheina, Y., Abdelmoumène, D., Salem, A.: Automatic generation control using second generation of FACTS interline power flow controller. In: 1st International Conference on Communications, Control Systems and Signal Processing (CCSSP), pp. 470–473. EL OUED, Algeria (2020)
14. Rahman, A., Saikia, L.C., Sinha, N.: Maiden application of hybrid pattern search-biogeography based optimisation technique in automatic generation control of a multi-area system incorporating interline power flow controller. IET Gener. Transm. Distrib. **10**, 1654–1662 (2016)
15. Topno, P.N., and Chanana, S.: Tilt Integral derivative control for two-area load frequency control problem. In: 2nd International Conference on Recent Advances in Engineering & Computational Sciences, pp. 1–6 (2015)
16. Yogendra, A.: Impact of hydrogen aqua electrolyzer-fuel cell units on automatic generation control of power systems with a new optimal fuzzy TIDF-II controller. Renew. Energy **139**, 468–482 (2019)
17. Ram, B.N., Saikia, L.C.: Automatic generation control of a solar thermal and dish-stirling solar thermal system integrated multi-area system incorporating accurate HVDC link model using crow search algorithm optimised FOPI Minus FODF controller. IET Renew. Power Gener. **13**(12), 2221–2231 (2019)
18. Mohamed, E.A., Ahmed, E.M., Elmelegi, A., Aly, M., Elbaksawi, O., Mohamed, A.: An optimized hybrid fractional order controller for frequency regulation in multi-area power systems. IEEE Access **8**, 213899–213915 (2020)
19. Jagatheesan, K., Anand, B., Samanta, S., Dey, N., Ashour, A.S., Balas, V.E.: Design of a proportional-integral-derivative controller for an automatic generation control of multi-area power thermal systems using firefly algorithm. IEEE/CAA J. Automatica Sinica **6**(2), 503–515 (2019)
20. Ram Babu, N., Bhagat, S.K., Saikia, L.C., Chiranjeevi, T.: Application of hybrid crow-search with particle swarm optimization algorithm in AGC studies of multi-area systems. J. Discrete Math. Sci. Cryptogr. **23**(2), 429–439 (2020)
21. Ram Babu, N., Saikia, L.C., Narrisetty V.: Maiden application of coyote optimizer algorithm with TIDN controller in AGC of a multi-area multi-source system. In: 2019 IEEE 16th India Council International Conference (INDICON), pp. 1–4. Rajkot, India (2019)

22. Ram Babu, N., Saikia, L.C., and Raju, D. K.: Maiden application of hybrid crow search algorithm with pattern search algorithm in LFC studies of a multi-area system using cascade FOPI-PDN Controller. In: Soft Computing for Problem Solving, pp. 337–351 (2020)
23. Bhagat, S K., Babu N. R., Saikia, L.C., Raju, D. K.: Maiden application of meta-heuristic techniques with optimized integral minus tilt-derivative controller for AGC of multi-area multi-source system. EAI Endorsed Trans. Scalable Inf. Syst. 1–9 (2020). https://doi.org/10.4108/eai.13-7-2018.164557
24. Ramoji, S.K., Saikia, L.C., Dekaraja, B., Behera, M.K., Bhagat, S K., Babu N.R.: Optimal unified frequency and voltage control of multi-area multi-source power system using the cascaded PIDN-TIDF controller. In: 2020 IEEE 17th India Council International Conference (INDICON), pp. 1–7 (2020)
25. Dekaraja, B., Saikia, L.C., Ramoji, S.K., Babu, N.R., Bhagat, S K., Behera, M.K.: Coordinated control of voltage and frequency in a three-area multisource system integrated with SMES Using FOI-FOPDF controller. In: 2020 IEEE 17th India Council International Conference (INDICON), pp. 1–7 (2020)
26. Askarzadeh, A.: A novel metaheuristic method for solving constrained engineering optimization problems: crow search algorithm. J. Comput. Struct. **169**, 1–12 (2016)

Redox Flow Battery Support for Combined ALFC-AVR Control of Multiarea Thermal System Incorporating Renewable Energy Sources

Biswanath Dekaraja, Lalit Chandra Saikia, and Naladi Ram Babu

Abstract The significance of a redox flow battery (RFB) in three-area systems with combined automatic load frequency control (ALFC) and automatic voltage regulator is demonstrated in this article. The system consists of the thermal, solar thermal power plant, dish-Stirling solar thermal plant, and hydro plant. A new cascade fractional-order (FO) proportional-integral (FOPI) and FOPD with filter coefficient (N) name as CFOPI-FOPDN controller is proposed as a secondary controller for the ALFC loop. The controller parameters are optimized using an artificial flora optimization algorithm. Analysis of dynamics responses reveals that the CFOPI-FOPDN controller provides better system dynamics than the PIDN and PI controllers. Overshoot, undershoot, and oscillations are reduced as the RFB is integrated into all areas of the system. The frequency deviation is a better input to RFB than the area control error signal for dynamic system performance improvement. Sensitivity analysis reflects that the optimized proposed controller parameters at nominal condition are robust and need not be reset repeatedly for a wide variation of system loading.

Keywords ALFC · AVR · Artificial flora algorithm · CFOPI-FOPDN controller · RFB

1 Introduction

The integrated electricity grid becomes increasingly complex as the number of consumers increases. The load in the power grid is fluctuating. The load-generation imbalance causes the frequency, tie-line power, and voltage to deviate from their schedule value. Hence, they need to control within specific limits. The governor function is used to restore frequency, and to maintain constant voltage via generator excitation, an automated voltage regulator (AVR) loop is used. Many works of literature [1–6] were found on automatic load frequency control (ALFC) studies where the

B. Dekaraja (✉) · L. C. Saikia · N. R. Babu
Department of Electrical Engineering, NIT Silchar, Assam, India

K. N. Das et al. (eds.), *Proceedings of the International Conference on Computational Intelligence and Sustainable Technologies*, Algorithms for Intelligent Systems,
https://doi.org/10.1007/978-981-16-6893-7_10

AVR loop was not considered. The ALFC loop and AVR loop have weak coupling [7, 8] although the AVR loop is faster. Hence, it needs further investigation.

In the literature, the authors of [9–16] have explored the combined ALFC-AVR loop. It is initially confined to an isolated thermal system [9, 10], two-area thermal–thermal systems [11], and two hydrothermal systems[12]. The isolated thermal system investigated in [9, 10] without considering physical limitations like generation rate constraint (GRC), communication delay (CD), and governor dead-band (GDB). The physical limitations make the system more practical. The authors of [11] presented two-area thermal systems neglecting GRC and GDB for thermal plant. The two-area hydrothermal systems demonstrated in [12] and analyzed system frequency response at different system prospects considering GDB for thermal plant. Renewable energy sources (RESs) are integrated into the interconnected power system to reduce global warming while addressing the load-generation mismatch[3]. The authors of [13] extended in this combined ALFC-AVR to multiarea multisource. The three-area thermal-diesel plant systems incorporating RESs like solar thermal power plant (STPP) are studied in [13]. The authors of [14–16] have explored three-area systems incorporating various RESs. It has been seen that the multiarea multisource system based on hydro, thermal, STPP, and dish-Stirling solar thermal power plant (DSTP) has not yet been investigated.

The energy storage devices (ESDs) such as capacitive energy storage (CES), superconducting magnetic energy storage (SMES), and redox flow battery (RFB) play a key role to reduce the system oscillations [6] in the ALFC studies. The three areas of the thermal system incorporating STPP, wind, and geothermal plant investigate having RFB [14]. However, the application of RFB is limited to a few studies [14] only. When RFB is used in various combinations of sources in the mutual ALFC-AVR system, the system dynamics may be different. Hence, it needs further investigation.

Researchers have been mainly focused on designing a supplementary controller for the ALFC loop in the joint ALFC-AVR model. In the past literature, the secondary controller such as conventional integral [9, 10], integral double derivative with filter (IDDF) [13], PID [11, 12], cascade fractional-order PDN and FOPIDN (CFOPDN-FOPIDN) [14], two degree of freedom tilt-integral-derivative (2DOFTID) [15], cascade FO integral, and FO PD with filter (CFOI-FOPDN) [16] has been successfully implemented in this combined scenario. A new cascade FO proportional-integral (FOPI)—FO proportional-derivative with filter (FOPDN) (CFOPI-FOPDN) controller is proposed and needs further study.

When the secondary controller's parameters are set optimally, it works better. The considered system presences nonlinearity constraints; hence, the conventional optimization method is not feasible due to the suboptimal solution and time-consuming. In order to overcome these issues, metaheuristics algorithms have been used. In the literature, the light searching algorithm (LSA) [13] and artificial flora algorithm [14, 16] have been fruitfully implemented in this amalgamated ALFC-AVR system. In this work, the AFA is used.

The sensitivity analysis is carried out to test the robustness of the proposed controller. In the past literature, changes in SLD magnitude [13], changes in SLD position[13], and various system conditions have been performed. The system loading is varied by ± 25% from its nominal states in this study to examine the controller's robustness.

The main objectives of this work are as follows:

(a) To build up in the merged ALFC-AVR loop of three-area systems based on the thermal, hydro, dish-Stirling solar thermal, and solar thermal power plant with appropriate physical constraints for hydro and thermal plants. An ESD-named RFB is integrated into all areas.
(b) To study the performance of the PI, PIDN, and CFOPI-FOPDN controller to find the best one.
(c) Investigate the effect of frequency deviation and ACE as an input signal to RFB on the system dynamics with the best controller obtained in (b). Also, analyze the effect of nonlinearity constraints on the system dynamics.
(d) Sensitivity analysis is used to verify the robustness of the best controller found in the previous step (b).

This work is prepared as follows: Sect. 2 describes the methodology containing system study, the relations between ALFC and AVR, the linearized model of RFB, and the projected controller. Section 3 included the optimization algorithm and results, and analysis presented in Sect. 4. The conclusion describes in Sect. 5, followed by References.

2 Methodology

2.1 Investigated System

The amalgamated ALFC-AVR loop of three-area systems comprises a thermal-STPP in Area-1, DSTP-thermal in Area-2, and hydrothermal Area-3. The GRC is taken as 3% per minute for thermal, 270% per minute lowering, and 360% per minute raising for hydro generation. The same governor speed regulation (4% p.u) is considered for all plants. The proposed system's detailed transfer function model is illustrated in Fig. 1a. The rated capacity for Area-1, Area-2, and Area-3 is 1000 MW, 2000 MW, and 3000 MW, respectively. The GDB for a thermal plant is 0.06% (0.036 Hz) considered. The isolation of the STPP and DSTP is taken as 0.30p.u. In Area-1, 1% of step load disturbance (SLD) is introduced for 1 s to study system performance. In order to determine the controller parameters, the integral square error (ISE) is used. At nominal system loading, the objective function (ISE) is minimized under the controller parameters' inequality restrictions. The cost function is given in (1)

$$J = \int_0^T \left(\Delta f_i^2 + \Delta P_{\text{tie},i-j}^2 + \Delta V_i^2\right) \mathrm{d}t \tag{1}$$

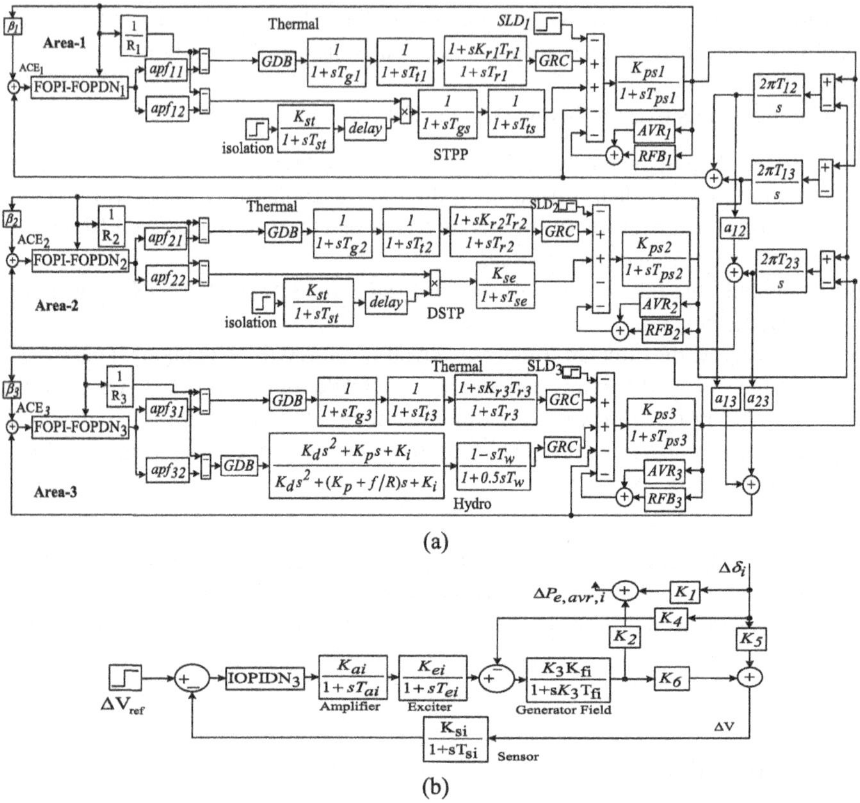

Fig. 1 **a** Investigated system model. **b** Detailed model of AVR

where $T = 120$ s (Simulation time), $J =$ objective function, and $i, j = 1,..,3$ $(i \neq j)$.

2.2 *Modeling of Cross-Coupling Coefficient for Combined ALFC-AVR Loops*

In the amalgamated ALFC-AVR model, the ALFC loop regulates the frequency via the governor mechanism. The AVR loop is used to maintain a constant voltage by varying the synchronous generator's field excitation (SG). The AVR controls the SG's terminal voltage as it builds up in an SG €, and therefore, the real power generation is written as (2)

$$P_e = \frac{V \times E'}{X_s} \sin(\delta) \tag{2}$$

where X_s = synchronous reactance of the SG, δ = rotor angle. Under small load perturbations, the frequency deviation can be regulated via the change in rotor angle ($\Delta\delta$) and induce voltage ($E^{'}$), which regulate the active power generation (ΔP_e) is written as (3) [8]

$$\Delta P_e = P_s \Delta\delta + K_1 E' \quad (3)$$

where P_s = synchronizing power coefficient. Moreover, the q-axis (V_q) and d-axis (V_d) components of the terminal voltage (V) are influenced by δ. The change in terminal voltage is written as (4) [7]

$$\Delta V = K_2 \Delta\delta + K_3 E' \quad (4)$$

The induced voltage in the SG is given in (5) [8]

$$E' = \frac{K_5 K_g}{1 + s K_5 T_f}(V - K_4 \Delta\delta) \quad (5)$$

where P_s, K_1, K_2, and K_3 are the cross-coupling coefficient between ALFC and AVR loops.

2.3 *Dish-Stirling Solar Thermal Plant (DSTP) Model*

In the late 1970s, dish-Stirling technology advancement started as the US. The Energy Department has funded the Jet Propulsion Laboratory's Solar Dish Electric Program. The technology is currently being established worldwide in many countries. This technology's special feature is the prime mover, known as a Stirling engine. Due to the high performance and emission-free operation of Stirling engines, Stirling heat engines are particularly suitable to transform solar thermal energy into electricity. The technology has been tested and has achieved a world record at the most efficient solar-to-grid energy conversion. As a result, new and larger megawatt plants have been built, and many more are being planned [18].

Since dish-Stirling technology will be a dominant force in the electric-powered house, user simulation, and computer model generation have become increasingly important to anticipate system response and functionality. Perhaps among the diverse concentrating technologies, the dish-Stirling concentrating solar thermal power generation system has unique technical characteristics: A parabolic-reflector unit focuses the sun's light on a smaller spot which the receiver unit then receives. A Stirling engine is coupled with an electric generator when in operation, and this generates electricity.

In this study, the solar energy converter (Φ) into heat is modeled as a first-order transfer function with time constant T_{st} is given in (6)

$$\Delta\text{heat}(s) = \frac{K_{\text{st}}}{1 + \text{sT}_{\text{st}}}\Delta\phi(s) \tag{6}$$

The out of the heat converts into mechanical torque with the help of Stirling engine (prime mover) is considered as a first-order transfer function given in (7)

$$\Delta P_m(s) = \frac{K_{\text{se}}}{1 + sT_{\text{se}}}\Delta\text{heat}(s) \tag{7}$$

where ΔP_m (s) and Δheat (s) are the change in mechanical output power and temparature changed, respectively.

2.4 Redox Flow Battery (RFB) Model

RFB is one of the most recent stationary energy storage systems. A key concern for RFB's wide-ranging acceptance is that they have high power, can be changed easily, perform well at low temperatures, are safer than other kinds of batteries, are flexible and take up less space. RFB is ideal for applications with power levels ranging from 10 KW to 10 MW and storage times ranging from 2 to 10 h [2]. During the conversion of chemical energy to electrical energy, one of the cells in the RFB has chemical energy, and the other cell is storing energy. An acid solution of vanadium ions is used as electrolytes to the battery cell using a pair of pumps. The design enables the two tanks to be sized by shape for RFB's power and energy capacities to be scaled more effectively than conventional sealed batteries. Since the RFB is a strongly non-linear model, its ALFC-AVR linearized model is shown in Fig. 2. [14]. During normal power system operation, RFB charges and rapidly discharge whenever the system has peak/sudden load demand. A dual converter performs AC/DC or DC/AC conversions.

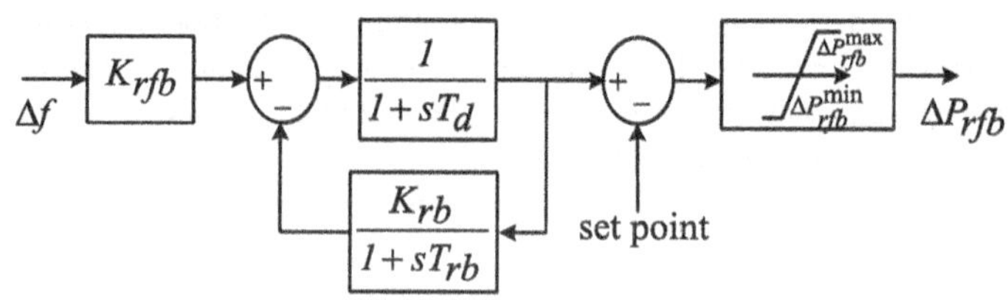

Fig. 2 Linearized model of RFB

2.5 Proposed CFOPI-FOPDN Controller

A cascade control strategy can accommodate several disturbances effectively, thereby improving control efficiency for set-point changes compared to a single-loop feedback control strategy. It is not possible to use a traditional single-loop control scheme for extreme disruptions and long-term delays. On the other side, a cascade control technique requires two loops, main or outer loop, and secondary or inner loop. The inner loop response to any system disruption is much quicker than the outer loop response. Figure 3 depicts the proposed controller structure.

The FO controllers are more flexible than the integer-order controller due to more tunable parameters [14]. The output of the proposed controller is expressed as (6)

$$\Delta P_{\mathrm{ci}} = \left[ACE_i \times \left(K_{\mathrm{Pi}} + \frac{K_{\mathrm{Ii}}}{s^{\mu i}} \right) - \Delta f_i \right] \left(K_{\mathrm{pi}} + \frac{s^{\lambda i} N_i K_{di}}{s^{\lambda i} + N_i} \right) \quad (8)$$

The proposed controller has seven tuneable knobs such as K_P, K_I, μ, K_p, K_d, λ, and N and optimized by using AFA. The derivative filter removes the unwanted noise of the input signal ΔP_{ref}.

The upper and lower limits of the controller gain $K_{\mathrm{Pi}}, K_{\mathrm{Ii}}, K_{\mathrm{pi}}, K_{\mathrm{di}}$ are 1 and 0 of the Area-*i*. The fractional-order integral (μ_i) and derivative (λ_i) operators' minimum and maximum values are 0.1 and 0.999. The filter coefficient of the FOPDN controller upper and lower bound is 100 and 1. These limits are chosen based on [14].

3 Artificial Flora Algorithm

This AFA [17] is inspired by the mechanism of flora migration and reproduction. A plant can not travel; it has the ability to spread seeds over a wide area, allowing offspring to find the most suitable habitat. First, the first plant, including its size and location, is generated at random by the algorithm. The original plant's location and spread distance as parameters are then replaced by the propagation function to produce the offspring plants. Finally, via the selection function, the perfect offspring is chosen as the new. The earlier plant appears to be the previous original plant. The

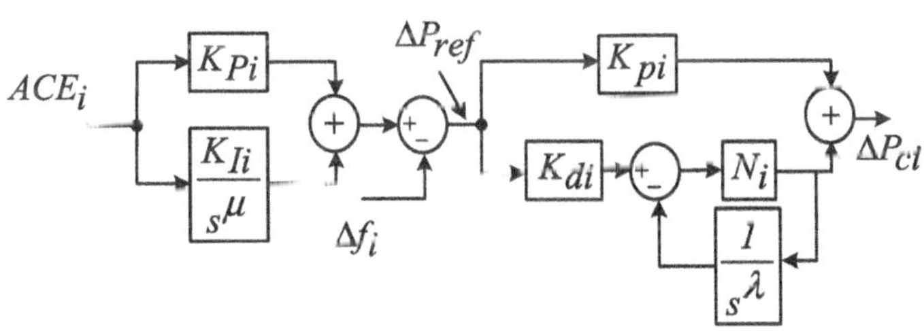

Fig. 3 Transfer function block diagram of CFOPI-FOPDN controller

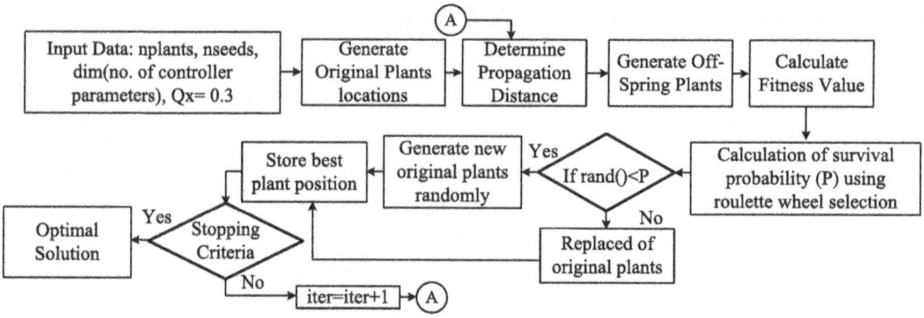

Fig. 4 Flowchart of AFA

iteration process continues until we find the best answer. The detailed mathematics of the AFA is available in [16]. Here, only a flowchart is given in Fig. 4.

4 Result and Analysis

4.1 Performance Analysis of PI, PIDN, and CFOPI-FOPDN Controller

The amalgamated ALFC-AVR loops of three-area thermal systems including STPP, DSTP, and hydro in Area-1, Area-2, and Area-3, respectively with proper nonlinearities constraints for hydro and thermal plants. The PI, PIDN, and CFOPI-FOPDN controllers investigate the dynamic responses in the considered system. The controllers gain and other parameters are optimized using the AFA at 1% of SLD in Area-1 at time $t = 1$ s. The optimal values are displayed in Table 1 and obtained dynamic responses accordingly. Only four responses are portrayed in Fig. 5a-d. The figures show that the proposed CFOPI-FOPDN controller gives better dynamic responses, viz. the magnitude of peak deviation and the number of oscillations. Hence, further studies are carried out CFOPI-FOPDN controller.

4.2 Effect of RFB on the System Dynamics

The combined AFLC-AVR system of three-area hydrothermal systems incorporating STPP in Area-1, Area-3, and DSTP in Area-2 was investigated in Sect. 4.1. Now, the system is investigated with the integration of RFB into all control areas. The RFB receives its input signal from the frequency deviation in each control area. The AFA subjected to 1% SLD in Area-1 at time $t = 1$ s is employed to optimize the proposed CFOPI-FOPDN controller parameters. The optimized values are not

Table 1 Optimal values of the different controllers' parameters

Controller	Parameters	Area-1	Area-2	Area-3
PI	${K_{Pi}}^*$	0.2981	0.3698	0.5847
	${K_{Ii}}^*$	0.0998	0.8894	0.6554
	K_{pavi*}	0.0323	0.5541	0.0365
	${K_{iavi}}^*$	0.3216	0.2219	0.1254
	K_{davi*}	0.5642	0.8774	0.2398
	${N_{avi}}^*$	23.541	62.231	12.652
PIDN	${K_{Pi}}^*$	0.8554	0.2415	0.8742
	${K_{Di}}^*$	0.2654	0.3012	0.8541
	${K_{Ii}}^*$	0.5421	0.0254	0.5612
	${N_i}^*$	10.256	36.584	24.351
	K_{pavi*}	0.5581	0.5243	0.2981
	${K_{iavi}}^*$	0.2018	0.3956	0.0321
	K_{davi*}	0.1124	0.0215	0.0874
	${N_{avi}}^*$	15.321	52.123	29.541
CFOPI-FOPDN	K_{Pi*}	0.8755	0.6965	0.5421
	${K_{Ii}}^*$	0.0251	0.0514	0.0214
	${\mu_i}^*$	0.9652	0.9678	0.9998
	${K_{pi}}^*$	0.2114	0.3128	0.9975
	K_{di*}	0.2335	0.4216	0.8513
	${\lambda_i}^*$	0.6991	0.2554	0.3561
	${N_i}^*$	31.225	69.661	0.3229
	K_{pavi*}	0.9995	0.8894	0.2154
	${K_{iavi}}^*$	0.2014	0.1022	0.6654
	K_{davi*}	0.5549	0.2779	0.3321
	${N_{avi}}^*$	26.599	54.012	69.011

shown and obtained the dynamic response. Figure 6a–c shows the comparisons of system responses. From the figures, it is seen that the integration of RFB into all control areas improves the system damping drastically in terms of peak deviation and settling time along with the fewer oscillations.

From the Fig. 6c, the capacity of RFB is calculated in each area are Area-1 = 9.9924 MW, Area-2 = 4.0794 MW, Area-3 = 4.429 MW, respectively. It is seen that RFB capacity in Area-1 is required about two times of Area-1 and Area-2 to mitigate the load-generation imbalance during the transient period.

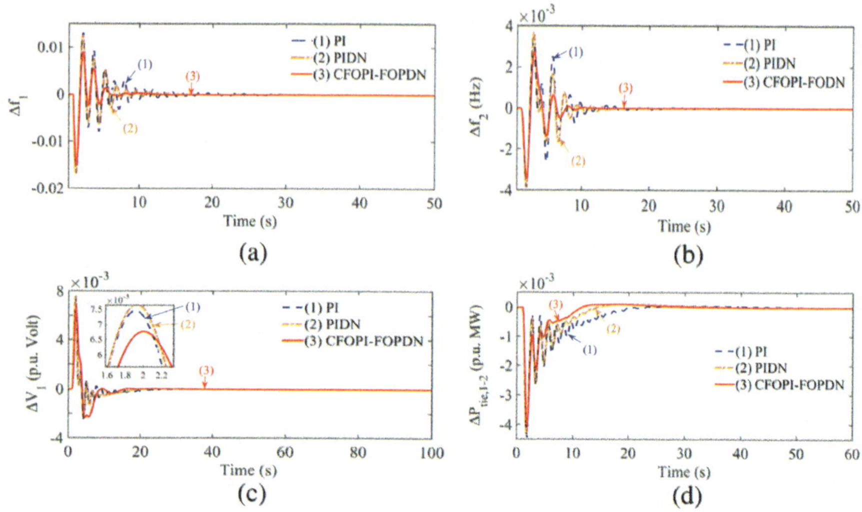

Fig.5 Comparisons of dynamic responses. **a** Δf_1. b Δf_2. **c** ΔV_1. **d** $\Delta P_{\text{tie},1-2}$

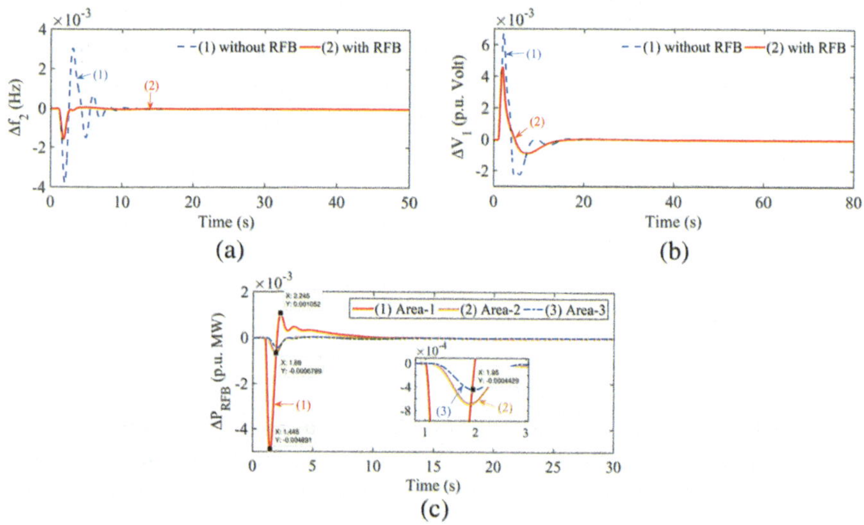

Fig. 6 Time-domain responses. **a** Δf_1. **b** ΔV_1. **c** Power deviation of RFB in all areas

4.3 Significance of Input Signal of RFB on System Dynamics

In order to analysis different input signals of the RFB charging/discharging impact on the system performance, two signals are used. One frequency deviation and another one is the area control error signal. In both cases, the proposed CFOPI-FOPDN

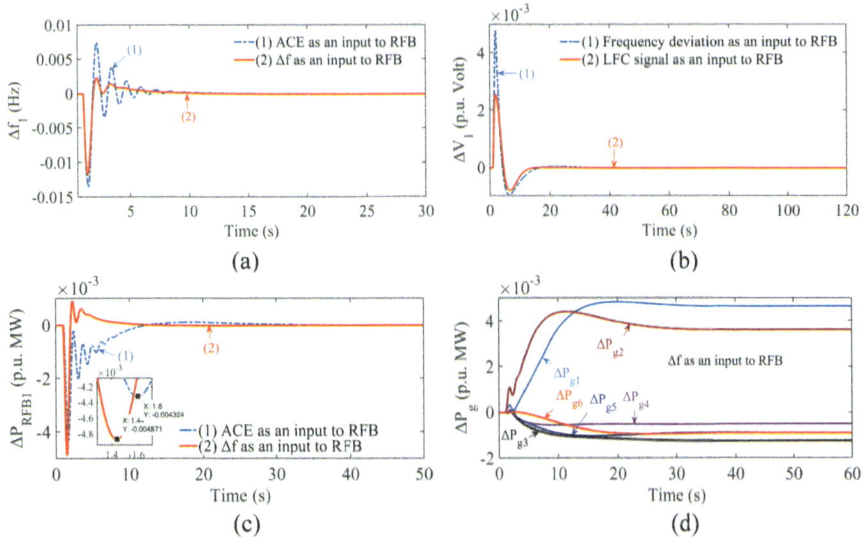

Fig. 7 Dynamic response comparisons. **a** Δf_1. **b** ΔV_1. **c** change of RFB power at different input signal in Area-1. **d** change of power generation of each unit

controller parameters are optimized using AFA one at a time. The optimal values of the CFOPI-FOPDN are not displayed. The dynamics performance comparisons in both cases are shown in Fig. 7a–d. It is examined that the frequency deviation as an input to RFB provides outstanding performance than area control error signal in terms of undershot and overshoot, whereas settling time almost the same.

Figure 7d shows the changes in generation responses with 1% SLD in Area-1 at $t = 1$se. It is clearly seen that the maximum power generation rate of generating unit-1 (thermal) in Area-1 is 1.5% per minute; unit-2 is 2.45% per minute, whereas Area-2 and Area-3 units are very less.

4.4 *Sensitivity Analysis of the CFOPI-FOPDN Controller*

The ALFC-AVR three-area thermal system model incorporates STPP, DSTP, and hydro units in Area-1, Area-2, and Area-3, respectively, and RFB in all three areas analyzed in Sect. 4.2. The time-domain responses of the proposed CFOPI-FOPDN controller at nominal system loading condition have been shown in Fig. 6a–d. The AFA was used to optimize the suggested controller parameters while the system loading changed by around ± 25% from its nominal loading. The optimal values are not displayed and obtained the dynamic responses accordingly. Figure 8a–d shows that the system dynamic responses.

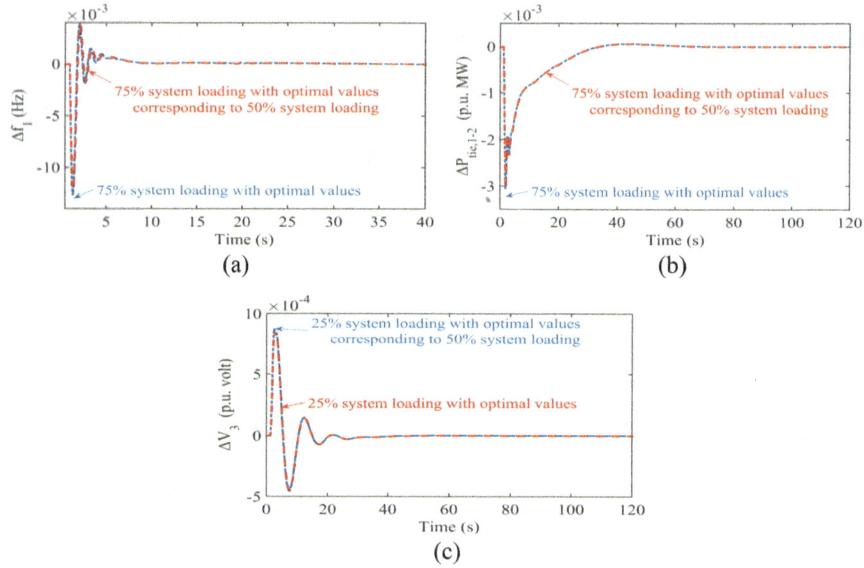

Fig. 8 Time-domain responses at diverse system loading. **a** Δf_1. **b** $\Delta P_{\text{tie},1\text{-}2}$. **c** ΔV_1

5 Conclusion

The amalgamated ALFC-AVR system of three-area thermal systems integrates STPP into Area-1, DSTP into Area-3, and hydro into Area-2 with appropriate GDB and GRC for both hydro and thermal plants. The integration of RFB into all the control areas is considered for the first time. A maiden attempt has been made to apply a new CFOPI-FOPDN controller as a secondary controller in the ALFC loop in this amalgamated ALFC-AVR model. AFA is used to optimize the controller parameters subjected to 1% SLD in Area-1 at time $t = 1$ s. According to the analysis relating to peak deviation, settling time, and the oscillations number, the suggested controller outperforms conventional controllers like PI and PIDN. The incorporation of RFB significantly enhances the performance of the system into the system. The frequency deviation is better than the ACE signal as an input to RFB. According to the sensitivity analysis, the suggested CFOPI-FOPDN controller is more resilient to a wide system loading change.

Appendix

Nominal parameters of the investigated system.

$i, j = 1, 2, 3, f = 60$ Hz, $T_{\text{ri}} = 10$ s, $K_{\text{ri}} = 0.5$, $T_{\text{ti}} = 0.3$ s, $T_{\text{gi}} = 0.08$ s, $R_i = 4\%$ p.u., $\beta_i = 0.425$, $T_{\text{ij}} = 0.0867$ p.u. MW/Hz, $T_{\text{wi}} = 1$ s, $K_{\text{phi}} = 4$, $K_{\text{dhi}} = 0.27$, K_{ihi}

$= 4$, $K_{psi} = 120$, $T_{psi} = 20$ s, $T_{spi} = 1.78$ s, $T_{sgi} = 0.1$ s, $T_{sti} = 3$ s. $T_{rbi} = 1.5$ s, $K_{rbi} = 0.4$ p.u. MW/Hz, $T_{di} = 0$ s, $K_{se} = 1$, $T_{se} = 1.2$ s, $K_{st} = 1.7$, $T_{st} = 1.7$ s. isolation $= 0.005$ p.u MW/m^2.

References

1. Saha, A., Saikia, L.C.: Performance analysis of combination of ultra-capacitor and supe conducting magnetic energy storage in a thermal-gas AGC system with utilization of whale optimization algorithm optimized cascade controller. J. Renew. Sustain. Energy **10**, 014103 (2018)
2. Arya, Y.: AGC of restructured multiarea multisource hydrothermal power systems incorporating energy storage units via optimal fractional-order fuzzy PID controller. Neural Comput. Applic. **31**, 851–872 (2019)
3. Ram Babu, N., Saikia, L.C.: Automatic generation control of a solar thermal and dish-stirling solar thermal system integrated multiarea system incorporating accurate HVDC link model using crow search algorithm optimized FOPI Minus FODF controller. IET Renew. Power Gener. **13**, 2221–2231 (2019)
4. Debbarma, S., Saikia, L.C., Sinha, N.: Automatic generation control using two degree of freedom fractional order PID controller. Int. J. Electric. Power Energy Syst. **58**, 120–129 (2014)
5. Sahu, P.C., Prusty, R.C., Panda, S.: Frequency regulation of an electric vehicle-operated micro-grid under WOA-tuned fuzzy cascade controller. Int. J. Ambient Energy 0143–0750 (2020)
6. Tasnin, W., Saikia, L.C.: Comparative performance of different energy storage devices in AGC of multisource system including geothermal power plant. J. Renew. Sustain. Energy **10**, 024101 (2018)
7. Elgerd O.I.: Energy systems theory. In: An introduction, 3rd edn. McGraw-Hill New Delhi (1983)
8. Saadat H.: Power system analysis. 3rd end. Tata McGraw Hill New Delhi (1994)
9. Rakhshani, E., Rouzbehi, K., Sadeh, S.: A new combined model for simulation of mutual Effects between LFC and AVR Loops. In: Asia-Pacific Power and Energy Engineering Conference, pp. 1–5. Wuhan, China. (2009)
10. Gupta, A., Chauhan, A., Khanna, R.: Design of AVR and ALFC for single area power system including damping control. In: Recent Advances in Engineering and Computational Sciences (RAECS), pp. 1–5. Chandigarh, India. (2014)
11. Sambariya, D.K., Nath, V.: Optimal control of automatic generation with automatic voltage regulator using particle swarm optimization. Univ. J. Control Autom. **3**, 63–71 (2015)
12. Vijaya Chandrakala, K.R.M., Balamurugan, S.: Simulated annealing based optimal frequency and terminal voltage control of multi source multi area system. Int. J. Electric. Power Energy Syst. **78**, 823–829 (2016)
13. Rajbongshi, R., Saikia, L.C.: Combined control of voltage and frequency of multi-area multisource system incorporating solar thermal power plant using LSA optimized classical controllers. IET Gener. Transm. Distrib. **11**, 2489–2498 (2017)
14. Dekaraja, B., Saikia, L.C.: Performance of redox flow battery in combined frequency and voltage control of multiarea multisource system using CFOPDN-FOPIDN controller. Int. Trans. Electric. Energy Syst. **31**, e12782 (2021)
15. Ramoji, S.K., Saikia, L.C.: Utilization of Electric Vehicles in combined voltage-frequency control of multiarea thermal-CCGT system using 2DOFTID controller. Energy Storage **3**, e234 (2021)
16. Dekaraja, B., Saikia, L.C., Ramoji, S.K., Babu, N.R., Bhagat, S K., Behera, M.K.: Coordinated control of voltage and frequency in a three-area multisource system integrated with SMES

using FOI-FOPDF controller. In: 2020 IEEE 17th India Council International Conference (INDICON), pp. 1–7 (2020)
17. Cheng, L., Wu, X.H., Wang, Y.: Artificial flora (AF) optimization algorithm. Appl. Sci. **8**, 329 (2018)
18. Babu, N.R., Saikia, L.C.: Load frequency control of a multi-area system incorporating dish-stirling solar thermal system and coyote optimized PI minus DF controller. In: 2020 IEEE International Conference on Power Electronics, Smart Grid and Renewable Energy (PESGRE2020), pp. 1–6. Cochin, India (2020)

Cascaded Neural Network Approach for Template-Based Array Synthesis

Sagar Dutta, Banani Basu, and Fazal Ahmed Talukdar

Abstract In this paper, we present a neural network approach to estimate the excitation required to generate the template-based shaped beam pattern using radial basis neural network (RBFNN). We have employed a 16-element linear array of isotropic sources with inter-element distance of 0.5λ. Using sets of known input/output pairs, a neural network is trained to construct a relationship between the input and output pair of the linear array. The model efficiently predicted the set of excitation voltages required to produce the desired template-based shaped beam pattern.

Keywords Excitation · Neural network · Radial basis function · Linear array · Radiation pattern

1 Introduction

Antenna arrays are widely used in applications such as radar and wireless communications. There are range of techniques that have been developed for the synthesis of arrays. Synthesis of radiation pattern consists of careful selection of the antenna parameters such as side-lobe level, null positions, and beam-width of the pattern, so that the radiation pattern reaches the desired specifications. In literature, over the past years, many beam forming techniques have been reported. In [1], author has reported the synthesis of the beam pattern in large arrays with unequal spacing. Pattern synthesis of circular antenna array is proposed in [2] by using nonlinear least square method. In the near- and far-field region, an optimization method is proposed for the flat-top pattern [3]. Because of the nonlinearity related to antenna radiation patterns, neural network approach has been adapted to get the same accuracy without

S. Dutta (✉) · B. Basu · F. A. Talukdar
Department of Electronics and Communication, National Institute of Technology Silchar, Silchar, Assam, India
e-mail: banani@ece.nits.ac.in

F. A. Talukdar
e-mail: fazal@ece.nits.ac.in

K. N. Das et al. (eds.), *Proceedings of the International Conference on Computational Intelligence and Sustainable Technologies*, Algorithms for Intelligent Systems,
https://doi.org/10.1007/978-981-16-6893-7_11

involving any stringent mathematical complexity. NN have the capability to assess a model to adapt its parameters after a training phase with known input/output pairs. The increased use of the neural network in various electromagnetic applications influenced its use in optimization and synthesis problems.

The synthesis of radiation patterns of an antenna array by changing the excitation is a nonlinear problem, and machine learning has the capability to handle nonlinear problems with the help of training. In [4], neural network has been used for the synthesis of periodic phased array antenna for any configuration. The problem of a non-uniform array synthesis has been solved by a neural network [5]. Preprocess coding neural network is used in [6] for antenna array synthesis in the desired direction. In [7, 8], pattern synthesis using hybrid Fourier neural network is reported, which utilized Fourier synthesis of the phases to predict the excitation. In [9–11], the author used an external MoM analysis tool to find out the phase to estimate the excitation for the desired pattern.

A neural network technique is used in this paper to estimate the excitation for three different beams (flat-top, pencil, and cosecant) using a linear 16-element array of isotropic antennas with 0.5λ inter-element distance. Unlike the strategies stated above, which use the Fourier method and an external MoM analysis method to calculate the excitation, our approach uses a neural network to predict the excitation required for the desired pattern. A suitable excitation data set which corresponds to a specific template is chosen. The data set consisting of templates and voltage sets is used to train the neural network. Given a specific template, the NN can predict the excitation voltage to produce the desired radiation pattern confining the template boundary. The synthesized beam and the desired template are observed to have a good agreement.

2 Neural Network

Artificial neural networks [12] ascribe their origin to the human brain's biological neuron. Their operation is based on the use of mathematical tools for basic general purposes and parallel processing that is capable of solving complex problems. Complex numerical methods provide conventional approaches for radiation pattern syntheses. Given their usefulness, these numerical methods takes time. The neural network, on the other hand, is faster because it does not require complex numerical procedures. ANNs do not depend on the complexity of the way the training data set is generated. This makes the calculation process in ANNs easier and quicker than many traditional methods. Multilayer perceptrons take longer to train than RBF neurons. Because of the RBNN's local neural response, they are better suited for synthesis.

Figure 1 shows the topology of a radial basis function neural network (RBFNN). It has an input layer, a hidden layer, and an output layer, with a radial basis function as the activation function. RBF neuron can be trained faster than MLP neuron, with both supervised and unsupervised learning, due to its two-stage training approach. The fundamental function parameters are in the first layer, whereas the activation function

Fig. 1 Radial basis neural network structure

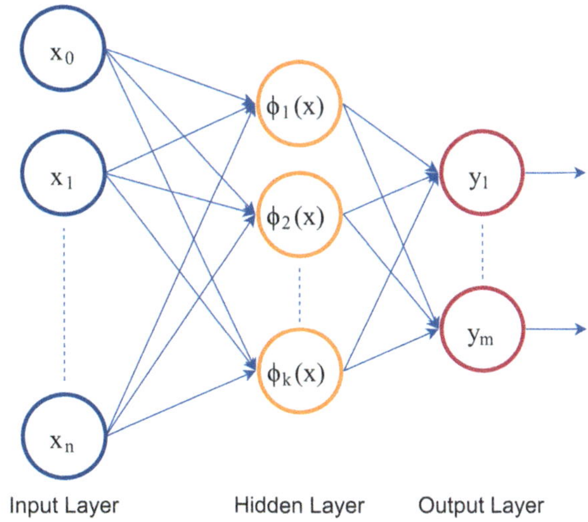

is in the hidden layer. RBFNN is a curve fitting network, unlike back propagation networks, which may be thought of as an implementation of an optimization problem.

Let us consider a series of N data points in which all n-dimensional input vectors $\mathbf{x} = x_i : i = 1, ..., n$ are assigned to its corresponding outputs y, and expressed as

$$f(x) = \sum_{p=1}^{P} w_p \varphi \left(\left\| \mathbf{x} - \mathbf{x^p} \right\| \right) \tag{1}$$

where w_p and $\mathbf{x^p}$ are the weight and center vectors, respectively, and $\varphi\left(\|\mathbf{x} - \mathbf{x^p}\|\right)$ is a Gaussian function given by,

$$\varphi \left(\left\| \mathbf{x} - \mathbf{x^p} \right\| \right) = exp \left(\frac{- \|\mathbf{x} - \mathbf{x^p}\|^2}{2\sigma^2} \right) \tag{2}$$

where σ controls the Gaussian width. RBFNN aims to identify the weights in order to allow the function to go through data points.

3 Array Model

Let us consider an N isotropic element linear array placed along the y-axis as shown in Fig. 2. The radiation from the individual antenna array elements contributes to the overall radiation pattern of the antenna array. The amplitude, phase, and inter-element spacing of the antenna elements are all taken into account while generating the required pattern.

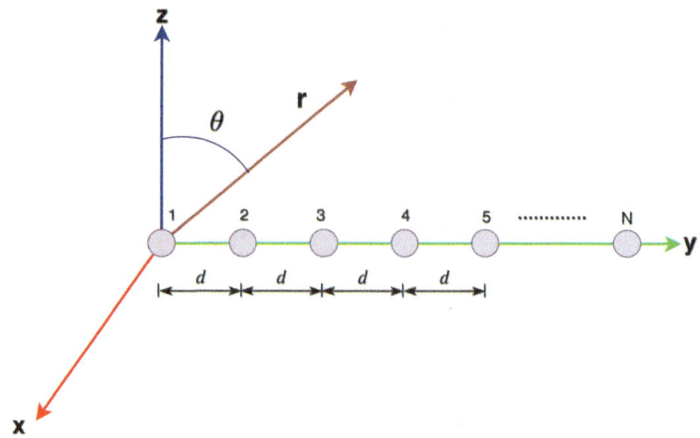

Fig. 2 Linear array configuration

The entire array's total electric field is given by,

$$\text{E(total)} = \text{E(Single Element)} \times \text{Array factor} \tag{3}$$

$$\mathbf{E}(\theta, \phi) = \mathbf{S}(\theta, \phi) \times AF(\theta, \phi) \tag{4}$$

where the array factor for the entire linear array can be written as

$$AF(\theta) = \sum_{n=1}^{N} I_n e^{j(n-1)(\mathbf{k}d\cos\theta + \beta_n)} \tag{5}$$

where N, d, I_n, β_n, and **k** are the number of antenna elements, the inter-element spacing, excitation coefficient, excitation phase, and the propagation vector for the n-th element. The expression for calculating the far-field magnitude and phase is given by,

$$|\mathbf{E}(\theta, \phi)| = \sqrt{Re\{\mathbf{E}(\theta, \phi)\}^2 + Img\{\mathbf{E}(\theta, \phi)\}^2} \tag{6}$$

$$\theta_E = \arctan \frac{Re\{\mathbf{E}(\theta, \phi)\}}{Img\{\mathbf{E}(\theta, \phi)\}} \tag{7}$$

where $Re\{\mathbf{E}(\theta, \phi)\}$ and $Img\{\mathbf{E}(\theta, \phi)\}$ are the real and imaginary parts, respectively.

Consider a column vector **x** containing of N excitation values applied to the antenna array. At a distance r from the origin, the electric field's co-polar component in the plane $\phi = 0$ and in the direction θ, can be denoted as

$$e(\theta) = \mathbf{x}^{\mathrm{T}}\mathbf{a}(\theta) \tag{8}$$

where $\mathbf{a}(\theta) = (a_0(\theta), a_1(\theta), ..., a_{N-1}(\theta)$ and it denotes relation between the voltages (x) applied to array elements in the direction θ and the radiation pattern $e(\theta)$.

A matrix $\mathbf{A}$ can be defined as $\mathbf{A} = (\mathbf{a}(\theta_1), \mathbf{a}(\theta_2)), ..., \mathbf{a}(\theta_N))$ and is used to calculate the electric field in the direction $\theta_i (i = 1...N)$, as

$$\mathbf{E}^{\mathrm{T}} = \mathbf{x}^{\mathrm{T}}\mathbf{A} \tag{9}$$

where $\mathbf{E} = (e(\theta_1), e(\theta_2), ..., e(\theta_N))^T$.

The suitable voltage set can be directly computed using the expression below when the amplitude and phase are known.

$$\mathbf{x}^{\mathrm{T}} = \mathbf{A}^{-1}\mathbf{E}^{\mathrm{T}} \tag{10}$$

where $\mathbf{A}^{-1}$ is the inverse of matrix $\mathbf{A}$. The neural network can predict the output with variety of inputs once it's trained. It is worth noting that, as in (5), the antenna configuration determines the function that associates the distribution of the radiated field with the excitation added to the array.

4 Synthesis of Radiation Pattern from Template

In most cases, only magnitude data is known and the phase information is not available. Without phase data, the neural network will be unable to associate the input-output vectors. To correctly approximate the necessary voltages, it is therefore critical to use both amplitude and phase data. Equation (5) has been used to calculate the array factor for the mentioned planar array having isotropic elements using MATLAB.

The entire electric field is represented as a complex variable, with the real and imaginary parts utilized to determine the absolute electric field distribution and phase distribution (θ_E), which serve as the neural network's input and output training data sets, respectively. The range of magnitude and phase of the excitation are kept within a range of 0 to 1 and -$\pi/2$ to $\pi/2$, respectively.

MATLAB is used to create the excitation voltage and its related electric field. A template is then used to represent the electric field. The training data set for the neural network is made up of these templates and their corresponding voltages. After training, the neural network can relate the electric field to the excitation voltage.

Three desired radiation patterns, flat-top beam, pencil beam, and cosecant beam have been chosen, and their corresponding template data-set is prepared. The phase for a desired pattern is estimated using the neural network. The calculated phase as well as the template data are fed into a trained neural network that estimates the excitation voltages. Figure 3 shows the complete procedure. The first neural network takes in the magnitude of desired electric field pattern and estimates corresponding phase. This estimated phase along with the magnitude of the desired pattern is then fed to the second neural network which estimates the corresponding excitation voltage.

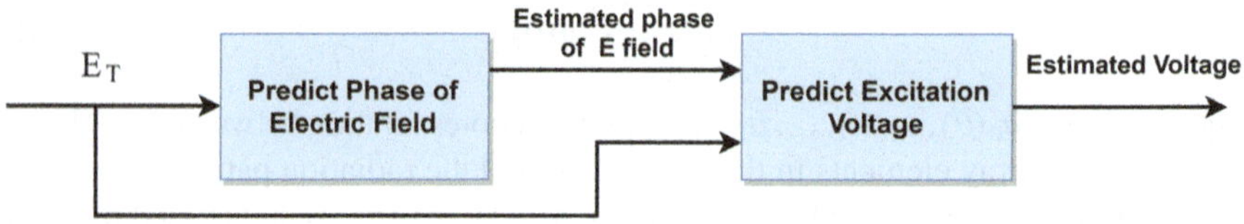

Fig. 3 Training procedure

5 Results

The simulations for our study were carried out using a 16-element linear array of isotropic elements with a 0.5λ inter-element spacing. For the synthesis of the radiation pattern, RBFNN with 250 neurons in the hidden layer and the k-means algorithm is used. The performance goal is set to 10^{-6} and learning rate at 0.01. For each beam pattern 5000 training data sets were created using MATLAB to train the neural network. The synthesis of the desired radiation patterns have been successfully produced by using neural network for flat-top beam, pencil beam, and cosecant beam.

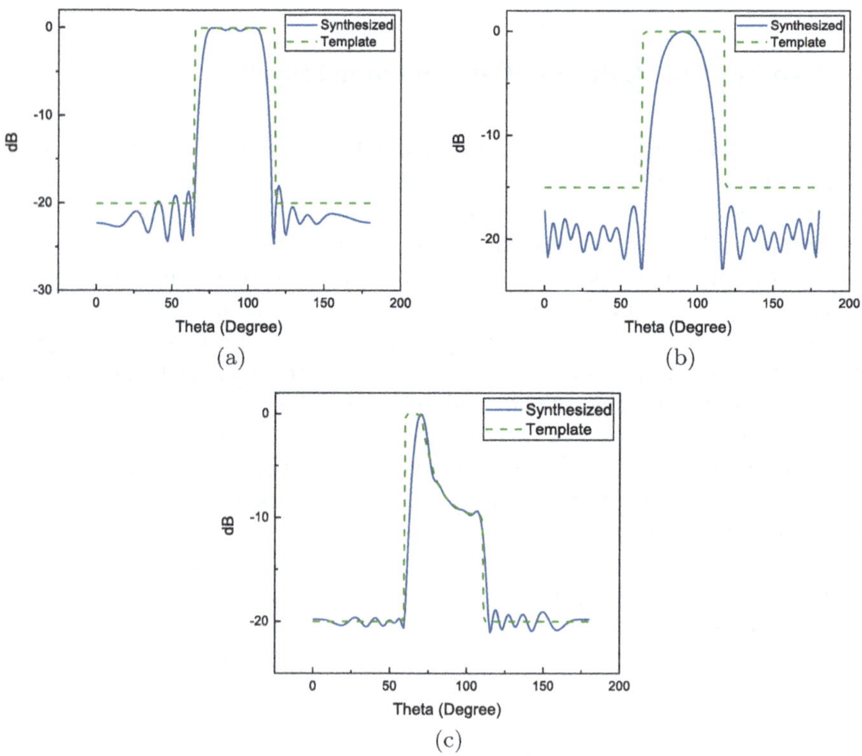

Fig. 4 Neural network synthesized **a** Flat-top beam, **b** pencil beam, **c** cosecant beam

We have trained NN separately with three sets of data corresponding to three beams mentioned above. Although the template is similar in Fig. 4a, b, the NN is trained to produce flat-top and pencil beam in each case. The goal of the training process is to obtain the excitation voltages which would generate the desired pattern within the boundaries of the template. The synthesized flat-top beam, pencil beam, and cosecant beam are shown in Fig. 4a–c respectively. Here, the beam width is kept fixed for all the templates from 60° to 120°. It is seen that the results show a very good agreement with the template specification. The synthesized patterns are within the boundary of the template with a sidelobe level of −20 dB. Moreover, Tables 1, 2, and 3 show the estimated magnitude and phase of the excitation voltages required to produce the different beams. Compared to the strategies stated above, which use the Fourier method and an external MoM analysis method to calculate the excitation, our approach successfully utilized a RBF neural network to predict the excitation required for the desired pattern.

Table 1 Synthesized voltages for flat-top beam

Element #	Magnitude	Phase (deg.)	Element #	Magnitude	Phase (deg.)
1	0.2247	−75.2539	9	0.1599	83.1266
2	0.2345	86.1334	10	0.2725	−88.8122
3	0.0552	−74.3395	11	0.5312	90
4	0.1784	−78.1620	12	0.3043	90
5	0.3783	−81.3220	13	0.0325	90
6	0.7989	−86.1994	14	0.2069	80.6443
7	1	−87.2404	15	0.1414	−76.2755
8	0.7717	−82.9394	16	0	0

Table 2 Synthesized voltages for pencil beam

Element #	Magnitude	Phase (deg.)	Element #	Magnitude	Phase (deg.)
1	0.1067	21.9297	9	0	0
2	0.6544	63.0608	10	0.5625	−22.6276
3	0.7993	−39.4296	11	0.6572	−1.7775
4	0.1676	36.1898	12	0.2828	25.1696
5	0.5758	86.7658	13	0.0873	−10.4148
6	0.5916	−38.0036	14	0.1911	−33.4571
7	0.4364	−26.2102	15	0.3533	−11.7556
8	0.6397	1.7863	16	0.4136	29.2913

Table 3 Synthesized voltages for cosecant beam

Element #	Magnitude	Phase (deg.)	Element #	Magnitude	Phase (deg.)
1	0.5026	−39.4769	9	0.3997	13.3233
2	0.2207	−51.0075	10	0.2508	90
3	0.1956	−39.1626	11	0.3530	8.3337
4	0.5796	−15.3727	12	0.3072	65.4895
5	1	−2.2625	13	0.2854	−26.7542
6	0.7548	32.6175	14	0.2665	21.4915
7	0.6458	85.3311	15	0.0891	−90
8	0.5277	−48.7357	16	0.0906	−11.6256

6 Conclusion

Cascaded neural network has been used successfully for synthesizing flat-top, pencil, and cosecant beams for a 16-element linear array with non uniform excitation. For the synthesis, the template was fed to the input of the cascaded neural network which estimated the phase of the electric field to estimate the excitation of the desired beam pattern. The produced templates and it is found that all the beam patterns are well within the confining boundaries.

Conflict of interest The authors declare that they have no conflict of interest.

References

1. Yang, K., Zhao, Z., Liu, Q.H.: IEEE Trans. Antennas Propag. **61**(2), 627 (2012)
2. Guo, H., Guo, C.J., Qu, Y., Ding, J.: Progr. Electromagnetics Res. **50**, 331 (2013)
3. Cai, X., Geyi, W.: IEEE Trans. Antennas Propag. z **67**(2), 980 (2018)
4. Bilel, H., Selma, L., Taoufik, A.: 2016 17th International Symposium on Antenna Technology and Applied Electromagnetics (ANTEM) (IEEE, 2016), pp. 1–6
5. Ayestarán, R., Las-Heras, F., Martinez, J.: J. Electromagnetic Waves Appl. **21**(8), 1001 (2007)
6. Zhao, Z., Zhao, H., Zheng, M.: Electronics Lett. **55**(2), 67 (2018)
7. Ghayoula, E., Bouallegue, A., Ghayoula, R., Fattahi, J., Pricop, E., Chouinard, J.Y.: 2017 IEEE International Conference on Systems, Man, and Cybernetics (SMC) (IEEE, 2017), pp. 3290–3295
8. Ghayoula, E., Ghayoula, R., Haj-Taieb, M., Chouinard, J.Y., Bouallegue, A.: Progr. Electromagnetics Res. **67**, 45 (2016)
9. Ayestaran, R.G., Las-Heras, F., Herran, L.F.: IEEE Antennas Wireless Propag. Lett. **5**, 45 (2006)
10. Ayestaran, R.G., Las-Heras, F., Herrán, L.F.: IEEE Trans. Antennas Propag. **55**(3), 832 (2007)
11. Ayestaran, R.G., Las-Heras, F.: IEEE Trans. Antennas Propag. **54**(8), 2420 (2006)
12. Kim, P.: (2017)

Deep Learning Assisted Technology for MIMO OFDM 5G Application

Shovon Nandi, Arnab Nandi, and Narendra Nath Pathak

Abstract 5G network complexity has come to a noticeable point with the help of the deep learning method. Deep learning also can optimize the performance parameter values and reduce the network cost as well. This paper delineates the deep neural network (DNN) model for the improvement of signal-to-noise ratio (SNR). Also, the channel estimation problem is focused and the BER; i.e., bit error rate is simulated by the subcarrier signals. The channel state information (CSI) is also portrayed using the robustness of the DNN-based working model. The DNN toolbox of MATLAB 2020b and Python 4.2.3 are used to construct deep learning models.

Keywords 5G · Beamforming · BER · Signal-to-noise ratio · Channel state information · Maximal combining ratio · Deep neural network

1 Introduction

The era of the Internet boom of today's world changes the technical field in a very fast manner. At the same time, we are very much concerned about accuracy, reliability, security, and efficient data communication. Many areas are highly benefited by the machine learning applications like intelligent signal processing, image processing, speech processing, video processing, and data analysis. The next generation is mainly dealing with big data. Lots of analytical tools are very much helpful in this regard. Dissimilar to the logic-dependent master systems that were predominant in the prior field of machine learning, the trust in information-guided strategies is propelled by the triumphs of pattern recognition tools dependent on that [1, 2]. An average artificial intelligence (AI) 'agent' sees and models its current circumstance and processes

S. Nandi (✉)
Bengal Institute of Technology, Kolkata, India

A. Nandi
National Institute of Technology, Silchar, India

N. N. Pathak
Dr BC Roy Engineering College, Durgapur, India

K. N. Das et al. (eds.), *Proceedings of the International Conference on Computational Intelligence and Sustainable Technologies*, Algorithms for Intelligent Systems,
https://doi.org/10.1007/978-981-16-6893-7_12

suitable activities, by expecting their outcomes [3]. These tools depend on many years old algorithms, for example, back-propagation, the expectation–maximization (EM) calculation [4] what's more, Q-learning, with various current algorithmic improvements, counting novel regularization strategies, and versatile learning rate plans [5]. The AI approach performs with a huge number of gathered instances. In each case, the model is trained with the set of related data and learned according to the fulfillment of the target job. As per the resolution of the trained model considered as a machine, the learning is made possible by the hypothesis class. The optimization of the performance parameters is truly depending on how well the machine is trained with a bulk amount of datasets [6, 7]. To achieve high-level accuracy, computational intelligence techniques like classification tools, deep learning tools, evolutionary algorithms, fuzzy sets, and rough sets have been used enormously [8, 9]. Also in the autonomous vehicle industry, the most attractive development is seen in self-driving cars. The major contribution of the application of reinforcement learning creates this achievement [10].

The future generation wireless networks get fuel from the massive multi-input multi-output technology. Here, in the 'frequency division duplex (FDD)' or 'time division duplex (TDD)', the base stations (BSs) are equipped with large arrays of cooperative antennas to spread the aroma of spatial multiplexing of several user equipment's (UEs) within the same frequency range. In this massive MIMO technology, the downlink of signal facilitates large degrees of freedom as the number of antennas in the BS is typically higher than the user counts. Hence, it is possible to fully control the transmitted signal in a proper direction and reduce the interference, which yields a beamforming gain and improves the effective SNR [11]. In a time-varying channel, the channel state information (CSI) helps in the analysis of decoded data at the receiving end. It has been observed that machine learning (ML)-based techniques can able to reduce feedback overhead and better CSI estimation in a massive MIMO system. Figure 1 shows the channel state information feedback route

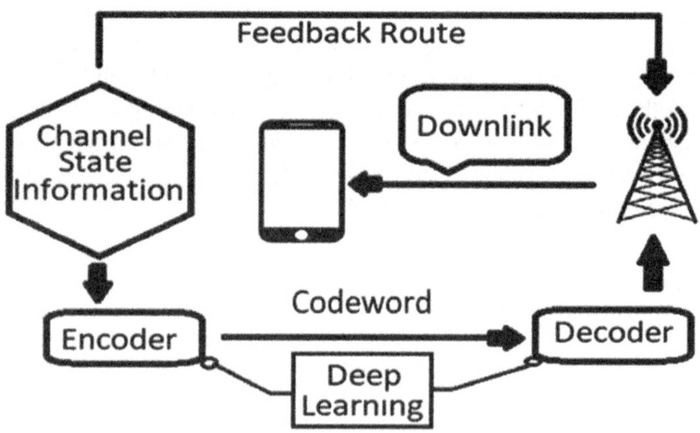

Fig. 1 Channel state information feedback in deep neural network (DNN) model

in a deep neural network (DNN) model. Deep neural network (DNN) or deep learning (DL) consists of knowledge-based deep layers. This DNN can be viewed as a subset of machine learning and depends on a special domain of neural network called artificial neural network—ANN [12–14]. The essential advantage of deep neural network (DNN) is mechanized component acknowledgment: the capacity to get includes and learn them naturally. The training process of the CSI at the receiving end undergoes pilot communication and CSI estimation. In [12], a new convolutional neural network (CNN)-based technique CSiNet was instigated to provide better sensing and recovery mechanism getting the feedback information of training data. The precoding matrix of CSI through encoding and convolutional neural network model generates codewords. The recovery of original CSI data is possible by the connected network.

2 ML Overview

In general, machine learning stipulates the changes with the help of artificial intelligence within the system which can carry out the tasks. There are some tasks like recognition, planning, robot, diagnosis, control, prediction. ML models are applied in these tasks of an intelligent agent with an environment. The architecture of a typical AI 'agent' is shown in Fig. 2. A trained model can efficiently perform the task provided to it using some primary arithmetic calculations.

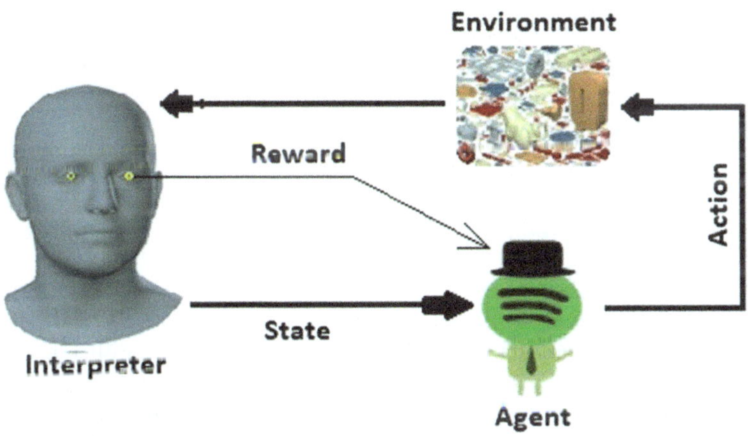

Fig. 2 A typical AI agent-based reinforcement learning system

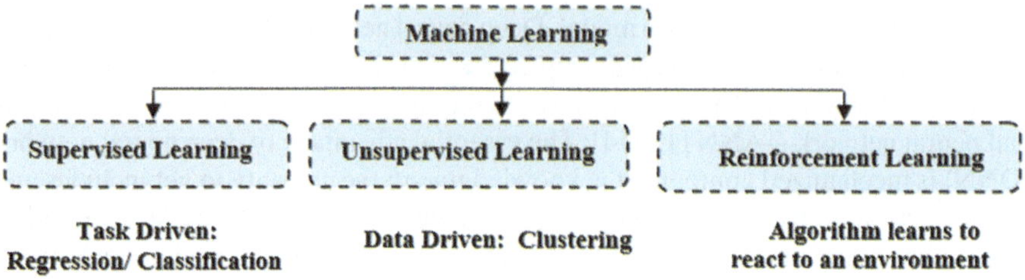

Fig. 3 Types of ML

2.1 Types of ML

The convention of domain knowledge is very much required in the field of conventional engineering; machine learning tools for driving the information can do a variety of works where design cycle, cost of usage, complexity decrease are included.

It can use the accessibility of information and computing resources in many engineering areas. Different kinds of machine learning algorithm models loan themselves to various tasks relying upon the accessibility of instances of wanted nature or feedback [15–17]. Machine learning is broadly categorized into different types of learning as shown in Fig. 3.

In the regression method, the data will come with a specified label and the algorithm will work on this defined dataset. This kind of algorithm lays out an answer key that can be used to estimate the data validation. Sometimes, the dataset is not defined with the label. The algorithms which deal with such data, predict the output after drawing out its characteristics and pattern. No structured data can be associated with an unsupervised algorithm. In the case of the reinforcement algorithm, no dataset is provided. Mostly it works on human behavior [18].

2.2 Role of ML in Network Layers

ML algorithms should be deployed and trained at different levels of the network which includes mathematical models, decision-making models as well as networking models. Because of sending data from its source to destiny, the necessary steps usually have to be followed. Similarly, ML algorithms follow such kind of networking to deploy the data finally as shown in Fig. 4.

2.3 Impacts of ML Methods on Physical (PHY) Layer

Different 5G PHY layer schemes like Massive MIMO, mmWave MIMO, Non-Orthogonal Multiple Access (NOMA), Cooperative Communication, and Cognitive

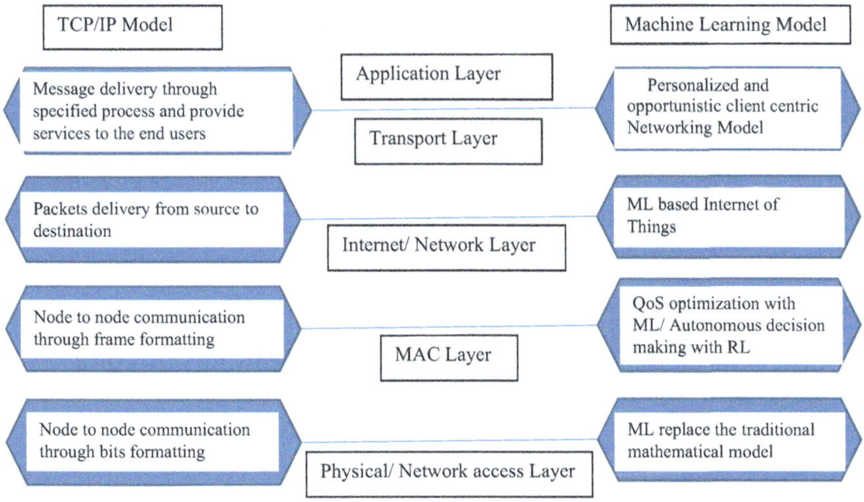

Fig. 4 Comparative view: TCP/IP model and machine learning model

Radio are mostly benefitted from ML application. The focus of this work is to fulfill the expanding need for fast and efficient communication and the prerequisites of 5G and more advanced networks. The old radio access policies are revived by the recent research on machine learning-based technology. The situation shows each cellular user equipment (UEs) can communicate the information all the while by utilizing a similar carrier frequency [19–21]. The superposition theorem of the additive white Gaussian noise (AWGN) along with the complex modulated signals coming from each UEs is applied at the base station [22, 23].

3 Simulation Model

The channel state information for both transmitting and receiving antenna is a vector of channel coefficients of OFDM subcarriers. The OFDM symbols are separated by the cyclic prefix in the time domain. As a consequence, we achieve the robustness of the system and are able to overcome the interference between two successive symbols [24, 25]. The transmission channel is considered the Rayleigh fading channel [26]. Multiple UEs are received proper signal interference noise ratio (SINR), with the coherent power allocation technique. This also helps in the performance improvement of joint symbol decoding at the receiver side. The simulated symbols obey the fading channel properties which are also incorporated during the creation of datasets in an offline method using DNN. This process is performed for mapping the obtained signals with the transmitted ones online. Five unique layers are considered to construct the algorithms; the ground layer is the source layer which consists of 400 real-valued feature vectors; the second layer is a recurrent neural network (RNN) layer which consists of 138 hidden layers; next layer is the feed-forward layer of a neural network whose output size is 25, multiple class neural network layer is the fourth one; the fifth layer is the classification layer whose output size is same as the previous one. The estimated values are achieved from the classification layer. The simulation process was conducted with the raining symbol of 500,000 along with the 120 epochs in the simulation process. The UEA and UEB are considered as the high data rate-driven equipment with higher constellation and the lower data rate-driven equipment with lower constellation, respectively.

4 Analysis of Beamforming and BER of Multiple Antenna System

The massive MIMO scheme results in huge spatial multiplexing gains, and hence, the capacity of the cellular network increases several fold. Further adding more antennas leads to reduced radiated power and greater simplicity in signal processing [27–30]. The main idea of beamforming in a wireless communication system is that it improves the energy efficiency of the system by focusing energy in a particular direction. The formation of signal beam is happening in a way so that it can able to maximize the reception or transmission capacity in a particular direction and minimize in other directions. It suppresses the signal in the direction of the interfering user or unintended user [31–34]. So, this is required to maximize the SNR for each user and suppress the interference. As we know that wireless channel experiences deep fade which can be overcome by the diversity scheme. Figure 5 shows the simple model of beamforming in single-input multiple-output (SIMO) system.

Assuming that the multiple antenna channel consists of L antennas, follows the equation $y = hx + n$, where T and R denote transmitting and receiving antenna, respectively; x and y_i (i is the integer) represent transmitted and received symbol; $h_1, h_2 \ldots h_L$ are the channel fading coefficient; n_i is the noise on ith receive antenna; so the L dimensional received vector is mentioned as,

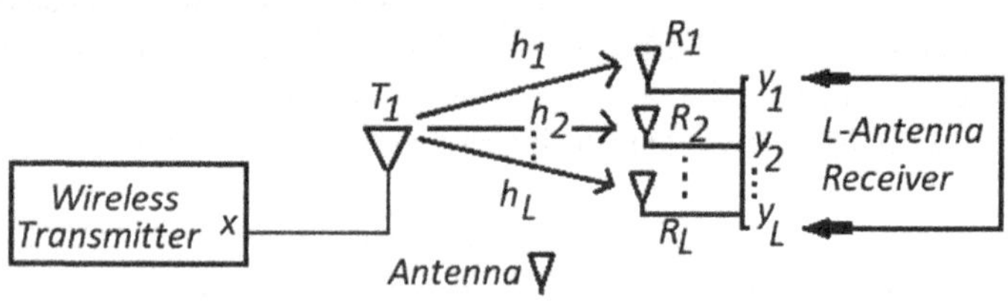

Fig. 5 Beamforming process in single-input multiple-output system

$$\begin{bmatrix} y_1 \\ y_2 \\ . \\ . \\ y_L \end{bmatrix} = \begin{bmatrix} h_1 \\ h_2 \\ . \\ . \\ h_L \end{bmatrix} x + \begin{bmatrix} n_1 \\ n_2 \\ . \\ . \\ n_L \end{bmatrix}$$

The beamformer of the receiver side is expressed as below: Consider w denotes the receive combiner or beamformer. Then the weighted combination of the received symbol or the output of the beamformer is,

$$\tilde{y} = [w_1^* w_2^* \ldots w_L^*] \begin{bmatrix} y_1 \\ y_2 \\ . \\ . \\ y_L \end{bmatrix}$$

The beamformer that maximizes the SNR is, $W = \frac{h}{\|h\|}$ is termed as the maximal ratio combiner (MRC). The BER for the multiple antenna system is $\approx \frac{1}{2^L} C_L^{2L-1} \left(\frac{1}{SNR}\right)^L$. It has been observed that as the number of antenna increases, BER decreases in the order of $\left(\frac{1}{SNR}\right)^L$. This is the advantage of diversity scheme. In MIMO system, the multiple antennas are incorporated over the system then the beamforming increases the reliability. Here the increased data rates can be achieved by transmitting multiple information streams in parallel, which is popularly known as spatial multiplexing. So it can be concluded that the maximize SINR is possible through the beamforming process.

In our work, we consider BPSK symbols with power P for the modulated symbols, $0 \rightarrow +\sqrt{P}, 1 \rightarrow -\sqrt{P}$. Also, the BER, P_e in AWGN channel is $P_e = Q\left(\sqrt{SNR}\right) \approx \frac{1}{2} e^{-\frac{1}{2} SNR}$. So it is concluded that BER decreases exponentially in AWGN channel as shown in Fig. 6. The average BER with L receives antennas after MRC combining as expressed as BER $= \left(\frac{1-\lambda}{2}\right)^L \sum_{l=0}^{L-1} (L + l + 1) C_l \left(\frac{1+\lambda}{2}\right)^l$ where λ is defined as $\sqrt{\frac{SNR}{SNR+2}}$. The variation of BER with respect to SNR is portrayed in Fig. 7 for MRC. Figure 8 shows SINR for MRC with CSI estimation.

The simulated output of Figs. 9 and 10 shows the BER versus SNR curve for user equipment A (UEA) and user equipment B (UEB) by varying the cyclic prefix (CP) of several schemes, where ML denotes maximum likelihood, SIC denotes successive interference cancelation, LS denotes least square, and DNN denotes deep neural network. Because of the power of the DNN, it can hinder the ISI viably and it can without much of a stretch get the channel state data dependent on the accessible training information. In any case, even in the ideal channel condition DNN-based analysis provides better execution. The minimum mean square and least square estimators are considered to get the powerful impact of pilot carriers for discovering the channel impulse response at the receiving end. It can be concluded from Figs. 9 and

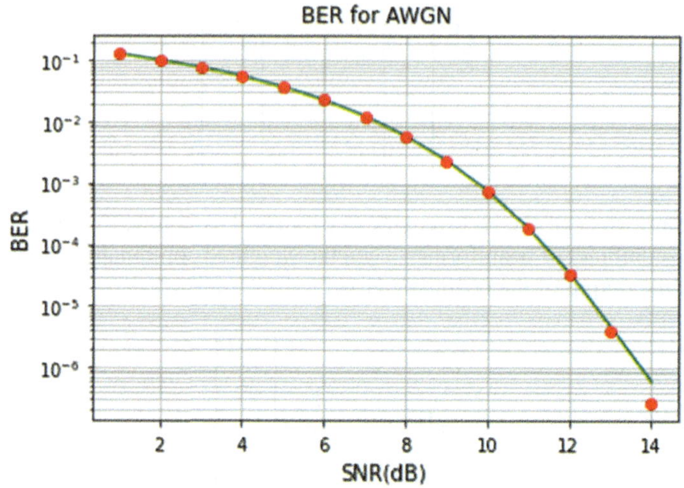

Fig. 6 BER versus SNR (dB) for AWGN channel

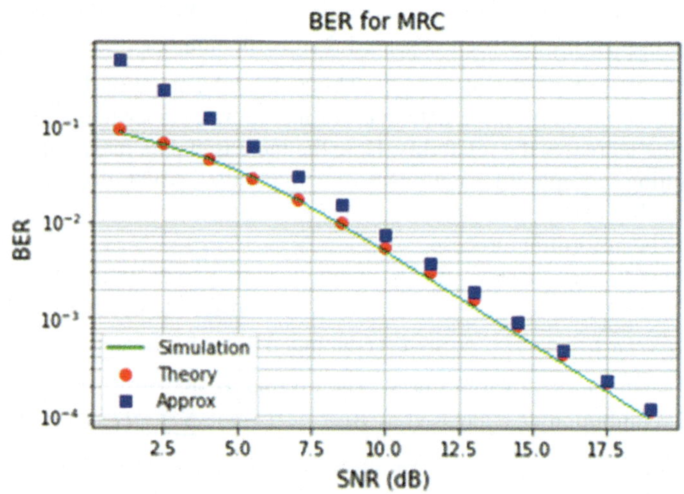

Fig. 7 BER versus SNR (dB) for MRC

10 that the DNN-based configuration can give optimal performance than maximum likelihood receivers for both the UEs.

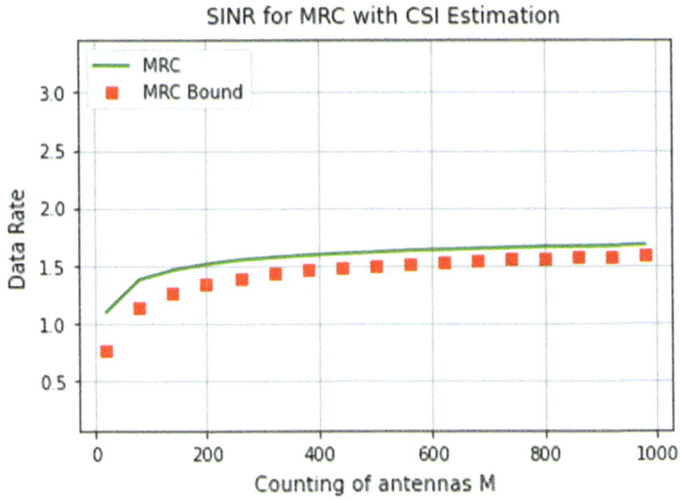

Fig. 8 SINR for MRC with CSI estimation

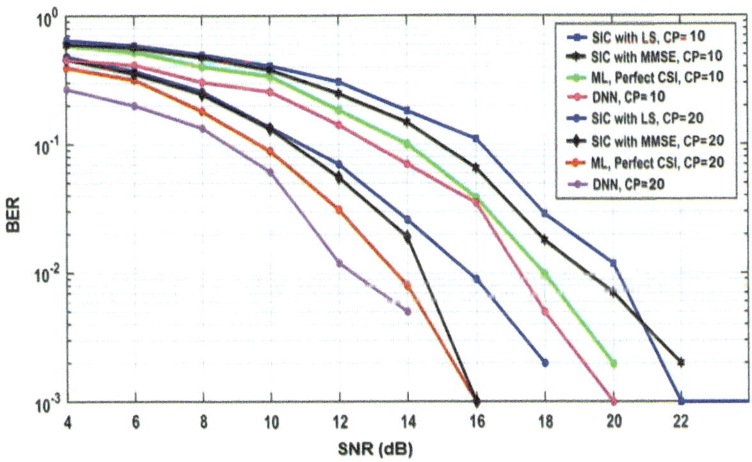

Fig. 9 BER value versus SNR curve for UE_A

5 Conclusion

It is seen that over the fading channel, the framework execution corrupts with the utilization of regular assessors. However, the deep neural network-assisted receiver provides performance comparable with the maximum likelihood receiver's performance. Indeed, even within the sight of obstruction, DNN offers prevalent execution, and it is particularly undifferentiated from the ideal channel conditions. The maximal

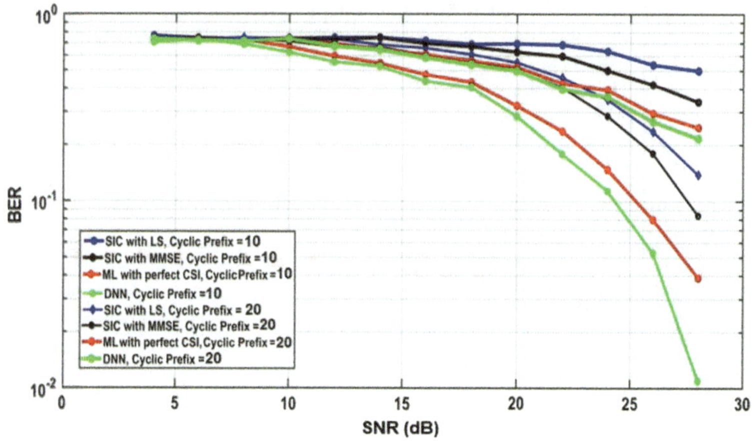

Fig. 10 BER value versus SNR curve for UE_B

likelihood estimator expects the zero-channel assessment problem, and it is created as a norm for examining the exactness of DNN-based beneficiaries.

References

1. Levesque, H.J.: In: Common Sense, the Turing Test, and the Quest for Real AI: Reflections on Natural and Artificial Intelligence. MIT Press (2017)
2. Das, S., Sarkar, M., Nandi, S.: A deterministic study of delta modulation (DM) using machine learning approach. In: 4th International Conference on Electronics, Materials Engineering and Nano-Technology (IEMENTech 2020). IEEE, (2020). ISBN: 978–1–7281–9286–4
3. Dey, A., Jain, S., Nandi, S.: New method of POS based on artificial intelligence and cloud computing. In: International Conference on Recent Advances in Energy-efficient Computing and Communication (ICRAECC), India, pp. 1–6. (2019). https://doi.org/10.1109/ICRAECC43874.2019.8995078
4. Dempster, A.P., Laird, N.M., Rubin, D.B.: Maximum likelihood from incomplete data via the EM algorithm. Journal of the royal statistical society. Series B (methodological), pp. 1–38. (1977)
5. Goodfellow, I., Bengio, Y., Courville, A., Bengio, Y.: In: Deep Learning. vol. 1, MIT Press Cambridge (2016)
6. Simeone, O.: A brief introduction to machine learning for engineers. Found. Trends in Signal Process. **12**(3–4), 200–431 (2018)
7. Kannan, S., Kim, H., Oh, S.: In: Deep learning and information theory: an emerging interface. IEEE ISIT Tutorial (2018)
8. Toet, A.: Computational versus psychophysical image saliency: a comparative evaluation study. IEEE Transactions on Pattern Anal. Machine Intell. **33**(11) (2011)
9. Sarkar, S., Das, S., Chaudhuri, S.S.: Hyper-spectral image segmentation using Rényi entropy based multi-level thresholding aided with differential evolution. Expert Syst. Appl. **50**, 120–129 (2016)
10. Mohammadi, M., Al-Fuqaha, A., Sorour, S., Guizani, M.: Deep learning for IoT big data and streaming analytics: a survey. IEEE Commun. Surveys Tutorials **200**(4), 0 2923–2960

(2018). https://doi.org/10.1109/COMST.2018.2844341. https://ieeexplore.ieee.org/document/8373692
11. Nandi, S., Pathak, N.N., Nandi, A.: Implementation of fast, adaptive, optimized blind channel estimation for multimodal MIMO-OFDM systems using MFPA. In: Intelligent Multi-Modal Data Processing. Wiley Publication, (2020). ISBN:9781119571384
12. Wen, C.-K., Shih, W.-T., Jin, S.: Deep learning for massive MIMO CSI feedback. IEEE Wireless Comm. Lett **7**(5), 748–751 (2018)
13. Lu, C., Xu, W., Shen, H., Zhu, J., Wang, K.: MIMO channel information feedback using deep recurrent network. IEEE Comm. Lett **23**(1), 188–191 (2019)
14. Wang, T., Wen, C.-K., Jin, S., Li, G.Y.: Deep learning-based CSI feedback approach for time-varying massive MIMO channels. IEEE Wireless Commun. Lett **8**(2), 416–419 (2019)
15. Hastie, T., Tibshirani, R., Friedman, J.: Unsupervised learning. In: The Elements of Statistical Learning. pp. 485–585. Springer (2009)
16. Sutton, R.S., Barto, A.G. et al.: In: Reinforcement Learning: An Introduction. MIT press (2018)
17. Wang, Y., Martonosi, M., Peh, L.S.: A supervised learning approach for routing optimizations in wireless sensor networks. In: Proceedings international workshop on Multi-hop ad hoc networks. ACM, pp. 79–86. (2006)
18. Bennis, M., Perlaza, S.M., Blasco, P., Han, Z., Poor, H.V.: Self organization in small cell networks: A reinforcement learning approach. IEEE Trans. Wireless Commun. **12**(7), 3202–3212 (2013)
19. Saeed, K., Korneylian, S.: Deploying an OFDM physical layer security with high rate data for 5G wireless networks. In: IEEE Canadian Conference on Electrical and Computer Engineering (CCECE), pp. 1–7. IEEE (2020)
20. Dey, A., Nandi, S., Sarkar, M.: Security measures in IOT based 5G networks. In: 3rd International Conference on Inventive Computation Technologies (ICICT), Coimbatore, India, pp. 561–566. (2018). https://doi.org/10.1109/ICICT43934.2018.9034365
21. Nandi, S., Sarkar, M., Nandi, A., Pathak, N.N.: Performance analysis of CO-OFDM system in a CR network. In: Chakraborty, G., Dutta (eds) Computer, Communication and Electrical Technology, ©Taylor & Francis Group, (2017). ISBN: 978–1–138–03157–9
22. Avinash, V., Narayanan, K.R., Chamberland, J.-F., Cheng, J.: A user-independent successive interference cancellation based coding scheme for the unsourced random access Gaussian channel. IEEE Trans. Commun. **67**(12), 8258–8272 (2019)
23. Nandi, S., Pathak, N.N., Nandi A.: Analysis of hard decision and soft decision decoding mechanism using viterbi decoder in presence of different adaptive modulations. Int. J. Future Gener. Commun. Netw. **13**(3), 3002–3012 (2020)
24. Nandi, S., Nandi, A., Pathak, N.N., Sarkar, M.: Performance analysis of cyclic prefix OFDM using adaptive modulation techniques. IJEECS **6**(8), 214–220 (2017). [ISSN: 2348–117]
25. Nandi, S., Nandi, A., Pathak, N.N.: Performance analysis of Alamouti STBC MIMO OFDM for different transceiver system. In: IEEE Conference on ICISS, pp: 883 – 887. (2017). https://doi.org/10.1109/ISS1.2017.8389305
26. Nandi, A., Kundu, S.: Energy level performance of packet delivery schemes in wireless sensor networks over fading channels. Int. J. Future Gener. Commun. Netw. **4**(2), 91–104 (2011)
27. Zheng, B., Zhang, R.: Intelligent reflecting surface enhanced OFDM: channel estimation and reflection optimization. IEEE Wireless Commun. Lett. **9**(4), 518–522 (2019)
28. Nandi, S., Pathak, N.N., Nandi, A.: A novel adaptive optimized fast blind channel estimation for cyclic prefix assisted space-time block coded MIMO OFDM systems. Wireless Personal Commun. (2020). https://doi.org/10.1007/s11277-020-07629-z
29. Nandi, S., Pathak, N.N., Nandi, A.: Efficacy of channel estimation and efficient use of spectrum using optimised cyclic prefix (CP) in MIMO-OFDM. IJEAT **9**(2), (2019). ISSN: 2249–895.8 https://doi.org/10.35940/ijeat.B4093.129219
30. Nandi, S., Chaudhuri, A., Bhar, J.: The performance comparison of different adaptive modulation schemes in presence of binary convolution encoder and viterbi decoder. IIJEC **6**(4), 001–009 (2018). [ISSN: 2321–5984]

31. Nandi, S., Pathak, N.N., Nandi, A.: Avenues to improve channel estimation using optimized CP in STBC coded MIMO-OFDM system—a global optimization approach. In: 5th International Conference on Microelectronics, Computing and Communication Systems (MCCS-2020), Ranchi, India (2020)
32. Trivedi V.K., Kumar, P.: Low complexity interference compensation for DFRFT-based OFDM system with CFO. IET Commun. **14**(14), 2270–2281 (2020)
33. You, L., Gao, X., Swindlehurst, A.L., Zhong, W.: Channel acquisition for massive MIMO-OFDM with adjustable phase shift pilots. IEEE Trans. Signal Process. **64**(6), 1461–1476 (2015). https://doi.org/10.1109/TSP.2015.2502550
34. Sarkar, M., Ghosh, S.: Development of a secured optical code-division multiple-access systems by implementing hybrid 2D-modified Walsh code. Opt. Eng. (2020). https://doi.org/10.1117/1.OE.59.10.106107

Multi-objective Hydro-Thermal-Wind Scheduling Applying PSO

P. Sowmith, N. Goutham Kumar, and B. Varun Kumar

Abstract Multi-objective particle swarm optimization (PSO) technique for solving scheduling of thermal-hydro-wind problems by considering the weighted sum method. There is an arrangement of Pareto means beside optimal result alone. To dominance, the relationship is the criteria behind the selection of various solutions to attain Pareto optimal solutions. The planning issue of fixed head present moment hydro-warm producing plants can be considered as minimization of complete fuel cost and emission related with all thermal units with synchronous fulfillment of balance and disparity requirements.

Keywords Weighted sum method · Particle swarm optimization · Optimization techniques · Multi-objective

1 Introduction

Multi-objective hydro-thermal-wind scheduling [1] According to engineering applications and science research, multi-objective optimization (MOP) has gained its fabulous substance. For the most part, MOP has several goals and even these destinations are conflictive. In the streamlining procedure, every one of the targets is upgraded at the same time, and finally, there is an arrangement of Pareto optimal solutions by choice of a lone optimal solution. Dominance relationship is the criteria behind the selection of various solutions to attain Pareto optimal solutions. Moving further, various concepts related to MOP are highlighted such as;

1. Formulation of multi-objective optimization (MOP)
2. Dominance relationship
3. Pareto optimal solutions.

P. Sowmith (✉) · B. Varun Kumar
Department of EEE, V R S E C, Kanuru, Vijayawada, A.P 520007, India
e-mail: sowmith@vrsiddhartha.ac.in

N. Goutham Kumar
National Fire Service College, Nagpur 440013, India

K. N. Das et al. (eds.), *Proceedings of the International Conference on Computational Intelligence and Sustainable Technologies*, Algorithms for Intelligent Systems,
https://doi.org/10.1007/978-981-16-6893-7_13

Hydro-thermal scheduling [2] is reached out to a MOPSO with considering emission and fuel cost simultaneously. Various traditional multi-objective methods have been proposed to tackle short-term environmental and economic hydro-wind-thermal scheduling problems, namely say

- Particle swarm optimization [3]
- GA—Genetic algorithm [4]
- Quadratic differential evolution
- Fuzzy satisfying evolutionary programming.

Where every one of the strategies referenced above convert the multi-target streamlining issue into a solitary target issue by utilizing loads, value punishment factors, or compromises. MOEAs can viably settle the deformities existing in the weight technique and imperative strategy and get the entire Pareto front through one run.

2 Formulation of Problem

2.1 *Objective Function*

Ideal planning of a force plant age can be resolved as complete age acquired from each creating plant when summarized bringing about insignificant expense by fulfilling the base and greatest constraints. The fundamental goal of aqueous booking depends on the arrival of a measure of water from every supply to create an adequate force to such an extent that costs of the fuel of warm plants are diminished. Planning of the fixed head present moment aqueous plants related to cost work which incorporates the quadratic capacity of thermal energy. Mathematically, the operating cost of a thermal unit is communicate as,

$$\text{Minimize } F = \sum_{t=1}^{T}\sum_{i=1}^{N_s} f_i(\mathrm{P}_{si}^t) \tag{1}$$

where N_s represents unit of thermal plants, T represents the total interludes of considered horizon scheduling, and ${P_{si}}^t$ expresses the power generation of ith thermal plants at tth interval time. The cost function of fuel is expressed as,

$$f_i(\mathrm{P}_{si}^t) = \mathrm{a}_i + b_i P_{si} + c_i P_{si}^2 \tag{2}$$

$$f = \sum_{i=1}^{N} \left(\alpha_i P_{si}^2 + \beta_i P_{si} + \delta_i\right) \tag{3}$$

The fuel cost co-efficient are $a_{i,}$ b_i, and c_i and $\alpha_{i,}$ β_{i}, and δ_i are the emission co-efficient

2.2 Constraints of Problem

2.2.1 Power Balanced Constraints

The thermal, hydro, and wind elements should reach total demand of load at every particular interval during the whole scheduling period.

$$\sum_{i=1}^{N_s} P_{si}^t + \sum_{j=1}^{N_h} P_{hj}^t + \sum_{k=1}^{N_w} P_{wk}^t = P_D^t \tag{4}$$

where $P_{s,i}^t$ is out-term power of ith thermal plant unit at tth interval; $P_{h,j}^t$ is the out-term power of jth hydro plant unit at tth interval; $P_{w,k}^t$ represents out-term power of k_{th} wind power plant at tth interval and P_D^t is total load demand.

2.2.2 Hydro Reservoir Continuity Equation Network

$$V_{hj}{}^t = V_{hj}{}^{t-1} + I_{hj}{}^t - Q_{hj}{}^t \tag{5}$$

where $V_{h,j}^t$ represents final repository volume of a reservoir i_{th} at tth interval; $I_{h,j}^t$ expressed as inflow; $Q_{h,j}^t$ represents eject; N_h represents hydro power plant s, $t =$ 1,2, …, T; $i =$ 1,2, …, N_h.

.

2.2.3 Operating Limits

i. Thermal power plant

$$P_{si,\min} \le P_{si}^t \le P_{si,\max} \tag{6}$$

The maximum and minimum generation limits are $P_{si,\max}$ and $P_{si,\min}$ of ith a thermal power plant, $i = 1, 2, \ldots$, Ns; $t = 1, 2, \ldots, T$

ii. Hydropower plant

$$P_{hj,\min} \le P_{hj}^t \le P_{hj,\max} \tag{7}$$

Maximum and minimum generation limits are $P_{hj,\max}$ and $P_{hj,\min}$ at jth hydro power plant, $t = 1, 2, \ldots, T; j = 1, 2, \ldots, Nh$

iii. Wind power plant

$$P_{wk,\min} \le P_{wk}^{t} \le P_{wk,\max} \tag{8}$$

where $P_{wk,\min}$ $P_{wk,\max}$ represents the minimum–maximum values of a wind power plant, $w = 1, 2, \ldots, Nw; t = 1, 2, \ldots, T$.

2.2.4 Discharge Limits of Hydro Plant

$$Q_{hj,\min} \le Q_{hj}^{t} \le Q_{hj,\max} \tag{9}$$

where $Q_{hj,min}$ $Q_{hj,max}$ represents a minimum–maximum discharges of jth hydro plant, $j = 1, 2, \ldots, Nh; t = 1, 2, \ldots, T$.

2.2.5 Volume Limits of Reservoir Storage

$$V_{hj,\min} \le V_{hj}^{t} \le V_{hj,\max} \tag{10}$$

where $V_{hj,\min}$, $V_{hj,\max}$ represents a minimum–maximum volume of $j_{th}\, j_{th}$ hydro plant, $j = 1, 2, \ldots, Ns; t = 1, 2, \ldots, T$.

2.2.6 Inceptive and Final Storage of Reservoir

$$V_{hj}^{o} = V_{hj,ini}^{t};\ V_{hj}^{T} = V_{hj,fin} \tag{11}$$

where V_{hj}^{0} V_{hj}^{T} are initial and final storages.

3 Methodology

There are two types of methods; they are (i) classical methods and (ii) heuristic methods

(i) Classical methods:
- Weighted sum method
- Weighted metric method
- Rotated weighted metric method
- Bensons
- Constraint method

(ii) Heuristic methods:

- non-dominated sorting GA
- MOPSO [5]
- MODE—Multi-objective DE
- Enhanced strength Pareto evolutionary algorithm.

3.1 Multi-objective Hydrothermal Scheduling

3.1.1 Weighted Sum Method

Consolidating a lot of goals into a solitary goal by adding every target to pre-increase by a client-provided weight [1].

$$\text{Min } F(x) = \sum_{m=1}^{M} w_m f_m(x) \tag{12}$$

$$\begin{aligned} \text{St } & g_{j(x)\geq 0,} && j = 1, 2, \ldots .j \\ & h_{k(x)\geq 0,} && k = 1, 2, \ldots .k \\ & x_i^{(L)} \leq x_i \leq x_i^{(U)} && \text{for } i = 1, 2, \ldots .n \end{aligned}$$

3.1.2 Pareto Optimal Front

It is a state where one cannot critic one solution as best or poorer unless one of the solutions in a group of solutions is fixed to be best or poor for the objective which is selected for a particular problem. It is used for studying income distribution and economics and life sciences. A compromise solution is obtained between cost function and emission; by using this, Pareto optimal, we will get the compromise solution; for an individual weight, we get different compromise solutions; by changing the weights, we select the best position where it is nearer to the origin. Pareto efficiency is a term that indicates the minimum efficiency and need not be accepted by all desired demands, and it does not guarantee the equality of all the needs. In power system, the Pareto optimal front is used, and it gives several control variables, and each solution is better or poor depends on the objective of the user [6].

4 Simulation Results

TEST CASE: For a thermal and hydro plant, a load is provided whose qualities are expressed here [2]

Table 1 Test case load sample [7]

Operation of time	Demand of load		
	First day (MW)	Second day (MW)	Third day (MW)
24.00–12.00	1200	1100	950
12.00–24.00	1500	1800	1300

$$\text{Steam plant}: \mathrm{H} = 500 + 8.0\mathrm{P_s} + 0.0016\mathrm{P_s^2}(\mathrm{MB\ tu/h})$$

$$\text{The cost of fuel is } 1.15\ \mathrm{R/MB\ tu}$$

$$150\ \mathrm{MW} \leq \mathrm{Ps} \leq 1500\ \mathrm{MW}.$$

$$\text{Hydro system}: q = 300 + 4.97\mathrm{P_H}\mathrm{acre} - \mathrm{ft/h}$$

$$0 \leq \mathrm{P_H} 1000\ \mathrm{MW}$$

$$q = 5300 + 12(\mathrm{P_H} - 1000) + 0.05(\mathrm{P_H} - 1000)^2 \mathrm{acre} - \mathrm{ft/h}$$

$$1000 < \mathrm{P_H} < 1100\ \mathrm{MW}$$

A wind constraint is added which is in the range of

$$0 \leq P_w \leq 10\ \mathrm{MW}$$

Emission objective function is also included

$$J = 0.006483 P_1^2 - 0.79027 P_1 + 28.82488$$

See Table 1.
Hydro-reservoir:

- 100,000 section of land ft. toward the beginning.
- Should have 60,000 sections of land ft. toward the finish of the timetable.
- Repository volume is restricted: 60,000 sections of land ft. $\leq$ v 120,000 sections of land ft.
- A consistent invasion into supply of 2000 sections of land ft./h over a whole 3-days' time frame (Tables 2 and 3; Fig. 1).

From the above Fig. 2, Table 4 Fuel Cost = 695,540 Rs/h, Emission Cost = 289,570 kg/h, and the weights are taken w1 = 0.1 and w2 = 0.9, then the compromise solution is 352070 with computational time 0.766 s, though obtained $\mathrm{P_H}$, $\mathrm{P_S}$, and $\mathrm{P_w}$ are the results of random variables. The demand of load and constraints are equals with the generation with least number of iterations (Table 5).

Table 2 Result for cost minimization of HTWS

Interval	PH	PS	Q	V	Pw	PG	Cost ($/h)	Computational time (s)
P1	435.65	758.71	2495.2	94,058	5.6412	1200	702,540	7.79254
P2	987.72	504.3	3748	73,082	7.9814	1500		
P3	169.65	921.71	1173.2	83,004	8.6427	1100		
P4	697.84	1096.9	3798.3	61,425	5.2413	1800		
P5	162.23	778.64	1136.3	71,790	9.1327	950		
P6	533.7	756.4	2982.5	60,000	9.9	1300		

Table 3 Result for emission minimization of HTWS

Interval	PH	PS	q	v	Pw	PG	Emission (kg/h)	Computational time (s)
P1	371.73	821.91	2177.5	97,870	6.3571	1200	296,120	4.951988
P2	625.35	867.23	3438	80,614	7.4247	1500		
P3	156.13	941.82	1106	91,343	2.0535	1100		
P4	934.17	861.11	4972.8	55,669	4.7218	1800		
P5	188.8	758.53	1268.3	64,449	2.6675	950		
P6	410.61	879.49	2370.7	60,000	9.9	1300		

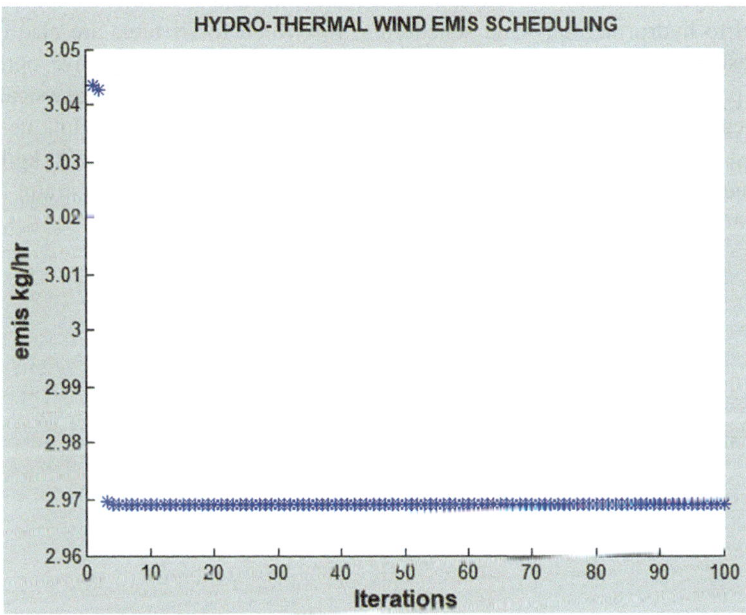

Fig. 1 Emission minimization of HTWS

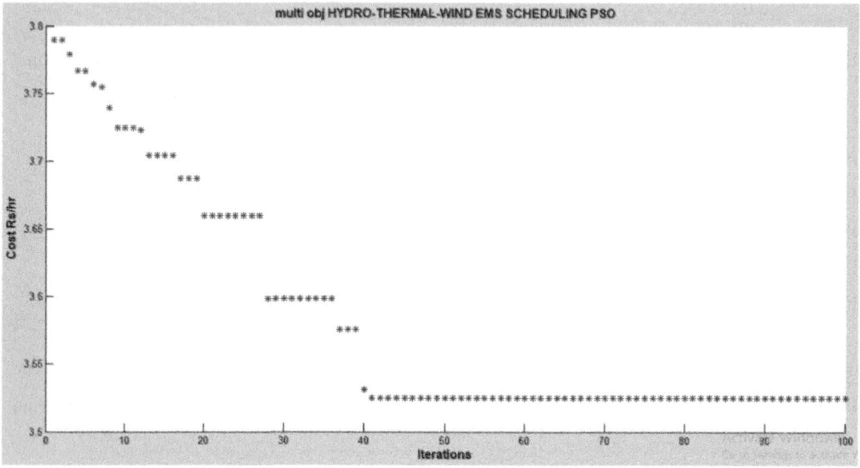

Fig. 2 Multi-objective HTWS using PSO

5 Conclusion

This analysis of paper presents a technique to diminish cost of the fuel for thermal plants a hydro-thermal-wind system is considered, and a heuristic method PSO applied to hydro-thermal-wind scheduling, for which fossil fuels are also one of the reasons to lessen the pollution caused by them. A multi-objective optimization is performed on hydro-thermal-wind scheduling problem using a classical technique called weighted sum applied to PSO. Compared to other techniques PSO got best solutions Fuel Cost = 695,540 Rs/h, Emission Cost = 289,570 kg/h with computational time 0.766 s. In future, we can add a solar plant constraint to this problem.

Table 4 Results for multi-objective HTS

Interval	P_H	P_S	q	V	P_w	P_G	Cost ($/h)	Emission (kg/h)	Compromise solution
P_1	241.64	949.58	1530.9	90,563	8.794	1200	695,540	289,570	352,070
P_2	666.08	823.91	3640.4	85,944	10	1500			
P_3	189.37	901.01	1271.1	94,690	9.626	1100			
P_4	1067.7	722.29	5636.5	51,051	10	1800			
P_5	55.079	887.02	603.74	67,807	7.899	950			
P_6	466.91	823.19	2650.5	60,000	9.9	1300			

Table 5 Comparison of PSO with other techniques

Algorithm	Minimum Cost ($)	Simulation time (s)	Emission (kg/h)
ORCSA-levy flight	709,862.048	0.18	–
ORCSA-cauchy	709,862.048	0.18	–
FIPSO	723,550.0	–	–
NSGSA [2]	709,652.4	0.899	306,970
WEIGHTED SUM (PSO)	695,540	0.766	289,570

References

1. Liu, Y., Zhan, X., Lyu, H., Qin, H.: Region search evolutionary algorithm with constraint handling for multi-objective short-term wind-solar-hydro-thermal scheduling. In: E3S Web of Conferences, vol. 233, no. 010 (2021). IAECST (2020)
2. Sowmith, P., Madhusudhanrao, R., Gouthamkumar, N.: Optimal scheduling of hydrothermal plant using particle swarm optimization. Lecture Notes in Mechanical Engineering. Springer, Singapore (2020)
3. Mandal, K.K., Basu, M.: Particle swarm optimization-technique based short-term hydrothermal scheduling. Energy Conv. Manage. **8**, 1392–1399 (2007)
4. Sowmith, P., Vamsi Krishna, N., Varun Kumar, B.: Conventional and heuristic optimization techniques comparison for economic load dispatch. In: Conference on Smart and Intelligent Systems. Springer
5. Benerjee, S.: Short Term hydro-thermal-wind scheduling based on particle swarm optimization technique. Electr. Power Energy Syst. **81**, 275–288 (2016)
6. Yu, B., Yuan, X.: Short-Term hydro-thermal scheduling using particle swarm optimization method. Energy Conv. Manage. **48**, 1902–1908 (2007)
7. Wood, A.J., Wollenberg, B.F.: Power Generation Operation and Control, 2nd edn. Wiley India, New Delhi (2012)

Design and Implementation of Recommendation System Using Sentiment Analysis in Social Media

Suchandra Dutta, Siddhartha Mondal, and Dhrubasish Sarkar

Abstract In this current era, sentiment analysis and also recommendation systems are some of the most popular subjects for the majority of the researchers, and on the other hand, these are the most important things for our regular life. The aim is to make valid recommendations and the most relevant tweets to the user of this model by extracting and analyzing random tweets based on a particular user address or hashtag. Nowadays, people value time over money, so we aim to save the user his or her most precious time and recommend the pertinent tweets. This model will filter out all the irrelevant as well as the inappropriate tweets and try to provide the gravest and valued tweets from Twitter. This filtration will happen based on the tweet's public features as well as the attributes of the respective tweeter such as his or her follower count. The most initial concepts of machine learning, recommendation system, and sentiment analysis are understood and implemented to be able to propose the mentioned model. The tools to work with the mentioned topics are implemented to evaluate prominent output with utmost effort throughout. Let alone being used in product recommendation systems, our proposed method is also capable enough to maintain a consistent and effective performance in other fields.

Keywords Sentiment analysis · Subjectivity · Polarity · Recommendation system · Machine learning · Social network analysis

1 Introduction

One of the chart-topping social media platforms, Twitter, is the focus of this project. Twitter is considered a decent social media platform with many prominent and respected public figures appearing on it alongside common people. Usually, it contains less inappropriate content instead of its strict terms and conditions set through the platform's creator and administrators. Considering people's value of time, the main goal of this project is to save the same as time is a human's most

S. Dutta · S. Mondal · D. Sarkar (✉)
Amity Institute of Information Technology, Amity University, Kolkata, India
e-mail: dhrubasish@inbox.com

K. N. Das et al. (eds.), *Proceedings of the International Conference on Computational Intelligence and Sustainable Technologies*, Algorithms for Intelligent Systems,
https://doi.org/10.1007/978-981-16-6893-7_14

precious entity. The goal is to decompose a hefty amount of information in the form of tweets and pick or extract the most pertinent data whose ideology revolves close to the search topic for the user. This model also separates the tweets considering their reachability; i.e., a tweet posted by a tweeter possessing 70 million followers is way more reachable than the same posted by a tweeter possessing 7000 followers which is exactly the same ideology followed in social media marketing. So using this research gap, we have tried to plan our model such that it can filter out the low-reachable tweets as well as recommend the ones which create a greater impact. Now, there are some tweets on the platform which spread like fire and are indicated as the 'viral' tweets, here is where likes and comments come into play.

A recommendation systems framework has proved to be an effective platform for providing consumers with useful suggestions. It is a large and tough job necessitates a variety of consumer preferences and other factors. As a result, a variety of suggestion methods have been suggested to address the issues such as content-based recommendation system, collaborative filtering-based recommendation system, and hybrid recommendation system.

So each approach, as well as the methodology, has its own set of benefits when it comes to solving specific problems. Collaborative filtering is the most common as well as commonly used methodology in recommender systems which relies on the use of data on the internet and user-generated data. The collaborative filtering system proposes objects based on user similarities.

Twitter opinion analysis helps you to monitor how consumers are reacting to your product or service on social media, and it will help you spot dissatisfied consumers or unfavorable reviews before they become a huge crisis.

Researchers have also worked on many projects related to sentiment analysis and recommendation system. However, the big issue is that we find that when user extract the tweets from Twitter, as per their need as like any '#' of any product name of something or '@' of any user name, then Twitter gives the user a CSV file where the details of polarity, subjectivity, follower count, name of the user, retweets, etc. Each detail is available but the main problem is that, if any user has 10 or 0 number followers in their account and another side if any user has a million number of followers in their Twitter account, so obviously the value or the weightage of the opinion of the user who has millions of followers is high the other side the weightage of the opinion of the other user who has 0–10 followers is low. So, in this paper, we are attempting to solve the issue by assisting the user to connect with the pertinent information he or she will be looking for.

1.1 Recommendation System

Recommendation system is a smart computer-based methodology that forecasts consumer rejection and uses and assists them in selecting products from a large pool of online items. Any Internet users have undoubtedly come across a Recommendation System. For example, Facebook recommends suitable friends, YouTube suggests appropriate videos, Glassdoor suggests suitable careers, TripAdvisor suggests suitable vacation destinations, Goodreads suggests fascinating books, and so on [1].

The recommendation of users or contents based on similarities is a naive recommendation approach that can be applied to Twitter. Based on the cosine similarity between their friends and tweet messages, TF-IDF weighting is used to suggest users follow as well as tweet messages to post [2]. Other applications that apply RSS include social networks, web news websites, entertainment pages, and other information management applications, in addition to online businesses [1]. Recommendation systems are widely used in various domains [3–5].

1.2 Different Types of Recommendation System

There are three different types of recommendation system –

a. Content-based recommendation system
b. Collaborative filtering-based recommendation system
c. Hybrid recommendation system

a. Content-based recommendation system—This filtering is done depending on the description of the commodity as well as the product or something else. Based on the meaning or definition of the product, the system determines its resemblance. The user's experience is used to search for related items that the user might like [2]. CBRS uses CBF to suggest items based on the user's profile and the definition of the object. His past quest or purchasing history may be included in the user profile. The machine learns to suggest objects that are close to those that the consumer has previously appreciated [1].
b. Collaborative filtering-based recommendation system—collaborative filtering is one of the most usable recommendation systems. 'A man is known by the company he keeps,' according to CFRS. That is, if CFRS assumes that two or more users' preferences have already aligned, then their interests are likely to match in the future [1].

   The suggestions are made depending on the user's activities. The user's past plays an important part [6].

   There are two types of collaborative filtering recommendation system, these two are,

   - User-User collaborative filtering
   - Item -Item collaborative filtering.

c. Hybrid Recommendation System—Hybrid recommendation system is a combination of a content-based recommendation system and a collaborative filtering-based recommendation system. The aim of integrating various filtering methods is to increase recommendation accuracy alongside removing the shortcomings of each filtering approach.

Twitter Sentiment Analysis—Sentiment analysis is a text mining as well as opinion mining method that is commonly used. Sentiment analysis on Twitter, on the other hand, refers to the use of sophisticated text mining tools to assess the sentiment of the tweets in terms of positive, negative, and neutral. In the case of Twitter sentiment analysis, maximum people use python language more than R language as python is easy to use. 'Tweepy' and 'Text Blob' are the two major libraries that have been used during the extraction of tweets from Twitter. Tokenization is one of the leading methods for extracting tweets from Twitter.

Why Twitter Is The Most Relevant Social Media Platform For This Project: We felt Twitter is our social media platform for the project because nowadays Internet is available all over the world and a majority of the people in this world have a Twitter account and millions of people tweet their opinion or any type of post regularly via their Twitter account. Companies use Twitter sentiment analysis to improve their business plans, analyze buyers' attitudes about goods products, or brands, see how people react to campaigns or product releases, and figure out why those products aren't selling.

2 Related Work

In this paper, researchers propose a lexicon-based approach to sentiment analysis of Twitter content. The technique solves normalization of the sentiment to provide an intensity of the sentiment rather than a positive/negative mark [7]. This sentiment analysis framework for SemEval2015 Task 10 Subtask B and E is defined in this article. After extracting several classes of features, including lexical, syntactic, lexicon-based, Z ranking, and semantic features, a logistic regression classifier was trained for subtask B [8]. In this paper, researchers provide a summary of popular Twitter features that are used to characterize certain accounts and their behavior and demonstrate how ineffective they are in this context. The number of retweets and fans, as well as the Klout ratings, are irrelevant to our research [9]. The results of this study would allow for a finer-grained analysis of Twitter data using the location variable, resulting in a more detailed analysis of customer sentiment and behavior [10]. To extract emotional insight from student forums, they use sentiment analysis. Polarity and emotion in interactions are measured and branded as favorable, negative, or neutral. Messages are categorized and visualized as six different feelings as part of a multiclass approach to understanding students' written thoughts [11]. The two case studies' ultimate goal is to compare the mood of location-based tweets to on-the-ground public opinion as seen in election results. In both scenarios, we find that Twitter's opinion about place correlates with election results. Regardless of the process, we see related trends in Twitter sentiment [12] against political parties and candidates [13]. In social network analysis, centrality is one of the most discussed principles. There is a lot of research on using centrality metrics to find the most important people or influential nodes [14, 15] in a social network. The challenge is to identify metrics that can be computed quickly and that can define consumers based

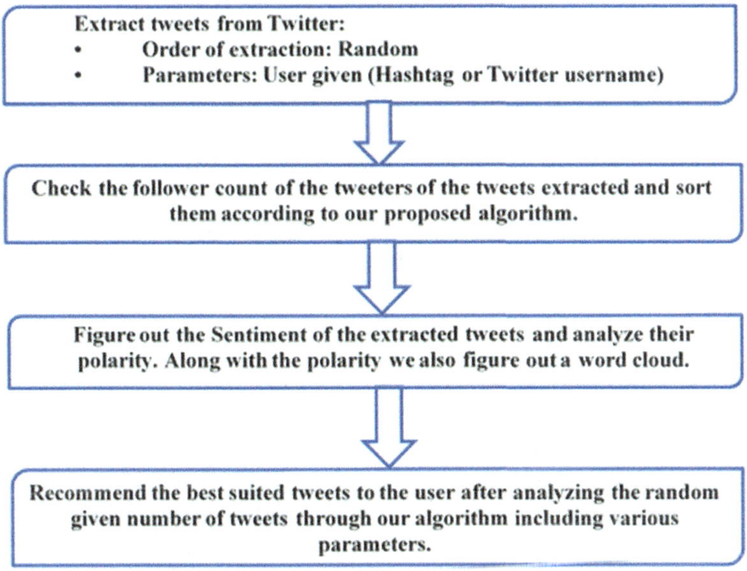

Fig. 1 Block diagram of the proposed model

on validity parameters that are as similar to the fact as possible [16]. By observing and classifying behavior, the model used subjectivity and polarity to create a customized recommender scheme. This gave the model two new ways to suggest, one based on the user's feedback and the other based on the subjects they covered [17].

3 Proposed Model

So here is the proposed model (Fig. 1) for this project the name of this model is the weightage-recommendation model from sentiment analysis. The above model is a working model for this project, and the process of this project is shown in this model.

4 Proposed Algorithm and Techniques

The above model (Fig. 2) shows the step-by-step process or can say the proposed algorithm of this project. So first, we collect the consumer key or API key as well as the consumer key secret and others and put it into the python program, after writing the program we get the CSV file as per as our '#' & '@' where all the extracted tweets or other relevant information are stored such as subjectivity, polarity, number

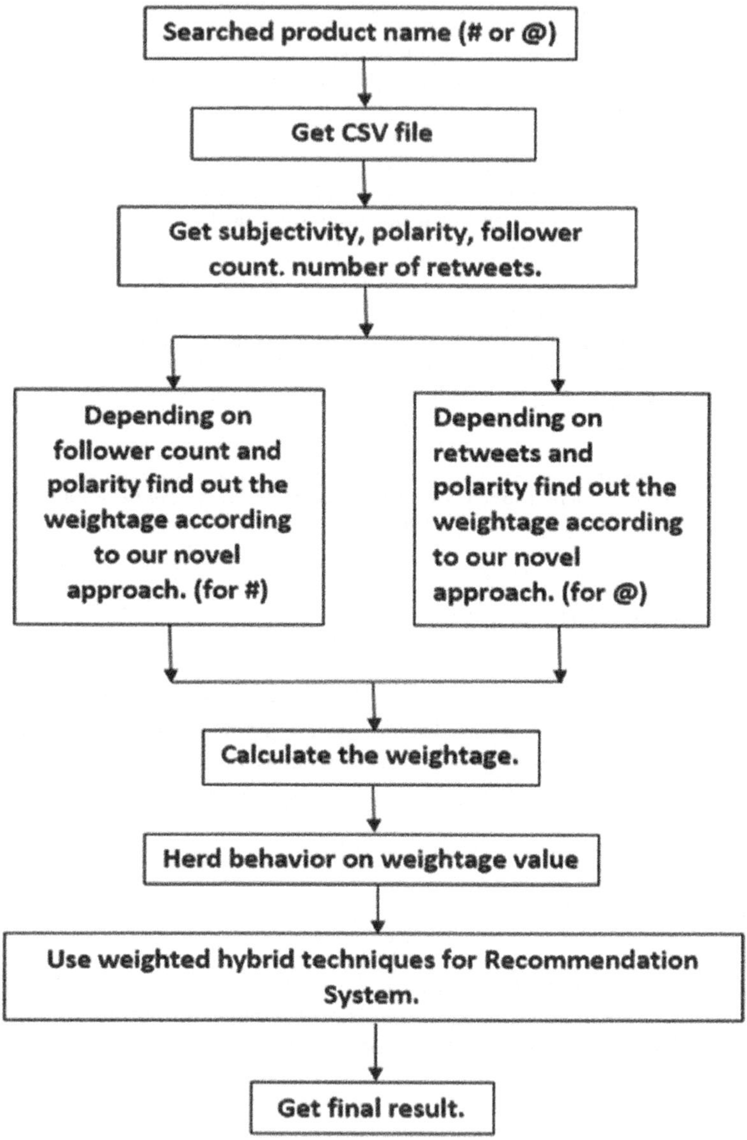

Fig. 2 Step-by-step process of the project

of follower count, and number of retweets. By using this information as the number of followers and the number of retweets, it is easy for us to explore weightage value.

The main issue is when a person has many numbers of followers then obviously the value of his/her opinion is more valuable so for solving this problem we can use our algorithm.

So, after getting the number of follower count from the CSV file it is easy for us to explore the weightage value according to the length of the number of followers, but this is only applicable for '#' tweets and for '@' tweets we can consider the number of retweets of this individual's account for find out the weightage value that we are capable of applying on the weighted average formula and then apply the recommendation system on it.

Proposed algorithm:

The idea is to allow appropriate weights to tweets based on the follower count of the user, which is calculated via the following algorithm,

If a person's Twitter account has less than 100 followers, then his or her single tweet will be valued at 2.

If a person's Twitter account has more than 100 followers but less than 1000, then his or her single tweet will be valued at 3.

If a person's Twitter account has more than 1000 followers but less than 10,000, then his or her single tweet will be valued at 4.

Like this, every time a user's follower count increases by the power of 10, his or her tweet's value increases by 1. That means, the number of the digit is equal to the weighted value.

For example, a tweet made by a tweeter who has 500,000 followers will be valued at 6 while the same tweet made by a tweeter who has 5,000 followers will be valued at 4.

Now, suppose there are 5 tweets and assuming arbitrary follower count for each of them, the calculations made for priority value would be like.

Tweet 1: 50,000 followers ———— 5————Neg.
Tweet 2: 600,000 followers ———— 6————Pos.
Tweet 3: 8,000 followers ———— 4————Pos.
Tweet 4: 110,000 followers ———— 6————Neg.
Tweet 5: 400 followers ———— 3————Neg

Priority Value: Neg = 5 + 6 + 3 = 14, Pos = 4 + 6 = 10, therefore overall polarity is Neg.

Pos = 35.71, Neg = 58.33% Neutral = null.

The technique used:

Here we use the weighted average formula for building a recommendation system,

$$W = \frac{Rv + Cm}{v + m} \tag{1}$$

where

W = weighted rating.

R = Rating (1–10).

v = number of positive sentiment or voting.

m = minimum of total number of positive sentiments.

C = average number of vote or positive sentiment.

In the above picture [eq. 1] [18], multiply each number by its weight and apply the results to obtain a weighted average. Find the number of all the factors multiplied by their weight and divide by the sum of the weights if the weights do not add up to one.

5 Python Script—Import Libraries

For this project, we can use several libraries in our python program for this particular project, so the libraries are TextBlob, sys, tweepy, matplotlib. pyplot, pandas, NumPy, os, nltk, by country, re, string, word cloud, PIL, nltk. sentiment.vander, Langdetect, nltk.stem, nltk.sentiment.Vader, etc. All the libraries mentioned here have different characteristics for different uses.

6 Result Analysis and Output

The above picture (Fig. 3) is a screenshot of the output of the project that helps to show the output for #IPLwin, as we can show in this picture the weightage value is clearly shown that is 4. Since we searched for 450 tweets and get the 304 numbers of positive tweets, 24 negative tweets, and 122 numbers of neutral tweets.

The above picture (Fig. 4) is the screenshot of the CSV file in EXCEL format, in which the weightage value is shown, as well as the polarity of the tweets and subjectivity of the tweets are also shown. The searched topic for this CSV file is '#Tesla'.

The above picture (Fig. 5) is shown the column diagram of the searched result that means '#Tesla', which helps to show the polarity of the extracted tweets, here in this graph or we can say diagram, −1.00 refers to that the negative polarity or sentiment, 0 refers to the neutral sentiment and + 1.00 refers to the positive polarity.

The above picture (Fig. 6) is shown the above WordCloud for the search topic ('#Tesla'). The word cloud is a unique visualization of text data that is often used to display keyword metadata on websites or to visualize free-from text.

7 Conclusion

Sentiment analysis, being the interdisciplinary field in today's era, crosses natural language, AI, and data, and even opinion mining. Sentiment analysis, particularly in the context of microblogging, is still in its initial phases and still being done. As a result, we suggest a couple of concepts that we believe are worth pursuing in the future, and it has the potential to be a leader to even better results.

Right now, we have just collaborated with a few other polarity parts and the followers count part. Apart from this in this paper, we are concentrating on accurate opinion mining from Twitter [12], and we can work on the weightage of the user

```
Please enter keyword or hashtag to search: #IPLwin
Please enter how many tweets to analyze: 450
Total number:  450
Positive number:  304
Negative number:  24
Neutral number:  122
```

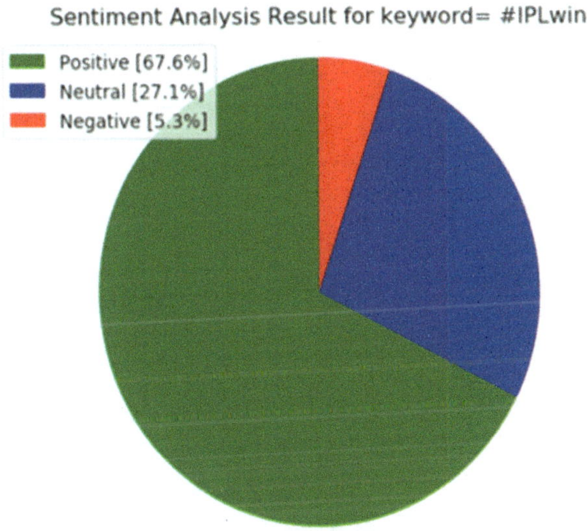


```
                                                    0
0    Right now @KKRiders doing what @RCBTweets woul...
1    How .@KKRiders ??\nUnbelievable!!!\n\n#ipl #IP...
2    #BREAKING #TWIST\n#MUMBAI WINS FINALLY with 11...
3    Woahhhhh!!!!!! WE WON!!!\nYAYYYY PALTANNNN🧿💙\n...
4    Only champion teams can win from these\n hopel...
..                                                ...
445  Appreciation post RUSSELL contribution 💜 he di...
446  sanjana ganesan(kkr) v/s sara tendulkar(MI)\nj...
447  RT @Ammu222105: Ipl is most entertainment leag...
448  RT @Ammu222105: Watch Live match to get bumper...
449  आंद्रे रसेल जैसे लोंडो को हमारे गांव में \nखरब...

[450 rows x 1 columns]
Jeetegi toh @mipaltan he 😁😁😁😁
#IPLWin #IPL2021 #KKRvsMI #MIvsKKR
The number of followers of the user are : 1757
1757
Weight = 4
```

Fig. 3 Output of weightage value

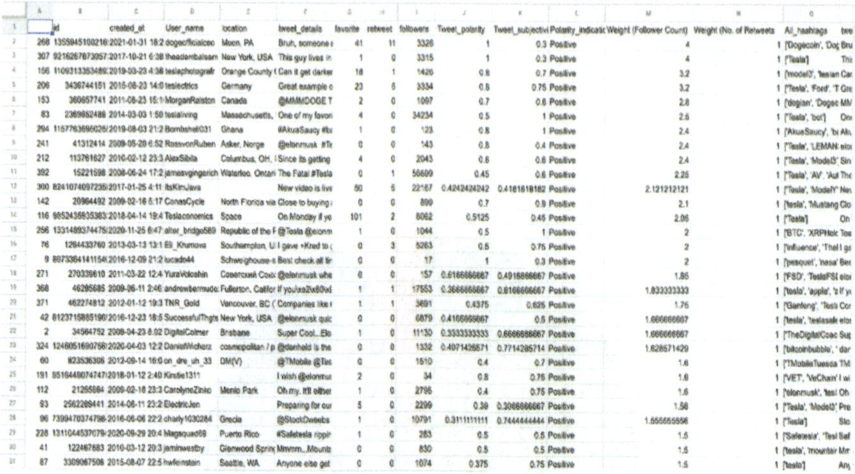

	id	created_at	User_name	location	tweet_details	favorite	retweet	followers	Tweet_polarity	Tweet_subjectivi	Polarity_indicatic	Weight (Follower Count)	Weight (No. of Retweets	All_hashtags	twe
268	1355945100216	2021-01-31 18:2	dogeofficialceo	Moon, PA	Bruh, someone	41	11	3326	1	0.3	Positive	4	1	['Dogecoin', 'Dog	Bru
307	9216267873057	2017-10-21 6:38	theadenbalsam	New York, USA	This guy lives in	1	0	3315	1	0.3	Positive	4	1	['Tesla']	Thi
156	1109313353489	2019-03-23 4:38	teslaphotograh	Orange County (	Can it get darker	18	1	1426	0.8	0.7	Positive	3.2	1	['model3', 'teslan	Car
206	3436744151	2015-08-23 14:0	teslectrics	Germany	Great example o	23	6	3334	0.8	0.75	Positive	3.2	1	['Tesla', 'Ford', 'T	Gre
153	360657741	2011-08-23 15:1	MorganRalston	Canada	@MMMDOGE T	2	0	1097	0.7	0.8	Positive	2.8	1	['dogian', 'Dogec	MM
83	2369652488	2014-03-03 1:50	teslaliving	Massachusetts,	One of my favori	4	0	34234	0.5	1	Positive	2.6	1	['Tesla', 'bot']	One
294	1157763696026	2019-08-03 21:2	Bombshell031	Ghana	#AkuaSaucy #b	1	0	123	0.8	1	Positive	2.4	1	['AkuaSaucy', 'b	Ak
241	41312414	2009-05-20 6:52	RossvonRuben	Asker, Norge	@elonmusk #T	0	0	143	0.8	0.4	Positive	2.4	1	['Tesla', 'LEMAN	elo
212	113761627	2010-02-12 23:3	AlexSibila	Columbus, OH,	Since its getting	4	0	2043	0.8	0.8	Positive	2.4	1	['Tesla', 'Model3'	Sin
392	15221598	2008-06-24 17:2	jamesvgingerich	Waterloo, Ontari	The Fatal #Tesla	0	1	56609	0.45	0.6	Positive	2.25	1	['Tesla', 'AV', 'Aut	The
300	8241074097235	2017-01-25 4:11	ItsKimJava		New video is live	50	6	22167	0.4242424242	0.4181818182	Positive	2.121212121	1	['Tesla', 'ModelY'	New
142	20964492	2009-02-16 5:17	ConasCycle	North Florida via	Close to buying	0	0	899	0.7	0.9	Positive	2.1	1	['tesla', 'Mustang	Clo
116	9852435835383	2018-04-14 19:4	Teslaconomics	Space	On Monday if yo	101	2	8062	0.5125	0.45	Positive	2.05	1	['Tesla']	On
256	1331489374475	2020-11-25 6:47	alter_bridge589	Republic of the F	@Tesla @elonm	1	0	1044	0.5	1	Positive	2	1	['BTC', 'XRPHolc	Tes
76	1264433760	2013-03-13 13:1	Eli_Krumova	Southampton, U	I gave +Kred to (	0	3	5263	0.5	0.75	Positive	2	1	['influence', 'That	I ga
9	8073364141154	2016-12-09 21:2	lucado44	Schweighouse-s	Best check all ti	0	0	17	1	0.3	Positive	2	1	['pesquet', 'nesa'	Bes
271	270339610	2011-03-22 12:4	YuraVoloshin	Советский Союз	@elonmusk whe	0	0	157	0.6166666667	0.4916666667	Positive	1.85	1	['FSD', 'TeslaFSI	elo
368	46295685	2009-06-11 2:46	andrewbermude	Fullerton, Califor	If you\xe2\x80\x	1	1	17553	0.3666666667	0.8166666667	Positive	1.833333333	1	['tesla', 'apple', 's	If y
371	462274812	2012-01-12 19:3	TNR_Gold	Vancouver, BC (	Companies like	1	1	3691	0.4375	0.625	Positive	1.75	1	['Ganfeng', 'Tesl	Co
42	8123715885196	2016-12-23 18:5	SuccessfulThgts	New York, USA	@elonmusk quic	0	0	6879	0.4166666667	0.5	Positive	1.666666667	1	['tesla', 'teslasafe	elo
2	34564752	2009-04-23 8:02	DigitalColmer	Brisbane	Super Cool...Elo	1	0	11130	0.3333333333	0.6666666667	Positive	1.666666667	1	['TheDigitalCoac	Sup
324	1246051690756	2020-04-03 12:2	DanielWicherz	cosmopolitan / p	@danheld is the	0	0	1332	0.4071428571	0.7714285714	Positive	1.628571429	1	['bitcoinbubble', '	dar
60	823536308	2012-09-14 16:0	on_dre_uh_33	DM(V)	@TMobile @Tes	0	0	1510	0.4	0.7	Positive	1.6	1	['TMobileTuesda	TM
191	9516449074747	2018-01-12 2:40	Kirstie1311		I wish @elonmu	2	0	34	0.8	0.75	Positive	1.6	1	['VET', 'VeChain'	I wi
112	21265884	2009-02-18 23:3	CarolyneZinko	Menlo Park	Oh my. It'll either	1	0	2795	0.4	0.75	Positive	1.6	1	['elonmusk', 'tesl	Oh
93	2562289441	2014-06-11 23:2	ElectricJen		Preparing for our	5	0	2299	0.39	0.3066666667	Positive	1.56	1	['Tesla', 'Model3'	Pre
96	7399470374798	2016-06-06 22:2	charly1030284	Grecia	@StockDweebs	1	0	10791	0.3111111111	0.7444444444	Positive	1.555555556	1	['Tesla']	Sto
228	1311044537079	2020-09-29 20:4	Magsquad69	Puerto Rico	#Safetesla rippi	1	0	283	0.5	0.5	Positive	1.5	1	['Safetesla', 'Tesl	Saf
41	122467683	2010-03-12 20:3	jaminwestby	Glenwood Sprin	Mmmm...Mount	0	0	830	0.5	0.5	Positive	1.5	1	['tesla', 'mountair	Mr
87	3309067506	2015-08-07 22:5	hwfeinstein	Seattle, WA	Anyone else get	0	0	1074	0.375	0.75	Positive	1.5	1	['tesla']	Any

Fig. 4 CSV file for '#Tesla'

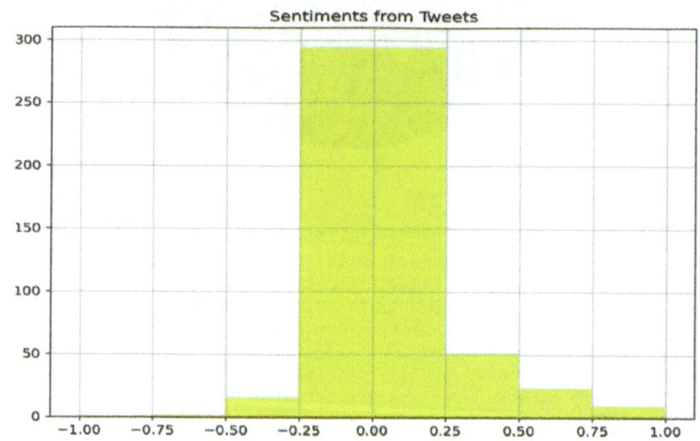

Fig. 5 Column diagram for #Tesla

Fig. 6 Word cloud for '#Tesla'

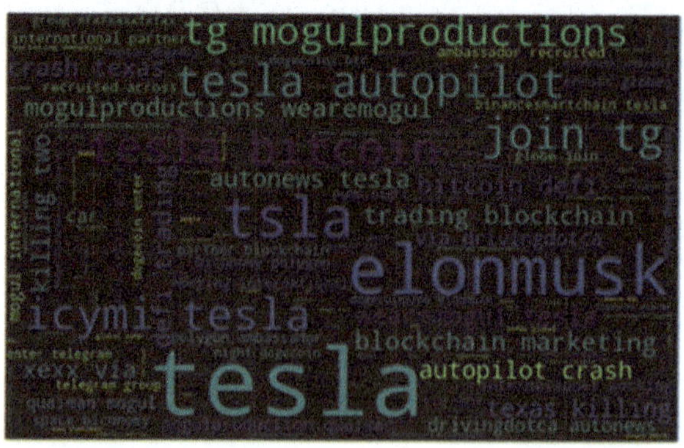

account which can help us to predict accurate sentiment with the help of sentiment analysis from Twitter.

References

1. Singh, P.K., Dutta Pramanik, P.K., Dey, A.K., Choudhury, P.: Recommender systems: an overview, research trends, and future directions, Int. J. Business Syst. Res. **15**(1), (2021)
2. Kim, Y., Shim, K., TWILITE: a recommendation system for Twitter using a probabilistic model based on latent Dirichlet allocation. Inf. Syst. **42**, 59–77 (2014). Elsevier https://doi.org/10.1016/j.is.2013.11.003
3. Sarkar, D., Gupta, M., Jana, P. and Kole, D.: Recommender system in healthcare: an overview. In: Srivastava, R., Kumar Mallick, P., Swarup Rautaray, S. and Pandey, M. ed. Computational Intelligence for Machine Learning and Healthcare Informatics. Berlin, Boston: De Gruyter, pp. 199–216. (2020). https://doi.org/10.1515/9783110648195-010
4. Roy S., Kundu S., Sarkar D., Giri C., Jana P.: Community Detection and Design of Recommendation System Based on Criminal Incidents. In: Bhattacharjee D., Kole D.K., Dey N., Basu S., Plewczynski D. (eds) Proceedings of International Conference on Frontiers in Computing and Systems. Advances in Intelligent Systems and Computing, vol 1255. Springer, Singapore. (2021). https://doi.org/10.1007/978-981-15-7834-2_7
5. Kundu S.S., Desai K., Ghosh S., Sarkar D.:Personalized Word Recommendation System Using Sentiment Analysis. In: Bhattacharjee D., Kole D.K., Dey N., Basu S., Plewczynski D. (eds) Proceedings of International Conference on Frontiers in Computing and Systems. Advances in Intelligent Systems and Computing, vol 1255. Springer, Singapore. (2021). https://doi.org/10.1007/978-981-15-7834-2_8
6. Doshi, S.: In: Brief on Recommender Systems (2019). b86a1068a4dd
7. Jurek, A., Bi, Y., Mulvenna, M.: Twitter sentiment analysis for security-related information gathering. (2014). https://ieeexplore.ieee.org/xpl/conhome/6968676/proceeding https://doi.org/10.1109/JISIC.2014.17
8. Hamdan, H., Bellot, P., Bechet, F., Lsislif: Feature extraction and label weighting for sentiment analysis in Twitter. In: Proceedings of the 9th International Workshop on Semantic Evaluation (SemEval 2015), pp. 568–573. Denver, (2015). https://www.aclweb.org/anthology/S15-2095.pdf
9. Cossu, J.V., Dugué, N., Labatut, V.: Detecting Real-World Influence Through Twitter. (2018). https://ieeexplore.ieee.org/xpl/conhome/7312803/proceeding , https://arxiv.org/pdf/1506.05903.pd
10. Yaqub, U., Sharma, N., Pabreja, R., Chun, S.A., Atluri, V.A., Vaidya, H.J.: Analysis and visualization of subjectivity and polarity of Twitter Location data (2018). https://dl.acm.org/doi/proceedings/https://doi.org/10.1145/3209281. https://doi.org/10.1145/3209281.3209313
11. Gkontzis, A.F., Karachristos, C.V., Panagiotakopoulos, C.T., Stavropoulos, E.C., Verykios, V.S.: Sentiment analysis to track emotion and polarity in student for a. ACM (2017) ISBN 978–1–4503–5355–7/17/09 https://doi.org/10.1145/3339909
12. Dutta, S., Sarkar, D., Roy, S., Kole, D. K., Jana, P.:A Study on Herd Behavior Using Sentiment Analysis in Online Social Network. In: Proceedings of 2021 International Conference on Communication, Control and Information Sciences (ICCISc), pp. 1–6. (2021). https://doi.org/10.1109/ICCISc52257.2021.9484918
13. Yaqub, U., Sharma, N., Pabreja, R., Chun, S.A., Atluri, V., Vaidya, J.: Location-based sentiment analysis and visualisation of Twitter election data. Digit. Gov.: Res. Pract. 1, 2, **14**, 19 (2020). https://doi.org/10.1145/3339909
14. Sarkar, D., Kole, D. K., Jana, P.: Survey of influential nodes identification in online social networks. Int. J. Virtual Commun. Soc. Netw. **8**(4), 57–67 (2016)

15. Sarkar, D., Debnath, S., Kole, D.K., Jana, P.: Influential nodes identification based on activity behaviors and network structure with personality analysis in egocentric online social networks. Int. J. Ambient Comput. Intell. **10**(4), 1–24 (2019)
16. Riquelme, F., Gonzalez-Cantergiani, P.: Measuring user influence on Twitter: a survey. Inform. Process. Manage. **52**(5), 949–975 (2016). https://doi.org/10.1016/j.ipm.2016.04.003
17. Kundu, S., Desai, K., Ghosh, S., Sarkar, D.: Personalized word recommendation system using sentiment analysis. AISC, In: Proceedings of International Conference on Frontiers in Computing and Systems, vol. 1255, pp. 81–93. (2021)
18. http://trailerpark.weebly.com/imdb-rating.html?source=post_page Accessed on 30 Apr 2021

Content-Based Movie Recommendation System with Sentiment Evaluation of Viewer's Reviews

Meet Singh and Dhrubasish Sarkar

Abstract Everyone loves watching movies, it provides relaxation, entertainment and motivation. And the most fascinating part about movies is that while at one end it unites people as a group together irrespective of their cast, religion, ethnicity, but on the contrary, it also differentiates each individual on the basis of their unique choices and preferences they have for the movies they might like. Humans are social beings, who depend on each other's opinions and recommendations to choose which movie to watch. This creates a demand for a robust recommender system which can recommend movies with not just high accuracy but also one which can save time. This paper proposes a movie recommendation system based on content-based filtering (CBF) Metcren and Someren (Proceedings Machanics Learning New Information Age, Workshop. pp 47–56, 2000 [1]) and sentiment analysis (SA) which uses correlation technique to find similarity between movie's features and reviews and analyses the sentiments on the reviews given by the user for that movie. TMDB and IMDB movie dataset was used to perform this experiment. Machine learning and deep learning algorithms were used for performing sentiment classification, and a comparative evaluation of performance metrics is performed resulting in the selection of the best model.

Keywords Recommendation systems · Content-based filtering · Sentiment analysis · Movie reviews

1 Introduction

Recommendation systems have become a necessary part of today's busy lifestyle. People are running out of time to complete the never-ending tasks within the limited 24-h time frame that they need a system which can facilitate and expedite their decision-making while saving not just time but also cognitive resources. A recommendation system recommends content to the user on the basis of their preferences

M. Singh · D. Sarkar (✉)
Amity Institute of Information Technology, Amity University, Kolkata, India
e-mail: dhrubasish@inbox.com

K. N. Das et al. (eds.), *Proceedings of the International Conference on Computational Intelligence and Sustainable Technologies*, Algorithms for Intelligent Systems,
https://doi.org/10.1007/978-981-16-6893-7_15

and other deciding factors. It uses a class of filtering techniques and pattern recognition algorithms to predict future based on past data. In today's era, there are thousands of choices to choose from for any given product, if its left upon the people to choose from, they will take forever to decide. A good recommendation system will provide personalized content to each person based on user preference and browser history. It has the capability to learn on itself by watching patterns and provide relevant suggestions, this made it very popular in online digital platforms like Netflix, Amazon, YouTube and Spotify. Recommendation systems are a powerful tool designed to provide recommendations on various items like advertising messages, movies, books, TV programs, music tracks, news articles, investment choices, restaurants, future friends in social network sites, courses in e-learning, jobs, research papers and clothes, though the most used application of recommendation systems is to recommend movies. Based on the recent advances in academic and industrial researches, analysing human sentiments has been proven to be a very effective medium for finding optimized recommendations. For each product, reviews are collected and sentiment analysis is performed to create a feature that reflects how positive or negative those reviews are. Then, this feature or value is used to perform matrix factorization in the recommender system.

Everyone loves watching movies but what we love more is to get recommended with movies similar to our taste. For this purpose, we have always been relying on the opinions and recommendations of our peer, friends and family members. We also explore various movie blogging sites which provide us with reviews about the movies. On the basis of recommendations, we get from friends or families combining them with the sentiments we analyse from the movie reviews we decide which movie to watch next. This was the motivation behind developing this recommendation system, it does not just recommend movies on the basis of our preferences but also provides sentiments attached with those movies whether positive or negative.

In this paper, the proposed content-based recommendation system is traversed through all possible suggestions and creates a customized list of recommendations that the user might find relevant/interesting. This is achieved through predictive modelling and heuristics with the available data. When there are lots of choices to choose from, content-based approach is very useful. In content-based methods, various factors like the ratings and buying behaviour of users are combined with the content information available for the items, e.g. Jane likes Spiderman movie. So, based on the similar genre keywords other science fiction movies, such as The Dark Knight will be recommended. To build the content-based recommender system pairwise similarity scores was computed for all movies based on their reviews, resulting in movies being recommended based on that similarity score and parallelly sentiments of the movie reviews were analysed for each recommended movie to provide a comprehensive list to the user for easy and effective decision-making.

The structure of this paper is as follows: Next section is the related work section where previous works are done in recommendation system and sentiment analysis field are discussed. Then there is the proposed methodology used to implement sentiment analysis and perform movie recommendations. The following section will be result and discussion, and then the last section is conclusion and future works.

2 Related Work

Debnath et al. [2] proposed a hybrid recommender system approach where movie recommendation was computed through a combination of both content-based and collaborative filtering methods. They assigned weights to those features which were used for content-based recommendations on the basis of their importance to the user. For collaborative filtering, they used the social network to perform regression analysis resulted in the values which were used for calculating the weights of the features. This experiment was performed on IMDB dataset.

Nazimuddin et al. [3] proposed a hybrid recommender system which has the elements of both content-based filtering and collaborative filtering in order to avoid the limitations caused by either of them. They proposed a diverse item selection algorithm where a diverse metric is used in order to extract the items which are dissimilar from each other through collaborative filtering method and then finally feeding the output of this method into the content-based filtering method. To perform this experiment, Movie Lens [4] dataset was used. They achieved better results when compared to content-based filtering and naive hybrid approach. Their proposed approach is provided high accuracy while also providing the benefit of diverse items being recommended. They claimed that the proposed model performs better than any rating algorithm.

Singh et al. [5] presented an experimental work on sentiment analysis performed on movie reviews and blog posts. Their proposed method was named SentiWordNet and used a variety of linguistic features, aggregation metric and scoring schemes to perform the sentiment analysis. They used two popular machine learning algorithms, naive Bayes and support vector machine (SVM), for the purpose of sentiment classification.

Uluyagmus et al. [6] proposed an approach where movie recommendation was based on content-based filtering approach where items were neural network architecture to user through using different feature sets from the past experience data of the user. They assigned weights to each feature in the feature set based on user's preference to calculate a ranking mechanism based on the duration spent on the movie the user spends on. This resulted in the creation of an extra feature which was a combination of various other features like actor, genre, director and keyword. In the end, all the features were used independently along with the combined feature to recommend movies to the users based on the rating generated on the data.

Son and Kim [7] proposed an approach where movie recommendation which was based on content-based filtering which recommended items to user by calculating correlations between them. For this, a multi-attribute network was used to reflect several changes effectively. The similarity between the items was found between both directly and indirectly linked items. To find the mutual relationships and to find patterns between linked items, centrality and clustering techniques were used. This enabled the proposed method to recommend a variety of items, helping the model to improve the performance. For dataset, MovieLens was used to perform the research resulting in high accuracy and robustness. Their proposed method solved the sparsity problem while also solving the over-specialization problem that happened in a lot of recommendation systems. They only used attributes to recommend similar items through the proposed method was the ratings attribute.

Kumar et al. [8] proposed a hybrid recommender system approach where movie recommendation was performed using a combination of both content-based and collaborative filtering methods. Sentiment analysis of tweets was additionally performed on collected data from the micro-blogging sites. Their proposed method used results of analysed sentiment of the Twitter data along with movie metadata and a social graph to recommend movies. Sentiment analysis was performed to find out the response of the user towards the movie. To implement the recommendation system, a weighted score fusion concept was used. Their proposed model provided better results compared to other generic models.

Reddy et al. [9] proposed an approach where movie recommendation was based on content-based filtering which recommended items to user by using genre correlations. Their proposed method was very simple, they used a rating mechanism where when a user provides a rating to a movie its genre is used to find other movies having similar genre through calculating the correlation between the user provided movie with other movies from the dataset. MovieLens dataset was used for this experiment.

Mehta and Gupta [10] proposed an approach where movie recommendation was performed using sentiment analysis and cosine similarity. They proposed a hybrid recommender system using cosine similarity score while improving the recommendations through sentiment analysis. Two different datasets were used: the first dataset was movie rating of top 5000 movies while for sentiment analysis they used movie tweets database which contains reviews of multiple movies. Their proposed method produced better results as compared to other generic models.

3 Proposed Methodology

The objective behind developing this movie recommendation system was intended to create a scalar and robust model. To achieve this, the movies dataset will pass through two parallel pipelines. First pipeline performed movie recommendation, it used a combination mapped for each movie's metadata and its description, whereas the second pipeline with perform sentiment analysis on users' reviews about the movie. Benefits of using content-based filtering algorithm are to provide a method which will not provide the same recommendations for every user, and rather it will provide personalized recommendations for each user. Let us say if user A likes to watch romantic movies, then similar movies would be recommended on the basis of similarity score between the romantic movie and other recommended movies. Sentiment analysis adds a major advantage by backing recommendations from the proposed system through providing individual classification of movie reviews from different uses whether its positive or negative. The combined output of both the pipelines together comprises the recommendation system. Proposed model's architecture can be understood from the flowchart in Fig. 1.

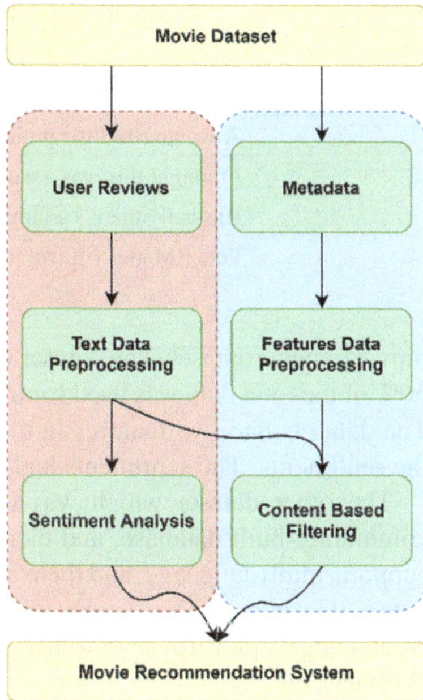

Fig. 1 Flowchart for the proposed movie recommendation system

The movie data is divided into two parts: user reviews and movie's metadata. User reviews are passed through pipeline one, movie review data is passed through a series of data pre-processing natural language processing (NLP) techniques resulting in the development of a word corpus which is further used for both sentiment analysis and as a parameter in content-based filtering, whereas the movies meta-data is also processed by passing through a series of data pre-processing techniques in order to clean the data and perform exploratory data analysis (EDA) to figure out which of the features are essential and useful for the recommendation systems. Feature engineering and feature selection are performed to create the master dataset. Now this master dataset along with the word corpus created by NLP is together used for finding out similarity score between different movies through the content-based filtering method. The combined output of both the sentiment analysis and content-based filtering recommendations is together presented to represent the complete movie recommendation system.

3.1 Data Acquisition

Two datasets have been used for this experiment: one is movie reviews dataset from IMDB [11] while the second data is the movies meta-data from TMDB [12]. The IMDB dataset has 50,000 movie reviews. This dataset is generally used for binary sentiment classification, and it has substantially more data than any of the previous

Table 1 Overview of the IMDB movie review dataset

Heading level	Example	Font size and style
0	One of the other reviewers has mentioned that ...	Positive
1	A wonderful little production. < br / > < br / > The...	Positive
2	I thought this was a wonderful way to spend ti...	Positive
3	Basically there's a family where a little boy ...	Negative
4	Petter Mattei's 'Love in the Time of Money' is...	Positive

movie reviews dataset. The dataset was required to be divided into 50/50 split where 50% of the total data was used for training while the latter 50% was used for testing. The datasets had two features in the dataset shown in Table 1, i.e. the reviews and the sentiments. The sentiments had two unique values: positive and negative.

The other dataset which was used is the TMDB movie database. This is a community-built database, and there are movie data from the year 2017. This data supports multi-language, and there are 45,000 movies along with each movies' meta-data in this dataset. The features which are provided in the dataset are data about cast, crew, budget, plot, revenue, release date, language, production company, country, overview, etc.

3.2 *Data Processing and Exploratory Data Analysis*

It was found through bi-variate analysis that counts of both positive and negative sentiments were equal and the count value was 25,000 words. The dataset was balanced as it contained equal count of semantics for reviews of both polarities: positive and negative. The dataset contained redundant words in HTML syntaxes, punctuations and stopwords in equal distribution; thus, a lot of data cleaning was required to be performed. Regex was used to remove HTML codes, URLs, emojis, stopwords, abbreviations and punctuations. Once the data is cleaned, now it can be used for transformation.

Data pre-processing of the dataset is performed using the natural language tool kit (NLTK) library [13]. Some words were pruned back to their base forms so that the length of the sentences is reduced. This is one of the most important steps because when machine learning modelling will be applied then two different words having the same base word would be encoded and tokenized in different manners. For example, 'watched' and 'watching' have the same root word 'watch'; however, they would be included separately with respect to any tokenizer to fix this issue. Lemmatization is performed on the text corpus in order to reduce the word to the root semantic word. Next was performed to retrieve additional information from the text corpus. Next the dataset was vectorized which allowed for the data to be converted into higher dimensional matrices. Vectorization made the word corpus properly suitable for advanced semantic analysis. For vectorizing, a dataset TFIDF and count vectorization methods were used. Next the labels were converted from text values like positive and negative to binary values like 1 and 0, it will be helpful when tensor compression or dimensionality reduction algorithm will be applied. Cloud visualization was used to show the most significant words present in positive and negative reviews as seen in Figs. 2 and 3.

IMDB movies metadata dataset had two files, and in one file, there were four columns: movie id, title, cast and crew as shown in Fig. 4, and in the second column, there were 20 columns including budget, genre, id, keyword, etc., as shown in Fig. 5. Both the dataframes were merged into a master dataframe on which various kinds of pre-processing techniques were applied in order to clean the data and make the data useful so that insights can be drawn from it. The types of data pre-processing techniques which were applied are as follows:

- Handling missing values
- Performing feature selection
- Typecasting the features
- Normalization and standardization.

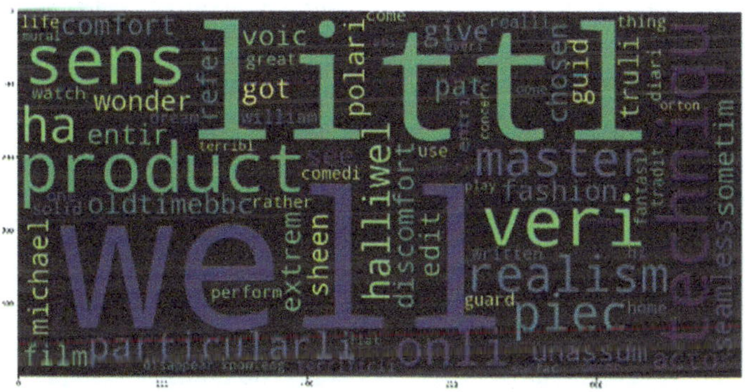

Fig. 2 Word cloud visualization plot for positive sentiments

Fig. 3 Word cloud visualization plot for negative sentiments

```
<class 'pandas.core.frame.DataFrame'>
RangeIndex: 4803 entries, 0 to 4802
Data columns (total 4 columns):
 #   Column    Non-Null Count  Dtype
---  ------    --------------  -----
 0   movie_id  4803 non-null   int64
 1   title     4803 non-null   object
 2   cast      4803 non-null   object
 3   crew      4803 non-null   object
dtypes: int64(1), object(3)
memory usage: 150.2+ KB
```

Fig. 4 TMDB movies metadata features overview for file 1

```
<class 'pandas.core.frame.DataFrame'>
RangeIndex: 4803 entries, 0 to 4802
Data columns (total 20 columns):
 #   Column                Non-Null Count  Dtype
---  ------                --------------  -----
 0   budget                4803 non-null   int64
 1   genres                4803 non-null   object
 2   homepage              1712 non-null   object
 3   id                    4803 non-null   int64
 4   keywords              4803 non-null   object
 5   original_language     4803 non-null   object
 6   original_title        4803 non-null   object
 7   overview              4800 non-null   object
 8   popularity            4803 non-null   float64
 9   production_companies  4803 non-null   object
 10  production_countries  4803 non-null   object
 11  release_date          4802 non-null   object
 12  revenue               4803 non-null   int64
 13  runtime               4801 non-null   float64
 14  spoken_languages      4803 non-null   object
 15  status                4803 non-null   object
 16  tagline               3959 non-null   object
 17  title                 4803 non-null   object
 18  vote_average          4803 non-null   float64
 19  vote_count            4803 non-null   int64
dtypes: float64(3), int64(4), object(13)
memory usage: 750.6+ KB
```

Fig. 5 TMDB movies metadata features overview for file 2

	director_name	actor_1_name	actor_2_name	actor_3_name	genres	movie_title	comb
0	John Lasseter	Tom Hanks	Tim Allen	Don Rickles	Animation Comedy Family	toy story	Tom Hanks Tim Allen Don Rickles John Lasseter ...
1	Joe Johnston	Robin Williams	Jonathan Hyde	Kirsten Dunst	Adventure Fantasy Family	jumanji	Robin Williams Jonathan Hyde Kirsten Dunst Joe...
2	Howard Deutch	Walter Matthau	Jack Lemmon	Ann-Margret	Romance Comedy	grumpier old men	Walter Matthau Jack Lemmon Ann-Margret Howard ...
3	Forest Whitaker	Whitney Houston	Angela Bassett	Loretta Devine	Comedy Drama Romance	waiting to exhale	Whitney Houston Angela Bassett Loretta Devine ...
4	Charles Shyer	Steve Martin	Diane Keaton	Martin Short	Comedy	father of the bride part ii	Steve Martin Diane Keaton Martin Short Charles...
...	...	...	...	...	...	...	...
36982	Joseph Kosinski	Tom Cruise	Miles Teller	Jennifer Connelly	Action Drama	top gun: maverick	Tom Cruise Miles Teller Jennifer Connelly Jose...
36983	Joel Crawford	Nicolas Cage	Emma Stone	Ryan Reynolds	Animation Adventure Family	the croods 2	Nicolas Cage Emma Stone Ryan Reynolds Joel Cra...
36984	Liesl Tommy	Jennifer Hudson	Forest Whitaker	Marlon Wayans	Music Drama	respect	Jennifer Hudson Forest Whitaker Marlon Wayans ...
36985	Ridley Scott	Matt Damon	Adam Driver	Jodie Comer	Drama	the last duel	Matt Damon Adam Driver Jodie Comer Ridley Scot...
36986	Paul Greengrass	Tom Hanks	Helena Zengel	Neil Sandilands	Drama Western	news of the world	Tom Hanks Helena Zengel Neil Sandilands Paul G...

Fig. 6 Overview of final pre-processed TMDB dataset

As a result, the final dataset had important features like director names, actor names, genres and movie title, as shown in Fig. 6.

3.3 *Proposed Model*

Sentiment Analysis: For modelling the dataset, multiple machine learning algorithms when implemented for the classification of the movie review into either positive or negative values. Logistic regression, linear support vector machine, multinomial naive Bayes and BERT were used.

BERT stands for bidirectional encoder representations from transformers, and it is a state-of-the-art machine learning model used for NLP tasks. It was developed by Devlin et al. at Google [14]. The proposed model was trained upon English language from the Wikipedia and BookCorpus data consisting of 2500 and 800 million words, respectively, and achieved state-of-the-art accuracy as of 2018. There are two variants of pre-trained general BERT model: the base model neural network architecture has a 12-layer, 768-hidden, 12-heads and 110 M parameter, whereas the large model has a 24-layer, 1024-hidden, 16-heads, 340 M parameter neural network architecture. To work with BERT model, Hugging Face transformer library was used.

Logistic regression model gave an accuracy score of 75% to the tfidf features. From the classification report from Fig. 7, it can be observed that weightage average of all three precision, recall and F1-score has given score of 0.75. From the confusion matrix, it was observed that 3663 values were true positives (TP), 1344 values were false-positive (FP), 1156 values were false negative (FN) and 3837 values were true negative (TN).

Linear support vector machine gave an accuracy score of 51% to the tfidf features. From the classification report from Fig. 8, it can be observed that weightage average of precision, recall and F1-score has given score of 0.75, 0.51 and 0.36. From the confusion matrix, it was observed that 5007 values were true positive (TP), 0 values were false-positive (FP), 4888 values were false negative (FN), and 105 values were true negative (TN).

Fig. 7 Classification report for logical regression classifier

	precision	recall	f1-score	support
Positive	0.74	0.77	0.75	4993
Negative	0.76	0.73	0.75	5007
accuracy			0.75	10000
macro avg	0.75	0.75	0.75	10000
weighted avg	0.75	0.75	0.75	10000

Fig. 8 Classification report for linear support vector machine classifier

	precision	recall	f1-score	support
Positive	1.00	0.02	0.04	4993
Negative	0.51	1.00	0.67	5007
accuracy			0.51	10000
macro avg	0.75	0.51	0.36	10000
weighted avg	0.75	0.51	0.36	10000

Multinomial naive Bayes model gave an accuracy score of 75% to the tfidf features. From the classification report from Fig. 9, it is observed that weightage average of all three precision, recall and F1-score has given score of 0.75. From the confusion matrix, it was observed that 3729 values were true positive (TP), 1278 values were false-positive (FP), 1213 values were false negative (FN), and 3780 values were true negative (TN).

BERT is based on the transformer architecture. Adam was used as optimizer, CategoricalCrossentropy was used as loss function, and SparseCategoricalAccuracy

Fig. 9 Classification report for multinomial naïve Bayes classifier

	precision	recall	f1-score	support
Positive	0.75	0.76	0.75	4993
Negative	0.75	0.74	0.75	5007
accuracy			0.75	10000
macro avg	0.75	0.75	0.75	10000
weighted avg	0.75	0.75	0.75	10000

Fig. 10 Classification report for BERT

	precision	recall	f1-score	support
Positive	0.90	0.91	0.90	4993
Negative	0.91	0.90	0.90	5007
accuracy			0.90	10000
macro avg	0.90	0.90	0.90	10000
weighted avg	0.90	0.90	0.90	10000

as accuracy metric. This deep learning model ran for 8 epochs and gave an accuracy score of 88% to the tfidf features, it also reached 0.04 loss value shown in Fig. 9. From the classification report from Fig. 10, it is observed that all three precision, recall and F1 score has given score of 0.90.

Figure 11 shows how BERT gave the best scores and was selected as the final model. And it gave accurate predictions as shown in Fig. 12.

Recommendation System: Leveraging the corpus created from the movie reviews data with the movie metadata, content-based filtering method is proposed. To find which item is most similar to the item which the user might like similarity score has been used. The similarity score is a number which ranges between 0 and 1, and it helps to determine the similarity; i.e. how much two items are similar to each other on a scale of 0 to 1. Similarity score is obtained through measuring the similarity amongst the text data of the different items. There are various methods to apply similarity score but for this experiment cosine similarity has been used. Cosine similarity is the metric which is used to measure how similar two textual data are irrespective of their size it mathematically measures the cosine of the angle between the two vectors projected in a multidimensional space as seen in Fig. 13. It is independent of magnitude and relatively easy and fast to calculate. It gives a great advantage over Euclidean distance as when two similar data are far apart then this distance measure suffers from the size of the data, but chances are that they may still be oriented closer together from cosine similarity. The smaller the angle, the higher the cosine similarity.

```
1250/1250 [==============================] - 1131s 863ms/step - loss: 0.2653 - accuracy: 0.8873 - val_loss: 0.3430 - val_accuracy: 0.8824
Epoch 2/2
1250/1250 [==============================] - 1076s 861ms/step - loss: 0.0766 - accuracy: 0.9733 - val_loss: 0.4837 - val_accuracy: 0.8820
<tensorflow.python.keras.callbacks.History at 0x7f38abeed090>
```

Fig. 11 Accuracy. val accuracy and loss score for the BERT classifier

```
This was an awesome movie. I watch it twice my time watching this beautiful movie if I have known it was this good :
 Positive
One of the worst movies of all time. I cannot believe I wasted two hours of my life for this movie :
 Negative
```

Fig. 12 Accurate predictions for positive or negative sentiments by BERT classifier

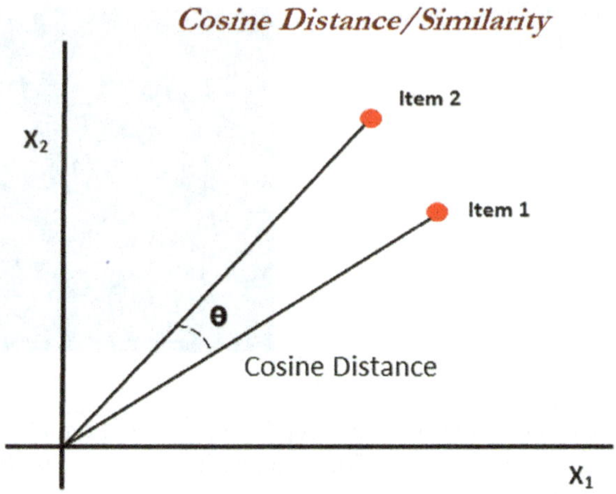

Fig. 13 Plot representing cosine distance similarity

Cosine similarity gave very accurate recommendations as seen in Fig. 14. When the movie 'JFK' was provided as input, the proposed recommendation calculated the cosine similarity between the movie and all other movies and as result recommended movies like 'Zero Dark Thirsty', 'Criminal', 'World Trade Center', 'Jack Ryan: Shadow Recruit', etc., which were all similar to each other. This validates that the proposed system is performing accurately.

Fig. 14 Generated recommendations by the proposed algorithm

```
get_recommendations('JFK')

2507                 Slow Burn
879         Law Abiding Citizen
2020               The Rookie
2193       Secret in Their Eyes
2697                     Bobby
753               The Sentinel
1202              Legal Eagles
817           American Wedding
65             The Dark Knight
3        The Dark Knight Rises
Name: title_x, dtype: object
```

4 Result Analysis and Discussion

An interactive web application is developed to perform the movie recommendation system. The home page will provide the users the option to provide the name of the movie. While typing the name of the movie, users will be getting recommendations by the auto complete engine of the web application to help them provide the name of the movie in a quicker manner.

When the users have been provided with the name of the movie, they need to select the enter button which will now send a request to the movie database API to collect meta-data along with user reviews about the movie. Once the meta-data and reviews are collected. Data pre-processing methods are being implemented to clean the data and to transform the data in the same manner in which the training dataset was pre-processed. This pre-process data is then used to display all the basic information about the movie like title, poster, description, release date, genre, rating, run-time and status. It is also used for providing information regarding the movie's cast, i.e. the directors, actors, producers, etc.

Next, the pre-trained BERT model will be used to perform sentiment analysis on the data collected about the movie from the API to display a positive and negative sentiment polarity for the given reviews.

Finally, the proposed content filtering-based recommendation approach will find similarities between the movie user provided and the rest of the other movies from the master dataset. All the movies which have a similar cosine score will be recommended to the users in the form of cards, displaying the title and poster of the movie.

5 Conclusion and Future Scope

Now days, recommendation systems are widely used in various domains including healthcare [15–17] where sentiment analysis [18, 19] plays a major role in building the model. The goal of this paper was to provide a low-cost yet efficient functionality method for recommending movies along with sentiments attached to the reviews of each movie. Brief literature survey was performed on various movie recommendation systems. Our proposed model is based on prior research performed on sentiment analysis and recommendation system using machine learning, deep learning and natural language processing. The proposed model uses state-of-the-art algorithm BERT of deep neural networks which resulted in improved accuracy compared to previous work. The proposed recommendation system is not just fast but also very informative in nature. It provides the dual functionality by not just simply recommending movies but also their movie sentiments, IMDB and TMDB datasets were used for model training and development. Various data pre-processing techniques have been applied like vectorization lemmatization, tfidf and stemming to transform the data into functional matrices in order to minimize computational restraint and to improve the performance. The proposed recommendation system does not just provide generic recommendations but rather the suggestions are more catered towards individual choices and preferences of each user this improves the arts of being selected by the user.

Future work can be done on the deployment of the model as an API service making the system available for public use. The API can also be used as an online model and be incorporated into any web application or software application development.

References

1. Meteren, R.V., Someren, M.V.: Using content-based filtering for recommendation. In: Proceedings Machines Learning New Information Age, Workshop, pp. 47–56. (2000)
2. Debnath, S., Ganguly, N., Mitra, P.: Feature weighting in content based recommendation system using social network analysis. In: Proceedings of the 17th international conference on World Wide Web (WWW '08). Association for Computing Machinery, New York, NY, USA, pp. 1041–1042. (2008). https://doi.org/10.1145/1367497.1367646
3. Uddin, M.N., Shrestha, J., Jo, G.: Enhanced content-based filtering using diverse collaborative prediction for movie recommendation. In: 2009 First Asian Conference on Intelligent Information and Database Systems, pp. 132–137. (2009)
4. Harper, F.M., Konstan, J.A.: The movielens datasets: history and context. ACM Trans. Interact. Intell. Syst. (TiiS) 5, 4, **19**, 19 (2015). https://doi.org/10.1145/2827872
5. Singh, V.K., Piryani, R., Uddin, A., Waila, P.: Sentiment analysis of movie reviews: a new feature-based heuristic for aspect-level sentiment classification, In: International Mutli-Conference on Automation, Computing, Communication, Control and Compressed Sensing (iMac4s), pp. 712–717. (2013). https://doi.org/10.1109/iMac4s.2013.6526500
6. Uluyagmur, M., Cataltepe, Z., Tayfur, E.: Content-based movie recommendation using different feature sets. In: World Congress on Engineering and Computer Science, vol. 1, San Francisco, USA (2012)
7. Son, J., Kim, S.: Content-based filtering for recommendation systems using multiattribute networks. Expert Syst. Appl. **89**, 404–412 (2017)
8. Kumar, S., De, K., Roy, P.: Movie recommendation system using sentiment analysis from microblogging data. IEEE Trans. Comput. Social Syst. **7**, 915–923 (2020)
9. Reddy, S., Nalluri, S., Kunisetti, S., Ashok, S., Venkatesh, B.: Content-based movie recommendation system using genre correlation. In: Smart Intelligent Computing and Applications. Smart Innovation, Systems and Technologies, vol. 105, Springer, Singapore (2019)
10. Mehta, R., Gupta, S.: Movie recommendation systems using sentiment analysis and cosine similarity. Int. J. Modern Trends Sci. Technol. **7**(01), 16–22 (2021) Copyright © 2021 ISSN: 2455–3778 Online (2020)
11. Maas, A.L., Daly, R.E., Pham, P.T., Huang, D., Ng, A.Y., Potts, C.: Learning word vectors for sentiment analysis. In: The 49th Annual Meeting of the Association for Computational Linguistics (ACL 2011)
12. The Movie Database (2021). Available at www.themoviedb.org/. Accessed from Jan 2021 to April 2021
13. Natural Language Toolkit (2021). Available at https://www.nltk.org/. Accessed from Jan 2021 to April 2021
14. Devlin, J., Chang, M., Lee, K., Toutanova, K: BERT: pre-training of deep bidirectional transformers for language understanding. In: Proceedings of the 2019 Conference of the North American Chapter of the Association for Computational Linguistics: Human Language Technologies, Volume 1 (Long and Short Papers), pp. 4171–4186. (2019)
15. Sarkar, D., Gupta, M., Jana, P. and Kole, D.:10. Recommender system in healthcare: an overview. In: Srivastava, R., Kumar Mallick, P., Swarup Rautaray, S. and Pandey, M. ed.

Computational Intelligence for Machine Learning and Healthcare Informatics. Berlin, Boston: De Gruyter, pp. 199-216. (2020). https://doi.org/10.1515/9783110648195-010

16. Roy S., Kundu S., Sarkar D., Giri C., Jana P.:Community Detection and Design of Recommendation System Based on Criminal Incidents. In: Bhattacharjee D., Kole D.K., Dey N., Basu S., Plewczynski D. (eds) Proceedings of International Conference on Frontiers in Computing and Systems. Advances in Intelligent Systems and Computing, vol 1255. Springer, Singapore.(2021) https://doi.org/10.1007/978-981-15-7834-2_7
17. Kundu S.S., Desai K., Ghosh S., Sarkar D.:Personalized Word Recommendation System Using Sentiment Analysis. In: Bhattacharjee D., Kole D.K., Dey N., Basu S., Plewczynski D. (eds) Proceedings of International Conference on Frontiers in Computing and Systems. Advances in Intelligent Systems and Computing, vol 1255. Springer, Singapore. (2021). https://doi.org/10.1007/978-981-15-7834-2_8
18. Dutta, S., Sarkar, D., Roy, S., Kole, D. K., Jana, P.:A Study on Herd Behavior Using Sentiment Analysis in Online Social Network. In: Proceedings of 2021 International Conference on Communication, Control and Information Sciences (ICCISc), pp. 1-6. (2021). https://doi.org/10.1109/ICCISc52257.2021.9484918
19. Parasrampuria, R., Ghosh, A., Dutta, S., Sarkar, D.: Recommending Insurance products by using Users' Sentiments.(2021). ArXiv, abs/2108.06210

New Characterization of Electrode of Supercapacitor with Its Application as a Backup Power Supply

Sambit Satpathy, Prakash Mohan, Chandra Shekhar Yadav, and Neeraj Kumar Mishra

Abstract Day by day, the energy storage system (ESS) is critical due to its massive applications in real time, like a hybrid car, wireless sensor network (WSN) and a Capa bus. The supercapacitor is gaining popularity in this relation because of high power density, low effective series resistance (ESR) and rapid charging with a long-life cycle. This makes supercapacitor a hot subject for research. This paper deals with the on electrochemical characterization of the supercapacitor with a short debate concerning supercapacitor as a backup power supply for different systems. Here we have considered a buck converter system which is used to generate and supply a 2–3 Amp current requirement in the industrial system with the backup power by the supercapacitor. The fluctuation of the topology is fully controlled by CMOS logic gate circuits. The whole system is working with a frequency of 1 MHz and a current output draw of 2.5A.

Keywords Supercapacitor · Constant current charging · Voltage doubler · Backup power supply

S. Satpathy (✉)
GLA University, Department of CEA, Mathura, Uttar Pradesh, India
e-mail: sambit.satpathy@suiit.ac.in

P. Mohan
Karpagam College of Engineering, CSE, Coimbatore, Tamilnadu, India

C. S. Yadav
Noida Institute of Engineering and Technology, CSE, Greater Noida, Uttar Pradesh, India
e-mail: hodcs@niet.co.in

N. K. Mishra
Bharat Institute of Engineering and Technology, ECE, Hydrabad, India

K. N. Das et al. (eds.), *Proceedings of the International Conference on Computational Intelligence and Sustainable Technologies*, Algorithms for Intelligent Systems,
https://doi.org/10.1007/978-981-16-6893-7_16

1 Introduction

Supercapacitors are energy storing devices [1] with high energy density. A number of circuits have been devised for efficient charging of supercapacitors [2]. Supercapacitors are environment-friendly as well as have a greater life cycle compared to batteries. Apart from this, supercapacitor is capable of pumping high current at low voltages due to which it is being used in a wide range of industrial applications. DC-DC converters also have a wide range of industrial applications owing to the property that it can be used to supply a constant voltage at a constant current and also to step up to step down DC voltages to the desired level. Supercapacitors have high power density due to which supercapacitors can be used to pump high power that too at lower voltages [3]. The only problem with supercapacitors is that supercapacitors with high capacitance have low working voltage as a result of which boost converters need to be used in cascade with supercapacitors in order to power industrial applications. Nowadays, batteries and supercapacitors are used as hybrid system for industrial applications like in hybrid vehicle [4–6].

In this work, a DC-DC buck converter system is being used to supply an industrial automotive application. So, the output has current draw of 2A–3A at 5 V. A control loop [3] has been designed for this buck converter that varies the duty ratio proportionately with changes in input voltage in order to yield a constant output. If the input voltage falls below the desired output voltage, the input voltage is cut off and the output load gets supplied by a supercapacitor module via a boost converter circuit. This boost converter circuit also has a control loop that yields a constant voltage irrespective of the decaying voltage across the supercapacitor module. The supercapacitor module is created by parallel combination of supercapacitors of 2.85 V and 3400F in order to store higher energy.

2 Prior Related Work

2.1 Electrochemical Characteristics

Here, we discuss the electrochemical behaviour of various energy storage systems additionally the specification and advancement of supercapacitor technology [7]. We addressed different energy storage system in Table 2 [8], such as lithium-ion, plum acid, vanadium redox and NaS battery (sodium and sulphur). Power, energy, lifetime, discharging time and efficiency have been considered as four parameters to evaluate their advantages and utility. Table 2 in the first and second columns displays the energy storage system's power and capacity. It has found more the power, is smaller in capacity. So that the lithium-ion battery has high energy, and the NaS battery capacity is higher than other systems. But the life of the battery system with lead acid is relatively longer than other systems, i.e. 20 years. It is important to mention any energy storage device listed has certain characteristics with both advantages

Table 1 Different electrochemical energy storage systems in modern power system [8]

Energy storage system	Power (MW)	Capacity (MWh)	Life time (Years)	Discharge time (hours)	Efficiency (%)	Advantage
Lithium-ion	≤ 100	≤ 0.25–25	≤ 15	≤ 1	≤ 90	Long-life cycle with high storage capability
Lead acid	≤ 100	≤ 0.25–25	≤ 20	≤ 4	≤ 85	Low cost and high recyclability
Vanadium Redox	≤ 50	≤ 250	≤ 10	≤ 8	≤ 80	Compatible and used in various renewable energy sources
NaS	≤ 50	≤ 300	≤ 15	≤ 6	≤ 80	Low cost and high recyclability

Table 2 Gate operation with on and off conditions

NOT gate	ON condition	OFF condition
N1	Input voltage > = 2.85 V	Input voltage < 2.85 V
N2	Input voltage > = 2.85 V	Input voltage < 1.43 V

and disadvantages. Researchers absorbed the benefits and sought to minimize the downside. So here we defined one more energy storage system, i.e. supercapacitor, that can combine with the above-mentioned storage systems (Table 1) for a more effective system in any application in real time [9, 10].

Supercapacitor is structurally and functionally identical to a regular capacitor. However, ultracapacitors with the value of large capacitance (kilofarad) [11] with a high power density are considered supercapacitors. The supercapacitor's specific power is about 1000–2000 W/kg with an efficiency of 95.5%, and life cycle is nearly 10 years which is the longest among the ESSs. The supercapacitor's effective series resistance (ESR) is very less so that the fast charging and discharging profile can be achieved. This helps supercapacitor to be used in hybrid cars [4], smart phones [2], WSNs [12] and other consumer electronics applications.

Figure 1 reveals the schematic view of the general supercapacitor or double-layer electrical condenser. A supercapacitor varies in two ways from a standard capacitor. Firstly, the parallel plates have a wider surface; secondly, there is a much narrower space between them. The electrodes are coated with activated carbon powder, and an electrical double film forms at the interface between activated carbon and electrolytes. As we charge the supercapacitor, the + ve ions are arranged around the

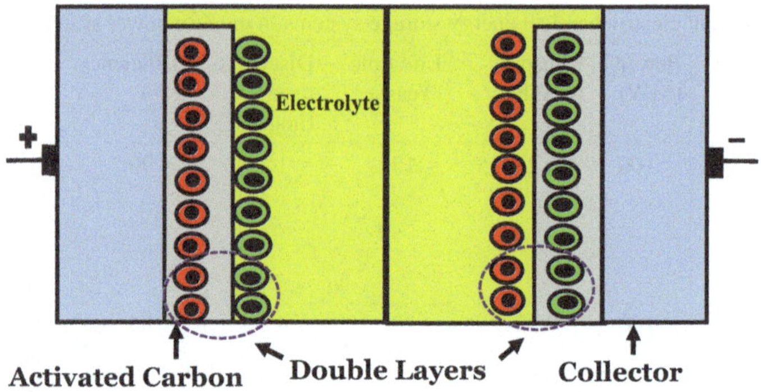

Fig. 1 Different electrochemical energy storage systems in modern power system [8]

negative electrode and the -ve ions on the positive electrode line, as seen in Fig. 1. This form of cation and anion structure is known as an 'electric double layer'. The purpose of 'activated carbon' raises the surface area of the electrode. The higher the surface area, the larger the charge storage, the greater the capacitance. This layer is triggered by physical movement of electrons, and there are no chemical reactions, like a traditional lithium-ion battery. It improves the charging and discharge (life cycle) of the supercapacitor [9] than other battery technology. The double-layer configuration contributes to lower ESR (series resistance equivalent), allowing fast charge of the supercapacitor compared to traditional lithium-ion batteries [13].

2.2 *Electrochemical Characteristics*

Supercapacitor electrodes and electrolytes are provided by a large variety of materials to determine the energy storage mechanisms in it. For understanding and the efficient use of supercapacitor's, the awareness of these processes is important. There are two major experimental studies to describe the supercapacitor electrode, and they are convenient methods for understanding the various mechanism of energy storage. Out of these two methods, one is cyclic voltammetry (see Fig. 2a, b) and other one is galvanostatic discharge as seen in Fig. 2c and d. In the cyclic voltammetry method, we need to measure current vs. applied voltage which is getting from cyclic voltammogram trace. The galvanostatic discharge, on the other hand, consists of a constant current discharge of electrode. The three basic behaviour forms which can be demonstrated by supercapacitor electrodes are explained below. Their electrical characteristics are compared and contrasted based on the basis of cyclical voltammetry and galvanostatic discharge methods.

2.3 *Pseudocapacitance Behaviour*

Pseudocapacitive electrodes exhibit a capacitor-like behaviour as their cyclic voltammogram is similar to their double-layer rectangular form (see Fig. 3a) and also have a linear galvanostatic discharge, which can observe from Fig. 3c. Pseudocapacitive electrodes can give a considerably greater specific-capacitance compared to double-layered electrodes which makes them attractive for applications requiring high energy density. Still they suffer two major problems, i.e. lesser cyclability and lower power density that are associated with the electrochemical reactions. This electrochemical reaction involves with irreversibility and which allows the electrode to age more rapidly as the dynamics of the reaction decrease its strength. The development of pseudocapacitive electrodes is a result of RuO2's peculiar electrochemical behaviour. Several cheap metal oxides have also been tested, for pseudocapacitors as possible electroactive materials [14]. The metal oxide materials and hydroxides are of considerable importance because of their abundance, high value of specific capacitance and low toxicity materials currently being studied.

Metal carbides and conducting polymers are inherently shown pseudocapacitive behaviour, in comparison with the metal oxides. Compared to inorganic materials, the major downside of conductor polymers is that although all of them have zero volumes for ions to cross, the conductive polymers that do not have an adequate amount of spaces to carry their ions and the intercalation creates extreme changes in thickness of the electrode. One of the pioneering conduction materials is polyaniline (PANI). Due to the progress made in technology in integrating PANI with other electroactive materials to improve the electrochemical efficiency of the supercapacitor, this material has been paying more attention in the past few years. The plurality of available supercapacitor's electrode must be taken into account, as the existence of functional groups contributing to a pseudocapacitive secondary response, which is a marginal part of the pseudocapacitive category.

3 Supercapacitor as Backup Power Supply

In this paper, the designing of a closed-loop DC-DC converter system [14] having feed-forward [15] path has been discussed. The feed-forward path varies the peak of triangular wave [16] or sawtooth wave with the fluctuation in input voltage. As a result of which the duty ratio of the switching pulse adjusts accordingly in order to yield a constant or regulated output. The DC-DC converter system comprises of a primary power source powered buck converter system supplying constant power to an application at a constant voltage, and a secondary or backup boost converter system powered by supercapacitor module for proving backup power to the application in case of failure of primary power source. A logic circuit has been designed to switch the power source from primary buck converter to the supercapacitor module powered boost converter system in case of primary system failure. Many control strategies for buck converter have been reviewed for varying the duty ratio proportionately with the fluctuations in input and output. The PWM control loop needs a sawtooth wave generator [15] which will vary the peak of triangular wave with fluctuations in input and a way to increase the reference voltage for PWM with increase in load current beyond the desired limit. The generation of triangular or sawtooth wave could be rather cumbersome for a wide range of input voltage fluctuations. These repeating periodic triangular waves could generate using ICs but not for a wide range of voltage or voltages exceeding 5 V. Another control strategy involves comparing the inductor current of the dc-dc converter [17] with a constant current at some point say at the output or through the output filter capacitor. In CCM, the inductor current waveform is triangular in nature, whose peak and slope would vary with changes in input voltage. But this system is not feasible for a wide range of voltage fluctuations and due to issues like converter operating in DCM.

3.1 Proposed System Circuit Diagram

In Fig. 2a, DC-DC converter system has been shown that is designed to supply load drawing a current of 2.5A at 5 V. This is the specification of industrial automotive load. The circuit system shown in Fig. 2 comprises of three parts. The part of the circuit bordered by green is the main muck converter system supplying the load. The part of the circuit bordered by blue is the supercapacitor powered boost

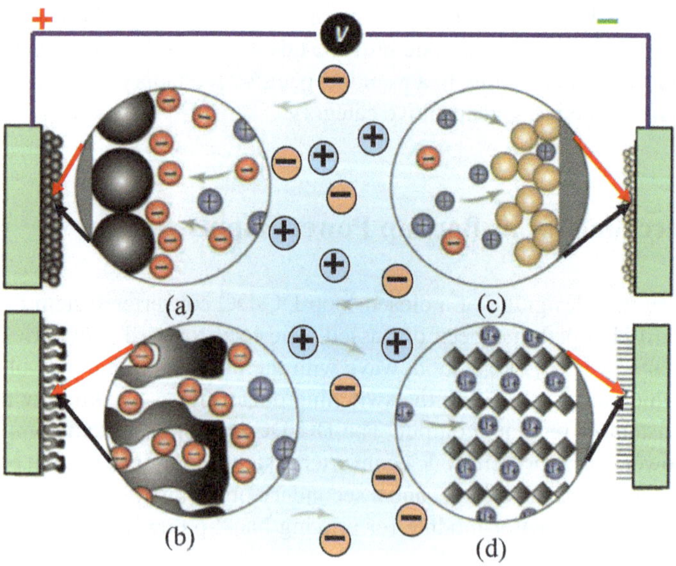

Fig. 2 Numerous mechanisms for the capacitive energy storage are shown: **a** Carbon particles, **b** porous carbon, **c** Redox pseudocapacitance, **d** Intercalation pseudocapacitance

converter system providing backup power to the load in case of failure of the main buck converter system. And the part of the circuit bordered by red is the sequential switching circuit responsible for cutting off the main power source and the buck converter system in case of failure and simultaneously connecting the supercapacitor module powered boost converter system with the load so that the load can function without interruption. The part of the circuit bordered by red is a digital circuit and is responsible for sequential switching. 'N_1' and 'N_2' are NOT gates, 'AN_1' and 'AN_2' are AND gates, and OR is an OR gate. In this circuit, 'N_1' has been designed to handle a maximum voltage of 2.85 V gate while 'N_2' is designed to handle a maximum voltage of 5.7. These NOT gates operate in off condition if input is below half of their maximum operating voltage and has been shown in Table 1. In Fig. 1, 'V_i' is the input voltage of the main power source 'S_1', 'M_1', 'M_2', 'S_3', 'M_3', 'M_4', 'S_6' and 'S_9' are MOSFETS, out of which 'M_1', 'M_2', 'M_3' and 'M_4'are controlled by the sequential circuit bordered by red for switching between the main power source and backup power source with the load. S_2, S_4, S_7 and S_8 are Schottky diodes. Schottky are preferred to general-purpose diodes due to lower voltage drop and fast recovery. 'L' and 'C' are the inductance and capacitance of the buck converter, and 'R_L' and 'R_C' are their equivalent series resistances. 'L_1' and 'C_1' are the inductance and capacitance of the boost converter, and 'RL_1' and 'RC_1' are their equivalent series resistances. All the 'R' are resistances of same value, and 'R_T' and 'R_{T1}' are also resistances of same value.

The 'Z', 'Z_1' and 'Z_3' in Fig. 2 represent Zener diodes that are used for voltage regulation. Their main purpose is to rid the circuit of any unwanted overshoot in the output voltage. 'V_{CS}' denotes the voltage of the supercapacitor module. 'V_0' denotes the output voltage of the entire circuit system while 'I_0' is the current drawn by industrial application. The value of 'V_t' is chosen to be twice the output voltage 'V_0'. 'A_1', 'A_2', 'A_3' and 'A_4' in the circuit are Op-amps out of which 'A_1' and 'A_4' in the circuit behaving as comparator. Till the input voltage 'Vi' remains above the output voltage any change in 'Vi' will lead to change in the peak of sawtooth wave at point 'C' which in turn will adjust the duty ratio proportionately, thereby keeping the output voltage constant. Similarly, when 'Vi' falls below 'V_0', 'V_{CS}' acts as the power source for the load, the duty ratio of boost converter changes accordingly to keep the output constant. The circuit operation has been shown in Tables 2 and 3.

4 Simulation Result

The components used for simulation are listed in Table 4. All the results are shown for the output current draw of 2.5A, although the design can sustain up to a current draw of 7A.

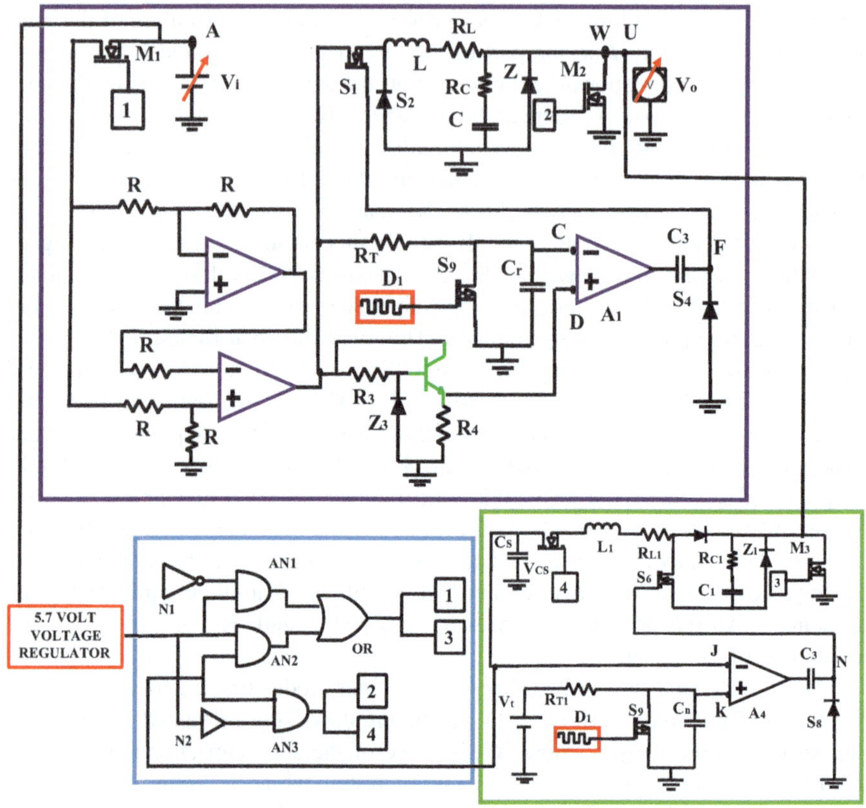

Fig. 3 Buck converter system with automated supercapacitor backup system [13]

Table 3 Operation of the circuit (see Fig. 3) shown

Condition	M1	M2	M3	M4	Input
$V_i > = V_O$	ON	OFF	ON	OFF	V_i
$V_i > = V_O$	OFF	ON	OFF	ON	V_{SC}

Table 4 List of the components used for simulation for the circuit shown in Fig. 3

Component	Specification
MOSFETS 'M1', 'M2', 'S3', 'M3', 'S6', 'S9'	2N7000
MOSFETS 'S1', 'M4'	Si2308BD
Op-amps A1, A4	LT1360
S5	BC547
A2	LM358P
A3	OPA37
Z, Z1, Z3	IN4732A
S2, S4, S7, S8	1N5819

4.1 Simulation Results for 270 s

In Fig. 3a, the blue curve represents the piecewise linear voltage or fluctuating input voltage obtained at node ‘A’ of Fig. 2 and the red dashed curve is the duty ratio that changes proportionately with changing input voltage in order to yield a constant output voltage which is shown in Fig. 5b. The duty ratio curve of Fig. 5a is obtained at node ‘E’, and the output voltage waveform shown in Fig. 5b is obtained at node ‘Y’ of Fig. 3. The waveforms shown in Fig. 5a and b are for the first 270 s till $V_i >= V_0$.

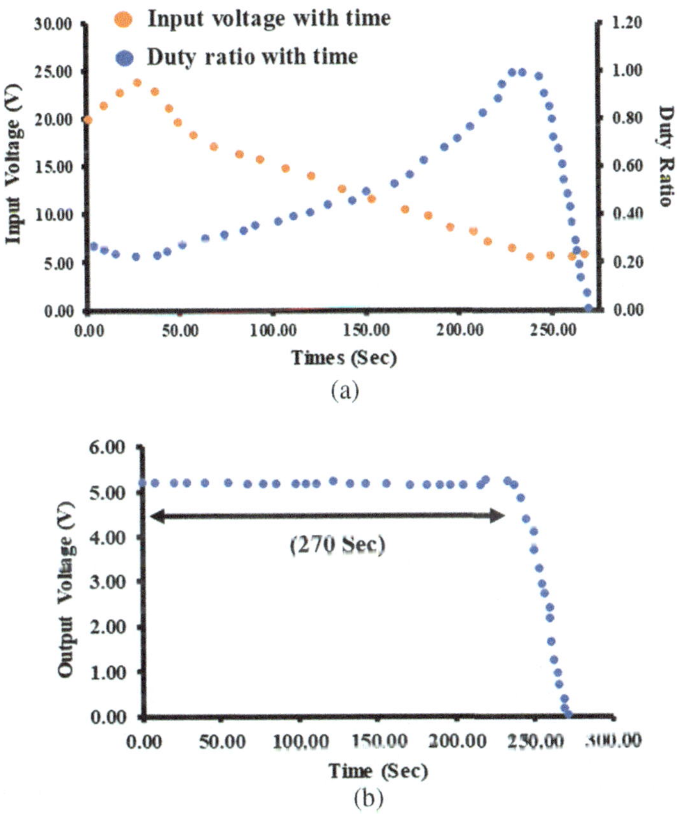

Fig. 4 **a** Input voltage across supercapacitor till 270 s, **b**output voltage source till 270 s

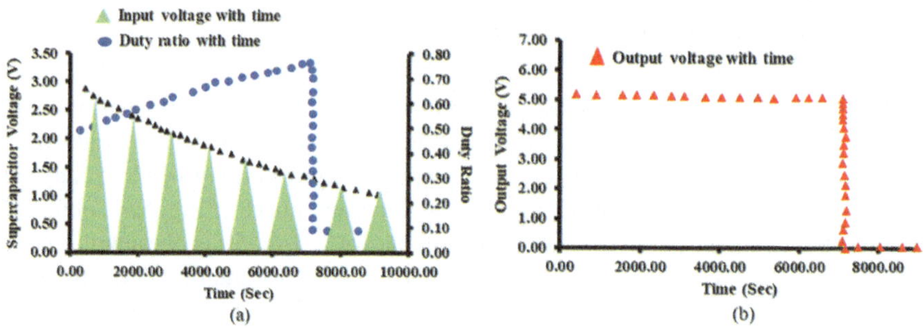

Fig. 5 **a** Input voltage across supercapacitor 270 s onwards, **b** output voltage source 270 onwards

4.2 Simulation Results 270 s Onwards

In Fig. 4a, the blue curve represents the piecewise linear voltage or fluctuating input voltage obtained at node 'F' of Fig. 4 and the red dashed curve is the duty ratio that is changing proportionately with the changing input voltage in order to yield a constant output voltage and this output voltage has been shown in Fig. 4b. The duty ratio curve of Fig. 4a is obtained at node 'N', and the output voltage waveform shown in Fig. 4b is obtained at node 'Y' of Fig. 4. The waveforms shown in Fig. 6a and b are the waveforms obtained 270 s onwards, i.e. when $Vi < Vo$.

5 Conclusion

An innovative method is developed, which consists of an analog buck circuit and a digital circuit. The objective of the digital circuit is to make sure that whenever the input voltage drops below a certain voltage then the bank of supercapacitors or supercapacitor module kicks in to maintain the constant output voltage, which in our design is about 5 V with an output current draw of 2.5A. This is evident from simulation results or waveforms shown in Figs. 3b and 4b. This module can supply from 12.5 W to 35-W power for the load to maintain the constant voltage at the output. This circuit is unique because it can handle high output current (2–5 Amp) at constant output voltage, which is 5 V. The digital design was an innovative in our case to switch the input power supply from wall power or any power module like battery to the charged supercapacitor bank. This part is mostly done using logic gates, comparators and reference signals, which are normally difficult in practical applications when all the voltages are fluctuation. We have also tested the circuit through simulation and found that switching occurs when the voltage drops below 5 V.

References

1. Satpathy, S., Das, S., Bhattacharyya, B.K.: How and where to use super-capacitors effectively, an integration of review of past and new characterization works on super-capacitors. J. Energy Storage **27**(101044), 1 (2020)
2. Winter, M., Brodd, R.J.: What are batteries, fuel cells, and supercapacitors? Chem. Rev. **104**(10), 4245–4270 (2004)
3. Camara, M.B., Gualous, H., Gustin, F., Berthon, A., Dakyo, B.: DC/DC converter design for supercapacitor and battery power management in hybrid vehicle applications—polynomial control strategy. IEEE Trans. Industr. Electron. **57**(2), 587–597 (2009)
4. Zuo, W., Li, R., Zhou, C., Li, Y., Xia, J., Liu, J.: Battery-supercapacitor hybrid devices: recent progress and future prospects. Adv. Sci. **4**(7), 1600539 (2017)
5. Jayananda, D., Kularatna, N., Steyn-Ross, D.A.: Supercapacitor-assisted LED (SCALED) technique for renewable energy systems: a very low frequency design approach with short-term DC-UPS capability eliminating battery banks. IET Renew. Power Gener. **14**(9), 1559–1570 (2020)
6. Ruiz, V., Blanco, C., Raymundo-Piñero, E., Khomenko, V., Béguin, F., Santamaría, R.: Effects of thermal treatment of activated carbon on the electrochemical behaviour in supercapacitors. Electrochim. Acta **52**(15), 4969–4973 (2007)
7. Faisal, M., Hannan, M.A., Ker, P.J., Hussain, A., Mansor, M. B., Blaabjerg, F.: Review of energy storage system technologies in microgrid applications: issues and challenges. IEEE Access **6**, 35143–35164 (2018)
8. Ahuja, P., Ujjain, S.K., Kanojia, R.: Electrochemical behaviour of manganese and ruthenium mixed oxide@ reduced graphene oxide nanoribbon composite in symmetric and asymmetric supercapacitor. Appl. Surf. Sci. **427**, 102–111 (2018)
9. Sengupta, A.S., Satpathy, S., Mohanty, S.P., Baral, D., Bhattacharyya, B.K.: Supercapacitors outperform conventional batteries [energy and security]. IEEE Consumer Electron. Magazine **7**(5), 50–53 (2018)
10. Borenstein, A., Hanna, O., Attias, R., Luski, S., Brousse, T., Aurbach, D.: Carbon-based composite materials for supercapacitor electrodes: a review. J. Mater. Chem. A **5**(25), 12653–12672 (2017)
11. Weddell, A.S., Merrett, G.V., Kazmierski, T.J., Al-Hashimi, B.M.: Accurate supercapacitor modeling for energy harvesting wireless sensor nodes. IEEE Trans. Circuits Syst. II Expr. Briefs **58**(12), 911–915 (2011)
12. Raza, W., Ali, F., Raza, N., Luo, Y., Kim, K.H., Yang, J., Kwon, E.E.: Recent advancements in supercapacitor technology. Nano Energy **52**, 441–473 (2018)
13. Vekslender, T., Ezra, O., Bezdenezhnykh, Y., Peretz, M.M.: Closed-loop design and time-optimal control for a series-capacitor buck converter. In: 2016 IEEE Applied Power Electronics Conference and Exposition (APEC) March, pp. 308–314. IEEE, (2016)
14. Liu, Y., Zhao, J., Xia, M., Qu, K.: Adaptive hysteresis band controller for buck converter with indirect input voltage feed-forward. In: Intelec 2013; 35th International Telecommunications Energy Conference, Smart Power and Efficiency October, pp. 1–6. VDE, (2013)
15. Satpathy, S., Bhattacharyya, B.K., Trivedi, M.C., Sengupta, A.S.: Buck topology powered exclusively by supercapacitor modules—a battery-less design concept. IEEE Access **9**, 25758–25767 (2021)
16. Cheng, K.Y., Yu, F., Lee, F.C., Mattavelli, P.: Digital enhanced V 2-type constant on-time control using inductor current ramp estimation for a buck converter with low-ESR capacitors. IEEE Trans. Power Electron. **28**(3), 1241–1252 (2012)
17. Smithson, S.C., Williamson, S.S.: A unified state-space model of constant-frequency current-mode-controlled power converters in continuous conduction mode. IEEE Trans. Industr. Electron. **62**(7), 4514–4524 (2015)

Classification of Indian Classical Dance Hand Gestures: A Dense SIFT Based Approach

R. Jisha Raj, Smitha Dharan, and T. T. Sunil

Abstract Computer vision techniques can improve the understanding of classical dance gestures and also open up oppurtunities for automatically annotating videos. In this paper, we study Bharatanatyam, an Indian classical dance, which uses hand gestures(mudras), facial expressions, and whole body movements to communicate the intended meaning to the audience. Open datasets on hand gestures of Bharatanatyam are not presently available. An exhaustive Bharatanatyam mudra dataset consisting of 15,396 distinct single hand gesture images of 29 classes and 13,035 distinct double hand gesture images of 21 classes was created for this. This paper deals with performance comparison of machine learning algorithms like Support Vector Machines, Multilayer Perceptron, Decision Tree, and Random Forest based on SIFT and dense SIFT features extracted from the images of the dataset. Dense SIFT descriptors which are 128 dimensional were further dimensionally reduced to 64 and 32 using Principal Component Analysis (PCA) and their performance was also evaluated.

Keywords Bharatanatyam mudra dataset · Dense SIFT · Feature descriptors · Hand gestures · Principal Component Analysis · SIFT

1 Introduction

Dance in India has an unbroken tradition since ancient times. Its themes are derived from mythology, legends, and classical literature [11]. There are two main streams of dance in India. They are classical and folk [11]. Bharatanatyam is an Indian classical dance which is practised mainly in South India. The performer uses graceful hand gestures (mudras), facial expressions, and body movements to communicate with the audience [21].

R. Jisha Raj (✉) · S. Dharan
College of Engineering, Chengannur, Kerala, India
e-mail: jisharaj@ceconline.edu

T. T. Sunil
College of Engineering, Attingal, Kerala, India

K. N. Das et al. (eds.), *Proceedings of the International Conference on Computational Intelligence and Sustainable Technologies*, Algorithms for Intelligent Systems,
https://doi.org/10.1007/978-981-16-6893-7_17

According to Natyashastra, a classical text on Indian dance, there are 28 Asamyukta Hastas (single hand gestures) and 23 Samyukta Hastas (Double hand gestures) [5, 8, 20] in Bharatanatyam. Open datasets on Bharatanatyam dance gestures are not presently available. An exhaustive open dataset[1] on Bharatanatyam dance gestures was created. The dataset contains 15,396 single hand gesture images and 13,035 double hand gesture images.

In this paper, SIFT and dense SIFT descriptors are used for feature extraction from mudra images. Dimensionally reduced dense SIFT versions are also experimented upon. The feature descriptors extracted are coded using bag of visual words. Each image is represented as histogram of codewords and are classified using conventional machine learning algorithms such as Support Vector Machines, Multilayer perceptron, Decision Tree, and Random Forest. Through the investigations, it is found that dense SIFT can be advantageous than SIFT with the correct choice of step size. Dimensionally reduced versions, 64 dimensional and 32 dimensional, were also experimented upon but found to be less accurate than dense SIFT.

2 Literature Review

Feature extraction is an important step in image classification. Feature descriptors describes relevant and distinct characteristics of images like edges, corners, and ridges [22].

A comparitive study of several descriptors has been made by authors for various applications like image matching [15], image retrieval, image classification, face identification [14], etc.

Scale-invariant feature transform (SIFT) [13] is one of the most powerful descriptors available. Ever since the origin of SIFT descriptors, they have been extensively used by researchers in image classification. SIFT descriptors have been used in the field of remote sensing [30], action recognition [23], leaf image classification [29], Batik fabric image classification [2], and also in large-scale image classification [24].

Some authors have also reported that using some features in addition to SIFT, like edge-SIFT [6] and color-SIFT [1, 26], improved image classification accuracy. Dense SIFT has been used by researchers in image fusion [12], image matching and face image classification [27, 28], wheat image classification [18], etc. Application of dimensionality reduction techniques on dense SIFT features has also been investigated for improved head pose detection [10].

In the context of Bharatanatyam mudras, studies based on very limited datasets were made by few authors. Mozarkar et al. [17] used 68 samples of 13 static double hand gestures of *Bharatanatyam* to develop a recognition system, and Sriparna Saha et al. [21] used 28 single hand gestures of Bharatanatyam to develop such a system. A rotation and scale-invariant gesture recognition system which could be applied for

[1] Full dataset can be downloaded from https://github.com/jisharajr/Bharatanatyam-Mudra-Dataset.git..

Bharatanatyam dance mudra recognition was experimented upon by Divya et al. [9]. A CNN based classifier for pose and hand gesture classification developed by Aparna Mohanty et al. [16] used 1400 samples of 10 single hand gestures and 840 samples of 14 double hand gestures of Bharatanatyam.

In this paper, we are comparing SIFT, dense SIFT, and dimensionally reduced versions of dense SIFT feature descriptors for classification of mudra images in the Bharatanatyam mudra dataset. Several classifier algorithms such as Support Vector Machines, Multilayer perceptron, Decision Tree, and Random Forest are used.

It may be seen from the above survey that there are very few serious studies on Indian classical dance gesture from a computer vision perspective. Some of the existing studies rely on a small dataset based on a limited number of mudras. This paper proposes an extensive dataset and analysis based on current focus areas of machine learning.

3 Dataset Description

Bharatanatyam has 28 single hand gestures (Asamyukta Hastas) and 23 double hand gestures (Samyukta Hastas) [5, 8]. The data collection was done with the help of volunteers who were trained in Bharatanatyam for more than 5 years. 15,396 single hand gesture images and 13,035 double hand gesture images were collected.

Samdamsam Mudra, one of the single hand mudras, is dynamic in nature and thus was excluded. *Katakamukha mudra* has three variations, and each of them was taken as a seperate class. This made up to 29 classes of static single hand mudras in the dataset. Two mudras in double hand, *Dola Mudra* and *Utsanga Mudra*, were excluded from the dataset. This was due to inclusion of large part of dancer's body in the captured images. Thus, double hand mudras in the dataset comprised of 21 different classes.

The dataset includes different views of the same mudra as perceived by a spectator. As the dancer moves on stage, a spectator sees the mudras at different angles. Considering this, 50–60 views of each mudra were obtained in succession. As an example, six different views of *Aralam mudra* and *Matsya mudra* are shown in Fig. 1.

The images were captured in a studio environment using Apple iPhone 6S 12 megapixel camera. The original images were in RGB format of size 3000 × 4000. The camera was mounted on tripod stand and kept at one meter distance from screen. A green screen served as the permanent background. The data collection environment is shown in Fig. 2.

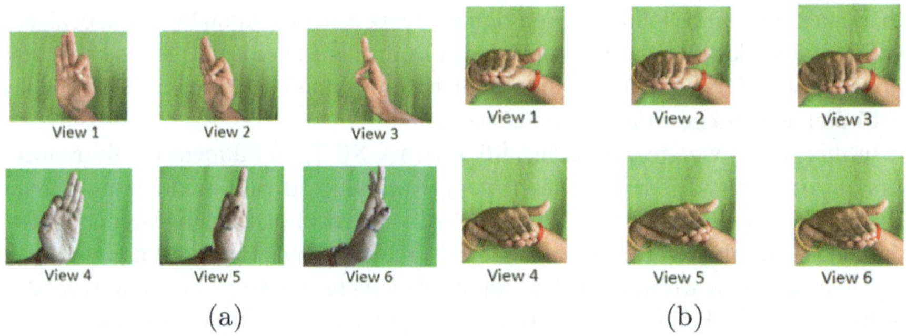

Fig. 1 Six different views of **a** *Aralam mudra*, one of the single hand mudras **b** *Matsya Mudra*, one of the double hand mudras

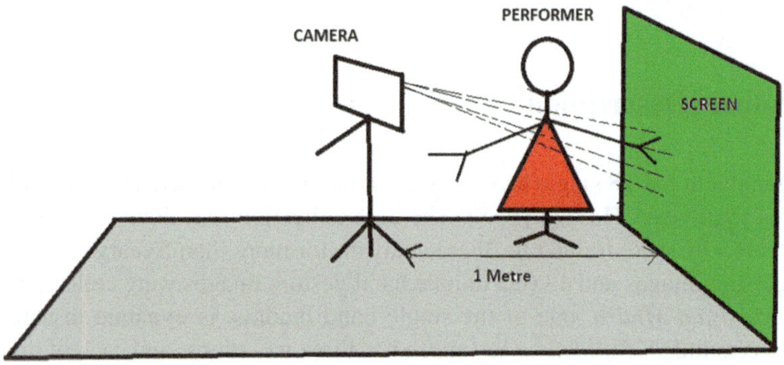

Fig. 2 Data collection environment

4 Mudra Classification

In this paper, a mudra classification system is implemented using feature descriptors and classifiers. Investigations were performed with SIFT and dense SIFT descriptors. Dimensionally reduced versions of dense SIFT descriptors were also explored. Thus, the feature descriptors used were

1. SIFT
2. Dense SIFT
3. Dimensionally reduced dense SIFT using PCA

The steps in the proposed system is shown in Fig. 3. All the 15,396 single hand gesture images of 29 classes and 13,035 double hand gesture images of 21 classes were resized and gray scale converted. The feature descriptors (SIFT and dense SIFT) extracted from the images were stacked to form a combined descriptor. These descriptors were first clustered using k-means, and codebook was generated. Features were then extracted from each image and mapped to the index of nearest codevector. Thus, each image was represented as a histogram of codewords. These were classified

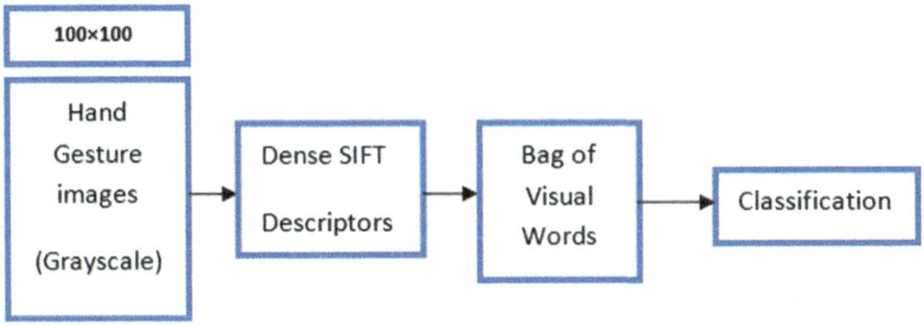

Fig. 3 Steps in the proposed mudra classification system

using various classifier algorithms like Support vector machines (SVM), Multilayer perceptron (MLP), Decision tree (DT), and Random forest(RF).

Trials were then made with dimensionally reduced dense SIFT descriptors. The dense SIFT feature descriptors were extracted and stacked. The combined descriptors which were originally 128 dimensional vectors were reduced to 64 and 32 dimensional vectors using PCA before being encoded using bag of visual words method. These encoded descriptors were then classified using the four classifier algorithms mentioned above.

5 Feature Detectors and Descriptors

Features are relevant and distinct points in an image. Feature points are also called keypoints. Keypoints are detected using keypoint detection algorithms. Once keypoints are obtained, they are represented using feature descriptors

5.1 SIFT

Scale-invariant feature transform [13] is a very powerful feature descriptor algorithm. Here, the image features are invariant to scale and rotation changes and partially invariant to changes in illumination and 3D viewpoint changes. Since it is localized in space and frequency, it is also robust to occlusion, clutter, and noise. These properties are of great relevance when dealing with dance gesture images. The points are also highly distinct, and many features can be obtained from even smaller objects[13].

The steps for extracting keypoints are: *Scale Space Peak Selection*: The first stage of keypoint detection is to find the location and scale which can be assigned repeatedly under different views of the same object. To find a scale-invariant point, a keypoint which is stable in all possible scales are to be located. For this, a continuous scale space function, a Gaussian function is used [3, 13].

Local Extrema Detection: For keypoint detection, each sample point is compared with eight of its neighbors and nine points each above and below the scale. The point qualifies as a keypoint if it is larger or smaller than all the other 26 points with which it is compared [13].

Keypoint Localization: The keypoints that are having low contrast and those that are poorly localized are to be removed by using Taylor expansion of scale space function [13].

Orientation Assignment: A consistent orientation is assigned to the keypoint based on its local image properties, so that keypoint descriptor can be represented relative to this direction and thus be scale invariant [13].

A region around the keypoint is chosen and an orientation histogram of the gradient directions of the sample points in this region is obtained. A 36 bin histogram covering 360 degree orientations of the gradients is used. Each sample in histogram is weighted by its gradient magnitude. A keypoint is created with the orientation as that of the highest peak in the histogram[13].

Extraction of Descriptor at Keypoint: A 16×16 region around a keypoint is considered. Relative orientation and magnitude of sample points in the region is computed. A weighted histogram for the 4×4 subregions are computed. It is an 8 bin histogram, and hence, the descriptor will be a vector of $4 \times 4 \times 8=128$ dimensions [13].

5.2 Dense SIFT

In dense SIFT instead of extracting descriptors at the keypoints alone, local feature descriptors are extracted at regular image grid points yielding a dense description of the images. The descriptors at the grid points of an image are concatenated into a vector to represent the image [27] (Fig. 4).

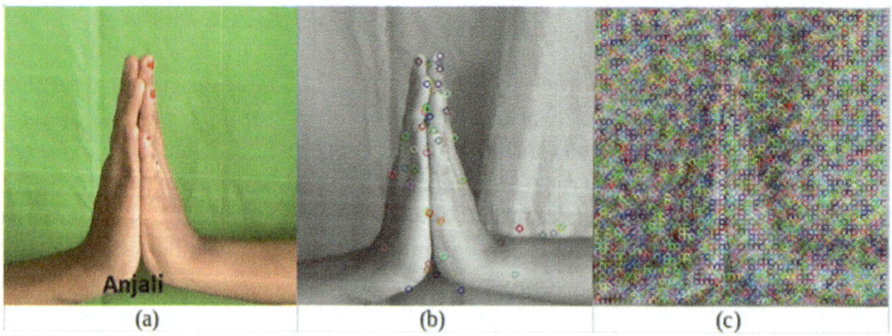

Fig. 4 Keypoint descriptors. **a** Original double hand mudra image, *Anjali Mudra* **b** SIFT keypoints **c** Dense SIFT keyoints

6 Principal Component Analysis

Dimensionality reduction techniques are used to reduce the dimensions or features of a dataset. Curse of dimensionality can make the training extremely slow and also will be difficult to find a solution [7].

Principal Component Analysis is a dimensionality reduction technique. First, it identifies the hyperplane that lies close to it and projects data onto it. The hyperplane onto which data is projected should be the one along which variance is maximum. It is also the hyperplane that minimizes the mean squared distance between the original dataset and its projection onto that hyperplane [7].

PCA identifies the hyperplanes along which the variance is maximum. These components are called the principal components. The second principal component will have the second highest variance and will be orthogonal to the first component. The third component will be orthogonal to the first two components and so on.

The dataset **X** contains the features along the columns. The covariance matrix of this data matrix **X** is then obtained as[7]

$$\mathrm{Cov} - \mathbf{X}\mathbf{X}^T \tag{1}$$

The covariance matrix is a square symmetric matrix, and hence, the eigen values are distinct and eigen vectors are linearly independent and orthogonal to each others. The projection matrix **W** is obtained with eigen vectors arranged in the decreasing order of their eigen values. This gives the principal components in the order of their significance. Next, decision on the number of components required is made. The component with lower eigen values can be discarded without much loss of information.

If the data originally has N dimensions, there will be N eigen values and N eigen vectors. Only P (P < N) eigen vectors are chosen, and these form the columns of a projection matrix, **W**. The new data matrix is obtained as [7]

$$\widehat{X} = \mathbf{X}^T\mathbf{W} \tag{2}$$

7 Bag of Visual Words(BOVW)

In BOVW, number of occurrences of image features are counted, and each image is represented as histogram of these features [25]. BOVW is obtained in four steps

Feature Extraction: Features of the images are first extracted. For this, any of the feature extraction method is used. In this paper, SIFT and dense SIFT descriptors are used [25].

Formation of Visual Vocabulary: The extracted feature vectors are clustered using k-means clustering. Cluster centers are obtained which form the codevectors or visual words. A visual vocabulary or codebook is thus formed from the images in the dataset. If the dataset is sufficiently representative, codebook becomes universal [25].

Quantize features using Visual Vocabulary: The codebook is used for quantizing features. A vector quantizer takes a feature vector and maps it to the index of the nearest codevector in a codebook[25].

Image representation: Each image is represented as a histogram of visual words. The histograms of images can be treated as feature vector for training classifiers[25].

8 Classifiers

The classifiers are used in the paper are
Support Vector Machines: They[4, 19] are also known as maximum marginal hyperplanes. They find a hyperplane that maximize the margin of separation between various classes. They can classify nonlinear data by using kernels. Kernels map the nonlinear data into a high-dimensional feature space where it becomes linear and easily separable.

They are originally binary classifiers. But they can be used for multiclass classification by using one-to-one approach and one-to-rest approach.

Here, trials were performed with linear, polynomial, and rbf kernels. Both one-vs-one and one-vs-rest approaches were attempted. The accuracy was poor, but among the three kernels, linear showed better performance with one-vs-rest approach.
Multilayer Perceptron: Multilayer [19] perceptron consists of feed forward network with an input layer, hidden layers, and output layer. Input layer consists of nodes equal to the features in the input. The number of hidden layers required and the number of nodes in each layer depend on the dataset and is chosen on trial and error basis. The nodes in the output layer equal the number of classes.

The input nodes equalled the cluster size and the output nodes the number of classes. Attempts were made with activation functions like rectified linear unit (ReLU) and Sigmoid. The optimizers attempted were Adam and stochastic gradient descent (SGD) with various learning rates like 0.01 and 0.001. ReLU activation function and Adam optimizer with learning rate 0.001 gave the accurate output.
Decision Tree: Decision [4] trees are powerful, nonparametric, supervised learning algorithms for fitting complex datasets. They create a model that predict the target variable by learning decision rules from the features of the dataset.

Trials were made by varying the number of samples reqired to split internal node from 2 to 5. Maximum accuracy was obtained when samples required to split the node was 2.

Random Forest: Random [19] forest classifier is an ensemble classifier with decision trees. It uses bagging method for training. Here, trials were made with 100, 200, and 300 estimators. RF with 100 estimators showed best performance.

9 Results and Discussions

15,396 single hand gesture images of 29 classes and 13,035 double hand gesture images of 21 classes were resized to 100× 100. They were grayscale converted and features extracted using SIFT and dense SIFT algorithms. The dense SIFT descriptors, originally 128 dimensional, were reduced to 64 and 32 dimensions. All experiments were done on a machine with Core i9 processor, 32GB RAM, and RTX 2080 GPU.

SIFT

For SIFT descriptors, only sparse keypoints were obtained, and it depends on the image characteristics. SIFT descriptors showed low accuracy with all the classifier algorithms for both single and double hand gesture images compared to dense SIFT. The highest accuracy was obtained with Random Forest with 78% for single hand gesture images and 91% with double hand gesture images.

Dense SIFT with 128 dimensional descriptors

For dense SIFT, the number of keypoints extracted per image varied based on the step size. Trials were made with many different step sizes. With a step size of 15 pixels, 49 keypoints were obtained for each image. These dense SIFT descriptors were most accurate with every classifier and were computationally more efficient. Best accuracy was obtained with Random Forest classifier with 94% for single hand gesture images and 97% for double hand gesture images.

Dense SIFT with 64 dimensional descriptors In an attempt to obtain a computationally efficient alternative, PCA was used to dimensionally reduce dense SIFT descriptors to 64. Dimensionality reduction by PCA was possible only when the images have a minimum of 64 keypoints. Hence, the step size of dense SIFT was chosen to be 14, so that the number of keypoints per image was 64. Best accuracy was with Random Forest classifier. For single hand gestures, it was 65% and for double hand gesture images it was 70%. The accuracy was decreased, and it was computationally inefficient.

Dense SIFT with 32 dimensional descriptors The dense SIFT descriptors were further reduced to 32 dimensions. The step size of dense SIFT was chosen to be 18, and the number of keypoints per image was 36. This was done to reduce computational overhead. The accuracy was 63% for single hand gestures and 68% for double hand gestures.

9.1 Summary

In dense SIFT, the number of keypoints for each image was decided by the step size. Table 1 shows the step sizes attempted and accuracy obtained for single hand gestures. Most accuracy was obtained with step size of 15. Similar trials were done for double hand gestures, and step size of 15 was found most accurate.

For each descriptor, the keypoints of all images were stacked and a combined descriptor was obtained. To encode this vector using bag of visual words method, k-means clustering was used. The cluster size was seen to influence the accuracy of the classifier, and so various cluster sizes were attempted. The best cluster size in all cases was found to be the squareroot of the feature vector (combined descriptor). Each image was represented as histogram of features and was used to train and test four different classifiers.

All classifier algorithms were tenfold cross-validated, and fine tuning was done. Parameters varied during fine tuning of classifiers and optimum parameter chosen is given in Table 2. Classification report was generated with the optimum parameters. Seventy percent of data was used for training and rest for validation purposes. Tenfold cross-validation of four classifiers with dense SIFT(128 D) features is shown in Fig. 5. The accuracy of Random classifier with all the feature descriptors are shown in Fig. 6. Computaional time for dense SIFT feature clustering (128 dimensional, 64 dimensional, and 32 dimensional) is given in Table 3.

Comparitive results for various classifiers based on precision, recall, F1-score, and accuracy for single hand gesture images and double hand gesture images are given in Tables 4 and 5 respectively.

Table 1 Influence of step sizes on accuracy of a Random Forest classifier using dense SIFT features from single hand gesture images

Step sizes of dense SIFT	Number of keypoints per image	Combined descriptor	Cluster size	Computation time (min)	Highest accuracy for single hand mudras (%)
12	81	(1247238,128)	1116	30	92.5
15	49	(754502,128)	870	15	94
18	36	(554328,128)	740	12	91

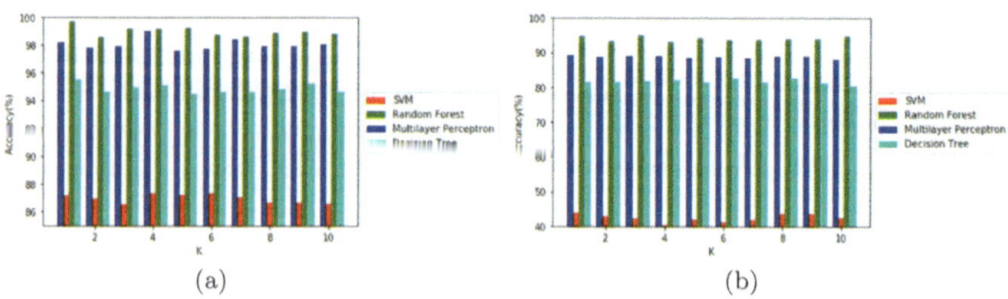

Fig. 5 Tenfold cross-validation of four classifiers with dense SIFT features from **a** Single hand gesture images **b** Double hand gesture images

Table 2 Parameters varied during fine tuning of classifiers and optimum parameter chosen

Classifier	Parameters varied for different trials	Optimum parameters based on cross-validation
SVM	Kernel = linear, polynomial, rbf; multiclass-classification = one-vs-one and one-vs-rest	Kernel = linear; multiclass-classification = one-vs-rest
MLP	Activation functions = ReLU, Sigmoid; Optimizers = Adam and SGD; Learning rate = 0.01, 0.001	Activation function = ReLU; Optimizer = adam; Learning rate = 0.001
DT	Min. samples at node for split = 2–5	Min. samples at node for split = 2
RF	No. of estimators = 100, 200, 300	No. of estimators = 100

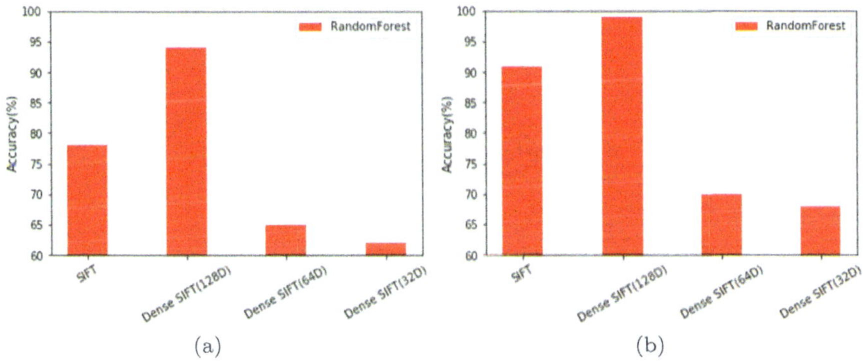

Fig. 6 Random Forest classifier accuracy with various feature descriptors extracted from **a** Single hand gesture images **b** Double hand gesture images

Table 3 Computation time for single and double hand gesture images when dense SIFT descriptors of different dimensions are used

Feature descriptors	Step size	Number of keypoints per image	Computation time for single hand gesture images (min)	Computation time for double hand gesture images (min)
Dense SIFT(128D)	15	49	15	12
Dense SIFT(64D)	14	64	20	18
Dense SIFT(32D)	18	36	12	8

Table 4 Classification report using feature descriptors from single hand gesture images

(a) Classification report of SVM and MLP

Single hand mudras

Descriptors	Combined descriptors	Clusters	SVM				MLP			
			Prec. (%)	Rec. (%)	F1-sco. (%)	Accu.(%)	Prec. (%)	Rec. (%)	F1-sco. (%)	Accu. (%)
SIFT	(537848,128)	700	58	12	13	12	62	60	60	60
Dense SIFT (128 D)	(754502,128)	870	57	43	43	43	89	89	89	89
Dense SIFT (64 D)	(985472,64)	990	21	26	26	21	60	60	60	60
Dense SIFT (32 D)	(554328,32)	740	20	25	25	20	62	62	61	61

(b) Classification report of Decision tree and Random forest

Single Hand Mudras

Descriptors	Combined descriptors	Clusters	Decision tree				Random forest			
			Prec. (%)	Rec. (%)	F1-sco. (%)	Accu. (%)	Prec. (%)	Rec. (%)	F1-sco. (%)	Accu.(%)
SIFT	(537848,128)	700	52	52	52	52	78	78	78	78
Dense SIFT (128 D	(754502,128)	870	82	82	82	82	94	94	94	94
Dense SIFT (64 D)	(985472,64)	990	61	61	61	61	65	65	65	65
Dense SIFT (32 D)	(554328,32)	740	62	62	62	62	63	63	63	63

Table 5 Classification report using feature descriptors from double hand gesture images

(a) Classification report of SVM and MLP

Double hand mudras

Descriptors	Combined descriptors	Clusters	SVM				MLP			
			Prec (%)	Rec. (%)	F1-sco. (%)	Accu.(%)	Prec. (%)	Rec. (%)	F1-sco. (%)	Accu.(%)
SIFT	(670674,128)	820	44	44	44	44	83	83	83	83
Dense SIFT (128 D)	(738715,128)	860	89	87	87	87	96	96	96	96
Dense SIFT (64 D)	(834240,64)	920	20	25	25	20	68	68	68	68
Dense SIFT (32 D)	(469260,32)	690	22	23	26	21	66	66	66	66

(b) Classification report of decision tree and random forest

Double hand mudras

Descriptors	Combined descriptors	Clusters	Decision tree				Random forest			
			Prec. (%)	Rec. (%)	F1-sco. (%)	Accu.(%)	Prec. (%)	Rec. (%)	F1-sco. (%)	Accu.(%)
SIFT	(670674,128)	820	60	60	60	60	91	91	91	91
Dense SIFT (128 D)	(733715,128)	860	95	95	95	95	97	97	97	97
Dense SIFT (64 D)	(834240,64)	920	65	65	65	65	70	70	70	70
Dense SIFT (32 D)	(469260,32)	690	65	65	65	65	68	68	68	68

10 Conclusion

In this paper, a comparison of SIFT and dense SIFT descriptors in the context of classification of Bharatanatyam mudra dataset is made. SIFT and dense SIFT descriptors were extracted from the images. They were combined and encoded using bag of visual words. The encoded descriptors were then classified using Support Vector Machines, Multilayer Perceptron, Random Forest, and Decision Trees. Dense SIFT descriptors were showing more precision, recall, F1-score, and accuracy than SIFT descriptors with all four classifiers. The most accuracy was obtained with Random Forest classifier.

Having obtained a dense SIFT descriptor which was more accurate than SIFT, experiments were performed with dimensionally reduced dense SIFT descriptors. Dense SIFT descriptors of 64 dimensions were computationally inefficient apart from being less accurate. And 32 dimensional dense SIFT descriptors were computationally efficient but showed decreased accuracy. So it is concluded that dense SIFT descriptors of 128 dimensions as input to a Random Forest classifier provided accurate classification of Bharatanatyam mudra dataset. For further enhancement of the work, other feature descriptors and convolutional neural networks may be explored.

References

1. Ai, D., Han, X., Ruan, X., Chen, Y.-W.: Adaptive color independent components based sift descriptors for image classification. In: 20th International Conference on Pattern Recognition, pp 2436–2439, IEEE (2010)
2. Azhar, R., Tuwohingide, D., Kamudi, D., Suciati, N., et al.: Batik image classification using sift feature extraction, bag of features and support vector machine. Procedia Comput. Sci. **72**, 24–30 (2015)
3. Babaud, J., Witkin, A.P., Baudin, M., Duda, R.O.: Uniqueness of the gaussian kernel for scale-space filtering. IEEE Trans. Pattern Anal. Mach. Intell. **1**, 26–33 (1986)
4. Bishop, C.M.: Pattern Recognition and Machine Learning, pp. 38–172, Springer (2006)
5. Coomaraswamy, A., Duggirala, G.K.: The mirror of gestures: being the Abhinayadarpana of Nandikeswara(English translation). pp. 25–51, Harvard University Press, London (1917)
6. Fidalgo, E., Alegre, E., González-Castro, V., Fernández-Robles, L.: Compass radius estimation for improved image classification using edge-sift. Neurocomputing **197**, 119–135 (2016)
7. Geron, A.: Hands-on machine learning with Scikit-Learn and TensorFlow: concepts, tools, and techniques to build intelligent systems, pp. 259–280, O'Reilly Media, Sebastopol, California (2017)
8. Ghosh, M.: Natyasastra(English Translation). pp. 170–190, Manisha Granthalaya Private Limited, Kolkota, India (1956)
9. Hariharan, D., Acharya, T., Mitra, S.: Recognizing hand gestures of a dancer. In: International Conference on Pattern Recognition and Machine Intelligence, pp. 186–192, Springer (2011)
10. Ho, H.T., Chellappa, R.: Automatic head pose estimation using randomly projected dense sift descriptors. In: 2012 19th IEEE International Conference on Image Processing, pp. 153–156, IEEE (2012)
11. Know India: National Portal Of India. Culture and Heritage: Performing Arts. https://knowindia.gov.in/culture-and-heritage/performing-arts.php. Accessed on 27 April 2021

12. Liu, Y., Liu, S., Wang, Z.: Multi-focus image fusion with dense sift. Inf. Fusion. **23**, 139–155 (2015)
13. Lowe, D.G.: Distinctive image features from scale-invariant keypoints. Int. J. Comput. vis. **60**(2), 91–110 (2004)
14. Mandal, B., Wang, Z., Li, L., Kassim, A.A.: Performance evaluation of local descriptors and distance measures on benchmarks and first-person-view videos for face identification. Neurocomputing **184**, 107–116 (2016)
15. Mikolajczyk, K., Schmid, C.: A performance evaluation of local descriptors. IEEE Tran. Patt Analys. Mach. Intell. **27**(10), 1615–1630 (2005)
16. Mohanty, A., Vaishnavi, P., Jana, P., Majumdar, A., Ahmed, A., Goswami, T., Sahay, R.R.: Nrityabodha: towards understanding indian classical dance using a deep learning approach. Signal Processing: Image Communication. **47**, 529–548 (2016)
17. Mozarkar, S., Warnekar, C.S.: Recognizing bharatnatyam mudra using principles of gesture recognition gesture recognition. Int. J. Comput. Sci. Netw. **2**(2), 46–52 (2013)
18. Olgun, M., Onarcan, A.O., Özkan, K., Işik, Ş, Sezer, O., Özgişi, K., Ayter, N.G., Başçiftçi, Z.B., Ardiç, M., Koyuncu, O.: Wheat grain classification by using dense sift features with svm classifier. Comput. Electron. Agric. **122**, 185–190 (2016)
19. Pedregosa, F., Varoquaux, G., Gramfort, A., Michel, V., Thirion, B., Grisel, O., Blondel, M., Prettenhofer, P., Weiss, R., Dubourg, V., Vanderplas, J., Passos, A., Cournapeau, D., Brucher, M., Perrot, M., Duchesnay, E.: Scikit-learn: Machine learning in Python. J. Mach. Learn. Res. **12**, 2825–2830 (2011)
20. Ramachandrasekhar, P.: Abhinayadarpanam. pp. 71–126, Giri Trading Agency Private Limited, Mumbai, India (2013)
21. Saha, S., Ghosh, L., Konar, A., Janarthanan, R.: Fuzzy l membership function based hand gesture recognition for bharatanatyam dance. In: 5th International Conference on Computational Intelligence and Communication Networks (CICN). pp. 331–335, IEEE (2013)
22. Salahat, E., Qasaimeh, M.: Recent advances in features extraction and description algorithms: a comprehensive survey. In: 2017 IEEE International Conference on Industrial Technology (ICIT), pp. 1059–1063, IEEE (2017)
23. Scovanner, P., Ali, S., Shah, M.: A 3-dimensional sift descriptor and its application to action recognition. In: Proceedings of the 15th ACM International Conference on Multimedia, pp. 357–360 (2007)
24. Simonyan, K., Vedaldi, A., Zisserman, A.: Deep fisher networks for large-scale image classification. In: Advances in Neural Information Processing Systems, pp. 163–171 (2013)
25. Tsai, C.-F.: Bag-of-words representation in image annotation: a review. Int. Schol. Res, Not (2012)
26. Verma, A., Liu, C., Jia, J.: New colour sift descriptors for image classification with applications to biometrics. Int. J. Biometr. **3**(1), 56–75 (2011)
27. Wang, J.-G., Li, J., Lee, C.Y., Yau, W.-Y.: Dense sift and gabor descriptors-based face representation with applications to gender recognition. In: 2010 11th International Conference on Control Automation Robotics and Vision, pp. 1860–1864, IEEE (2010)
28. Wang, J.-G., Li, J., Yau, W.-Y., Sung, E.: Boosting dense sift descriptors and shape contexts of face images for gender recognition. In: 2010 IEEE Computer Society Conference on Computer Vision and Pattern Recognition-Workshops, pp. 96–102, IEEE (2010)
29. Wang, Z., Lu, B., Chi, Z., Feng, D.: Leaf image classification with shape context and sift descriptors. In: 2011 International Conference on Digital Image Computing: Techniques and Applications, pp. 650–654, IEEE (2011)
30. Yang, Y., Newsam, S.: Comparing sift descriptors and gabor texture features for classification of remote sensed imagery. In: 2008 15th IEEE International Conference on Image Processing, pp. 1852–1855. IEEE (2008)

A Grid-Connecting Control Scheme for Reactive Power Compensation of PV Inverter

Jaya Chandra Panigrahi, Narmadha Bejjupalli, Yatindra Gopal, and Y. N. Vijaya Kumar

Abstract Production of power at unity level is possible to handle the active power production only. Fulfillment of the phantom power to the load is generated by grid only. The requirement of phantom power in count to active power reduced when PV inverter with grid tied make smart. Hence, grid requirements of reactive power reduced and lesser supply of reactive power. In this manuscript, two different approaches are analyzed such as limiting the delivered active power or oversizing of the PV inverter. Static reactive power maintain is used to manage voltage levels within a satisfactory range. Result analysis of the numerous static phantom power compensation control strategies has been performed at the MATLAB software platform.

Keywords Reactive power compensation · Photovoltaic · Three phase inverter · FACTS devices · Phantom power

1 Introduction

Grid-tied solar-based inverter (GTI) are customarily intended to work at unity power factor (UPF) which implies that they have capacity to create true power as it were [1–3]. As motor loads which run on electrical supply are dominatingly inductive, they will in general devour more amount of inductive power which is oscillating in nature [1, 4]. Presently, this phantom power necessity is catered by network as it were [4, 5]. With the expansion in infiltration of enormous number of distributed energy resources (DERs) which pump just real power into the grid, the plant power factor gets poor according to utility perspective, influencing the performance of the grid [6]. Some time ago, the oscillatory power infusion/retention is made by using FACTS devices for moderating power quality issues. There are various kinds of FACTS devices like series, shunt, series–series, series–shunt which are utilized dependent on the prerequisite. Be that as it may, these devices experience the ill

J. C. Panigrahi · N. Bejjupalli · Y. Gopal (✉) · Y. N. Vijaya Kumar
Department of Electrical Engineering, SVCET (A), Chittoor, A.P, India

K. N. Das et al. (eds.), *Proceedings of the International Conference on Computational Intelligence and Sustainable Technologies*, Algorithms for Intelligent Systems,
https://doi.org/10.1007/978-981-16-6893-7_18

effects of the downside of enormous measuring, significant expense, and huge space of establishment and so on.

By virtue of this scenario presented above, there is a need to manage the phantom power stream in both the transmission and distribution side as this may influence the voltage regulation moreover [6]. In this paper, control techniques for GTI to supply phantom power in addition to the true power are examined. As the quantity of GTI expands, their use as oscillatory power curtailment devices will help in grid voltage regulation and lessen the need of costly capacitor banks. Three-phase inverter circuit is considering for power conversion process. The PWM technique used for gate pulses is SPWM technique at 10 kHz switching frequency [7].

To utilize the unused power capacity of the smart inverters, there are two principle choices: oversizing the inverter or real power reduction. Oversizing the inverter has an additional expense that should be considered. Active power decrease can be carried out in various manners, for example, fixing the greatest power highlight say 30% less than the full rating capacity of power or dependent on point of common connection voltage as the reference voltage. This strategy prompts generation decrement from the installed solar PV panels and subsequently diminished acquiring to the solar energy system proprietors [8–10].

2 Photovoltaic Modeling

The PV panel is the heart of the total circuit through which the solar energy will be converted into the electrical energy (DC form) [7, 9]. There exists a diode and a series resistance and a parallel resistance in modeling the PV panel. The output of solar cell depends on the open-circuit voltage and short-circuit current, mathematical equations of the solar PV panel model are given Eqs. (1–5). The equivalent figure of a solar cell is shown in Fig. 1.

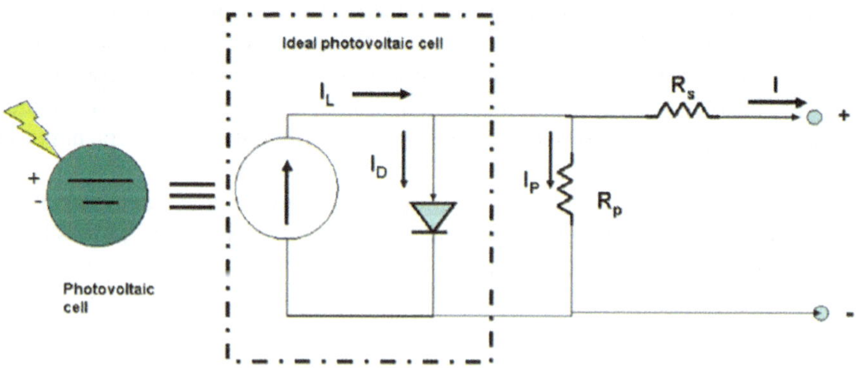

Fig. 1 Equivalent network of solar cell

$$Voc = Voc + Kv \cdot (Tc - 25) \tag{1}$$

$$Isc = \frac{Iph - Id - Vpv/Rp}{1 + Rs/Rp} \tag{2}$$

$$\text{Iarray} = Ipv\,.\,\text{Npar} \tag{3}$$

$$\text{Varray} = Vpv\,.\,\text{Nser} \tag{4}$$

$$\text{Parray} = \text{Iarray}\,.\,\text{Varray} \tag{5}$$

3 Inverter Circuit with PV System

The inverter circuit which has been under study is basically a voltage source inverter (VSI). This is responsible for carrying the whole operation of exchanging the apparent power between AC side and DC side. The inverter circuit is placed after the solar panel to convert the unidirectional supply to the bidirectional supply. There are different types of switches in use for the conversion process such as diodes, mosfets, and IGBTs; of these, IGBT has better switching performance than others, so they are used as switching devices for conversion process.

The PWM technique used for gate pulses is SPWM technique at 10 kHz switching frequency. A three-phase IGBT inverter circuit with appropriate filters is used. Before inverter circuit, a boost converter is used to boost the level of DC voltage produced by the panel to the required level. A DC link (capacitor) is used between the boost converter and inverter circuit to maintain the voltage which in turn maintains the integrity of the total mechanism. The circuit is shown in Fig. 2.

4 Capability Curve

The capability curve shows the capacity of the inverter for injecting or absorbing the reactive power without causing any fatality to the equipment [11, 12]. In this case, the capacity is shown for both real and reactive components of power injection using the electric potential, current, and power limits.

(a) **Voltage Limits:** The variation of DC voltage can be from the initial to maximum limit so as to extract maximum active power from the solar panel. But there are some limitations for minimum amount of voltage of the panel which allows the permissible value of AC voltage into the inverter as per IEC codes.

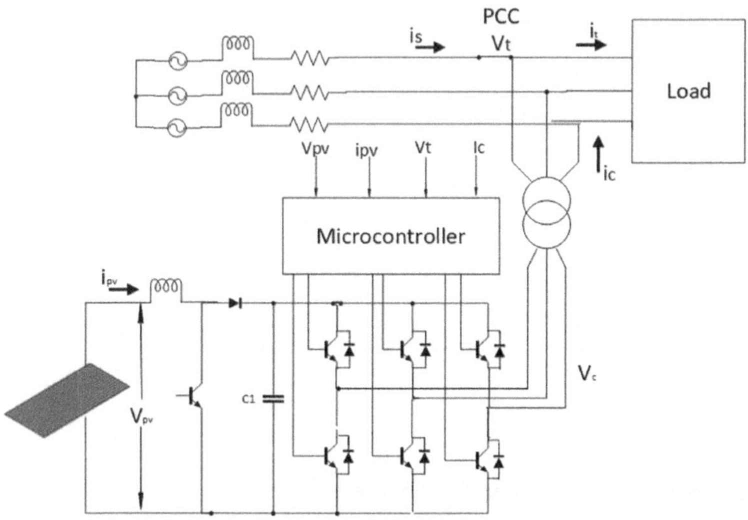

Fig. 2 Three-phase inverter circuit

$$V\min = 2\sqrt{2V} \tag{6}$$

when the inverter is working at a maximum power point of solar panel, the minimum solar insolation required to maintain the voltage range at the inverter side is calculated from the formula

$$G\min = \left(\frac{Vdc\min - Vocstc}{\mathrm{Kv}} + 25 - Ta\right) \times \frac{800}{\mathrm{Noct} - 20} \tag{7}$$

Besides this limit, the maximum voltage range should also be taken into consideration as it is depending on the open-circuit voltage of solar cells connected in series. The maximum open-circuit voltage depends on minimum temperature limit chosen at the location. Equation (8) determines the limit for maximum open-circuit voltage.

$$Voc\max = Vocstc + Kv \cdot \left(Ta\min + G\min \cdot \frac{Noct - 20}{800} - 25\right) \tag{8}$$

$$V\max = Voc\max\ .\mathrm{Nser} \tag{9}$$

The maximum voltage at the inverter terminals from the solar panel has an impact on delivering capability of the active and reactive power.

$$P^2 + (Q + Vg^2/X)^2 = (Vg\,.\,Vi)/X \tag{10}$$

(b) **Current Limits:** The maximum inverter current (Ii) injection by the inverter under the circumstances imposes the restrictions on active and oscillatory power feed-in capability of the inverter. The limits of power injection are expressed in Eq. (11).

$$P^2 + Q^2 = (Vg.li)^2 \tag{11}$$

$$Ii = \frac{\sqrt{P^2 + Q^2}}{\text{Vg}} \tag{12}$$

From the above equations, active and reactive power injections at the grid side are shown in Fig. 3.

Considering the variation in the DC voltage at the inverter input terminals, reactive power can be referred as

$$Qmpp(G, T, Vdc) = \frac{Vgrid.Vmpp(G, T), M}{X} \tag{13}$$

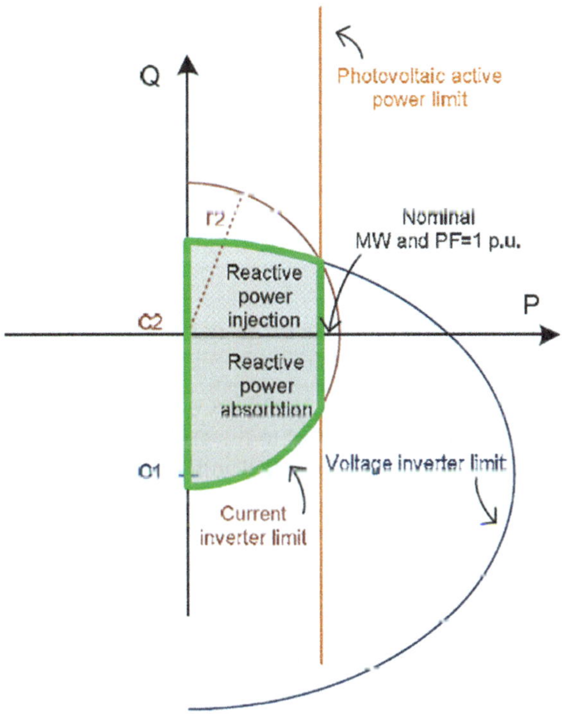

Fig. 3 Capability Curve of an Inverter

As the DC voltage tends to alter the modulation index (MI), for keeping MI constant, DC voltage should be constant; therefore, reactive power is also constant for a given DC voltage. In order to deliver the reactive power to the grid, MI can be imposed which sets the output voltage value of inverter (Vac).

5 Reactive Power Compensation Techniques

For controlling the reactive power, many power electronic devices came into force due to the technological developments from late 1900s. But the disadvantages such as lack of space for installation and some other constrain the power engineers thought to use the solar inverter as a reactive power controlling device which has an advantage of using the inverter at its rated capacity under normal working conditions (when the inverter is not delivering the rated active power) [13, 14].

As the inverters are current limiting devices, there are two possible ways to design the inverter [15–17]. One is by making the full rating of inverter to be available for real power generation or by keeping some current for reactive power generation at its rated real power. The main objective of the reactive power principle is to improve system performance and balance the system which can be obtained by load compensation and regulating the grid-side voltage. The reactive power injection is a bit complex compared to the active power injection as the reference for reactive component cannot be achieved in a simple way.

Some of the reactive power compensation methods include constant reactive power injection, constant power factor operation at inverter, power factor depending on the real power generation at solar panel, and voltage as a function of reactive power. In this paper, reactive power control using oscillatory power as a function of electric potential method has been discussed in detail.

5.1 Oscillatory Power as a Function of Electric Potential

The other name for this method is volt-var function $Q(V)$. The inverter provides the required oscillatory power to the grid depending on the voltage profile of the grid. The inverter will absorb or deliver the phantom power depending on the voltage level programmed in the inverter control strategy [18, 10]. The volt-var characteristics are shown in the below figure for which if the voltage at the grid side is +4% outside the nominal value, and then the inverter tends to deliver or absorb the reactive power depending on the value of voltage.

The $Q(V)$ function is a local control function influenced by the variables of inverter and grid. If an appropriate communication system is used, then this can be implemented at the global level also.

The controller consists of mainly two components of control units, namely voltage controller and current controller (Figs. 4 and 5).

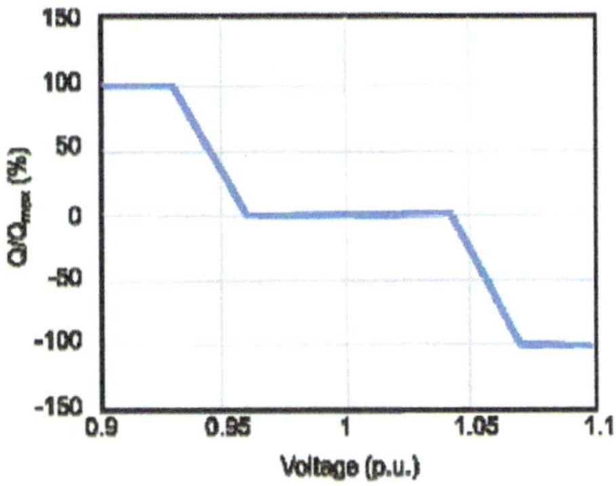

Fig. 4 Volt-var profile at grid side for $Q(V)$ control

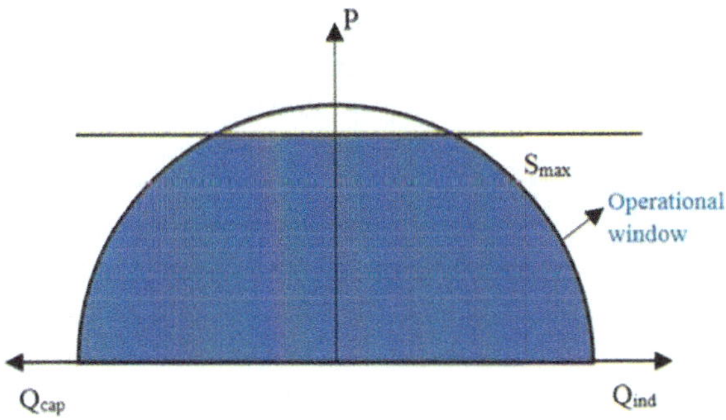

Fig. 5 Characteristic curve for $Q(V)$ control

5.2 *Voltage Controller*

The DC voltage is controlled using the reference value obtained from the MPPT technique [19]. The PI controller is fed with an error obtained by comparing the DC link voltage with a certain value for Idref generation which controls the active power.

5.3 *Current Controller*

The grid currents are transformed using the transformation techniques (Park's transformation). Refer (14) and (15)

$$Vd = V\alpha * \cos(\theta) + V\beta * \sin(\theta) \tag{14}$$

$$Vq = -V\alpha * \sin(\theta) + V\beta * \cos(\theta) \tag{15}$$

The obtained transformations are compared to the output of voltage controller which is the set value, and the error obtained is fed to an appropriate controller. The

Qref value is taken from the comparison of actual value of voltage to the reference value set, through which the Iqref is obtained, which further controls the reactive power at the inverter side. The voltage profile for which the inverter should inject or absorb the reactive power is programed by the user. By using the command signal provided to the inverter, it will operate by comparing the actual voltage profile at the grid side to the given reference value and decides the value of phantom power to be injected or absorbed by the inverter.

6 Simulation Results and Discussions

Simulations are carried out for a 50 KW three-phase solar inverter. The inverter has IGBT/diode as a switching device with sinusoidal PWM technique with a carrier frequency of 10 kHz. The simulation was carried out for a variable irradiance from 1000 to 200 W/m^2 in steps of 200 units per second. The results obtained are shown.

Figure 6 shows the reference current generation for controlling the reactive power either by injecting or absorbing the power depending on the grid voltage. In the initial state, the current reference is zero means that the voltage profile is equal to the reference value and no error has been generated. After certain time interval (at $t = 1.1$ s), the I_{qref} is maximum at the negative side which means the total reactive power required by the load is supplied by the inverter itself. After some time (at $t = 1.5$), the I_{qref} is near to -0.5 which means, it is delivering a 30% of the reactive power required by the load, and the remaining reactive power will be provided by the grid.

Figure 7 shows the power injection/absorption by the inverter to the grid side depending on the load requirement. The active power and reactive power injections will follow the reference signals I_d and I_q which will be obtained by the error values of DC voltages and AC voltages.

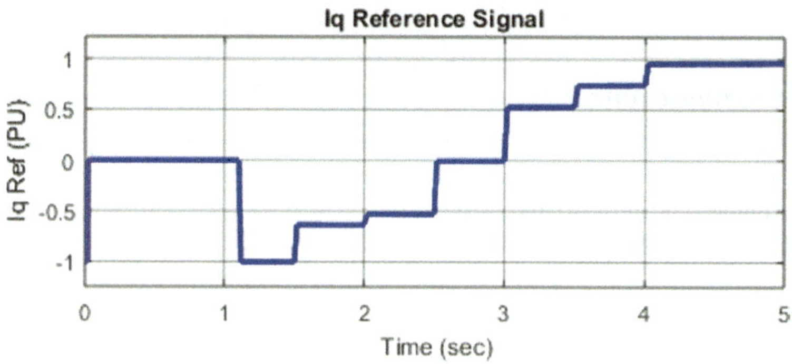

Fig. 6 I_{qref} Generation for a step change in the grid voltage profile

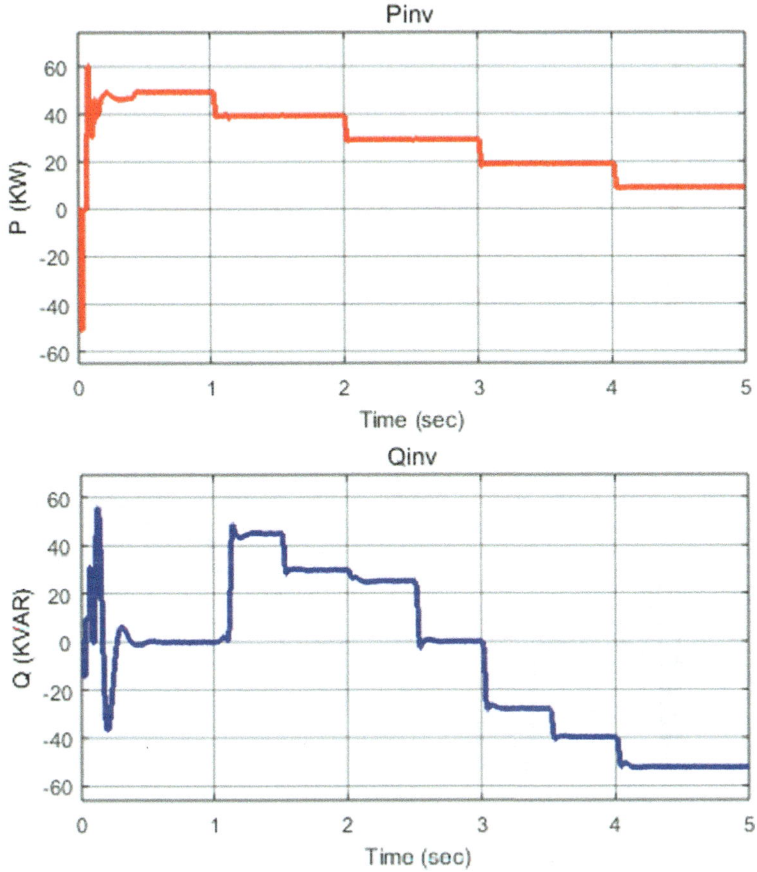

Fig. 7 Active and reactive component of power at inverter side

Figure 8 shows the active power taken by the load and the reactive power absorbed by the inverter after a certain time (at $t = 4$ s). Depending on the load requirement after the power injection by the inverter circuit, the grid provides the remaining amount of the demand. If the voltage profile of the grid exceeds the maximum permissible limit, then the inverter also tends to absorb some amount of reactive power in order to stabilize the voltage profile of the grid; in that case, the solar inverter will act as a sink.

From the above waveforms, it is evident that whenever there is a change in the voltage other than the actual value/reference value, there is either pumping of oscillating power along with the true power at day time. Table 1 depicts the values of real and oscillatory power with time duration at inverter side.

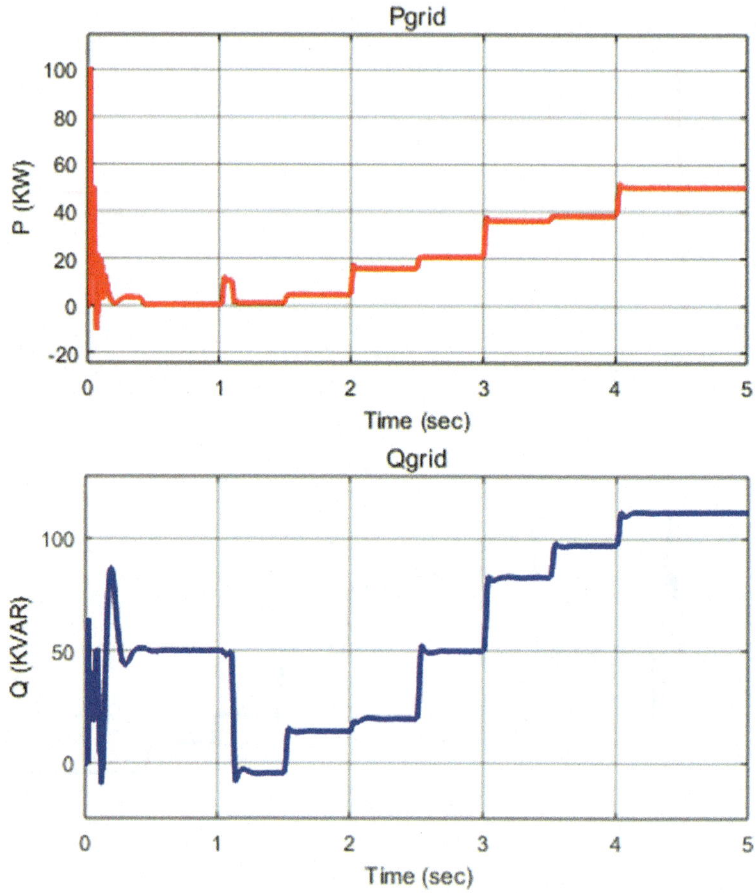

Fig. 8 Active and reactive component of power at grid side

Table 1 True power and oscillatory power values for the step change in voltage

Time step	True power (KW)	Reactive power (KVAR)	Voltage change
0–1	49.5	0	1
1–1.5	39.5	46	0.9
1.5–2	39.5	40	0.94
2–2.5	29.5	30	0.95
2.5–3	29.5	0	1
3–3.5	19.5	–30	1.05
3.5–4	19.5	–40	1.06
4–5	10	–50	1.09

7 Conclusions

This manuscript performed work based on grid connected 3-ϕ PV inverter system with reactive power compensation component. 3-ϕ PV inverter gives the strongly built reactive power and active power regulation competence. Hence, generate the fewer harmonics current with faster response of MPPT technique and smooth transition. The reactive power curtailing and controlling feature should be made available in PV plant to develop the efficient operation of the system when the PV irradiance is not abundant. The performance analysis of grid-connected 3-ϕ PV inverter system is developed at MATLAB software platform. A simulation result of different static reactive power control approach has been presented, and the results are discussed in detail.

References

1. Hunter, G., Andrade, I., Riedemann, J., Blasco-Gimenez, R., Pena, R.: Active and reactive power control during unbalanced grid voltage in PV systems. In: IEEE (2016), pp. 3012–3017
2. Taghikhani, M.A., Farahani, A.D.: A reactive power based reference model for adaptive control strategy in a SEIG. Eng. Technol. Appl. Sci. Res. **8**, 2477–2484 (2017)
3. Yang, Y., Blaabjerg, F., Wang, H., Simoes, M.G.: Power control flexibilities for grid-connected multi-functional photovoltaic inverters. IET Renew. Power Gener. Res. Article 1–10 (2015)
4. Keawthai, S., Po-Ngam, S.: Simplified active power and reactive power control with MPPT and islanding detection for three-phase grid-connected photovoltaic inverters. IEEE (2015)
5. Arellanes, A, Rodriguez, E., Orosco, R., Perez, J., Beristain, J.: Three-Phase grid-tied photovoltaic inverter with reactive power compensation capability. In: IEEE International Autumn Meeting on Power, Electronics and Computing (2017)
6. Severin, N., Malcolm, M.S., Wilson, E., Liwci, W.: Comparison of voltage control methods in distribution systems using Q-V based PI and droop controls of solar inverters. In: IEEE (2017), pp. 1–5
7. Gopal, Y., Panda, K.P., Birla, D., Lalwani, M.: Swarm optimization-based modified selective harmonic elimination PWM technique application in symmetrical H-bridge type multilevel inverters. Eng. Technol. Appl. Sci. Res. **9**, 3836–3845 (2019)
8. Chakraborty, S., Kumar, R., Haldkar, A.K., Ranjan, S.: Mathematical method to find best suited PV technology for different climatic zones of India. Int. J. Energy Environ. Eng. **8**, 153–166 (2017)
9. Dhaked, D.K., Gopal, Y., Birla, D,: Battery charging optimization of solar energy based telecom sites in India. Eng. Technol. Appl. Sci. Res. **9**, 5041–5046 (2019)
10. Gopal, Y., Birla, D., Lalwani, M.: Reduced switches multilevel inverter integration with boost converters in photovoltaic system. SN Appl. Sci. **2**, 1–15 (2020)
11. Rajeshwari, P.N.: D-STATCOM for reactive power compensation. Int. J. Electr. Electron. Res. **3**, 111–118 (2015)
12. Li, H., Wen, C., Chao, K.-H., and Li, L.-L.: Research on inverter integrated reactive power control strategy in the grid-connected PV systems. Energies 1 22 (2017)
13. Ashutosh, B.: A simulation model for reactive power compensation. Int. J. Res. Appl. Sci. Eng. Tcchnol. **3**, 536–540 (2015)
14. Harirchi, F., Simoes, M.G., AlDurra, A.: Multi-functional double mode inverter for power quality enhancement in smart-grid applications. IEEE, pp1–8 (2016)

15. Gomez-Gonzalez, J.F., Canadillas-Ramallo, D., Gonzalez-Díaz, B., Mendez-Perez, J.A., Rodríguez, J., Sanchez, J., Guerrero-Lemus, R.: Reactive power management in photovoltaic installations connected to low-voltage grids to avoid active power curtailment. In: International Conference on Renewable Energies and Power Quality, 5–11 (2018)
16. Spring, A., Wirth, G.B., Pardatscher, G.R., Witzmann, R.: Grid influences from reactive power flow of photovoltaic inverters with a power factor specification of one. IEEE Trans. Smart Grid **7**, 1222–1229 (2016)
17. Avdhoota, T., Falguni, B., Ajay, B.: Novel control strategy for grid-connected Pv system with reactive power compensation and MPPT control. Int. J. Innov. Res. Electr. Electron. Instrum. Control Eng. **2**, 1472–1476 (2014)
18. Varma, R.K., Siavashi, E.M.: PV-STATCOM: a new smart inverter for voltage control in distribution systems. IEEE Trans. Sustain. Energy 1–12 (2018)
19. Ozbay, H., Oncu, S., Kesler, M.: SMC-DPC based active and reactive power control of grid-tied three phase inverter for PV systems. Int. J. Hydrogen Energy 1–1 (2017)

Design and Operation of Drip Irrigation System to Orchard by Using Stored Rain Water Through Solar Photovoltaic Water Pump

Uddipta Ghosh, Ranajit Kumar Biswas, Debasish Das, Debargha Banerjee, and Baibaswata Das

Abstract The experiment was conducted to examine the prospect of water storage structure and drip systems along with the solar photovoltaic pump which were used for the purpose of irrigation to orchard. The estimated cost of storage water was found as Rs. 0.67/lit. It was found that the solar-operated battery pump had discharged 1971.52 L of water in an uninterrupted run of 2.90 h. By using this discharge, the volume of water can irrigate the 0.33 ha area of land of orchard. The findings of this experiment may reduce the problem of the water deficiency of the western part of the West Bengal, especially in the summer season to some extent.

Keywords Water storage · Solar photovoltaic pump · Drip system · Discharge

1 Introduction

Birbhum, Purulia, Bankura are more precisely in the western part of the West Bengal is suffering by the deficit of water in hot season. This region has major problem of low humidity and high temperature [2]. Due to less organic matter in soil of this region and uneven topography, most of the rainfall flow away as a runoff. There are no such provisions of storing of this runoff water in this region, and also, evaporation and seepage loss also play a major role for creating drought situation. As there is deficit of water, agricultural production is not up to the mark so it made impediment for the betterment of the economic condition of the farmers of this particular region [4]. Generally, check dam, dug cum bore well, open well, farm pond, etc., are used

U. Ghosh (✉) · D. Das · D. Banerjee
Department of Civil Engineering, IIS College of Engineering, Kalyani, India

R. K. Biswas
Department of Soil and Water Engineering, Bidhan Chandra Krishi Viswavidyalaya, Haringhata, India

B. Das
Department of Civil Engineering, Dr. Sudhir Chandra Sur Institute of Technology and Sports Complex, Kolkata, India

K. N. Das et al. (eds.), *Proceedings of the International Conference on Computational Intelligence and Sustainable Technologies*, Algorithms for Intelligent Systems,
https://doi.org/10.1007/978-981-16-6893-7_19

for storage of water. But here is also a problem because there are not so much soil erosion control practices adopted in this region and so siltation of dam also created problem [3]. Among the irrigation system, drip irrigation is found the most effective in maintaining the optimum soil moisture in soil root zone with improved yield and water use efficiency with high return value. It helps not only saving in water but also saving in labour cost and fertilizer cost. But every irrigation system needs pump set which help to operate that system, but people of this region is so poor that they cannot afford the cost of electricity or diesel operated pump. So, there should be a suitable alternative way to operate the pump without expensing the huge money. Keeping this in view, the experiment was undertaken to evaluate the performance of solar operated drip irrigation system by using the stored water of water silo to orchard.

2 Previous Studies

Drip irrigation delivers water (and fertilizer) directly to the roots of plants, thereby improving soil moisture conditions; in some studies, this has resulted in yield gains of up to 100%, water savings of up to 40–80%, and associated fertilizer, pesticide, and labour savings over conventional irrigation systems [6]. Due to water scarcity, the available water resources should be utilized very effectively through water saving irrigation technologies to maximize the yield. Furrow and basin irrigation methods are commonly adopted to irrigate the crop, which causes adverse effects of water excess and water stress [5, 11]. Drip irrigation is the most efficient and new technology of irrigation in India to supply precise amounts of water directly into the vicinity of root zone at right time, matching with the consumptive water demand of plant for optimum growth, improved yield, and quality of produce with substantial water saving [12]. The drip method of irrigation also helps in reducing overexploitation of groundwater that partly occurs in surface irrigation. Water saving through drip irrigation system is estimated to be in the range of 12–84% in different crops, besides its beneficial impact on crop yields [9, 10]. Drip irrigation have various advantages over the conventional irrigation in terms of economics, yield etc. Water saving and the water use efficiency of different crops cultivated under drip irrigation are significantly higher when compared to those under flood irrigation. Hence, there is strong basis to encourage the farmers to adopt the drip irrigation method. Solar photovoltaic water pumping (SPVWP) is a cost-effective application in remote locations in developed countries [8]. The economy and reliability of solar electric power made it an excellent choice for remote water pumping [7]. Advantages of PV pumping systems include low operating cost, unattended operation, low maintenance, easy installation, and long life. These are all important in remote locations where electricity may be unavailable [1].

3 Study Area

Due to some unavoidable circumstances, the experiment was conducted at the Instructional Farm of Faculty of Agricultural Engineering, Bidhan Chandra Krishi Viswavidyalaya. The experimental setup was made on the similar environment, topographical and soil condition with the targeted study area. The total area of 45 m × 45 m was taken under consideration. The soil of this farm was silty loam. It is suitable for sloppy land where the runoff water of higher elevation land can be collected at lower elevation land by constructing water silo. But in the present study, the farm pond was constructed, and for immediate study, the farm pond was filled with rain water and partially with tube well water.

4 Methodologies

4.1 Crop Water Requirement

Lemons (Orchard) have been cultivated as test crop with the spacing of 4.5 m × 4.5 m for conduction of this experiment. Total crop water requirement was calculated by using Eq. 1

$$\mathrm{ET_{crop}} = E_{\mathrm{pan}} \times K_p \times K_c \times k_w \tag{1}$$

where $\mathrm{ET_{crop}}$ = crop water requirement, E_{pan} = pan evaporation, K_p = pan factor, k_w = wetted area and K_c = crop factor.

After calculation of the crop water requirement in mm, the depth of the water requirement was multiplied by the spacing between the crops to determine the volume of water required per crop per day.

4.2 Construction of Water Storage Structure

Water storage structure was made with the following dimensions:

Length of the structure = 30 m, top width of the structure = 5 m, depth of the structure = 2 m, side slope of the structure = 1:1, bottom length of the structure = 26 m, and bottom width of the structure = 1 m (Fig. 1).

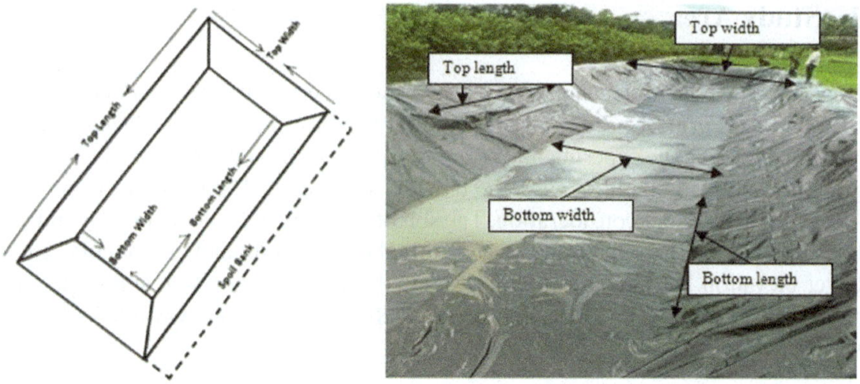

Fig. 1 Water storage structure

4.3 Lining of the Water Storage

Lining was used to prevent the seepage loss from the water storage structure. Here, polythene lining was used to check the seepage loss. By using the heat sealing method, small polythene film pieces were jointed to get the require width and depth of the constructed water storage structure. After placing the polythene sheets, which were tightly fixed with the ground, surface by using the rope and boulders.

4.4 Arrangement of the Solar Operated Pump

A solar photovoltaic pump with the capacity of 0.25 horse power with the connection of 12 V recharge rechargeable has been used to run the drip irrigation system, and the data of head and discharge rate from the pump was taken after every 15 min of interval when the pump was operated. The performance of the pump was evaluated based on the full utilizations of the solar chargeable battery during continuous operating until it was worn out.

5 Results and Discussions

5.1 Crop Water Requirement

It was estimated from the data which was collected from the Department of Agricultural Meteorology and Physics, BCKV, that the average pan evaporation of the month of December, January, February, March, April and May has been shown in

Table 1 Crop water requirement in liter per plant per day

Month	Pan factor	Crop factor	E_{pan} (mm)	ET_{crop}
December	0.8	0.9	2.00	8.75
January	0.8	0.9	1.75	7.64
February	0.8	0.9	2.00	8.75
March	0.8	0.9	4.00	17.20
April	0.8	0.9	5.00	21.87
May	0.8	0.9	5.00	21.87

Table 1. By using Eq. 1, the crop water requirement (ET_{crop}) was estimated for the respective months and the results have been shown in Table 1. As per the recommendation made by the Vermeiren and Jabling, 1980, the pan factor, crop coefficient and percent wetted area for orchard crops were taken as 0.8, 0.9 and 30%, respectively [13].

5.2 *Expulsion Rate of Water Through Solar Photovoltaic Water Pump*

It was observed from the experiment that the solar pump has given a discharge of almost 0.09 L/s (675.20 L/h) with the capacity of running duration of 2.92 h at continuous operation. It was measured that the suction and delivery head were 0.3 m and 0.6 m, respectively, which is standard for operating the drip system efficiently.

5.3 *Water Storage Capacity and Cost of Construction of the Water Silo Structure*

Total capacity of the water storage structure was estimated approximately 176 m^3. For making this storage capacity, 176 m^3 volume of earthwork was done at Rs. 212.98/m^3 (Rs. 107/m^3 for earthwork excavation and Rs. 107/m^3 for scattering). So, total expenditure was Rs. 37,484/-. For lining and covering purpose of water silo, 245 m^2 polythene was required and the cost was Rs. 11805/. The total cost for making the water storage was =(11,805 + 3391 + miscellaneous cost 9595/-) = Rs. 58884/-.

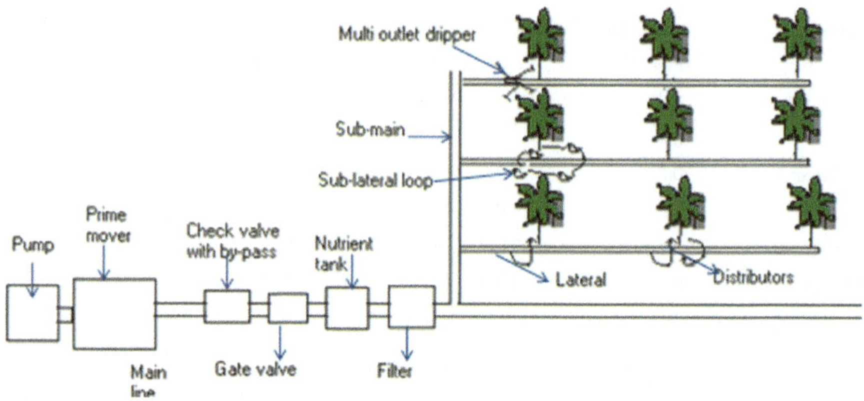

Fig. 2 Design of drip irrigation system

5.4 Drip System

The drip system consisted of main line, lateral pipe, dripper, filter and other accessories. The diameter and length of the main line and lateral were 25 mm, 45 m and 12 mm, 450 m, respectively. The discharge capacity of the dripper was 4lph, and total 400 numbers of drippers were used. To install the whole drip irrigation system, the cost was required as Rs. 6500/- approximately (Fig. 2).

5.5 Estimation of Water Price

Service life of the water silo = 10 years, expenditure of water silo = Rs. 58,885/-, annual investment = Rs. 58,885/10 = 5888.5/-, interest on investment at 10% = 58,885 × 0.1 = 5888.8/-, total amount cost = (5888.5 + 5888.5) = Rs. 11,777/-, cost of water = Rs. 11777/(176 × 1000)L = Rs. 0.70/L This cost of water is within the permissible limit of the poor people of the western part of the West Bengal.

6 Summary and Conclusion

So, from the findings of the experiment, it have been observed that the solar pump discharged approximately 1972 L of water with no further charging. It has also observed that the average water requirement of lemon crop was maximum 22 L per plant per day during the summer. Assume the water requirement of the crop was 30 L per plant per day by including the various factor during the environment condition. It was also observed that the solar pump was operated total 5.90 h at a stretch (2.90 h by using solar battery +3 h by using the sunshine). By conducting the experiment, it was observed that the average release of water from the solar-operated pump was 675.20 L/h (Art.3.2). So, it was observed that almost 3989 L of water have discharged per day by a solar pump which can irrigate the 133 numbers of lemon crop per day or 0.33 ha area. Most of the farmers of the targeted region are marginal farmers so this irrigated area could be sufficient for them.

The water silo with the combination of solar pump-operated drip irrigation system possibly an economy fit and possible way to mitigate the problem of water scarcity of the targeted region of the current experiment. In the light of the experience gained during the investigation and result obtained, it is felt that the following points should be given due consideration in future studies. To reach a specific conclusion and recommendation, the same experiment needs to be repeated and more research work should be done. To realize the effect of drip irrigation, new work has to be started on fertigation technique. The laterals of drip irrigation needs to be placed below covering field by polythene strips.

References

1. Abu-Aligah, M.: Design of photovoltaic water pumping system and compare it with diesel powered pump. Jordan J. Mech. Industr. Eng. **5**(3), 273–280 (2011)
2. Biswas, R.K., Ghosh, U.: Water harvesting, land and water use planning towards sustainable agriculture in water scarce-poverty struck western districts of West Bengal. Int. J. Biomath. Syst. Biol. **3**(1), 1–11 (2016)
3. Biswas, R.K., Rana, S.K., Mallick, S.: Performance of drip irrigation in papaya cultivation in new alluvium agro-climatic zone of West Bengal. Ann. Agric. Res. **20**(1), 116–117 (1999)
4. Ghosh, U., Biswas, R.K.: Design and operation of drip irrigation system to orchard by using stored rain water (Jaller gola) through chargeable battery and solar photovoltaic water pump. Int. J. Sci. Environ. Technol. **6**(4), 2501–2507 (2017)
5. Hedge, D.M., Srinivas, K.: Growth, productivity and wateruse of banana under drip and basin irrigation in relation to evaporation replenishment. Indian J. Agron. **35**(1–2), 106–112 (1990)
6. Maisiri, N., Senzanje, A., Rockstrom, J., Twomlow, S.: On farm evaluation of the effect of low cost drip irrigation on water and crop productivity compared to conventional surface irrigation system. Phys. Chem. Earth A B C **30**, 783–791 (2005)
7. Meah, K., Ula, S., Barrett, S.: Solar photovoltaic water pumping: opportunities and challenges. Renew. Sustain. Energ. Rev. **12**(4), 1162–1175 (2008)
8. Narayanamoorthy, A.: Efficiency of irrigation: a case of drip irrigation. In: Occasional Paper National Bank for Agriculture and Rural Development, Mumbai, vol. 45, p. 38 (2006)
9. Narayanamoorthy, A.: Evaluation of drip irrigation system in Maharashtra, Mimeograph Series No.42. Gokhale Institute of Politics and Economics, Pune (1999)
10. Naryanamoorthy, A.: Beneficial impact of drip irrigation. A study based on western India. Water Resour. J. **195**, 17–25 (1997)
11. Raina, J.N., Thakur, B.C., Verma, M.L.: Effect of drip irrigation and polyethylene mulch on yield, quality and water-use efficiency of tomato (Lycopersicon esculentum). Indian J. Agric. Sci. **69**, 430–433 (1999)
12. Thangaselvabal, T., Suresh, S., Prem Joshua, J., Sudha, K.R.: Banana nutrition—a review. Agric. Rev. **30**, 24–31 (2009)
13. Vermeiren, I., Jobling, G.A.: Localized irrigation: design installation, operation, evaluation. Irrigation and Drainage paper 36, p. 203. FAO, Rome (1980)

Quantification of Urinary Bladder for Early Detection of Hazard in Oliguric Patient Under Dialysis Using Embedded System

V. M. Senthil Kumar, Ravindrakumar Selvaraj, S. Sasikanth, and Ekkirala Kranti Kumar

Abstract Patients suffering from acute kidney injury (AKI) are oliguric or nonoliguric which increases the creatinine levels, and the urine solute concentrations differs. The daily urine excretion has worse prognosis if urine output (UO) is less than 0.5 ml/kg/h. In this paper, an embedded based portable device with IoT support is used to determine and quantify urinary bladder UO. The bladder pressure is measured using the proposed setup which measures the intra-abdominal pressure. Quantification is done using IAP in different positions and stages. The measurements are validated using hospital data of oliguric patient under dialysis. The embedded based device uses piezoelectric phenomenon is portable, less cost and is designed for bedside use. The wrapped belt varies if bladder put pressure on the belt. Interdigitated capacitor (IDC) structures is to be used in future.

Keywords AKI · Oliguric · Urinary bladder · Embedded system · Dialysis · Sensor · IoT

1 Introduction

In India, about 18% of the total population lost their lives due to deficiency of healthcare system. In modern era, the need of advanced technology enhances the quality of health services. Remote health monitoring will be improved through IoT based design. The data analysis software should have effective methods for diagnosis. The usage of swarm intelligence will improve the diagnosis process. Major kidney failures lead to death if unnoticed. Suitable early detection methodologies for diseases like azotemia and kidney damage can avoid the loss. Urinary

V. M. S. Kumar · S. Sasikanth · E. K. Kumar
Malla Reddy College of Engineering and Technology, Secundrabad, Telangana, India

R. Selvaraj (✉)
Biomedical Engineering, Sri Shakthi Institute of Engineering and Technology, Coimbatore, Tamil Nadu, India

K. N. Das et al. (eds.), *Proceedings of the International Conference on Computational Intelligence and Sustainable Technologies*, Algorithms for Intelligent Systems,
https://doi.org/10.1007/978-981-16-6893-7_20

incontinence (UI) is the urinary bladder control/sensation problem in autistic children, diabetic patient, and the elderly sometimes due to excessive alcohol intake, pregnancy, enlarged prostate and tumor/cancer [1]. The quantification of urinary bladder through multisignal processing through multielectrodes, intravesical pressure, electrical impedance plethysmography methods through strain gauge, imaging, electromagnetic and nerve electroneurographic recordings are used but not suitable for bedside or portable units. In current clinical settings, catheterization and catheter-based sensors are employed mostly due to its accuracy and precision but lags in other aspects like expert usage, need of computers and fixed posture [2]. In ultrasound imaging, a catheter is inserted through the urethra, electrical impedance tomography (EIT) with multiple electrodes are also used [3]. The work focuses to eliminate increase in bladder cancer, avoid second cancers and increase survival rates. Screening and early protection help in avoiding cancer. The main objective of the proposed system is to design an efficient embedded based system for detection the bladder for dialysis patients and perform early detection of hazard of oliguric patient. The mortality rate can be reduced for intra-abdominal hypertension (IAH) and the abdominal compartment syndrome (ACS) patients. Low levels of pressure are absorbed in this study for the IAH and ACS patients. The high-thoracic pressure can be quantified by proper data recorded for large periods of time.

2 Literature Survey

For the bladder quantification, several methods and systems are implemented and investigated in literature. In past, the bladder wall abnormal region was detected using magnetic resonance (MR) cystography [4]. The measurement was based on wall thickness and bent rate for detecting the path from each voxel on the inner border to the outer border. The path and length can be found by thickness mapping through electric field tracing [5]. The data can be transferred using wireless modules [6]. Ultrasound was used to detect the cancerous growth when designed with point-of-care (POC) assembly which are found safe, low cost and better in long term use [7]. Large volume of high-resolution images for bladder cancer is recorded for analysis. The other measurement method was using a wireless infrared spectroscopy [8] ultra-wideband (UWB) device with radar and classifier found moderate in performance due to equipment complexity [9]. Along with the bladder fullness, the pressure was also measured using Liquid pressure sensor [10]. Artificial neural network estimates the bladder status from the electroneurogram (ENG) signal of pudendal nerve in real time [11]. Using situ sensor, the bladder activity can be monitored using closed-loop control [12]. The measurement can be done through a chip implanted in the muscle of the bladder, and the information is transmitted through wireless for real-time measurement [13]. The simulated circuit uses an integrated circuit with micro-electro-mechanical (MEMS) pressure sensor. For transmission, RF antennas were fabricated. Similar type of chip with bladder pressure data compression was used for efficient measurement and diagnosis [14]. Apart from this, other works have been carried out in literature for bladder detection and closed-loop control

[15]. Water accumulation detection using impulse radio ultra-wideband (IR-UWB) radar. The implantable sensor based on hall effect with magnetometer reads the pressure exerted through contraction and expansion [16]. Non-invasive methods using near-infrared spectroscopy are efficient due to its wearable structure with timely alerts [17]. The sensors are fabricated on a flexible PCB and tested with artificial bladder prototype. A similar implantable sensor for bladder measurement was done using multilevel resistor ladder which measures the bladder discrete resistance values [18]. EIT sensors were used for stimulation and evaluation in a ring and semicircle electrode formation [19]. Any design should manage the drift, hysteresis and material damage due to long term use. The biocompatible polypyrrole/agarose hydrogel composite found better for the sensor designs. Nathalia et al. [20] investigated the bladder fullness influence on urinary tract obstruction. The detection was done using dynamic renal scintigraphy. Internet of Things based projects were developed in literature for other applications [21]. Mohd. Hamim et al. [22] coined the work on IoT for biomedical application with wearable electrodes and android apps. Alwan and Prahald Rao [23] used Zigbee technology for healthcare system, while Zulfikar Fajar et al. addressed on SMS systems. The works on urinary bladder detection and diagnosis is not done in national level [24]. Jeyalakshmi presented a non-invasive NIRS wearable sensor for bladder fullness sensation and alerting system using simulation [25]. A real-time product was not developed. Internet of Things based projects were developed in literature for other applications [21, 26]. Shivam Trivedi and Alice. N. Cheeran presented an integrated system for heart rate analysis [27]. In literature, several works are carried out for the implementation of biomedical application through system of chip architecture [28, 29] and new devices like FinFET and CNTFET. The authors are aiming to develop similar urinary bladder monitoring unit using the FinFET and CNTFET devices in future [30–32].

3 Background Methodology

The urinary bladder monitoring unit shown in Fig. 1 uses a magnetic field sensor to detect the fields emitted from the coin shaped permanent magnet embedded with a silicon membrane. The magnetic bar is placed at the median sagittal plane, while the lower external abdominal contains the sensor. The unit contains a central fulcrum and a side fulcrum.

Figure 2 shows the hardware block diagram of the continuous bladder monitoring system. It is categorized into three different unit (i.e.,) sensing unit, controlling unit and display unit. The system is designed with arduino uno microcontroller which takes all controlling activities in the proposed design. It receives the signals from the sensing unit which is mounded on the abdominal belt. The filling status of the bladder is measured by means of pressure. Once the belt sensors produce high pressure, it will transmit to the controlling unit. The controller receive it and display the command "bladder is full" on the serial monitor which is connected to the controlling unit. The hardware block consists of the microcontroller, sensors, 16 MHz ceramic resonator, Liquid Crystal Display (LCD) and IoT module.

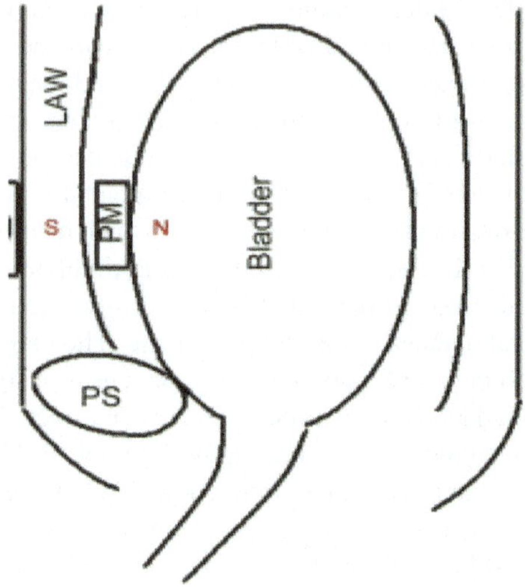

Fig. 1 Urinary bladder monitor

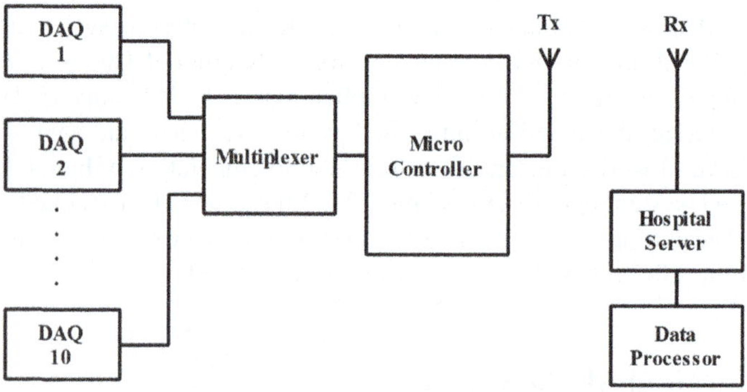

Fig. 2 Hardware block diagram of the project

4 Proposed Methodology

Major kidney failures lead to death if unnoticed. Suitable early detection methodologies for diseases like azotemia and kidney damage can avoid the loss. The work is based on the design and development of a different prototype device for monitoring the urinary bladder status for the patent willing for dialysis. The embedded based device is portable, less cost and is designed for bedside use. The piezoelectric phenomenon was used to detect the level of urine through pressure. The belt is wrapped around the stomach. The increase in bladder size will put pressure on the belt. For patients with critical pathology interdigitated capacitor (IDC) structures is to be used so the volume is measured through capacitance. For patients having mobility problems, wireless belt is to be used. The measured information is processed through the data acquisition unit which is connected to multiple sensors and microcontroller kit. The other sensors add to the device are the pressure, heart rate and body temperature. This information of the patient is sent to the server. The normal and abnormal conditions are classified, and diagnosis is done at the server end. For feature extraction, wavelet transform and non-negative matrix factorization (NMF) technique are to be used. For classification, neural network is to be combined with swarm intelligence. Using large set of real-world data collected from medical laboratories and hospitals is to be used. Through multipoint connections many patients are to be connected with the device. So, in a hospital ward with multiple beds, all patients can be monitored through the single-window system using IoT. The abnormal variations in the heart beat and pressure can be detected, and death can be avoided. In current status, only the analysis through manual intervention is carried out. The bladder detection was done only on the analysis of databases obtained. No solid framework is presented in literature. The proposed project helps the physician to diagnose the critical stages in kidney failures and give alerts and warning during emergency. The needs of a biomedical device for intimation of critical stages in urinary track diseases are higher. The proposed device is safety in its use and protects from dangerous situation. The device is design in the belt form with a miniature hardware kit embedded in the belt. The product is design flexible and can be used for different conditions of urinary problems. Based on the disease range and condition, the architecture of the product is varied. The price range also changes. The device uses a battery which is rechargeable. It can be used more than 6 h. The main block diagram of the product is given in Fig. 2. Multiple data acquisition units are used to get the sensor information of multiple patients. The information collected are temperature, pulse and pressure. The information of individual patients are multiplexed and only one channel information is given in microcontroller. The microcontroller send the data to the hospital; server where the data processing is carried out. The process is given in Fig. 4.

5 Results

The steps followed are given in Fig. 3. The results are obtained for embedded system. Sensors used temperature Sensor LM35, Pulse Sensor plug-and-play heart rate sensor for embedded units. High-intrathoracic pressure is monitored using the sensor data. The bladder pressure is measured using the proposed setup which measure the intra-abdominal pressure and is measured via the indwelling urinary catheter. It is expressed in mmHg. Intra-abdominal pressure (IAP) is the pressure within the abdominal cavity. Normal IAP in a well child is 0 mmHg and in a child on positive pressure ventilation is 1–8 mmHg.

The outcome of the results

IAP varies after every 4–6 h in patients with risk factors and changes in semi-recumbent position. The IAP measurement is indeterminate in prone positioning

Fig. 3 Process flow diagram

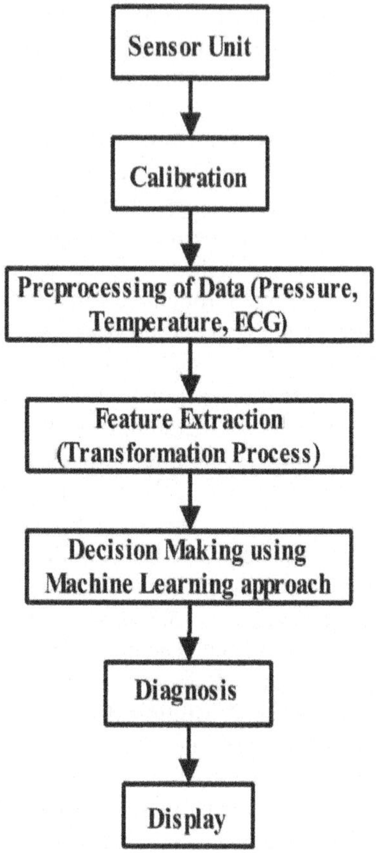

in IAH unstable patients. The mean IAP in critically patients shows that the quantification need minimum optimal volume for proper measurement. Compared to intravesical method the elevation in quantification is less in the proposed setup. The quantification shows that the semi-recumbent position and body mass percentiles effects the voltage obtained at the sensor output which affects the IAP output. For oliguria patients with pneumonia complication, the sensor output is varying by 10% for each elevation 5°. Tables 1 and 2 show the data obtained in the sensors for patient with and without semi recumbent position. The patients with the need of dialysis have IAP greater than 20 mm Hg if the reports show that the new organ failure or dysfunction happened (Record only, No human trials made: Mr. Sugumar, Erode kidneys center, Age 47, Diagnosis: **end-stage renal disease (ESRD)**.

The quantification is modeled to replicate the ESRD diagnosis using the setup by adjusting the belt. In Fig. 6, the normal patient belt is adjusted in full sitting position to model the ESRD problem. Drastic variation is found in IAP. The study shows that the head of bed >30°, Paracentesis, Percutaneous drainage, Nasogastric decompression, Rectal decompression and Hemodialysis/ultrafiltration are followed

Fig. 4 Pin configuration of LM35, pulse and ESP sensor

Table 1 Sensor output with respect to IAP (ADC 10 bit, reference voltage –5 V for 30 mmHg) data: male age 53

S. No	Equivalent intra-abdominal pressure mmHg	Change in IAP	Sensor output relative with 5 V
1	12	14	3.4
2	22	25	4.2
3	24	29	4.4
4	27	31	4.7

for treatment. The vasopressors and optimized ventilation can provide temporary pressure control.

Communication of Data:

The Wi-Fi module for IOT application is ESP8285 with 1 MiB of built-in flash. Currently, these sensors are implemented with respect to the block diagram Fig. 4.

Table 2 Sensor output in semi recumbent position respect to IAP (ADC 10 bit, reference voltage =5 V for 30 mmHg)

S. No	Intra-abdominal pressure mmHg	Condition diagnoses	Sensor output relative with 5 V
1	o	Child-Normal IAP	0.2 V
2	2	Grade I	0.31
3	5“‘	Grade I	1
4	12	Grade II	2.1
5	22	Grade III	3.8
6	24	Grade III	4.1
7	27	Grade IV	4.5
8	29	Grade IV	4.7

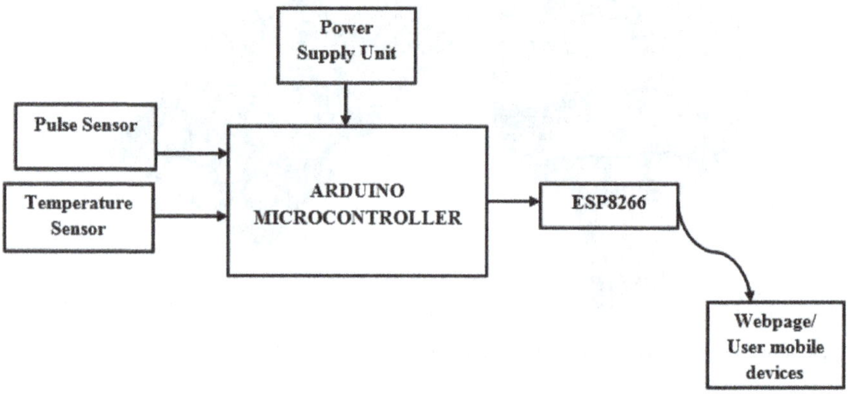

Fig. 5 The sensor interfacing with IoT

The sensor interfacing with IoT is shown in Fig. 5. The bladder unit is tested in our laboratory. The pictures are given Fig. 6 when the testing was done.

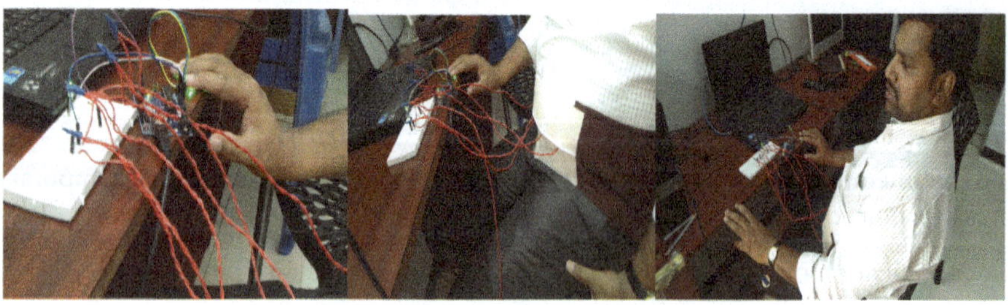

Fig. 6 Prototype developed with the bladder detection unit with sensors—pulse and temperature

The sensor data are transmitted to the IoT module which consists of raspberry pi. The program for machine learning approach is written in python. This decision-making module analysis the received data applies the machine learning algorithm and finds the abnormality. The abnormality identified is send to the web server. Through multiplexing all patient data are acquired, analyzed and stored in the server for future correspondence. The proposed method uses swarm intelligence to diagnose. The IoT and swarm intelligence based prototype will be efficient for the problem.

The paper presents the quantification of urinary bladder for early detection of hazard in oliguric patient under dialysis using embedded system. IoT support is used to determine and quantify the status of the urinary bladder. The designed product uses an ATMEL128 microcontroller with IOT support. The data is transmitted through GSM and to IoT server. The implementation was done and the prototype is tested. The embedded based device is portable, less cost and is designed for bedside use. The piezoelectric phenomenon was used to detect the level of urine through pressure. The belt is wrapped around the stomach. The increase in bladder size will put pressure on the belt. For patients with critical pathology, interdigitated capacitor (IDC) structures is to be used so the volume is measured through capacitance.

6 Conclusion

The paper presents a quantification of urinary bladder for early detection of hazard in oliguric patient under dialysis using embedded system. The detection helps in determining the condition and daily urine excretion. The urine output (UO) can be detected using the pressure monitoring embedded based portable device with IoT support. The piezoelectric phenomenon is used in the design embedded based device which is portable, less cost and is designed for bedside use. The wrapped belt varies it bladder put pressure on the belt. In future, interdigitated capacitor (IDC) structures are to be used in future.

References

1. Dakurah, M.N., Koo, C., Choi, W., Joung, Y.-H.: Implantable bladder sensors: a methodological review. Int. Neurourol. J. **19**(3), 133–141 (2015)
2. Niestoruk, L., Beuth, T., Petry, K., Balzer, M., Stork, W., Mueller-Glaser, K.: A concept for wearable long-term urinary bladder monitoring with ultrasound. Feasibility study. In: 5th European DSP Education and Research Conference (EDERC). Amsterdam, pp. 134–138 (2012)
3. Shin, S., Moon, J., Kye, S., Lee, K., Lee, Y.S., Kang, H.: Continuous bladder volume monitoring system for wearable applications. In: 39th Annual International Conference of the IEEE Engineering in Medicine and Biology Society (EMBC), Seogwipo, pp. 4435–4438 (2017)
4. Duan, C., Yuan, K., Liu, F., Xiao, P., Lv, G., Liang, Z.: Volume-based features for detection of bladder wall abnormal regions via MR cystography. IEEE Trans. Biomed. Eng. **58**(9), 2506–2512 (2011)

5. Zhao, Y., Zhu, H., Duan, C., Gu, X., Liang, Z.: A precise calculation of bladder wall thickness for detection of bladder abnormalities via MR cystography. In: 2011 IEEE Nuclear Science Symposium Conference Record, Valencia, pp. 3153–3157 (2011)
6. Cao, H., et al.: A wireless strain sensor system for bladder volume monitoring. In: 2011 IEEE MTT-S International Microwave Symposium, Baltimore, MD, pp. 1–1 (2011)
7. Basak, A., Ranganathan, V., Bhunia, S.: A wearable ultrasonic assembly for point-of-care autonomous diagnostics of malignant growth. In: 2013 IEEE Point-of-Care Healthcare Technologies (PHT), Bangalore, pp. 128–131 (2013)
8. Molavi, B., Shadgan, B., Macnab, A.J., Dumont, G.A.: Noninvasive optical monitoring of bladder filling to capacity using a wireless near infrared spectroscopy device. IEEE Trans. Biomed. Circ. Syst. 1932–4545 (2013)
9. Krewer, F., Morgan, F., O'Halloran, M.: Incontinence management for the elderly: development of a radar-based bladder volume monitor. In: 2014 IEEE Antennas and Propagation Society International Symposium (APSURSI), Memphis, TN, 1137–1138 (2014)
10. Wenlong, D., Wenyuan, L., Zhigong, W., Xiaoying, L.: Implanted electronic siren to alarm for bladder full with urine. In: 2014 36th Annual International Conference of the IEEE Engineering in Medicine and Biology Society, Chicago, IL, pp. 630–633 (2014)
11. Geramipour, A., Makki, S., Erfanian, A.: Neural network based forward prediction of bladder pressure using pudendal nerve electrical activity. In: 2015 37th Annual International Conference of the IEEE Engineering in Medicine and Biology Society (EMBC), Milan, pp. 4745–4748 (2015)
12. Karam, R., et al.: Real-time classification of bladder events for effective diagnosis and treatment of urinary incontinence. IEEE Trans. Biomed. Eng. **63**(4), 721–729 (2016)
13. Majerus. S. et al.: Wireless bladder pressure monitor for closed-loop bladder neuromodulation. In: 2016 IEEE SENSORS, Orlando, FL, pp. 1–3 (2016)
14. Karam, R., Majerus, S.J.A., Bourbeau, D.J., Damaser, M.S., Bhunia, S.: Tunable, lightweight on-chip event detection for implantable bladder pressure monitoring devices. IEEE Trans. Biomed. Circ. Syst. **11**(6), 1303–1312 (2017)
15. Karam, R., Bhunia, S., Majerus, S., Brose, S.W., Damaser, M.S., Bourbeau, D.: Real-time, autonomous bladder event classification and closed-loop control from single-channel pressure data. In: 2016 38th Annual International Conference of the IEEE Engineering in Medicine and Biology Society (EMBC), Orlando, FL, pp. 5789–5792 (2016)
16. Milani, A., Mazzocchi, T., Iacovacci, V., Pinzi, N., Ricotti, L., Menciassi, A.: Magnetic sensing system for monitoring the volume of an artificial bladder. In: 2018 7th IEEE International Conference on Biomedical Robotics and Biomechatronics (Biorob), Enschede pp. 877–882 (2018)
17. Fong, D., Alcantar, A.V., Gupta, P., Kurzrock, E., Ghiasi, S.: Non-invasive bladder volume sensing for neurogenic bladder dysfunction management. In: IEEE 15th International Conference on Wearable and Implantable Body Sensor Networks (BSN), Las Vegas, NV, pp. 82–85 (2018)
18. Kim, M.K., et al.: Implantable bladder volume sensor based on resistor ladder network composed of conductive hydrogel composite. In: 39th Annual International Conference of the IEEE Engineering in Medicine and Biology Society (EMBC), Seogwipo, pp. 1732–1735 (2017)
19. Liang, X., Xu, L., Tian, W., Xie, Y., Sun, J.: Effect of stimulation patterns on bladder volume measurement based on fringe effect of EIT sensors. In: IEEE International Conference on Imaging Systems and Techniques (IST), Abu Dhabi, United Arab Emirates, pp. 1–5 (2019)
20. Cosenza, N.N., Lau, F., Lima, M.C.L., Amorim, B.J., Mosci, C., Lima, M.L., Ramos, C.D.: Influence of bladder fullness on the detection of urinary tract obstruction by dynamic renal scintigraphy. Radiol. Bras. **50**(4), 237–243 (2017)
21. Patii, N., Iyer, B.: Health monitoring and tracking system for soldiers using Internet of Things (IoT). In: International Conference on Computing, Communication and Automation (ICCCA), pp. 1347–1352 (2017)

22. Hamim, M., Paul, S., Hoque, S.I., Rahman, Md.N., Ifat-Al Baqee.: IoT based remote health monitoring system for patients and elderly people. In: International Conference on Robotics, Electrical and Signal Processing Techniques (ICREST), pp. 533–538 (2019)
23. Alwan, O.S., Prahald Rao, K.: Dedicated real-time monitoring system for health care using ZigBee. Healthcare Technol. Lett. **4**(4), 142–144 (2017)
24. Fajar, M.R.Z., Hadiyoso, S., Rizal, A.: An interfacing digital blood pressure meter with arduino-GSM module for real-time monitoring. In: International Conference on Control, Electronics, Renewable Energy and Communications (ICCREC), pp. 98–102 (2017)
25. Jeyalakshmi, M.S.: Wearable system design for detecting urinary bladder fullness. Int. J. Pure Appl. Math. **119**(15), 2183–2189 (2018)
26. Kaur, A., Jasuja, A.: Health monitoring based on IoT using Raspberry PI. In: International Conference on Computing, Communication and Automation (ICCCA), pp. 1335–1340 (2017)
27. Trivedi, S., Alice, N., Cheeran: Android based health parameter monitoring. In: International Conference on Intelligent Computing and Control Systems (ICICCS), pp. 1145–1149 (2017)
28. Jeyaseelan, J.P., Selvaraj, R.: Design of new implantable stimulator chip (SoC) for non-invasive/minimally invasive biomedical application. In: International Conference on Communication and Signal Processing (2014)
29. Priyadarshini, J.J., Ravindrakumar, S.: Design of low-power blink detector for minimally invasive implantable stimulator (SoC) using 180 nm technology. In: Suresh, L., Dash, S., Panigrahi, B. (eds.) Artificial Intelligence and Evolutionary Algorithms in Engineering Systems. Advances in Intelligent Systems and Computing, vol. 324. Springer, New Delhi (2015)
30. Senthilkumar, V.M., Muruganandham, A., Ravindrakumar, S., Gowri Ganesh, N.S.: FINFET operational amplifier with low offset noise and high immunity to electromagnetic interference. Microprocess. Microsyst. **71**, 102887 (2019)
31. Senthilkumar, V.M., Ravindrakumar, S., Nithya, D., Kousik, N.V: A vedic mathematics based processor core for discrete wavelet transform using FinFET and CNTFET technology for biomedical signal processing. Microprocess. Microsyst, **71**, 102875 (2019)
32. Senthil Kumar, V.M., Ravindrakumar, S.: Design of an area-efficient FinFET-based approximate multiplier in 32 nm technology for low-power application. In: Wang, J., Reddy, G., Prasad, V., Reddy, V. (Eds.) Soft Computing and Signal Processing. Advances in Intelligent Systems and Computing, vol. 898. Springer, Singapore

Shorted Non-radiating Edges Integrated 2 × 1 Rectangular Microstrip Antenna for Concurrent Improvement of Gain and Polarization Purity

Himanshu Deka, L. Lolit Kumar Singh, Sudipta Chattopadhyay, and Abhijyoti Ghosh

Abstract A 2 × 1 array of rectangular microstrip antenna (RMA) with fully shorted non-radiating edges mounted on two different types of mostly available and popular substrates namely RT-Duroid, and FR4 have been proposed for simultaneous improvement of co-polarization gain and polarization purity. The mutual coupling effects between the two radiating elements have also been studied by varying the gap between the radiating patch and documented in this paper. The simulated studies have been carried out using the high-frequency structure simulator (v. 14). The array structure with FR4 substrate provides a co-polarization gain of 5.3 dBi and polarization purity of 29.77 dB when the gap between two radiating patches is λ/4. On the other hand, the array structure with RT-Duroid substrate provides a co-polarization gain of 5.3 dBi and polarization purity of 29.77 dB when the gap between two radiating patches is λ/4. Both the results are quite good as compared to the conventional rectangular patch antenna. Both the proposed structure also show good impedance matching and stable radiation pattern with in the operational frequency range.

Keywords Rectangular microstrip antenna · Antenna array · Co-polarization gain · Polarization purity · Shorted non-radiating edges

1 Introduction

In the wireless radio communication system, the patch antenna is the most traditional and desired contender because of its small size, multifunctionality, and easy fabrication methods, etc. Despite all the advantages typical microstrip antenna have some constraint like only 3–4 dBi gain, 3–4% input impedance bandwidth, poor polarization purity (co-polarization (CP) to cross-polarization (XP) separation), etc. [1, 2]. RMA is one of the most dynamic candidates among the different microstrip structure which provides linearly polarized broadside radiation in its fundamental mode, i.e.,

H. Deka · L. L. K. Singh · S. Chattopadhyay · A. Ghosh (✉)
Department of Electronics and Communication Engineering, Mizoram University, Tanhril, Aizawl, Mizoram 796004, India

K. N. Das et al. (eds.), *Proceedings of the International Conference on Computational Intelligence and Sustainable Technologies*, Algorithms for Intelligent Systems,
https://doi.org/10.1007/978-981-16-6893-7_21

TM_{10}. However, a point of orthogonally polarized radiation, referred to as cross-polarized (XP) radiation, takes place in the fundamental mode also [3]. Among the different feeding techniques XP radiation is most significant in probe feed structures [4]. Different techniques like modifying the antenna structure [5–7], different feeding techniques [8, 9], defected ground structure (DGS) [10–13], defected patch structure (DPS) [14, 15] are proposed by different researchers for the betterment of radiation characteristics of rectangular microstrip antenna. All these structures provide good amount of polarization purity (CP-XP isolation) but fails to provide higher gain.

The concept of implementing the fully and partially shorting walls at the non-radiating edges to improve polarization purity has been reported in [16, 17]. Shorting posts with defected patch surface (DPS) to improve radiation characteristics have been reported in [18], where single-element radiating structure has been considered for study. However, because of low gain issue of single-element RMA, it is undesirable that the array structure must be implemented for practical applications. Shorting walls and posts have been considered for array structure mainly to improve scanning range [19], reduction of mutual coupling [20], gain enhancement [21] only but not simultaneously.

In this present study, two 2 × 1 array rectangular microstrip antenna with completely shorted non-radiating edges, one with RT-Duroid substrate ($\varepsilon_r = 2.33$), and another one is with FR4 substrate ($\varepsilon_r = 4.4$) have been investigated for improvement of co-polarization gain and polarization purity simultaneously (Fig. 1a (top view), Fig. 1b (3D view), Fig. 1c (side view)). The mutual coupling effects between the radiating elements have been studied by varying the gap between two patches.

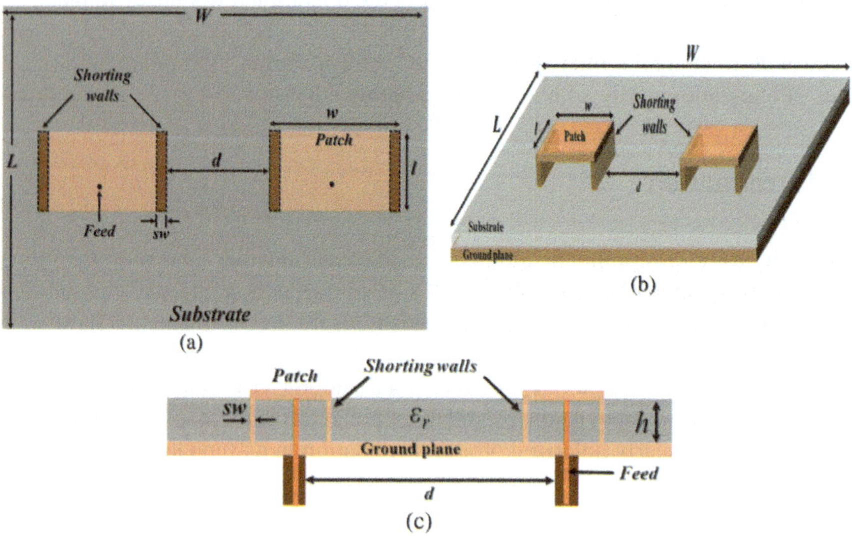

Fig. 1 The proposed 2 × 1 array of rectangular patch antenna with shorted non-radiating edges **a** top view, **b** 3D view, and **c** side view

Table 1 Parameters used in the proposed RMA ($h = 1.575$ mm)

Substrate	ε_r	L (mm)	W (mm)	l (mm)	w (mm)	Thickness of the shorting plate (sw) (mm)
Fr4-Epoxy	4.4	100	100	12	18	0.5
RT-Duroid	2.33					

The current antenna structure with RT-Duroid dielectric provides the CP gain (9 dBi) and polarization purity (29.77 dB) when the gap between the radiating elements is λ/4. On the other hand, the proposed structure with FR4 substrate provides the best performance of CP gain (5.3 dBi) and polarization purity (22.34 dB) when the gap between the radiating elements is λ/4.

2 Proposed Structure

At first two conventional rectangular microstrip antennas using thin (0.1 mm) copper sheet with length (l) 12 mm and width (w) 18 mm with two different substrates, i.e., FR4 ($\varepsilon_r = 4.4$) and RT-Duroid ($\varepsilon_r = 2.33$) with substrate dimension of 100 mm^2 × 100 mm^2 and thickness (h) 1.575 mm and have been designed using HFSS (V.14). Then, both the non-radiating sides of the RMA are shorted with the ground plane by placing two copper strip of thickness (sw) 0.1 mm and height (sh) 1.575 mm by making groves of same thickness on the substrate. This conventional patch with shorted non-radiating walls has upgraded to 2 × 1 array along the H-plane. All the antenna structures have been optimized for proper matching. In the array configuration, three models have been studied by changing the distance between the patch (λ, λ/2, λ/4) along the H-plane. The detail parameters that are considered in the present design have been shown in Table 1.

3 Results and Discussion

The simulated [22] results regarding input characteristics as well as radiation characteristics obtains from the different proposed structure have been documented in this section. The reflection coefficient (S_{11}) profile of the different structures with Fr4 and RT-Duroid dielectrics is shown in Fig. 2. Both the figures show a good impedance matching of all the structures. The present structures with RT-Duroid resonates around 9 GHz, while the proposed structures with FR$ substrate resonates around 6.7 GHz.

For array structure, the mutual coupling between the array elements is also an important parameter to study. Figure 3 shows the isolation in the H-plane. For both, the structures shows quite significant isolation (S_{12}) between the ports. The

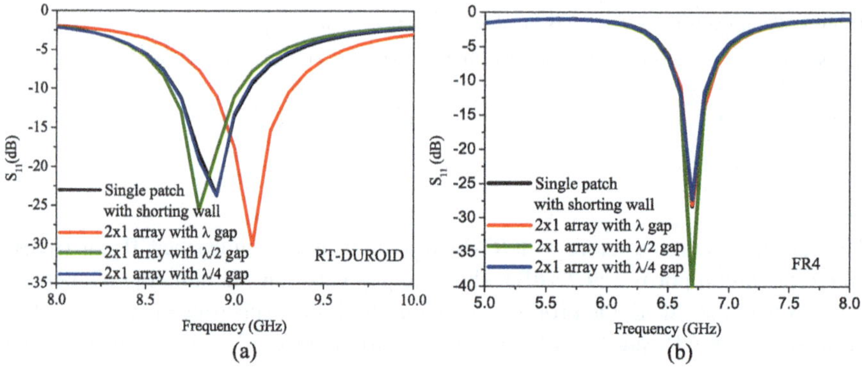

Fig. 2 Reflection coefficient profile (S_{11}) of the conventional rectangular patch antenna with **a** RT-Duroid, **b** Fr4 substrate

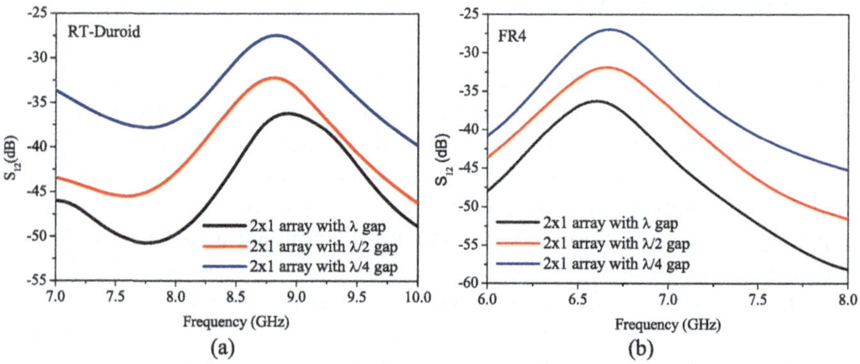

Fig. 3 Reflection coefficient profile (S_{12}) of the conventional rectangular patch antenna with **a** RT-Duroid, **b** Fr4 substrate

figure reveals that the structure with λ gap between the two radiating elements is having highest isolation, i.e., lowest mutual coupling and the isolation between ports decreases as the gap between the radiating elements decreases.

The complete radiation profile of all the proposed antennas with RT-Duroid substrate have been shown Fig. 4. The detailed outcome of the different input and radiation parameters have been documented in Table 2.

It can be seen that single-element patch with shorting wall provides maximum polarization purity (CP-XP isolation) with moderate increment in gain. The array structure with a gap of λ between the radiating elements provides highest gain of 9.1 dBi with 24.51 dB of polarization purity, while the array structure with a gap of λ/4 between the radiating elements provides co-polarization gain of 9 dBi with 29.77 dB of polarization purity which is much better as compared to the conventional rectangular patch antenna. The above results can be verified from the radiation plot also.

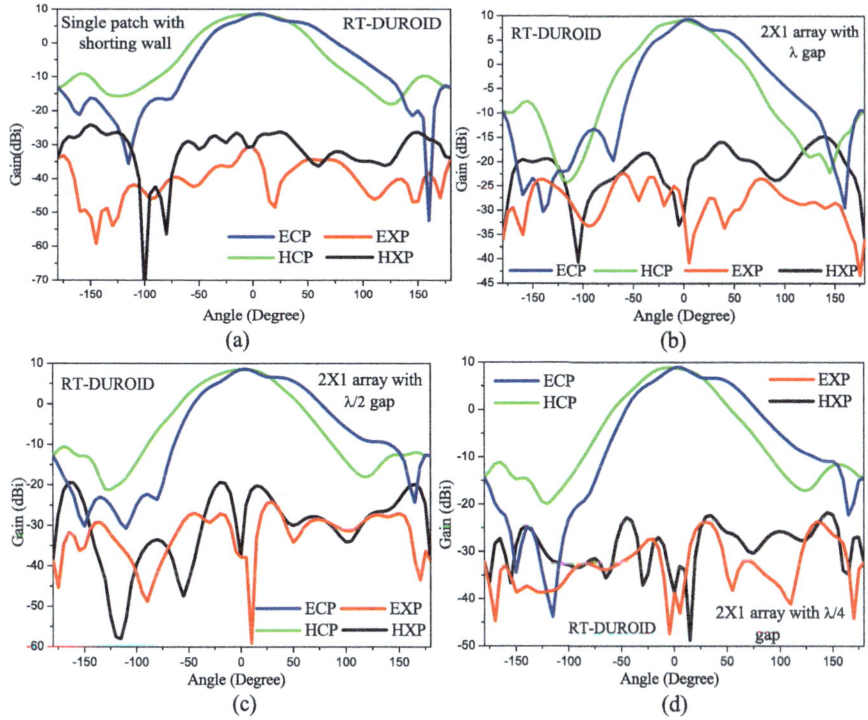

Fig. 4 Radiation profile of present structure with RT-Duroid substrate **a** single patch with shorting wall, **b** 2 × 1 array with shorting wall with λ gap, **c** 2 × 1 array with shorting wall with λ/2 gap, **d** 2 × 1 array with shorting wall with λ/4 gap between radiating elements

Table 2 Output of different characteristics of proposed structure with RT-Duroid substrate

S. No	Structure	Gap between the patches (*d*)	Resonance Frequency (GHz)	Gain (dBi)	H-Plane CP-XP isolation (dB)
1	Conventional RMA	–	7.2	6.9	15.64
2	Single-element shorted RMA	–	8.9	8.2	31.68
3	2 × 1 array shorted RMA	λ	9.1	9.1	24.51
4	2 × 1 array shorted RMA	λ/2	8.8	8.6	28.77
5	2 × 1 array shorted RMA	λ/4	8.9	9.0	29.77

The entire radiation profile of all the structures with FR4 substrate have been shown Fig. 5 and the detailed outcome of the different input and radiation parameters like resonance frequency, co-polarization gain, and polarization purity have been documented in Table 3.

The results in the Table 3 for FR4 substrate is also comparable for the structure with RT-Duroid also. The array structure with different gap between the radiating elements provides higher gain (5.3 dBi for array with λ/4 gap between the radiating elements) as compared to the conventional rectangular patch antenna (3.05 dBi). The polarization purity provides by the proposed structures is also better than the conventional rectangular patch structure. The radiation plots also verify the above results.

A comparative study of proposed antennas with other recently published papers has been shown in Table 4 which clearly shows the superiority of the propose structures.

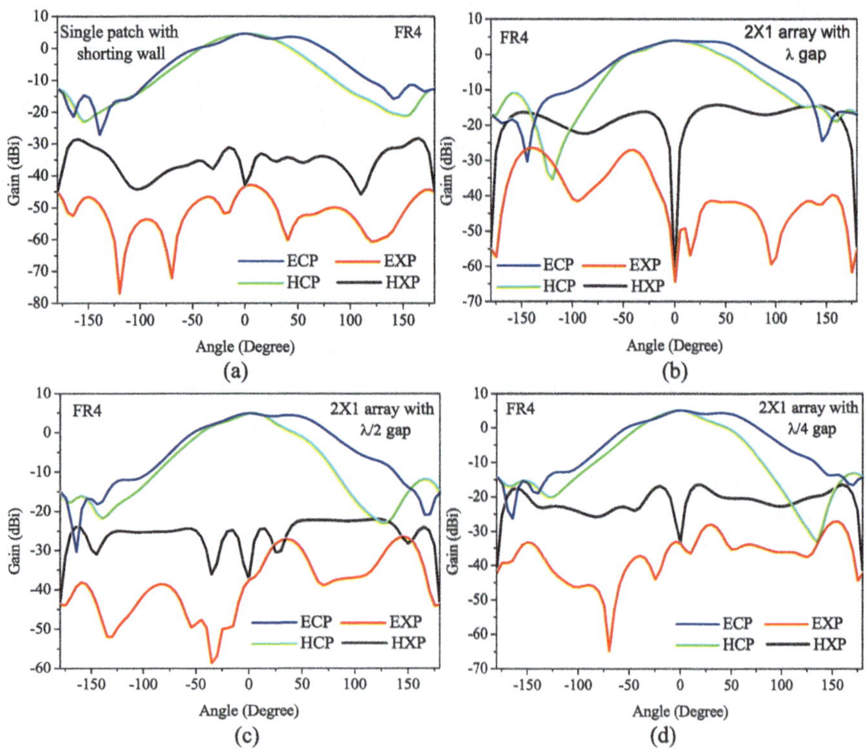

Fig. 5 Radiation profile of present structures with FR4 substrate **a** single patch with shorting wall, **b** 2 × 1 array with shorting wall with λ gap, **c** 2 × 1 array with shorting wall with λ/2 gap, **d** 2 × 1 array with shorting wall with λ/4 gap between radiating elements

Table 3 Input and radiation characteristics output of different proposed structure with FR4 substrate

S. No	Structure	Gap between the patches (d)	Resonance Frequency (GHz)	Gain (dBi)	H-Plane CP-XP isolation (dB)
1	Conventional RMA	–	5.4	3.05	18.97
2	Single element shorted RMA	–	6.7	3.5	18.81
3	2 × 1 array shorted RMA	λ	6.64	4.5	19.8
4	2 × 1 array shorted RMA	λ/2	6.64	5	26.86
5	2 × 1 array shorted RMA	λ/4	6.7	5.3	22.34

Table 4 Comparison of the proposed work with recently published works

Ref. No	Gain (dBi)	H-Plane CP-XP isolation (dB)
[6]	7	30
[7]	7.9	13
[12]	7	28
[13]	8.5	22
[14]	6.13	30
[16]	6.1	25
[17]	7.5	30
[18]	6.3	22.5
Present work with RT-Duroid substrate with a gap of λ/4 between radiating elements	9	29.77
Present work with FR4 substrate with a gap of λ/2 between radiating elements	5	26.86

4 Conclusion

In the present investigation, 2 × 1 element array of shorted RMA has been studied thoroughly to achieve high gain, low XP radiation with appreciably good mutual coupling effect. Different input and output characteristics such as impedance bandwidth, co-polarization gain, polarization purity have been studied through high-frequency structure simulator (HFSS V14). Two different type of mostly available and popular substrate namely FR4 and RT-Duroid have been used. Further 2 × 1 array of RMA with the different gap between the radiating elements has also been investigated. This is done with the outlook to investigate the effect of shorted non-radiating edges on the different input and radiation characteristics of an antenna array structure. The study concludes that the separation between adjacent elements ($\approx \lambda/4$) yield better performance if the array elements are placed in H-plane.

References

1. Garg, R., Bhartia, P., Ittipiboon, A., Bahl, I.: Microstrip Antenna Design Handbook. Artech House, Norwood U.S.A (2001)
2. Kraus, J.D., Marhefka, R.J., Khan, A.S.: Antenna Wave Propagation. McGraw Hill Education, New Delhi India (2015)
3. Huynh, T., Lee, K.F., Lee, R.Q.: Cross-polarization characteristics of rectangular patch antennas. Electron. Lett. **24**(8), 463–464 (1988)
4. Petosa, A., Ittipiboon, A., Gagnon, N.: Suppression of unwanted probe radiation in wide band probe-fed microstrip patches. Electron. Lett. **35**, 355–357 (1999)
5. Loffler, D., Wiesbeck, W.: Low-cost X-polarised broadband PCS antenna with low cross-polarisation. Electron. Lett. **35**, 1689–1691 (1999)
6. Granholm, J., Woelders, K.: Dual polarization stacked microstrip patch antenna array with very low cross-polarization. IEEE Trans. Antennas Propag. **49**, 1393–1402 (2001)
7. Baligar, J.S., Revankar, U.K., Acharya, K.V.: Broadband two-layer shorted patch antenna with low crosspolarisation. Electron. Lett. **37**, 547–548 (2001)
8. Chin, C.H.K., Xue, Q., Wong, H., Zhang, X.Y.: Broadband patch antenna with low cross-polarisation. Electron. Lett. **43**, 137–138 (2007)
9. Sim, C.-Y.-D., Chang, C.C., Row, J.S.: Dual-feed dual-polarized patch antenna with low cross polarization and high isolation. IEEE Trans. Antennas Propag. **57**, 3321–3324 (2009)
10. Ghosh, A., Ghosh, D., Chattopadhyay, S., Singh, L.L.K.: Rectangular microstrip antenna on slot type defected ground for reduced cross polarized radiation. IEEE Antennas Wirel. Propag. Lett. **14**, 324–328 (2015)
11. Ghosh, A., Chakraborty, S., Chattopadhyay, S., Nandi, A., Basu, B.: Rectangular microstrip antenna with dumbbell shaped defected ground structure for improved cross polarized radiation in wide elevation angle and its theoretical analysis. IET Microw. Antennas Propag. **10**, 68–78 (2016)
12. Ghosh, A., Chattopadhyay, S., Chakraborty, S., Basu, B.: Cross type defected ground structure integrated microstrip antenna: a novel perspective for broad banding and augmenting polarization purity. J. Electromagn. Waves Appl. **31**, 461–476 (2017)
13. Ghosh, A., Basu, B.: Triangular slotted ground plane: a key to realize high gain, cross-polarization free microstrip antenna with improved bandwidth. Turk. J. Electr. Eng. Comput. Sci. **27**, 1559–1570 (2019)
14. Chakraborty, S., Ghosh, A., Chattopadhyay, S., Singh, L.L.K.: Improved cross polarized radiation and wide impedance bandwidth from rectangular microstrip antenna with dumbbell shaped defected patch surface. IEEE Antennas Wirel. Propag. Lett. **15**, 84–88 (2016)
15. Ghosh, A., Ghosh, S.K., Ghosh, D., Chattopadhyay, S.: Improved polarization purity for circular microstrip antenna with defected patch surface. Int. J. Microw. Wirel. Technol. **8**, 89–94 (2016)
16. Poddar, R., Chakraborty, S., Chattopadhyay, S.: Improved cross polarization and broad impedance bandwidth from simple single element shorted rectangular microstrip patch: theory and experiment. Frequenz **70**, 1–9 (2016)
17. Ghosh, D., Ghosh, S.K., Chattopadhyay, S., Nandi, S., Chakraborty, D., Anand, R., Raj, R., Ghosh, A.: Physical and quantitative analysis of compact rectangular microstrip antenna with shorted non-radiating edges for reduced cross-polarized radiation using modified cavity model. IEEE Antennas Propag. Mag. **56**, 61–72 (2014)
18. Ghosh, A., Singh, L.L.K., Basu, B.: Wide bandwidth microstrip antenna with defected patch surface for low cross polarization applications. Int. J. RF Microwave Comput. Aided Eng. **27**, 1–10 (2017)

19. Waterhouse, R.B.: The use of shorting posts to improve the scanning range of probe-fed microstrip patch phase array. IEEE Trans. Antennas Propag. **44**, 302–309 (1996)
20. Tunio, I.A., Mahe, Y., Razban, T., Froppier, B.: Mutual coupling reduction in patch antenna array using combination of shorting pins and metallic walls. Prog. Electromagn. Res. C **107**, 157–171 (2021)
21. Ou, J.-H., Huang, J., Liu, J., Tang, J., Zhang, X.Y.: High-gain circular patch antenna and array with introduction of multiple shorting pins. IEEE Trans. Antennas Propag. **68**, 6506–6515 (2020)
22. HFSS: High-frequency structure simulator. Version 14, Ansoft Corp

A Mutual Authentication and Key Agreement Protocol for Smart Grid Environment Using Lattice

Daya Sagar Gupta

Abstract Internet of Things (IoT) has been experienced in various communication technologies. In recent years, as an application of IoT, smart grid dragged great attention from industry and academia. Smart grid can establish the sharing of information among smart meters and the service provider via the Internet protocol (IP)-based communication. This information sharing among various entities in smart grid system may lead to make the system vulnerable to various security and privacy issues. Various security protocols are existed in the literature which provides the secure communication of smart meters with the service provider by establishing a secure connection. However, many of them fail to propose a security protocol where the service provider can share the information with different smart meters and can negotiate a session key with each smart meters for further communication. Further, the existed schemes are exposed to quantum attack as they have used the traditional cryptographic tool in their implementation. Considering these issues, we design a lattice-based mutual authentication and key agreement protocol for smart grid which permits a secure communication among the service provider and the smart meter. The proposed protocol also allows the service provider to initiate a communication with smart meters in the smart grid system. The security analysis shows the security resistance of the proposed scheme with existed as well as quantum attack.

Keywords Security · Privacy · Smart grid · Authentication · Lattice-based cryptography

1 Introduction

Recently, Internet of things (IoT) has prolifically brought in drastic revolution in the field of wireless communication and technology. IoT is a network that connects different things such as sensors, actuators, and smart devices through the Internet.

D. S. Gupta (✉)
Department of Computer Science and Engineeirng, Rajiv Gandhi Institute of Petroleum Technology, Jais, Amethi, Uttar Pradesh 229304, India

K. N. Das et al. (eds.), *Proceedings of the International Conference on Computational Intelligence and Sustainable Technologies*, Algorithms for Intelligent Systems,
https://doi.org/10.1007/978-981-16-6893-7_22

Rapid development of smart devices for different smart applications has opened doors in the improvement of human life. Smart grid (SG) has been introduced as one of an smart applications of IoT. SG is capable for energy management in the smart homes and has the potential that can capture the whole advanced electric power grid. SG is a huge network of millions of connected power equipment which could be considered as the *things* in the design of IoT network. Since SG and IoT have the same nature of working, thus IoT could be used in the implementation of SG. The IoT integrates the features of SG and is used as a tool to collect and share the energy-based information in the network. In modern days, the conventional power grid is replaced by SG due to its smart forming features like reliability, interactive, efficiency, and intelligent. SG as a smart network of electrical things has a distributed control over different power suppliers and exhibits the current as well as future demands of the customers [1]. In recent days, SG has dragged the attention from power industry engineers and researchers in the area of communication and electric power generation. Some of them also referred SG as the future grid as it integrates the digital communication technologies to enhance the traditional electrical power grid [2].

A number of different entities such as smart devices, control centers and communication modules are being found in SG network. Being a smart device, smart meter plays a significant role in SG. Smart meters are used to monitor the electricity consumption and regular flow of power in the network on a regular basis [3]. The power resources are supplied to the end-users by various service providers according to the information transferred by the smart meter fixed at user end. The information exchanged among the smart meters and the service provider is routed through an insecure network and hence lead to be different security issues. SG networks may face a big harm to the security threats and thus must implement the hard security mechanisms so that the different security goals such as authentication, confidentiality, and integrity can be achieved in SG environment [4]. Various security mechanisms for SG have been found in the literature till date [5–8]. Many researchers have designed different security protocols using traditional security approaches such as elliptic curve cryptography (ECC) and bilinear pairing. These systems of cryptographic tools are costly in their implementation. Besides, most of the existed security protocols for SG have used traditional systems like public-key cryptography (PKC), identity-based cryptography (IBC), bilinear Diffie–Hellman problem (BDHP), Chinese remainder theorem (CRT), Diffie–Hellman problem(DH) computational Diffie–Hellman problem(CDH), and other DH-type related techniques to design their security protocols. These traditional security protocols suffer from different security attacks and incur a high communication, storage, and computational cost. Moreover, the earlier cryptographic methods including PKC, CRT, IBC, BDH, ECC, BP, etc., will be exposed to the future quantum computer technologies. Shor [9] showed that many of these cryptographic techniques are insecure against quantum computing environments. It has been noted that the cryptographic designs based on the lattice resist the quantum computers and mitigate the vulnerabilities persisted in traditional cryptographic methods [10, 11]. To deal with the aspects above, we design a lattice-based mutual authentication and key agreement protocol for SG environment. The main contributions of the proposed protocol are summarized as follows.

- It is designed on the concept of lattice cryptography. It uses the lattice hard assumptions small integer solution (SIS) and inhomogeneous small integer solution (ISIS) problems which confirm the post-quantum defended nature of the proposal.
- A security analysis is presented which is used to proof the security of the proposed protocol.
- It assures the security requirement for SG network including anonymity, replay attack, mutual authentication, data privacy, quantum security, etc.
- The proposed protocol uses only addition and multiplication among the matrices and vectors and thus is very efficient and simple and can be easily implemented in IoMT scenario.

1.1 Structure of Paper

Further, the paper is organized in the following manner. Section 2 presents literature survey. In Sect. 3, some definitions and concepts are to be used in the protocol design. In Sect. 4, the proposed protocol is described. Section 5 tells about the security analysis of proposed protocol. Section 6 concludes the paper.

2 Related Work

In the year 2011, Fouda et al. [12] proposed a message authentication protocol as a security equipment in SG for the secure communication. In their proposal, they used DH key exchange [13] to negotiate a secret key and for mutual authentication among smart meters. Nicanfar et al. [8] in their proposal designed a mutual authentication protocol between smart meters of SG network and a cloud server using an initial password. They also proposed a key agreement protocol for secure communications in SG communications. However, they have used PKI technique. Garg et al. [14] developed a secure authentication scheme for smart meters in SG environment with the security requirements such as anonymity, trust, and mutual authentication. They claimed that their protocol reduces energy consumption, computation, and communication overheads. Their designed protocol exhibits key agreement protocol using ECC and functions. Mahmood [3] designed an ECC-based authentication scheme for SG systems. The proposal claimed that it satisfies all the security requirements along with the mutual authentication and costs low computation and communication overheads. Zhang et al. [15] presented an anonymous authentication and key exchange scheme for SG environment. In their protocol, the service provider and the smart meter firstly authenticate each other and then share a common secret key for their further communication. Kumar et al. [16] proposed an ECC-based authentication protocol in SG to show demand response preservation. In their protocol, they also designed a protocol for session key agreement between a device in SG and an utility center for secure communication of confidential information. They claimed that the security analysis proofs the resistance of various attacks.

3 Lattice

Lattices are become a powerful tool for the future cryptography due to its robustness nature. The hard assumptions on a lattice can strengthen the cryptographic designs in the presence of quantum computers. A lattice is a regular structure of any form. The mathematical definition of a lattice can be structured as follows:

Definition 1 Given a set of vectors $\mathbf{x_1}, \mathbf{x_2} \ldots \mathbf{x_n} \in \mathcal{R}^m$, a lattice denoted by $\mathcal{L}$ can be defined as:

$$\mathcal{L}(\mathbf{x_1}, \mathbf{x_2} \ldots \mathbf{x_n}) = \left\{ \sum_{i=1}^{n} a_i \mathbf{x_i} : a_i \in Z^+ \right\} \quad (1)$$

These vectors $\mathbf{x_1}, \mathbf{x_2} \ldots \mathbf{x_n}$ must be a linearly independent set and are know as basis vectors. The integers m and n are dimension and rank of the lattice, respectively [17].

The shortest nonzero vector in $\mathcal{L}$ is known as *minimum distance* of $\mathcal{L}$ and can be calculated by the following equation:

$$\mathcal{D}_{\min}(\mathcal{L}) = \min_{\mathbf{x} \in \mathcal{L} \setminus \{0\}} \| \mathbf{x} \| \quad (2)$$

Moreover, a lattice $\mathcal{L}$ must have a basis and may have different sets of basis. However, each such set has equal number of elements in it.

Lemma 1 *A lattice $\mathcal{L}$ must have at least one basis to generate $\mathcal{L}$.*

The basis of $\mathcal{L}$ can be represented by a matrix $\mathbf{X} = [\mathbf{x_1}, \mathbf{x_2} \ldots \mathbf{x_n}] \in Z^{m \times n}$. The matrix $\mathbf{X}$ is called *basis matrix* in which the columns are set of basis vectors. Now, a lattice $\mathcal{L}$ can be further defined by the following equation:

Definition 2 Let a basis matrix be $\mathbf{X} = [\mathbf{x_1}, \mathbf{x_2} \ldots \mathbf{x_n}] \in Z^{m \times n}$, the generated lattice $\mathcal{L}$ by $\mathbf{X}$ in $\mathcal{R}^m$ can be expressed as $\mathcal{L}(\mathbf{X}) = [\mathbf{Xa} : \mathbf{a} \in Z^n]$ where $\mathbf{Xa}$ represents a usual matrix–vector multiplication.

The basic hard assumptions in a lattice $\mathcal{L}$ include the shortest vector problem (SVP) and closest vector problem (CVP). In SVP, the assumption is to find a short nonzero vector (whose Euclidian norm is minimum) on a given lattice $\mathcal{L}$ and in CVP, the assumption is to find the closest vector to a given vector in $\mathcal{L}$. The another extension of SVP and CVP is *appr*SVP and *appr*CVP, respectively [18].

Definition 3 (Shortest Vector Problem (SVP)). Given a lattice $\mathcal{L}(\mathbf{X})$ and its basis matrix $\mathbf{X} \in Z^{m \times n}$, it is hard in finding a nonzero vector $\mathbf{a} \in \mathcal{L}(\mathbf{X})$ for $\| \mathbf{a} \| = \mathcal{D}_{\min}(\mathcal{L})$.

Definition 4 (Closest Vector Problem (CVP)). Given a lattice $\mathcal{L}(\mathbf{X})$, its basis matrix $\mathbf{X} \in Z^{m\times n}$ and a vector $\mathbf{b} \notin \mathcal{L}$, it is hard in finding a nonzero vector $\mathbf{a} \in \mathcal{L}(\mathbf{X})$ such that $\| \mathbf{b} - \mathbf{a} \| = \mathcal{D}_{\min}(\mathcal{L})$.

3.1 *q-Ary Lattice*

For an integer modular $q \approx poly(m)$, a lattice $\mathcal{L}$ which satisfies $Z_q^m \subseteq \mathcal{L} \subseteq Z^m$ is said to be *q-ary* lattice. The definition of a *q-ary* lattice is laid here:

Definition 5 Given a modular matrix $\mathbf{X} \in Z_q^{m\times n}$ for $(m < n)$ and $n \approx poly(m)$, two *q-ary* lattices $\Lambda_q^{\perp}$ and Λ_q can be written as:

$$\Lambda_q^{\perp} = \{\mathbf{a} \in Z^n : \mathbf{X}\mathbf{a} = \mathbf{0} \bmod q\} \tag{3}$$

and

$$\Lambda_q = \{\mathbf{a} \in Z^n : \mathbf{a} = \mathbf{X}^{\mathrm{T}}\mathbf{b} \bmod q, \forall\, \mathbf{b} \in Z^m\} \tag{4}$$

In the literature, many lattice-based developments had used *q-ary* lattices and their hard assumptions to design a cryptographic protocol. There are two hard problems on *q-ary* lattices which are used in designing of our PiLike protocol. These problems are expressed in the following manner.

Definition 6 (Small Integer Solution (SIS) problem). Given a modular matrix $\mathbf{X} \in Z_q^{m\times n}$ and $\alpha \in Z^+$, it is hard to obtain a vector $\mathbf{a} \in Z^n \backslash \{0\}$ such that $\mathbf{X}\mathbf{a} = \mathbf{0} \bmod q$ with $\| \mathbf{a} \| \le \alpha$.

Definition 7 (Inhomogeneous Small Integer Solution (ISIS) problem). Given a modular matrix $\mathbf{X} \in Z_q^{m\times n}$, $\alpha \in Z^+$ and a random vector $\mathbf{b} \in Z_q^m$ be a random vector, it is hard to obtain a vector $\mathbf{a} \in Z^n \backslash \{0\}$ such that $\mathbf{X}\mathbf{a} = \mathbf{b} \bmod q$ with $\| \mathbf{a} \| \le \alpha$.

The SIS and ISIS problems are secure against quantum attack according to the authors in [19, 20]. Further, the improvements in [19] by [21] showed that SIS/ISIS problems are at least as hard as SVP, shortest independent vector problem (SIVP), etc.

4 Proposed Mutual Authentication and Key Agreement Protocol

This section describes the proposed protocol in detail. Two entities are mainly involved: (i) the service provider SP, which has capability to initiate the communication (ii) a number of smart meters SM_i. The proposed protocol mainly includes three different phases, namely *Setup, Registration, Login, and Authentication,* and *Key agreement* which are discussed below. For the sake of clarity, the nomenclatures used are mentioned in Table 1.

Table 1 Notations

Notation	Meaning
m	Security parameter
n, q	An integer and a prime number
$\mathbf{X}$	A matrix from $Z_q^{n\times n}$
$\mathbf{x}$	Secret-key of SP
$\mathbf{P}$	Public-key of SP
$H_1(\cdot)$	$Z_q^n \times Z_q^n \times Z_q^n \times \{0,1\}^* \rightarrow \{0,1\}^m$
$H_2(\cdot)$	$Z_q^n \times \times\{0,1\}^* \rightarrow \{0,1\}^m$
a_i	Secret value of SM_i
ID_i	Identity of SM_i
N_i	ith Node
$\mathbf{sk_i}$	Private-key of N_i

4.1 Setup Phase

This phase of the proposed protocol is needed to create a list of system parameters and it is executed by SP. It takes 1^m as input and produces the system parameters in the following manner.

(1) SP picks a prime modulus q and an integers n.
(2) SP picks a matrix $\mathbf{X} \in Z_q^{n\times n}$, where all the operations on matrix are done using modulus q.
(3) SP picks two cryptographic hash functions $H_1 : Z_q^n \times Z_q^n \times Z_q^n \times \{0,1\}^* \rightarrow \{0,1\}^m$ and $H_2 : Z_q^n \times \times\{0,1\}^* \rightarrow \{0,1\}^m$.

4.2 Registration Phase

This phase is executed by the smart meter SP. The steps of this phase can be described as follows:

(1) SP selects a random vector $\mathbf{x} \in Z_q^n$ as its secret key and computes $\mathbf{P} = \mathbf{x}^\mathbf{T} \cdot \mathbf{X}$ as its public key.
(2) SP picks a random integer $a_i \in Z_q$ as a secret value of smart meter SM_i.
(3) SP selects a random string ID_i for the identity of SM_i.
(4) SP stores ID_i and a_i of each SM_i in its database with some other essential information like name, service number, etc.
(5) At last, SP stores $\langle \text{ID}_i, a_i, \mathbf{X}, \mathbf{P}, H_1, H_2\rangle$ in SM_i's memory for $\{i = 1, 2, \ldots, k\}$.

4.3 Login and Authentication

SP initiates this phase for the mutual authentication with the corresponding SM_i. SP and SM_i cooperatively execute the following steps:

(1) SP chooses a vector $\mathbf{y} \in Z_q^n$ randomly and then calculates $\mathbf{A} = \mathbf{X} \cdot \mathbf{y}$ and broadcasts $\langle \mathbf{A}, T_1 \rangle$ to all smart meters. Here, T_1 is current timestamp.
(2) On receiving $\langle \mathbf{A}, T_1 \rangle$, SM_i chooses a random vector $\mathbf{r_i} \in Z_q^n$ and computes $\mathbf{B_i} = \mathbf{X} \cdot \mathbf{r_i}$, $\mathbf{C_i} = \mathbf{r_i^T} \cdot \mathbf{P^T}$, $D_i = H_1(\mathbf{A}||\mathbf{B_i}||\mathbf{C_i}||T_2)$, $E_i = ID_i \oplus D_i$ and $F_i = H_2(\mathbf{C_i}||ID_i||a_i)$. Here, Here, T_2 denotes current timestamp at SM_i.
(3) Now, SM_i sends the tuple $\langle \mathbf{B_i}, E_i, F_i, T_2 \rangle$ to SP.
(4) On receipt of $\langle \mathbf{B_i}, E_i, F_i, T_2 \rangle$, SP checks whether $(T_2 - T_1) < \Delta T$. After successful verification of the stimulated time period, it computes $\mathbf{C_i'} = \mathbf{B_i^T} \cdot \mathbf{x}$.
(5) SP computes $D_i' = H_1(\mathbf{A}||\mathbf{B_i}||\mathbf{C_i'}||T_2)$ and $ID_i' = D_i' \oplus E_i$.
(6) SP then, checks its database for ID_i' and extract corresponding secret value a_i.
(7) Now, SP computes $F_i' = H_2(\mathbf{C_i'}||ID_i'||a_i)$. If $F_i' = F_i$, SP successfully authenticates SM_i else aborts the process.
(8) SP calculates $G_i = H_2(\mathbf{C_i'}||ID_i'||T_3)$ for the current timestamp T_3 and sends $\langle G_i, T_3 \rangle$ to SM_i as a response.
(9) On receiving $\langle G_i, T_3 \rangle$, SM_i checks the validity of time period as $(T_3 - T_2) < \Delta T$. If it is true, SM_i calculates $G_i' = H_2(\mathbf{C_i}||ID_i||T_3)$ and checks whether $G_i' = G_i$. If it is true, SM_i authenticates SP as a legal service provider else aborts the communication.

4.4 Key Agreement Phase

Once the service provider SP and smart meter SM_i are mutually authenticated, both of them construct a session key for their further communication. In the protocol development for mutual authentication, the secret parameter D_i combines the random secrets $\mathbf{y}$ and $\mathbf{r_i}$ of SP and SM_i, respectively. Thus, D_i must be involved in the development of session key. Another secret parameter $\mathbf{C_i}$ is also known to both SP and SM_i. Thus, SP and SM_i negotiate a common session key as $K = H_1(\mathbf{C_i}||D_i||T_1)$

5 Security Analysis

This section considers the security analysis of the proposed protocol. The different security requirements and attacks on the proposed protocol are laid here.

1. *Anonymity*: The service provider SP has stored the identity ID_i of a particular smart meter SM_i in its database securely. Thus, an adversary $\mathbb{A}$ can not fetch ID_i either from database of SP or from memory of SM_i. Further, if $\mathbb{A}$ intercepts the

login message $\langle \mathbf{B_i}, E_i, F_i, T_2 \rangle$ and tries to fetch ID_i form it, he/she fails since ID_i is included in E_i as $E_i = \text{ID}_i \oplus D_i$ and D_i is unknown to $\mathbb{A}$. Therefore, the proposed protocol satisfies anonymity property.

2. *Untraceability*: To avoid the traceability of smart meters, the proposed protocol uses random ephemeral secret keys. The login message $\langle \mathbf{B_i}, E_i, F_i, T_2 \rangle$ is changed in every transmission as the random secrets $\mathbf{y}$ and $\mathbf{r_i}$ changes. After intercepting a number of login messages from SP intended to different smart meters, the adversary $\mathbb{A}$ will be unable to differentiate the login messages because of the random secrets. Thus, $\mathbb{A}$ cannot trace the login messages intended for a particular smart meter SM_i. Thus, the proposed exhibits untraceability requirement.
3. *Replay attack* In the proposed protocol, the timestamps and randomized values are used to prevent the replay attack. In the broadcast message $\langle \mathbf{A}, T_1 \rangle$ and login message $\langle \mathbf{B_i}, E_i, F_i, T_2 \rangle$, timestamps T_1 and T_2 are used. At verification end, the validity of timestamp is checked to prevent the replay attacks. Similarly, smart meters also include a time stamp T_3 which prevent the replay attack. Further, these timestamps are also inserted in the computation of different entities as $D_i = H_1(\mathbf{A}||\mathbf{B_i}||\mathbf{C_i}||T_2)$ and $G_i = H_2(\mathbf{C_i}||\text{ID}_i||T_3)$ to check the validity of timestamps. Thus, the receiver of any messages checks the freshness of timestamp and then verifies the validity of that timestamp in the present message. Hence, the proposed protocol denies the possibilities of replay attack.
4. *Session key security:* In this attack, an adversary $\mathcal{A}$ tries to find the session key K negotiated between SP and SM_i. $\mathcal{A}$ could try it either by attacking the messages exchanged during the login and authentication phase or could use the public parameter. In our proposed protocol, the session key is generated as $K = H_1(\mathbf{C_i}||D_i||T_1)$, which contains $\mathbf{C_i}$ and D_i as private parameters and T_1 is used for freshness of the session key. Thus, $\mathcal{A}$ is unable to compute the correct session key.
5. *Impersonation attack:* Let a registered smart meter SM_i somehow gets the identity ID_j of another registered smart meter SM_j and tries to impersonate SM_j. In this case, SM_i after getting $\langle \mathbf{A}, T_1 \rangle$ from SP, chooses a random vector $\mathbf{r_i} \in Z_q^n$ and computes $\mathbf{B_i} = \mathbf{X} \cdot \mathbf{r_i}$, $\mathbf{C_i} = \mathbf{r_i^T} \cdot \mathbf{P^T}$, $D_i = H_1(\mathbf{A}||\mathbf{B_i}||\mathbf{C_i}||T_2)$, $E_i = \text{ID}_k \oplus D_i$ and $F_i = H_2(\mathbf{C_i}||\text{ID}_i||a_i)$. SM_i then sends $\langle \mathbf{B_i}, E_i, F_i, T_2 \rangle$ to SP for authentication. On the other hand, SP checks freshness of timestamp and computes $\mathbf{C_i'} = \mathbf{B_i^T} \cdot \mathbf{x}$ and $D_i' = H_1(\mathbf{A}||\mathbf{B_i}||\mathbf{C_i'}||T_2)$. SP extract $\text{ID}_k' = D_i' \oplus E_i$ and finds $F_i' = H_2(\mathbf{C_i'}||\text{ID}_k'||a_i)$. It can be easily seen that $F_i' \neq F_i$ because SP uses a_i instead of a_k. Thus, the proposed provides security against impersonation attack.
6. *Man-in-the-Middle Attack:* Let the adversary $\mathbb{A}$ intercepts the tuple $\langle \mathbf{A}, T_1 \rangle$ and a login tuple $\langle \mathbf{B_i}, E_i, F_i, T_2 \rangle$. $\mathbb{A}$ now, tries to generate a new login tuple. It can be noted that the generation of new login tuple requires the secret parameters $\mathbf{r_i}$, ID_i and a_i of SM_i. Similarly, the tuple $\langle G_i, T_3 \rangle$ cannot be constructed by $\mathbb{A}$ because $\mathbb{A}$ does not know the secret parameters. Hence, the proposed scheme resists the man-in-the-middle attack.

7. *Quantum Attack:* The proposed protocol uses SIS/ISIS lattice problems to design its security. Thus, the proposed protocol is secure in a post-quantum environment, i.e., the scheme resists quantum attacks.

6 Conclusion

The recent technologies enhance the conventional power grids into smart grid. In this paper, we have proposed a novel lattice-based mutual authentication and session key negotiation protocol among service provider and smart meters in SG. The use of lattice-based hard problems ensures the withstand of the proposed scheme against existing attacks and futuristic quantum attacks as well. The proposed scheme is simple and efficient to be implemented in SG as it only uses lightweight matrix operations. The security analysis has been also presented which proves the security of the proposed protocol.

References

1. Fang, X., Misra, S., Xue, G., Yang, D.: Smart grid-the new and improved power grid: a survey. IEEE Commun. Surv. Tutor. **14**(4), 944–980 (2011)
2. Farhangi, H.: The path of the smart grid. IEEE Power Energy Mag. **8**(1), 18–28 (2009)
3. Mahmood, K., Chaudhry, S.A., Naqvi, H., Kumari, S., Li, X., Sangaiah, A.K.: An elliptic curve cryptography based lightweight authentication scheme for smart grid communication. Fut. Gener. Comput. Syst. **81**, 557–565 (2018)
4. Tsai, J.L., Lo, N.W.: Secure anonymous key distribution scheme for smart grid. IEEE Trans. Smart Grid **7**(2), 906–914 (2015)
5. Gupta, D.S., Biswas, G.: On securing bi and tri-partite session key agreement protocol using ibe framework. Wirel. Person. Commun. **96**(3), 4505–4524 (2017)
6. Gupta, D.S., Islam, S.H., Obaidat, M.S., Hsiao, K.F.: A novel identity-based deniable authentication protocol using bilinear pairings for mobile ad hoc networks. Adhoc Sens. Wirel. Netw. **47** (2020a)
7. Gupta, D.S., Islam, S.H., Obaidat, M.S., Vijayakumar, P., Kumar, N., Park, Y.: A provably secure and lightweight identity-based two-party authenticated key agreement protocol for IIOT environments. IEEE Syst. J. (2020c)
8. Nicanfar, H., Jokar, P., Beznosov, K., Leung, V.C.: Efficient authentication and key management mechanisms for smart grid communications. IEEE Syst. J. **8**(2), 629–640 (2013)
9. Shor, P.W.: Polynomial-time algorithms for prime factorization and discrete logarithms on a quantum computer. SIAM Rev. **41**(2), 303–332 (1999)
10. Gupta, D.S., Biswas, G.: Cryptanalysis of Wang et al'.s lattice-based key exchange protocol. Perspect. Sci. **8**, 228–230 (2016)
11. Gupta, D.S., Islam, S.H., Obaidat, M.S., Karati, A., Sadoun, B.: Laac: lightweight lattice-based authentication and access control protocol for e-health systems in iot environments. IEEE Syst. J. (2020b)
12. Fouda, M.M., Fadlullah, Z.M., Kato, N., Lu, R., Shen, X.S.: A lightweight message authentication scheme for smart grid communications. IEEE Trans. Smart Grid **2**(4), 675–685 (2011)
13. Diffie, W., Hellman, M.: New directions in cryptography. IEEE Trans. Inf. Theor. **22**(6), 644–654 (1976)

14. Garg, S., Kaur, K., Kaddoum, G., Rodrigues, J.J., Guizani, M.: Secure and lightweight authentication scheme for smart metering infrastructure in smart grid. IEEE Trans. Indus. Inf. **16**(5), 3548–3557 (2019)
15. Zhang, L., Zhao, L., Yin, S., Chi, C.H., Liu, R., Zhang, Y.: A lightweight authentication scheme with privacy protection for smart grid communications. Fut. Gener. Comput. Syst. **100**, 770–778 (2019)
16. Kumar, N., Aujla, G.S., Das, A.K., Conti, M.: Eccauth: a secure authentication protocol for demand response management in a smart grid system. IEEE Trans. Indus. Inf. **15**(12), 6572–6582 (2019)
17. Gupta, D.S., Biswas, G.: A novel and efficient lattice-based authenticated key exchange protocol in c-k model. Int. J. Commun. Syst. **31**(3), e3473 (2018)
18. Gupta, D.S., Biswas, G.: Design of lattice-based elgamal encryption and signature schemes using sis problem. Trans. Emerg. Telecommun. Technol. **29**(6), e3255 (2018)
19. Ajtai, M.: Generating hard instances of lattice problems. In: Proceedings of the twenty-eighth annual ACM symposium on Theory of computing, pp 99–108 (1996)
20. Mukherjee, S., Gupta, D.S., Biswas, G.: An efficient and batch verifiable conditional privacy-preserving authentication scheme for vanets using lattice. Computing **101**(12), 1763–1788 (2019)
21. Micciancio, D., Regev, O.: Worst-case to average-case reductions based on gaussian measures. SIAM J. Comput. **37**(1), 267–302 (2007)

Review on High-Performance Concrete by Replacing Conventional Cement with Fly Ash and Silica Fume

Suvrajit Banerjee, Baibaswata Das, Debargha Banerjee, Debasish Das, and Uddipta Ghosh

Abstract High-performance concrete has become a necessary topic in nowadays construction engineering. Now, how the proper amalgamation of fly ash, silica fume can make an effect in the performance of the concrete is the matter of discussion of this study. High-performance concrete at a lower cost by replacing conventional cement is important. Now for improving basic properties of concrete high-performance concrete (HPC) with a lesser W/C ratio and utilization of superplasticizers in conjunction with mineral admixtures are necessary. Cement is substituted by mineral admixture such as fly ash (FA) at several percentages and silica fume (SF) is added in a proportion of 10% of the total weight of cement. It is also noted that the amount of workability and strength required is achieved at optimum percentage. Furthermore, analysis of important parameters such as durability, consistency, strength in terms of various strength values such as—design compressive strength, tensile strength (split tensile), and also flexural strength for a higher grade of the concrete mix such as M80 has also been discussed in this study.

Keywords High-performance concrete · Mineral admixture · Fly ash (FA) · Silica fume (SF) · Workability · Durability · Compressive strength · Tensile strength (split tensile) · Flexural strength

1 Introduction

Undoubtedly, one of the most prime materials for construction worldwide is concrete. It is the need of modern days to design a high-strength concrete, which can be used for

---

S. Banerjee · B. Das (✉)
Civil Engineering Department, Sur Institute of Technology (SurTech), Surermath, 540, Dumdum, Kolkata 700074, India
e-mail: baiba.tanan@dsec.ac.in

D. Banerjee · D. Das · U. Ghosh
Department of Civil Engineering, JIS College of Engineering, Barrackpore-Kalyani Expy, Phase III, Block A, Kalyani, West Bengal 741235, India
e-mail: debasish.das@jiscollege.ac.in

K. N. Das et al. (eds.), *Proceedings of the International Conference on Computational Intelligence and Sustainable Technologies*, Algorithms for Intelligent Systems,
https://doi.org/10.1007/978-981-16-6893-7_23

several purposes in multi-weather conditions. In different weather, conditions where conventional concrete faces issues related to durability and strength for dealing with this kind of problem appropriate use of admixtures is required to adjust the properties of conventional concrete to make it more productive for multi-weather situations.

High-performance concrete: In current years, the term "high-performance concrete" has been set in motion in the construction industry. Concrete is generally termed as high-performance concrete when certain parameters such as high workability, high density, high strength, high durability, high-elastic modulus, etc., are satisfied. HPC should also exhibit lower permeability and more resistance to the chemical attacks as compared to normal concrete. Certain mineral admixtures like fly ash, gram granulated blast furnace slag (GGBFS), metakaolin, silica fume, in one hand, and certain chemical admixtures such as plasticizers and superplasticizers on the other hand an accurate proportion is required along with the usual ingredients for the production of high strength concrete. Lower water-cement ratio, the use of some admixtures like fly ash and silica fume and, super-plasticizer in proper proportion differentiate the high strength concrete from convention concrete. High-performance concrete has several edges over conventional concrete properties like high strength, high durability, low permeability, and compaction without the tendency of segregation, comparatively higher mechanical properties, higher toughness, and ease of placing, resistance to impact, and much lesser effects of microcracking. In modern construction technology, effective use of admixtures will lead to a curtailment in cement content which in turn reduces the rate of CO_2 emission and decrease the environmental pollution.

2 Experimental Study

Ingredients: OPC of 53 grade adhering to IS: 12269 also zone II sand according to IS: 383. Size of coarse aggregates of 12.5 mm and 20 mm are selected for this particular study. 1710 kg/m^2 is figured as the bulk density of coarse aggregate. Producing silicon metal or ferrosilicon allows silica fume comes as a byproduct. At the early stages of concrete silica fume plays an important role in gaining strength. From coal-fired power plants as principal byproduct fly ash is found. Mineral admixtures such as fly ash, gram granulated blast furnace slag (GGBFS), and silica fume are industrial waste materials and sustainable to some extent. BFSA chemicals are the main source of water-reducing admixture Glenium B233. Hydrochric acid (HCL) and sulphuric acid (H_2SO_4) are the main ingredients to test durability (Table 1).

3 Experimental Process

From ACI 211.1-91 (reapproved 2009), the guidelines for mix design are taken (Table 2).

Table 1 Factors taken for the design mix

Ingredients	Relative density	Fineness modulus
Cement	3.135	283.94(m^2/kg)
Fine aggregate	2.528	2.61
Coarse aggregate	2.74	7.95
Silica fume	2.35	
Fly ash	2.05	

Table 2 M80 grade concrete for mix proportion

Specification	Mix ratio	Water-cement ratio
M80	1:0.8:1.66	0.33

Table 3 Mix proportions for concrete

Name	Description	Cement weight (kg)	Coarse aggregate weight (kg)	Fine aggregate weight (kg)	Silica fume weight (kg)	Fly ash weight (kg)
CM	Control mix	600.2	1059.99	502.3	60	0
HPC-1	SF 10.3% + FA 4.9%	569.99	1059.99	502.3	58.7	28
HPC-2	SF10.3% + FA 10.3%	539.8	1059.99	502.3	56	56
HPC-3	SF10.3% + FA 15.3%	510.3	1059.99	502.3	52.56	78
HPC-4	SF10.3% + FA 20.3%	480.1	1059.99	502.3	49.45	97.46
HPC-5	SF10.3% + FA 25.3%	449.9	1059.99	502.3	46.34	114
HPC-6	SF10.3% + FA 30.3%	420.2	1059.99	502.3	43.28	127.32

The addition of silica fume by the weight of 10% of cement is typical for all experimental mixes. Cement substitution with fly ash can vary up to thirty percent concerning the weight of the cement (Table 3).

3.1 Experimental Procedure

The cubes of measurement 150 mm × 150 mm × 150 mm were cast to know the compressive strength after 7 days and 28 days, respectively. The beams of measurement 500 mm × 100 mm × 100 mm were cast to know the flexural strength after 7 and 28 days. For finding the tensile strength (split tensile) after 7 and 28 days,

Table 4 Compressive strength test data

Name	Compressive strength (N/mm^2)		Slump (mm)
	M80 grade		
	7 days	28 days	28 days
CM	54.44	80.1	71
HPC 1	56.33	82.12	70
HPC 2	60.32	85.03	67
HPC 3	62.57	91.2	67
HPC 4	57.63	87.3	66
HPC 5	52.52	83.1	62
HPC 6	52.1	78.9	64

cylindrical specimens having a measurement of 300 mm × 150 mm were cast. Specimens were cast and then removed from the mounds after a difference of 1 day for every examination. 7 and 28 days strength test by water curing of the specimen is done. Also after the removal of specimens from water curing, they were allowed to dry under shade, and subsequently tested for various factors such as—compressive strength, split tensile strength, and flexural strength, respectively. The cubes of measurement 100 mm × 100 mm × 100 mm were cast to know the concrete durability. After completion of 28 days water curing period, the specimens were also tested for 60 days durability test by immersing the specimens in 5% H_2SO_4 and 5% concentrated HCL, respectively. Weight loss occurred due to acid immersion, and finally, the compressive strength was checked after acid immersion.

4 Results and Analysis

4.1 Result of the Compressive Strength Test

In this test, silica fume inclusion by weight of cement is 10% is done. Substitution of fly ash by weight of cement is 15% gives required strengths (Table 4 and Fig. 1).

4.2 Tensile Strength (Split Tensile)

See Table 5 and Fig. 2.

Fig. 1 Compressive strength for SF and FA

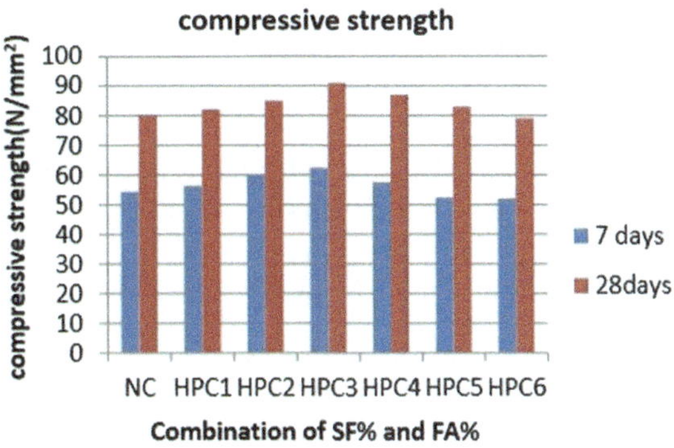

Table 5 Tensile strength (split-tensile) test data

Name	Tensile strength (split tensile) (N/mm^2)	
	M 80 grade	
	7 days	28 days
CM	2.4	2.99
HPC1	2.7	3.55
HPC2	2.9	3.8
HPC3	3.2	4.23
HPC4	2.7	4.03
HPC5	2.6	3.69

Fig. 2 Tensile strength (split tensile) strength

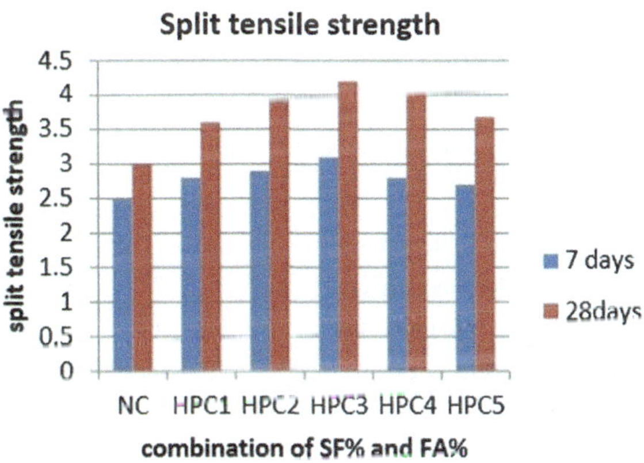

4.3 *Flexural Strength*

See Table 6 and Fig. 3.

Table 6 Flexural strength test data

Name	Flexural strength (N/mm^2)	
	M 80 grade	
	7 days	28 days
CM	4.66	6.1
HPC1	5.3	6.7
HPC2	6.2	7.8
HPC3	6.65	8.6
HPC4	6.09	7.4
HPC5	5.23	6.5

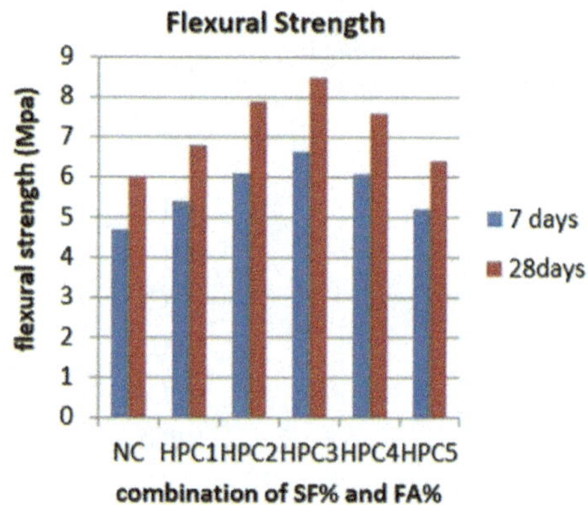

Fig. 3 Flexural strength

4.4 *Durability Test*

See Tables 7, 8 and Figs. 4, 5.

Table 7 When immersed in HCL

Name	Specimen weight (N/mm^2)		Compressive strength (N/mm^2)	
	Before immersing	60 days later	Before immersing	60 days later
HPC1	26.2	2.47	78	72
HPC2	2.76	2.55	84	78
HPC3	2.74	2.56	87	82
HPC4	2.69	2.42	84.99	77
HPC5	2.72	2.53	83	74.99
HPC6	2.74	2.48	79.99	73

Table 8 Durability test when immersed in H_2SO_4

Mix name	Specimen weight (N/mm^2)		Compressive strength (N/mm^2)	
	Before immersing	60 days later	Before immersing	60 days later
HPC1	2.67	2.52	82.1	76.2
HPC2	2.68	2.55	82.9	78.1
HPC3	2.72	2.62	87	82.1
HPC4	2.74	2.64	80.1	76.2
HPC5	2.66	2.51	76.9	72.3
HPC6	2.73	2.64	72.2	67.2

Fig. 4 Test on compressive strength (before and after immersing in HCL)

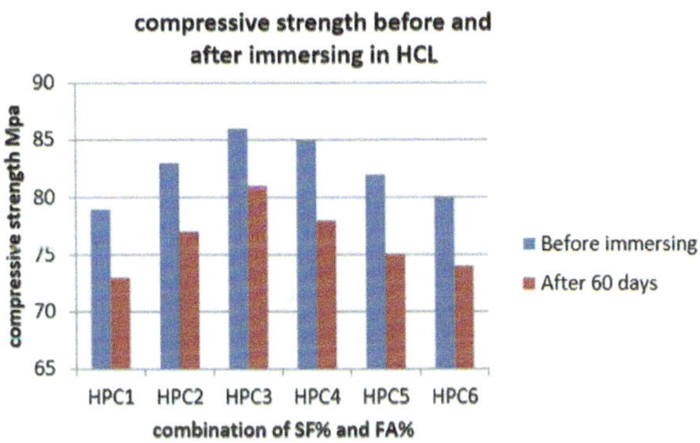

Fig. 5 Test on compressive strength (before and after immersing in H_2SO_4)

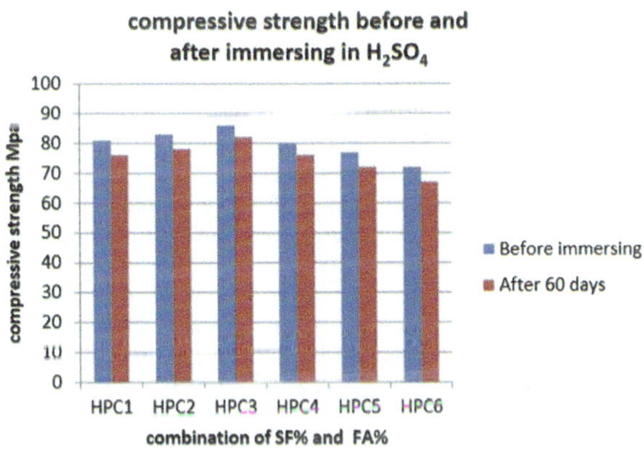

5 Conclusions

Cementitious materials are considered an essential product for making high-performance concrete. Fly ash, granulated slag, silica fume, metakaolin, and different types of mineral admixtures are nowadays used to enrich various concrete properties. This paper has shown us that the utilization of mineral admixtures such as silica fume and fly ash has improved the various properties of the test sample. Properties like design compressive strength (28 days), tensile Strength (split tensile), flexural strength as well as durability in the presence of H_2SO_4 and HCL have improved a lot. For example, 11% increment in compressive strength is observed after involving fly ash as 15% replacement of cement and silica fume as 10% replacement of cement. Also, it has been observed that 8.5 N/mm^2 is the measured flexural strength value after the period of 28 days and 4.2 N/mm^2 is the measured tensile strength value after the period of 28 days. The substitution of cement with mineral admixture has improved the durability criteria. For example, the rate of weight loss happened due to HCL and H_2SO_4 solutions for the combination of silica fume 10% as the addition of cement and 15% fly ash as substitution of cement is observed as minimum compare to all trails.

References

1. Faouzi, A.Z.I.M., Belachia: Supplementary cementitious materials. High-performance concrete in Algeria for a more economical and more durable concrete (state of art report) (2007)
2. Magudeaswaran, P., Eswaramoorthi, P.: Experimental investigation of mechanical properties on silica fume and fly ash as partial cement replacement of high-performance concrete. IOSR J. Mech. Civil Eng. (IOSR-JMCE) **6**(4), 57–63 (2013)
3. Hariharan, A.R., Santhi, A.S., Mohan Ganesh, G.: Study on strength development of high strength concrete containing fly ash and silica fume. Int. J. Eng. Sci. Technol. **3**, 2955–2961 (2011)
4. Duval, R.: Influence of silica fume on the workability and the compressive strength of high-performance concretes. Cem. Concr. Res. **28**, 533–547 (1998)
5. Chinnaraju, K., Subramanian, K., Senthil Kumar, S.R.R.: Role of fly ash and silica fume on durability characteristics of high-performance concrete. Int. J. Adv. Civil Eng. 17–27
6. Perumal, K., Sundararajan, R.: Effect of partial replacement of cement with silica fume on the strength and durability characteristics of high-performance concrete. In: Our World in Concrete and Structures, pp. 397–404
7. Pitroda, J., Zala, L.B., Umrigae, F.S.: Experimental investigations on partial replacement of cement with fly ash in design mix concrete. Int. J. Adv. Eng. Technol. 126–129 (2012)
8. Kiran, T.G.S., Ratnam, M.K.M.V.: Fly ash as a partial replacement of cement in concrete and durability study of fly ash in acidic (H_2SO_4) environment. Int. J. Eng. Res. Dev. **10**(12), 01–13 (2014)
9. Sironiya, S., Jamie, S., Verma, M.P.: Experimental investigation on fly ash and glass powder as partial replacement of cement for M-25 grade concrete. Int. J. Sci. Adv. Res. Technol. **3**(5), 322–324 (2017)
10. Shabab, M.E., Shahzada, K., Gencturk, B.: Synergistic effect of fly ash and bentonite as partial replacement of cement in mass concrete. KSCE J. Civil Eng. 1–9 (2015)

Study of Various Methods and Technologies Used in Wastewater Treatment to Control Water Pollution

Debargha Banerjee, Debasish Das, Uddipta Ghosh, Suvrajit Banerjee, and Baibaswata Das

Abstract Due to industrial or factory disposal as well as domestic effluents, this become more and more dangerous day by day. Hence, various technologies or techniques are used for better purification of water. Pollution in water is mainly caused due to organic, inorganic and biological contaminates amongst these pesticides are very usual and occurred due to agricultural source, which creates an environmental problem. Several methods of wastewater treatment exist as activated carbon adsorption, chemical oxidation, biological treatment, etc. This activated charcoal involves phases of pollutants without any decomposition into pollution problem, whereas chemical oxidation mineralizes all substances and it is economically suitable for removal of all particles from the water and makes it pure again. It has been observed that wastewater pollutants from various industries, factories and households to be infused in surface water and underground wells. In order to achieve potable safe water or to safeguard the existing water resources, these water contaminating pollutants need to be removed. Over the past few years, numerous methods are put forward to remove the unwanted substances from water bodies which are detrimental in nature.

Keywords Biological contaminants · Activated carbon adsorption · Agricultural source · Chemical oxidation

D. Banerjee (✉) · D. Das · U. Ghosh
Department of Civil Engineering, JISCE, Kalyani 741235, India

D. Das
e-mail: debasish.das@jiscollege.ac.in

S. Banerjee · B. Das
Department of Civil Engineering, Dr. Sudhir Chandra Sur Institute of Technology and Sports Complex, 540 Dumdum Road, Kolkata 700074, India
e-mail: baiba.tanan@dsec.ac.inx

K. N. Das et al. (eds.), *Proceedings of the International Conference on Computational Intelligence and Sustainable Technologies*, Algorithms for Intelligent Systems,
https://doi.org/10.1007/978-981-16-6893-7_24

1 Introduction

Water is one of the most vital and crucial natural resources of the earth which is required for sustainability of all forms of life on the surface of the earth. The quality of life and adaptation of living beings which include both humans and wildlife are greatly influenced by the quality, quantity and availability of fresh water. Water use is basically categorized as commercial, domestic, industrial, irrigation, mining and public supply water.

For the successful establishment and maintenance of varied human activities, a constant supply of fresh and clean water is very essential. However, human settlements and the industrial activities are producing liquid and solid waste in huge quantity and thereby polluting the water sources all over the world. Due to the massive rise in pollution throughout the world, water is going to be one of the scariest natural resources by the end of twenty-first century. With the advancing modern civilization, rapidly growing technology and drastically developing economy, it is creating a massive treat of water pollution and scarcity through its own activities. There is going to be worldwide impact from the problem of fresh water deficiency [1].

2 Literature Review

As we all know that the population on the earth is increasing day by day, so the daily needs are also increasing. Water is one of the most important needs of everyone's life. The earth is covered by 96.5% of water, still we are suffering for water because only 0.7% of water on the earth is useable and treated as fresh water. In those stages, various units and processes are there such as physical units, chemical units and biological units. Under these, there are various processes are there such as screening, sedimentation, flocculation, aeration, disinfection, etc [2].

3 Water Deficiency

Water and sustainable development are interconnected with each other. In the absence of fresh and salt water, there is no development between social and economic development. Increasing population day by day causes bacteria, sewage which contaminated water adversely.

3.1 Some of the Major Reasons for Water Deficiency

- There is an excessive use of water in industries, agriculture where it causes a major shortage of freshwater. Due to excessive use in agriculture, pollution occurs for run off of pollutants like fertilizers and pesticides into lakes, rivers, streams [3].
- Climatic change due to deforestation is one of the main reasons for scarcity of freshwater all over the world.

3.2 Effects of Water Deficiency

- World Health Organization director general said that all over the world scarcity of water effect approx. four people amongst ten people.
- Scarcity of water directly effects the life of people.
- Shortage of water affects farmer to produce product which effects earning. Water scarcity causes many diseases, unhygienic issues.
- Economic development increases demand of water in industries, which causes shortage of freshwater for people.

3.3 Solutions to Prevent Water Deficiency

- After using water, close the tap properly. Examples: Whilst brushing the teeth, close the tap.
- Wastewater treatment plant, sewage treatment plant and industrial treatment plant are required to protect water from wasting and control pollution [4].

4 Characteristics of Water

4.1 Physical Characteristics

Temperature—It had a hard effect on some of the properties and characteristics of water like density, thermal capacity, salinity, specific weight, etc. The temperature increased with increasing rates of chemical and biological reactions [5].

Colour—This property has an impact on our life. Colour in water identifies the good or bad quality in water. It gives the identification of being unfit to drink even if it is safe or for public use.

Turbidity—It is a measure of the light-transmitting properties of water and is comprised of suspended and colloidal material. It is important for health issues and aesthetic reasons. Generally, water is affected by human activities, decaying plant matter, plant nutrients, suspended sediments, etc. It provides an estimate of total suspended solids (TSS).

Taste and Odour—Water is always tasteless and odourness. If any kind of smell is found, then it is considered as water pollution. Water pollution may cause huge effects to human life. It may lead to great damage to animals as well. Water quality is based on this property.

4.2 Chemical Characteristics

PH—Acidic and basic property which are present in water is identified by PH of water. It is defined as the negative log of the hydrogen in concentration. PH scale varies from 0 (acidic) to 14 (alkaline). PH of freshwater is 4.5. Water becomes acidic due to the presence of dissolved acids and CO_2. The permissible limit of PH is 6.5–8.5 (for drinkable water) [5].

Hardness—Hardness is defined as the concentration of multivalent metallic cation in solution. The hardness of water often occurs due to the presence of a salt of calcium and magnesium. Temporary hardness is occurred due to carbonate and bicarbonate of calcium and magnesium, whilst permanent hardness occurs due to chloride and sulphate of calcium and magnesium. The permissible value of total hardness is 75–200 ppm [6].

Alkalinity—Alkalinity is defined as the number of ions in water that will react with neutralizing hydrogen ions (H^+). It is a measure of the ability of water to neutralize acids. It comes in water due to minerals or it may be produced due to atmospheric CO_2 mixed in water or due to microbial decomposition of organic matter.

BOD—Biological oxygen demand indicates the amount of O_2 required by the living organism for their oxidation of biodegradable organic matter which is present in water under aerobic conditions. If organic matter is very high, then loss of dissolved O_2 occurs which causes an anaerobic environment in water. If the BOD value is higher, than it causes water pollution.

COD—Chemical oxygen demand is required for the oxidation of organic matter presents in water by a strong chemical oxidizing agent like K_2CrO_7. It is usually measured to determine the presence of the amount of organic matter in water that contains toxic or antimicrobial chemicals.

4.3 Biological Characteristics

Bacteriological aspects of water pollution: Human beings and other animals discharge a large number of intestinal bacteria into stool and urine. That is why bacteria presents in drinking water when the water source is contaminated with stool.

Algae: Growth of algae may affect the quality of water as well may cause water diseases. It provides colour odour characteristics to water. It produces toxic components. **Coliforms**: It is a harmless aerobic lactose fermenters organism.

5 Causes of Water Pollution

There are many types of water pollutants, which include contamination due to chemical wastes, volatile organic compounds (VOCs), insecticides and herbicides, domestic wastes, food processing wastes, etc.

Industrial effluents and sewages into the water bodies cause different organic and inorganic pollutants. Some of inorganic materials are arsenic, boron, barium, chloride, calcium, copper, cobalt, lead, iron, potassium, sodium, silver, tin, zinc and many more.

5.1 Major Types of Water Pollutants

- **Industrial wastages**: Industries produce a large amount of wastes including toxic chemicals. This contains lead, mercury, sulphur, nitrates, etc. We can see some of the industries do not have proper waste management, which is effecting the rivers, lakes. The chemicals may increase the amount of minerals present in the water and can change the temperature of the water.
- **Waste sewages**: Sewages are mainly produced from households and discharged into the river or lake water. Then, this water will carry pathogens, which are a harmful pollutant, and also carries some bacteria and cause some of health issues and diseases [7].
- **Household wastes**: The wastes produced from houses like as rubber, food, glass, plastic, paper, etc., can be mixed into the marine waters through canal water. And it is very difficult to decompose those wastes; it can take 200 years to decompose all of these wastes. But these wastes not only polluted the water but also harm animal under the water.
- **Oil leakage**: When a large quantity of oil leaks from different sources, it can be a threat to marine life which will never dissolve. If the leakage of oil is high, then the toxicity level of pollutants also very dangerous.

- **Burning of fossil fuels**: Different types of fossil fuels like as oil and coal after burning the ashes can mix in the atmosphere. This particles can be harmful after mixing with rain water which is called as acid rain. Also burning of the coal discharge carbon dioxide which causes global warming.
- **Radioactive wastes**: Nuclear fusions can produce nuclear energy which contains a toxic chemical uranium. The nuclear is very serious pollutants if they are not disposed properly to prevent any type of nuclear accident. We have some examples of nuclear accident in Russia and Japan [8].

6 Steps of Wastewater Treatment

- **Preliminary treatment**—It is the very first procedure of wastewater treatment, where the wastewater enters to the treatment plant through a bar screen in which the larger waste particles such as plastics, broken branches, papers, etc., are removed. After that the water enters to the grit tank, where the small solid particles are removed with the help of physical procedures.
- **Primary treatment**—In primary treatment, it involves the physical separation of solids from the water such as the suspended solids, floated solids (oil particles) and settleable solids. The wastewater allows to flow from the primary settling tank so that the particles can be removed after several hours, where the solid particles are settled down in the bottom of the tank and the oil particles float in the top of the surface.
- **Secondary treatment**—After primary treatment, secondary treatment is done, it involves the biological processes to break the chain of organic matters present in water such as human waste, kitchen waste and dead animal waste. By this process, microorganisms are converted in nonsettleable solid to settleable solid.
- **Tertiary treatment**—After primary treatment and secondary treatment, tertiary treatment is done, it involves the chemical process in which disinfecting compounds are added to water such as chlorine and alum. It is done to remove the harmful organisms which cannot be removed in biological process and to neutralize the hazardous chemicals present in the water [9].

6.1 Units and Their Operation of Wastewater Treatment Plants

6.1.1 Physical Units

- **Screening**—It is the first unit operation used in a wastewater treatment plants. Screening is used to remove the papers, plastics, metals, etc. So that it should not affect and clog in the equipment in downstream. Nowadays, in modern wastewater treatment, plants both coarse screens and fine screens are used.

- **Comminution**—It is the process in which the solids flowing along the water is cut up, to reduce their size and to increase their surface area. So that it should not affect the subsequent treatment plant equipment. In modern wastewater treatment, plants pressurized comminutor is used so that the required fineness can be achieved.
- **Flow equalization**—In this process, the hydraulic velocity of flow of water is controlled so that it can flow in a laminar way to the downstream. Equalization of flow retains the high volume of incoming flow and the fluctuations of flow, so that the aeration and the mixing which are required can be done properly.
- **Sedimentation**—It is the process in which the suspended dirt particles present in water allows to settle down at the bottom of a tank. In this process, no equipment is required, the whole procedure is done by the means of gravity. It helps to pure the water at some degree.
- **Flotation**—Flotation is a separation process in which gas bubbles are introduced in water as a transport medium so that the suspended particles such as fats, oil, grease and industrial wastes can be removed. In this process, the suspended particles attach to the bubbles and come to the surface of the water and then removed. This process is also known as dissolved air flotation.
- **Flocculation**—It is also a separation process where fine solid particles create floc, when chemical agents like alum, aluminium sulphate, etc., are added to water then the solid particles bind together and form heavy floc. Then, it floats on the surface or settled down to the bottom so that they can be easily removed.
- **Granular-medium filtration**—In this process, water is allowed to flow through a filtration chamber in which the remaining suspended or the colloidal solids are removed. The filtration chambers are generally made up of sands and gravels which collect the impurities present in water.

6.1.2 Biological Units

- **Activated sludge process**—This process is done for both the municipal and industrial wastewater. In this process, highly concentrated microorganisms are used to remove the nutrients by increasing the quality of water.
- **Aerated lagoon**—In this process, wastewater is treated by using physical, biological as well as chemical process. Lagoon is a pond like body which can hold wastewater and treat it for a certain period of time. Almost the whole process occurs naturally, but sometimes aeration devices are used for addition of oxygen in the wastewater so that the microbial action reduces the pollutants present in water. Aeration is done for making the treatment more efficient, so that less area is required.
- **Trickling filter**—It consists of rocks, coke, gravel, slag, polyurethane foam, sphagnum peat moss and plastic media through which the water flows downward. Trickling filter is an aerobic system which uses microorganisms as a medium. The organic matters attach to the microorganisms help to remove the organic matters and the pathogens from the wastewater.

- **Stabilization ponds**—It is a man-made large water body in which the colloid water is allowed to be treat naturally by the influence of sunlight, wind, algae and microorganisms. It also helps to remove the organic matters and the pathogens from the wastewater.
- **Anaerobic digestion**—It is the process in which microorganisms or the organic matters are breakdown into bio degradable materials such as animal manure, food waste and biogas. The whole process occurs in the absence of oxygen. Generally, anaerobic digestion can be used as wastewater sludge treatment, agriculture manure, etc. One of its end product is biogas, which is combusted for generating electricity and heat or can be used as renewal gas.
- **Biological nutrient removal**—In this process, the total amount of nitrogen and phosphorus are removed by the help of microorganisms at different environmental conditions. It is done before the discharge of water to the surface or ground water.

6.1.3 Chemical Units

- **Chemical precipitation**—It is the process in which dissolved metals present in wastewater changes into solid or insoluble particles. Chemical precipitation is used to remove the toxic ionic constituents from wastewater with the help of some counter ions.
- **Adsorption**—It is the process in which water is made pure by removing its objectionable taste and odour, including chlorine. Adsorption is done when molecules present in water binds themselves with the surface of the solid particles. Typically, carbon or activated carbon is used to remove toxic chemicals and gases, sometimes microorganisms also.
- **Disinfection**—The pathogenic microorganisms which are harmful to health are removed in this process. If the microorganisms are not removed from the drinking water, than consuming it will cause illness. So that, these are either totally removed or reduced to permissible limit by using chlorination, ultraviolet light and ozone.
- **Dechlorination**—It is generally done when the treated water is used for industries. In this process, the excessive chlorine is removed after chlorination because it deposits on the surface of the equipment which can cause health issues also may lead to the corrosion also.

7 Conclusion

This paper contains a review of various methods and techniques that may be used for the purpose of recycling treatment, recovery and reuse of the wastewater to obtain clean and fresh water. It is being observed that there are numerous options that are feasible and readily available in the rapidly developing and advancing world. Various research activities in this topic have led to significant advancement and development in the diverse processes and methods that are employed for the purpose of wastewater treatment, and all the emergent issues and technological options are being considered and examined in this report.

References

1. Anderson, J., et al.: Climbing the ladder: a step by step approach to international guidelines for water recycling. Water Sci. Technol. **43**(10), 1–8 (2001)
2. Armon, R., et al.: Residual contamination of crops irrigated with effl uent of different qualities: a fi eld study. Water Sci. Technol. **30**(9), 239–248 (1994)
3. Mara, D.D.: Domestic Wastewater Treatment in Developing Countries. Earthscan Publications, London (2004)
4. Ludwig, H.F., et al.: Textbook of Appropriate Sewerage Technologies for Developing Countries. South Asian Publishers Pvt. Ltd., New Delhi (2005)
5. Metcalf & Eddy, Inc.: Wastewater Engineering: Treatment, Disposal and Reuse, 3rd ed. New York, McGraw-Hill (1991)
6. Metcalf & Eddy, Inc.: Wastewater Engineering: Treatment, Disposal and Reuse, 4th ed. New York, McGraw-Hill (2003)
7. Oron, G., et al.: Effect of effl uent quality and application method on agricultural productivity and environmental control. Water Sci. Technol. **26**(7–8), 1593–2160 (1992)
8. Sobsey, M.: Inactivation of health-related micro-organisms in water by disinfection processes. Water Sci. Technol. **21**(3), 179–195 (1989)
9. Stephenson, T., et al.: Membrane Bioreactors for Wastewater Treatment. IWA Publishing, London (2000)

Identification of Malignant Lymphoblast Cell in Bone Marrow Using Machine Learning

Jyoti Rawat, Sachin Rawat, Indrajeet Kumar, and J. S. Devgun

Abstract Leukemia and lymphoma are the neoplastic proliferative disorders of white blood cells (WBCs), which increases the malignant lymphoblast cells in the blood. The objective of this work is to perform segmentation of white blood cell image into dissimilar region and then analyze those reasons for particular application in hematology to localize the lymphoblast cells. The population of lymphoblast cells is enormously grows in bone marrow. Acute leukemia often causes erroneous diagnosis due to its imprecise nature. Moreover, due to the rapid growth of the disease, it may damage the normal blood cells and cause death. Early prediction and diagnosis are very significant and conscientious factor for the hematologist. In this work, an unsupervised learning algorithm, namely k-means is used for segmenting the nucleus and cytoplasm. Various color and geometric features are extracted to analyze the malignant and healthy leukocyte cells. These features are then subjected to the feature elimination to reduce the high dimensional feature space into optimal sized features set using principal component analysis. Furthermore, various binary classifiers like SVM, PNN, k-NN, and SSVM are used for classifying the extracted regions into nucleus, cytoplasm, and background cells. The results yielded through the experimentations establish that the proposed system successfully eradicate the extraneous features, effectively classify the malignant lymphoblast cells from the healthy lymphocytes and preserves the maximum accuracy in comparison to that of other methods.

J. Rawat (✉)
DIT University, Dehradun, India

S. Rawat
Patch Infotech Pvt Ltd, Gurugram, Haryana, India

I. Kumar
GEHU University, Dehradun, India

J. S. Devgun
M. M. Institute of Medical Sciences an Research, Solan, HP, India

K. N. Das et al. (eds.), *Proceedings of the International Conference on Computational Intelligence and Sustainable Technologies*, Algorithms for Intelligent Systems,
https://doi.org/10.1007/978-981-16-6893-7_25

Keywords White blood cells (WBCs) · Hematological analysis · Lymphoblast cell localization · Acute leukemia · Geometric features · Color features · k-means clustering

1 Introduction

In the area of hematological analysis, image processing has been widely used to assist pathologist in learning verity of microscopic images. In machine learning, image is considered as an important aspect for conveying information, and extracting this information such that it can be used for particular application. In the medical application, it is feasible to assemble the diagnosis system to aid medical experts due to advanced image processing techniques. There have been many studies to automate analysis and detection of various hematological disorders. To diagnosis the patient's, pathologist needs a lot of information from their blood cells [1]. Specific knowledge is required for executing this work that can be automatically analyzed using the image processing techniques, so hematology expert can get more objective stastical information. In this paper, a method is designed that is able to evaluate and discriminate the human being's peripheral blood into leukemia affected or healthy cells by localization of lymphoblast cells [2, 3].

Leukemia is regarded as an anomalous propagation of immature WBCs, called progenitor cells in the bone marrow and lymphoid tissues. Over the time, leukemia infected cells can outnumber the normal red blood cells (RBCs), white blood cells (WBCs), and even platelets, which leads to serious trouble of anemia or thrombocytopenia [4, 5]. It can be very delicate task to analyze the leukocytes (white blood cells) in random area of stem cells. The author [6] presents a method for discriminating normal cells to abnormal white cell using multilayer perceptron (MLP) neural network and extracted 17 features like shape, color, and size. Bayesian regulation and Levenberg–Marquardt algorithms were implemented to train the classifier and achieve good classification performance. Ravikumar [7] performed white cell segmentation and classify into their five different types using fuzzy c means, acquired high accuracy for 91%. Several authors [8–11] have reported the classification of the leukocytes by using the color space transformation and support vector machine classifier with linear discriminant analysis for feature selection. Lina et al. [9] segment the leukocyte with different size of images as 47×47 and 57×57 pixels and recognize the white cell with Euclidean distance method.

Various authors segmented the white cell and performed classification of white blood cell by using different segmentation algorithms and extracted various texture and shape features to recognizing lymphocyte cells [10, 12–21]. Kulkarni and Bhosale [20] derived a novel approach for analysis of leukemic cells by employing color conversion algorithms. Deshmukh and Panditrao [22] presents technique that focuses on automation using median filtering, automatic thresholding for removing outliers. Geometric descriptors of nucleus are examined, and then, contour signature is applied to evaluate boundary irregularity that can be used for association of cancer cells.

An earlier work done by the present author [23–26], a technique based on image fusion for segmenting the nucleus of white cell and various other segmentation and classification technique are presented. Nevertheless, the algorithm depicted did not confine the whole leukocyte but only the nucleus region for detection of leukocytes.

Hence, in this paper, work is performed on microscopic images of peripheral blood cell; the most important thing is to separate the erythrocytes, leukocytes, thrombocytes, and plasma based on their asymmetry. Before localizing, lymphoblast cell need to divide the leukocyte cells into different region. In this work, segmentation is performed to differentiate the lymphocyte into two regions as nucleus and cytoplasm. Then, classification is performed to localize the malignant cell using various classifiers to assess the performance of system.

2 Methods

The first and most important step for an image processing application is to acquire an image with high quality, precision, and noiseless that is painstaking as a problematical task due to inexperienced operators. For this test, all the microscopic images of WBCs were obtained from the benchmark dataset ALL-IDB [27]. A detailed analysis of various steps involved in this study is provided in the sections that follow, and it has been graphically depicted in Fig. 1.

2.1 Segmentation Module

Image is acquired from database and then subjected to various image processing related steps to increase their feasibility for the application. The resultant images are then implemented with k-means clustering in order to extract the nucleus and cytoplasm region from the cell background as shown in Figs. 2 and 3.

In segmentation, first step is to extract the required region containing cytoplasm and nucleus of WBC from its microscopic images through k-means clustering without damaging the color content of the image. k-means algorithm helps us to cluster or categorize the data as per the similarities between the different classes of data [28]. In the study [29], the region containing the WBC's nucleus has been extracted using k-means clustering, while in [30] this technique is used for WBCs' segmentation. The k-means algorithm has also been implemented for distinguishing the regions present in the input image into $L * a * b$ color space with three different regions, i.e., background, cytoplasm, and nucleus [31].

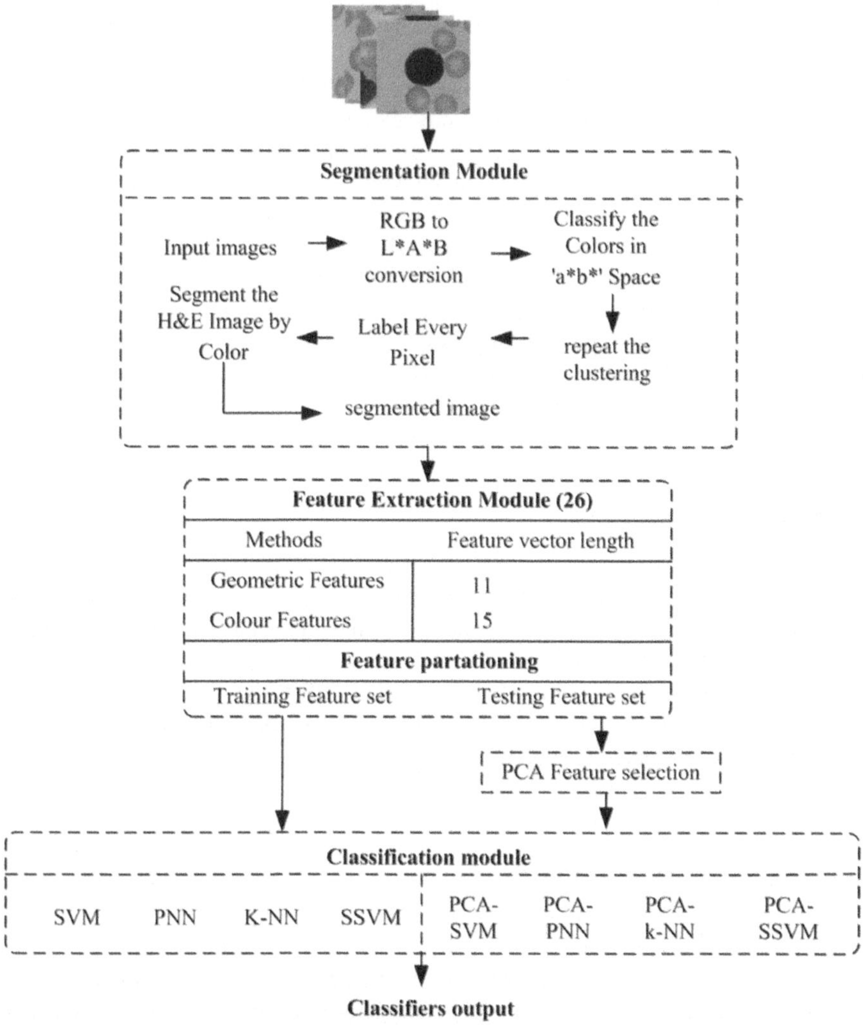

Fig. 1 Description of proposed classification system

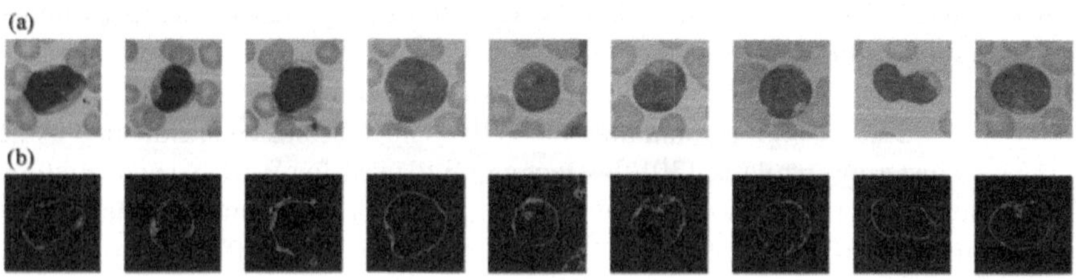

Fig. 2 Cytoplasm extraction through k-means clustering, **a** input images, **b** extracted cytoplasm region from input image

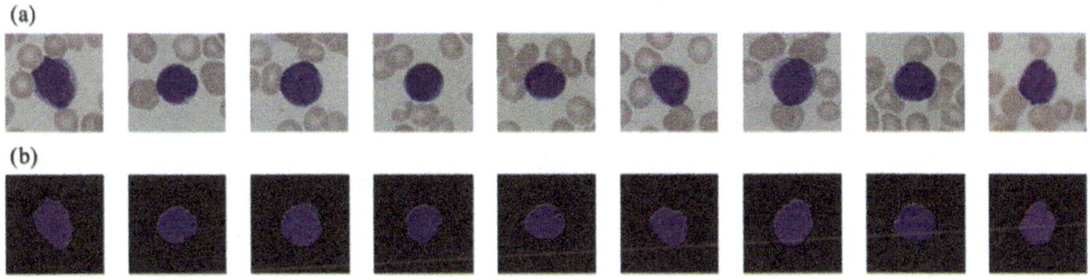

Fig. 3 Extraction of region containing nucleus through k-means clustering, **a** input images, **b** final extracted nucleus

2.2 *Feature Extraction and Selection Module*

The selective prominent features of microscopic elements are generally used for their cell classification. Generally, the recent literature is observed to be more focused on texture and shape features extracted from particular blood cell [32–34]. However, this study concentrates on geometric and color-based features of nucleus and cytoplasmic region of white cell stated in Fig. 4.

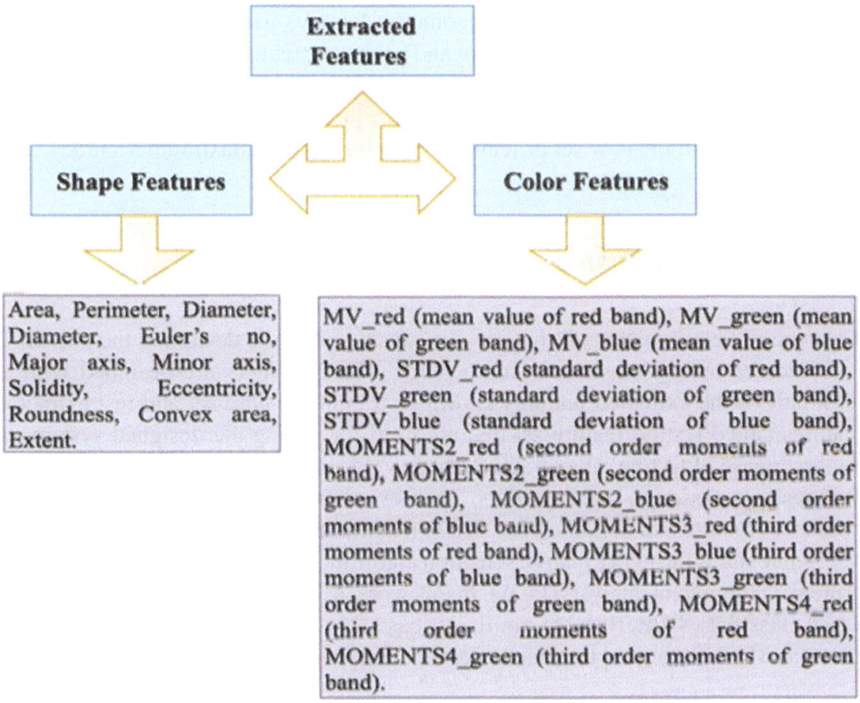

Fig. 4 List of extracted shape and color features

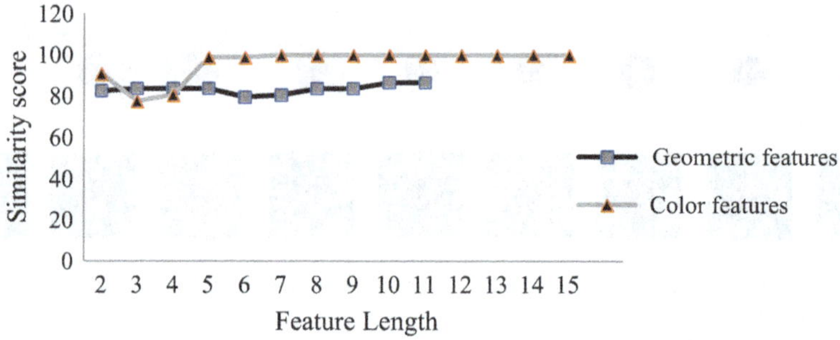

Fig. 5 Co-relation graph for selective color and geometric features using PCA feature selection

As mentioned in Fig. 4, there are various morphological dissimilarities between the lymphoblast and lymphocytes. For this study, principal component analysis (PCA) is utilized. If there are some normal cells, even though they look the same, it is suspected that there might be differences or variations in the type of cells. It is required to overserve the inverse correlation among the cells [35].

To improve the accuracy and performance of the model, classification is preceded by feature selection which is applied on all the feature subsets extracted from set of features containing color as well as geometric features irrespective of their length. The co-relation graph for selective color and geometric features is shown in Fig. 5. For a reduced feature set, PCA selects ten feature descriptors for geometric features, and six feature descriptors are selected for color features. At last, calculated the overall accuracy got with the new set of features, which gives the maximum accuracy.

2.3 Classification Module

The feature vector instances, that are not considered while designing the classifier, are used for testing classifier's competency. The classifiers are designed with at least half of the extracted features known as training feature set and rest extracted features called testing feature set that are used for testing the designed system. In the first phase of testing, all extracted features (11 geometrical features and 15 color features) are passed to the classifiers. The next phase deals with using the reduced number of features for classification. These features are reduced by selecting the most relevant features through a feature selection algorithm based on similarity score called principal component analysis (PCA). In this work, a decision boundary or geometric-based classifier (SVM, SSVM), similarity-based classifier (k-NN), and probabilistic classifier (PNN) are used to achieve best accuracy of classification.

3 Experimental Evaluation and Discussion

In this work, an online benchmark database ALL-IDB [27] is used. Moreover, different authors have been utilized images taken locally from the hospitals which are not available publically [12–14, 18]. Thus, this work could not directly compare with the other methods [6, 20, 22, 26]. In the early stage of leukocyte analysis, leukocyte identification is done by segmentation process. Blood cells segmentation performed in ALL-IDB 2 datasets. Segmentation results shown in Figs. 2 and 3. The aim of feature extraction phase is to convert the microscopic images to quantitative data that can correctly represent the visual information and then to simultaneously extract the feature descriptors (color and shape) that are most significant to consequent classification process. Confusion matrix and performance metrics are given in Fig. 6.

Experiment 1

In the experiment 1, the proposed system has been tested for color feature set using different classifiers (smooth support vector machine (SSVM), k-nearest neighbor (k-NN), probabilistic neural network (PNN), and support vector machine (SVM)). The achieved performance for feature based on color symmetry is tabulated in Table 1 without applying any feature reduction algorithm.

In Table 2, classification performance is reported after applying PCA feature dimension reduction techniques, and the computation time is also reported in Table 3.

For total extracted color feature with feature length 15, the highest classification accuracy and sensitivity are 97.95 and 96.96% is achieved for SVM classifier as well as the highest specificity rate is 100% gained by PNN as per Table 3. The highest classification accuracy for reduced color feature set with feature length six is 98.97% achieved by SVM. The maximum obtained sensitivity and specificity is 100% by using PCA-SVM and PCA-PNN, PCA-k-NN, respectively as shown in Table 2.

	Positive (ALL infected)	Negative (normal)
Positive (ALL infected)	TP (true positive) Correctly identified (infected) lymphocyte	FP (false positive) Misclassified healthy objects
Negative (normal)	FN (false negative) Healthy identified cells	TN (true negative) Misclassified ALL infected objects

$$specificity = \frac{TN}{TN + FP} \times 100\%$$

$$sensitivity = \frac{TP}{TP + FN} \times 100\%$$

$$accuracy = \frac{TP + TN}{TP + TN + FP + FN} \times 100\%$$

Fig. 6 Confusion matrix and performance metrics

Table 1 Classification accuracy using different classifiers for combined nucleus and cytoplasm color features without feature selection

Classifier	CM_{CN}			Acc (%)	Sen (%)	Spec (%)
SVM		P	N	**97.95**	**96.96**	98.46
	P	32	1			
	N	1	64			
PNN	P	28	5	94.90	84.84	**100**
	N	0	65			
k-NN	P	27	6	66.32	81.81	98.46
	N	1	64			
SSVM	P	5	28	68.37	15.15	95.38
	N	3	62			

Note FV: Feature vector, P: Positive, N: Negative, L: Length of feature vector, CM_{CN}: Combined confusion matrix for performance evaluation, Positive: ALL-infected white blood cells, Negative: Healthy white blood cells.

Table 2 Classification accuracy using PCA-based feature selection [FV (L)] 6

Classifier	CM_{CN}			Acc (%)	Sens (%)	Spec (%)
		P	N			
PCA-SVM	P	33	0	**98.97**	**100**	98.46
	N	1	64			
PCA-PNN	P	31	2	97.96	93.93	**100**
	N	0	65			
PCA-k-NN	P	29	4	95.92	87.87	**100**
	N	0	65			
PCA-SSVM	P	10	23	75.51	30.30	98.46
	N	1	64			

For the evaluation of the designed system, computation time as well as its accuracy both are crucial factors. The computation time and classification accuracy are listed in Table 3 with respect to classifiers.

Experiment 2

In this experiment, all the extracted geometric features are used for the testing the designed system. The achieved highest accuracy, max-sensitivity, and max-specificity are 87.75%, 81.81%, and 90.76%, respectively by SVM without using any feature dimension reduction techniques as shown in Table 4. After applying feature dimension reduction techniques for geometric features, the highest classification accuracy is 89.80%, and max-specificity is 93.84% accomplished by PCA-k-NN.

Similarly, the max-sensitivity is 87.87% attained by two different classifiers, i.e., PCA-SVM, PCA-PNN (Table 5).

Table 3 Comparative analysis with respect to accuracy and computation time for characterization of lymphocyte into two classes of normal and ALL-infected cells for color and geometric features

Classifiers	Total computation time (seconds) for color features	Total computation time (seconds) for geometric features	Classification accuracy for color features	Classification accuracy for geometric features
	$TCT_{Nucleus-cytoplasm}$	$TCT_{Nucleus-cytoplasm}$	Nucleus-cytoplasm	Nucleus-cytoplasm
SVM	3.9	3.9	97.95	87.75
PNN	5.6	25.4	94.90	81.63
k-NN	11.2	11.2	66.32	62.24
SSVM	2.8	2.9	68.37	67.35
PCA-SVM	3.3	3.7	**98.97**	86.73
PCA-PNN	3.2	5.6	97.96	86.73
PCA-k-NN	**1.3**	8.0	95.92	**89.80**
PCA-SSVM	2.7	**2.6**	75.51	78.57

Table 4 Classification accuracy using different classifiers for combined nucleus and cytoplasm shape features without feature selection

Classifier	CM_{CN}			Acc (%)	Sens (%)	Spec (%)
SVM		P	N	87.75	81.81	90.76
	P	27	6			
	N	6	59			
PNN	P	24	9	81.63	72.72	86.15
	N	9	56			
k-NN	P	19	14	62.24	57.57	64.61
	N	23	42			
SSVM	P	7	26	67.35	21.21	90.76
	N	6	59			

Table 5 Classification accuracy using PCA feature selection with different classifiers for combined nucleus and cytoplasm shape features for feature length [FV (L)] 10

Classifier	CM_{CN}			Acc (%)	Sens (%)	Spec (%)
		P	N			
PCA-SVM	P	29	4	86.73	**87.87**	87.69
	N	9	57			
PCA-PNN	P	29	4	86.73	**87.87**	87.69
	N	9	57			
PCA-k-NN	P	27	6	**89.80**	81.81	**93.84**
	N	4	61			
PCA-SSVM	P	24	93	78.57	72.72	81.53
	N	12	53			

After the successful evaluation of different experiments performed by different classifiers, it has been observed that color features as well as geometric features both are relevant for localization of lymphoblast (cancerous cell) for peripheral blood images. It can assist hematology expert to analyze the microscopic images by providing the necessary information about shapes of the leukocyte.

4 Conclusion

In this work, the malignant lymphoblast cell is identified using microscopic blood smear. The segmentation module is implemented by k-mean clustering method achieved a precise solution. The proposed classification system is more reliable and yet yields comparable classification rate in range of 89.80–98.97% for different classifiers based on geometric and color feature set. In general, it can be said that the identification of the malignant lymphoblast cell with the help of microscopic images of WBCs is achievable through machine learning and can yield noteworthy accuracy. The method has proved to be effective, efficient, and simpler enough to be considered for application in the medical field.

References

1. Kim, K., Song, J., Golshani, F., Panchanathan, S.: Automatic classification of cells using morphological shape in peripheral blood images. In: Internet Multimedia Management Systems, 2000 Oct 11, vol. 4210, pp. 290–298. International Society for Optics and Photonics
2. Bain, B.J.: A Beginner's Guide to Blood Cells, 2nd ed. Department of Hematology, St Mary's Hospital Campus, Imperial College, London (2004). ISBN 1-4051-2175-0
3. Houwen, B.: Blood film preparation and staining procedures. Lab hematol. **6**, 1–7 (2000)
4. Leukaemia and Lymphoma Society: Blood and Marrow Stem Cell Transplantation, Revised 2013
5. American Cancer Society: "Facts Spring 2015" Leukaemia Lymphoma Society: Fighting Blood Cancer, Revised April 2015
6. Nasir, A.A., Mashor, M.Y., Hassan, R.: Classification of acute leukaemia cells using multilayer perceptron and simplified fuzzy ARTMAP neural networks. Int. Arab J. Inform. Technol. **10**(4), 1–9 (2013)
7. Ravikumar, S., Shanmugam, A.: WBC image segmentation and classification using RVM. Appl. Math. Sci. **8**(45), 2227–2237 (2014)
8. Ramesh, N., Dangott, B., Salama, M.E., Tasdizen, T.: Isolation and two-step classification of normal white blood cells in peripheral blood smears. J. Pathol. Inform. **3** (2012)
9. Na, L., Chris, A., Mulyawan, B.: A combination of feature selection and co-occurrence matrix methods for leukocyte recognition system. J. Softw. Eng. Appl. **5**(12), 101–106 (2013)
10. Ushizima, D.M., Lorena, A.C., De Carvalho, A.C.: Support vector machines applied to white blood cell recognition. In: Fifth International Conference on Hybrid Intelligent Systems (HIS'05), 2005 Nov 6, p. 6. IEEE
11. Sarrafzadeh, O., Rabbani, H., Talebi, A., Banaem, H.U.: Selection of the best features for leukocytes classification in blood smear microscopic images. In: Medical Imaging 2014: Digital Pathology, 2014 Mar 20, vol. 9041, p. 90410P. International Society for Optics and Photonics

12. Huang, D.C., Hung, K.D., Chan, Y.K.: A computer assisted method for leukocyte nucleus segmentation and recognition in blood smear images. J. Syst. Softw. **85**(9), 2104–2118 (2012)
13. Stadelmann, J.V., Spiridonov, I.N.: Automated classification of leukocytes in blood smear images. Biomed. Eng. **46**(2), 83–87 (2012)
14. Mircic, S., Jorgovanovic, N.: Application of neural network for automatic classification of leukocytes. In: 2006 8th Seminar on Neural Network Applications in Electrical Engineering, 2006 Sep 25, pp. 141–144. IEEE
15. Ongun, G., Halici, U., Leblebicioglu, K., Atalay, V., Beksaç, M., Beksaç, S.: Feature extraction and classification of blood cells for an automated differential blood count system. In: IJCNN'01. International Joint Conference on Neural Networks. Proceedings (Cat. No. 01CH37222), 2001 Jul 15, vol. 4, pp. 2461–2466. IEEE
16. Theera-Umpon, N.: White blood cell segmentation and classification in microscopic bone marrow images. In: International Conference on Fuzzy Systems and Knowledge Discovery, 2005 Aug 27, pp. 787–796. Springer, Berlin, Heidelberg
17. Shivhare, S., Shrivastava, R.: Morphological granulometric feature of nucleus in automatic bone marrow white blood cell classification. Int. J. Sci. Res. Publ. **2**(5), 2250–3153 (2012)
18. Tabrizi, P.R., Rezatofighi, S.H., Yazdanpanah, M.J.: Using PCA and LVQ neural network for automatic recognition of five types of white blood cells. In: 2010 Annual International Conference of the IEEE Engineering in Medicine and Biology, 2010 Sep, pp. 5593–5596. IEEE
19. Young, I.T.: The classification of white blood cells. IEEE Trans. Biomed. Eng. **4**, 291–298 (1972)
20. Kulkarni, T.A., Bhosale, D.S.: A robust morphological analysis of normal and abnormal leukemic cells populations in acute lymphoblastic leukemia. Int. J. Electr. Electron. Data Commun. **2**(8) (2014)
21. Bikhet, S.F., Darwish, A.M., Tolba, H.A., Shaheen, S.I.: Segmentation and classification of white blood cells. In: 2000 IEEE International Conference on Acoustics, Speech, and Signal Processing. Proceedings (Cat. No. 00CH37100), 2000 Jun 5, vol. 4, pp. 2259–2261. IEEE
22. Deshmukh, A., Panditrao, A.: Automated leukemia detection using contour signature in blood microscopic images. Int. J. Res. Comput. Commun. Technol. 304–8 (2014)
23. Wurflinger, T., Stockhausen, J., Meyer-Ebrecht, D., Böcking, A.: Robust automatic coregistration, segmentation, and classification of cell nuclei in multimodal cytopathological microscopic images. Comput. Med. Imaging Graph. **28**(1–2), 87–98 (2004)
24. Abbas, N., Saba, T., Mehmood, Z., Rehman, A., Islam, N., Ahmed, K.T.: An automated nuclei segmentation of leukocytes from microscopic digital images. Pakistan J. Pharm. Sci. **32**(5) (2019)
25. Rawat, J., Bhadauria, H.S., Singh, A., Virmani, J.: Review of leukocyte classification techniques for microscopic blood images. In: 2015 2nd International Conference on Computing for Sustainable Global Development (INDIACom), 2015 Mar 11, pp. 1948–1954. IEEE
26. Rawat, J., Singh, A., Bhadauria, H.S., Kumar, I.: Comparative analysis of segmentation algorithms for leukocyte extraction in the acute Lymphoblastic Leukemia images. In: 2014 International Conference on Parallel, Distributed and Grid Computing, 2014 Dec 11, pp. 245–250. IEEE
27. Labati, R.D., Piuri, V., Scotti, F.: All-IDB: the acute lymphoblastic leukemia image database for image processing. In: 2011 18th IEEE International Conference on Image Processing, 2011 Sep 11, pp. 2045–2048. IEEE
28. Tatiraju, S., Mehta, A.: Image Segmentation using k-means clustering, EM and normalized cuts. Department of EECS, vol. 1, pp. 1–7 (2008)
29. Sinha, N., Ramakrishnan, A.G.: Automation of differential blood count. In: TENCON 2003. Conference on Convergent Technologies for Asia-Pacific Region, 2003 Oct 15, vol. 2, pp. 547–551. IEEE
30. Rajendran, S., Arof, H., Mokhtar, N., Mubin, M., Yegappan, S., Ibrahim, F.: Dual modality search and retrieval technique analysis for leukemic information system. Sci. Res. Essays **6**(2), 247–255 (2011)

31. Mohapatra, S., Patra, D., Satpathy, S.: Automated leukemia detection in blood microscopic images using statistical texture analysis. In: Proceedings of the 2011 International Conference on Communication, Computing and Security, 2011 Feb 12, pp. 184–187
32. Saraswat, M., Arya, K.V.: Automated microscopic image analysis for leukocytes identification: a survey. Micron **1**(65), 20–33 (2014)
33. Putzu, L., Caocci, G., Di Ruberto, C.: Leucocyte classification for leukaemia detection using image processing techniques. Artif. Intell. Med. **62**(3), 179–191 (2014)
34. Nixon, M., Aguado, A.: Feature Extraction and Image Processing for Computer Vision. Academic Press (2019)
35. Guo, Q., Wu, W., Massart, D.L., Boucon, C., De Jong, S.: Feature selection in principal component analysis of analytical data. Chemom. Intell. Lab. Syst. **61**(1–2), 123–132 (2002)

Predicting Tamil Nadu Election 2021 Results Using Sentimental Analysis Before Counting

K. Devasena, M. Sarika, and J. Shana

Abstract Sentiment analysis (SA) could be a way to examine people's suppositions from composed dialects. These days, due to the development of social media, people's conclusions are exceptionally pivotal for decision-making in each viewpoint. Apart from counting the individual votes, techniques such as sentimental machine learning models can be used to classify the text to find whether the given statement is in favor or not to the particular political party from collecting the data from social media like Facebook and Twitter. In this work, an attempt is made to apply different classifiers like Naive Bayes (NB) and support vector machine (SVM) in predicting the result of the Tamil Nadu election 2021 from Twitter and Facebook posts. The model was built using the tweets obtained before the counting of votes began on May 2nd, 2021. Using the model, the results in Tamil Nadu could be predicted easily.

Keywords Sentimental analysis · Multinomial Naive Bayes · Support vector machine · Election results

1 Introduction

Over a long time, there has been a tremendous development within the utilization of estimation investigation in social media for following the feeling of the open approximately a specific item or point such that the companies, media organizations are progressively working on ways to extract people's opinions from social media since it generates extensive sentiment-rich datasets in various genres such as tweets, comments, and blog posts. At the same time, there has been significant research on how sentiments are expressed and plays a vital role in various fields like marketing, political, and entertainment. Two major techniques can be used in SA—"Approach based on Knowledge" and "Approach based on Machine learning" [2]. In this paper,

K. Devasena · M. Sarika (✉) · J. Shana
Coimbatore Institute of Technology, Coimbatore, Tamil Nadu 641014, India

J. Shana
e-mail: shana@cit.edu.in

K. N. Das et al. (eds.), *Proceedings of the International Conference on Computational Intelligence and Sustainable Technologies*, Algorithms for Intelligent Systems,
https://doi.org/10.1007/978-981-16-6893-7_26

the machine learning approach was primarily used for the preparation set that builds the classifier to evaluate the sentiments or opinions of the people. Named information is sent to the machine learning show, and the calculation gives the specified predefined categories such as whether positive or negative. In text classification, named information is utilized as preparing information to derive a classification show, and after that, it can consequently classify unlabeled content information utilizing the inferred demonstration. The goal of the model is to predict the results of the Tamil Nadu election 2021 by analyzing the sentiment of the text (i.e., positive or negative) to a particular political party. The tweets and posts collected from Twitter and Facebook during the period from January 2021 to March 2021 were based on people's opinions about the two major parties in Tamil Nadu, Anna Dravida Munnetra Kazhagam (ADMK) and Dravida Munnetra Kazhagam (DMK). The sentiments of these social media posts were analyzed using machine learning models. The analysis results could summarize the polarity of the users' opinion thereby suggesting which party is favored by the people the most. The collected texts were then analyzed using machine learning models such as supervised learning algorithms, namely Naive Bayes and support vector machine.

2 Related Work

In [1], the authors have investigated the potential to model political sentiment through the mining of social media and combine sentiment analysis using supervised learning and volume-based measures, and the author evaluates the model's result against the conventional election polls and the final election result. In [2], various kinds of classifiers such as Naive Bayes, SVM, and greatest entropy were used to analyze the posts of Twitter related to a wide range of electronic products such as mobiles, laptops, video games, and television sets. In [3], the author has classified Malayalam tweets into positive and negative statements using machine learning models like NB, SVM, and RF. In [4] the authors have analyzed an outsized set of manually labeled tweets in several languages, used them as training data, and constructed automated classification models. The models have classified the text as negative, neutral, and positive. In [5], the author has tested some neural network architectures (CNN, LSTM, BiLSTM) and BERT architectures on previous Russian sentiment evaluation datasets, and the model has practically achieved the human level in this. In [6], the author has investigated the task of automatically predicting the overall recommendation (accept, reject, or sometimes borderline) and also identifying the sentences with positive and negative sentiment polarities from the review. In [7], the authors have collected the data from Twitter based on their relevance with the topic searched, and various classifiers like Naive Bayes classifier, support vector machine, the data have been classified into positive and negative statements based on their sentiment. The work in [8] analyzed Twitter data and classified the dataset into negative, neutral, positive, irrelative, average classes using an SVM classifier. In [9] the author has used Ekran's model to identify emotions in large datasets generated on Twitter and for modeling intelligent agents and has used Naive Bayes, support vector machines, decision trees, logistic regression, and neural networks. The author in [10] uses AdaBoost. MH models with 500 rounds of boosting for train data and interprets F-measure for baseline and four combinations of features: n-grams and lexicon features (n-gram + lex), n-grams, and part-of-speech features (n-gram + POS), n-grams, lexicon features and microblogging features (n-grams + lex + twit), and finally all the features combined.

3 Methodology

The process of building the classifier model, in general, is depicted in Fig. 1. The steps are elaborated in Sects. 3.1 through 3.6.

3.1 Dataset Collection

Here, primary data collection has been used to collect the dataset, i.e., people's opinions collected directly from Facebook and Twitter.

3.1.1 Facebook

The data have been gathered from the post's comments containing people's opinions about both the political parties with a sampling rate of 100 per comment per week during active election months (January 2021–March 2021). Around 300 data per

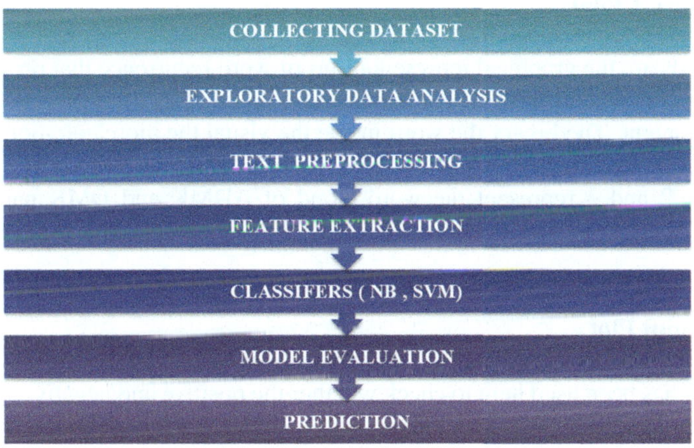

Fig. 1 Methodology used in text classification

party have been gathered from the comments of the post using APIs like App secret, App ID, access token, and also with the post's link. However, about 400 entries containing both parties have been removed while cleaning the text, which results in a final 100 entries per party. Here, there are two sets of datasets, i.e., each per party—ADMK_dataset and DMK_dataset. For each set of datasets, there are 100 posts, altogether 200 data for training the model.

3.1.2 Twitter

The data have been gathered from public tweets containing the candidates' names with a sampling rate of 100 tweets per week per candidate during active election months (about 4 months for TN Election). It should be mentioned that the method was also applied to a dataset of 100 tweets per week per candidate, around 900 tweets per party are gathered for the given latitude and longitude (11.1271 °N, 78.6569 °E) for Tamil Nadu. However, about 600 repetitious tweets contain both candidates' names which are then removed, resulting in a final 250 tweet dataset per party. Here, there are two sets of datasets, i.e., each per party—ADMK, DMK. For each set of datasets, it has 250 data, altogether 200 data for training the model.

The attributes of the dataset are as follows:

- Text—It talks about how people feel about the particular party, i.e., people's opinion about the party.
- Labels—It talks about whether the opinion is in favor of the party or not, i.e., either positive or negative.

3.2 Exploratory Data Analysis

3.2.1 Word Cloud

Word clouds are the graphical representations of word recurrence that provide more prominent conspicuousness to words that show up more as often as possible amid a source content. The bigger the word inside the visual the more common the word was inside the document(s).

Figures 2 and 3 represent the word cloud of ADMK and DMK parties from Facebook posts and Twitter tweets.

3.2.2 Count Plot

Count plot has been used here to check whether the positive and negative statements are equal in numbers to proceed further.

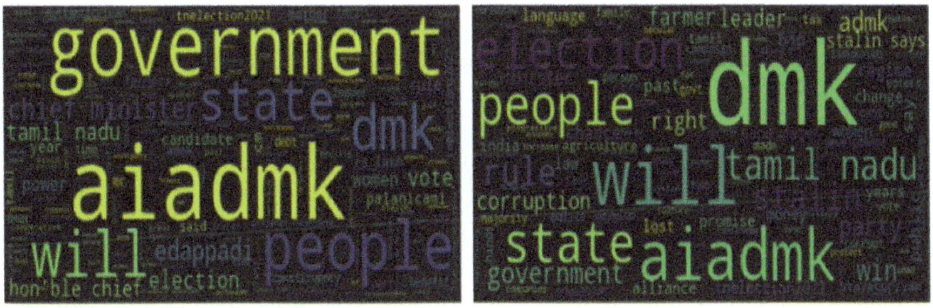

Fig. 2 Word cloud of ADMK and DMK from Facebook

Fig. 3 Word cloud of ADMK and DMK from Twitter

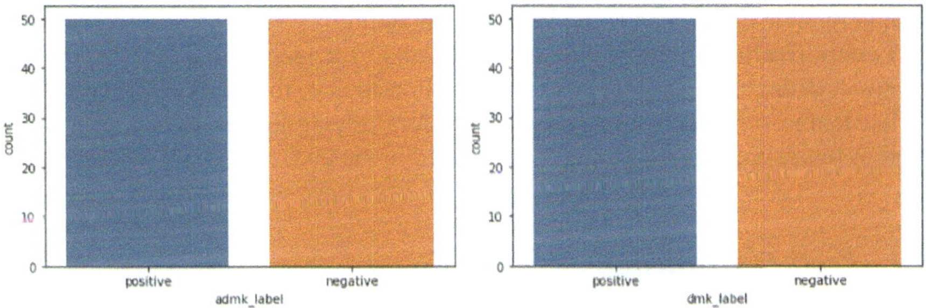

Fig. 4 Count plot of ADMK and DMK data from Facebook

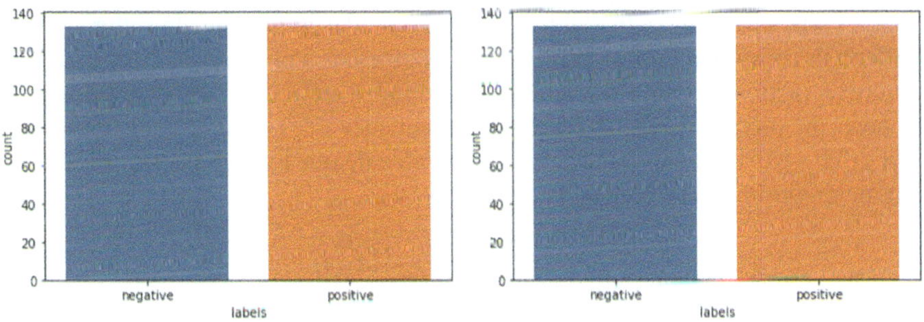

Fig. 5 Count plot of ADMK and DMK from Twitter

Figures 4 and 5 represent the dataset containing an equal number of positive and negative statements in the ADMK and the DMK from Facebook posts and Twitter tweets.

3.3 Text Preprocessing

In text classification, preprocessing the information is as imperative as show building, and when it comes to unstructured information like content, this preparation is indeed more critical. The preprocessing steps utilized here are recorded below,

- **Removal of punctuations**—Punctuations are those unhelpful parts of the text or noise. Removing noise will be helpful in text analysis on pieces of data like tweets.
- **Removal of stop words**—Stop words are those which occur frequently like "a," "the," "is," "are." These words also don't carry much information as compared to the other texts. Removing stop words are necessary from the given text so that more importance will be given to those meaningful texts.
- **Tokenizing**—Tokenization is a way of splitting text data into smaller units like ["Jayalalithaa," "memorial," "corrupted"] at the same time it removes certain characters such as punctuation; hence, it is a final string that is detached from the given text. The main aim of tokenization is to transform unstructured text material into reliable data which is used in model building
- **Lemmatization**—Lemmatization is a method for identifying, structuring, and investigating the total set of possible relationships of the word in the given text, like lemma for studies is study and lemma for studying studies. Hence, the process is to figure out the basic form of each word in the text.

3.4 Feature Extraction

Machine learning algorithms learn from a predefined set of highlights from the preparing information to supply yield for the test information. But, the matter that emerges when working with dialect handling (content) is that the algorithms cannot work on the raw content straightforwardly. So, there is a prerequisite for a couple of feature extraction methods to change over text into a matrix (or vector) of highlights. So here, the procedures utilized are, term frequency (TF)–inverse dense frequency (IDF) (TF−IDF). Equation (1) is the measurable degree to gage the significance of a particular term amid a corpus.

$$tf = idf = t\,ft * id\,ft \tag{1}$$

Equation (1) shows is that the term frequency and is that the inverse document frequency.

3.5 Classifier Model

3.5.1 Multinomial Naive Bayes

The Naive Bayes algorithm has been used because it is used to find the probabilities of classes assigned to texts by using the joint probabilities of words and classes. The elemental thought of Naive Bayes is to discover the probabilities of classes assigned to texts. That Naive Bayes accepts that the nearness of a specific feature in a class is irrelevant to the presence of any other feature additionally accept each feature is given rise to importance. Naive Bayes needs less training data, less training time, and less memory consumption.

Given that they computed a database of probabilities for terms appearing in the dataset (positive or negative). It can proceed to the last step of the Naive Bayes classifier, which is the classification.

$$y = \arg\max k \in \{Positive, Negative\} p(C_k) i = 1\, n\ p(x_i|C_k) \tag{2}$$

Equation (2) shows y is the outcome; $p(C_k)$ is the prior probability of the class; $p(x_i|C_k)$ is the likelihood, and for every incoming text, it has to compute the probability of such text being positive and negative (i.e., for each class), and the final veridic will be given by the largest probability.

3.5.2 Support Vector Machines

Support vector machine (SVM) is a supervised learning algorithm that is useful for all regression, analysis, and classification problems. One of the main reasons why SVM is preferred is that it can be applied to any kind of vector which encodes any kind of data. In order to leverage the power of SVM text classification, texts have to be transformed into vectors. Hence, support vector machines (SVM) could be a non-probabilistic binary linear classifier geometrically prepared SVM model can be seen as a hyperplane that isolates positive and negative mathematically hyperplanes can be characterized as within the Eq. (3).

$$w.x + b = 0 \tag{3}$$

This can also be written in vector form $\langle w.x \rangle + b = 0$, where w is known as weight vector, consequently compares to normal vector to the isolating hyperplane, x is an input vector or text, b indicates the perpendicular distance from the hyperplane; support vector is given by Eq. (4) for separating hyperplane (w, b).

$$\text{class}(x) = \text{sign}(\langle w.x \rangle + b) \tag{4}$$

The distance from the hyperplane to the closest point is known as the margin of somewhere it finds the linear separator with greatest minimal separate utilizing back vectors in high dimensional space in text classification linear kernels have been utilized. By Eq. (5), it can be illuminated as the optimization problem.

$$\text{minimize}\|w\|^2/2 \tag{5}$$

Subject to: $y_i(\langle w.x\rangle + b) = 1$, where $i = 1, 2, \ldots n$, where X_i is training example with label y_i and $\|w\|$ is the L2 norm of the weight.

3.6 Model Evaluation and Prediction

In model evaluation, the model is evaluated using different performance metrics such as accuracy, confusion matrix, classification report, roc_auc score. A model with good performance metrics is considered the best deployment model. A confusion matrix is an N × N matrix that tells about the performance of the classification model; N is the number of classes. The matrix compares the particular target values with those predicted by the machine learning model.

The model is then used to predict the winning party in the Tamil Nadu elections 2021. Prediction for unseen data tells us how this model learned the training data, and how it will predict the sentiment for new data entry.

4 Experimental Results

4.1 Facebook Data

Around 200 unseen data were extracted through the same process as the training dataset. After cleaning, filtering, and preprocessing, nearly, 120 data entries have remained for prediction. Using the unseen data, the model has predicted DMK will win against ADMK in the Tamil Nadu election results 2021.

Figure 6 represents the number of positive (1) and negative (0) statements about ADMK and DMK parties from the Facebook posts.

Consider the case of the ADMK posts, in Table 1, the support vector machine has performed better than multinomial Naive Bayes. Its accuracy is 80% compared to the DMK post with an accuracy of 66.6%.

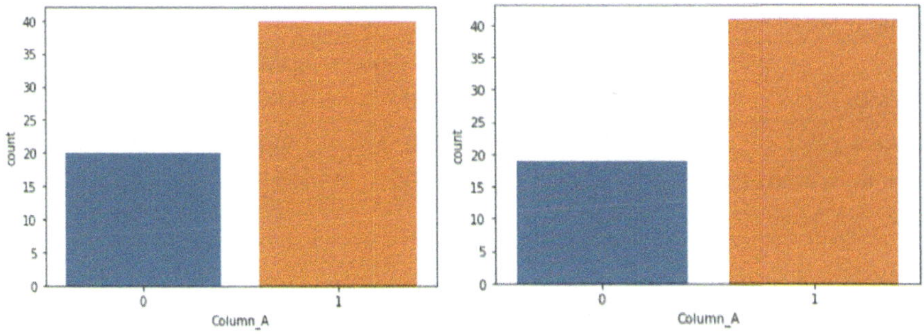

Fig. 6 Count plot of unseen ADMK posts and DMK posts from Facebook

Table 1 Model performance of Facebook post

	ADMK (%)		DMK (%)	
Metrics	MNB	SVM	MNB	SVM
Accuracy (%)	66	80	80	73.3
Confusion matrix (%)	66.67	80	40	33.3
ROC_AUC score (%)	70	80	38.88	20.83

4.2 Twitter Data

Around 300 unseen data have been extracted through the same process as the training dataset. After cleaning, filtering, and preprocessing, nearly, 100 data entries have remained for prediction. Using the unseen data, the model has predicted DMK will win against ADMK in the Tamil Nadu election results 2021.

Figure 7 represents the number of positive (1) and negative (0) statements about ADMK and DMK parties from Twitter tweets.

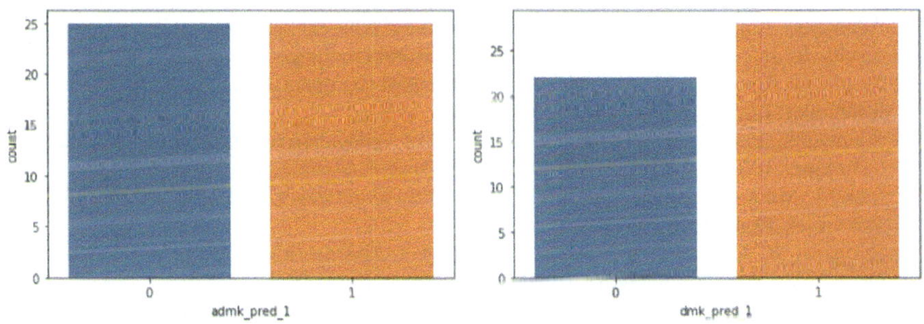

Fig. 7 Count plot of unseen ADMK and DMK from Twitter

Table 2 Model performance of Twitter data

	ADMK		DMK	
Metrics	MNB	SVM	MNB	SVM
Accuracy (%)	62.16	60.81	56.71	52.23
Confusion matrix (%)	62.16	60.81	56.72	52.24
Roc_AUC score (%)	64	60	56	51

Consider the case of the ADMK post, from Table 2, the multinomial Naive Bayes has performed better than support vector machine. Its accuracy is 62.16% compared to the DMK post with an accuracy of 56.71%.

5 Conclusion

This paper concludes that Facebook posts and Twitter tweets can be expressed as mirrors reflecting the people's opinion in political elections and helps to do an analysis. In this case, the model predicts that the DMK party will gain a majority over the ADMK party in Tamil Nadu with a margin of less than 5% based on the social media tweets and posts collected before the counting of votes. The model summarizes the results with the analysis report that the party DMK would win a majority of seats in the Tamil Nadu election 2021. The work could not be published before the counting of votes however the counting results matched with the model's analysis. The future scope of this paper is to detect irrelevant content in social media automatically and alert the user that it is irrelevant content for them.

References

1. Adam, B., Alan F.S.: On using Twitter to monitor political sentiment and predict election results. In: International Joint Conference for Natural Language Processing (IJCNLP) (2012)
2. Surbhi, C., Romil, A., Soumay, G.A.J.: An analysis of sentimental data using machine learning techniques. Int. J. Comput. Appl. (2017)
3. Soumya, S., Pramod, K.V.: Sentimental analysis of Malayalam tweets using machine learning techniques. ICT Exp. (2020)
4. Igor, M., Miha, G., Jasmina, S.: Multilingual Twitter sentiment classification: the role of human annotators. PLoS ONE (2016)
5. Rubtsova, Y.: Constructing a corpus for sentiment classification training. In: Conference of Open Innovations Association (2015)
6. Wang, K., Wan, X.: Sentiment analysis of peer review texts for scholarly papers. In: International ACM SIGIR Conference on Research and Development in Information Retrieval (2018)
7. Ankur, G., Jyoti, G., Sitesh, V.: Real-time sentiment analysis of tweets using Naive Bayes. In: International Conference on Next Generation Computing Technologies (NGCT) (2016)

8. Munir, A., Shabib, A., Iftikhar, A.: Sentimental analysis of tweets using SVM. Int. J. Comput. Appl. (2017)
9. Rodrigo, S.-A., David, V.-C.: Facebook impact and sentiment analysis on political campaigns. In: The 19th Annual International Conference (2018)
10. Efthymios, K., Theresa, W., Johanna, M.: Twitter sentiment analysis: the good the bad and the OMG! In: Fifth International AAAI Conference on Weblogs and Social Media (2011)

Ticket Dispensation Using Face Detection and Classification

V. L. Harisudhan, P. Narean Karthick, and J. Shana

Abstract Face detection in recent years attracted more developers and researchers in various domains in creating an intelligent system for a purpose. This paper introduces an approach to automatically dispense tickets to customers by face detection and classification. Dispensation is done by detecting the face of the people at the ticket counter and classifying them into three different categories to issue tickets. Face detection and classification methods depend on certain features within the face. The machine learning algorithms used are a pre-trained model namely Haar feature-based cascade classifier for face detection, MobileNetV2, and Support Vector Machine (SVM) for classification. The dataset has been manually collected, which has the images of people residing in different parts of Tamil Nadu, India under various age groups. The performance of MobileNetV2 and SVM models are analyzed. Hence MobileNetV2 performs better than SVM.

Keywords Face detection · Classification · Haar feature-based Cascade classifier · Convolutional Neural Network · Transfer learning · Support Vector Machine · MobileNetV2 · Computer vision · Image processing

1 Introduction

Face detection was once an arduous problem in the area of image processing. But in recent years researchers and developers are investing their valuable time in this field. Now the face detection systems are used in various fields with enormous applications like face recognition, photography, enhancing security, marketing, etc. It has evolved and developed beyond our imagination. This paper introduces a way of using face detection to solve a common problem. In most parts of the world, tickets at the entry section in parks, tourist spots, theme parks, etc. are hand-dispensed. In this computer

V. L. Harisudhan (✉) · P. Narean Karthick · J. Shana
Coimbatore Institute of Technology, Coimbatore, Tamil Nadu 641014, India

J. Shana
e-mail: shana@cit.edu.in

K. N. Das et al. (eds.), *Proceedings of the International Conference on Computational Intelligence and Sustainable Technologies*, Algorithms for Intelligent Systems,
https://doi.org/10.1007/978-981-16-6893-7_27

world, the tickets are not to be issued manually. So this can be approached and implemented in an intelligent way by using machine learning models. In machine learning, the system learns on its own by surfing through the data and finding the patterns. In general, ticket fare differs with the age and is divided into three categories,

0–3 years (baby), 3–13 years (child), and 13 and above (adolescent/adult). The aim is to analyze and build an efficient machine learning model to detect faces from the photo images of the people waiting at the ticket counter to buy tickets. Then classify the detected faces into the above-mentioned categories and dispense the appropriate ticket.

2 Existing Work

In [1] the authors explain the theory on face recognition using the Haar cascade method, neural network architecture, implementation of convolution neural network (CNN) with preprocessing, and analysis of the model. In [2] the authors compare various face recognition algorithms to detect the face and classify the recognized faces into human faces and non-human faces using CNN, SVM, and K-Nearest Neighbor (KNN) algorithms. In [3], the authors estimate humans' real-time age, gender, and emotion by analyzing facial images using methods and techniques such as Viola-Jones, Principal Component Analysis (PCA), SVM, Histogram of Oriented Gradients (HOG), and CNN. In [4] the authors proposed image classification on the MNIST dataset using the CNN algorithm. In [5] the authors reviewed some classification techniques which can handle medical data and mainly focused on SVM. In [6] the authors proposed web data extraction from Instagram to detect the faces using the Haar cascade algorithm and classified them into images with selfies and images without selfies. In [7] the authors proposed automatic age and gender classification with limited data using simple CNN architecture. In [8] the authors proposed facial expression classification using techniques such as the Viola-Jones algorithm, PCA, KNN, SVM, HOG, Multi-Layer Perceptron Neural Network (MLPNN). In [9] the author proposed face detection using Haar cascade algorithm and real-time face emotion classification using CNN. In [10] the author's proposed face recognition based on CNN to train the model with faces. In [11] the authors proposed modified convolutional neural networks for face recognition. In [12] the authors have proposed image classification using CNN on CIFAR-10 dataset implemented with MatconvNet. In [13] the author proposed fruit image classification using MobileNetV2 and compared it with mobileNetV1.

3 Methodology

Figure 1 represents the steps involved in the process of building the machine learning models. It comprises two main phases namely face detection and preprocessing

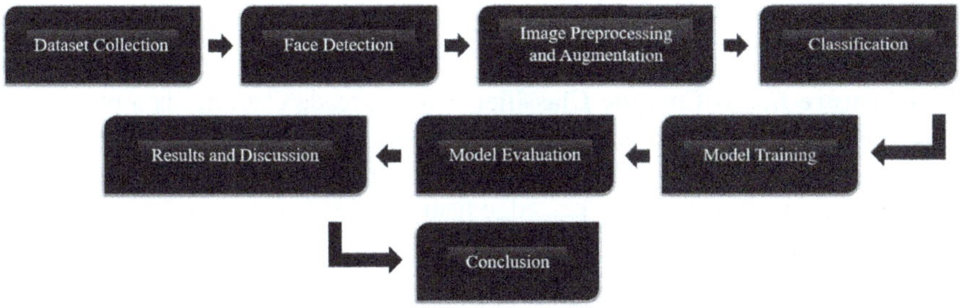

Fig. 1 The proposed work of this paper

as first phase and classifier, model building using the training dataset, and model evaluation as the second phase.

3.1 Dataset Collection

The dataset has images of people from various parts of the state of Tamil Nadu, India. The scope of data is restricted within Tamil Nadu because people from other states and other countries have different facial features.

Figure 2 depicts the images in the data set of classes that are mentioned in Table 1.

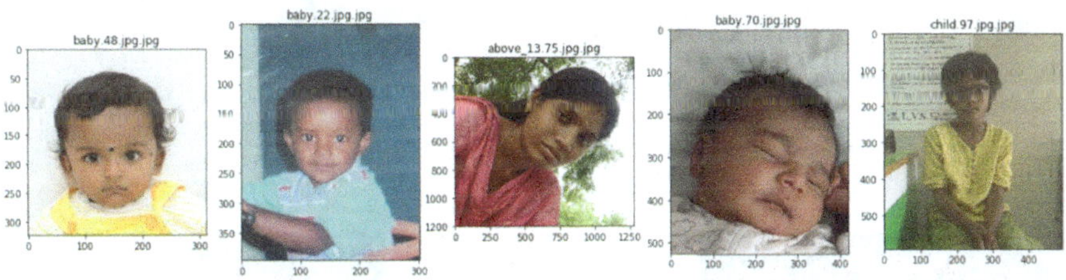

Fig. 2 Images from the dataset

Table 1 Annotation of data set

Classes/Categories	Number of images
0–3 years old (baby)	150
3–13 years old (child)	182
13 and above years old (adolescents/adults)	180
Total	512

3.2 Face Detection

Haar Feature-Based Cascade Classifier. Haar cascade classifier is a pre-trained model that is used to detect objects by extracting features from the image. In this case, the object is the face that is to be detected from the images. The features can be detected using Haar features, i.e., edge features, line features, and four-rectangle features.

In Fig. 3 the features are detected. The horizontal black rectangle represents darker eyes compared to the lighter forehead (white rectangle). The vertical black and white rectangles are the features that focus on the same eyes, where the eye is darker than the part between the eyes.

In computer vision, the library named Open source Computer Vision (OpenCV) has Haar cascade algorithm as a pre-trained model to detect the face from the image.

The red-lined square boxes in Fig. 4 illustrate the region of interest detected from the images using the Haar feature-based cascade classifier. These detected faces are extracted from the image and given to the model for training, validation, and prediction.

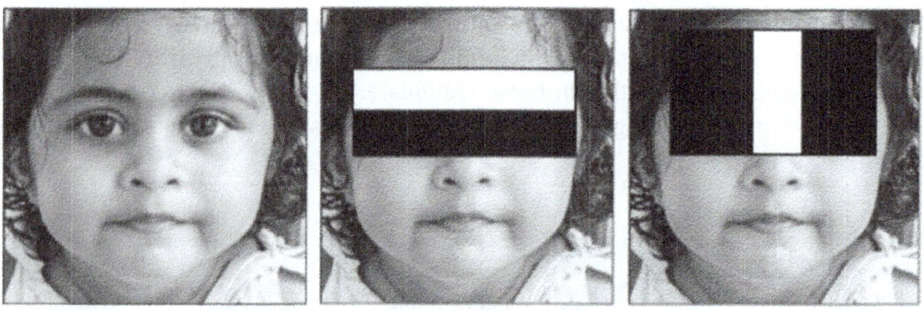

Fig. 3 Features identified, after elimination of irrelevant features

Fig. 4 Face detected by Haar feature-based cascade classifier

Table 2 MobileNetV2 model summary

Layer	Output_shape	Parameters
mobilenetv2_1.00_224 (Model)	(None, 7, 7, 1280)	2,257,984
global_average_pooling2d	(None, 1280)	0
Dropout	(None, 1280)	0
Dense	(None, 3)	3843

3.3 *Image Preprocessing and Augmentation*

The red-lined square boxes Fig. 4 are extracted from the images and are preprocessed.

Resizing. The detected faces are in different sizes, so a base size is established for all the images before feeding into the machine learning model. The base size established for the images is 224 × 224 for MobileNetV2, and 200 × 200 for SVM.

Data Augmentation. Artificially the training dataset is modified to increase the image data to increase the model's performance.

Images as NumPy array. Directly images cannot be fed into the machine learning model. So the images are converted into grayscale images. The array holds the greyscale value of preprocessed images. The greyscale value lies between 0 and 255.

3.4 *Classification*

3.4.1 MobileNetV2

MobileNetV2 has 154 layers of convolutional neural network with convolutional layers, bottleneck layer, and pooling layers. First, a baseline model is created and pre-trained with the ImageNet data set, commonly known as transfer learning. Upon that this model architecture has a global average pooling layer, a dropout layer, and a dense layer with a rectified linear activation function (ReLU). Table 2 depicts the detailed view of the MobileNetV2 model.

3.4.2 Support Vector Machine

SVM is a classification and regression algorithm in which each data point is a coordinate in n-dimensional space. Classification happens by deriving a hyperplane that makes the difference between the classes.

Many planes can be separated, but only one plane must be selected based on the closest points of both classes, in Fig. 5 the maximum distance between each class, such a plane is called hyperplane ($w \cdot x + b = 0$). Those closest points are known

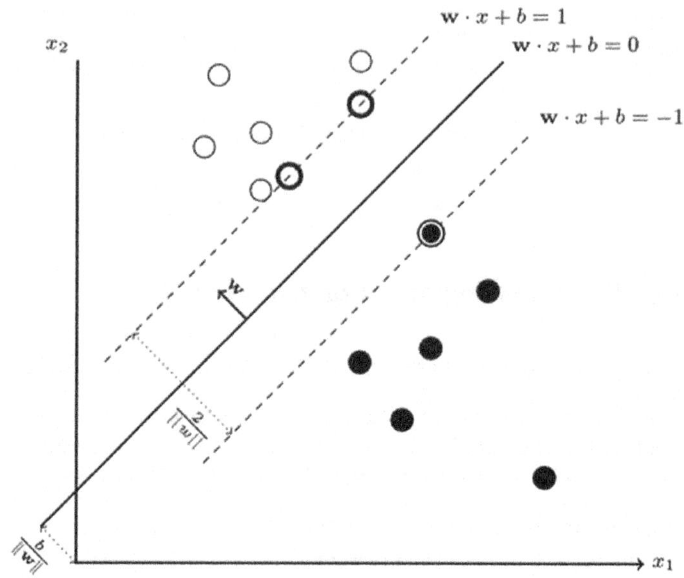

Fig. 5 An illustration of a support vector machine model

as support vectors. The line passes through the vector that is presently closer to the hyperplane in Fig. 5. The distance between the line w·x + b = 1 and w·x + b = −1 from the hyperplane is called margin.

According to the model. Data are images of size 200 × 200 were converted into arrays then flatten and along with class label value it forms a data set. As the model is based on images there will be non-linearly separable points so that soft margin was used and then kernel trick, were used to convert into a higher dimension. After that linear planes were passed to classify into three classes.

3.5 *Model Training*

MobileNetV2 model is trained with 410 images and the SVM model is trained with 435 images. The preprocessed training images are fed into the models for training.

3.5.1 MobileNetV2 Model

The sparse categorical cross-entropy is used as the loss function to calculate the error. The loss function is optimized by Adam optimizer with a learning rate of 1×10^{-4}. The model is trained with 269 epochs.

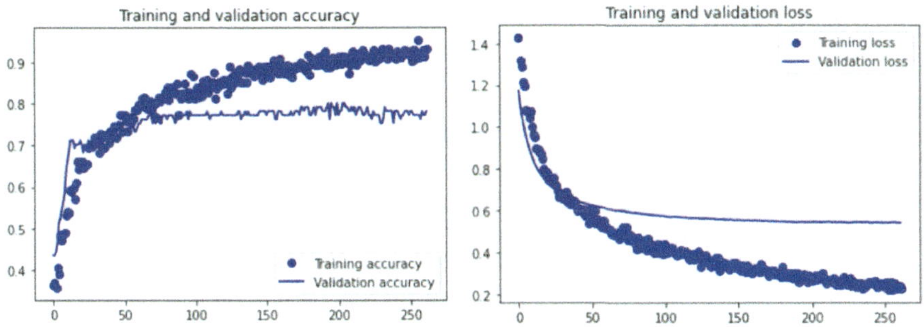

Fig. 6 Train and validation curve from the MobileNetV2 model

In Fig. 6 the training and validation curve depicts the accuracy and loss during training and validating the MobileNetV2 model.

3.5.2 SVM Model

The hyperparameter C, gamma, and kernel (Radial bias function, linear, polynomial, and sigmoid) are initialized. The gridsearchCV method is utilized to tune the hyper-parameters that are mentioned above. The model is trained with 5 folds for each of 80 images.

Figure 7 depicts the maximum training score and the cross-validation score deviates from the training score because of less training data.

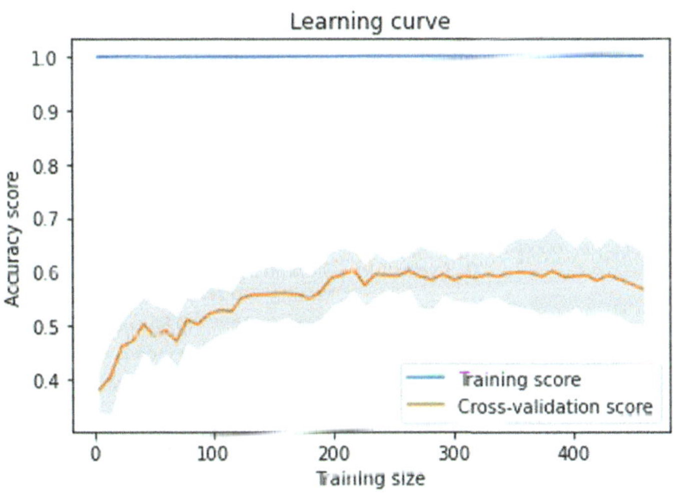

Fig. 7 Train and validation curve from the SVM model

3.6 Model Evaluation

Evaluating the model is to analyze the performance of the algorithm. The model can be evaluated using the validation data which is fed to the model. The models predict the classes for validation data and compare it with actual values. Accuracy, classification report, and confusion matrix are the metrics used to analyze and evaluate the machine learning models.

3.7 Results and Discussion

The unseen data, i.e., the validation data is fed to the model for prediction. The validation data size for MobileNetV2 is 20% of total data, and for SVM it is 15%.

Table 3 depicts the training accuracy and validation accuracy of the MobileNetV2 model and SVM model. Table 4 shows the results obtained from MobileNetV2 and SVM models on validation data. It is observed that the MobileNetV2 model performs better than the SVM model.

Figure 8 depicts the number of correctly classified images in validation data, i.e., 79 images out of 101 are correctly classified by the MobileNetV2 model and 50 images out of 77 are correctly classified by SVM. This model is used to predict the class for the unseen images given in Fig. 9.

Figure 9 depicts the prediction by the MobileNetV2 model. The classes with probabilistic values are found for each image which predicts the category of the image. The ticket fare can be allocated according to these categories.

Table 3 Train and validation accuracy from SVM and MobileNetV2 models

Model	Train accuracy (%)	Validation accuracy (%)
SVM	100	64.93
MobileNetV2	95.34	80.19

Table 4 Model comparison based on classification report

Class	Precision	Recall	F1-Score
0–3 years old	SVM: 0.52 MobileNeV2: 0.85	SVM: 0.67 MobileNeV2: 0.85	SVM: 0.59 MobileNeV2: 0.85
3–13 years old	SVM: 0.61 MobileNeV2: 0.72	SVM: 0.57 MobileNeV2: 0.70	SVM: 0.59 MobileNeV2: 0.71
13 and above years old	SVM: 0.81 MobileNeV2: 0.79	SVM: 0.72 MobileNeV2: 0.83	SVM: 0.76 MobileNeV2: 0.81

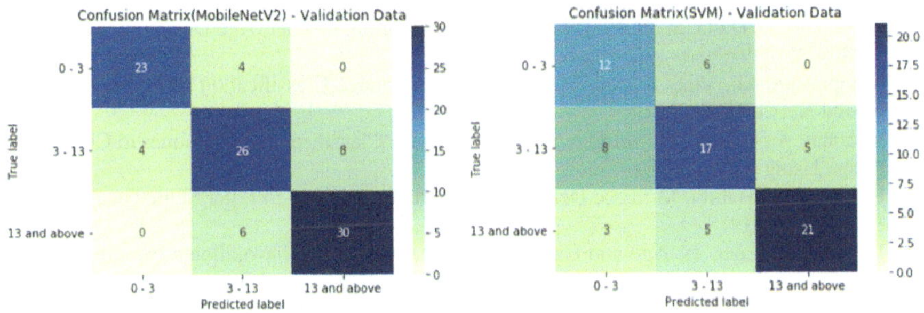

Fig. 8 Confusion matrix of MobileNetV2 model and SVM model

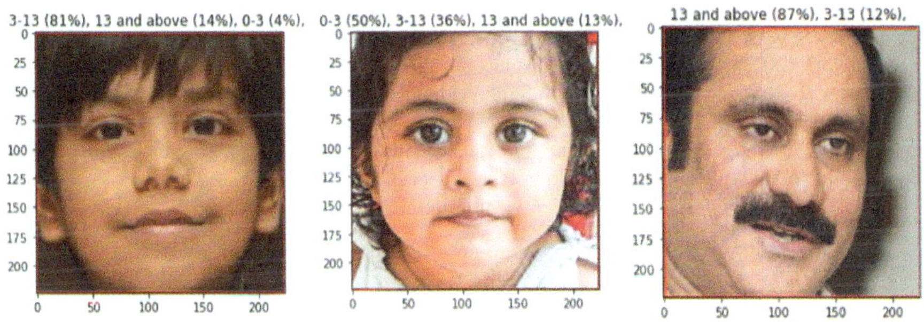

Fig. 9 Classes predicted by the model on unseen data

3.8 *Conclusion*

In most parts of the world, tickets are hand-dispensed at the entry section, the intended work is aimed to build a system that would dispense tickets in the counter by face detection and classification using machine learning. The work also stated how the faces are detected from an image using the Haar cascade algorithm. The two classifiers SVM and MobilNetV2 were implemented to perform the classification of face images. The detection and classification speed reduces manual work in the counters. This method of ticket dispensation eliminates discrepancies.

Future work may include capturing faces from video footage or live video streams and detecting multiple faces from a frame at a time, and classifying all on the whole. This system has various applications in different sectors which can make the process simple and efficient.

References

1. Habrman, D.: Face Recognition with Preprocessing and Neural Networks (2016)
2. Paul, S., Acharya, S.: A Comparative Study on Face Recognition Algorithms (2020)

3. Sharma, D.J., Dutta, S., Bora, D.J.: Real-Time Emotion, Gender, Age Detection Using CNN (2020)
4. Ramprasath, M., Vijay Anand, M., Hariharan, S.: Image Classification using Convolutional Neural Network (2018)
5. Dhabhai, A., Gupta, Y.K.: Empirical Study of Image Classification Techniques to Classify the Image Using SVM (2016)
6. Priadana, A., Habibi, M.: Face Detection using Haar Cascades to Filter Selfie Face Image on Instagram (2019)
7. Levi, G., Hassner, T.: Age and Gender Classification using Convolutional Neural Networks (2015)
8. Abdulrazzaq, M., Dino, H.: Facial Expression Classification Based on SVM, KNN and MLP Classifiers (2019)
9. Hussain, S.A., Al Balushi, A.S.A.: A Real-Time Face Emotion Classification and Recognition Using Deep Learning Model (2020)
10. Wang, J., Li, Z.: Research on Face Recognition Based on CNN (2018)
11. Coşkun, M., Uçar, Y., Yıldırım, Ö., Demir, Y.: Face Recognition Based on Convolutional Neural Network (2017)
12. Anwar Hossain, Md., Shahriar Alam Sajib, Md.: Classification of Images using Convolutional Neural Network (CNN) (2019)
13. Xiang, Q., Wang, X., Li, R., Zhang, G., Lai, J., Hu, Q.: Fruit Image Classification Based on MobileNetV2 with Transfer Learning Technique (2019)

Estimating the Effectiveness of Paratransit Service in Guwahati City

Saikat Deb, Debasish Das, and Mriganka Mazumdar

Abstract The Intermediate Public Transport (IPT) operates as feeder for the public transportation in larger cities. It acts as a key means of mass transportation in medium to small sized municipalities. The IPT contributes to mobility improvement, however the unchecked growth of IPT leads to the increasing traffic congestion and air pollution, and reduced safety. For a sustainable transport infrastructure, efficient IPT system is needed. To manage efficiently, the performance needs to be assessed by taking account of service perception, community perceptions and the level of satisfaction for passengers. In this study the performance of IPT in Guwahati, India was measured utilizing Benchmarks employing seven Key Performance Indicators (KPI). Level of service (LOS) scale spanning from LOS 1 (best service quality) to LOS 4 (worst service quality) is designed to evaluate the performance of the KPIs. IPT of Guwahati has been found to satisfactory journey speed and comfort with LOS1. Average time to wait, availability of IPT and GPS facilities fall under LOS 3, LOS 4 and LOS 4 categories are the performance areas that need to be addressed. With service level LOS 2 average number of passengers in IPT and safety of IPT performs satisfactorily.

Keywords Accessibility · KPI · Safety · Average number of passengers · Comfort

1 Introduction

As land use changes and the growing population of Indian cities increase, the demand for transportation is increasing. Lower utilisation of public transport is the consequence of inefficient transit facilities. This forces passengers to change to other modes, causing congestion and poor mobility. Therefore, the problem must be addressed by enhancing mass transit service quality. The mass transit or public

S. Deb (✉) · M. Mazumdar
Civil Engineering Programme, Assam Down Town University, Guwahati, Assam, India

D. Das
Department of Civil Engineering, JIS College of Engineering, Kalyani, West Bengal, India

K. N. Das et al. (eds.), *Proceedings of the International Conference on Computational Intelligence and Sustainable Technologies*, Algorithms for Intelligent Systems,
https://doi.org/10.1007/978-981-16-6893-7_28

transit comprises both conventional, fixed time and fixed route transits and non-conventional, fixed or flexible routes and flexible schedules [1]. The non-conventional transit is made up of cars, taxis, rickshaws, minibuses, cabs and more. Conventional transit usually travels under mixed conditions, leading to significant time to wait along with an increase of journey time. In contrast, bye its flexibility, connectivity, on-demand availability, comfort and easy operation, IPT offers increased access to narrow streets in which buses cannot run. In the lack of an organized transit system, IPT covers a substantial part of urban travel as an alternative public transport. In addition to constant performance monitoring and upgrading technology, IPT has the capability to ensure clean, low-emission and better safety [2]. Cities without a structured public transit are obviously more IPT passenger service cars in cities [3]. The performance of a city's public transit system can therefore not only be measured by traditional transport, but also IPT. The main aim of this study is therefore to evaluate the performance of the IPT system with a focus on future enhancements. Because evaluation of performance is a very important phase to update any system [4–6]. We may analyze the performance and establish the aim for upgrading the IPT system through the benchmarking procedure. Benchmarking includes systematic and consistent measurement and monitoring of performance. This helps to highlight where improvement has to be made immediately based on the present scenario assessment. Benchmarking is a complex procedure involving an identification and development of KPIs [7].

2 Methodology

Figure 1 shows the methods used in order to meet the goals of the study. The field of research is chosen to include IPT as one of the forms of public transit. Seven KPIs are assessed in order to evaluate the performance and to identify the areas of performance for improvement measures. The data are collected from the primary and secondary sources based on the need of the study. For each indicator, the methodology and service standard criteria are taken on the basis of conventional LOS categories from several manuals, with some being drawn-up by the authors. The approaches for evaluation indicate the areas where IPT is excellent and the places where improvements are needed.

2.1 Study Area Selection

IPT system and city buses plays an important role in the growing transportation demand of Guwahati city. Three-to-six-seater auto rickshaws, and maxi cabs are used as IPT vehicles in the city. In the city, IPT provides a door-to-door service and shared services throughout important corridors. A study area of 262 km^2 has been selected which fall under the Guwahati Metropolitan Development Authority (GMDA). As

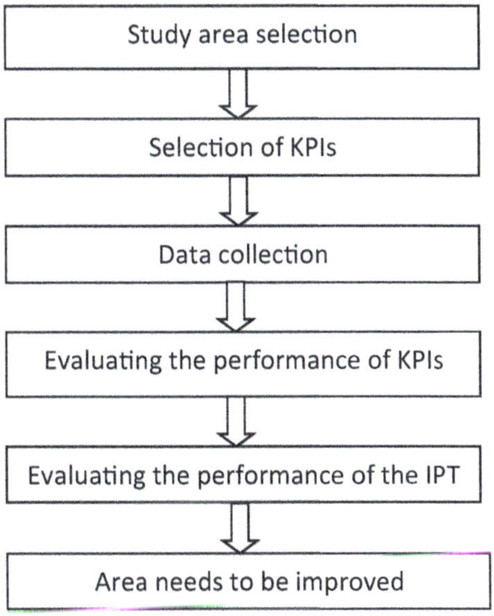

Fig. 1 Methodology

per the 2011 census data, 11.2 lakhs inhabited the Guwahati city. The current road network is classified under IRC: 86-1983 as arterial, sub arterial, collector and local road.

2.2 *Selection of KPIs and Service Attributes*

Benchmarking includes two steps: firstly, the identification of relevant KPIs and, secondly, an evaluation of the selected KPIs' performance based on specified standard values [8]. The KPIs are picked and defined to reflect the transit service [9]. Table 1 represents the selected KPIs for this study. In Table 1, column 3 represents the techniques adopted for the estimation of the particular KPI. The following is a brief summary of all KPIs:

Availability of IPT. The number of IPT vehicles per thousand inhabitants is assessed for availability. Unregulated distribution of IPT leads to traffic jams whereas, the lack of IPT means that IPT users are likely to transfer to private mode. For mobility and traffic management, therefore, availability is vital. Table 2 shows the LOS criterion for availability.

Average time to wait. Increased time to wait for an IPT vehicle imply insufficient capacity. The time to wait for the IPT is therefore an essential consideration. Time to wait is estimated by gathering the information about maximum acceptable waiting time of the passengers. Based on the acceptable time limitations, a cumulative distribution curve has been drawn. On the basis of percentile data, the waiting time service level is determined. Table 2 shows the LOS criterion for the indicator.

Table 1 Key performance indicators (KPI)

Broad areas of performance	KPI	Evaluation technique
Accessibility	Availability of IPT	[10]
Ease ofeuse	Average time to wait	Author
	Average number of passengers in IPT	Author
Comfort	Perception based analysis- driver and co-passengers' behavior, vehicle cleanliness, seating proximate to the driver	Author
Mobility	Average speed of the vehicles	[10]
Security	Rate of accidents for IPT	Author
ITS Facility	GPS facility	[10]

Table 2 Criteria of LOS for all the KPIs

LOS values	Availability of IPT/10,000 population	Average time to wait (Percentiles)	Average number of passengers in IPT (%)	Comfort (average of relative values)	Average speed of the vehicles (kmph)	Rate of accident for IPT (%)	GPS facilities (%)
1	<4	100	0	>0.75	>20	<10	≥75
2	5–6	75	<20	0.75–0.5	18–20	10–15	50–75
3	7–8	50	20–50	0.5–0.25	16–18	15–20	25–50
4	>8	25	>50	< 0.25	<16	>20	<25

Average number of passengers in IPT. For passengers, overloaded vehicles are risky. Vehicular occupancy survey was carried out at different locations of the city to gather the information about average number of passengers in IPT. The LOS criteria is computed based on the IPT carrying more passengers than their capacity.

Comfort. Comfort is the main KPI, which draws passengers to utilize IPT even when transit services are available to them. If a journey is painful or unpleasant by a mass transit mode, a person having an access to alternative mode will select that mode. While a person without option is very uncomfortable and less inclined to use transport as soon as another alternative is provided [9]. Comfort is measured in terms of users' perception. A questionnaire survey was carried out. Four comfort related criteria were presented in the questionnaire and the interviewees were requested to assess these criteria at a Likert 5-point scale with 1 indicating lowest acceptance and 5 indicating maximum acceptance. For each element, individual acceptance was then estimated using the following equation for all respondents:

$$\mathrm{AI}_r = \sum (S_j \times F_j) \div n \tag{1}$$

where

AI_a = Index of acceptability for the service attribute 'r',
F_j = frequency of respondents giving rating j to service attribute 'r',
S_j = scale value of the rating j,
n = summation of frequencies.
$= \sum F_j$.

For each criteria the relative values are obtained by dividing the index values with the maximum rating five. The overall performance of the comfort is indicated by the average of the relative values. The relative value ranges from 0 to 1. This range has been categorised into four groups for the purpose of defining LOS criteria as set out in Table 2.

Average speed of the vehicles. Average speed of the vehicles indicates mobility characteristics of the IPT. Table 2 shows the LOS requirements for the average speed of the vehicles.

Rate of accident for IPT. IPT accident rate is defined as the overall accident ratio to the overall recorded accident number of the calendar year in question. Table 2 indicates the LOS criteria.

GPS facility. The ratio of GPS enabled IPT vehicles to the total number of IPT vehicles is calculated to measure these criteria. Table 2 indicates the LOS criteria for this indicator.

3 Collection of Data

The data have been collected through on-field surveys and secondary sources. Table 3 shows data from on-field survey and secondary sources.

The speed and delay survey, average number of passenger survey in IPT and the questionnaire survey to gather users' perception data about average time to wait and comfort is conducted on-field. Field surveys consists of speed and delay surveys, vehicular occupancy survey (conducted to know average number of passengers/IPT

Table 3 Data collected from primary and secondary sources

Data	Characteristics	Data-source
Vehicle growth rate and number of vehicles	Secondary	GMDA
Accident-related data	Secondary	Head office of the city traffic police
Comfort and time to wait for IPT	Primary	Questionnaire survey
Speed of the vehicles	Primary	Speed and delay survey
Average number of passengers in IPT	Primary	IPT occupancy survey

Table 4 Attributes incorporated in questionnaire survey

Key performance indicator	Qualitative attributes	Remarks
Comfort	Behavior of the driver	A five-point Likert scale has been designed to collect the perception data
	Behavior of the co-passenger	
	Vehicle cleanliness	
	Seating proximate to the driver	
Waiting time	State your acceptable time to wait in minutes for IPT?	

vehicles) and questionnaire survey in order to know the perspective of the users' regarding waiting time and comfort. The first part of the questionnaire consists of questions related to comfort condition of the IPT vehicles and in the second part respondents were requested to state their maximum acceptable waiting time. Table 4 indicates the comfort related attributes and question related to acceptable waiting time.

In order to carry out an attitude survey, 300 passengers were interviewed on several sites across the city. The primary intention of the speed and delay study is to assess on the major corridors about the average travel speed and to determine the journey times between the origin and the destination (O-D). The journey time from the designated corridor to the fixed control locations is documented in this investigation. In accordance with corridors that had a common IPT system with defined routes, the speed and delay measuring corridors of IPT were assessed. During peak periods, two runs are done in either direction or average travel speeds are calculated. Travel times were also calculated in the similar way. This survey is carried out at 16 significant places throughout peak hours of morning and evening.

4 Results and Discussion

The overall journey length is 6 km found from the trip makers survey for IPT in the study area. It is noted that 50% of the journeys have a length of less than 4 km and 70% have a length of fewer than 6 km. For all income groups having a monthly income more than 5000 rupees, the average monthly expense for IPT is 8.12 percent of their monthly income. On the other hand, the users with a monthly income less than 5000 spend about 16% of the monthly income in using the IPT. The IPT occupancy study shows that auto rickshaws and maxi cabs are overloaded by school children during peak hours which is extremely harmful to them. In such case, the observed load factor was found to be more than 2. IPT drivers are found to flaunt traffic rules by stopping the vehicles abruptly and suddenly on the road to enable the passengers to board or alight which lead to traffic congestion and serious safety hazards.

Each KPI has a major impact on the overall IPT system performance. The Calculated Level of Service (CLOS) which indicate the present performance of the IPT are described below:

4.1 *Availability of IPT*

5330 auto rickshaws and 7534 maxi cabs are used as IPT vehicles in Guwahati, which make the total IPT vehicles as 12,864. 1,120,000 people reside within the city limit. Therefore,

$$\text{Availability of } IPT/1000 \text{ population} = \frac{12864}{1120000} \times 1000 = 11.48$$

As per Table 2, the calculated LOS (CLOS) is observed to be 4.

4.2 *Average Time to Wait for IPT*

Average time to wait for IPT is calculated based on users' perception data as discussed in earlier sections. The declared acceptable time to wait for IPT are given in Fig. 2, according to the cumulative proportion of passengers. The LOS criteria for the 100, 75, 50 and 25 percentile values as described in the methodology were assessed based on acceptable time to wait. Acceptable time to wait are <3, 3–6, 6–10, or 10 min for specified percentile values. The average waiting time for travelers is 6.44 min which fall under LOS 3 as per Table 2.

Fig. 2 Maximum acceptable waiting time

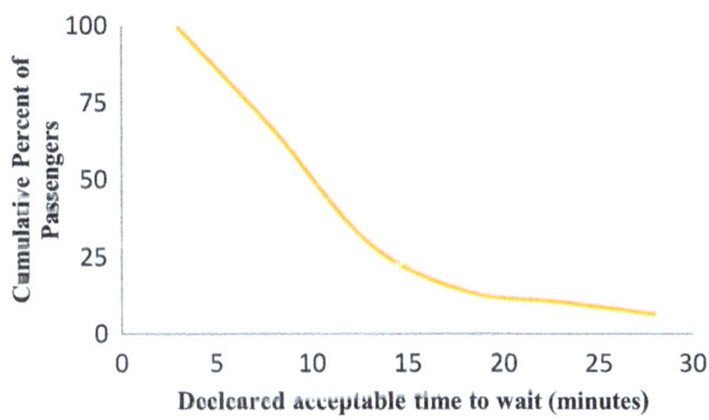

Table 5 Cumulative distribution of LOS frequency

LOS	Location frequency	Frequency percentage	Cumulative percentage
1	0	0	0
2	9	56.25	56.25
3	7	43.75	100
4	0	0	100

Table 6 Index value of comfort

Indicators	Weights given by number of users					AI_r	Relative value ($AI_r/5$)
	5	4	3	2	1		
Behavior of driver	129	95	58	5	13	4.07	0.81
Behavior of co-passengers	118	110	38	25	9	4.01	0.80
Vehicle cleanliness	82	123	95	0	0	3.96	0.79
Seating proximate to driver	112	68	52	14	54	3.57	0.71

4.3 Average Number of Passengers in IPT

By executing an IPT vehicle occupancy inspection at 16 distinct locations, the average number of passengers at IPT is computed. LOS is calculated for each location as per the values given in Table 2. The calculated LOS for the city is estimated by finding out the median value of the cumulative frequency distribution of LOS. The cumulative distribution values are shown in Table 5 and the CLOS was found to be 2.

4.4 Comfort

For each attribute, the index values are calculated using Eq. 1 and for all indicators, the relative values are identified and reported in Table 6.

For the four comfort indicators, the average relative value is 0.78. As per Table 2, the CLOS value is found to be 1.

4.5 Speed of the Vehicles

The mean speed of the vehicles is estimated from the data obtained through primary surveys in main corridors during the busy hours. From the mean speed values, the LOS for each of the routes are estimated. The mean speed of vehicles is found to be

21 km/h for all the routes. Therefore, calculated LOS for the IPT system is found to be 1.

4.6 *Rate of Accident for IPT*

The LOS of rate of accident criterion is developed on Guwahati city's 2015–2018 accident statistics. Total accidents number for the year 2018 is considered for the estimation of CLOS value. IPT accounted for nearly 153 accidents out of 987 total accidents within the city limit. The rate of accident $\left(\frac{(143\times 100)}{987}\right)$ is calculated to be 14.48%. As per Table 2, the CLOS value is found to be 2.

4.7 *GPS Facility*

The operators can regulate IPT movement with the GPS integrated in the vehicles to make sure services operate smoother. The IPT has no GPS facilities for the study area. Therefore CLOS value for this indicator is 4.

The actual service level is the average CLOS of all KPIs. The average LOS is 2.4, which shows that the IPT system has low performance. Table 7 summarizes and indicates the IPT performance report for the area of research.

5 Relevance of the Work

Public transportation is very important for a sustainable growth of an urban area. Public transport helps make a city sustainable by: Improving air quality. Reducing the need for many separate trips by private vehicles in dense urban areas. Replacing many separate emissions-producing vehicles with fewer transit vehicles that generally emit less pollution on a per person basis. But public transportation system in maximum of the cities are limited to some particular routes. Moreover, it would not be practical also to implement public transport system in all the routes. Therefore, a more flexible transport system is needed to increase the accessibility of public transport modes. IPT system, by its flexibility, connectivity, on-demand availability, comfort and easy operation, offers increased access to narrow streets in which buses or public transport cannot run. In the lack of an organized transit system, IPT covers a substantial part of urban travel demand which in turn will reduce the use of personal vehicle and will help to make the city traffic sustainable.

Table 7 Report on IPT performance

KPIs	CLOS	LOS targeted	Suggestions to achieve target
GPS facilities	4	3	Vehicle should be updated with GPS facilities for better monitoring and control
Availability of IPT	4	2	New IPT vehicles should not be permitted to be registered. IPT vehicles should be confined to operate zone wise. They should act as a feeder trip to mass transit
Average time to wait for IPT	3	2	Frequency could be increased during the morning and evening peak hours
Rate of accident for IPT	2	1	Drivers of the IPT should be instructed to follow traffic rules. The practice of stopping the vehicle abruptly and haphazardly should be completely stopped
Average number of passengers in IPT	2	1	Passengers should be encouraged to use city bus transit whenever possible
Average speed of the vehicles	1	1	Journey speed can be further boosted by avoiding the overloading
Comfort	1	1	Comfort of the users can be further improved by avoiding the overloading

6 Conclusion

By means of benchmarking it is observed that in the comfort and average speed of vehicles the IPT system in Guwahati performs better. In these two KPIs, the CLOS values are found to be 1. In the categories of accessibility, ease of use and ITS facilities, IPT performance is seen to be quite bad. Under accessibility, level of service of availability is found to be 4. Under ease of use area, the level of service of average time to wait, and average number of passengers in IPT are found to be 3, and 2, respectively. The accident rate CLOS is found to be 2 under the performance area of safety. The overall service level of IPT is 2.4, which implies that performance has to be improved. Table 7 included suggestions for improving the IPT system.

The IPT service can be systematized and highly coordinated in terms of routes and frequencies with the city bus system. Continuous review of the IPT system will help enhance the mobility and reduce Guwahati's private motorization.

References

1. Ministry of Urban Development (MoUD), Service Level Benchmark's for Urban Transport at a Glance, Urban MOBILITY INDIA CONFERENCE, New Delhi, India (2009)
2. Ministry of Urban Development (MoUD), Final Report, National Transport Development Policy Committee, New Delhi, India (2012)
3. Ministry of Urban Development (MoUD), Study on Traffic and Transportation Policies and Strategies in Urban Areas in India, New Delhi, India (2008)
4. Narasipuram, R.P., Somu, C., Yadlapalli, R.T., Simhadri, L.S.: Efficiency analysis of maximum power point tracking techniques for photovoltaic systems under variable conditions'. Int. J. Innov. Comput. Appl. **9**, 230–240 (2018)
5. Narasipuram, R.P.: Optimal design and analysis of hybrid photovoltaic-fuel cell power generation system for an advanced converter technologies. Int. J. Math. Modelling Numerical Optimisation **8**, 245–276 (2018)
6. Deb, S., Ahmed, M.A.: Variability of the perception about safety, comfort, accessibility and reliability of city bus service across different users' groups. Int. J. Civil Eng. Technol. (IJCIET) **9**, 188–204 (2018)
7. Transportation Research Board, Transit Capacity and Quality of Service Manual, 2nd edn (2003)
8. Kanuganti, S., Arkatkar, S.S., Singh, A.P., Sarkar, A.K.: Quantification of level-of-service index for bus routes in developing countries: a case study in India. J. Eastern Asia Soc. Transp. Stud. **10**, 1347–1366 (2013)
9. Kesten, S.: A new passenger oriented performance measurement framework for public rail transportation systems. Promet Traffic Transp. **26**, 299–311 (2014)
10. The Centre of Excellence in Urban Transport (CoE), Service Level Benchmarking in Urban Transportation for Indian Cities, CEPT University, Ahmedabad, India (2013)

A State-of-the-Art Review on Multi-criteria Decision Making Approaches for Micro-grid Planning

Nishant Thakkar and Priyanka Paliwal

Abstract Micro-grid (MG) comprises of small-scale distributed generation system which includes conventional as well as non-conventional sources. Micro-grid offers an effective solution for meeting growing energy demand. With the increasing penetration of micro-grid, the micro-grid planning has gained attention. The micro-grid planning is focused on multiple objectives comprising of economic, technical and environmental concerns. Multi criteria decision making (MCDM) is an operation research tool which can efficiently deal with such kind of complex planning scenarios. This paper presents a comprehensive literature survey on MCDM techniques for micro-grid planning. A classification of different techniques with a brief discussion has been presented. The techniques have been discussed with respect to their criteria and objectives. The objective of this papers is to develop an understanding of MCDM techniques for their effective utilization in micro-grid planning.

Keywords Micro-grid · Planning · Multi-objective problem · MCDM · Optimization

1 Introduction

Increasing population has led to an increase in world-wide energy consumption. Traditionally used generating sources based on fossil such as coal, natural gas and petroleum-based fuel have contributed in increasing environmental pollution. In order to address climate concerns, major countries around the world have moved towards distributed generation (DG) sources based on renewable energy [1]. DGs cover various energy generation technologies, such as solar, gas turbines, wind, fuel cell, microturbine and energy storage system (ESS) [2]. However, integration of

N. Thakkar (✉) · P. Paliwal
Department of Electrical Engineering, Maulana Azad National Institute of Technology, Bhopal, India

P. Paliwal
e-mail: priyankapaliwal@manit.ac.in

K. N. Das et al. (eds.), *Proceedings of the International Conference on Computational Intelligence and Sustainable Technologies*, Algorithms for Intelligent Systems,
https://doi.org/10.1007/978-981-16-6893-7_29

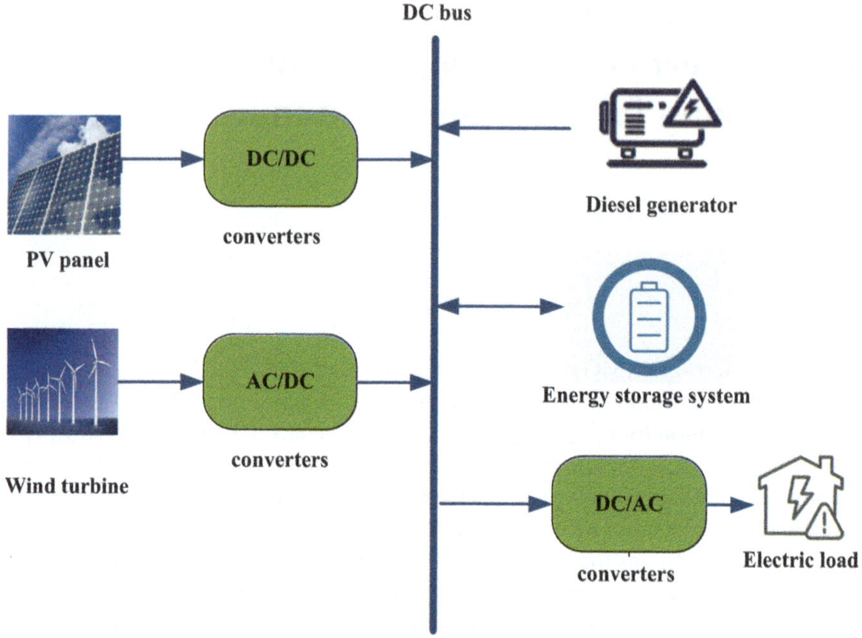

Fig. 1 Islanded micro-grid architecture

renewable DGs is not free from challenges. One significant issue is the intermittent and randomized behaviour of solar and wind energy. These problems severely affect system power quality and reliability. To address this problem, the concept of micro-grid has been introduced [1]. The micro-grid comprises of a combination of low voltage, small-scale complementary generation technologies. It is a small-scale system which is located near to consumer end. Figure 1, presents a general architecture of islanded micro-grid.

Micro-grid delivers reliable and flexible operation in both grid and islanded connected mode [3]. The planning of micro-grid affects system reliability, security, stability and economy. Hence, planning and technology evaluation plays a significant role in micro-grid deployment [4]. The micro-grid planning is a complex multi-objective problem [5]. Therefore, in order to solve such problems, MCDM has emerged as an effective and efficient tool for micro-grid planning. In this paper a brief review on existing literature based on MCDM in context of micro-grid planning is presented. This work also discusses various MCDM method and their merits over other planning methods.

The remaining paper is ordered as follows: Section 2 presents micro-grid planning and need of MCDM framework. Section 3 highlights the classification of MCDM and presents a discussion on commonly used MCDM techniques. Section 4 presents the conclusion drawn from this review.

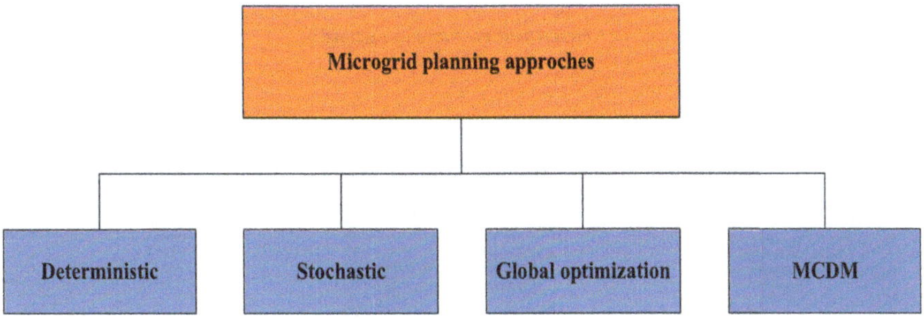

Fig. 2 Classification of micro-grid planning techniques

2 Micro-grid Planning

The micro-grid planning has to deal with the determination of optimal size and site of available DGs and energy storage systems (ESS). The MG planning comprises of various objective such as power reliability, system stability, economic and technical parameters [6]. Many literature have reported micro-grid planning with different objectives and optimization techniques [7]. Both, single as well as multi-objective formulations have been discussed in literature [8]. Generally, single objective problem is based on minimization and maximization of specific element. However, with the advancement of new DGs, and their involvement in micro-grid, the focus of planning has shifted towards addressing multiple objectives. The multiple objectives comprises of various criteria and sub-criteria [9]. Thus, in order to manage such problem, MCDM has been adopted by many researchers.

Figure 2 presents a classification of different approaches used for micro-grid planning. These approaches comprise of deterministic, stochastic and heuristic optimization-based techniques. Amongst these techniques, this paper is focused on MCDM method due to various advantage over other techniques as follows [10]:

Suitable for multi-dimensional problem.
Can be adopted to solve different problems
Method is quite as fast compared to other methods

3 Multi-criteria Decision Making

In MCDM method, problem comprises of goal, set of criteria and set of alternatives [11, 12]. The decision maker ranks the available criteria-based on priority of objectives. Generally, MCDM comprises of five elements viz. objective, conditions, alternatives, preferences and outcomes [8]. A classification of different MCDM techniques [9] has been presented in Fig. 3.

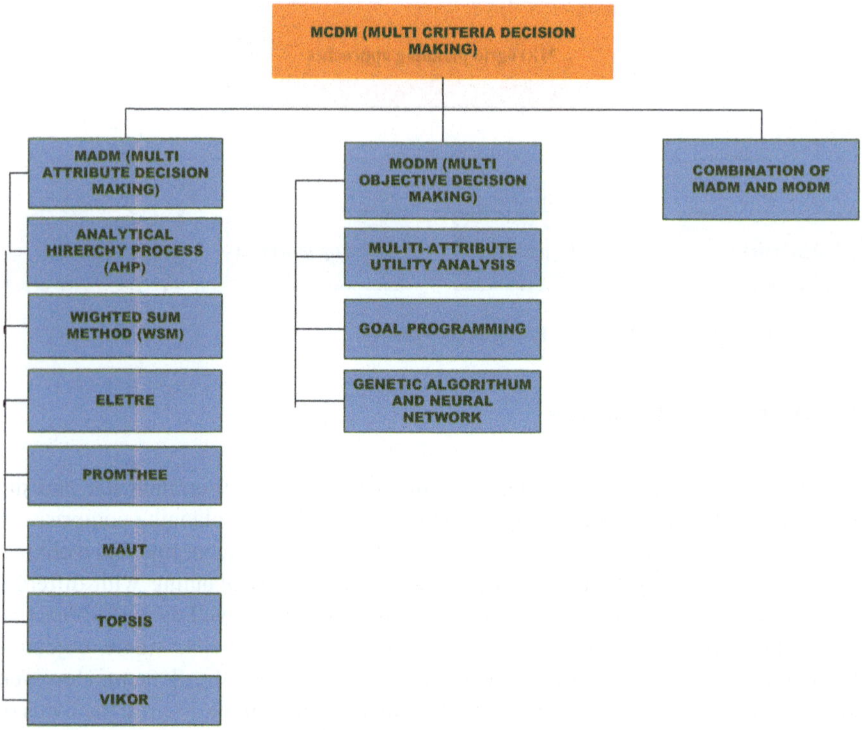

Fig. 3 Classification of MCDM techniques

3.1 Weighted Sum Method (WSM)

The weight sum method was first presented by Fishburn in 1967 [9]. This method use the weight vector to change all individual objective functions into single objective function [10]. It is one of the simplest MCDM method. In WSM, additive hypothesis is used, which indicates that overall value of each alternative is equal to the total sum of product.

3.2 Analytic Hierarchy Process (AHP)

AHP is one of the most important and highly used technique in MCDM. This method was primarily proposed by Saaty in 1970s [13]. AHP converts complex decision-making problem into small problems. Thus, criteria are converted into sub-criteria. This considerably simplifies the problem [12]. AHP is based on hierarchy process. It is mainly used in determining the score of alternative and relative weights of decided criteria [11]. In AHP, problem is categorized into three levels: goal or main objective,

Table 1 Saaty's performance scale [12]

Weight	Description
1	Equal preference
3	Moderate preference
5	Strong preference
7	Very strong preference
9	Extremely strong preference
2, 4, 6, 8	Intermediate values

criteria and alternative. The AHP method use Saaty's scale for assigning weightage in order to compare different criteria [13]. The Satty's preference scale is presented in Table 1 [10]. In Ref. [14], authors used the AHP method for sustainable development and RES selection of an electric generation system. AHP builds a matrix based on i and j element and coefficient m_{ij} which is the result of pairwise comparison for each level of hierarchy. The coefficient is called as preference weight. As show in Table 1, all first nine integers judge the performance of connected element [12].

3.3 Topsis

The TOPSIS method was first presented by Hwang and Yoon in 1981 [9]. The TOPSIS computes the distance of solution form negative and positive ideal solution. Based on this, it selects the best alternative [10]. Both positive and negative solution are hypothetically derived for the method. The procedure of TOPSIS method is depicted in Fig. 4 [15].

Carolina Marcelino et al. [16], used the TOPSIS and AHP method for finding best alternative for hybrid micro-grid. Here, five different battery technologies with three different criteria such as maximization of locally available RES, minimization of electric cost and loss of load probability are considered. Lazzerini [12], applied the TOPSIS and AHP method for ranking of four criteria such as cost of energy, environmental impact and power line load level.

3.4 ELECTRE

ELECTRE (Elimination Et Choix Traduisant la Realite) method was made accessible by Roy [9]. The ELECTRE method has various variations such as ELECTRE I, II, III, IV, IS and TRI. All ELECTRE extension have different procedures. The ELECTRE I and ELECTRE IS are used for assigning kernel for preferred alternative in single or group of option. ELECTRE II, III and IV are used for classification of option and ELECTRE TRI are used for assigning the predefined category to all options [15]. In Ref. [17], the ELECTRE III method was used for decentralized

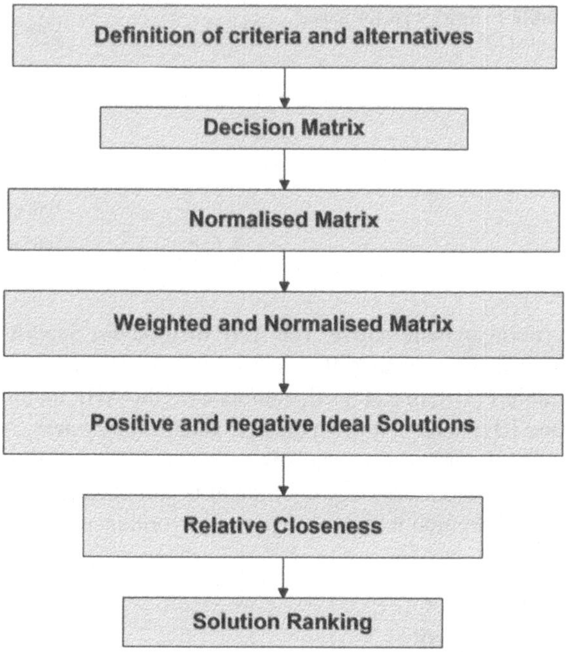

Fig. 4 Procedure of TOPSIS method

energy system optimization. In reference [30], the comprehensive review of fuzzy-MCDM for energy policy is studied. Here fuzzy AHP, fuzzy ELECTRE and fuzzy PROMETHEE have been analysed.

3.5 Promethee

PROMETHEE method was presented by Brans and vincke in 1985 [9]. It is one of the most efficient methods of MCDM. The PROMETHEE method has six different versions namely PROMETHEE I, II, III, IV, V and VI. The PROMETHEE method comprises of following steps [15].

- Independent preference between two variables is presented by decision maker.
- The presented alternative is compared based on the preference function.
- Alternatives are compared and measure value are fill in matrix.
- Partial ranking is providing through PROMETHEE I approach.
- Final ranking of alternative is done through PROMETHEE II.

In reference [18], PROMETHEE method has been used for selection of suitable location for micro-grid placement.

3.6 MAUT

The MAUT method was presented by Edwards and Newman in 1982 [9]. In this method utility function for each decision criterion is defined. Here, utility for each criterion is generated in the interval of 1 and 0 [19]. In Ref. [20], MAUT and PROMRTHEE method have been discussed and compared.

3.7 Vikor

VIKOR method was introduced by Opricovic in 1988 [9]. This method finds solution, which is closer to ideal one and it also ranks the alternative [21]. It is basically based on ideal point solution, which is minimizing individual regret and maximize the group utility [22]. In Ref. [15], VIKOR method has been used for integration and comprehensive evaluation of various micro-grid schemes. In [21], author used VIKOR method to select the best RESs.

4 MCDM in Micro-grid Planning

A wide variety of MCDM techniques have been applied by researchers for micro-grid planning with different criteria and methodology. Table 2 presents summary of literature survey on application of MCDM method for micro-grid planning. The review has been done with respect to methods, criteria and objectives.

5 Conclusion

In order to fulfil energy requirement, several developed countries have moved towards deployment of renewable energy sources. For maximization of utility of RESs, the micro-grid plays an important role. Therefore, micro-grid planning plays a significant role in developing efficient utility framework. The micro-grid planning is based on several parameters such as social, environmental, technical, political and economic. With the study of several literatures, it has been concluded that MCDM provides an appropriate way of or dealing with such complex planning problems. With an objective to familiarize readers with MCDM, a discussion on different MCDM techniques has been presented in paper. Some of the prominent papers with results have been discussed. Based on review, it can be concluded that MCDM based planning can provide an effective way of designing micro-grid. The review on MCDM techniques presented in this paper is expected to assist system planners in selection of appropriate method.

Table 2 Summary of MCDM techniques

References	Method	Criteria	Summary	Reported results
[12]	Fuzzy AHP and TOPSIS assisted, LPDM	• Power reliability′ power quality • Proper handling of electricity demand • Cost reduction	Design of efficient energy system in micro-grid	Presents new approach and compared it with WS method
[23]	AHP-CRITIC and VIKOR	• Cost • Reduction of peak load • Reduction of emission	Comprehensive evaluation of micro-grid	Solar PV system is obtained as best system for micro-grid
[16]	AHP-TOPSIS	• Electricity cost • Loss of load probability • Maximize locally available RESs	Five different battery technologies are assessed for selecting best option in hybrid micro-grid	Among five different battery technologies, Li-ion battery scored highest rank
[11]	AHP-Assisted MCDM	• Economic • Technical • Structural • Social aspect	Energy resources planning in micro-grid using MCDM	Based on the weight assignment combination of solar PV and grid obtained highest score (0.136)
[24]	AHP (for weighting of different criteria) MULTIMOORA (for different plan polarization)	• Cost of energy • Energy not served • Profits for grid power supply during peak load • Capital costs • Total emissions	Different combination of RESs for micro-grid optimal planning	Combination of DG, biomass and microturbine scored highest rank
[4]	Combination of BWM (best–worst method), entropy weighting and grey cumulative prospect theory	• Electricity supply reliability • Environmental protection • Economy	Micro-grid optimal planning	Combination of solar PV (100 kW), wind turbine (200 kW) and BSS (600 kWh) are optimal combination
[25]	AHP (for optimization of problem) and non-dominated sorting genetic algorithm-II (for optimization of method)	• Optimal size • Location of DERs • Minimizing power mismatch in islanded mode • Investment and operation costs	Micro-grid planning for reliable and efficient operation	Proposed new approach and verified for different conditions
[26]	AHP	• Technical • Economical • Managerial • Social	Environment friendly and efficient system development	Role of MCDM in energy management

(continued)

Table 2 (continued)

References	Method	Criteria	Summary	Reported results
[27]	Integrated fuzzy-MCDM model with Fuzzy-Delphi approach	• Technological • Environmental • Economic • Social performance	Different MCDM techniques are used for assessment of optimal battery storage system	Li-ion batteries are selected as optimal storage system
[28]	AHP and ARAS	• Performance • Economics • Sustainable development • Ecological integrity • Socially responsible operation	Selection and analysis of different energy generation technologies	Based on criteria preferences nuclear energy is ranked as best alternative
[29]	WSM, WPM and TOPSIS (for order Preference if similar idea)	• Reduction in active Losses • Reduction in Reactive Losses • Cost of Power Loss	Three different MCDM technique are compared at two different power factor for evaluation of preference of loop configured micro-grid	The combination of two DG system of 0.82 power factor are considered as best configuration

References

1. Paliwal, P., Patidar, N.P., Nema, R.: Planning of grid integrated distributed generators: a review of technology, objectives and techniques. Renew. Sustain. Energy Rev. **40**, 557–570
2. Paliwal P.: Reliability constrained planning and sensitivity analysis for Solar-Wind-Battery based Isolated Power System. Int. J. Sustain. Energy Plann. Manage. **29**, 109–126 (2020). https://doi.org/10.5278/ijsepm.4599
3. Bhargava, S.K., Das, S.S., Paliwal, P.: Multi-objective optimization for sizing of solar-wind based hybrid power system: a review. Int. J. Innov. Res. Sci. Eng. Technol. **3**(3), 195–201 (2014)
4. Zhao, H., Guo, S.: Selecting the optimal micro-grid planning program using a novel ulti-criteria decision making model based on grey cumulative prospect theory. In: MDPI, Energies, vol. 11 (2018). https://doi.org/10.3390/en11071840
5. Tsoutsos, T., et al.: Sustainable energy planning by using multi-criteria analysis application in the island of Crete. Energy Policy **37**(5), 1587–1600 (2009). https://doi.org/10.1016/j.enpol.2008.12.011
6. Paliwal, P., Patidar, N.P., Nema, R.: Probabilistic indices for analyzing the impact of DER penetration on system reliability. IET Renew. Power Generation **14**(12, 07), 2154–2165 (2020). https://doi.org/10.1049/iet-rpg.2019.1214
7. Liu, Y., et al.: Modeling, planning, application and management of energy systems for isolated areas: a review. Renew. Sustain. Energy Rev. **82**, 460–470 (2018). https://doi.org/10.1016/j.rser.2017.09.063
8. Daim, T., et al.: Research and technology management in the electricity industry: methods, tools and case studies. Green Energy Technol. **60**, 17–31 (2013). https://doi.org/10.1007/978-1-4471-5097-8

9. Kumar, A., Sah, B., et al.: A review of multi criteria decision making (MCDM) towards sustainable renewable energy development. In: Renewable and Sustainable Energy Reviews, Elsevier, vol. 69(c), no, November 2016, pp. 596–609, https://doi.org/10.1016/j.rser.2016.11.191
10. Lazzerini, B., Pistolesi, F.: Efficient energy dispatching in smart micro-grids using an integration of fuzzy AHP and TOPSIS assisted by linear programming. In: Conf. of the European Society for Fuzzy Logic and Technology, Italy (2013)
11. Bohra, S.S.: AHP-assisted multi-criteria decision-making model for planning of micro-grids. IECON 2019–45th Annu Conf. IEEE Ind. Electron. Soc. **1**, 4557–4562 (2019)
12. Lazzerini, B., Pistolesi, F.: A linear programming-driven MCDM approach for multi-objective economic dispatch in smart grids. In: SAI Intelligent Systems Conference 2015 November 10–11, 2015, London, UK
13. Kumar, A., Sah, B., et al.: Application of multi-criteria decision analysis tool for design of a sustainable micro- grid for a remote village in the Himalayas. In: 6th Internatinal Conference of Renewable Power Generation, vol. 2017, issue 13, October 2017, pp. 2108–2113. https://doi.org/10.1049/joe.2017.0702
14. Ahmad, A., Tahar, R.M.: Selection of renewable energy sources for sustainable development of electricity generation system using analytic hierarchy process: a case of Malaysia. In: Renewable Energy, Elsevier (2013)
15. Kolios, A., et al.: A Comparative study of multiple-criteria decision-making methods under stochastic inputs. In: MDPI, Energies, July 2016, pp. 1–21. https://doi.org/10.3390/en9070566
16. Marcelino, C., et al.: A combined optimization and decision making approach for battery supported HMGS. J. Oper. Res. Soc. **71**(5) (2020)
17. Papadopoulos, A., Karagiannidis, A.: Application of the multi-criteria analysis method Electre III for the optimisation of decentralised energy systems. In: Omega, Science Direct, vol. 36 (2008), pp. 766–776. https://doi.org/10.1016/j.omega.2006.01.004
18. Akinyele, D., Ighravwe, D.: Evaluation of solar PV micro-grid deployment sustainability in rural areas: a fuzzy STEEP approach. In: 2019 IEEE PES/IAS PowerAfrica (2019), pp. 593–598
19. Fridgen, G., Rusic, M.: Energy cooperatives as an application of micro-grids: multi-criteria decision support for investment decisions. In: 36th International Conference on Information System, Fortworth, USA (2015)
20. Wang, M., Lin, S., Lo, Y.: The comparison between MAUT and PROMETHEE. In: Conf. IEEE, IEEM (2010), pp. 753–757
21. Kumar, M., et al.: Selection of best renewable energy source by using VIKOR method. In: Technol Econ Smart Grids Sustain Energy (2017), pp. 1–10. https://doi.org/10.1007/s40866-017-0024-7
22. Zhang, Y., Yang, J., et al.: Evaluation of multi-energy micro-grid based on prospect theory and VIKOR method. In: 12th IEEE PES Asia-Pacific Power and Energy Engineering Confrence, vol. 3 (2020)
23. Zhang, Y., et al.: Evaluation of multi-energy microgrid based on prospect theory and VIKOR method In: IEEE PES Asia-Pacific Power and Energy Engineering Conference (APPEEC) (2020), pp. 1–5. https://doi.org/10.1109/APPEEC48164.2020.9220545
24. Moravej, Z., Afshar, H.: Optimal planning of micro-grid using multi criteria decision analysis. Int J. Multidisciplinary Sci. Eng. **5**(8), August 2014, issn 2045-7057
25. Contreras, S.F., et al.: Optimal micro-grid planning for enhancing ancillary service provision. J. Mod. Power Syst. Clean Energy **7**(4), 862–875 (2019). https://doi.org/10.1007/s40565-019-0528-3
26. Gouvea, S.E., et al.: Application of MCDM method for technologies selection to support energy management. Procedia Manuf. **39**, 1289–1296 (2020). https://doi.org/10.1016/j.promfg.2020.01.340
27. Zhao, H., Guo, S.: Comprehensive assessment for battery energy storage systems based on fuzzy-MCDM considering risk preferences. Energy **168**, 450–461 (2019). https://doi.org/10.1016/j.energy.2018.11.129
28. Gedimi, V.: Analysis and choice of energy generation technologies: the multiple criteria assessment on the case study of Lithuania. In: Conf. ICSEEA2012, Energy Procedia, Elsevier, vol. 32 (2013), pp. 11–20. https://doi.org/10.1016/j.egypro.2013.05.003

29. Javaid, B., et al.: Analysis techniques for performance evaluation. In: Conf. iCoMET 2019, IEEE (2019)
30. Kaya, L., et al.: A comprehensive review of fuzzy multi criteria decision making methodologies for energy policy making. In: Energy Strategy Reviews (2019), pp. 207–228

Availability of Different Agricultural Biomass for Energy Applications in Villupuram and Cuddalore Districts

S. Rajasekaran and A. Saraswathi

Abstract Biomass briquette technology is one of the alternative sources of energy for conventional fuel; the biomass briquette can be used for both domestic and industrial applications. In India, Tamil Nadu state major agriculture production depends on paddy, sugarcane, cotton and groundnut in most of the districts. In northeastern part of Tamil Nadu, particularly in Villupuram and Cuddalore districts, the major agricultural wastes are sugarcane bagasse, groundnut shells, rice husk and cashew nut shell. The objective of this paper is to study the availability of agricultural waste in Villupuram and Cuddalore districts to produce biomass briquette. By varying the cashew nut shell and sugarcane bagasse ratio with different ratio of binding agent as sawdust and paper pulp, the investigation has conducted to find the highest calorific value of biomass briquette. In addition to the study, economic assessments have made for solid briquette production.

Keywords Cashew nut shell · Sugarcane bagasse · Sawdust · Renewable energy · Cost analysis

1 Introduction

All over the world, the commercial and industrial activities are increased due to increased population. Energy plays an important role, in developing countries like India. The global primary energy requirement depends on the fossil fuels such as coal, natural gas, petroleum products and nuclear fuel, as per global bio energy statistics In the year 2018, 81% of total primary energy supply depends on the fossil fuels [1]. The renewable energy technology shares of 13.8% include biomass [2]. Biomass briquette technology is a kind of unpolluted coal technology that will fascinate to

S. Rajasekaran (✉)
Department of Mechanical Engineering, University College of Engineering, Villupuram, Tamil Nadu 605103, India

A. Saraswathi
Department of Electrical and Electronics Engineering, University College of Engineering, Villupuram, Tamil Nadu 605103, India

K. N. Das et al. (eds.), *Proceedings of the International Conference on Computational Intelligence and Sustainable Technologies*, Algorithms for Intelligent Systems,
https://doi.org/10.1007/978-981-16-6893-7_30

prevent global warming and serve to conserve forest resources. The utilization of biomass is the third largest source of energy all the world. The biomass is one type of renewable energy [3] and sustainable source of energy, producing heat and power for commercial and industrial applications. Positive environmental impacts of biomass release very low sulfur content. The original form of biomass is very difficult for loading, unloading, transport and storage due to irregular shape and size because of low bulk density.

The biomass resources can be subdivided in terms of properties such as woody and non-woody [4]. Different types of biomass materials with combination of non-biomass material are used for briquette production from agriculture residue, fruit waste, woody biomass, pantry solid waste, textile solid waste, paper and cardboard waste, garden waste and plastic waste, etc. The various raw materials [5] for making biomass briquette from agriculture waste are: groundnut shell, sugarcane bagasse, cashew nut shell, custard shell, cottons stalk, rice husk, coffee husk, wheat straw, etc. The various binder materials used are: starch, waste paper, cassava starch, molasses, clay, lime, cow dung, etc. Biomass briquette is one among the methods for converting the low bulk density agricultural waste [6] into high-density fuel with 20% increase in combustion quality.

The density is one of the important parameters in briquette, and the highest density of briquette releases higher energy/volume ratio [7]. Hence, higher density briquettes are desirable for handling transportation and storage. The developing countries heavily depend on biomass [8] as alternative sources for their energy needs. Chandaka et al. [9] implemented a project in India, to promote conversion of waste agricultural biomass into energy in partnership with Birla Institute of Management and Technology. Ghosh [10] studied and reviewed the biomass and biowaste supply chain for bioenergy and biofuel production and investigated the cause of the major challenges and issues in India. Patil and Deshannavar [11] have investigated the composition and calorific value of dry sugar cane leaves and compared with other agricultural waste.

In India, Tamil Nadu state, plenty of agriculture wastes are available due to vast agricultural-based crop production such as groundnut, sugarcane, cashew nut and paddy. The net cultivated area [12] of agricultural land in Tamil Nadu is around 52,438.4 lakh ha. As per the Government of India, Department of Revenue states that, the net cultivate land [13] in Villupuram and Cuddalore district are 3307 lakh ha and 2199 lakh ha, respectively. In the year 2017–2018, nearly 6650 tons of cashew nut [14] are produced in Villupuram and Cuddalore districts (Formerly South Arcot District). In Tamil Nadu, nearly 34.5% cashew nut are produced in Villupuram and Cuddalore districts in land area of around 33,458 ha. In Tamil Nadu, Villupuram and Cuddalore districts are cultivating maximum sugarcane, among all districts of 30.67% and 12.28%, respectively. It is concluded that the availability of agricultural waste in Villupuram and Cuddalore districts is higher from cashew nut, sugarcane bagasse and rice husk compared to other agricultural waste.

Singh and Gu [15] have studied the feasibility of biomass conversion to energy in Indian scenario. They analyzed the utilization of agro-industry waste and leafy biomass, waste water, industrial waste, food industry waste, animal waste and municipal solid waste. Hiloidhari et al. [16] have studied various crops cultivated in India. They analyzed that 39 residues from 26 crops produce 686 MT on annual basis, in which 234 MT are estimated as surplus for bioenergy generation. They concluded that residue from sugar cane is the highest followed by rice. Balasubramani et al. [17] have analyzed the solid briquette manufacturing system model for environmental to improve the productivity. They concluded that the system could produce low carbon energy system and rural development. Afsal et al. [18] have investigated the combustion characteristics of solid briquette using vegetable market waste and sawdust. By varying the different ratio of vegetable market waste and sawdust using bentonite clay as a binding agent, the proximate analysis and ultimate analysis are carried out. They concluded that the highest value is recorded for 25:75 of vegetable market waste and sawdust.

The various literature studies show that the survey is based on availability of agricultural waste in a country or in a particular state. In this study, the availability of major agricultural waste is in a particular district, and utilization for making solid briquette with limited transportation of raw material is within the district. In this present study, the solid briquettes have produced from the various agricultural wastes [19] such as cashew nut shell, sugarcane bagasse and the binding agent as sawdust and paper waste. In addition to the study, a financial feasibility of solid briquette manufacturing process has to be carried out to find the net present value, payback period and benefit cost ratio.

2 Materials and Methods

2.1 Raw Materials—Agricultural Waste

The raw materials used for the production of solid briquettes are sugarcane bagasse and cashew nut shell which are available in plenty as agricultural waste in Villupuram and Cuddalore districts in Tamil Nadu. The required quantity of raw materials is collected from the farmer. Cashew nut shell is one of the most abundant forms of tropical biomass waste which can be used for energy generation. The cashew comprises of approximately 30% nut and 70% shell.

The cashew nut shell is very hard with shell thickness that varies from 1–2 mm is shown in Fig. 1. The sugarcane bagasse is shown in Fig. 2, is an agro-industrial residue, and is the source of raw material which contains cellulose, hemicelluloses and lignin [20]. The moisture content has removed while undergoing the pretreatment process. Sawdust is easily available biowaste from woods as it has higher specific heat [21] than the normal wood logs; it plays better replacement.

Figure 3 shows the sawdust biowaste used as a binding agent. It has the moisture content about 10%, and ash content is around 3.3%, which is lesser than coal. It is fibrous in nature and stands for longer heating period. The paper pulp used in

Fig. 1 Cashew shell

Fig. 2 Sugarcane bagasse

Fig. 3 Sawdust

Fig. 4 Resin made with starch

biomass briquettes as resin acts a better binding material for the components like (crushed sugarcane bagasse, sawdust and cashew nut shell). Thus, the pulp is formed with the help of paper, and starch binds those materials to attain a densified compact biomass briquettes. The rate of increased compactness increases the burning time of solid briquettes and also achieves the required shape and size.

Figure 4 shows that sawdust and paper waste (resin made with starch) are used as a binder agent with sugarcane bagasse and cashew nut shell. The paper waste is collected from municipal solid waste, and sawdust is collected from the saw mill. The calorific value of paper waste and sawdust is 3226 kcal/kg and 3629 kcal/kg, respectively.

2.2 Processing of Solid Briquette

The production process of solid briquette from agricultural waste is: cleaning, drying, size reduction and binder addition. The cashew nut shell and sugarcane bagasse are the primary raw material for briquette production and the sawdust, waste paper as binding agent. The collected raw material sorted and cleaned before sending to the feed stock is called sieving. The unwanted materials are removed, ensuring the recovered size before the feed stock. Sieves and magnetic conveyor are used as screening equipment to remove the impurities such as mud, soil, dirt and other unwanted materials. Washing of sugarcane bagasse and cashew shell can improve the combustion properties by applying water as a mild solvent.

Size reduction of sugarcane bagasse is an important process, prior to solid briquette production. It also increases the bulk density which improves the flow of biomass during densification. Different methods of size reduction [22] for briquette production are: chopping, chipping, hammer milling, grinding, shredding and crushing. The cashew shell can be chipped before to the feed stock. The hammer mill is common

Fig. 5 Sample of solid briquette

equipment used to reduce the size of biomass for densification. The production process of solid briquette consists of raw materials collection, processing [23], size reduction, preparation of binding agent and densification. Densification can be done by applying heat and pressure. The output of the densification process is a compressed block of solid briquette as shown in Fig. 5. The solid briquettes are produced from cylindrical briquette machine and the finished product size of 75 mm diameter and 270 mm length.

2.3 Economic Implication of Solid Briquette Technology

The economic analysis is intended to find the profitability of a solid briquettes production.

Agricultural wastes for the briquette process are the major determinant of the machinery. The production cost of the solid briquette includes capital cost, installation cost, operating cost and running and maintenance cost. The economic feasibility of this project is studied to find the net present value, benefit cost ratio and payback period [24].

2.3.1 Net Present Value (NPV)

The net present value is used to determine the probability of any project. The net present value is positive; it means that the project can be accepted, prohibited and so feasible. The discounted value of the net cash inflow represents the net present value of the project. In the present investigation, 12% discount factor is used to discount the net cash inflow representing the opportunity cost of capital.

2.3.2 Benefit Cost Ratio (BCR)

The BCR is a decision rule to accept or reject the project. If the BCR is greater than one, then the project is accepted. It measures the present valve of returns per rupee of invested and its relative measure.

2.3.3 Payback Period (PP)

Payback period represents the duration period of time required for the project to pay for itself. In the present economic feasibility study, payback period has been calculated by successively deducting the initial cost of investment from the net returns until the initial investment is fully recovered. According to the payback principle, payback period is shorter, then more desirable to the project.

3 Result and Discussion

3.1 Calorific Value of Biomass Component

Proximate analysis and higher heating value have estimated using bomb calorimeter. Table 1 shows the results from bomb calorimeter such as moisture content, volatile matter, ash content and fixed carbon for different raw materials. The cashew net shell is used as major % of raw material due to higher calorific value. The caloric value of sawdust is higher compared with paper waste. Table 2 shows different ratio of biomass raw material and its calorific value.

In this investigation, five samples are prepared with different ratio of cashew shell, sugarcane bagasse, sawdust and paper pulp. The calorific value of biomass components results shows that the cashew nut shell has higher calorific value followed by sawdust, paper waste and bagasse.

Table 1 Proximate analysis and calorific value of different biomass species

Properties	Moisture content	Volatile matter	Ash content	Fixed carbon	Calorific value
Unit	%	%	%	%	Kcal/kg
Cashew shell	6.47	72	1.05	20.48	4890
Bagasse	9.1	74.9	2.5	13.5	2270
Sawdust	8.3	1.4	14.1	76.2	3629
Paper waste	5.5	37.1	36.5	20.9	3226

Table 2 Calorific value of biomass waste for different samples

Biomass samples	Biomass	Cashew shell	Bagasse	Sawdust	Paper pulp	Total calorific value
Sample I	% of raw material	50	10	30	10	100
	Weight (g)	150	30	90	30	300
	Calorific value (kcal/kg)	733.5	68.1	326.6	96.7	1224.9
Sample II	% of raw material	55	5	30	10	100
	Weight (g)	165	15	90	30	300
	Calorific value (kcal/kg)	806.8	34.1	326.6	96.7	1264.2
Sample III	% of raw material	60	5	25	10	100
	Weight (g)	180	15	75	30	300
	Calorific value (kcal/kg)	880.2	34.1	272.1	96.7	1283.1
Sample IV	% of raw material	55	10	25	10	100
	Weight (g)	165	30	75	30	300
	Calorific value (kcal/kg)	806.8	68.1	272.1	96.7	1243.7
Sample V	% of raw material	60	5	20	15	100
	Weight (g)	180	15	60	45	300
	Calorific value (kcal/kg)	880.2	34.1	163.3	145.2	1222.8

3.2 *Economic Assessment of the Project*

The economic assessment of the project [25] includes the cost of raw material, equipment cost and operation cost. From the results of the economic analysis, the monthly gross returns of briquette and payback period are estimated. Table 3 is involved in the production cost, expenditure of raw materials and other cost in Indian rupees and its percentage. The requirements of agricultural waste for the production of solid briquette are calculated as per the result from the analysis of higher calorific value sample. Table 2 shows the higher calorific value of different % ratio of raw material. The solid briquette contains 60% of cashew nut shell, 5% of bagasse, with 25% of sawdust and 10% of paper pulp. By using the above data, 90 tons of cashew nut shell, 7.5 tons of sugarcane bagasse, 30 tons of sawdust and 22.5 tons of paper pulp are purchased per month to produce 150 tons of solid briquette production.

Table 3 Particulars of raw materials and operating cost

Group	Particulars	Amount (rupees)	Total percentage
A	Raw materials cost (agricultural waste)		75.7
	Cashew nut shell (90 T/month @ Rs. 5000/T)	450,000	
	Sugarcane bagasse (7.5 T/month @ Rs. 2500/T)	18,750	
	Sawdust (30 T/month @ Rs. 2400/T)	72,000	
	Municipal paper waste (22.5 T/month @ Rs. 1000/T)	225,000	
	Total Raw Materials Cost	765,750	
B	Operational Cost (Labor Cost)		12.5
	Skilled Operator (3 No./Month @ Rs 18,000)	54,000	
	Unskilled Labor (6 No./Month @ Rs 12,000)	72,000	
	Total Labor Cost	126,000	
C	Electricity Cost and Consumables	120,000	11.8
D	Grand Total	1,011,750	100
E	Monthly Gross Returns (Briquettes 150 T @ Rs 7500/T)	1,125,000	
F	Monthly Net returns (E – D)	113,250	

The output of biomass briquette has been calculated on monthly basis, the average cost and return using briquette machine for 8 h per day and 25 working days in a month. The output of solid briquette is 6 tons/day and 25 working days in a month to produce 150 tons. The total expenditure of the raw materials cost is shared by 75.7%, labor cost shared by 12.5% and electrical and consumables shared by 11.8%. The monthly gross return from the biomass briquette is of Rs. 113,250 by selling the briquette of Rs. 7500 per ton. The details of the annual gross returns and costs for the briquette machine for each year are presented in Table 4. It is observed that the aggregate present value of the net returns from the sale of solid briquette at the tenth year end is around Rs. 4,414,710 and the NPV of Rs. 1,014,710.

Average annual return = Total net return/Working life years of machine = 6,660,000/10 = Rs. 666,000.

Payback period = Initial Investment/Average annual return = 3,400,000/666,000 = 5.1 years.

Net present value = PV of net returns – Initial Cost = 4,414,710 – 3,400,000 = Rs. 1,014,710.

Benefit cost ratio = PV of net returns/Initial cost = 4,414,710 /3,400,000 = 1.29.

Table 4 Annual net returns and costs of briquette machine Rs. 3,400,000

Year	Gross return (Rs)	Operating cost (Rs)	Labor cost (Rs)	Total operation and maintenance cost (Rs)	Net return (Y) (Rs)	DF at 12%	Present value of net returns (Rs)
1	13,500,000	12,141,000	126,000	12,267,000	1,233,000	0.8929	1,100,893
2	13,500,000	12,141,000	252,000	12,393,000	1,107,000	0.7972	882,494
3	13,500,000	12,141,000	378,000	12,519,000	981,000	0.7118	698,256
4	13,500,000	12,141,000	504,000	12,645,000	855,000	0.6355	543,368
5	13,500,000	12,141,000	630,000	12,771,000	729,000	0.5674	413,654
6	13,500,000	12,141,000	756,000	12,897,000	603,000	0.5066	305,499
7	13,500,000	12,141,000	882,000	13,023,000	477,000	0.4523	215,771
8	13,500,000	12,141,000	1,008,000	13,149,000	351,000	0.4039	141,763
9	13,500,000	12,141,000	1,134,000	13,275,000	225,000	0.3606	81,137
10	13,500,000	12,141,000	1,260,000	13,401,000	99,000	0.322	31,875
S*					27,000	0.322	8693

4 Conclusion

In this study, the agricultural wastes available in Villupuram and Cuddalore districts are analyzed. The raw materials used for the study are cashew nut shell, sugar-cane bagasse, sawdust and paper waste. The availability of agricultural biomass in Villupuram and Cuddalore districts is maximum from cashew nut and sugar cane. Every year, nearly 6650 tons of cashew nut are produced from the agricultural land in Villupuram and Cuddalore districts. By varying the different ratio of raw material, the analysis is conducted in five samples. In conclusion, the results obtained from the higher heating value sample have been selected for economic analysis. From the investigation, it is concluded that Sample III of biomass briquette has the highest calorific value of 1283.1 kcal/kg.

Sample III contains 60% of cashew nut shell and 5% of sugar cane bagasse with 25% sawdust and 10% paper pulp as binding agent. By using Sample III data biomass compounds, the economic feasibility is studied. It is concluded that the average annual return for 10 years' service of the machine is found to be Rs. 666,000 with the payback of 5.1 years. By converting the agricultural biomass waste as a supplementary source of energy into solid briquette, the requirement of fossil fuel can be reduced. It can be effectively utilized for the production of 150 tons of biomass briquette per month with minimum transportation of agricultural waste within the district. The future scope of the project can be implicated; by analyzing the availability of agricultural biomass in the locality, the briquette can be produced.

References

1. Global Bioenergy Statistics 2020, World Bio Energy Association
2. Regional Wood Energy Development Programme in Asia, Proceedings of the International Workshop on Biomass Briquetting, New Delhi, India (3–6 April 1995)
3. International Energy Agency (IEA). World Energy Outlook, 1998 edn, www.iea.org (1998)
4. ISO 17225-1. Solid Bio Fuels—Fuel Specifications and Classes—Part. 1: General Requirements. ISO, Geneva, Switzerland (2014)
5. Kpalo, S.Y.: A review of technical and economic aspects of biomass briquetting. Sustainability **12**, 1–30 (2020)
6. Onukak, I.E.: Production and characterization of biomass briquettes from tannery solid waste. Recycling. **17**, 1–19 (2017)
7. Demirbas, A.: Physical properties of briquettes from waste paper and wheat straw mixtures. Energy Convers. Manage. **40**, 437–445 (1999)
8. Gunjanpatil: The possibility study of briquetting agricultural wastes for alternative energy. Indonesian J. Forestry Res. **6**, 133–139 (2019)
9. Chandak, S.P., Chari, K.R., Memon, M.A.: Converting waste agricultural biomass into energy: experiences and lessons learnt from a capacity building and technology demonstration project in India. J. Jpn. Inst. Energy **94**, 1129–1147 (2015)
10. Ghosh, S.K.: Biomass and bio-waste supply chain sustainability for bio-energy and bio-fuel production. Proc. Environ. Sci. **31**, 31–39 (2016)
11. Patil, R.A., Deshannavar, U.B.: Dry sugarcane leaves: renewable biomass resources for making briquettes. Int. J. Eng. Res. Technol. **10**, 232–235 (2017)
12. Government of India, Department of Revenue, Land resources
13. Government of Tamil nadu, Department of Economics and Statistics, Hand Book 2019
14. Kishore, V.V.N., Srinivas, S.N.: Biofuels of India. J. Sci. Ind. Res. **62**, 106–123 (2003)
15. Singh, J., Gu, S.: Biomass conversion to energy in India—a critique. Renew. Sustain. Energy Rev. **14**, 1367–1378 (2010)
16. Hiloidhari, M., Das, D., Baruah, D.C.: Bioenergy potential from crop residue biomass in India. Renew. Sustain. Energy Rev. **32**, 504–512 (2014)
17. Balasubramani, P., Anbumalar, V., Nagarajan, M.S., Maheandera Prabu, P.: Biomass briquette manufacturing system model for environment. J. Alloys Compounds **686**, 859–865 (2016)
18. David, A.A.R., Baiju, V., Muhammed Suhail, N., Parvathy, U., Rakhi, R.B.: Experimental investigations on combustion characteristics of fuel briquettes made from vegetable market waste and saw dust. Mater. Today: Proc. **33**, 3826–3831 (2020)
19. Mythili, R., Venkatachalam, P.: Cofiring of biomass with coal: a perspective. J. Sci. Ind. Res. **72**, 122–126 (2013)
20. Tursi, A.: A review on biomass: importance, chemistry, classification, and conversion. Bio Fuel Res. J. **22**, 962–979 (2019)
21. Antwi-Boasiako, C., Cheampong, A.B.B.: Strength properties and calorific values of sawdust-briquettes as wood-residue energy generation source from tropical hardwoods of different densities. Biomass Bio Energy **85**, 144–152 (2016)
22. Solano, D., Vinyes, P., Arranz, P.: Biomass Briquetting Process. UNDP-CEDRO Publication, Beirut, Lebanon (2016)
23. Tripathi, A.K. Iyer, Kandpal, T.C.: A techno-economic evaluation of biomass briquetting in India. Biomass and Bio Energy **14**, 479–488 (1998)
24. Abhilash Shahapur, N., Mamle Desai, R., Abdul Kader, Md.: Agricultural wastes as a supplementary source of energy: an economic assessment of a briquette making facility. J. South Pacific Agriculture **20**, 25–33 (2017)
25. Kaliyan, N., Vance Morey, R.: Factors affecting strength and durability of densified biomass products. Biomass Bio Energy **33**, 337–359 (2009)

Deep Wavelet-Based Compressive Sensing Data Reconstruction for Wireless Visual Sensor Networks

K. Sekar, K. Suganya Devi, P. Srinivasan, and T. Dheepa

Abstract In recent times, compressive sensing (CS) provides data robustness by projecting the raw signal into a different domain and transmitting the predicted signal rather than the source through sparse signal sampling. But the evidence from the physical world shows that the signals are not precisely sparse, which will slow down the performance of the existing state-of-the-art signal reconstruction procedures. Deep convolutional networks provide the better learning capability of standard autonomous feature extraction, thus achieving significant improvement in several sectors. In this paper, we proposed a deep wavelet-based compressive sensing signal recovery algorithm, which reconstructs the signal with better quality and running quicker than the existing state-of-the-art approaches and enhances the lifetime of wireless visual sensor networks (WVSNs).

Keywords Compressive sensing · Deep convolutional neural networks · Signal reconstruction · Discrete wavelet transform · Wireless visual sensor networks

1 Introduction

Wireless sensor network (WSN) has resource limitations such as power, bandwidth, and computational abilities [1]. WSN has typically used to capture information from the real-world environment. Wireless visual sensor networks (WVSNs) have been used to collect image or video information [2]. Visual applications need different views to gather the information from the application field, requiring more visual sensors. Usage of more vision sensors for a particular application provides great perspective or reconstruction of the landscape, but energy constraint, computing, and bandwidth become problems that need to be tackled. So, to minimize the consumption

K. Sekar (✉) · K. S. Devi · P. Srinivasan
National Institute of Technology Silchar, Silchar, Assam, India
e-mail: k_rs@cse.nits.ac.in

T. Dheepa
Konugnadu College of Engineering and Technology, Tholurpatti, Tamilnadu, India

K. N. Das et al. (eds.), *Proceedings of the International Conference on Computational Intelligence and Sustainable Technologies*, Algorithms for Intelligent Systems,
https://doi.org/10.1007/978-981-16-6893-7_31

of those resources, an image compression technique is required to generate virtual images with the smallest data transmission size.

Wireless sensor networks consume more energy to forward the data to the base station than the other computations. The compressive sensing technique [3, 4] has recovered the signal from fewer signal measurements. But it has two challenges, firstly, which recovers the sparse signal, and secondly, it demands immense challenge for signal recovery. A promising innovation combining CS with deep learning procedures eliminated the sparsity constraint [5–7], but recovery is still slow. While these neural pretrained models have not been optimized with CS, they showed impressive reconstruction quality to established methods.

In general, WSN consumes more energy at the time of data transmission than other processes such as capturing and computational processing. To minimize the energy utilization, (1) choose an image from a vision sensor with information variations from a predefined object, (2) on captured images, using decentralized compression, since the image captured is more prominent in size, with the high resource consumption rate.

From the above, the principal significance of this article is to provide less energy consumption-based solutions in deep learning, which means compression of images and selection of images that have better statistics to the information stored on every vision sensor.

2 Related Works

In this section, we review the implementation of different CNNs for image reconstruction processes. In particular, we are going discuss about implementing DWT into CNNs. Subsequently, studies on feature extraction are unleashed. The article [8] described deep learning-based structured signal reconstruction. The greatest challenge is that as the signal grows in size, so will the network. Mousavi and Baraniuk [9] introduced the deep inverse signal recovery procedure, and which has used the block measurement matrices. In [10], the authors discussed the fast CS recovery using generative models (F-CSRG) technique, in which they concentrate on the latent space variable to reconstruct the signal. Wei-Sheng Lai et al. [11] proposed laplacian pyramid super-resolution network (LapSRN). LapSRN is a cascaded CNNs and progressively recovers the signal from a low-resolution image

FFDNet [12] takes input as a noise-level map and speeds up the training and testing of the signal, and covers more area. LapCSNet [13] has been introduced by Wenxue Cui et al., which contains two different sub-networks for sampling and recovery. Sergey Ioffe and Christian Szegedy [14] proposed batch normalization technique, which performs normalization at each mini-batch training.

We propose an effective CNN-based compressive sensing methodology that aims to trade among performance and reliability to discuss earlier problems. The suggested down-sampling procedure utilizing discrete wavelet transform (DWT) for sampling image, and it does not lose any picture details or intermediate attributes due to its inversion capability [15–17]. DWT captures the brightness and location of feature coefficients of the image. We use the inverse wavelet transform (IWT) with a convolutional projection layer to recover feature vector observations in image extraction methods.

3 Theory of Compressive Sensing

In compressive sensing (CS) [3, 15], after observing a less number of linearly encoded samples, we attempt to restore a high-dimensional feature signal. Let the given signal vector $\mathbf{x} \in \mathbb{R}^N$, which should be recovered for the down-sampled measurements, $\mathbf{y} \in \mathbb{R}^M$

$$\mathbf{y} = \Phi\mathbf{x} + \epsilon \tag{1}$$

where Φ - $M \times N$ basis matrix and $M \ll N$. ϵ-is the distortion coefficient. The above measurement is not an accurate one, and to reconstruct the original data measurement $\mathbf{x}$ effectively, we need some structure, which performs dimensionality reduction of the signal without loss features. The compressed signal matrix $\mathbf{X}$ contains the $N \times 1$ columns of signal entries. Then, it is given as,

$$\mathbf{x} = \Psi\vartheta + \epsilon \tag{2}$$

where distortion $\epsilon \in \mathbb{R}^N$, $\Psi = [\Psi_1, \Psi_2, \ldots, \Psi_N]$, is a $N \times N$ orthonormal sparsifying basis matrix, ϵ is a $N \times 1$ noise vector, and $\vartheta = [\vartheta_1, \vartheta_2, \ldots, \vartheta_N]$ is a $N \times 1$ transform basis entries of signal vector of $\mathbf{x}$. Then, $\mathbf{x}$ is M-sparse if ϑ with M nonzero entries.

The sparse signal contains a small proportion of coefficients for value factors and a significant majority of zero factors. Measured signal matrix $\mathbf{X}$ has the vector coefficients with fewer high-valued elements and a large number of zero or low valued factors; it is then sparse and compressible in Ψ. Signal $\mathbf{x}$ has been regarded as K-compressible under some conditions on a Ψ orthogonal basis. Signal $\mathbf{x}$ could then be acquired from linear sample measurements ($K < M \leq N$) and reconstructed. A sparsely regularized convex optimization or greedy algorithm is used to recover the signal $\mathbf{x}$ from measurements $\mathbf{y}$.

4 Proposed Work

VSN uses camera sensors to provide view and resolution capabilities, as well as environmental monitoring. With resource availability, an image is constructed from all the data acquired by a camera. The sensor node performs some computations on the image before transmission in the WVSN application. WVSN naturally has fewer resources; as a response, a resource-saving method is needed by submitting the specified image based on specific criteria. Every visual sensor sends the captured

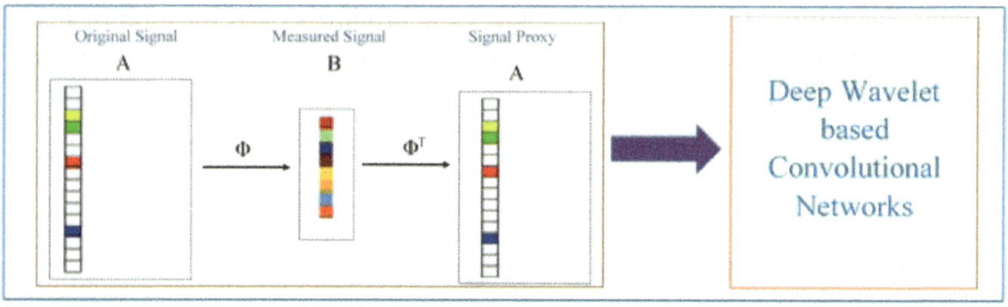

Fig. 1 Schematic diagram of compressed signal recovery using deep convolutional networks

image as a reference image to the processor. The sensor node will then capture new images, compress them, and compare new images with those referenced. The similarity measure is used as the basis for decisions when transmitting the image. The DWT method of image compression is performed after preprocessing the image. After preprocessing, the image has been given as an input source for blocks of the wavelet transform. DWT is a filter with 2-dimension and separable with its rows and columns, and it has to apply a low-pass filter (L) and a high-pass filter on the input image. In the next level, the output has given as input to the same filter and so on. In traditional approaches, the captured image is actually fed into the deep wavelet convolutional neural network (DWCNN). The proposed system uses compressive image data as input for the DWCNN network, enhancing the image reconstruction processes. Figure 1 illustrates the compressed sampling of image data and feeds this proxy signal to the deep wavelet-based convolutional network for further process.

4.1 Deep Wavelet-Based Convolutional Networks

Deep convolutional networks (DCNs) are composed of five main layers: convolutional layers, pooling, ReLU layers, fully connected layers, and loss layers. Convolutional layers are the center of such networks. They are added with series of convolutional filters with a small receptive range, repeated across the whole field of vision and feature vectors of the shape. Pooling layers are aggregating features in feature map patches under down-sampling, and nonlinearity of the ReLU layer causes nonlinearity in the total network decision function and activates the network function. All the inputs from one layer are coupled to every activation unit of the following layer which are fully connected layers. Finally, the loss layer then compares the estimated values of the fully connected layer to the actual values during training, intending to decrease the gap between the estimated and the actual value as much as feasible. A back-propagation algorithm is used to train convolutionary layers entire network and refine filters. All five layers perform significant roles in the main application of DCNs, which provides the classification of images.

4.2 Deep Wavelet-Based Compressive Sensing Signal Recovery

The proposed signal reconstruction DWCS network extracts the signal proxy **a** from measured signal vectors **b** applying IWT. The DWCS acquires knowledge from both signal vector **a** and IWT while training the images. DWCSNet takes signal proxy $\tilde{\mathbf{a}}$ as input from the measurement signal vectors **b**. Based on this, the dimension of the network should be changed and should improve the network dimensions from $\mathbb{R}^{\mathbb{M}}$ to $\mathbb{R}^{\mathbb{N}}$ and forwarded to the down-sampling operations. The measured signal matrix Φ is a fixed one in the general compressive sensing method. And for every b_r $(1 \leq r \leq M)$ is a combination of a_r signal vectors $(1 \leq r \leq N)$. The training data set $\mathcal{G}_{T_D} = [(b_1, a_1), (b_2, a_2), (b_3, a_3), \ldots, (b_p, a_p)]$ formed **p** pair of signal vector and measurement vectors, and the test set also contains pair of signal vectors and measurement vectors. In DCN, every other convolutional layer introduces the non-linearity of the ReLU towards its performance. The $(p, q)^{th}$ element of lth feature vector in the first of the convolutional layers receives a range of $\tilde{a}$ as its input, and delivers **a** then the result is,

$$[A^l{}_{Conv_1}] = \mathcal{H}(ReLU((\mathcal{W}^l{}_1 \times \tilde{\mathbf{x}}_{(p.q)}) + (T^l{}_1)_{(p,q)})) \quad (3)$$

Eventually, by avoiding the boundaries formed by nonexistent-padding the input, the down-sampling operator $\mathcal{H}()$ uses the output of ReLU(.) to the scale of the received signal.

Similarly, the feature maps of the 2nd , 3rd, 4th, and 5th convolutional layers are created. Although the filtering models in 2nd, 3rd, 4th, and 5th layers of the network may differ, the concepts in those layers are the same as 1st layer. Let $layer_1, layer_2, layer_3, layer_4$, and $layer_5$, using the channel size of 7×7 and the output of down-sampled image shapes are after every convolutional layers are 64×64, 32×32,16×16 ,8×8 , and 4×4, respectively. These filters are used to down-sampling the image data, and the deconvolutional layers reconstruct the image with better visual quality using the up-sampling. In particular, DWCSNet considers each low- and high-frequency sub-bands as inputs and seems to be secure to conduct down-sampling against loss of data. Figure 2 shows the architecture of the proposed deep wavelet-based CSnet.

4.3 Energy Model

Since the energy supply for sensor nodes in WVSN's is provided mainly from batteries with limited capacity. Thus, energy utilization becomes a crucial criterion and should implement a specific procedure or methodology for visual sensors to consume less energy while capturing images. Overall the following three activities concerning WVSN's energy intake, as data capturing, processing and transmission.

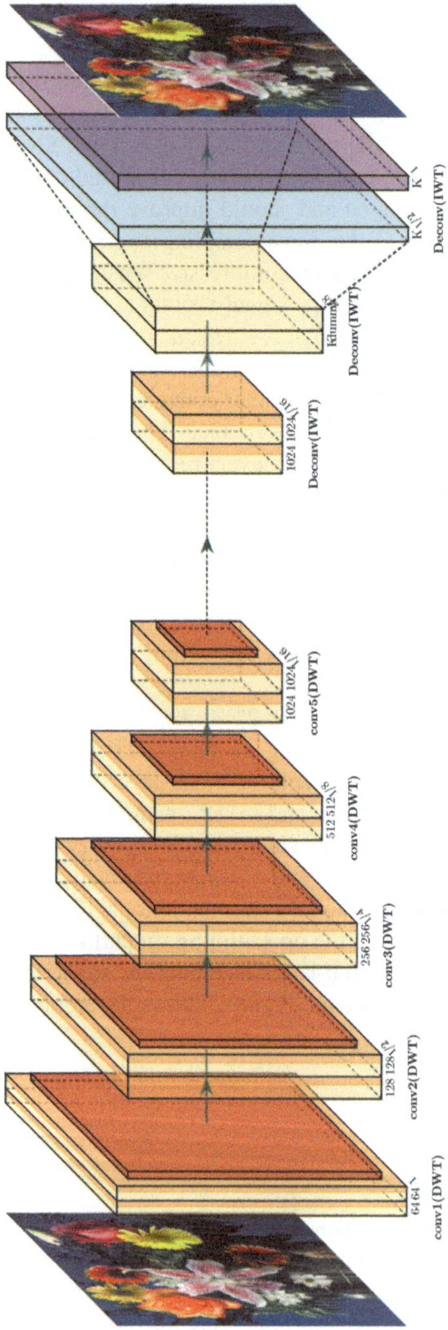

Fig. 2 Architecture of proposed deep wavelet-based CSNet

$$E_{Net} = E_{Capt} + E_{Proc} + E_{Trans} \quad (4)$$

where E_{Net} is the total energy consumed by the sensor and WVSN. E_{Capt}, E_{Proc}, and E_{Trans} are the energy consumption at capturing the image, processing the data, and data transmission to the base station. In this work, we consider the energy cost given by the ratio of total bytes of data captured, processed, and transmitted to the maximum possible data. The energy consumption is directly proportional to the number of bytes of data received at the base station. Equation(5) provides the total energy cost.

$$E_{Cost} = \frac{\sum_{i=1}^{N} E_i I_i}{\sum_{i=1}^{N} E_i} \quad (5)$$

where N represents the number of visual sensors present in the WVSN, the energy consumption of the data traveling distance through the visual sensor network has been E_i. $I_i(t)$ is an indicator function and $I_i = 1$ if the sensor i transmits data at a maximum level and < 1 means that the node i transmits data at a reduced rate, and I_i equal to 0 if i does not transmit data at t. The total energy consumption here is proportional to the amount of information transmitted by the network.

5 Results and Discussions

As we discussed earlier, the main objective of this work is to implement the machine learning frameworks to reconstruct the images with better visual quality from down-sampling of image. Using DWCSNet to get proxy input $\tilde{\mathbf{x}} = \Phi^T \times \mathbf{b}$, and has five layers with the following details. $layer_1$ has 64 filters, each having one channel of size 7×7, $layer_2$, has 32 filters, $layer_3$ with 16 filters , fourth with 8 filters and fifth with 1 filter and each having one channel of size 7×7. For training and testing purpose, we considered 64×64 size of sub-images from dataset used in LapCSNet [13]. This work performed with the support of MATLAB software and matconvnet tool [18].

Figures 3 and 4 show the visual information of reconstructed images after performing different deep convolutional networks for images of set 14. For validation, we considered BM3D [8], F-CSRG [10], FFD_Net [12], and LapCSNet [13] models for comparison with our model. For this experiment, we set the sampling rates of 0.01 and 0.02 with 256×256 patches with a batch size of 64 for two different images from set14, and the results are shown in Figs. 3 and 4, respectively. DWCSNet provides better results than the other methods.

PSNR in dB versus signal compression ratio details provided in Fig. 5, while compression ratio betters with an increase of PSNR, and Fig. 6 depicts the plot of structural similarity index (SSIM) versus compression ratio. In comparison, an increase of SSIM gives better compression of the signal. SSIM of images reconstructed using DWCSNet better than the other networks.

Fig. 3 Comparison of reconstructed image quality different CS methods at the sampling rate = 0.01

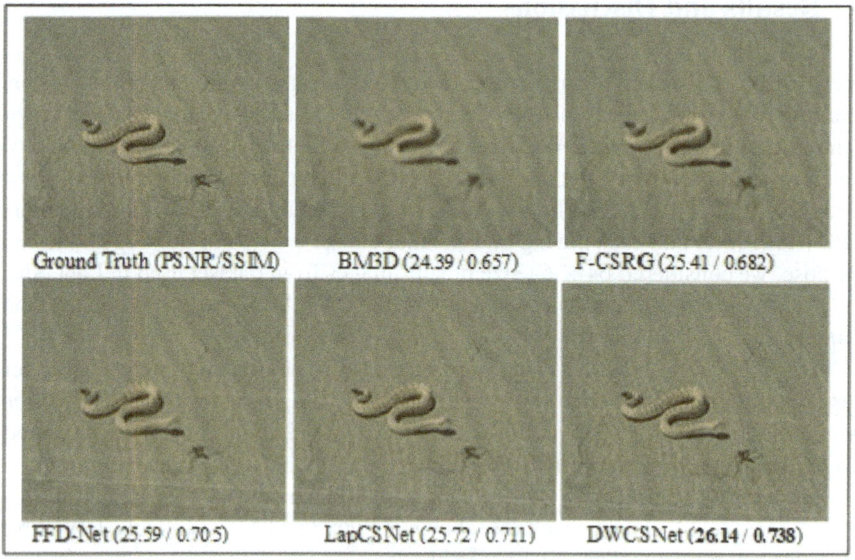

Fig. 4 Comparison of reconstructed image quality different CS methods at the sampling rate = 0.02

Fig. 5 PSNR (dB) versus compression ratio(%)

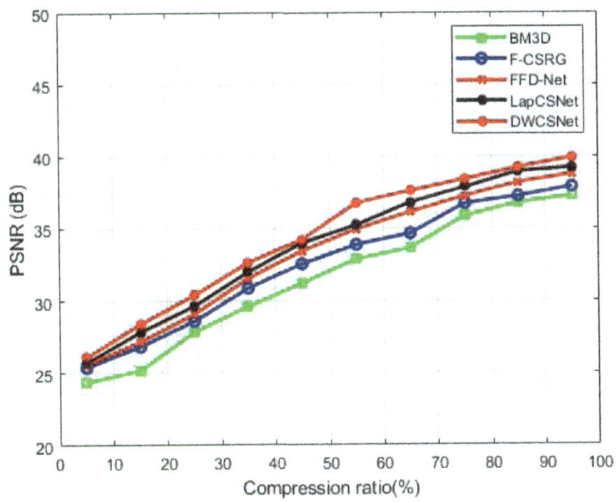

Fig. 6 SSIM versus compression ratio(%)

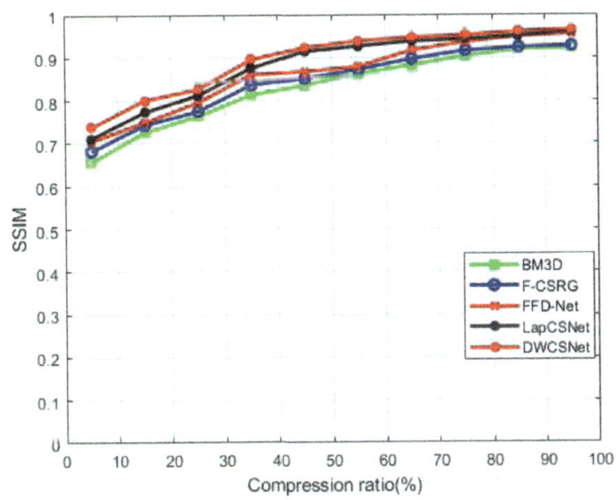

The overall energy consumption cost, while data transmission is provided by Fig. 7. DWCSNet proves that it reconstructs the compressed signal using a wavelet-based deep network and requires minimum samples to recover the signal with better visual quality. The above helps reduce energy utilization and increase the lifetime of optical sensors and WVSN. From the above results, the DWCSnet outperforms other state-of-the-art methods by a better margin. From the above, we observed that the proposed method provide better visual quality performance at the time of image reconstruction than the other methods

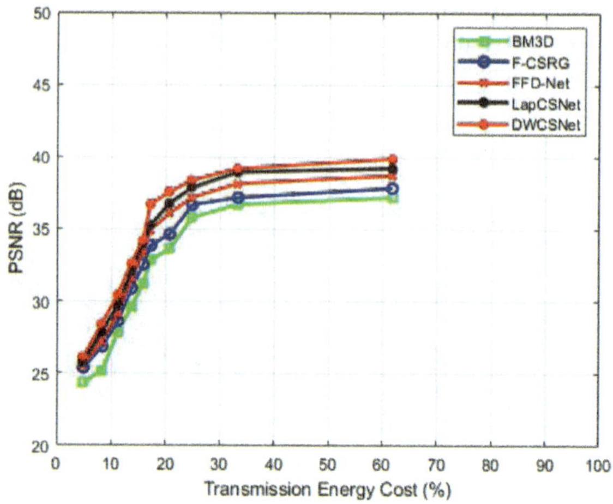

Fig. 7 Transmission energy cost (%) versus PSNR (dB)

6 Conclusions

In this work, the DWCSNet method provides a better trade-off between signal sampling and recovery. The DWT is implemented as a down-sampling procedure to minimize image quality and extend the dynamic range of the sample. It is incorporated with a pooling layer into any convolutional layers. In particular, DWCSNet considers each low- and high-frequency sub-bands as inputs and seems to be secure to conduct down-sampling against loss of data. The proposed method effectively recovers the image from down-sampling data. The superior predicting tasks such as identifying and segmenting the artifacts are performed mainly by pooling for down-sampling and then up-sampling for dense predictions. Therefore, the drawbacks of pooling activity in these tasks may still be undesirable. In the future, DWCSNet could be extended for the effective retention of characteristics on a smaller scale image data.

References

1. Akyildiz, I.F., Su, W., Sankarasubramaniam, Y., Cayirci, E.: A survey on sensor networks. Commun. Mag. **40**(8), 102–114 (2002)
2. Sastra, N.P., Wirawan, I., Hendrantoro, G.: Energy efficiency of image compression for virtual view image over wireless visual sensor network. J. Netw. **10**(6) (2015)
3. Donoho, D.L.: Compressed sensing. IEEE Trans. Information Theor. **52**(4), 1289–1306 (2006)
4. Sekar, K., Suganya Devi, K. Srinivasan, P.: Energy efficient data gathering using spatio-temporal compressive sensing for WSNs. Wireless Pers Commun. **117**, 1279–1295 (2021)
5. Wang, J., Chen, Y., Hao, S., Peng, X., Hu, L.: Deep learning for sensor-based activity recognition: a survey. Pattern Recogn. Lett. **119**, 3–11 (2019)
6. Wu,Y., Rosca, M., Lillicrap, T.P.: Deep compressed sensing. CoRR. abs/1905.06723 (2019)

7. Shi, W., Jiang, F., Zhang, S., Zhao, D.: Deep networks for compressed image sensing. CoRR. abs/1707.07119 (2017)
8. Mousavi, A., Patel, A.B., Baraniuk, R.G.: A deep learning approach to structured signal recovery. In: 2015 53rd Annual Allerton Conference on Communication, Control, and Computing (Allerton) (2015)
9. Mousavi, A., Baraniuk, R.G.: Learning to invert: signal recovery via deep convolutional networks. In: 2017 IEEE International Conference on Acoustics, Speech and Signal Processing (ICASSP), pp. 2272–2276 (2017)
10. Xu, S., Zeng, S., Romberg, J.: Fast compressive sensing recovery using generative models with structured latent variables. In: ICASSP 2019—2019 IEEE International Conference on Acoustics, Speech and Signal Processing (ICASSP) (2019)
11. Lai, W.-S., Huang, J.-B., Ahuja, N., Yang, M.-H.: Deep Laplacian pyramid networks for fast and accurate super-resolution. In: IEEE Conference on Computer Vision and Pattern Recognition (ICCVPR) (2017)
12. Zhang, K., Zuo, W., Zhang, L.: Ffdnet: toward a fast and flexible solution for cnn-based image denoising. IEEE Trans. Image Process. **27**(9), 4608–4622 (2018)
13. Cui, W., Xu, H., Gao, X., Zhang, S., Jiang, F., Zhao, D.: An efficient deep convolutional Laplacian pyramid architecture for cs reconstruction at low sampling ratios. In: IEEE International Conference on Acoustics, Speech and Signal Processing (ICASSP) (2018)
14. Ioffe, S., Szegedy, C.: Batch normalization: accelerating deep network training by reducing internal covariate shift. CoRR (2015)
15. Suganya Devi, S.K.K., Srinivasan, P., Dheepa, T., Arpita, B., Dolendro Singh, L.: Joint correlated compressive sensing based on predictive data recovery in wsns. In: 2020 International Conference on Emerging Trends in Information Technology and Engineering (ic-ETITE), pp. 1–5 (2020)
16. Sekar, K., Suganya Devi, K., Srinivasan, P., SenthilKumar, V.M.: Deep wavelet architecture for compressive sensing recovery. In: 2020 Seventh International Conference on Information Technology Trends (ITT), pp. 185–189 (2020)
17. Liu, P., Zhang, H., Lian, W., Zuo, W.: Multi-level wavelet convolutional neural networks. CoRR. abs/1907.03128 (2019)
18. Vedaldi, A., Lenc, K.: Matconvnet—convolutional neural networks for matlab. In: Proceeding of the ACM International Conference on Multimedia (2015)

Optimized 64-bit Reversible BCD Adder for Low-power Applications and Its Comparative Study

K. N. Hemalatha, S. Girija, and B. G. Sangeetha

Abstract Reversible logic has emerged its importance in the framework of recent technology as such optical computing and quantum computation. Reversible logic does not loose bits of information during computation. In the proposed work, a class of new design for reversible 4-bit and 64-bit BCD adder circuits is designed. Design of 64-bit BCD adder is first of its kind when related with the present reversible BCD adder in the literature. Proposed design uses 11 constant inputs, 22 garbage outputs, and the quantum cost are 72. Quantum cost of the circuit is reduced extremely compared to existing design. Design of 64-bit reversible BCD adder is first of its kind analyzed with all existing design available in the literature.

Keywords Reversible logic · BCD adder · Quantum cost · Garbage output · Constant inputs

1 Introduction

In low-power VLSI, power and speed are the significant terms in VLSI circuit design. Low power supply, reducing switching activity, and shrinking technology of the devices are some of the traditional methods to optimize in terms of area, power, and speed. These conventional methods are not sufficient for the criteria to meet its demand in power optimization. The conventional design approach uses irreversible gates. Conventional gates generate energy loss due to the loss in the information bit through computation. Information of bit loss occurs as the input and output vectors are not equal, i.e., not mapped. Hence, classical circuits dissipate heat for each loss of bit during computation. Landauer specified that each bit loss generates kTln2 joule

K. N. Hemalatha (✉) · S. Girija
Department of ECE, Dr. Ambedkar Institute of Technology, Bangalore, India

S. Girija
e-mail: girija.ec@drait.edu.in

B. G. Sangeetha
Department of ECE, RNS Institute of Technology, Bangalore, India

K. N. Das et al. (eds.), *Proceedings of the International Conference on Computational Intelligence and Sustainable Technologies*, Algorithms for Intelligent Systems,
https://doi.org/10.1007/978-981-16-6893-7_32

of energy, k the Boltzmann's constant, and T the temperature at which operations are carried [1]. In 1973, Bennett proved that in view to reduce the energy dissipation, logic operations must be carried using reversible logic [2]. Thus, state-of-the-art skill has to be used in reversible logic to lessen power and delay optimization. No loss of information bit in reversible circuits, and hence, the operation is reversible if it contains the reversible. The major constraints that focus on the designer's understanding of the reversible logic are for the reduction in the usage of gates and minimize the unwanted outputs generated in the design. Reversible circuit preserves input either 0 or 1 at a constant value, named as constant inputs. Garbage output referred to the output is not included in the function further [3]. Quantum cost is recognized as the cost of a primitive gate.

The proposed paper is planned as mentioned: Sect. 1 introduces the essential design of reversible circuits with the important design constraints that are discussed. In Sect. 2, the basics of reversible logic are discussed, and Sect. 3 discusses the significant reversible gates that are available with the block diagrams and functions. In Sect. 4, an irreversible BCD adder and the design issues involved are discussed. Also, the proposed reversible 4-bit and 64-bit BCD adder design issues are discussed. Section 5 discusses the result and assessment of the design proposed with the other important existing circuits available in the literature in terms of the parameters of optimization. Section 6 mentions the relevance of the paper to the theme of the conference, and Sect. 7 refers to the conclusions and future enhancement of the proposed work.

2 Reversible Logic

A reversible circuit avoids the information loss in logical computation with the inputs being made equal to the outputs as shown in Figure [1]. A reversible logic gate is an m*m gate, where m is no. of inputs or outputs and outputs are regenerated using the inputs. To reduce power, reversible circuits are used as an alternative to irreversible logic. In reversible logic, inputs are mapped one to one with the outputs by which will conserve power as there will not be an information loss (Fig. 1).

Vital design parameters used in the reversible circuit are:

Fig. 1 Block diagram of reversible logic gates

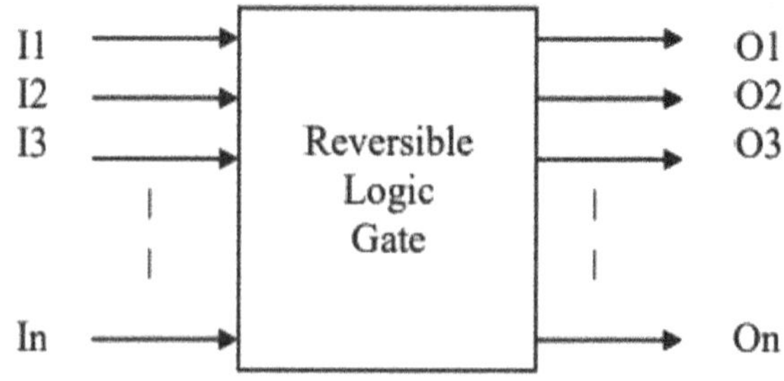

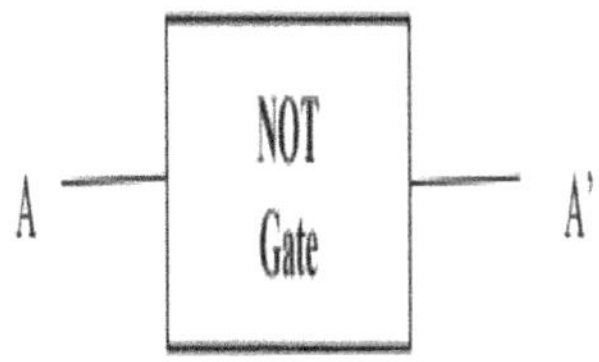

Fig. 2 NOT gate block diagram

1. Fan-outs are not permitted in reversible circuits.
2. Minimum quantum cost has to be maintained.
3. As for as possible, reduce the garbage outputs.
4. Usage of minimum Ancilla inputs.
5. Latency.

3 Reversible Logic Gates

The significant logic gates generally used in designing reversible logic are as discussed.

3.1 NOT Gate

Simple 1 × 1 reversible gate is NOT gate. The quantum cost is 1 for 1 × 1 gate, i.e., NOT gate represented in Fig. 2.

3.2 Feynman Gate

Feynman gate is a 2 * 2 reversible gate as represented in Fig. 3. Quantum cost of a Feynman gate is 1. Feynman gate (FG) is used for copying the output [4].

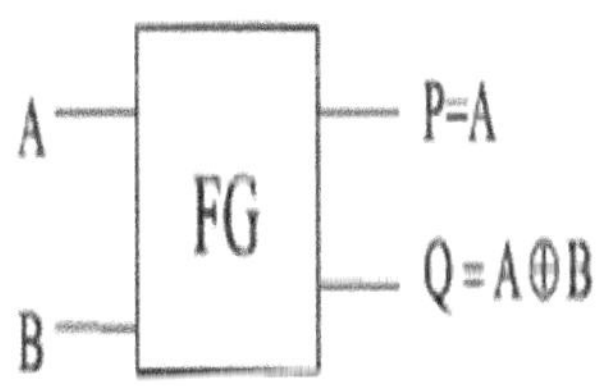

Fig. 3 Feynman gate block diagram

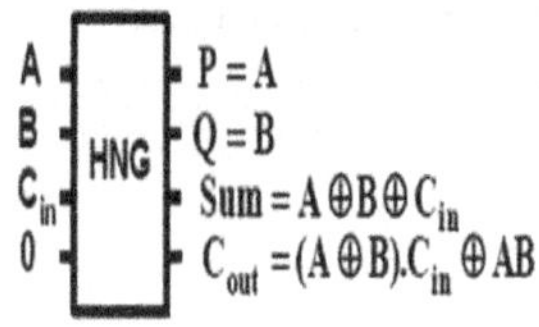

Fig. 4 HNG gate block diagram

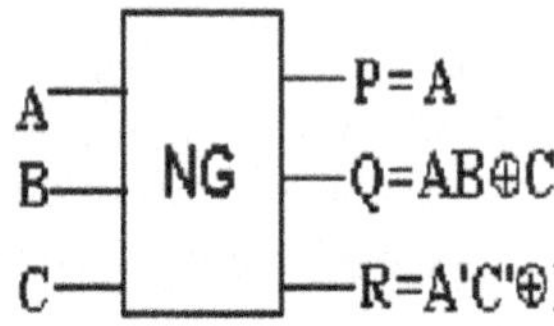

Fig. 5 NG gate block diagram

3.3 HNG Gate

HNG gate [5] represented in Fig. 4 is one of the important reversible gates that may be use independent reversible 1-bit full adder circuit with a quantum cost 6. Most designs in the literature use the HNG gate as a reversible 1-bit full adder. The proposed design uses an HNG gate as a reversible 1-bit full adder circuit for the realization of a 4-bit ripple carry adder reversible circuit.

3.4 NG Gate

NG gate [6] is a 3 * 3 gate as represented in Fig. 5. NG gate is having quantum cost of 8. Proposed design uses NG gate in the overflow correction logic circuit.

4 Proposed Work

4.1 BCD Adder

Digital circuits take care of the numbers represented in (BCD), i.e., binary coded decimal number. The circuit is capable of adding 2 BCD numbers and generates a result in BCD numbers. BCD uses ten digits from 0 to 9 represented in the form of binary 0 to 9; i.e., BCD number is denoted as a 4-bit binary number [7]. Each BCD number should not be greater than 9. BCD number like an example 526 is represented as.

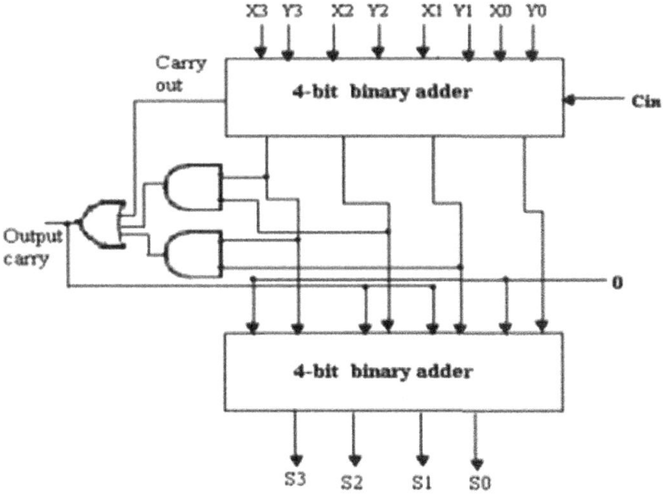

Fig. 6 Conventional BCD adder

5	2	6
↓	↓	↓
0 1 0 1	0 0 1 0	0 1 1 0

BCD numbers find their usage in the display of decimal numbers as in digital clocks and counters. Two BCD numbers are added by using the BCD adder circuit. The addition of BCD numbers is identical to the addition of binary numbers excluding the change if the sum of 2 BCD digits is greater than 9 or carry generation. A correction factor of six is added to change the invalid numeral into a valid numeral. Figure 6 represents the conventional BCD adder diagram.

4.2 BCD Reversible Adder

BCD reversible adder comprises of three blocks:

1. 4-bit ripple carry.
2. Overflow detection circuit
3. Overflow correction circuit.

Initially, 4-bit ripple carry adder adds the numbers and generates sum (S3-S0) in the form of binary. The generated output is verified for a carry out or a value greater than "9," by the correction circuit and then produces a 6-correction bit, "O" output using Eq. (1).

$$O = C_{out} + S_3(S_1 + S_2) \tag{1}$$

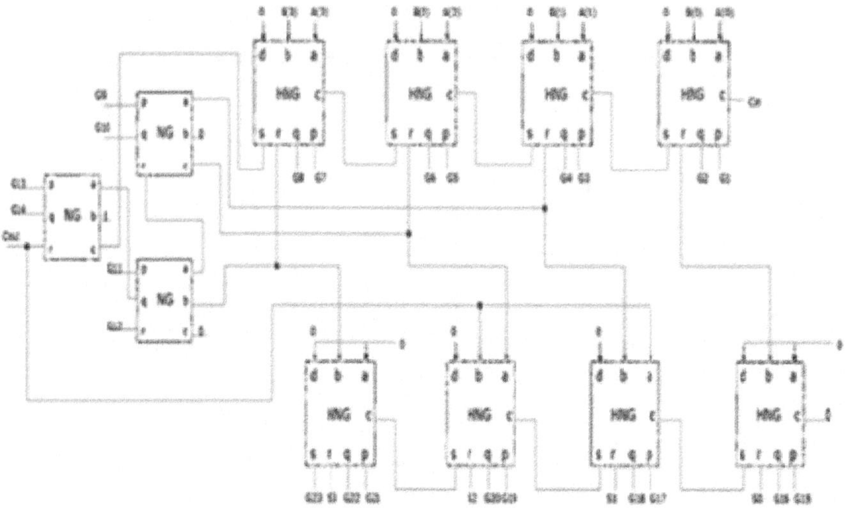

Fig. 7 Proposed 4-bit BCD reversible adder

Generally, the design of BCD reversible adder is as follows: 4-bit ripple carry reversible adder which produces initial sum in the form of binary addition of the two BCD digits. Four-bit parallel reversible adder is realized by the HNG gate. In the next part, detection of BCD overflow is done. NG gate is used in the detection circuit. In the firm, reversible circuits do not allow fan-out. In the correction circuit, a parallel reversible adder of 4-bit is taken into consideration to add the error correction. Then, the output generated by the 4-bit first reversible ripple carry adder gives the final result. The proposed design of reversible BCD adder is realized by using an HNG gate and NG gate. The proposed reversible 4-bit BCD reversible adder design is as represented in Fig. 7. Four-bit BCD reversible adder is constructed by using 11 reversible gates; hence, the gate count of the design is 11.

Proposed design is extended for the 64-bit reversible BCD adder and verified the working functionality using Xilinx tool.

5 Results and Implementation

Table 1 shows the evaluation parameters of recent 4-bit BCD reversible adder with the proposed circuit of 4-bit BCD reversible adder. Garbage outputs and quantum cost are improved in the design proposed with respect to other designs.

Considered the designs which are having less Ancilla inputs and garbage outputs are compared here in the existing designs. Figure. 8 shows the comparison graph of proposed reversible 4-bit BCD adder with other existing 4-bit reversible BCD adder in terms of reversible logic parameters.

Table 1 Analysis of 4-bit BCD reversible adder

	Constant input	Garbage output	Quantum cost
[4]	68	72	440
[8]	28	24	220
[9]	16	16	676
[10]	56	64	336
[5] Design 3*	8	24	412
[6] Design 3*	4	3	280
[11]	6	10	Not mentioned
[12]	17	22	Not mentioned
[13] Design 2*	10	6	90
Proposed design	11	22	72

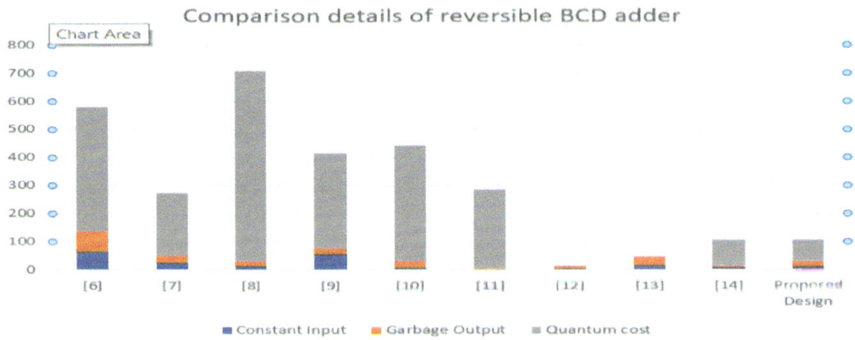

Fig. 8 Simulation output for proposed 4-bit BCD reversible adder

Figure 9 denotes the simulation output for the proposed 4-bit BCD reversible adder using NG and HNG gates. Figure 10 represents the RTL schematic of 4-bit proposed design. Functionality of the design is verified using Xilinx Vivado tool.

Proposed design is extended for 64-bit BCD reversible adder and verified its functionality using the simulation tool as in Fig. 11.

Figure 12 shows the top module RTL schematic for 64-bit reversible BCD adder, both input a and input b are 64 bits, and input carry Cin is single bit. Output D contains 64 bits sum, and output carry Cout is of single bit. Figure 13 shows the RTL schematic for optimized 64-bit proposed reversible BCD adder.

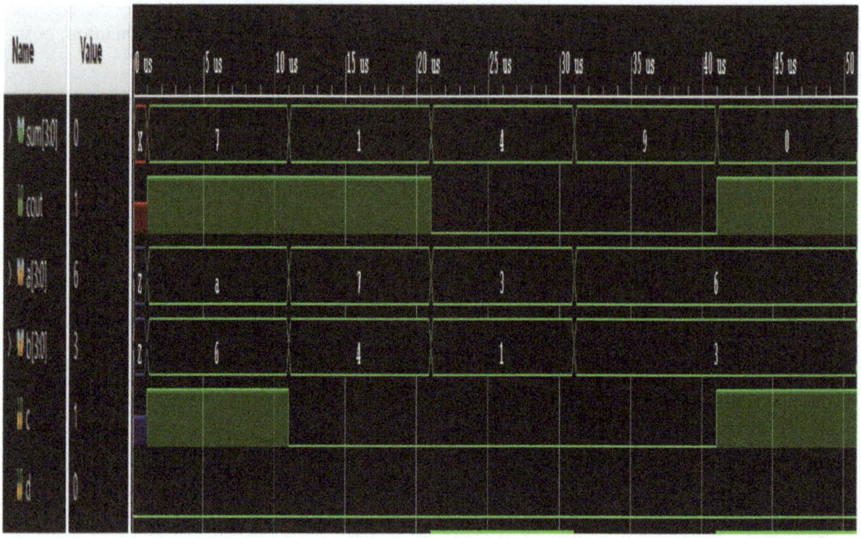

Fig. 9 Simulation output for proposed 4-bit BCD reversible adder

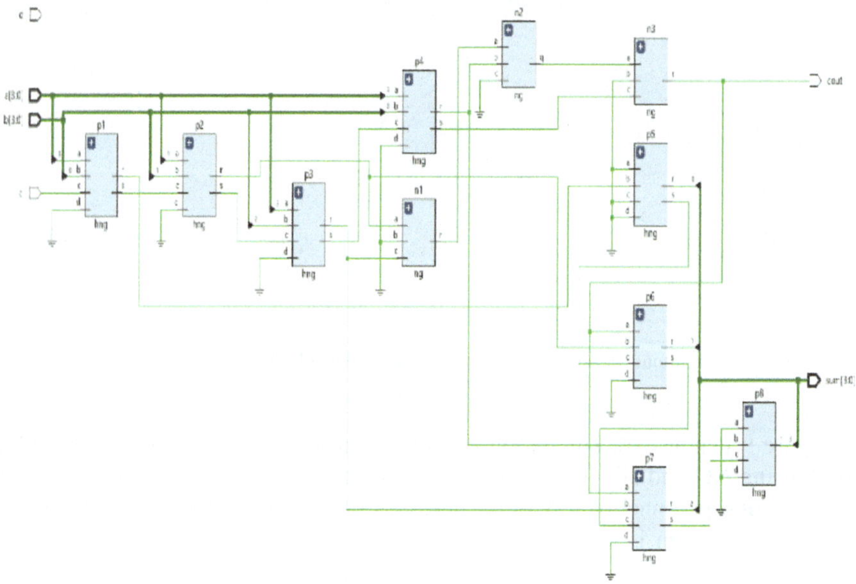

Fig. 10 RTL schematic for proposed 4-bit BCD reversible adder

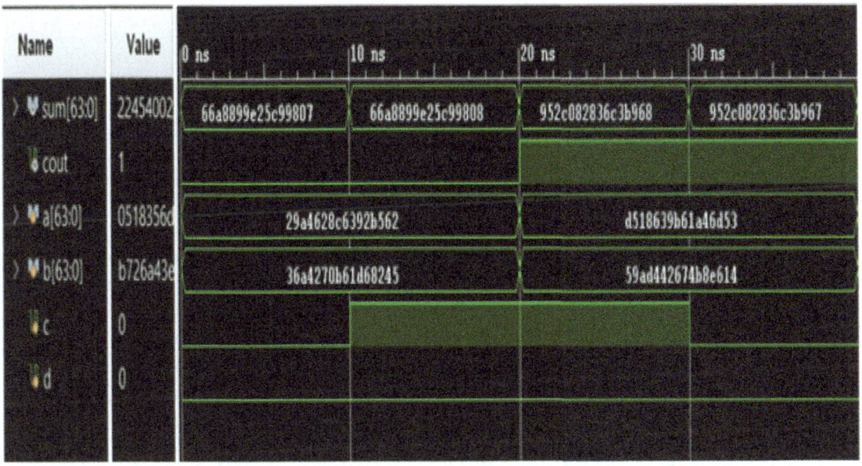

Fig. 11 Simulation output for proposed 64-bit BCD reversible adder

6 Relevance of the Paper

Power dissipation and heat generation are a severe problem in today's IC technology. Fundamental reason for power dissipation and heat generation is because of the bit erasure during irreversible operations. Since reversible logic operations are information lossless, it reduces the power dissipation. Hence, reversible logic is an emerging technology used in applications like quantum computing, optical computing. Reversible circuits designed can be used in Innovative Technology for Societal Development.

7 Conclusion

The reversible circuits are designed to lessen the power dissipation thereby eventually increase the speed and lifetime of the circuit. More research is attracted in the direction of reversible logic as it has huge applications in the developing technologies. This paper proposed an efficient optimized 4-bit and 64-bit BCD reversible adder improving the quantum cost and power dissipation. In the existing system, the quantum cost is 72. In the proposed work, the quantum cost and the power dissipation parameters are reduced. Design of 64-bit BCD adder is first of its kind analyzed with all existing design available in the literature.

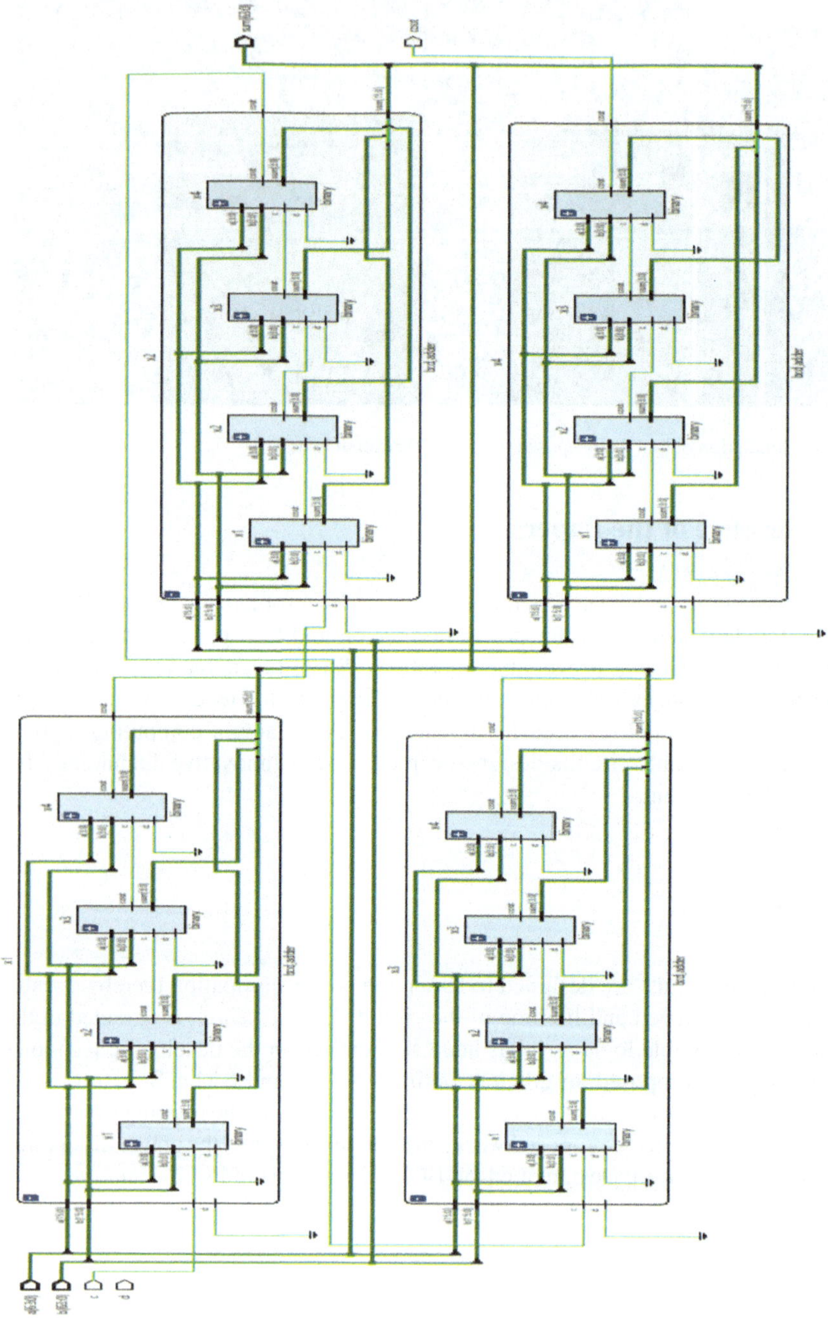

Fig. 12 RTL schematic for proposed reversible 64-bit BCD adder

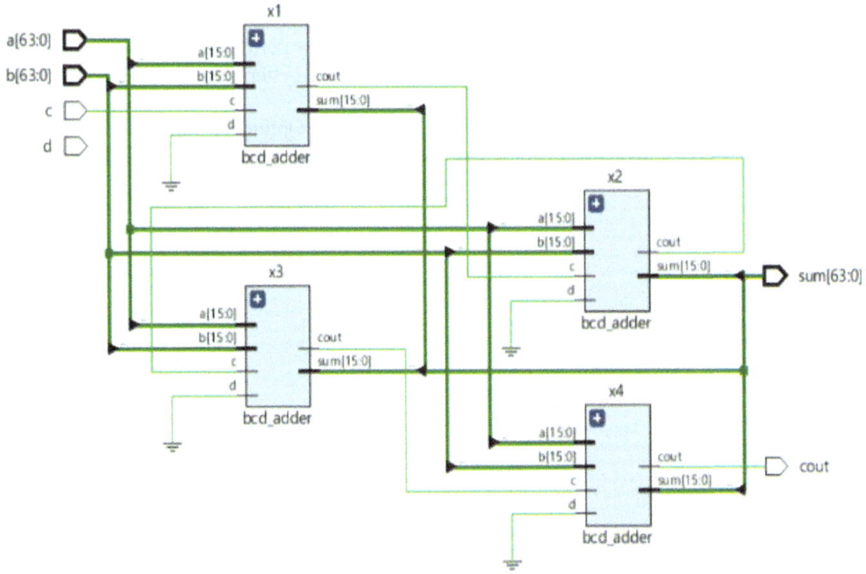

Fig. 13 Optimized RTL schematic for proposed reversible 64-bit BCD adder

References

1. Landauer, R.: Irreversibility and heat generation in the computational process, IBM1. Res. Develop. **5**, 183–191 (1961)
2. Bennet, C.H.: Logical reversibility of computation, IBM 1. Res. Develop. **17**(6), 525–532 (1973)
3. Bhuvana, B., Kanchana Bhaaskaran, V.S.: Design of reversible adders using a novel reversible BKG gate. In: Online international conference on green engineering and technologies (IC-GET), IEEE (2016)
4. Babu, H.M.H., Chowdhury, A.R.: Design of a compact reversible binary coded decimal adder circuit. Elsev. J.. Syst. Arch. **52**, 272–282 (2006)
5. Mohammadi, M., Haghparast, M., Eshghi, M., Navi, K.: Minimization optimization of reversible bcd-full adder/subtractor using genetic algorithm and don,t care concept. Int. J. Quantum Inf. **7**(5), 969–989 (2009)
6. Thapliyal, H., Ranganathan, N.: Design of Efficient Reversible Logic based Binary and BCD adder circuits. ACM J. Emerg. Technol. Comput. Syst. **9**(3) (2013)
7. Ankush, Sonam, M.: An improved design of low power BCD Adder using reversible computing. Int. J. Adv. Sci. Technol. **104**(2017):33–42 (2017)
8. Biswas, A.K., Mahmudul Hasan, M., Chowdhury, A.R., Hasan Babu, H.Md.: Efficient approaches for designing reversible binary coded decimal adders. Microelectron. **1**(39)(12), 1693–1703 (2008)
9. Thomsen, M.K., Gruck, R.; Optimized reversible binary-coded decimal adders. Syst. Archit. **54**(7), 697–706 (2008)
10. Mohammadi, M., Eshghi, M., Haghparast, M., Bahrololoom, A.: Design and optimization of reversible bcd adder/subtractor circuit for quantum and nanotechnology-based systems. World Appl. Sci. J. **4**(6), 787–792 (2008)
11. Bhagyalakshmi, H.R., Venkatesha, M.K.: Optimized reversible BCD adder using new reversible logic gates. J. Comput. 2(2) (2010) ISSN 2151-9617

12. Hasan Babu, H.Md., Chowdhury, A.R.: Design of a reversible binary coded decimal adder by using reversible 4-bit parallel adder. In: Proceedings of the 181h international conference on VLSI design and 4th international conference on embedded systems design, pp 1063–9667/05, IEEE (2005)
13. Nagamani, A.N., Ashwin, S., Agrawal, V.K.: Design of Optimized Reversible Binary and BCD Adders. In: International Conference on VLSI Systems, Architecture, Technology and Applications (VLSI-SATA), IEEE (2015)

Disease Prediction Using Various Data Mining Techniques

Nikita Sharma, Dhrubasish Sarkar, and Suchandra Dutta

Abstract The data mining methods are often beneficial in open domains like business, marketing, and retail. Healthcare is one of these areas that are still in its development phase. The healthcare industry is very rich in information, but sadly, not all data is carefully mined or discovered to draw out the effective, hidden patterns, and due to the lack of helpful analysis tools for discovering hidden relationships and trends in data, decision making is hampered (Dangare and Apte in Int J Comput Appl (0975–888) 47(10), 2012 [1]). There are a lot of advanced techniques in determining the domain that is used to discover knowledge from the healthcare database, particularly in diseases like heart disease, lung cancer, Parkinson's disease, and others. This paper analyzes the predictions of heart disease on a dataset with 13 attributes. On the heart disease dataset, decision trees, Naive Bayes, support vector machine, linear discriminant analysis, logistic regression, and KNN are evaluated as data mining approaches (classification algorithms). Usually, the algorithms are compared based on their performance measure, and the algorithm with the highest accuracy is implemented on the heart dataset for the further model prediction.

Keywords Random forest classifier · LDA · Logistic regression · KNN · Healthcare

1 Introduction

Data mining is the process of non-trivial extraction of implicit, previous unknown and useful information from large databases or data sets. Pretty much, it will in general be said that the entire instrument incorporates data assessment from different outlooks and a variety of data from it. The revealed information can be utilized for directing

N. Sharma
Novel Medicare Solutions Private Limited, Kolkata, India

D. Sarkar (✉) · S. Dutta
Amity Institute of Information Technology, Amity University Kolkata, Kolkata, India
e-mail: dhrubasish@inbox.com

K. N. Das et al. (eds.), *Proceedings of the International Conference on Computational Intelligence and Sustainable Technologies*, Algorithms for Intelligent Systems,
https://doi.org/10.1007/978-981-16-6893-7_33

various areas, for instance, the social insurance industry. As of now in time, the medicinal services area produces huge data about patients and infection acknowledgment. Information mining gives a lot of strategies to find concealed examples from information. The nature of administration remains as the critical test which is looked by the human services division. Standard of administration shows perceiving illness accurately and giving powerful medicines to patients. Worse than average conclusion can prompt annihilating outcomes, which are unseemly. Seventeen million absolute worldwide passing are because of respiratory failures and strokes, as said by a study of the World Health Organization. Work over-trouble, mental weight, and various issues are the reasons in light of which heart infirmities happen fundamentally [2, 3].

As a general rule, it is found as the principle purpose for death in grown-ups. Determination is a confounded and basic undertaking that should be executed correctly and productively. The end is as regularly as conceivable made subject to the ace's detects, comprehension, and information. This prompts bothersome results and addition in clinical costs of drugs provided for patients. Along these lines, a prearranged clinical identification framework is detailed that exploits the gathered database and choice emotionally supportive network. This machine can help in deciding illness with low clinical tests and powerful medicines.

Information technology (IT) driven network is inconceivably acknowledged; for its data noteworthiness, it ends up being an or more of any association. The job of information innovation in human services is entrenched. Proceeded with progression in IT applications inside the fields of care has raised individuals' desires for higher consideration with diminished use. The motorization of emergency clinic data frameworks has given partner in nursing access to significant clinical data just [4].

Clinical information handling has astounding potential for investigating the concealed examples inside the information sets of the clinical area. These examples are frequently utilized for clinical assignment. Be that as it may, the offered crude clinical information are heterogeneous in nature and voluminous. This information got the chance to be gathered in associate in nursing sorted out sort. This gathered information is frequently then coordinated to make an emergency clinic information framework. Information preparing innovation gives a client situated way to deal with novel and concealed examples inside the information.

1.1 Heart Disease

The heart is one of the most important organs of the human body organ system. It is only a siphon, which siphons blood through the body. In the case that the body's blood dispersion is poor, by then, the organs will, as a rule, as suffer, and if the heart stops working completely, passing occurs in a matter of seconds. Life is completely dependent upon the capable working of the heart. The word heart disease suggests an illness of heart and vein structure inside it [2].

Feature name	Description	Domain value
Age	Age in years	30<age <77
Gender	Male =1, Female = 0	0,1
Chest pain	1=atypical angina, 2= typical angina, 3= asymptomatic , 4=nonanginal pain	1,2,3,4
Resting blood pressure	Mm Hg admitted at the hospital	94-200
Serum cholesterol	In mg/dl	120-564
Fasting blood sugar > 120 mg/dl	Fasting blood sugar>120 mg/dl(1=true,0= false)	0,1
Resting electrocardiographic results	0=normal,1= having ST-T,2=hypertrophy	0,1,2
Max heart rate achieved	-	71-202
Exercise induced angina	1=yes,0=no	0,1
Old peak st depression induced by exercise relative to rest	-	0-6.2
Slope of peak exercise st segment	1= up sloping,2= flat, 3= down sloping	1,2,3
Number of major vessel(0-3) colored by fluoroscopy	-	0,1,2,3
Thallium scan	3=normal,6= fixed defect, 7= reversible defect	3,6,7

Fig. 1 Attributes of the heart dataset

In the diagram (Fig. 1), a few components have been depicted. The attributes of the heart dataset, Personal history as well as family history, smoking, poor eating habits, fast heartbeat, high blood cholesterol, overweight, physical inactivity, and hypertension are all factors that increase the risk of heart disease. These and other factors are used to establish whether a patient has heart disease. Much of the time, a determination is commonly founded on the patient's present test outcomes and the specialist's understanding. Along these lines, the conclusion is a perplexing errand that requires a lot of experience and high expertise.

2 Related Work

Data mining is the process of extracting hidden and previously discovered patterns, correlations, and information from huge databases. Heart and circulatory system diseases are the first leading cause of death in Australia, causing 33.7% of all deaths. Several data mining techniques are used in the diagnosis of heart disease such as Naïve Bayes, decision tree, and neural network [5]. Data mining is the process of selecting, discovering, and modeling huge amounts of data. Decision tree models are commonly used in data mining to examine data and induce the tree, and it contains the rules that will be applied to produce predictions. The accuracy of three data mining techniques is compared [6]. The paper [7] identifies key characteristics and data mining approaches that can improve the reliability of cardiac patient survival prediction. This study uses nine categorization models to predict patient survival:

decision tree (DT), adaptive boosting classifier (AdaBoost), logistic regression (LR), stochastic gradient classifier (SGD), random forest (RF), gradient boosting classifier (GBM), extra tree classifier (ETC), Gaussian Naive Bayes classifier (G-NB), and support vector machine (SVM). The technique of extracting hidden knowledge from massive amounts of raw data is known as data mining. The difficulty of describing the many data mining algorithms employed in the field of medical prediction is tackled in this survey article. Different algorithms and combinations of many target qualities, such as decision trees, Naive Bayes, and KNN, are the major emphasis [8]. This paper [9] presents a comparison of data mining approaches for diabetes early detection. We utilize a publicly available data collection with 520 instances and 17 characteristics apiece. The approaches of Naive Bayes, neural network, AdaBoost, k-nearest neighbors, random forest, and support vector machine were examined. This article [10] identifies the relationships between opposing factors and calculates diabetes based on its major characteristics. The feature range supplied by the recursive random forest elimination algorithm was considerable. The diabetes estimation was examined using the random forest classifier. This study [11] includes a discussion of machine learning and image fusion classification approaches that have been shown to aid healthcare practitioners in detecting cardiac disease. The authors in [12] use a taxonomic tree structure to offer a comprehensive evaluation of measles cases and death in this study.

3 Objective

The primary goal of this examination is to build up a shrewd cardiovascular sickness forecast framework or coronary illness expectation framework related to the assurance of the casualties of coronary illness with the assistance of weighted associative classifiers, which will be utilized in making educated call with the most precision. The system is applied in remote areas like rural regions or countryside, to replicate like human symptomatic experience for treatment of the ill heart. The system is simply updated as and once the new coaching knowledge set is available. Conjointly the system is reasonable and includes less variety of steps because the prediction rules are predetermined and are just a function/library call away throughout the heart disease prediction system [13].

We have to match different algorithmic rules that are typically utilized in prediction systems to develop a hybrid algorithm, which might be used for multidimensional knowledge analysis, multidimensional knowledge is that the knowledge that records the facts associated with the variable entities known as dimensions. The benefit of multidimensional knowledge is that it helps to generate more significant insights [14].

4 Methodology: Basic Terms and Concepts

AI is that segment of information science that alludes to the usage of man-made consciousness, which adds to frameworks the capacity to gain and improve as a matter of fact without being particularly customized consequently. PC based knowledge spins around the improvement of PC programs that can get to information and use it to get a handle on for themselves [14, 15].

Random Forest Classifiers: Random forest may be a variety of supervised machine learning formula supported ensemble learning.

Ensemble learning may be a variety of learning wherever differing kinds of formulas or the same algorithm multiple times to create a lot of powerful prediction models.

The random forest formula combines multiple formula of identical sort, i.e., multiple call trees, leading to a forest of trees, and therefore, the name "Random Forest." The random forest formula is used for each regression and classification tasks.

K-nearest neighbors (KNN): K-nearest neighbors (KNN) formula may be a variety of supervised formula which might be used for each classification still as regression prophetical issues. However, it is in the main used for classification prophetical issues in trade.

Logistic Regression: Logistic regression is employed once the dependent variable (the target) is categorical.

For example, after we need to spot that associate email is spam or not. Hence, it is not appropriate for classification problem]. Logistic regression is employed in biological sciences. The info is fitted into a simple regression model, that is then acted upon by a supply regression operate predicting the target categorical dependent variable.

LDA: Linear discriminant analysis is utilized as a spatiality decrease method inside the pre-preparing step for design arrangement.

Support vector machine: The support vector machine is a controlled learning method that sorts data into two characterizations. It is utilized for arrangement and relapse investigation. The yield of the help vector machine speaks to a guide of arranged information with the edges between the two focuses as far separated as could be expected under the circumstances. Support vector machines are utilized in text orders and picture arrangement. The principle undertaking of SVM is to figure out which classification another information has a place in. It is likewise a non-twofold straight arrangement.

Naïve Bayesian: Naïve Bayesian, as from the name recommends, utilizes Bayes' hypothesis to group objects. The gullible Bayes' calculation is definitely not a solitary term yet an assortment of calculations, which expect sound freedom between the qualities of the information focuses. A portion of the famous employments of gullible Bayes' classifiers incorporate spam channels, text examination, and clinical conclusion. The basic understanding of gullible Bayes' classifiers is the utilization of likelihood hypothesis and measurable autonomy.

5 Proposed Prediction System

At present, numerous clinics oversee social insurance knowledge using the human services data model, as the model contains a lot of measure of information used to remove shrouded data for making shrewd clinical findings. This examination means to build up an Intelligent Heart Disease Prediction System as shown in Fig. 2 that helps in perceiving of cardiac illness using a heart database. To grow such, clinical terms, for example, gender-wise orientation, circulatory strain, and cholesterol like 13 info properties are utilized. The information mining order strategies like K-closest neighbors, linear discriminant analysis, support vector machine, logistic regression, Naïve Bayesian, and random forest classifiers are utilized [2, 16].

Compact stages of data analysis can also be said as life cycle processes of data science.

The below stages are concisely understood to facilitate a better workflow of data analytics: -

1. Preparation phase: acquiring data and transforming it into a suitable manner for computation.
2. Analysis phase: using computer programs to draw insights from the data.
3. Reflection phase: it is concerned with critical thinking and communication with the client regarding their requirements.
4. Dissemination phase: presenting our reports in the form of paper or presentation.

Fig. 2 Proposed framework of the model

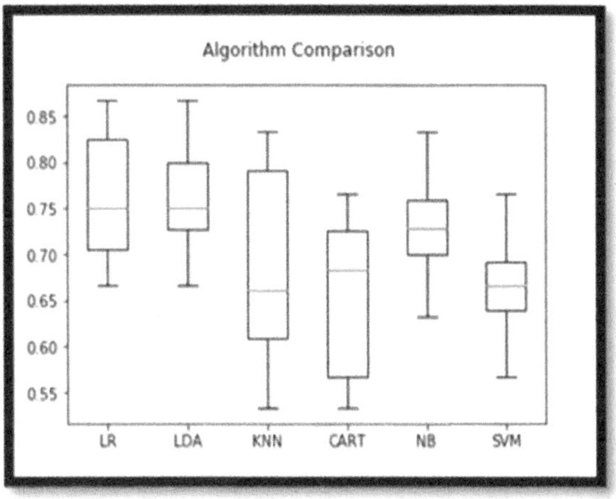

Fig. 3 Algorithm comparison dataset

6 Result Analysis

The heart dataset was obtained from the UCI repository. Various algorithm techniques (K-nearest neighbors, linear discriminant analysis, support vector machine, logistic regression, Naïve Bayesian, and random forest classifiers) were applied, and it was seen that logistic regression had the highest mean [16, 17] (Fig. 3).

The logistic regression model was applied on the dataset, and a descriptive analysis and an inferential analysis were drawn and found that out of 299 people, 139 has the disease, and 160 do not have the disease.

The spyder editor (3.7) of anaconda navigator was used for the execution of the codes, which was written in Python.

Figure 4 Descriptive analysis of the heart dataset shows the detailed and inferential results of the dataset. It clearly can be seen that which portion of the population is diagnosed with cardiovascular disease and which confusion Matrix: Confusion matrix, as shown in Fig. 5, is the summary of prediction results, basically a table that describes the performances of the classification model (classifier). It not only shows the errors but also shows the types of errors that occur while the classifier is run upon a dataset.

7 Conclusion

The purpose was to compare the algorithms that provide the best company with different performance measures using machine learning and knowledge discovery data techniques. All data are pre-processed and sent to the featuring section. Each algorithm works best in some situations and worse in some situations. The data mining classification techniques like K-nearest neighbors, random forest classifiers,

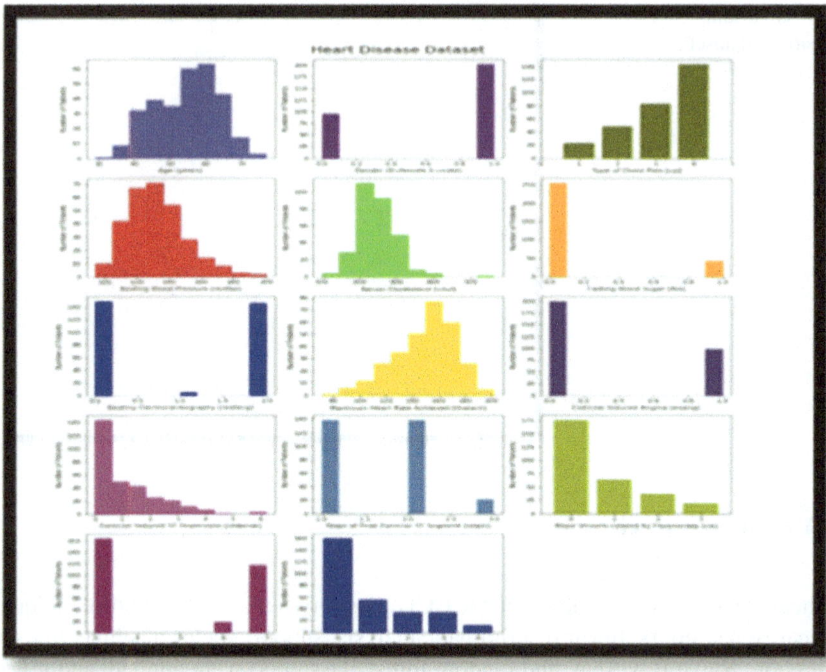

Fig. 4 Descriptive analysis of the heart

Fig. 5 Confusion matrix

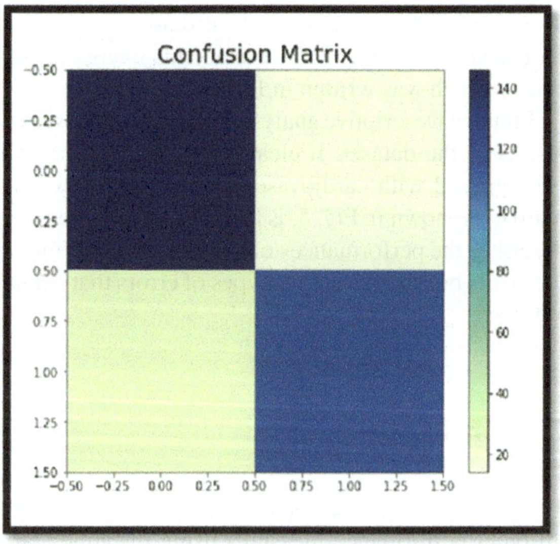

linear discriminant analysis, support vector machine, logistic regression, and Naïve Bayesian are used and are considered some of the best models to be fitted in this study [16]. Use of feature selection in high dimensional data, the risk of models getting choked is very high because of an increase in the training time with several features. With an increasing number of features, models have a high risk of overfitting. Therefore, feature selection helps to overcome these problems without much loss of the complete information. This paper is the initial step in learning the diagnosis of heart disease. Nevertheless, there are certain limitations like the author's knowledge in the domain, secondly, the computing power based on the tools and thirdly the time limit of the study. This study requires much accurate expertise in the field. In future, the work can be expanded and enhanced for the automation of various types of disease prediction. It also extended to find other types of diseases with the uses of these attributes.

References

1. Dangare, C.S, Apte, S.S.: Improved study of heart disease prediction system using data mining classification techniques. Int. J. Comput. Appl. (0975–888), **47**(10) (2012)
2. Gudadhe, M., Wankhade, K., Dongre, S.: Decision support system for heart disease based on support vector machine and artificial neural network. In: 2010 International Conference on Computer and Communication Technology (ICCCT), Allahabad, Uttar Pradesh, pp. 741–745 (2010)
3. Parthiban, L., Subramanian, R.: Intelligent heart disease prediction system using CANFIS and genetic algorithm. Int. J. Med. Health Sci. **1**(5) (2007)
4. Weng, S.F., Reps, J., Kai, J., Garibaldi, J.M., Qureshi, N.: Can machine-learning improve cardiovascular risk prediction using routine clinical data? PLoS ONE **12**(4), e0174944 (2017). https://doi.org/10.1371/journal.pone.0174944
5. Aparna, K., Reddy, C.S., Prabha, I.S, Srinivas, K.V.: Disease prediction in data mining techniques. IJCST **5**(2), 246–249 (2014)
6. Kamaraj, K.G., Priyaa, D.S.: Multi disease prediction using data mining techniques. Int. J. Syst. Soft. Eng. **4**(2):12–14 (2016)
7. Ishaq, S., Sadiq, M., Umer, S., Ullah, S., Mirjalili, V., Rupapara, M.: Nappi, improving the prediction of heart failure patients' survival using smote and effective data mining techniques,". IEEE Access **9**, 39707–39716 (2021). https://doi.org/10.1109/ACCESS.2021.3064084
8. Vijiyarani, S., Sudha, S.: Disease prediction in data mining technique—a survey. Int. J. Comput. Appl. Inf. Technol. **II**(I), 17–21 (2013)
9. Chaves, L., Marques, G.: Data mining techniques for early diagnosis of diabetes: a comparative study. Appl. Sci. **11**(5), 2218 (2021). https://doi.org/10.3390/app11105221
10. Rashid Abdulqadir, H., Mohsin Abdulazeez, A., Assad Zebari, D.: Data mining classification techniques for diabetes prediction. Qubahan Acad. J. 1(2), 125–133. https://doi.org/10.48161/qaj.v1n2a55
11. Diwakar, M., Tripathi, A., Joshi, K., et al.: Latest trends on heart disease prediction using machine learning and image fusion, Mater. Today. Proc. **37**(4). https://doi.org/10.1016/j.matpr.2020.09.078
12. Lashari, S.A., Ibrahim, R., Senan, N., Taujuddin, N.: Application of data mining techniques for medical data classification: a review. pp. 737–759 (2018). https://doi.org/10.1007/978-981-15-3514-7_56
13. Wasan, S.K., Bhatnagar, V., Kaur, H.: The impact of data mining techniques on medical diagnostics. Data Sci. J. **5**, 119–126 (2006). https://doi.org/10.2481/dsj.5.119

14. Palaniappan, S., Awang, R.: Intelligent heart disease prediction system using data mining techniques. In: 2008 IEEE/ACS International Conference on Computer Systems and Applications, Doha, pp. 108–115 (2008). https://doi.org/10.1109/AICCSA.2008.4493524.
15. Shao, Y. E., Hou, C.-D., Chiu, C.-C.: Hybrid intelligent modeling, schemes for heart disease classification. Appl. Soft Comput. **14**(Part A), 47–52 (2014)
16. Agrawal, R., Srikant, R.: Fast algorithms for mining association rules. In: Publication: VLDB '94: Proceedings of the 20th International Conference on Very Large Data Bases, pp. 487–499 (1994)
17. Ordonez, C.: Association rule discovery with the train and test approach for heart disease prediction. IEEE Trans. Inf. Technol. Biomed. **10**(2), 334–343 (2006). https://doi.org/10.1109/TITB.2006.864475

Classification of Medicinal Plant Species Using Neural Network Classifier: A Comparative Study

Anurag Protim Saikia, Rahul Roy, Subir Datta, Sadhan Gope, Subhasish Deb, and Ksh Robert Singh

Abstract There is a rising requirement that medicinal plants are properly classified and identified since they are extremely vital to our lives. However, it is quite difficult to identify such biological things because they are not governed by mathematical function. In this paper, the classification of six different species of medicinal plants is discussed using different feature extraction techniques, namely morphological, color, and texture. The image dataset is comprised of a total of 90 sample leaves images of six different species of medicinal plants that are acquired from suburbs of Assam, India. The performance of the above three features toward classification of medicinal plants is studied in this work. For effective leaf recognition, the backpropagation neural network (BPNN) classifier is used. The study reveals that the proposed classification model is capable of classifying six different species of medicinal plant with an overall recognition accuracy 98.88% using suitable feature combination.

Keywords Color model · Image · Morphological feature · Neural network · Segmentation · Texture feature

1 Introduction

Plants play a key role in the Earth's biome, providing support, shielding, and fostering a prosperous environment. A number of these plants possess some medicinal key essential properties. In computer, vision automated recognition of the plant leaves is a challenging task. Various sectors of academic, industrial as well as research areas such as medicinal and botanical research will immensely benefit from an efficient automated leaf recognition apparatus for medicinal plants. Processing of the plant leaf images thereafter employing pattern recognition techniques can smoothen the task of recognizing medical plant leaves [1]. In modern times, herbal medicines are becoming more popular, because they are both healthy and inexpensive for human health. According to the World Health Organization, herbal medicines are used by 80%

A. P. Saikia (✉) · R. Roy · S. Datta · S. Gope · S. Deb · K. R. Singh
Department of Electrical Engineering, Mizoram University, Aizawl, Mizoram 796004, India

K. N. Das et al. (eds.), *Proceedings of the International Conference on Computational Intelligence and Sustainable Technologies*, Algorithms for Intelligent Systems,
https://doi.org/10.1007/978-981-16-6893-7_34

percent of people in Asia and Africa for primary health care. In the lush green forests of southern and northeastern India, there are several medicinal plant varieties that are used as medicines [2]. As a consequence of deforestation and pollutants, several plant leaves have been extinct. This means that medicinal plants must be identified and preserved for future generations. Rising of illegal trafficking and maltreatment in the pharmacognosy and to not have adequate resource person, an electronic and reliable system for detecting and classifying the bulk of data is essential to reduce and eliminate malpractice [3].

1.1 Related Works

Works related to medicinal plants classification-based gray tone spatial dependency matrices (GTSDM), local binary pattern (LBP) operators and textures are discussed in [2], where they yielded a classification performance of 94.7%. Features like color and shape are presented in [3, 4]. According to the literature, gray level co-occurrence matrix (GLCM) and gray level run length matrix (GLRLM) are commonly used for extracting texture attributes for classification purposes. In this paper, implementation of GLCM and GLRLM to texture attribute classification of various medicinal plants is discussed [5–7]. Nguyen et al. [8] suggested SURF characteristics in combination with a bag of words model with SVM claissifier as a method for Android leaf-based plant recognition with 95.94% accuracy. SVM classifier is used for the classification of medicinal leaf images with different shape, multi-features, law marks features [9–11]. Applications of BPNN for different classification tasks are also presented in [12–15, 20]. The BPNN classification classifier used by Abirami et al. was 92.6% percent accuracy, and Amlekar et al. [15] used ANN to investigate the process of leaf detection, achieving a recognition rate of 96.53% percent. There are applications of GLCM-based texture characteristics for classification tasks in a variety of other discipline present in [16–19].

To sum up, there is still very less amount of works on medicinal plants classification method using neural network for various medicinal leaves and achieves high accuracy. We realized that a neural network should be added to the classification approach as much as feasible by referring to relevant literature and research experience. Backpropagation neural network is quite helpful too. In addition to achieve higher classification accuracy, we used morphological, color, and texture (GLCM) features classification data in backpropagation neural network. To put it another way, this removes a barrier to studying the identification and classification of different medicinal plants. The proposal in this paper was to a highly efficient medicinal plants species classification method using neural network classifier. The main steps of this study are the extraction and classification of functions. All characteristics have been extracted from each plants leaves of our own medicinal plants dataset. Morphological, color, and texture features are from each plants and leaves are given to the classifier. The suggested approach employs a backpropogation neural network as a classifier. The paper is organized accordingly: Sect. 1 discusses an overview of the importance of medicinal plant, need for the classification of medicinal plants and related works, Sect. 2 describes the method proposed with different feature extraction methods with BPNN classifier, and Sect. 3 discusses result and analysis part. Section 4 brings up the conclusion of the research work.

2 Proposed Work

2.1 Acquisition of Image Samples

The digital images of the herbal plant's leaves have been gathered from the countryside of Assam. For the collection work to be carried out, the mobile camera is being used to obtain the digital form of the digital medicinal plant leaf. After that, the computer attains the digital color images in the unprocessed JPEG format. The working data collection consists of splitting the images into one size 500 × 500 × 3 and resulting in a total of 90 pictures, 15 images from each 6 species and stored as JPEG format. The image dataset is used to create the testing and the training sets. The whole new methodology is displayed in Fig. 2.

2.2 Raw Dataset Preprocessing

Figure 1 shows representative sample photos for each of the six species in the database.

The dataset is preprocessed on three subtasks: The following are:

- Firstly, all images have been resized to 500 × 500 pixels because of the various dimensions of the dataset.
- Secondly, extraction of region of interest(RoI) from the resized leaf images, i.e., for further segmentation process.
- Thirdly, image segmentation algorithms are of various forms and the basic is of thresholding, i.e., discontinuity in intensity color values.

Fig. 1 Representation of sample images (only one for each species) from database

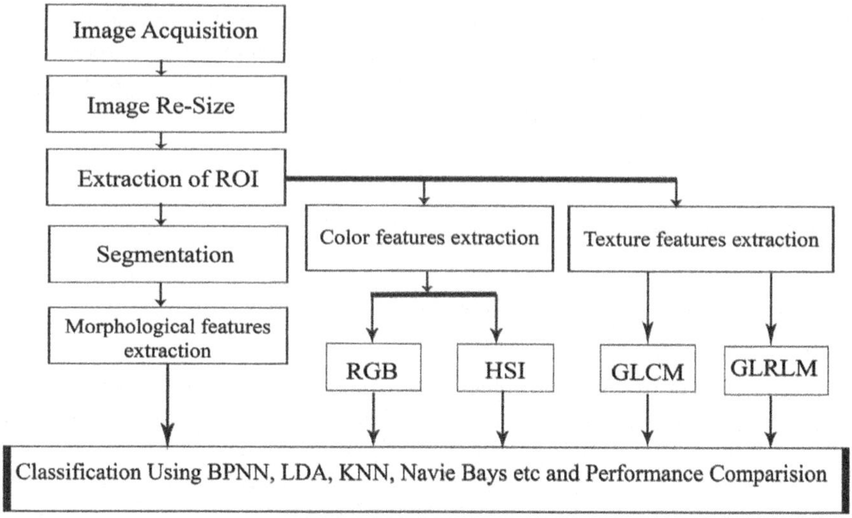

Fig. 2 Flow diagram of proposed scheme

2.3 *Morphological Feature Extraction*

The extracted morphological features are:

1. Area.
2. Length/Perimeter.
3. Convex Area.
4. Minor Axis length.
5. Major Axis length.
6. Eccentricity.

2.4 *RGB to HSI Color Model Conversion*

The following equations are used to calculate the H, S, and I components of each RGB pixel.

$$H = \begin{cases} \Theta & if\, B \leq G \\ 360 - \Theta & if\, B > G \end{cases} \tag{1}$$

$$S = 1 - \frac{3}{(R + G + B)}[\min(R, G, B)] \tag{2}$$

$$I = \frac{1}{3}(R + G + B) \tag{3}$$

The RGB to HSI conversion equation is defined in Eqs. (1–3).

where $\theta = \cos^{-1}\left\{\frac{\frac{1}{2}[(R-G)+(R+B)]}{[(R-G)^2+(R-G)(G-B)]^{\frac{1}{2}}}\right\}$ where H, S, I denotes hue, saturation, and intensity [21].

2.5 Color Features Extraction Method

The following six numbers of color features are extracted, and the mathematical expression for the color features is given in Eqs. (4–7).

$$\text{Mean}(\mu) = \frac{\sum\sum P(i, j)}{MN} \tag{4}$$

$$\text{Variance} = \sum_{i=0}^{G-1}\sum_{j=0}^{G-1}(i-\mu)^2 P(i, j) \tag{5}$$

$$\text{Range} = \max(\max(i)) - \min(\min(i)) \tag{6}$$

$$\text{Skewness } g_1 = \left(\frac{m_3}{m_2}\right)^{\frac{3}{2}} \tag{7}$$

where,

$$m_3 = \sum \frac{(x-\bar{x})^3}{n}$$

$$m_2 = \sum \frac{(x-\bar{x})^2}{n}$$

$$\text{Kurtosis}(\gamma) = \frac{1}{MN\sigma^4}\sum_{i=1}^{M}\sum_{j=1}^{N}(P(i, j) - \mu) \tag{8}$$

$$\text{Smoothness} = 1 - \frac{1}{(1+\sigma^2)} \tag{9}$$

where M, N is the image dimensions and $P(i, j)$ is the color values of I and J columns.

2.6 Texture Features Extraction Method

2.6.1 Gray Level Co-occurrence Matrix

The GLCM method is used to obtain an object's texture detail. The frequency of pixel pairs appearing can be expressed by a matrix in one direction. The input images are quantized on a grayscale of 64 for the purpose of minimize processing time. So a 64*64 GLCM is needed. Pixel pairs have a level of occurrence of 0°, 45°, 90°, and 135°. The extracted GLCM features are contrast, correlation, energy, homogeneity [22].

2.6.2 Gray Level Run Length Matrix (GLRLM)

The GLRLM is computed for a subset of orientations: 0°; 45°; 90°; and 135°. The input images are quantized to a grayscale of 64 in order to minimize processing time. These four GLRLMs have been measured on average, and 11 features were extracted from the resulting GLRLM, including short run high gray level, long run low gray level, long run, run length non-uniformity, low gray level run, high gray level run, gray level non-uniformity, run percentage, short run low gray level, short run, and long run high gray level [24].

2.7 Classification Using Backpropagation Neural Network (BPNN)

The primary basis for classification is the existing data from which one can learn the classification function, and also, classification model can be constructed from that existing data. With the help of this model, the record data can be made to a given category one in a database map and the later can be used to make predictions about the data. Classifiers have been used to identify samples in a variety of data mining approaches including decision trees, logistic regression, and neural network algorithms [20]. The suggested classification technique employs a multilayer neural network with backpropagation [23]. Figure 3 shows the general configuration of BPNN, which is nothing more than a multilayer feedforward network with the main component being the nonlinear conversion units, the feedforward network is made up of the input layer, the hidden layer, and the output layer. Under normal conditions, a three-layer BP network can complete an arbitrary n-dimensional to m-dimensional mapping with just one hidden layer. The input to the neural network when used as a classifier consists of n elements which correspond to the n characteristics of plant leaves. Equation (14) represents the number of nodes in the hidden layer.

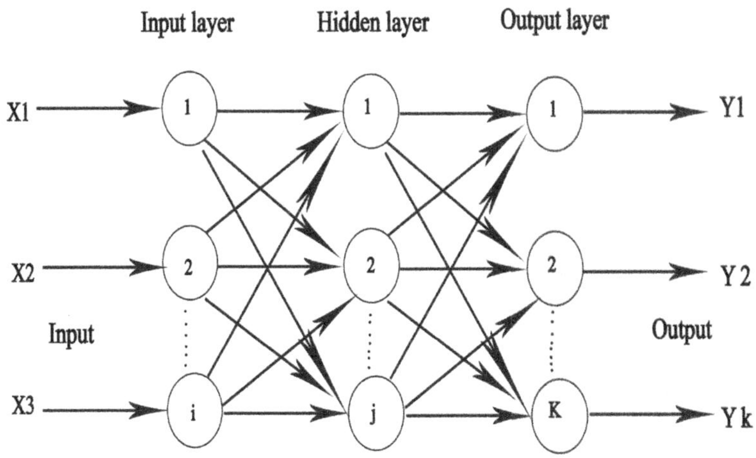

Fig. 3 A schmatic diagram of the BPNN model containing input, hidden, and output layer

$$N_{\text{nodes}} = \left[\frac{I}{T} \times 100\right] + \sqrt{T} \tag{10}$$

where the number of inputs and training data is denoted by I and T.

3 Result and Analysis

Morphological, color, and texture features were extracted from the sample images. The extracted morphological features are area, perimeter, convex area, major axis length, eccentricity, and minor axis length. We have extracted all six morphological features from total of 15 images of each 6 species; hence, morphological features were extracted from total 90 numbers of images, and color features were extracted from same dataset. Total six numbers of color features were extracted from each of the color plan, namely R, G, B, H, S, I color plan. For each plan of these two color models, every color features were extracted, and extracted color features are mean, variance, range, skewness, kurtosis, smoothness.

3.1 Results of Features Extraction

All 90 sample images were used to extract the texture features. These four GLCMs are averaged at different angles which are evaluated along with pixel pairs 0°, 45°, 90°, and 135°. Four main statistical characteristics are extracted, namely contrast, correlation, energy, and homogeneity. Total six number of morphological features are extracted, and color features are extracted from each color plan, namely R, G, B, H, S, I color plan.

3.2 Performance Plot and Confusion Matrix for BPNN

When the following conditions are met, the network immediately stops training:

1. If the validation error does not decrease after a period of 10 epochs, the network will be unable to generalize
2. When the maximum epoch has been achieved (1000)
3. When the mean square error reaches zero, the process is said to be complete [24].

Figure 4 shows performance plot for combine features (Morphological +Color+ GLCM), and the best validation performance of the network is 0.2211 MSE at 29th epoch for combined features (Morphological +Color+ GLCM). Figure 5 shows the confusion matrix test dataset on the neural networks created from the classification result of Combine (Morphological + Color+ GLCM)-based features, and the created artificial neural network system has classified 89 cases correctly out of 90 given input of total plant images. They have given 98.8% correct classification result.

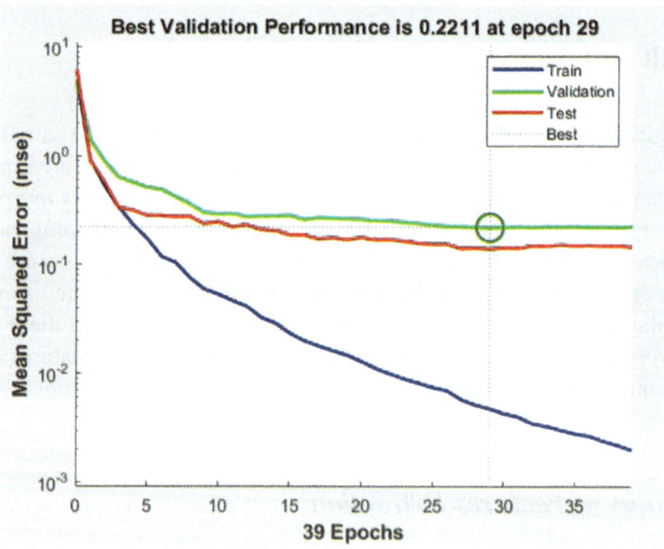

Fig. 4 Performance plot of training, validation, testing for combine features

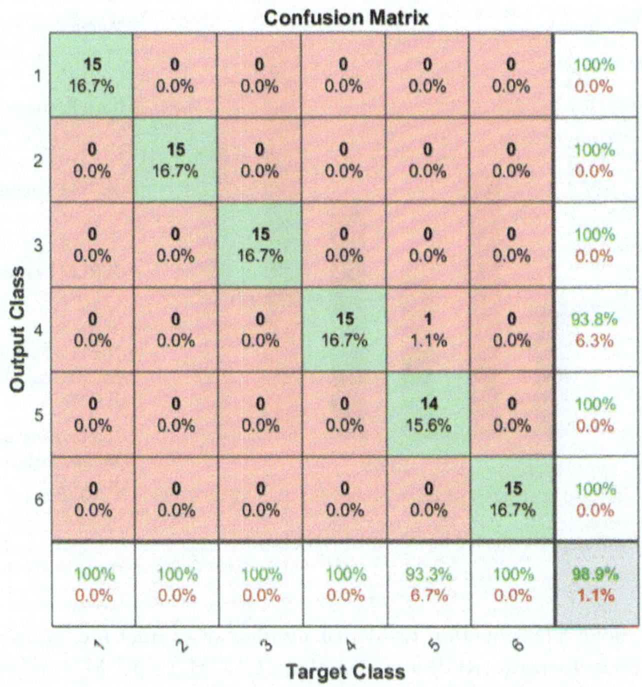

Fig. 5 Confusion matrix for combine (Morphological + Color+ GLCM)-based features

Table 1 Classification result of BPNN

Features	1	2	3	4	5	6
Morphological features	13	11	14	14	12	15
Color features	12	15	13	15	14	15
GLCM features	15	14	10	15	11	15
GLRLM features	15	15	11	15	13	14
Combine (Morphological+ Color+GLCM)	15	15	15	15	14	15
Combine (Morphological+ Color+GLRLM)	15	15	14	15	13	13

3.3 *Results of Classification*

The classification of six different varieties of medicinal plants is successfully implemented with a BPNN. Table 1 shows the performance of various classifiers on medicinal plant data depending on the proposed feature sets.

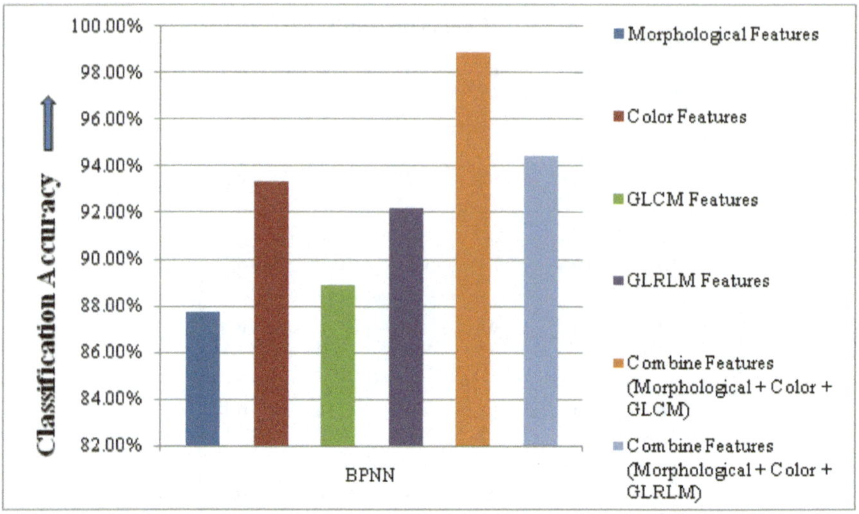

Fig. 6 Classification performance based on BPNN classifier

Figure 6 shows classification results of number of correct recognition for morphological, color features (RGB and HSI plan), GLCM, GLRLM, combine features (Morphological + Color+GLCM) and combine (Morphological + Color+ GLRLM) features. From morphological features, we achieve accuracy of 87.77%. For color features, the accuracy is 93.33%, and texture features (both GLCM and GLRLM) are 88.83% and 92.22 %, respectively, combine(Morphological + Color+ GLRLM) features 94.44% and combine features (Morphological + Color+GLCM) we were achieve overall accuracy about 98.88%.

3.4 Comparing Classification Accuracy Between Different

We carried out a comparison of performance study of three well-known classifiers with BPNN classifier accuracy, which are linear discriminant analysis (LDA), naive Bayes (NB), and the k-nearest neighbors method (k-NN) and presented in Table 2. In Fig. 7, the performance of different classifier is shown. The analysis of performance with default parameters in the dataset revealed that the BPNN usually outperforms the other methods in the combined features (Morphological + Color+ GLCM), where overall accuracy of 98.88% was achieved.

Table 2 Comparison of different classifier accuracies

Features	BPNN (%)	KNN (%)	LDA (%)	NB (%)
Morphological features	87.77	65.55	88.88	68.88
Color features	93.33	75.55	77.77	77.77
GLCM features	88.88	77.77	94.44	72.22
GLRLM features	92.22	70.00	76.66	68.88
Combine (Morphological + Color + GLCM)	98.88	80	85.55	70
Combine (Morphological + Color + GLRLM	94.44	84.40	75.55	76.66

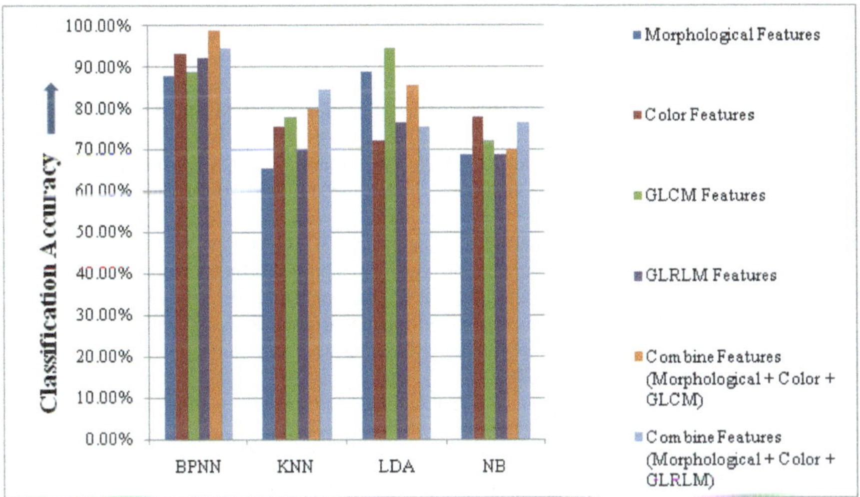

Fig. 7 Comparison of classification accuracy of BPNN classifier other statistical classifiers

4 Conclusion

The classification of six different types of medicinal plants was carried out in this work. Related works highlighted an overview of earlier works. It is found that most of the earlier works are based on morphological features that required various image preprocessing steps. This work incorporates neural networks to classify six different medicinal plant varieties on the basis of morphological, color, and texture features. The efficacy of the various different texture features, namely morphological, color, and combine feature, is studied and compared. It is found that color and GLCM features are able to yield better classification results as compared to morphological features. It is observed that the proposed classification task is able to attain a classification accuracy of about of 98.88% considering all features together. The following are some of the benefits of BPNN over other classifiers: ability to recognize complicated nonlinear correlations between dependent and independent variables without extensive statistical training but major drawbacks of the method are proneness to overfitting. Future work would involve research in various deep learning algorithms that can be used to compare results. As a result, future work may be driven in that direction to make implementation more efficient.

References

1. Haralock, R.M., Shapiro, L.G.: Computer and Robot Vision. Addison-Wesley Longman Publishing Co., Inc (1991)
2. Arun, C.H., Sam Emmanuel, W.R., Christopher Durairaj, D.: Texture feature extraction for identification of medicinal plants and comparison of different classifiers. Int. J. Comput. Appl. **62**(12), 1–9 (2013)
3. Gopal, A., Prudhveeswar Reddy, S., Gayatri, V.: Classification of selected medicinal plants leaf using image processing. In: 2012 International Conference on Machine Vision and Image Processing (MVIP). IEEE (2012)
4. Janani, R., Gopal, A.: Identification of selected medicinal plant leaves using image features and ANN. In: 2013 International Conference on Advanced Electronic Systems (ICAES). IEEE (2013)
5. Priya, C.A., Balasaravanan, T., Selvadoss Thanamani, A.: An efficient leaf recognition algorithm for plant classification using support vector machine. In: International Conference on Pattern Recognition, Informatics and Medical Engineering (PRIME-2012). IEEE (2012)
6. Sathwik, T., et al.: Classification of selected medicinal plant leaves using texture analysis. In: 2013 Fourth International Conference on Computing, Communications and Networking Technologies (ICCCNT). IEEE (2013)
7. Chaki, J., Parekh, R., Bhattacharya, S.: Plant leaf recognition using texture and shape features with neural classifiers. Pattern Recogn. Lett. **58**, 61–68 (2015)
8. Nguyen, Q.-K., Le, T.-L., Pham, N.-H.: Leaf based plant identification system for android using surf features in combination with bag of words model and supervised learning. In: 2013 International Conference on Advanced Technologies for Communications (ATC 2013). IEEE (2013)
9. Kan, H.X., Jin, L., Zhou, F.L.: Classification of medicinal plant leaf image based on multi-feature extraction. Patt. Recogn. Image Anal. **27**(3), 581–587 (2017)
10. Dahigaonkar, T.D., Kalyane, R.T.: Identification of ayurvedic medicinal plants by image processing of leaf samples. Int. Res. J. Eng. Technol. (IRJET) **5**(5), 351–355 (2018)
11. Puri, D., Kumar, A., Virmani, J., et al.: Classification of leaves of medicinal plants using laws' texture features. Int. J. Inf. Technol. 1–12 (2019)
12. Saitoh, T., Kaneko, T.: Automatic recognition of wild flowers. In: International Conference on Pattern Recognition, vol. 2. IEEE Computer Society (2000)
13. Al Bashish, D., Braik, M., Bani-Ahmad, S.: A framework for detection and classification of plant leaf and stem diseases. In: 2010 International Conference on Signal and Image Processing. IEEE (2010)
14. Vennila Ramalingam, A.S., Palanivel S.: Species classification of aquatic plants using GRNN and BPNN AI and society **29**(1), 45–52
15. Manisha, A., et al.: Leaf features based plant classification using artificial neural network. IBMRD's J. Manage. Res. **3**(1), 224–232 (2014)
16. Mall, P.K., Singh, P.K., Yadav, D.: Glcm based feature extraction and medical x-ray image classification using machine learning techniques. In: 2019 IEEE Conference on Information and Communication Technology. IEEE, 2019, pp. 1–6

17. Hashia, B., Mir, A.H.: Texture features' based classification of mr images of normal and herniated intervertebral discs. Multimedia Tools Appl. **79**(21), 15171–15190
18. Anami, B.S., Malvade, N.N., Palaiah, S.: Automated recognition and classification of adulteration levels from bulk paddy grain samples. Inf. Process. Agric. **6**(1), 47–60 (2019)
19. Chandraprabha, K., Akila, S.: Texture feature extraction for batik images using glcm and glrlm with neural network classification (2019)
20. Wang, Z., et al.: Review of plant identification based on image processing. Arch. Computa. Meth. Eng. **24**(3), 637–654
21. Gonzalez, R.C., Woods, R.E.: Digital image processing (2002)
22. Haralick, R.M., Shanmugam, K., Dinstein, I.H.: Textural features for image classification. IEEE Trans. Syst. Man, Cybern. **6**(1973), 610–621
23. Aakif, A., Faisal Khan, M.: Automatic classification of plants based on their leaves. Biosyst. Eng. **139**, 66–75 (2015)
24. Singh, K.R., Chaudhury, S.: Comparative analysis of texture feature extraction techniques for rice grain classification. IET Image Process. **14**(11), 2532–2540 (2020)

A State-of-the-Art of Stock Market Prediction Using Neural Network in Recent years

Mousumi Banerjee and Vanita Garg

Abstract Stock market has always been a complex, intriguing, and yet interesting topic for investors. Almost everyone is directly or indirectly connected with stock markets, and stock price movements have widespread and far-reaching implications. Stock prices are affected by several variables such as news, blogs, and historical details. It is very difficult to predict accurate price of buying and selling stocks with regards to yielding maximum profit. The traditional buy and sell strategies have been replaced by artificial neural network and machine learning technologies, and according to some studies, they are capable of producing great results. This paper provides a literature review using artificial neural network and machine learning methodology of stock market forecasting.

Keywords Stock prediction · Artificial neural network · Machine learning

1 Introduction

Stock market have always attracted people, and the last two decades have increased its popularity with the flexibility of online stock trading. It has always been a long-term endeavor of people to make sense of stock market movements and improve their predictions, and thus, the need for newer tools and technologies has always existed. Stock market participants want to increase their investment and earn a healthy return by buying and selling at the right time. Luo et al. [1] presented IGOA an improved grasshopper optimization algorithm IGOA that uses opposition-based learning, levy-flight, and Gaussian mutation to increase the global and local exploration capabilities of the original GOA. In comparison with Sca, PSO, and Ga, the suggested learning technique will result in a much more robust kernel gradient boosting machine model with better prediction. Yang et al. [2] a new turbo stock selection with stock forecasting system has been created. The focus of this article was on stock prediction and grading. Huang et al. [3] suggested a new variational-based decomposition model to

M. Banerjee · V. Garg (✉)
Department of Mathematics, School of Basic and Applied Sciences, Galgotias University, Greater Noida, Uttar Pradesh, India

K. N. Das et al. (eds.), *Proceedings of the International Conference on Computational Intelligence and Sustainable Technologies*, Algorithms for Intelligent Systems,
https://doi.org/10.1007/978-981-16-6893-7_35

forecast financial data. The genetic algorithm is used in this model to optimize the vmd parameter. The vmd breaks down the data into long short-term pattern. Chen et al. [4] to forecast stock prices, a hybrid model integrating XGBoost and improved firefly algorithm was suggested with IFA developing hyper parameter of the XGBoost and mean variance model for allocating investment proportion of the portfolio. MA et al. [5] used a trading simulation to compare the efficiency of various average and omega investment optimization models for regular trading without transaction fees. Finally, paper discusses the success of when the transaction fee is factored in, these models emerge. The long short-term network architecture and the deep recurrent neural network model are compared to associated network model supplied by Ding and Qin [6]. The associated model was found to be 95% accurate. In this paper, we have collected researches which applied neural network for stock prediction.

The paper is organized as following section given the detailed summary of papers published in 2016 to 2021. Starting from 2016, we have included the related paper and has given the critical/analytical approaches to all the literature.

1.1 Paper in (2016)

Chiang et al. [7] suggested a backpropagation neural network (BPN) approach to predict the stock prices. BPN model is tested on seven company's stocks performance and used seven technical indicators as input variable. The experiments performed in Taiwan stock exchange showed that three corporation had positive return and three corporation had negative return. The BPN methodology did not generate significant positive return after a period of training, but it is still superior to the buy and hold strategy according to the results. Limitation of the BPN model is that it gets slow after few iterations and cannot give accurate result. Choosing proper input variables is another reason for quality of result accuracy. Trading decisions are likely to become more productive if appropriate input variables are chosen.

Akita et al. [8] proposed innovative operation that transform newspaper article to paragraph vector using Long Short-Term Model to predict macroeconomic time series data. LSTM model is a sort of recurrent neural network that is an efficient method of learning long-term dependencies in data. Tokyo stock exchange sources had been used for the proposed model. In this paper, researcher's procedure is to predict ten companies closing price using regression from numerical and textual information and employ LSTM model. For this experiment, data from 2001 to 2008 is taken from Nikkei 225 Newspaper. The results shows that textual news gives better outcomes than numerical information, and LSTM model is more effective to represent time series data than other models. In this paper, input vector is combination of textual information and stock prices. Both have different dimensions, and if incorporating more input variables like moving average, moving average convergence divergence can make the prediction accuracy slow. So better prediction model can be used to avoid this problem.

1.2 Paper in (2017)

Samarawickrama et al. [9] presented paper on recurrent neural network to anticipate everyday stock return prices of Colombo stock exchange. The architecture of this model constructed feed forward neural network simple neural network, long short-term model, and GRU model. Three companies were selected from three different sectors namely banking, finance, manufacturing, and diversified holding sectors. Last two days data open, high and close value were taken as input variable to predict closing value of stocks. GRU generated highest error as compare to LSTM and feed forward neural network. If the number of last two days data were increased for input variables, only LSTM models produce best results. Model can be extended to perform other companies not for only to predict closing price but for open high and low price as well.

Khare et al. [10] suggested to forecast the short-term stock prices using two models namely multilayer perception model and long short-term model. Ten listed company's data from New York stock exchange is used over a duration of one year. Every 10 min data is considered for each stock which is observed for 85,000 to 90,000 points. In this paper, technical indicators are categorized in three different types namely trend, cyclic, and momentum type. Individual result is calculated for two different models—MLP and LSTM. Architecture is same for MLP model and LSTM. It is concluded that forecasting short-term prices, MLP outperformed the LSTM model. Major limitation of MLP is that it cannot solve time series problem. Stock market data are all time series data. In this model, yesterday's price is considered for forecasting today's prices. MLP model would not work where postpositional or fundamental trade is considered.

1.3 Paper in (2018)

Khanjarpanah and Pishvaee [11] paper is focused on financial planning assets and portfolio management. A novel robust versatile model and its updated methods are proposed and compared to deterministic model. The output of the proposed model is demonstrated using solution of stocks from the Tehran stock exchange market, which is the most important Iranian financial market. The result shows that the deterministic model outperform robust model at lower level of violation penalties, while robust models outperform deterministic models at higher levels of violation penalties.

Hamadameen et al. [12] the principal of dominated solution is used to find a pareto optimal solution (POS) for multi-objective linear programming (MLP) problem without transforming the MLP problem into its special linear programming problem Training method is proposed and compared with the result of the existing method and found that efficient results was performed by the proposed method.

Hossain et al. [13] paper proposed a hybrid approach to forecast the closing price of the stocks. Basically, the hybrid method is a mixture of networks of long short-term memory (LSTM) and gated recurrent unit (GRU). LSTM and GRU are efficient recurring system that can perform quickly and accurately in terms of regression-based predictive performance. The S&P 500 index historical data for 66 years from 1950 to 2016 was used. Adam optimizer is used to train the data. The hybrid network has showed prediction accuracy. This model is good for predicting closing price of the stocks. Since this is regression-based prediction model, it would be challenging to predict opening price for this model as sentiments involved with opening price of the stock.

Thomas Fischer et al. [14] used component stocks of the S&P 500 from December 1989 to September 2015 for an applied LSTM networks for estimation only in directional movement. Obtain end of each month component lists for the S&P 500 from December 1989 to September 2015 and first accessed the month end constituent lists for the S&P 500 by Thomas Reuters. Consolidated single binary matrix for these sections, all stocks have ever been a part of the index over the span of time. Downloaded daily return of all detail's dividend, stock split, and corporate action from January 1990 until October 2015. Concluded that LSTM network can derive useful information constructively from data from noisy financial time series.

Siyuan Liu et al. [15] used three-layer stacked LSTM model to predict stock prices and compared with single layer LSTM model. This experiment implemented on CSI index for the period of 18/05/2017 to 29/01/2019.It is found that three-layer LSTM gives more accurate result than single layer LSTM model.

Gosswami et al. [16] provided a robust method PCA with backpropagation is used for measuring the share price and evaluated how it varies with the actual price. Since stock prices data are very high-dimensional data to reduce dimensionality a statistical tool, principal component analysis is used. Data source taken from Dhaka Stock Exchange (Bangladesh). Researcher selected ten different company's last 6 month closing price data as a sample data for a training test and compared it with its actual price and evaluated their performances. Performance was not satisfactory due to small sample and if big size data is used to train the network, the performance would improve significantly. Backpropagation worked in a gradient descent way, eliminating the error between the real output and desired output, but adding more input variable and when increased data size gets the model slow. Using different neural network method, LSTM can be used. LSTM decrease the error with gate method.

Siddique et al. [17] proposed a hybrid model to support vector regression with particle swarm optimization (PSO) and compared the model with standard SVR. Tata Steel stock price data is considered for testing and training the model and data set from (24/07/2001 to 19/03/ 2018) used. Errors measured in training process with MAE, RMSE, and MAPE 2.7602, 5.7413, and 0.6899 percent (approx.), respectively, and errors in evaluation phase are 2.9291, 6.4949, 0.7085% (approx.), respectively. It was found that PSO-SVR surpassed the standard SVR. But limitation with PSO is that it easily falls into local optimum in high-dimensional space and gives a low convergence rate in iteration process. Genetic algorithm can be another method to optimize data as it has the ability to avoid being trapped in local optimal solution.

In Maiti and Roy [18], the purpose of the research is to develop a commutating technique for solving interval-valued and multi-choice bi-level programming for the stack berg game using an intuitionistic fuzzy method.

1.4 Paper in (2019)

Zhang et al. [19] introduced a new method for predicting stock closing price using a generative adverbial network (GAN) with a multilayer perceptron as the comparator and long short-term memory as the creator. Dataset is taken from S&P 500 for training and testing the data. LSTM model takes longer time to train the model and cannot demonstrate accurate result.

Garg and Deep [20] used two version of biogeography-based optimization-BBO and blended BBO to propose a novel technique to addressing the traditional portfolio optimization model. Portfolio optimization can be done in two ways using the mean and variance model: first with the expected return known and the optimization method applied directly to the problem and second with the expected return, investigated across predefined range then the optimization technique applied. The result of comprehensive studies shows that LX-BBO outperforms blended BBO when it comes to portfolio optimization.

Das et al. [21] proposed a method using ranking function incorporated with fuzzy slack and surplus variable to find the best solution of the problem. Result compared with fully fuzzy linear programming problem and found that the ranking function technique is more reliable and versatile.

Ramezanian et al. [22] proposed a genetic network programming model which is graph-based evolutionary algorithm and combined the GNP model with ARMA-GARCH model to predict the stock prices and determine the profit. Researcher could be add more technical tool and fuzzy logic to improve the model.

Zandieh and Mohaddesi [23] solved the portfolio rebalancing problem in this paper when security returns are measured by unknown variable, and transaction costs are taken into account. The suggested model's performance is examined using a rebalancing strategy. For greater performance, meta heuristics approach instead of a global solver is suggested. Portfolio restructuring is a term that has been used to describe the process of rebalancing portfolio.

Singh et al. [24] developed two models to forecast the pattern of stock market prices. The first version is an LSTM model, and the second version is a hybrid model that contains the SVM model, KNN model and a random forest classifier using the majority algorithm. Since stock prices often fluctuate with news and sentiments, so a sentiment analyzer was used by both models to verify the sentiment analysis on the company news. Natural language processing tool is used for sentimental analysis, data obtained from Twitter, and other news outlets. It was found that hybrid model gave an accurate 92 percent performance. Limitation of LSTM model is that it gives inconsistent result in each time used. Using different algorithm like stacking algorithm can be used to predict stock market trend.

Reddy et al. [25] proposed the Multi-Source Multiple Instance (M-MI) system to predict the stock price movement. Sentiments plays a very important role in stock prediction. Twitter may help us to predict the stock prices. This method only tells us that trend is positive or negative but for prediction movement implemented RNN based approach used and later combine these two approaches to predict the final value. We may call this as multi-source multiple instance model. For training the data, the BSE closing and opening price is used. Due to overfitting and underfitting data which might lead to noise in dataset, so that Multi-Source Multiple Instance model might not predict accurate result.

Eapen et al. [26] proposed combined multiple pipelines of convolution neural network and bi-directional LSTM model to forecast stock market index. Convolutional is a technique of extraction features for high-dimensional data as seen in stock prices. Data collected from S&P index of daily closing, opening, high and low price dated from 02/01/2008 to 27/11/2018 for a period of ten years duration. Tested several single pipelines deep-learning models using CNN and bi-directional LSTM. Mean square error (MSE) is a statistical tool considered to calculate error. Proposed model demonstrated the best result. Ad delta optimizer is used which worked in a gradient descent manner for correcting weight with each iteration in the model that causes possibility to slow down the model.

Zhang et al. [27] proposed an LSTM approach to predict the stock price and compare the model with conventional ARIMA model, and prediction ability is compared with GRU. Stock prices data relevant to multifactor so model is performed on single factor and multifactor model. Dataset considered 1990–2015 giving 6109 observations and separated into two subgroups. Training subset has 6009 records and testing subset has 100 records. For trained the data backpropagation through time (BPTT) is used. It is a gradient-based technique. Observed that multifactor recurrent neural network outperformed all other model. So the prediction ability of the LSTM model is superior than the GRU model. Limitation with this technique, it has a difficulty with local optima and cannot get accurate result to cope with this problem different method can be used.

Zhang et al. [28] proposed to forecast the highest price of a stock. In this paper researcher compared the PSO-LSTM model with standard LSTM model to forecast the prices. Experiments are applied in three stocks Amazon, Apple, and Microsoft from US market. Daily open, close, high, low, Adj close, and volume data had been taken The LSTM model optimized by PSO which increases the accuracy of prediction in all three stocks, providing better result than the standard LSTM model, but PSO has some limitations that it has slow convergence rate in every iteration process specially if problem is complex.

1.5 *Paper in (2020–2021)*

Latunde [29] formulated a multi-factor model in capital asset management under some uncertain criteria. The mathematical model was developed in response to investors desire to solve a particular problem in capital asset management involving multiple variables. In this problem taxation, deprecation considered as input variable. Objective is to maximize investors' assets value. The result represents that the model provided an adequate reflection of the system under consideration and estimated present value of the efficiency of the asset is optimized with minimal risk or liabilities.

Haq et al. [30] analysis used a deep generative model to forecast next price movement by merging feature selected using different feature selection technique to obtain subset of features. To begin, daily stock data was used to generate a set of 44 valuation metrics. A logistic regression model (support vector and random forest) that has been trained independently is used to calculate the worth of 88 stocks. The result reveals that combining feature obtained using different feature selection procedure and putting them into a deep generative model outperformed and state of methods (Fig. 1).

In the year 2016, backpropagation method is used most.

In the year 2017, LSTM method is used most.

In the year 2018, back propagation and LSTM method are used.

In the year 2019, 2020, and 2021, LSTM method is used most.

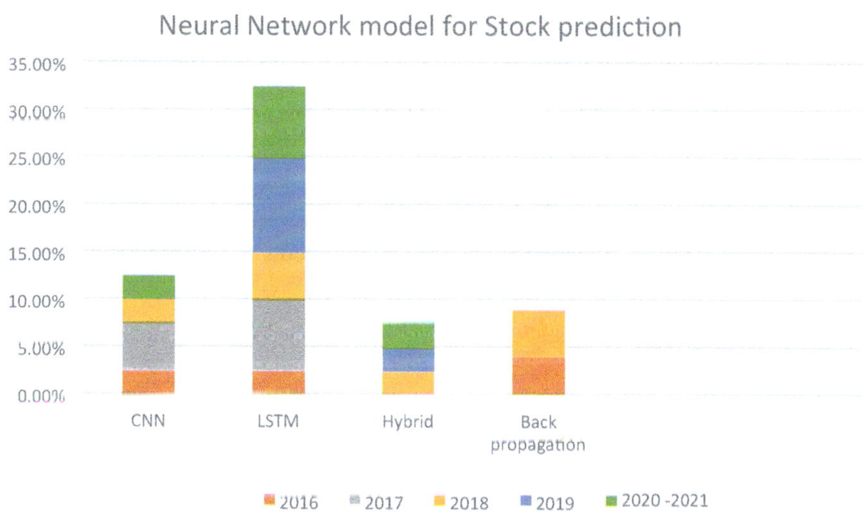

Fig. 1 Neural network model for stock prediction

2 Mathematical Formulation

Stock prices are time series data consists of regular opening value of transaction data stocks, high value, low value, and volume. Defining daily profit in stocks, when trader buy and sell the stock.

$$P_{i,t} = \text{closing price of the stocks}$$
$$P_{i,t-1} = \text{Previous day closing price of the stocks}$$
$$i = \text{No of stocks}(i = 1, 2, 3 - - - - n)$$
$$\text{Return of the stock over previous p day is calculated} =$$
$$\text{Profit} = R_{i,t} = \frac{p_{i,t} - p_{i,t-1}}{p_{it}}$$

The aim of the trader is to maximize the profit and minimize the loss.

3 Result and Discussion

This review paper presents solving stock prediction problem for the period 2016–2021.This paper gives a detailed summary of recent published articles in the topic of stock market prediction by using LSTM, RNN, CNN, and hybrid method. The objective of this paper is to understand which techniques have been used in given period (2016–2021) to solve stock prediction problem and observed that LSTM method is widely used in the problem. The study aims to highlight that stock prediction and portfolio optimization also requires a higher level of prediction and an algorithm-based technique can give a higher level of prediction. Modern algorithms have artificial intelligence and machine learning attributes which develops their own intelligence (machine intelligence). It is understood and expected that there will be multiple algorithms for specific stocks, portfolios and with regards to specific news or events (like outcome of an election) which may have the requirements to be further customized and developed as per specific requirements or outcomes. The predictions and result analysis will entail further customization and optimization scope.

4 Conclusion

This paper is focused on historical data as an input features for data optimization to predict stock prices and portfolio optimization problem. Market is also affected specially in Indian market widely affected by sentiments and socio-political factors. This requires further studies to explore how to retrieve more valuable and significant factors from news and optimize the system to learn the distribution of data more precisely. The advance markets like the USA use algorithm-based predictions more widely as compared to markets like India. This is because the advance markets are more matured and are less disturbed by factors like sentiments or political events. There cannot be one algorithm that suits all markets or all requirements and a phenomenon which is constantly increasing and will always remain an evolving phenomenon.

Acknowledgements Authors are thankful to the Galgotias University for providing the academic support, resources and the facilities in the department that enabled us to accomplish our work.

References

1. .Jie, L., Huiling, C., Qian, Z., Yueting, X., Hui, H., Xuehua, Z.: An improved grasshopper optimization algorithm with application to financial stress prediction. Appl. Math. Modell. **64**, 654–668 (2019)
2. Yang, F., Chen, Z., Li, J., Tang, L.: A novel hybrid stock selection method with stock prediction. Appl. Soft Comput. J. 1 27 (2019)
3. Huang, Y., Gao, Y., Gan, Y., Ye, M.: A new financial data forecasting model using genetic algorithm and long short-term memory network. Neurocomputing (2020)
4. Chen, W., Zhang, H., Mehlawat, M.K, Jia, L.: Mean–variance portfolio optimization using machine learning-based stock price prediction. Appl. Soft Comput. 106943 (2019)
5. Ma, Y., Han, R., Wang, W.: Portfolio optimization with return prediction using deep learning and machine learning. Expert Syst. Appl. **165**, 113973 (2019)
6. Ding, G., Qin, L.: Study on the prediction of stock price based on the associated network model of LSTM. Int J Mach Learn Cyber **11**, 1307–1317 (2020)
7. Chuan-Wen, C., Lin, J.-B., Chen, C., Lin, Y.T.: Backpropagation neural network model for stock trading points predictions. Int. Res. J. Appl. Fin. **VII**(10), 254–266 (2016)
8. Akita, R., Yoshinara, A., Matsubara, T., Uehara, K.: Deep learning for stock prediction using numerical and textual information. In: IEEE/ACIS 15th International Conference on Information and Communication Systems (2016)
9. Samarawickrama, A.J.P., Fermando, T.G.I.: A recurrent neural network approach in predicting daily stock prices. In: International Conference on Industrial and Information Systems (*ICIIS*) (2017), pp. 1–6 [9]
10. Khare, K., Darekar, O., Gupta, P., Attar, V.Z.: Short term stock price prediction using deep learning. (RTEICT). IEEEEE (2017), pp, 482–486
11. Khanjarpanah, H., Pishvaee, M.S.: A fuzzy robust programming approach to multi-objective portfolio optimization problem under uncertainty. Int. J. Math. Operat. Res. **12**(1) (2018)
12. Hamadameen, A.O., Hassan, N.: Pareto optimal solution for multi-objective stochastic linear programming problems with partial uncertainty. Int. J. Math. Operat. Res. **12**(2) (2018)
13. Hossain, M.A., Karin, R., Thulasiram, R., N.D. Bruce, Wang, Y.: Hybrid deep learning model for stock price prediction. In : IEEE Symposium Series on Computational Intelligence (2018)
14. Fischer, T., Krauss, C.: Deep learning with long short-term memory networks for financial market predictions. Eur. J. Operat. Res. 654–659 (2018)
15. Liu, S., Liao, G., Ding, Y.: Stock transaction prediction modelling and analysis based on LSTM. In: International Conference on Industrial Electronics and Applications, pp. 2287–2290 (2018)
16. Goawami, T., Kumar, S., Hasan, S.M.: Stock market data analysis and future stock prediction using neural network. Int. J. Comput. Sci. Inf Secur. **16** (2018)
17. Siddique, M., Mohanty, S., Panda, D.: A hybrid forecasting model for stock prediction of stock value of tata steel using support vector regression and particle swarm optimization. Int. J. Pure Appl. Math. **119**(14), 1719–1727 (2018)

18. Mait, S.K., Roy, S.K.: Analysing interval and multi-choice bi-level programming for Stackelberg game using intuitionistic fuzzy programming. Int. J. Math. Operat. Res. **16** (2018)
19. Zhang, X., Li, C., Morimoto, Y.: A multi-factor approach for stock price Prediction by using recurrent neural networks **8**(1), 9–13 (2019)
20. Garg, V., Deep, K.: Portfolio Optimization Using Laplacian Biogeography-based Optimization. Springer (2019)
21. Das, S.K., Mandal, T., Behera, D.: A new approach for solving fully fuzzy linear programming problem' Int. J. Math. Operat. Re. **15**(3), 201 (2019)
22. Ramezanian, R., Peymanfar, A., Ebrahimi, S.B.: An integrated framework of genetic network programming and multi-layer perceptron neural network for prediction of daily stock return: an application in Tehran stock exchange market. Appl. Soft Comput. J. (2019)
23. Zandieh, M., Mohaddesi, S.O.: Portfolio rebalancing under uncertainty using meta-heuristic algorithm'. Int. J. Operat. Res. **36**(1), 201 (2019)
24. Singh, S., Ahmad, M., Bhattacharya, A., Azhagiri, M.: Predicting stock market trends using hybrid SVM Model and LSTM with sentiment determination using natural language processing. Int. Journal of Eng. Adv. Technol. **9**(1) (2019)
25. Reddy, A.V.M., Krishna, Y.D.V., Miranam, S.: Stock market prediction using RNN and sentiment analysis. Int. J. Adv. Res. Ideas Innovat. Technol. **5**(3), 5–8 (2019)
26. Eapen, J., Verma, A.,Bein, D.: Novel deep learning model with CNN and Bi directional LSTM for improved stock market index prediction. IEEE, pp.0264–0270 (2019)
27. Zhang, Y., Yang, S.: Prediction on the highest price of the stock based on PSO-LSTM neural network, In: 3rd International Conference on Electronic Information Technology and Computer Engineering (2019)
28. Zhanga, K., Zhonga, G., Donga, J., Wanga, S., Wanga, Y.: Stock market prediction based on generative adversarial network. Procedia Comput. Sci. **147**, 400–440 (2019)
29. Tolulope, L.: Multifactor modelling in asset management. Int. J. Math. Operat. Res. **17**(3) (2020)
30. Haq, A.L., Zeb, A., Lei, Z., Zhang, D.: Forecasting daily stock trend using multi -filter feature selection and deep learning. Expert Syst. Appl. **168**, 114444 (2021)

Expanding Electricity Access in Rural Uttarakhand by Mobilization of Local Resources

Divya Asija, Rajkumar Viral, and Tarannum Bahar

Abstract Off-grid systems generate electricity without relying on the utility grid, and hence, they are referred as stand-alone installations. These are better suited to the most distant regions where the grid cannot reach, and no other energy source is available. Since they are frequently the most cost-effective and environmentally benign alternative for applications distant from the utility grid, these systems include the bulk of resources such as wind, geothermal, solar photovoltaic, biomass, and hydro potential accessible in distant parts of the world. Lights and other distant stations, supplemental energy units for emergency services or military uses, and rural and remote village electrification are only a few examples. Hydropower, particularly small hydropower, is perhaps the most significant of the "renewables" for electrical power generation on the globe, accounting for 45 percent of global energy output. The modular size hydro is often run-of-river, without dams or water storing, and is among the maximum profitable and eco sustainable green system to consider for remote village electrification in developing nations as well as future hydro development worldwide. Uttarakhand, in contrast, is one of the Indian states with a high hydro potential of nearly 20,000 MW, of which 16,500 MW is fully defined and technically viable at 1160 MW, and around 5525 MW is under progress (All India Regionwise Generating Installed Capacity of Power, CEA, Ministry of Power, Government of India, 2021 [1–3]. For rural electrification, small hydro power (SHP) projects primarily comprise small, medium (grid linked typically), and mini/micro/pico/watermills (off-grid connected usually). The study discusses the Indian power situation, the role of SHPs in rural electrification in Uttarakhand, the SHP principle, categorization, benefits, and drawbacks. The paper will explain

D. Asija (✉) · R. Viral
Department of Electrical and Electronics Engineering, ASET, Amity University Uttar Pradesh, Noida, India
e-mail: dasija@amity.edu

R. Viral
e-mail: rviral@amity.edu

T. Bahar
Galgotia University, Greater Noida, U.P., India

K. N. Das et al. (eds.), *Proceedings of the International Conference on Computational Intelligence and Sustainable Technologies*, Algorithms for Intelligent Systems,
https://doi.org/10.1007/978-981-16-6893-7_36

government support and strategies in Uttarakhand for the growth of SHPs. It also discusses the technological, economic, and environmental viability of SHPs.

Keywords Rural electrification · SHPs · Mini/Micro/Pico watermills

1 Introduction

The Indian electricity sector has a total installed capacity with all generation resources available of 382.730 GW as on April 30, 2021 [1], which has fifth position all over the world. This capacity also includes total of captive generation with an extra 95.012 GW. However, if we consider the non-conventional power generation, it is found 88.55% of the total installed volume and 11.45% of remaining is from renewable power generation [2]. During the fiscal year from 2011–12, India produced 855 BU (855,000 MU i.e., 855 TWh [3]) of electricity capacity.

As per analysis, it is observed that more than 1/3rd rural population of India has erratic supply and rest 6% of the town population are still deprived of electricity. The sector-wise installed capacity in MW as on July 31, 2020 [1–4], can be examined in Fig. 1, while Fig. 2 shows the per capita consumption of electricity (kWh) of different states in India in comparison with national average as on July 31, 2020.

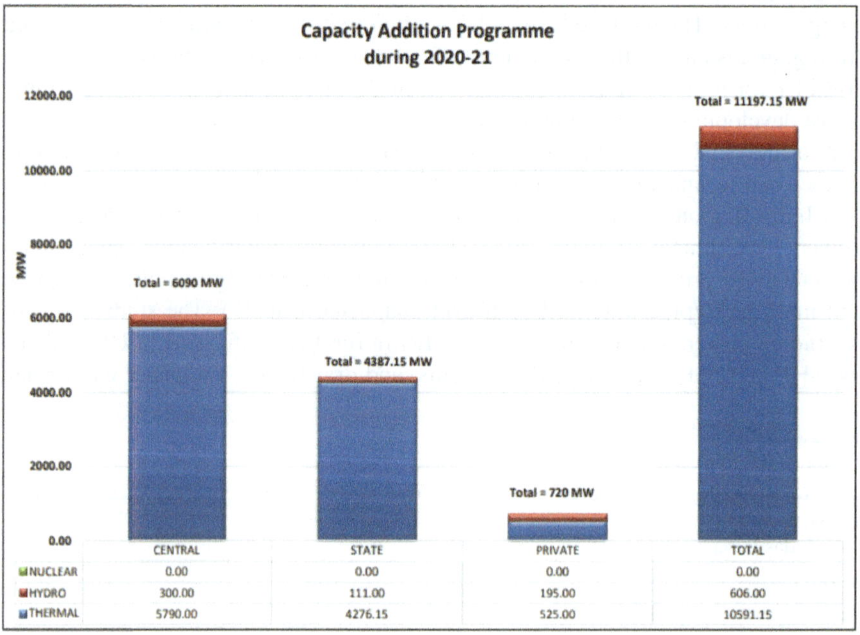

Fig. 1 Sector-wise installed capacity during 2020–21

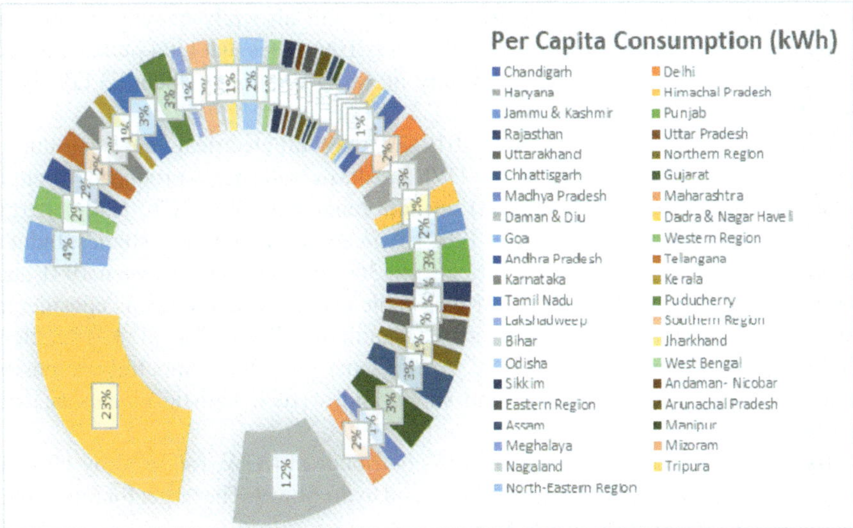

Fig. 2 Per capita consumption of electricity (kWh) of different states in India in comparison with national average as on July 31, 2020 [1–4]

In terms of concern to hydro power including large and small, the existing installed capacity as on April 30, 2021, is nearly 477.74 GW, which also include renewable power plants such as huge hydro power plants contributing approximately 36.80% of overall installed generation of India [5, 6]. As on April 2021, India has renewable resources-based generation of electricity, nearby 25% of its entire capacity including SHPs of 4786.81 installed capacities (MW). Nearly, 8846 remote/rural villages, with total installed capacity, biogas plants are 4.4 million, total of 1800 micro hydro plants as of August 2020 [3–7]. Even before to the worldwide epidemic, India's total electricity consumption is expected to rise about 50% from 2019 to 2030, nevertheless development is now expected to be nearer to 35.0% in the Stated Policies Scenario (STEPS) and 25.0% in the late recovery situation. Finally, it would endanger the few parts of India's where continuous attempts are made to boost the campaign against energy scarcity, since lesser-pay people would be enforced to rely upon significantly contaminating and amateurish energy resources. In STEPS India, scheme surpasses the goals to constitute in the Paris agreement on the Nationally Determined Contribution (NDC). From 2005 to 2030, India's GDP reduces its emissions intensity by 40%, above the current NDC's target of 33.5%. Furthermore, non-fossil fuels account for about 60% of energy generating capacity, much exceeding India's goal of 40%. By 2040, India's headship in vast distribution of renewable energy expertise will have increased major lead in the field of solar photovoltaic, wind energy system, and battery storage systems which crosses up to $40 billion/year. As a result, India will account for one out of every seven parts of outlay disbursed on such type of three categories of apparatus in year 2040, equated to one out of every twenty today. Over the next 10 years, India's renewable energy workforce will increase by one million people. Additionally, India's electrical security will also rely heavily on storage technology. The Central Electricity Authority (CEA) states that total pumped hydro storage system has total installed capacity of 4.8 GW among which 3.3 GW is in operational stage [8].

All through, the financial year of 2019–20, the total electrical energy produced through various utilities were 1383.50 TWh and the overall energy produced (through small and large utilities/non-utilities) in India remained 1598 TWh, while energy consumption in FY2019 remained at 1208 kWh/capita. According to studies, India's per capita power usage increased thrice from 360 to 1208 kWh between 1991 and 2021. As a result, the deployment of renewable energy technologies is considered vital. The Government of India's focus on attaining 'Power for all' has accelerated capacity addition in the country. At the same time, the competitive intensity is increasing at both the market and supply sides (fuel, logistics, finances, and manpower).

By 2022, solar energy is estimated to contribute 114 GW, followed by 67 GW from wind power and 15 GW from biomass and hydropower. The target for renewable energy has been increased to 227 GW by 2022.

In FY22, the total thermal installed capacity in the country stood at 234.72 GW. Installed capacity of renewable, hydro, and nuclear energy totaled 95.01 GW, 46.20 GW, and 6.78 GW, respectively.

2 Uttarakhand Energy Scenario

As far concerning to Uttarakhand, it is ironic in environmental resources specifically water and forestry with numerous glaciers, tributaries, thick woodlands, and snow-covered foothill mountaintops.

The topographical surroundings of the state-run diverge by a petite aloofness. Furthermore, due to forestry restrictions or the high cost of handling, a nested grid connection would be unlikely in these circumstances. This framework also includes the operation and maintenance (O&M) service. The state Uttarakhand has several large and small tributaries, watercourses and springs that might be accustomed to produce power for confronting the electricity necessities to isolated rural/remote inhabitants, small house landlords, minor societies, campsite locations, and inaccessible gatehouses [9].

The state Uttarakhand is the only foremost state in the republic of India in context to exploitation of renewable energy resources (RESs) which has highest hydro power-based generation including small and large hydro. Electrification is one of the most vital requirements for socio-economic development of rural areas of the state Uttarakhand. The linkages between RESs, sustainable livelihoods, and socio-economic development are well recognized. Uttarakhand is predicted to have a hydropower capacity of 20,236 MW, of which 3594.85 MW has been harnessed through hydroelectric projects till date. Nature has blessed the state with rivers like

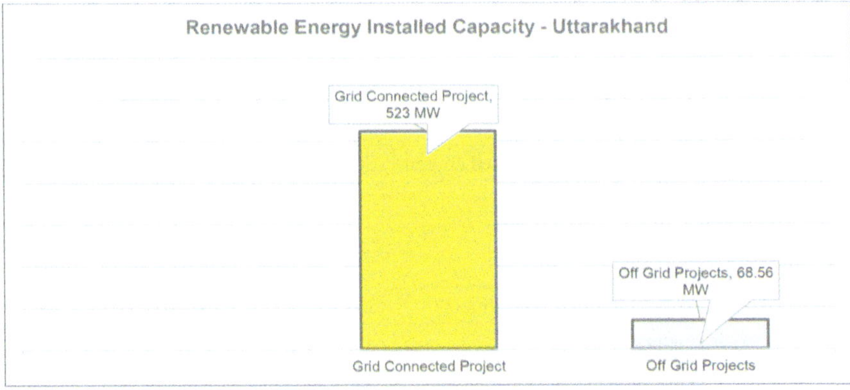

Fig. 3 RESs capacity in Uttarakhand as on Dec 31, 2020 [9]

as the Ganga, Yamuna, and Kosi. As a result, the area is perfect for hydropower development.

Figure 3 represents the RESs capacity in Uttarakhand as on Dec 31, 2020 [10]. In Uttarakhand, total energy consumption 7250.73-million-unit (MU) Watt in the year 2020–21 including domestic sectors (1484.86 MU watt), commercial (941.90MU watt), industrial (4197.73 MU watt), street lighting (53.88 MU watt), agriculture (295.98 MU watt), and water works/waste disposal (276.38 MU watt), respectively. Along with this in rural electrification, the total electrified inhabited revenue villages by Uttarakhand Power Corporation Ltd. (UPCL) are 15,298, by Uttarakhand renewable energy development agency (UREDA) 266 [11–13].

3 SHPs Status in Rural Electrification (RE) in Uttarakhand

The rural electrification (RE) schemes of the renewable ministry at state and center level intends to offer elementary lights/energy amenities using RESs in areas without no access electricity in isolated register/homesteads parishes, and where energy networks interconnections are not practicable or not economically feasible. India is one of the countries in the world that has a dedicated ministry for promoting renewable energy, the MNRE, and an exclusive public sector financial agency, the Indian RE Development Agency (IREDA). Similarly, in Uttarakhand, UREDA is the nodal state agency which dedicated to promoting all the RESs based technologies throughout the state. As on March 31, 2011, in Uttarakhand, installation of off-grid/decentralized RE systems/devices for rural electrification were biogas plants nos.-12,659, water pumping windmill nos.-0, SPV pumps nos.-26, solar photovoltaic (SPV): street lighting systems (SLS) nos.-568, home lighting system (HLS) nos.-91,307, solar lantern (SL) nos.-4,023, and power plants (PP) nos.-18,003. Among other RE technologies like, aero hybrid systems-0 MW, solar cooker nos.-10,534,

Table 1 Achievement in RE by UREDA in Uttarakhand as on Nov 31, 2020

S. no	Particulars	Capacity (MW)
A	Grid-connected project	523.55
1	Bio energy program	72.60
A	Bagasse and non-bagasse-based co-generation power plant	72.60
2	Solar energy program	254.55
A	Solar PV power plant	216.40
B	Solar rooftop and small plant	18.15
C	Solar canal bank/top power projects	20
3	Hydro energy program	196.40
A	Micro, mini, and small hydro project	196.40
B	Off-connected project	68.56
1	Bio energy program	59
A	Rice husk and biomass based captive power plant	59
2	Solar energy program	3.66
A	Solar decentralized PV power plant	3.66
3	Hydro energy program	5.90
A	Micro and mini hydro plants	5.90

biomass gasifiers (rural + industrial) nos.-350, and waste to energy −3.07 MW has been installed in total nos. of 472 villages and 34 hamlets as on Nov 31, 2020, under RE [13–17].

In addition to this, physical progress of RE projects under implementation in Uttarakhand as on Nov 31, 2020, was overall sanctioned village nos.-672, village completed-472, ongoing villages-142 including total hamlets-63, and completed hamlets-34 [3–6]. In the line, Table 1 represents the current position of Uttarakhand achievement in RVE by UREDA as on Nov 31, 2020 [10–12].

4 Small Hydro Power (up to 25 MW)

The energy available from the free-falling water stream easily can be changed in the machine-driven energy and ultimately can be converted into electrical/electricity energy. The total sum of useful energy that may be apprehended is proportionate to the perpendicular inaccessibility of the stream's water drop (called the total head) and the capacity/volume (known as flowrate) of the water falling. In this background many small-scale schemes, e.g., Mini (from 100 kW to 1 MW)/Micro (up to 100 kW)/Pico (<50 kW) [10]. Figure 4a and b show typical view of SHP scheme. SHPs are beneficial to electrify the remotely located rural place where grid connectivity not feasible. In Uttarakhand as on March 31, 2011, the total identified SHPs sites are 444 (1577.44 MW) out of which the installed sites are 95 (134.12 MW) and

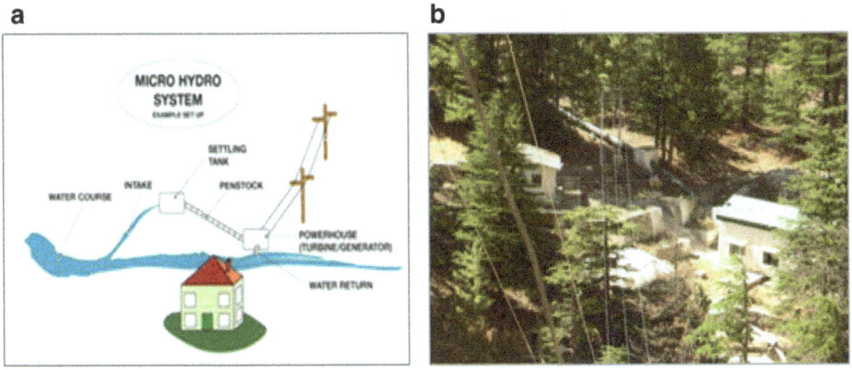

Fig. 4 Energy generation from small hydro power projects **a** Schematic of a small hydro project **b** MHP near Gangotri in operation (3 × 100 kW)

the project under implementation 55 (230.65 MW) [9]. (For more details of SHPs please see the reference [11–16]).

4.1 SHP Concept

In most of the common cases, any hydro turbine changes the water kinetic energy into mechanical energy that can rotate the shaft of coupled electric generator easily and generate electricity. Scientifically, the power available from water turbine, i.e., P (in Watt), generate the power at the shaft on turbine shat and can be expressed as:

$$P = \eta_t.\rho_w.g.Q.h_{\text{eff}} \tag{1}$$

where $\eta_t$represents the efficiency available of the hydraulic turbine, ρ_w water density (kg/m^3) of stream, the acceleration factor of gravity g in m/s^2, Q available water or the volume per second at the turbine shaft in m^3/s, and h_{eff} the operative gravity head of discharge athwart the mechanical turbine (m). The water turbine has hydraulic efficacies in between the range of 80 and 90% and generator can achieve the efficiency almost 90 and 98% [8].

4.2 Mini/Pico/Micro Hydro

UREDA is building MHPs in Uttarakhand for both RE and grid feeds. In excess of 250 communities and villages have been captivated because of these schemes, which have so far approved 34 MHPs with an aggregate capacity of 3.41 MW. In rural areas of the state where the nationwide or state grid cannot be long-drawn-out,

UREDA has installed a significant number of MHPs. Till now, 41 MHPs with a total capacity of 3.960 MW have been placed in Uttarakhand's remote villages, with another 18 MHPs with a total capacity of 1.965 MW under construction. UREDA commissioned 6 MHPs with a total capacity of 500 kW in 2020–21 (1MHP with a capacity of 100 kW in districts Tehri, 1MHP with a capacity of 25 kW in district Uttarkashi, 2 MHPs with capacities of 100 and 50 kW in Chamoli, and 2 MHPs with capacities of 200 and 25 kW in district Bageshwar) [15–19].

4.3 Watermill (Gharats)

According to the UREDA study, there are 15,449 traditional watermills (gharats) in operational or extinct state. With MNRE's financial help, UREDA is upgrading these historic watermills for electromechanical, mechanical, and electrical application. In isolated places, the modernized watermills of this mountainous terrain are a vital and crucial source of electricity. These watermills have been used to crush grains as well as supply energy to local regions. Modified watermills can generate up to 5 kW of electricity, which is enough to power 15–20 homes within a 500 m radius. The government of India's Small-Scales Industry Department (SSID) has chosen watermill (or Gharat) as an SSID government that has excluded watermill components and goods from VAT [18]. Apart from the MNRE subsidy, the state government is offering a grant of INR 6000/watermill for renovation of electric energy output that can be cost of Rs. 35,000/- for mechanical energy output for conventional watermills. Images of commissioned gharats in various parts of the state are shown in Fig. 5a, b [19, 20].

Fig. 5 Watermills/Gharats in operation for RVE **a** Gharat situated in Sunderdhunga Valley **b** Watermill Sahaspur, 30 km from Dehradun

5 State Government Policies and Incentives in Promoting of RE Programs

Realizing the facts, govt. has taken various steps to promote and illuminate such kind of rural and remote areas, mainly in border and Himalayan region. Under previous headings, we already mentioned RESs available in the state, center, and state govt. different incentives/schemes to electrify the remote areas under each RESs. Except this, Uttarakhand govt. is also providing subsidies in each technology, so that everyone can afford them and illuminate. *(For detailed policy please see website of the respective State like mentioned in refs* [17–22]).

6 Analysis of SHPs for Techno-Economic and Ecological Feasibility

The goal of this study is to evaluate four existing micro/pico-hydro projects (Gangotri 3 × 100 kW, Harsil 2 × 10 kW, Dalmir 2 × 5 kW, and Bhairon Gahti 1 × 10 kW) in Uttarakhand's Uttarakashi district (Fig. 6). Four factors may be considered to discover and analyze appropriate sites for pico-hydropower projects: technical, financial, environmental, and social impacts. Site selection and technical problem assessment are included in technical analysis. The overall investment costs and revenue from engineering viable projects are calculated as part of the financial study. CO_2 emissions from transmission lines and diesel engines are contrasted to those from a pico-hydropower system to measure the impact on the environment. The villagers' responses to surveys about their standard of living when using electricity from an already existing pico-hydropower plant are being used to assess the social impact.

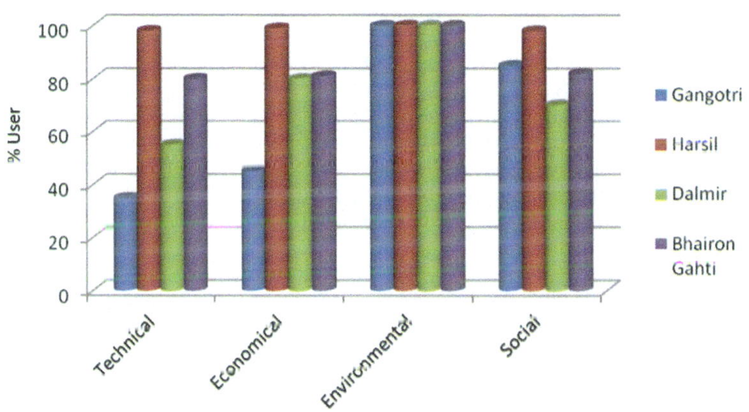

Fig. 6 Techno-economic and environmental feasibility analysis of SHPs

7 Conclusion

Electrification is a big problem in very remote rural areas, where lot of identified RE sites cannot be extended soon due to economic reasons for public and private developers. As an alternative to grid expansion or as a supplement to grid-provided electricity, SHPs offer significant potential to provide a stable and secure, eco-friendly energy source. As a result, an action has been taken in this study to discuss the multiple SHPs and their claim for electrifying rural communities in the area. The government of Uttarakhand's incentives, schemes, and policies to boost electrification initiatives are also highlighted. The research focuses primarily on the techno-economic and ecological viability of SHPs. If these schemes and programs implemented in such a way, then Uttarakhand soon become completely electrified state in all aspects.

References

1. All India Regionwise Generating Installed Capacity of Power, CEA, Ministry of Power, Government of India, December (2021)
2. http://www.renewindians.com/2021/12/Renewable-energy-contribution-in-india.html.
3. Power sector at a glance: all India data, Ministry of Power, Government of India, June (2021)
4. Remme, U., et al.: Technology development prospects for the Indian power sector. International Energy Agency France, OECD, February (2011).
5. Year End Review—2021 Press Information Bureau, Government of India, Dec. (2021)
6. New and Renewable Energy, Cumulative deployment of various Renewable Energy Systems as on 31/12/2021. Ministry of New and Renewable Energy, Government of India, Dec. (2021)
7. http://cea.nic.in, Accessed on Dec.2021.
8. https://iea.blob.core.windows.net/assets/1de6d91e-e23f-4e02-b1fb-51fdd6283b22/India_Energy_Outlook_2021.df
9. Energy scenario at Glance in India: Central Electricity Authority (CEA), Planning wing/Integrated Resource Planning Division's Annual Report, accessed on Sep.2021.
10. MNRE official website: http://mnes.nic.in (also exist http://mnre.gov.in.), accessed on June. 2021
11. Urja, A.: A bi-monthly newsletter of the MNRE, Government of India **5**(8) (2021)
12. Annual report 2011–12, Ministry of Power of India
13. Annual report 2020–21, MNRE, also available on: http://www.mnre.gov.in/annualreport/2020_21_English/Chapter%203/chapter%203.htm. Accessed on Dec. 2021.
14. http://rggvy.gov.in/rggvy/rggvyportal/plgsheet3.jsp, Rajeev Gandhi Vidyut Garamin Yojana, Official website, Ministry of Power, Govt. of India, 2021, accessed in June. 2021
15. http://www.powermin.nic.in. Accessed on June 2021
16. http://ahec.org.in 2011, Accessed on June 2021
17. MNRE's, Ministry's Annexure-I of circular No.15/6/2006-07-RVE 31-07-2020
18. Uttarakhand at glance 2020–21, Report prepared by Directorate of Economics and Statistics, Uttarakhand, also available on http://www.uk.gov.in/pages/display/116-at-a-glance
19. Kanase Patil AK, Saini RP, Sharma MP (2020) Integrated renewable energy systems for off grid rural electrification of remote area. Int. J. Renew. Energy **35**(6), 1342 1349
20. Raman, P., Murali, J., Sakthivadivel, D., Vigneswaran, V.S.: Opportunities and challenges in setting up solar photo voltaic based micro grids for electrification in rural areas of India. Renew. Sustain. Energy Rev. **16**(5), 3320–3325 (2012)

21. Annexure of G.O. No.263/I (2)/2008-04 (8)-96/2001, dated 29-1-2008, Policy for harnessing renewable energy sources in Uttarakhand with Private Sector/Community Participation. Government of Uttarakhand
22. Sharma Ramesh, C., Bisht, Y., Sharma, R., Singh, D.: Gharats (watermills): indigenous device for sustainable development of renewable hydro-energy in Uttrakhand Himalayas. Renew Energy **33**(10), 2199–2206 (2008)

Detection of Abnormalities in Mammograms Using Deep Convolutional Neural Networks

K. Suganya Devi, K. Sekar, Nivedita Singh, Subhra Jyoti Baroi, Dhiraj Kr. Sah, and Kaushik Borahi

Abstract Deep learning is a category of machine learning algorithms and has sparked a great deal of interest in its applicability to radiography challenges due to its tremendous improvement. It is using an end-to-end learning methodology, which effectively utilizes training datasets with comprehensive clinical annotations. In this paper, we propose modified CNN methods that can effectively identify breast cancer in the investigation of mammography. In this work, tumor descriptions are required through the early training phase, and later phases require only image-level descriptors, removing the need for hardly accessible tumor descriptions. In comparison with earlier approaches, all of our modified CNNs for categorizing the mammograms with promising result in accuracies analyzed through various color schemes.

Keywords Breast cancer · Preprocessing · Digitized mammograms · ResNet · MobileNet · EfficientNet

1 Introduction

Breast cancer is the most common risk factor for mortality among women globally, and it is the only kind of cancer that affects them. Several diagnostic strategies have been implemented to assist in the rapid diagnosis and therapeutic procedures and minimize the number of fatalities from the disease. Many assisted breast cancer diagnostic approaches were employed to improve prediction accuracy [1]. According to International Agency for Research on Cancer (IARC) press release GLOBOCAN-2020 [2], with 23 lakh cases detected in 2020, women breast cancer became the most frequently diagnosed cancer type in the globe. The rate of survival of breast cancer patients of the world varies significantly [3]. The low survival rates in developing countries are primarily leading to a shortage of early diagnosis techniques

K. Suganya Devi (✉) · K. Sekar · N. Singh · S. J. Baroi · D. Kr. Sah · K. Borahi
National Institute of Technology Silchar, Silchar, Assam, India
e-mail: suganya@cse.nits.ac.in

K. N. Das et al. (eds.), *Proceedings of the International Conference on Computational Intelligence and Sustainable Technologies*, Algorithms for Intelligent Systems,
https://doi.org/10.1007/978-981-16-6893-7_37

After developing computer-aided detection (CAD) systems, matters have changed, enabling better diagnostic results and thus overcoming human errors.

The early CAD systems are not achieving significant improvement in the detection of breast cancer[4]. The fast growth of artificial intelligence, particularly deep learning, stimulates radiologists' attention to use these approaches to increase cancer detection efficiency [5]. This work proposed modified CNN architectures of ResNet50 V2, ResNet101 V2, EfficientNet, and MobileNet V2 to detect and classify breast cancer from mammograms under three different color schemes, such as Black and White, Inferno, and Virdis.

The paper organized as follows: Sect. 2 provides the details of related works on breast cancer detection. Section 3 elaborates the proposed methodology: results and discussions and conclusions provided by Sects. 4 and 5.

2 Related Works

This section reviewed the development of CNNs for the classification and detection of breast cancer and its performances. Ozekes et al. [6], proposed a method for automated mass detection using templates, in which the pixels in the mammogram scanned in 8 directions, and region of interest(ROI) identified using the minimum and maximum intensity thresholds. Jen et al. [7] focused on Abnormal mammogram detection using abnormality detection classifiers (ADC), which provides the sensitivity of 86% on DDSM. Araujo et al. [8], investigated breast cancer through hematoxylin and eosin stains of images and performs image-based classifications.

In [9], the authors' proposed methods to remove multiplicative noise to enhance non-uniform illumination simultaneously normalize the brightness across an image and increase contrast using discrete Fourier transform (DFT). The paper [10] discussed a CAD system based on faster region Based CNN(R-CNN) on a mammogram; the device identifies and classifies malignant or benign lesions giving a sensitivity of 90%. Li et al. [11] proposed an improvised CNN model, known as the DenseNet-II, for classifying the benign and malignant mammography images and achieved 95.6%. Agarwal et al. [12] proposed a patch-based CNN method for automated mass detection in full-field digital mammograms (FFDM) also explored a transfer learning approach.

Li et al. [13] classifying the tumors based on H&E (Hematoxylin and Eosin) stained microscopic breast histology images also shed some light on patch extraction. In [14], the authors developed a mammography-based CNN breast cancer detection model that provides better results than the other benchmark models. Jeffers et al.[15] used clinical breast imaging-reporting and data system (BI-RADS) to compare three breast density metrics on full-field digital mammographic (FFDM) images.

3 Proposed Methodology

This section discussed the dataset used, followed by the preprocessing of mammography images and segmentation process under different color schemes, followed by a discussion on well-known CNN architectures and their respective modified architecture and the strategy for training the images using the architectures.

3.1 Dataset

We used the dataset of MIAS MiniMammographic database (which is smaller than the usual MIAS database of mammograms), and this has developed by the Mammographic Image Analysis Society (MIAS). The mammogram films are digitized to a 50-micron pixel edge using a Joyce-Loebl examining microdensitometer. There is an aggregate of 322 digitized pictures. It likewise incorporates the radiologist's "truth"—markings of the nearness of any anomalies. The images in the dataset are minimized pixel rates up to 200-micron and are padded to get 1024×1024 image.

3.2 Preprocessing

Digitized mammogram images are usually affected by a lot of noise. The task at hand is to eliminate noise without losing out on any crucial information by applying suitable noise removal techniques. Median Blur Filter is a nonlinear statistical filter is one such noise removal technique, which substitutes the gray values of every pixel in an image with the neighborhood of the pixels, often called the window, with the gray level median. The kernel size's median level is determined by the range of pixels covered by it. Preprocessing has been done with a 3×3 median filter kernel widely used to remove impulse, salt-and-pepper noises, and remove scratches present on the given images.

The Multi Otsu is used to set a thresholding intensity, helping differentiate the breast region from the background and pectoral muscles. It is by using this threshold intensity which the image will be binarized. The pixel intensity is more significant than the threshold intensity given by the value 1 and 0 otherwise. Post this the single seeded region growing algorithm. The algorithm begins with the first seed and grows with the neighboring homogeneous elements, and the seed element may be a single pixel or a region. We consider every image's top-left corner pixel as the seed pixel and set the right orientation for all the images. From this, the pectoral muscle region has been identified, so with simple bitwise-AND masking, the specified area is removed and generating the final preprocessed image. Figure 1 shows a mammogram input image and its preprocessed output image.

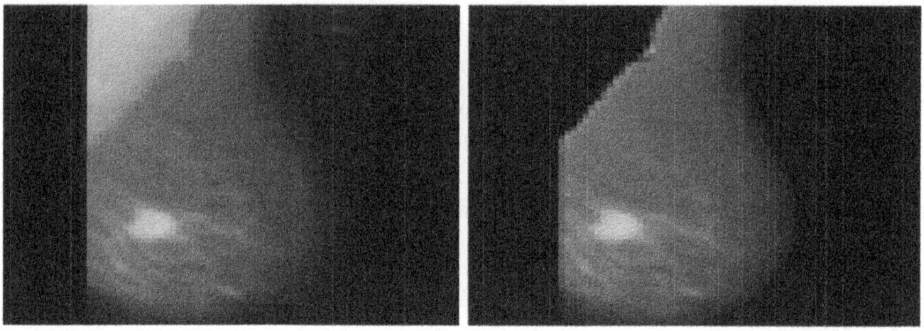

(a) (b)

Fig. 1 **a** Input image, **b** preprocessed output

3.3 Segmentation

The primary purpose of segmentation is to be able to identify regions of interest. Image segmentation is defined as the method of dividing the given image into non-intersecting homogeneous sub-images. The primary purpose of segmentation is to be able to identify regions of interest. In the segmentation stage, the preprocessed image passed through three phases: contrast limited adaptive histogram equalization (CLAHE), histogram equalization, and morphological enhancement. CLAHE enhances the contrast of the overall image; hence, the regions containing abnormality are roughly highlighted.

A graphical representation of the intensity distribution of an image is called a histogram. Histogram equalization is a technique to improve image contrast. In CLAHE, the contrast amplification is limited; therefore, it only focuses on the regions with very high density; it ignores the regions with density more than the average tissue density, which may or may not contain abnormality. Therefore, this work proposes using a custom-made histogram equalizer to improve the contrast and not miss out on abnormalities. The approach is helpful in background and foreground images which are both bright or dark.

Morphology enhancement is a technique of image processing depending on shapes. Top-hat filtering measures the morphological opening of the image. It then subtracts the output from the original image, while bottom-hat morphological operator subtracts the input image from the output of morphological closure on the input image.

We consider top-hat and bottom-hat filters used on dark backgrounds for light objects and bright backgrounds for dark objects. After the threshold value selection, the histogram is converted into regions. The histogram equalized images are binarized. After binarization, we could convolve the histogram equalized images then obtain the suspected regions of interest of an image. With an ROI mask and preprocessed image masking, the overlapped areas have increased intensity by 8%. In non-overlapping areas, the intensity is reduced by 4%, and the segmented image is created. Figure 2 shows the output images after every stage of the segmentation phase.

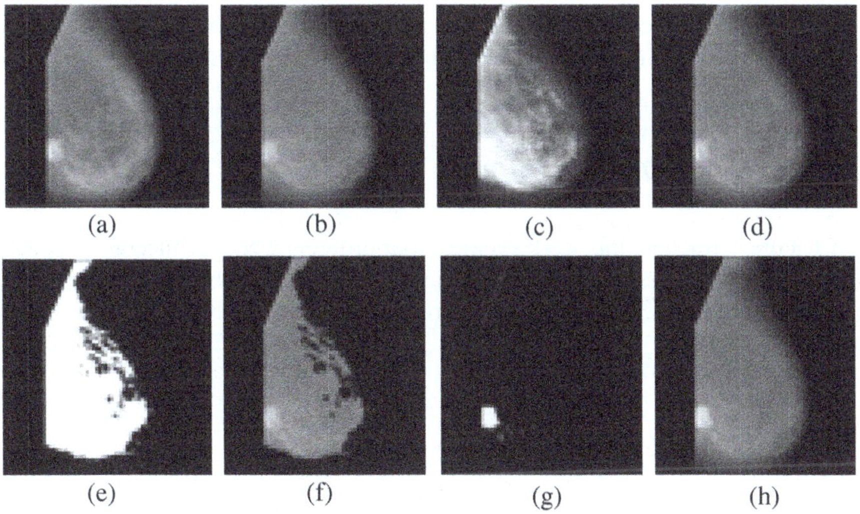

Fig. 2 **a** Input image, **b** CLAHE output, **c** histogram equalized output, **d** morphologically enhanced output, **e** binarized output, **f** convolved output, **g** ROI masked, **h** segmented output

3.4 Color Scheme

The Viridis package combines R color scales for the Python matplotlib library, which contains four different color scales: Viridis: the primary choice, and three alternatives, namely magma, plasma, and inferno. This work chooses Viridis and inferno color schemes. The simple reason is that these color schemes are chosen because they are perceptually uniform, meaning that values close to each other or colors look similar Then values far from each other have more different-appearing colors, consistently across the range of values, making it easy to differentiate between them. So the segmented black and white(B&W) images are converted into two different color

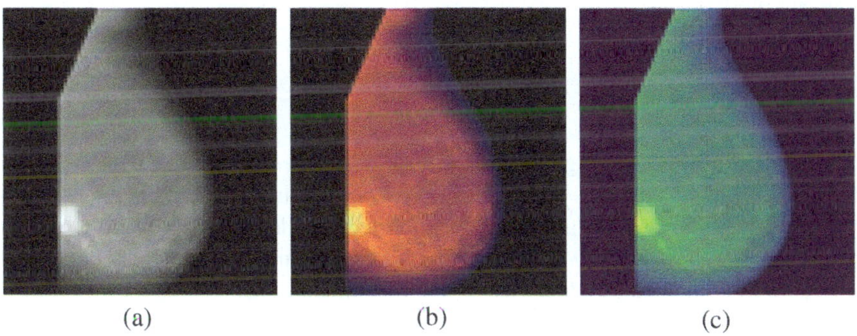

Fig. 3 Different color schemes: **a** B&W, **b** inferno, **c** viridis

schemes, namely Inferno and Viridis, using simple functions, namely cv2.cvtColor(). Figure. 3 shows the images under different color schemes.

3.5 CNN Architectures

For feature extraction, this work explores four different CNN architectures, namely ResNet50 V2 [16], ResNet101 V2, EfficientNet [17], and MobileNet V2 [18]. Both ResNet50 and ResNet101 are well-explored, but thats not the case with EfficientNet and MobileNet V2.

3.5.1 ResNet

Residual Network (ResNet) is a classic neural network. ResNet is a ConvNet that has residual blocks added with skip connections. This skip connection helps deal with the vanishing gradient problem; this is done between two layers where the activation layer output of a layer L is fed directly to the activation function(ReLU) of layer L+2. The ResNet50 consists of five stages; every stage has a set of convolution and identity blocks. Both convolution block and identity blocks have three convolution layers each. ResNet101 V2 is a convolutional neural network that is 101 layers.

3.5.2 EfficientNet

The EfficientNet model consists a set of eight different models as $B_0, B_1, \ldots, B_7$. The subsequent models vary with more parameters and higher accuracy. The CNN models employ a scaling method that uses a straightforward but extremely effective compound coefficient to equally scale all depth/width/resolution dimensions. EfficientNet group developed a new baseline network using a neural architecture approach and scaled up to create a set lower in size but quicker than many popular convNet models.

3.5.3 MobileNet V2

A highly recommended image classification architecture for mobile devices, embedded systems, and computers without GPU or low computational efficiency provides better results. It is based on a streamlined architecture that uses depthwise separable convolutions to form a lightweight deep neural network at its core layer is a depthwise separable filter, named as depthwise separable convolutions.

3.5.4 Modified CNN Architecture

In this work, we introduced modified versions of the four discussed CNN architectures and are named m_ResNet50V2, m_ResNet101 V2, m_EfficinetNet, and m_MobileNetV2. The architectures such that m_ResNet50V2, m_ResNet101 V2, m_EfficinetNet, and m_MobileNetV2 use the feature extractor of ResNet50V2, ResNet101 V2, EfficinetNet, and MobileNetV2, respectively. A flattening layer is introduced in the modified architecture after the feature extraction layer to convert the pooled feature map to a single column. Three dense layers are present in the architecture, and a dropout layer follows each dense layer. A dense layer is the most basic layer in a neural network. This feeds all its neurons from the previous layer, and each neuron supplies one output to the next layer. The units were chosen in this model, i.e., 512, 264, 2 are experimentally determined. The role of the dropout layer is to prevent the model from overfitting. In this architecture, we consider ReLU function for activation of convolutional layers. Figure 4 is the schematic representation of the modified architecture.

Both CNN and modified CNN's portrayed above are prepared on the Mini-MIAS dataset with input measurements 224 × 224. The dataset is split into training and testing sets with the end goal of an aggregate of 216 and 106 pictures in the training set and testing set individually. The modified CNNs are trained for 500 epochs using two different learning rates 10^{-2} and 10^{-6}. Higher learning rates are utilized because training a CNN without preparation requires more time to gain proficiency than the analyzed pictures. Computations were performed on Google Collab using Keras library with Tensorflow2 as the backend. Testing accuracy and time are taken for the model to converge taken as evaluation metrics for the proposed system.

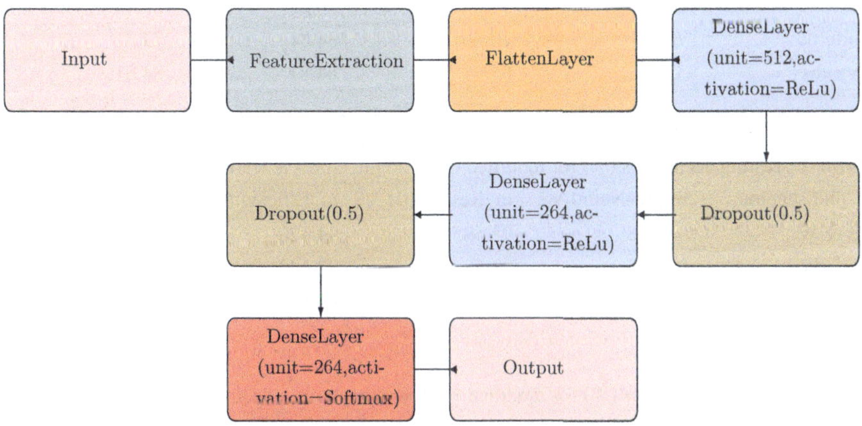

Fig. 4 Procedure for modified CNN architecture

4 Experimental Results and Discussions

In this paper, an end-to-end system for detecting abnormalities in mammograms is designed. With this motive, widely used CNN models were analyzed and proposed modified versions of each of them and compared their respective accuracies using the MIAS MiniMammographic database. The proposed methodology trains the mammogram images after preprocessing and segmentation techniques. We train 75% images and test 25% images to classify the bread cancer.

In trial #1, the EfficientNet overcomes other CNN architectures with a testing accuracy of 86.2% at the learning rate of 10^{-6} with B&W color scheme. The accuracy details of various CNN methods are provided by Table 1. Table 2 provides the trail #2 performances Viridis color scheme with m_ResNet101 V2 provided the better accuracy of 97.358% with the learning rate of 10^{-6}. At the learning rate of 10^{-2}, both B&W and Viridis color schemes perform well with m_EfficientNet and m_MovileNet V2, respectively, and provides the accuracy of 97.358% given by Table 3. It can be inferred that the modified CNNs are substantially enhanced performance compared with the prominent CNN architectures.

Table 4 gives the time convergence of different modified CNN methods. The method, m_EfficientNet with B&W color scheme, is the best since it takes the least time to converge, i.e., 52s. But based on the fact that models trained on lower learning rates are preferred, it can be affirmatively said based on the experimental values that the m_ResNet101 V2 in Viridis color scheme is the best-performing model.

Table 1 Testing accuracies (%) for CNN architectures *t* learning rate 10^{-6}

Color scheme	ResNet50 V2	ResNet101 V2	EfficientNet	MobileNetV2
B & W	84.35	35.65	86.2	58.43
Inferno	70.00	70.46	84.81	72.31
Viridis	55.65	55.65	84.81	84.81

Table 2 Testing accuracies (%) for modified CNN architectures at learning rate 10^{-6}

Color scheme	m_ResNet50 V2	m_ResNet101 V2	m_EfficientNet	m_MobileNetV2
B & W	94.528	93.585	94.528	94.528
Inferno	93.585	93.585	92.642	93.585
Viridis	94.528	97.358	94.528	93.585

Table 3 Testing accuracies (%) for modified CNN architectures at learning rate 10^{-2}

Color scheme	m_ResNet50 V2	m_ResNet101 V2	m_EfficientNet	m_MobileNetV2
B & W	93.585	94.528	97.358	93.588
Inferno	95.472	94.528	96.415	93.585
Viridis	94.528	93.585	96.415	97.358

Table 4 Comparison based on time convergence (s) of modified CNNs

Color scheme	CNN	Learning rate	Time (s)
B & W	m_EfficientNet	10^{-2}	52
Viridis	m_ResNet101 V2	10^{-6}	368
Viridis	m_MobileNet V2	10^{-2}	286

5 Conclusions

An essential aspect of the methodology proposed for recognizing abnormalities in mammograms is vividly discussed in this work. The proposed strategy utilizes definite preprocessing and segmentation procedures for image enhancement, before which were used to train the CNN. It has been indicated that the modified frameworks significantly upgraded the performance of the major CNNs architectures. The automated CAD framework developed with m_ResNet101 V2 provides the best outcomes on both the grounds of testing accuracy and a lower learning rate under the Viridis color scheme. In future work, the investigation will focus on how well the modified CNN architectures would perform on a larger dataset with transfer learning methods.

References

1. Global Health Estimates Technical Paper WHO/HIS/HSI/GHE/2013.4 (2013)
2. IARC Press Release No. 201 GLOBOCAN-2020 (2020)
3. Coleman, M.P., Quaresma, M., Berrino, F., Lutz. J.M., De Angelis, R., Capocaccia, R., Baili, P., Rachet, B., Gatta, G., Hakulinen, T., Micheli, A., Sant, M., Weir, H.K., Elwood, J.M., Tsukuma, H., Koifman, S., Silva, E.G.A., Francisci, S., Santaquilani, M., Verdecchia, A., Storm, H.H., Young, J.L.; CONCORD Working Group. Cancer survival in five continents: a worldwide population-based study (CONCORD). Lancet Oncol. **9**(8), 730–56.(2008) https://doi.org/10.1016/S1470-2045(08)70179-7
4. Lehman, C.D., et al.: Diagnostic accuracy of digital screening mammography with and without computer-aided detection. JAMA Intern. Med. **175**, 1828–1837 (2015)
5. Sekar, K., Suganya Devi, K., Srinivasan, P., SenthilKumar, V.M., Deep wavelet architecture for compressive sensing recovery. In: 2020 Seventh International Conference on Information Technology Trends (ITT), pp. 185–189 (2020)
6. Ozekes, S., Osman, O., Camurcu, A.Y.: Mammographic mass detection using a mass template. Korean J Radiol. **6**(4), 221–228 (2005). https://doi.org/10.3348/kjr.2005.6.4.221
7. Jen, C. C., Yu, S.-S. et al.: Automatic detection of abnormal mammograms in mammographic images. Exp. Syst. Appl. **42**(6), 3048–3055 (2015)
8. Araújo, T., Aresta, G., Castro, E., Rouco, J., Aguiar, P., Eloy, C., Polónia, A., Campilho, A.: Classification of breast cancer histology images using convolutional neural networks. PloS one **12**(6) (2017). https://doi.org/10.1371/journal.pone.0177544
9. Amel, H.A., Jamila, H.: Image enhancement by using homomorphic filtering model. Qalaai Zanist Sci. J. **2** (2017). https://doi.org/10.25212/lfu.qzj.2.2.32
10. Ribli, D., Horváth, A., Unger, Z. et al.: Detecting and classifying lesions in mammograms with deep learning. Sci. Rep. **8**, 4165 (2018). https://doi.org/10.1038/s41598-018-22437-z

11. Li, H., Zhuang, S., Li, D.-A., Zhao, J., Ma, Y.: Benign and malignant classification of mammogram images based on deep learning. Biomed. Sign. Process. Control **51**, 347–354 (2019)
12. Agarwal, R., Diaz, O., Lladó, X., Yap, M.H., Martí, R.: Automatic mass detection in mammograms using deep convolutional neural networks. J. Med. Imaging (Bellingham) **6**(3), 031409 (2019). https://doi.org/10.1117/1.JMI.6.3.031409
13. Li, Y., Wu, J., Wu, Q.: Classification of breast cancer histology images using multi-size and discriminative patches based on deep learning. IEEE Access pp. 1–1 (2019). https://doi.org/10.1109/ACCESS.2019.2898044
14. Yala, A., Lehman, C., Schuster, T., Portnoi, T., Barzilay, R.: A deep learning mammography-based model for improved breast cancer risk prediction. Radiology **292**(1), 60–66 (2019)
15. Jeffers, A.M., Sieh, W., Lipson, J.A., Rothstein, J.H., McGuire, V., Whittemore, A.S., Rubin, D.L.: Breast cancer risk and mammographic density assessed with semiautomated and fully automated methods and BI-RADS. Radiology **282**(2), 348–355 (2017)
16. Vesal S., Ravikumar N., Davari A., Ellmann S., Maier A., Classification of breast cancer histology images using transfer learning. In: Campilho A., Karray F., ter Haar Romeny B. (eds) Image Analysis and Recognition. ICIAR 2018. Lecture Notes in Computer Science, vol. 10882. Springer, Cham (2018)
17. Tan, M., Le, Q.V.: EfficientNet: rethinking model scaling for convolutional neural networks. CoRR abs/1905.11946 (2019)
18. Sandler, M., Howard, A., Zhu, M., Zhmoginov, A., Chen, L.C.: MobileNetV2: inverted residuals and linear bottlenecks. In: The IEEE Conference on Computer Vision and Pattern Recognition (CVPR), pp. 4510–4520 (2018)

Orange Fruit Recognition Using Neural Networks

Surya Pratap Singh, Shivam, and Indrajeet Kumar

Abstract For the completion of this work, recognition of orange fruit is achieved using multiple methods of convolutional neural network (CNN). For the achieving of image accession, the dataset upholds 820 sample images in total. The selected dataset has been divided in separate two classes, where the first class contains images of orange fruit, whereas the second class contains images of other fruits (apple, banana, mellon, etc.). The identification is done using self-created CNN, LeNet-5, VGG-19, and MobileNetV2, these are predefined CNN models that have a fixed layers. Architecture is specifically designed for LeNet-5 model, while the other used models are VGG-19 and MobileNetV2 been imported by taking the help of predefined libraries and furthermost have been implemented as per requirement of the authors. Once the execution of this experiment was achieved successfully, the observation came into highlight hat validation accuracy that was achieved for this work by using self-created CNN, LeNet-5, VGG-19, and MobileNetV2 was 96.80%, 89.60%, 87.50%, and 95.90%, respectively.

Keywords Orange fruit recognition · Deep learning · LeNet-5 · VGG-19 · MobileNetV2

1 Introduction

In present scenario and during COVID pandemic time, it was seen that vitamin C was an important source that acted as an immunity booster for the COVID patients as well as for every individual [1]. Because vitamin C acted as an immunity booster for the COVID patients, it was thus given approval by the medical administrator representatives for allowing vitamin C to be used as an antidrug immunity booster for the patients [2]. Vitamin C was either consumed in the form of tablets, powders, or by taking fruits that had citric acid in them [3]. Some of the vitamin C and citric acid sources of fruits are lemon, lime, orange, etc. [4]. These vitamin C sources

S. P. Singh · Shivam (✉) · I. Kumar
Graphic Era Hill University, Dehradun, Uttarakhand, India

K. N. Das et al. (eds.), *Proceedings of the International Conference on Computational Intelligence and Sustainable Technologies*, Algorithms for Intelligent Systems,
https://doi.org/10.1007/978-981-16-6893-7_38

are very effective and thus helps to intake natural vitamin. Today, orange fruit has become one of the widely preferred fruit for vitamin C intake and thus the production of orange fruit has taken a rise rapidly over the time [5]. The top producers of orange fruit worldwide during the period of 2010–2018 are Brazil, China, Eastern European, United States, Egypt, Mexico, Turkey, and South Africa [6]. The total amount of oranges produced by these countries each year in million metric tons is 15.6, 7.3, 6.2, 4.7, 3.0, 2.5, 1.8, and 1.6, respectively. From a period of ten years in last interval, Brazil has been the top producer of orange fruit worldwide by leading the table from last ten years in a row [7]. It is observed that whenever there is a rise in of production in any of the aspects, it not only becomes a topic of production but also becomes an important topic for the researchers to make any of the research in any field [8]. Today, majority of the population are dependent on the agricultural farming in different categories. Some people make their living and full day meals by cultivating apples and then selling them door to door, similarly some to the cultivation of orange fruits and then sell them door to door. What makes orange fruit recognition and important and impactful topic for the researchers is that there are many fruits that looks like an orange fruit, but actually is not an orange fruit. There are so many fruit like sweet lemon, lemon, etc., are look like the orange but they are not. Thus, the correct prediction of the orange fruit is a key point to watch out for. Thus, orange fruit recognition has also become an important topic today for the researchers [9]. On a regular basis, the research work on orange fruit recognition by applying different-different method such as by using neural networks, image segmentation, etc., is coming forward. Manual detection of orange fruits is easy but sometimes it becomes difficult to identify them individually when to be detected for multiple oranges at a single time as the time consumed is more [10]. But by taking the help of image processing and other models, this manual workforce can be weakened.

Thus, it has been observed and noticed that orange fruit recognition has become an important research that a researcher looks forward to, based on its shape, size, and different factors. Few of the similar works on orange fruit recognition have been discussed in studies [11–16]. In this study [11], a new and different method for bringing out recognition of orange fruit has been discussed. Under this work, different segmentations methods edge based and color based are applied on the orange images and further the observation and recognition of oranges present either on the tree or in bucket of fruits is done. Another co-related type of problem is carried out in study [12] using laser range finder. Under this study, for bringing out the recognition of the orange, a sensor is used that gives the range attenuation data of surface and an algorithm is used over that uses the data generated by the sensor to locate fruit.

In this study [13], a completely different concept was used for the recognition of orange fruit. In their experiment, different preprocessing techniques named reign labeling, size filtering, perimeter-based-detection, etc., are used to identify fruit and an algo is needed that can detect fruit present in cluster by using shape analysis approach. Thus, it was observed that although their approach was completely different, but the results achieved were impeccable.

Another method for the orange fruit recognition is performed in this study [14]. In completion for this experiment, they achieved the recognition of orange fruit by using different image processing methods. Their previously used method that was Naïve Bayes method was this replaced with the SVM classification to increase and achieve the better quality. The features were extracted using the GLCM algorithm and further compared their model with existing model by primarily focusing on determining the performance in view to sensitivity, model accuracy, total time taken for the execution, specificity, etc. New and different approach has been preferred for the orange fruit recognition in this study [15]. Using different strategies of computer vision, the four main features on the basis of intensity, color of the fruit, its shape, and texture were extracted, and further fusion of texture and color features were proposed that was considered as the backbone of the experiment. By taking the help of classifier of minimum distant, the recognition was performed and from the altered sub bands of wavelet, the concurrent features were derived. Similar type of experiment has been performed in this study [16]. With the help of CNN, recognition of fruits was performed. Initially, the main regions or the features of the image were extracted by taking the help of selective searching algorithms. Next by the means of entropy of the fruits, the regions were selected and at last, these selected regions were taken as an input for the CNN and the recognition was achieved.

2 Material and Methodology

2.1 Dataset Preparation

For completion of preferred work, dataset is being downloaded from Kaggle which contain total of 760 images. 420 images of orange and 340 images of fruits other than oranges. Moreover, for this research work, the number of samples available were less, thus images of orange and non-orange fruits were collected from another dataset and therefore, total number of images generated after combining all were 820 out of which 460 are orange fruit images and 360 are images of fruits other than orange, further these images have been divided into two different classes namely training class and the testing class respectively where total of 500 images being considered for training purpose and 320 for testing. The detailed clarification of selected dataset preparation and their bifurcation is precisely displayed in Table 1.

Table 1 Precise details of how dataset is prepared and its bifurcation

Preferred class	Samples	Data bifurcation		Total
		Train set	**Test set**	
Orange class	460	300	160	460
Other than orange class	360	200	160	360
Total	820	500	320	820

2.2 Proposed Work

The workflow chart of experiment made out for "orange fruit prediction using CNN methodology" is shown in reference to Fig. 1.

The entire workflow chart involves the segments of image preprocessing, dataset's detailed bifurcation, and also of deep learning model where different models are applied over the dataset, and the result section where prediction results are displayed weather the fruit is orange or not. The brief detail of each segment is described in upcoming phase of this manuscript.

2.2.1 Preprocessing Section

Considering image preprocessing as a most important step for this experiment, the results were successfully achieved. It ensures pixel values of image surely lies in a particular range, and thus ensure systems comprehensive processing is produced effectively. Resultant pixel value for preferred key pixel is defined according to the formula as per the Eq. (1).

$$Y_i = \frac{X_i - \text{least}(X)}{\text{maximal}(X) - \text{least}(X)} \tag{1}$$

where the intensity scale may differ from 0 to 255.

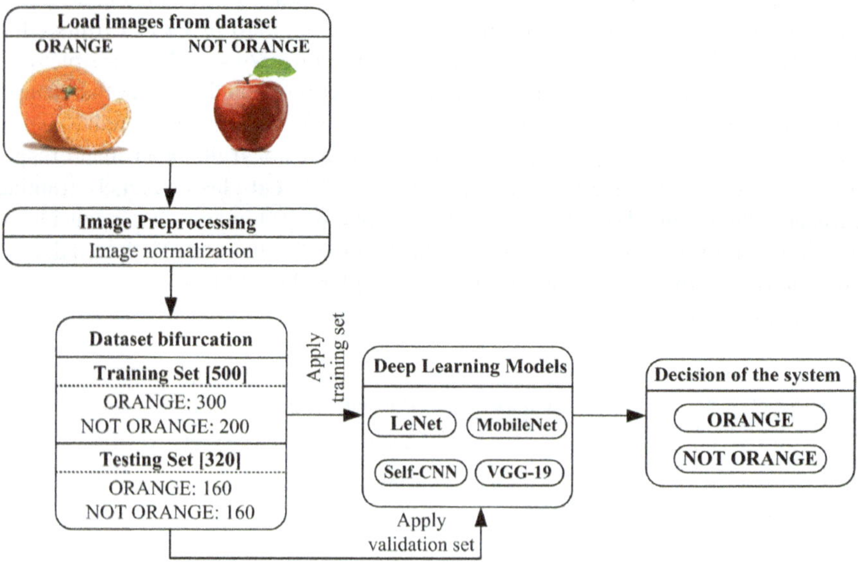

Fig. 1 Workflow chart for extensive experiment

2.2.2 Deep Learning Models

For this experiment, various types of deep neural (predefined) network models are used. Self-created CNN, MobileNetV2, LeNet-5, and VGG-19 are the model that have been taken into consideration for predicting whether the fruit is orange or not.

Here, authors have compared self-created CNN, LeNet-5, VGG-19, and MobileNetV2. Where LeNet-5 is an existing model been compared with the new proposed models mainly VGG-19 and MobileNetV2. Furthermost, on bringing out the comparison between above two mentioned models, it was specifically observed that the MobileNet model is opted as a network having lightweight, uses depth-case separable CNN layer, and thus the running complexity and parameters were able to be reduced to a great extent. The aim of using the mentioned different models together on similar dataset is to find difference between these deep learning models.

In this experiment, size of the image passes to input layer of LeNet-5 is considered as 64×64. The filters used to intensify the overall performance (models) of convolutional functioning on the specific used dataset containing images are 32, 64, and kernel size is considered as 3×3. Padding used is valid for all convolutional layers. The image size for VGG and MobileNet is 3×3 and it is all hidden layers are predefined hence they are called, and dataset is passed to them for training purpose. The formula due to which production of the wished and preferred dimension of the turnout assigned map or the turnout attribute been generated has been discussed below in form of Eq. (2).

$$w_2 = \left(\frac{w_1 - f + 2_p}{S} + 1\right),$$
$$h_2 = \left(\frac{h_1 - f + 2_p}{S} + 1\right) \tag{2}$$

Here, $w_1 \times h_1 \times d_1$ is preferred size of our input image whereas w_1 and h_1 represent width and height, where d_1 is exact quantity of band of color in our image. Next, all remaining entities that include the actual measurements of used filter are represented by f and similarly p talks about highest number of padding and oppositely s determines stride value. All precise details of each used model have been discussed next.

LeNet-5: This model contains many layers which primarily focus on the classification and recognition of numeric digits that are hand-written or are character numbers [17]. The architecture came first into sights in 2015 been proposed by LeCun. LeNet-5 includes a layer that is takes the input, and also layer that works for generating the output. Apart from input and output layers, it has two separate convolutional layers, two separate pooling layers, and mainly two entire connection layers. Over the period of time, this model has been widely preferred and suggested when it comes for character digits recognition.

VGG-19: Visual geometry group short termed as VGG has two variant layer versions—one has 16 layers in total referred as VGG-16 and the other one has 19 layers in total that is referred as VGG-19. The architecture first came into spotlight in 2014 been proposed by minds of Simonyan and Zisserman [18]. Contest which was held in area of localizing and classifying tracks, the VGG had grabbed both first and also second rankings, respectively. It contains five segments of the convolutional (CNN) layers at initial phase, further headed by three fully and exactly connected layers. On brief, comparing of both variants of VGG model that is of 19 layers and 16 layers, respectively, it was found that VGG-19 has more expensive and deeper structure as compared to VGG-16 [19].

MobileNetV2: The name of the model itself contains the word "mobile" that simply justifies that MobileNet model is a well-known CNN architecture primarily focusing to achieve the mobile versions classification and image. The model was proposed in 2017 by Google and involves the in-depth (deep) convolutions under the helps to filter the inputs without assembling new filters and also includes the point-to-point (exact) convolutions, due to which intensive generating of the newly features according to two different core and important cables which are present inside this model architecture [20]. In reference to activation functions, ReLU and SoftMax are the two different and separate functions used for increasing the non-linearity for completion of this work.

Sigmoid: An activation function, named sigmoid function, is also known or taken as the logistic function. Sigmoid function is a function that mathematically expresses sigmoid curve also known as a curve that is S-shaped and works for all the input values that are real and does involves non-negative derivate exactly once and after every point. The practical representation of sigmoid function is given in Eq. (3).

Sigmoid function (g) is depicted by taking help of an exponential formula that is demonstrated in Eq. (3).

$$g = 1./(1.0 + \exp(-z)) \tag{3}$$

where g denotes the sigmoid function, exp refers to the exponential, and z is the range ideally 0 and 1.

ReLU: It is a widely accepted activation function, where ReLU is a short term used for rectified linear unit. Over the period, it has come into observation that ReLU has turned out to be an emerging activation function and has achieved a better popularity all across when supposed to state a part in CNN model [21]. Its nature of action is considered as a functional method for the model and thus is preferably used in order to break the linearity of image by increasing the non-linearity as images are distinctly and majorly non-linear. The working formula for ReLU function has been discussed in Eq. (4).

$$f(x) = \text{maximum}(X, 0)$$

$$f(x) = \begin{cases} x & \text{if } x > 0, \\ ax & \text{otherwise.} \end{cases} \tag{4}$$

To be noted that for $a \leq 1$, this equals to

$$f(x) = \max(a, ax)$$

3 Experiment and Results

3.1 Environmental Setup

The entire demonstration of the experiment is performed on system that has specs of a Intel® Core™ i7-6500U Processor, CPU@ 2.60 GHz, total RAM of 12 GB, whereas two Gigabyte Intel HD Graphics 520 and a 512 Gigabyte hard drive. All images used in this experiment have been stored in system that is mentioned above. Spyder (Python platform) is preferred, and the complete experiment is accomplished on same platform.

3.2 Result Discussion

The work has highlighted that when different models (LeNet-5, MobileNetV2, VGG-19, and self-created CNN) are applied over the same dataset was found that CNN created and MobileNet have performed better than the other models, where difference between accuracy of CNN and MobileNet was very close to each other. For accuracy analysis, refer to Table 2. Reason why CNN performed better is because the model was designed as per dataset, whereas MobileNet and VGG are the predefined models, mainly designed for big dataset. CNN is designed of this specific dataset with limited

Table 2 Attained results of considerable experiments achieved for this experiment

Model	Train phase			Test phase		
	Total cases	Accuracy (%)	Overall loss	Total cases	Accuracy (%)	Overall loss
LeNet-5	500	97.9	0.0825	320	89.6	0.8969
MobileNetV2	500	98.6	0.0068	320	95.9	0.1375
VGG-19	500	93.7	0.1551	320	87.5	0.2815
CNN	500	99.3	0.0324	320	96.8	0.1144

number of layers, and filter size which is suitable for dataset is provided and that is the main the reason that CNN performed better than all other predefined models.

Kindly, refer to Fig. 2 for the experiment analysis of all the models. Graphs were generated based on training and validation accuracy and training and validation loss which were observed at the time of training when being trained on 25 epochs and for each epoch, all the values were recorded and represented in graph in order to compare values for all the models at each step. Figure 3 shows the prediction.

3.3 Comparative Analysis

This current performed experiment has been compared with earlier published experiment [13]. In this work [13], authors are interested in automation harvesting and classification is done using various preprocessing techniques named as region labeling, size filtering, and perimeter-base-detection whereas some have determined quality of fruits using Naïve Bayes classifier and SVM and have found that SVM performs better than Naïve Bayes. In the present study, CNN is used for classification purpose and limited images are used and accuracy is impressive. It is also shadowed how proposed system is comparatively easier for implementation and generally fast in reference to study [14]. Therefore, proposed work can also be preferred when it comes for the problem like recognizing orange and different fruits.

4 Conclusion

For the completion of this work, advanced and well-known CNN models mainly self-created CNN, MobileNetV2, LeNet-5, and VGG-19 were used, all producing different training and testing result on single dataset. During the completion of this experiment, authors made a conclusion that self-created CNN and MobileNet performed almost equally and same on single dataset and the accuracy of training and testing of both the models are typically of same range. Oppositely, VGG performs poor as compared to other models and the main reason for this is out dataset consists of very a smaller number of images and VGG is designed to work more efficiently on big datasets; on the other hand, MobileNet is designed to work on mobile phones and can perform better on small datasets whereas self-created CNN is designed as per the dataset, so it has performed typically better than other models.

This experiment can likely to be extended further in order to achieve comparatively better results and modes by additionally applying better image preprocessing methods or approaches. Point to note is that these methods should be applied pre-training of our model.

Another important aspect to note can be the increase in value of the RGB images. Apart from this, another deep learning model named GoogleNet can also be used and brought to an extent to achieve better accuracy and also bring out the comparison of

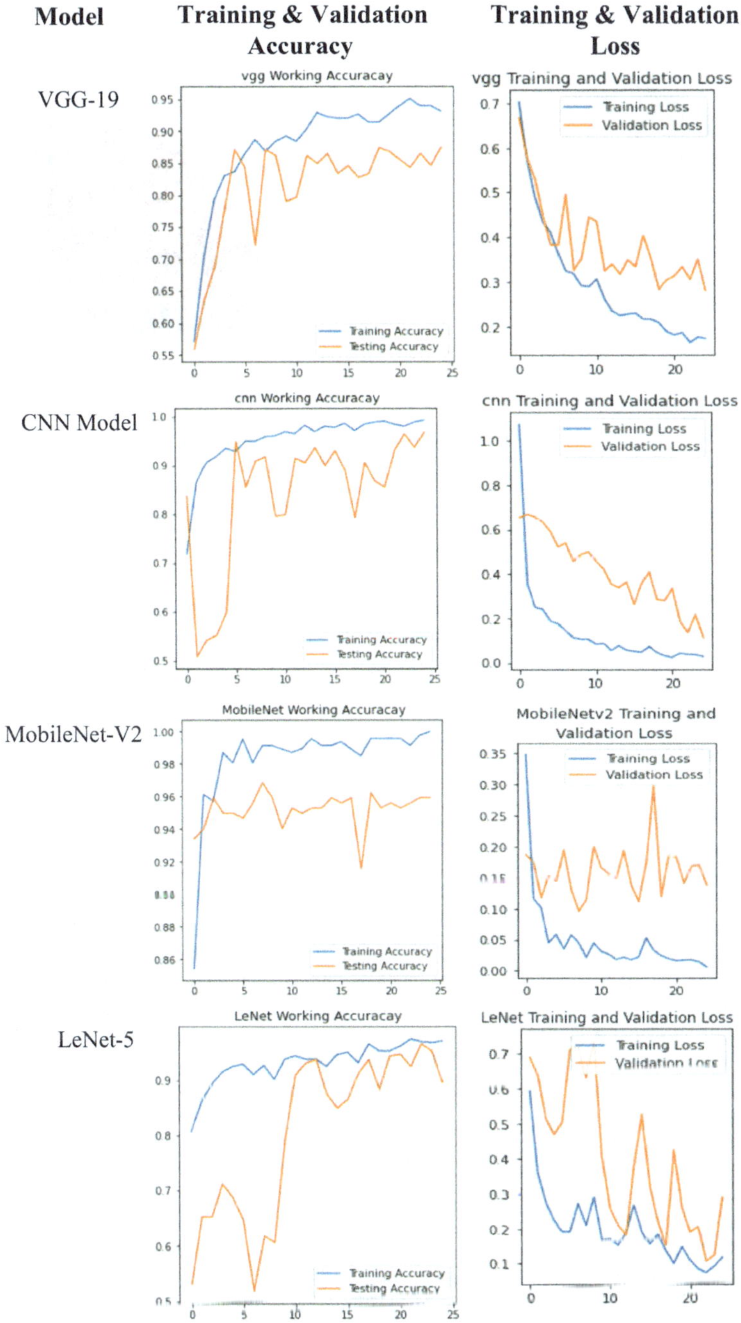

Fig. 2 Learning curve representation—VGG-19, CNN, MobileNetV2, LeNet-5

Orange Fruit

Fruit Other than orange

Fig. 3 Prediction by models: orange versus not orange

all the models. GoogleNet is generally considered as one of the advanced versions of all the models, thus expected to achieve good results of the experiment if been performed using the GoogleNet model. Also, data augmentation can also be done and thus is believed to be a better work for this experiment in future.

References

1. Erol, A., High-dose intravenous vitamin C treatment for COVID-19 (2020)
2. Abobaker, A., Alzwi, A., Alraied, A.H.A.: Overview of the possible role of vitamin C in management of COVID-19. Pharmacol. Rep. 1–12 (2020)
3. Al-, Z.M.J., Hussain, Y.A., Ali, A.A., Al-, M.D.: The influence of vitamin-C intake on blood glucose measurements in COVID-19 pandemic. J. Infect. Dev. Count. **15**(02), 209–213 (2021)
4. Hernández, Y., Lobo, M.G., González, M.: Determination of vitamin C in tropical fruits: a comparative evaluation of methods. Food Chem. **96**(4), 654–664. 2015 (2006)
5. Coutinho, T.C., Ferreira, M.C., Rosa, L.H., de Oliveira, A.M., de Oliveira Júnior, E.N. Penicillium citrinum and Penicillium mallochii: new phytopathogens of orange fruit and their control using chitosan. Carbohyd. Polym. **234**, 115918 (2020)
6. Mgonja, C. T., & Utou, F. E. Assessment of Fruits Handling in Tanzania-Case of Orange Fruit. Assessment, (2017) 6(6).
7. Costa, G.V.D., Neves, C.S.V.J., Telles, T.S.: Spatial dynamics of orange production in the state of Paraná, Brazil. Revista Brasileira de Fruticultura **42**(2) (2020)
8. Grenz, J., Thalmann, C., Stämpfli, A., Studer, C., Häni, F.: RISE–a method for assessing the sustainability of agricultural production at farm level. Rural Dev. News **1**(2009), 5–9 (2009)
9. Adisa, V.A.: The influence of molds and some storage factors on the ascorbic acid content of orange and pineapple fruits. Food Chem. **22**(2), 139–146 (1986)
10. PMohanty, P., Hughes, D.P., Salathe, M.: Using deep learning for image-based plant disease detection. Front Plant Sci. **7**(2016), 1–10 (1419)
11. Thendral, R., Suhasini, A., & Senthil, N. A comparative analysis of edge and color-based segmentation for orange fruit recognition. In 2014 International Conference on Communication and Signal Processing (2014) (pp. 463–466). IEEE.

12. Jiménez, A.R., Jain, A.K., Ceres, R., Pons, J.L.: Automatic fruit recognition: a survey and new results using range/attenuation images. Pattern Recogn. **32**(10), 1719–1736 (1999)
13. Hannan, M.W., Burks, T.F., Bulanon, D.M.: A machine vision algorithm combining adaptive segmentation and shape analysis for orange fruit detection. Agricult. Eng. (2009)
14. Komal, K., Sonia, D.: GLCM algorithm and SVM classification method for orange fruit quality assessment. Int. J. Eng. Res. Technol. (IJERT) **8**(9), 697–703 (2019)
15. Arivazhagan, S., Shebiah, R.N., Nidhyanandhan, S.S., Ganesan, L.: Fruit recognition using color and texture features. J. Emerg. Trends Comput. Inf. Sci. **1**(2), 90–94 (2010)
16. Hou, L., Wu, Q., Sun, Q., Yang, H., & Li, P. (2016, August). Fruit recognition based on convolution neural network. In 2016 12th International Conference on Natural Computation, Fuzzy Systems and Knowledge Discovery (ICNC-FSKD), pp. 18–22. IEEE.
17. Guo, T., Dong, J., Li, H., Gao, Y.: Simple convolutional neural network on image classification. In: 2017 IEEE 2nd International Conference on Big Data Analysis (ICBDA) (2017), pp. 721–724. IEEE
18. Simonyan, K., Zisserman, A.: Very deep convolutional networks for large-scale image recognition. arXiv preprint arXiv: (2014). 1409.1556
19. Hemdan, E.E.D, Shouman, M.A., Karar. M.E.: Covidx-net: a framework of deep learning classifiers to diagnose covid-19 in x-ray images. arXiv preprint arXiv:2003.11055. (2020)
20. Sae-Lim, W., Wettayaprasit, W., Aiyarak, P.: Convolutional neural networks using MobileNet for skin lesion classification. In: 2019 16th International Joint Conference on Computer Science and Software Engineering (JCSSE) (2019), pp. 242–247. IEEE
21. Shivam, Singh, P.S., Kumar, I.: Rice plant infection recognition using deep neural network systems, (2021) CEUR **2786**, 384–393

Machine Learning-Based Method for Recognition of Paddy Leaf Diseases

Gulivindala Suresh, Narla Venkata Lalitha, and Aditya Kumar Sahu

Abstract Agriculture sector has been facing extensive challenges such as climate instability, cropping pattern, inadequate use of fertilizers, disease identification, and promoting new technologies. Owing to this, recognizing plant disease is one of the leading concerns in boosting productivity. In this paper, an efficient image processing technique has been utilized to classify diseases that occur in rice plant. Initially, background portion of an RGB rice plant image is removed in preprocessing phase. Next, three different clusters are obtained from the image through K-means clustering algorithm. Later, the diseased portions from these clusters of image are retrieved using histogram and color values. Color and texture features are calculated on diseased images. Finally, these obtained features are subjected to the classification phase using a support vector machine (SVM) classifier. Experimentation validates that the proposed detection method through SVM achieves maximal accuracy of 83.3% by outperforming other existing methods.

Keywords Disease detection · Leaf disease · Paddy leaf · Precision farming · SVM classifier

1 Introduction

Agriculture is one among the prominent sources of income in many countries and around 70% of people depend on agriculture in India [1]. The farmers cultivate different crops owing to the environmental circumstances of land and need. Nevertheless, farmers face various problems such as natural disasters, water shortages, and plant diseases, [2]. With the advent of technology, these problems can be reduced to some extent. This leverages on time recognition and prevention of the disease

G. Suresh (✉) · N. V. Lalitha
Department of ECE, GMR Institute of Technology, Rajam, Andhra Pradesh, India
e-mail: suresh.g@gmrit.edu.in

A. K. Sahu
Department of Computer Science and Engineering, Vignan's Foundation for Science, Technology and Research (Deemed to be University), Guntur, Andhra Pradesh 522202, India

K. N. Das et al. (eds.), *Proceedings of the International Conference on Computational Intelligence and Sustainable Technologies*, Algorithms for Intelligent Systems,
https://doi.org/10.1007/978-981-16-6893-7_39

which enhances the productivity of yield. Hence, recognition of plant disease is a demanding task, and therefore, it is a hot research topic in the agriculture sector. The manual monitoring process for plant diseases is quite difficult as it entails more handling time, high volume of work, and knowledge on the plant diseases. Therefore, digital image processing (DIP) methods are being explored to recognize the plant diseases.

Recognition of plant diseases using DIP technology involves various stages viz., image acquisition, preprocessing, segmentation, building the features, and classification [3]. These steps are carried on the exterior portion of the diseased plants [4]. In general, leaves are critical sources to detect plant diseases. In paddy crops, bacterial leaf blight (BLB), leaf smut (LS), and brown spot (BS) are the most common diseases [5]. However, every disease of paddy leaf has its features and differs in color, size, and shape [6]. To suffice this, several researchers developed various feature extraction methods and detection techniques to recognize paddy leaf disease. Still, there is possibility to leverage the detection accuracy of paddy leaf diseases. Henceforth, in our work, color features are combined with texture features to describe the disease in a better manner to improve accuracy. The major contributions of proposed work are given below:

1. A combination of color and texture features is explored to describe the disease characteristics.
2. An algorithm is designed for the automatic detection of disease through an SVM classifier.
3. A rigorous evaluation is carried out, wherein all the classifiers viz., KNN, Naïve Bayes (NB), decision tree (DT), and logistic regression (LR) have been tested for this method.

The remaining paper is structured as follows: Sect. 2 presents the review of related work of plant disease classification, Sect. 3 discusses materials and method, Sect. 4 comprises of proposed work. Experimentation and outcomes of the method are dealt in Sect. 5, conclusion of proposed work is elucidated in Sect. 6.

2 Related Work

Texture feature-based healthy and unhealthy plant leaf classification and detection are presented in [7]. Using thresholding and segmentation, the green portion of the image is removed and five statistical texture features are calculated for the selected segment in an image. These features are given to the multi-class SVM classifier and got 94% overall accuracy. This method is tested on ten different plant species. In [8], an SVM-based classification to detect the tomato leaf is healthy or disease has been proposed. The gray level co-occurrence matrix is considered for texture feature extraction. Experimentation is performed on 400 healthy, 400 infected image dataset, and achieved 99.83% accuracy using SVM with a linear kernel function. Six different types of tomato plant diseases are detected in [9]. Otsu's segmentation is used for image preprocessing. For feature extraction, ten intensity-based statistical features are calculated and given to the DT classifier for classification. The average accuracy of 78% is achieved in this work.

An image segmentation-based plant leaf recognition is elucidated in [10]. Feature extraction comprises of color and texture attributes. Four different plants with five different diseases are taken to evaluate their algorithm. SVM classification is used and attained an accuracy of 97.6%. Prajapati et al. [5] developed a method to classify rice leaf diseases, namely BLB, BS, and LS. Saturation of the original image is used to remove background, and segmentation is carried through K-means clustering. Three models are designed with different textures, colors, and shape features. The model is trained and tested using SVM and got a test accuracy of 73.33%. Detection and classification of the fungal rust disease in the pea plant are presented in [11]. To detect the disease in the early stage, microscopic images are considered. Using DWT, features are extracted and fed to SVM classifier. A dataset with 500 images has been tested and 89.6% accuracy is achieved.

Scale invariant feature transform (SIFT)-based rice leaf disease detection and identification are proposed in [12]. Bag of words (BoW) is used to convert the features that are generated by SIFT to words for classification and then divided into clusters using K-means. This system is trained by using SVM and evaluated using precision, recall, and accuracy. The accuracy of this model is 94.16%. Rice blast is one of the rice plant diseases and to detect this disease spots, an improved KNN algorithm by K-means has been proposed in [13]. Color rice leaf images were prepared using DIP technique in Lab color space, and the KNN algorithm has been used to detect disease spots. Image histogram is performed using Otsu method, and the Euclidean distance-based squared classification is followed. This process results in to reduce the gray level in binary images. This model is evaluated and achieved 94% accuracy.

To detect and classify the five different rice leaf disease images viz., sheath blight, BS, LS, false smut, and stem rot. NB classifier and minimum distance classifier are used and compared their performance [14]. To compare the performance of these two classifiers, 200 rice leaf images are considered. Color and texture features along with morphology and structure attributes were used. Minimum distance classifier gave high accuracy of nearly 81% to that of NB classifier's 69%. In [15], the author presented a classification of plant leaf diseases using directional local quinary patterns (DLQPs). DLQP is used for texture feature descriptor to detect diseased plant leaf, and SVM is used as a classifier. The authors focused on six different diseases of a tomato plant leaf, three different diseases of potato plant leaf, and three different diseases of an apple plant leaf. Compared the performance of three feature descriptors, namely directional local quinary patterns, local binary pattern, and local ternary patterns, and achieved the highest average accuracy for DLQP with 96.5%.

3 Materials and Method

The proposed method comprises of four stages such as initial processing, segmenting the image, feature extraction, and classification. Methods related to feature extraction and classification are elucidated in the following section.

3.1 Local Binary Pattern

Local binary pattern (LBP) is a spatial operator that provides a spatial structure of the gray images [16]. It concentrates on local information or texture content in a gray image. This local information is computed as the difference between intensities of the center pixel g_{c} with its neighbors g_{n} as shown in Fig. 1. Negative difference is represented with "0" and positive difference with "1" which are denoted as binary vector given in Eq. (1).

$$\mathrm{LBP}_{N,R} = \sum_{n=0}^{N-1} f(g_{\mathrm{n}} - g_{\mathrm{c}})2^{p} \tag{1}$$

where N is the number of neighboring pixels, R is the radius, and p is the binary weight.

Uniform LBPs are regular patterns that convey 90% of textural information and can be rotation invariant [17]. All the rotation invariant uniform transitions up to two are considered and accommodated in 58 bins. Similarly, other non-uniform transitions into one bin, accounting for a total of 59 bins as given in (2).

$$\mathrm{LBP}_{N,R}^{\mathrm{riu}} = \begin{cases} \sum_{n=0}^{N-1} f(g_n - g_c)2^{p}, & \text{if } \mathrm{UT}\left(\mathrm{LBP}_{N,R}\right) \leq 2 \\ N+1, & \text{otherwise} \end{cases} \tag{2}$$

where $\mathrm{LBP}_{N,R}^{\mathrm{riu}}$ is the rotation invariant uniform LBP with transitions up to two.

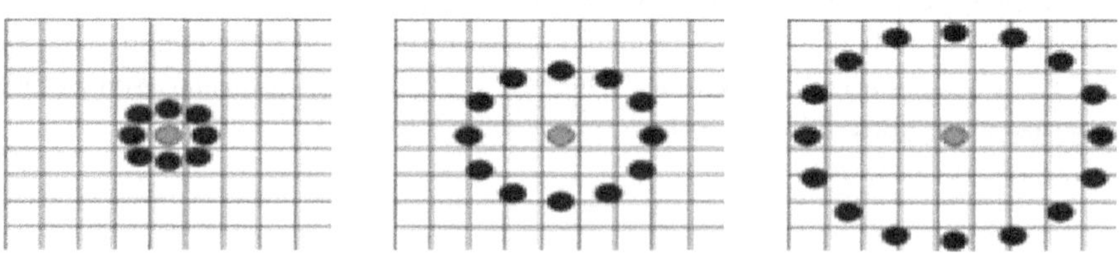

a.N=8, R=1 b. N=16, R=2, c. N=24, R=3

Fig. 1 LBP with various radii and neighbors

For each image, the rotation invariant uniform LBP of 59 bins is obtained and is treated as a texture feature vector (FV).

3.2 Color Moments

Color is a strong descriptor. Color moments are used to partition the images based on their color characteristics. Orengo defined three central moments viz., mean M_c, standard deviation σ_c, and skewness S_c for a color image [18]. Fourth-moment kurtosis K_c is included to capture the image details effectively. These four moments are considered for every channel of an RGB color image as given in Eqs. (3)–(6), to obtain a set of 12 color features for the image.

$$M_c = \frac{1}{N}\sum_{n}^{N} P_{nc} \tag{3}$$

$$\sigma_c = \sqrt{\left(\frac{1}{N}\sum_{n}^{N}(P_{nc} - M_c)^2\right)} \tag{4}$$

$$S_c = \frac{\left(\frac{1}{N}\sum_{n}^{N}(P_{nc} - M_c)^3\right)}{\sigma_c^3} \tag{5}$$

$$K_c = \frac{\left(\frac{1}{N}\sum_{n}^{N}(P_{nc} - M_c)^4\right)}{\sigma_c^4} \tag{6}$$

where c indicates one of the three color channels, i.e., R, G, and B, and n ranges from 1 to N, the amount of pixels available in the image. P_{nc} is the pixel intensity of the nth pixel in cth channel.

3.3 Support Vector Machine

SVM is used to define a hyperplane that maximizes the shortest distance between any data point [19]. Optimal hyperplane can be achieved by minimizing the below objective function

$$\left\{\frac{1}{2}|w|^2 + C\sum_{i=1}^{k}\varepsilon_i\right\}, \varepsilon_i \geq 0 \tag{7}$$

$$\text{Subject to } s_i\left(w^T r_i + b\right) \geq 1 - \varepsilon_i, \quad i = 1, 2, \ldots k$$

where w is a normal vector, C is trade-off margin width and misclassifications, ε_i is a slack variable that allows data points to slip off the margin but penalizes them. Linear SVM can be defined by Eq. (7). In certain cases, the data are not linearly separable, so the kernel function is used to translate the data into another space for linear separation. Various kernel functions are available for this purpose such as RBF and polynomial kernel. Moreover, feature data in this work are not linearly separable. Hence, RBF and polynomial kernels have been explored in this work.

4 Proposed Work

Color images of paddy leaf with various diseases are considered. These images are preprocessed to obtain a binary image and further used to produce a color image without any background. Segmentation is carried to identify leaf portion, diseased portion, and background. Texture and color features are obtained from the infected portion of the leaf. These FVs are different for various diseases of paddy leaf. The classifier is modeled, trained, and evaluated to classify the paddy leaf diseases based on these color and texture features. Process flow of proposed work is shown in Fig. 2 and is detailed in the following steps.

4.1 Preprocessing

Input color images of different sizes are represented with fixed size $M \times N$. These color images are converted to the HSV model, and component "S" is considered

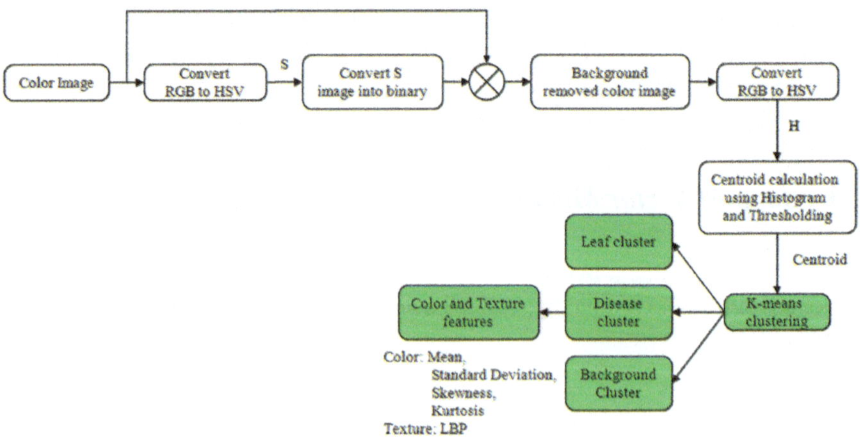

Fig. 2 Proposed method for classification of rice leaf disease

for the binary image. This binary image is used as a mask for input color image to remove the background.

4.2 Segmentation

Image segmentation is a critical step in clustering diseased portions in the images. Earlier [5] explored three different approaches for segmentation and shown that the HSV approach has performed well. Therefore, this approach has been adopted for segmentation to obtain the three clusters. HSV is applied to the background removed the color image, and segmentation is performed based on hue component. K-means clustering is performed on hue component to cluster disease portion and the green region as another cluster which is a healthy portion of the leaf. The diseased portion is identified among three clusters using its histogram and color values. This green portion is removed, and the diseased portion is considered for further processing to extract features.

4.3 Feature Extraction

Texture and color features play a substantial role in object detection, image recognition, and retrieval. In this work, LBP is explored to extract texture features. All the non-uniform and uniform patterns account for 59 patterns with unity radius which acts as texture features. Color moments are calculated from each channel of the diseased portion in the RGB model. This results in 12 color features. By fusing texture and color features, a FV of 71 dimensions is obtained. These features are used to train the model for classification.

4.4 Classification

A classification model is built to classify three diseases based on the features. SVM with RBF and polynomial kernels are explored to perform the identification of three diseases. Later, other classifiers viz., KNN, NB, DT, and LR are used to carry out this task. Further, the performance of these models has been investigated with regard to metrics viz., precision, recall, and F_1 Score.

5 Experimental Results and Discussion

Experimentation is performed to assess the performance of various classifiers in identifying paddy leaf diseases through texture and color features. Dataset of 120 paddy leaf images with three diseases viz., BLB, BS, and LS is considered [5]. A set of experiments is performed to endorse the proposed approach. Initially, segmentation is performed on diseased leaf images to obtain the features. Qualitative results of leaf portion, segmented, and diseased portions for three diseases are shown in Fig. 3. Later, the evaluation of the SVM classifier with different kernels has been explored. Finally, the performance of the SVM classifier is compared with other classifiers viz., KNN, LR, DT, and NB.

Texture and color features are obtained from the diseased portion of the segmented image to form a FV of size 71 and are used to train the SVM classifier. All diseased images in the dataset are split into training and testing group in the ratio of 80:20. Based on this, 96 images are allocated for training and 24 images for testing. The model has been tested and assessed using precision, recall, F_1 score, and accuracy using Eqs. (8)–(11).

$$\text{precision} = \frac{\text{TP}}{\text{TP} + \text{FP}} \tag{8}$$

$$\text{recall} = \frac{\text{TP}}{\text{TP} + \text{FN}} \tag{9}$$

$$F_1\ \text{Score} = \frac{2 \times \text{precision} \times \text{recall}}{\text{precision} + \text{recall}} \tag{10}$$

$$\text{accuracy} = \frac{\text{TP} + \text{TN}}{\text{TP} + \text{FP} + \text{TN} + \text{FN}} \tag{11}$$

where true positive (TP) is correctly predicted positive values, true negative (TN) is correctly predicted negative values, false positive (FP) and false negative (FN) occur when actual class contradicts with predicted class.

The method is evaluated using precision, recall, and F_1 score metrics with regard to various classifiers. The outcome of various classifiers on the classification of BLB, BS, LS diseases is plotted in Fig. 4. The results advocate that SVM-RBF classifier achieved an F_1 score of 0.86, 0.83, and 0.73 in classifying BLB, BS, and LS diseases, respectively, and its performance is better when compared with other classifiers.

Further, the overall performance of classification with various models is assessed in terms of accuracy and is shown in Table 1. Accuracy quoted here is the average classification accuracy of all three diseases. It is apparent from Table 1 that SVM with RBF has outperformed other classifiers with an average classification accuracy of 83.3%. Researchers in [5, 14, 20] worked on rice leaf disease classification through

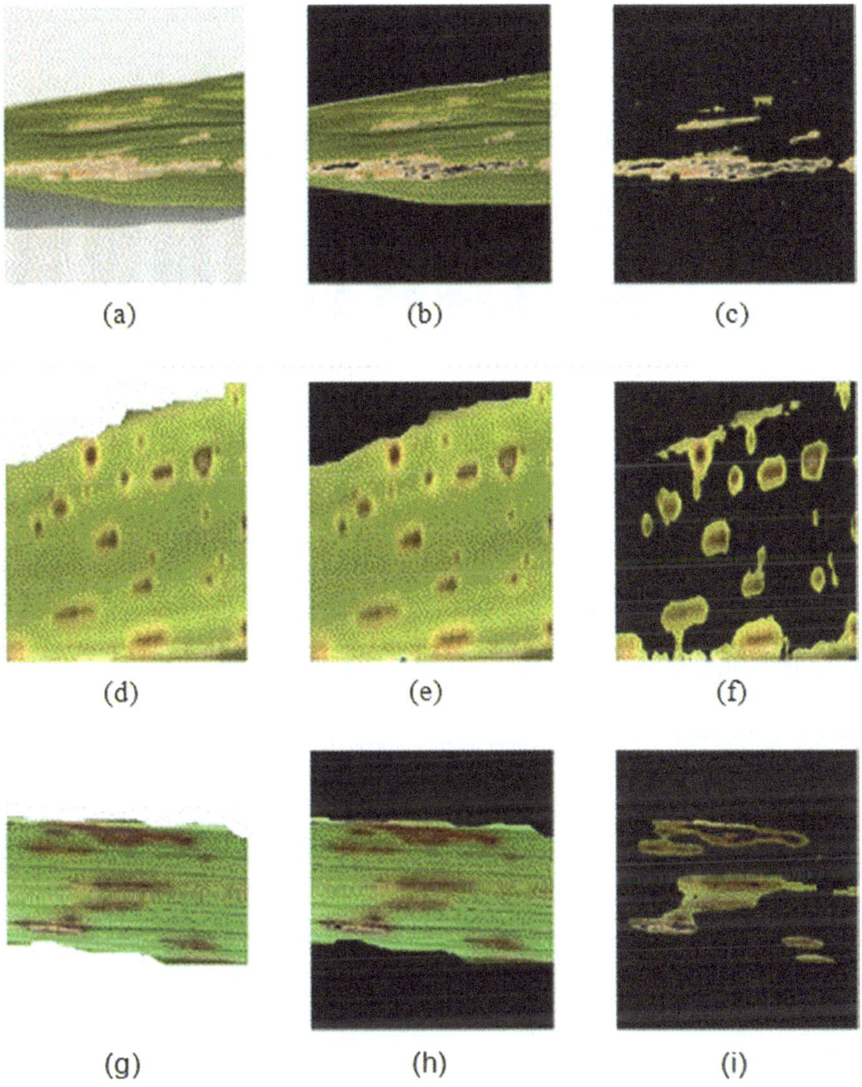

Fig. 3 Qualitative results of segmentation: **a** Leaf image with bacterial leaf blight disease, **b** background removed leaf portion, **c** diseased portion, **d** leaf image with brown spot disease, **e** background removed image, **f** diseased portion, **g** leaf image with leaf smut disease, **h** background removed image, **i** diseased portion

various classifiers viz., NB, SVM, and minimum distance. Classification accuracy of the proposed method using SVM-RBF is superior to other works as given in Table 2.

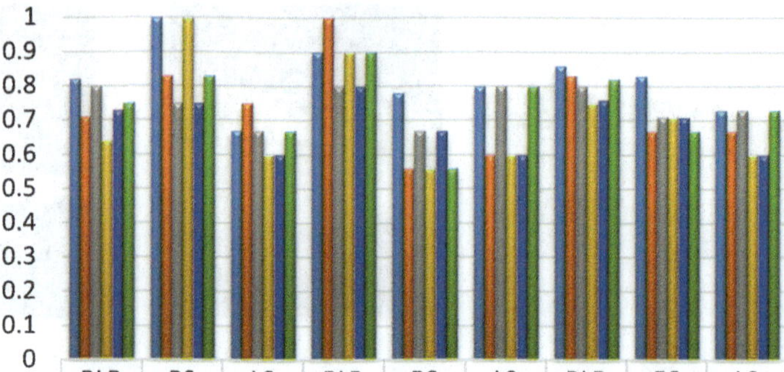

	Precision			Recall			F1		
	BLB	BS	LS	BLB	BS	LS	BLB	BS	LS
SVM-RBF	0.82	1	0.67	0.9	0.78	0.8	0.86	0.83	0.73
SVM-Poly	0.71	0.83	0.75	1	0.56	0.6	0.83	0.67	0.67
KNN	0.8	0.75	0.67	0.8	0.67	0.8	0.8	0.71	0.73
Naïve Bayes	0.64	1	0.6	0.9	0.56	0.6	0.75	0.71	0.6
Decision Tree	0.73	0.75	0.6	0.8	0.67	0.6	0.76	0.71	0.6
Logistic regression	0.75	0.83	0.67	0.9	0.56	0.8	0.82	0.67	0.73

Fig. 4 Performance of various classifiers on classification of BLB, BS, LS diseases

Table 1 Performance analysis of various classifiers

Classifier used in proposed method	Accuracy in %
SVM—RBF kernel	83.3
SVM—Polynomial kernel	75
Naïve Bayes	70.83
KNN	75
Logistic regression	75
Decision tree	70.83

Table 2 Comparative analysis of classification accuracy

Method	Classifier used	Accuracy in %
Phadikar [20]	Bayes classifier	79.5
	SVM	68.1
Prajapati et al. [5]	SVM	73.3
Sharma et al. [14]	Bayes classifier	69
	Minimum distance classifier	81
Proposed method	SVM–RBF	83.3

6 Conclusion

In this digital era, technology has leaped and image processing has been extensively explored in agriculture for various activities like weed detection, crop grading, yield estimation, and disease identification in the agricultural sector. In this work, the classification of rice plant disease among three different types of diseases has been developed. Color rice leaf images are preprocessed to remove the background portion. Background removed image is segmented into three clusters viz., leaf, diseased, and background using K-means clustering. The diseased portion is identified among three clusters using its histogram and color values. Color and texture features are calculated for diseased images, and SVM is used for classification purpose. Other classifiers such as KNN, NB, DT, and LR are considered for evaluation which validated that SVM with RBF kernel performed well. Comparative analysis of the developed method has shown that SVM with RBF kernel performed well in comparison with other state-of-art methods.

References

1. Khirade, S.D., Patil, A.: Plant disease detection using image processing. In: International Conference on Computing Communication Control and Automation, pp. 1–4 (2015)
2. Pantazi, X.E., Moshou, D., Tamouridou, A.A.: Automated leaf disease detection in different crop species through image features analysis and one class classifiers. Comput. Electron. Agric. **156**, 96–104 (2019)
3. Kamal, M., Nor, A., Masazhar, I., Rahman, F.A.: Classification of leaf disease from image processing technique. Indones. J. Electr. Eng. Comput. Sci. **10**(1), 191–200 (2018)
4. Inácio, D., Rieder, R.: Computer vision and artificial intelligence in precision agriculture for grain crops : a systematic review. Comput. Electron. Agric. **153**, 69–81 (2018)
5. Prajapati, H.B., Shah, J.P., Dabhi, V.K.: Detection and classification of rice plant diseases. Intell. Decis. Technol. **11**(3), 357–373 (2017)
6. Barbedo, J.G.A., Koenigkan, L.V., Santos, T.T.: Identifying multiple plant diseases using digital image processing. Biosyst. Eng. **147**, 104–116 (2016)
7. Arivazhagan, S., Shebiah, R.N., Ananthi, S., Vishnu Varthini, S.: Detection of unhealthy region of plant leaves and classification of plant leaf diseases using texture features. Agric. Eng. Int. CIGR J. **15**(1), 211–217 (2013)
8. Mokhtar, U., et al.: SVM-based detection of tomato leaves diseases. Adv. Intell. Syst. Comput. **323**, 641–652 (2015)
9. Sabrol, H., Kumar, S.: Intensity based feature extraction for tomato plant disease recognition by classification using decision tree. Int. J. Comput. Sci. Inf. Secur. **14**(9), 622–626 (2016)
10. Singh, V., Misra, A.K.: Detection of plant leaf diseases using image segmentation and soft computing techniques. Inf. Process. Agric. **4**(1), 41–49 (2017)
11. Singh, K., Kumar, S., Kaur, P.: Support vector machine classifier based detection of fungal rust disease in Pea Plant (Pisam sativam). Int. J. Inf. Technol. **11**(3), 485–492 (2019)
12. Bashir, K., Rehman, M., Bari, M.: Detection and classification of rice diseases: an automated approach using textural features. Mehran Univ. Res. J. Eng. Technol. **38**(1), 239–250 (2019)
13. Larijani, M.R., Asli-Ardeh, E.A., Kozegar, E., Loni, R.: Evaluation of image processing technique in identifying rice blast disease in field conditions based on KNN algorithm improvement by K-means. Food Sci. Nutr. **7**(12), 3922–3930 (2019)

14. Sharma, V., Mir, A.A., Sarwr, D.A.: Detection of rice disease using Bayes' classifier and minimum distance classifier. J. Multimed. Inf. Syst. **7**(1), 17–24 (2020)
15. Ahmad, W., Shah, S.M.A., Irtaza, A.: Plants disease phenotyping using quinary patterns as texture descriptor. KSII Trans. Internet Inf. Syst. **14**(8), 3312–3327 (2020)
16. Ojala, T., Pietikäinen, M., Mäenpää, T.: Multiresolution gray-scale and rotation invariant texture classification with local binary patterns. IEEE Trans. Pattern Anal. Mach. Intell. **24**(7), 971–987 (2002)
17. Li, Z., Liu, G., Yang, Y., You, J.: Scale- and rotation-invariant local binary pattern using scale-adaptive texton and subuniform-based circular shift. IEEE Trans. Image Process. **21**(4), 2130–2140 (2012)
18. Stricker, M., Orengo, M.: Similarity of color images. In: SPIE Conference on Storage and Retrieval for Image and Video databases, pp. 381–392 (1995)
19. Hastie, T., Tibshirani, R., Friedman, J.: The elements of statistical learning: data mining. Infer. Prediction (2013)
20. Phadikar, S.: Classification of rice leaf diseases based on morphological changes. Int. J. Inf. Electron. Eng. **2**(3), 460–463 (2012)

Data Path Designing in Multi-voltage Domain

Ratnesh Kumar Gupta, Ajay Somkuwar, and Lalita Gupta

Abstract Multi-voltage design is a well-known design technique in core design for reduction in power dissipation of core. The different components in core operate at different voltage and at different speed. These different components are provided with different voltages based on its performance requirements. Data path in core has functional blocks which handle the computation to be performed. These blocks are designed and handled in such a way that the required operation is performed efficiently. As technology nodes are going smaller and smaller, meeting the required expectation becomes critical. This paper shows the different techniques that were used to reduce the power and to meet the timing requirements of the circuit. The dynamic power was reduced by 1.1%, and leakage was reduced by 0.4%. The timing results showed increase in positive paths from 252 to 281 greater than 0.2 ps in setup and hold negative paths were reduced from 32 to 0. The results were converged for 7 nm and 10 nm.

Keywords Power dissipation · Power reduction · Data path · Timing violation

1 Introduction

As the technology nodes are getting lower and lower, meeting all the design parameters while assuring the performance and quality is getting complex. The frequency at which the processor work is getting higher to meet the expectation. The reduction in cycle time impacts the timing as well as power for these processors. The data path mostly being on high speed paths have criticality in meeting the timing and power. The timing includes meeting all the paths without having any setup or hold violation. The power being directly dependent on frequency ($P = CV^2f$) increases proportionally. Since the demand for low power products is increasing day by day, the power reduction is one of the biggest challenges. Several low power techniques such as clock gating [1], power gating [2], multi-bit sequential [3], and low leakage

R. K. Gupta (✉) · A. Somkuwar · L. Gupta
Department of Electronics and Communication Engineering, Maulana Azad National Institute of Technology, Bhopal 462003, India

K. N. Das et al. (eds.), *Proceedings of the International Conference on Computational Intelligence and Sustainable Technologies*, Algorithms for Intelligent Systems,
https://doi.org/10.1007/978-981-16-6893-7_40

devices are mostly being followed wherever possible. The power contribution from interconnect is still a challenge to handle and accurate modeling of real power dissipation in the simulation tools, and power measuring tools are something that can be worked on to minimize and accurately work on the power minimization technique. In this paper, the data path designing was done to meet the timing and power at 7 and 10 nm technology nodes.

2 Data Path Designing

Data path design refers to the designing of functional blocks which are involved in all the computation that is carried out. These functional blocks can be registers, adders, multiplexers, counters, etc., as shown in Fig. 1. These blocks are so designed such that critical requirements such as timing, power and clock synchronization in the design are met. For meeting the power budget for these blocks, the blocks are designed to consume minimum power. One of the well-known techniques to reduce power is to cut down the supply voltage (power dissipation is square of voltage supplied). The higher voltage is provided to the critical path blocks with high speed requirement, and lower voltage is provided to the non-critical path with low speed requirement.

Setup Time
It is the minimum amount of time required for the data to be stable before the arrival of active edge of the clock such that data is sampled correctly by the clock.

$$T_{\mathrm{clk2}}Q + T_{\mathrm{Comb.}} + T_{\mathrm{set-up}} \leq T_{\mathrm{clk}} + T_{\mathrm{skew}} \tag{1}$$

In Eq. (1), $T_{\mathrm{clk2}}Q$ represents the time taken by FF_1 to generate Q after arrival of active edge of clock, T_{Comb} is the combinational delay, T_{setup} is the required setup time, T_{clk} is clock period, and T_{skew} represents delay of two buffer in Fig. 2.

Hold Time
It is the minimum amount of time required for the data to be held stable after the arrival of active edge of the clock such that the data is sampled correctly by the clock.

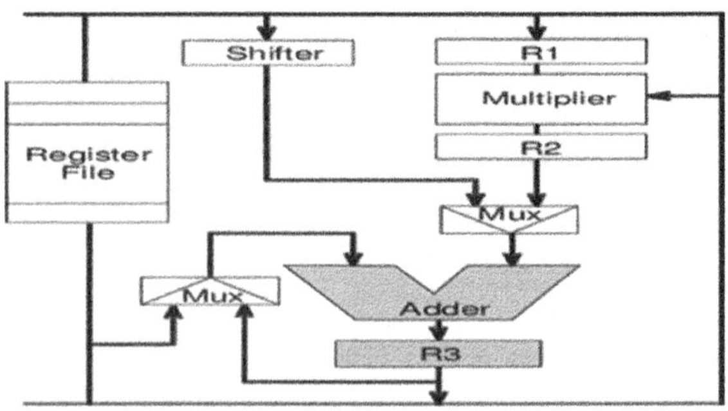

Fig. 1 Basic data path design architecture

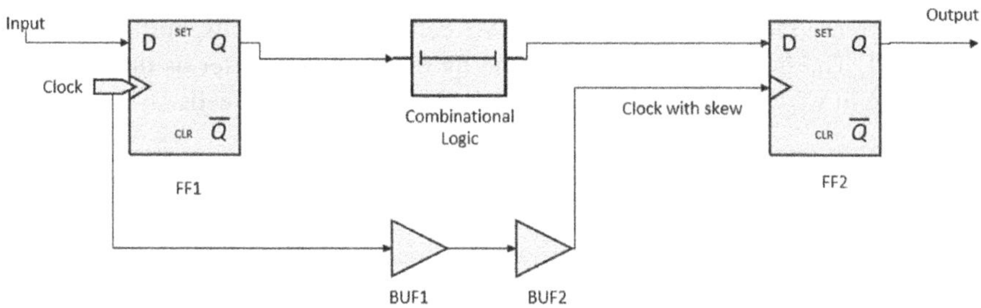

Fig. 2 Two flop scenario

$$T_{\mathrm{clk}2Q} + T_{\mathrm{Comb.}} \geq T_{\mathrm{hold}} + T_{\mathrm{skew}} \tag{2}$$

In Eq. (2), T_{clk2Q} represents the time taken by FF_2 to generate Q after arrival of active edge of clock, $T_{\mathrm{Comb.}}$ is the combinational circuit delay, T_{hold} is the required hold time, and T_{skew} is the delay of two buffer in Fig. 2.

2.1 *Techniques to Fix Timing Violations*

Setup fixing majorly depends on data path optimization, input transition and output capacitance, clock latency and skew, clock DRVs (max transition and max capacitance), crosstalk. Generally, hold fixing will start after clock tree synthesis (CTS) stage, because before CTS stage clock tree will not build, so we will not see the exact clock skew and latency in tool. Hold fixing depends majorly on clock skew, clock and data drivers, cross talk, etc.

Setup Fix in Data Path

For fixing the setup time violation, from Eq. (1), the priority is on reducing the combinational path delay. The combinational circuit delay includes cell delay, net delay, clock tuning, placement, etc. The placement of the blocks was done such that driver and receiver are close to each other. The cell delay was decreased by reducing the load, it has to drive and by upsizing it if the space was available in layout. The net delay was improved by moving some critical nets to higher metal layer and optimizing the routing. The clock tuning was done at last as there was no scope to fix the set violation in data path.

Hold Fix in Data Path

The hold violation arises when the data captured is not stable for the minimum time (hold time) required after the arrival of active edge of clock. From Eq. (2), if the combinational delay is increased, the hold violation can be fixed. Other parameters in above equation are sometimes difficult to vary requires extra effort in changing those parameters to change for fixing the hold violation. The combinational delay can be increased in following ways high VT cells, min delay cells, buffer insertion, clock pull. The hold fix was mainly done by inserting the buffer as the sufficient setup margin was available and swapping the LVT cell to HVT cells.

3 Power Analysis

The power dissipated in processor decides the overall performance of SOC or devices on which it is mounted to. The power dissipated should be as minimum as possible while improving the other parameters such as performance and area.

As the technology nodes are going smaller, the higher power dissipated will create the temperature to rise in the smaller area (power density = *P*/*A*) which may completely disable the functioning of transistors.

Total Power Dissipated: Leakage + Short Circuit + Switching

3.1 Leakage Power: VCC (Leakage Current)*

Figure 3 shows the leakage current in transistor. The different leakage currents explained below.

Subthreshold Leakage Current
Subthreshold leakage is the most important contributor to static power in CMOS.

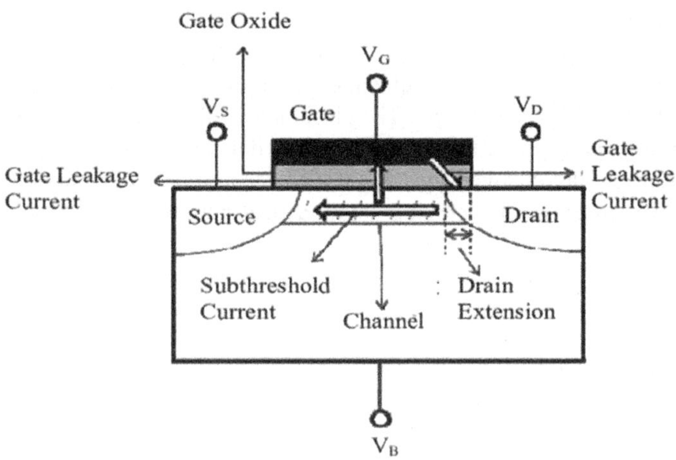

Fig. 3 Transistors leakage currents

$$I_{\text{sub}} = I_{\text{s}}.e^{\frac{q(V_{\text{GS}}-V_{\text{T}}-V_{\text{offset}})}{n\text{KT}}}\left(1 - e\frac{-qV_{\text{DS}}}{kt}\right) \quad (3)$$

In Eq. (3) [4], I_{sub} represents subthreshold current, I_{s} is saturation current, V_{T} is the threshold voltage, n represents emission coefficient, K is Boltzman constant, T is the absolute temperature, VDS is drain to source voltage, and V_{Offset} represent the offset voltage.

From Eq. (3), I_{sub} depends on primarily on threshold voltage (V_{T}). So, as the V_{T} increases, the subthreshold current decreases exponentially. The primary reason for keeping high threshold devices is to suppress the leakage currents.

Source to Drain Leakage
When transistor is off, there is non-zero current flowing from source to drain.

Gate Leakage
Even though the gate is insulted, there is some amount of current always leaking from gate terminal.

$$I_{\text{Gate}} = W_{\text{eff}}L_{\text{eff}}A\left(\frac{V_{\text{OX}}}{T_{\text{OX}}}\right)^2 \exp\left[\frac{-B(1-\left(1-V_{\text{OX}}/\text{Ø}_{\text{OX}})^{\frac{3}{2}}\right)}{\frac{V_{\text{OX}}}{\text{Ø}_{\text{OX}}}}\right] \quad (4)$$

In Eq. (4) [5], W_{eff} represents the effective width of the device, L_{eff} is the effective length of the device, Ø_{OX} is barrier height for tunneling particles, A is area, V_{OX} represents the potential drop across the thin oxide layer, T_{OX} represents the oxide thickness, and A and B are physical parameters. As can be seen from Eq. (4), the gate current depends inversely on T_{OX}. As oxide thickness increases, the gate leakage current reduces. Leakage current can be reduced by use of high VT cells, use of stack devices, power gating, etc.

*3.2 Short Circuit Power = VCC * (Short Circuit Current)*

$$P_{\text{SC}} = T_{\text{sc}}V_{\text{DD}}I_{\text{Peak}}f \quad (5)$$

In Eq. (5), P_{sc} represents short circuit power dissipation, T_{SC} is the duration and slope of input signal, I_{Peak} is peak current, and F represents the frequency of operation. The short circuit current flows for a small duration when the input and output are transitioning as both the transistor NMOS as well as PMOS are on allowing current to flow to from VDD to VSS. The situation is depicted in Fig. 4.

Possible fix for short circuit power reduction
The supply voltage given to circuit can be reduced to a point where it is less than the sum of the threshold voltage value of the transistors PMOS and NMOS such that both transistor will not be on at the same time avoiding any power dissipation due

Fig. 4 Short circuit current path

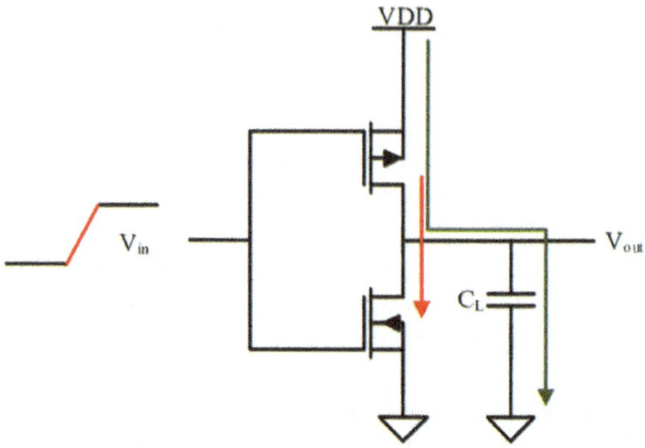

to short circuit current. By matching the rise and fall times of the input and output signals, short circuit power can be reduced.

3.3 *Switching Power*

Switching power is also known as dynamic power. This is the power dissipated when the device switches from one state to another.

$$\text{Switching Power} = \text{AF} * (\text{Diffusion Cap of Driver} + \text{Load Cap} + \text{Gate Cap of Receiver}) * V^2 * F \quad (6)$$

In Eq. 6 [6], V represents the voltage supplied, AF is the activity factor, and F represents frequency of operation. The switching power depends on the dynamic capacitance, voltage, and frequency. For reducing the dynamic capacitance, the load cap can be reduced which will reduce the overall dynamic power. The activity factor (AF) is the number of times the signal is making transitions divided by the number of phases. The clock has an activity factor of one as it is taken as reference. Methods to reduce dynamic power include reducing the cap on the high activity factor nodes, clock gating, voltage if possible can be reduced (Fig. 5).

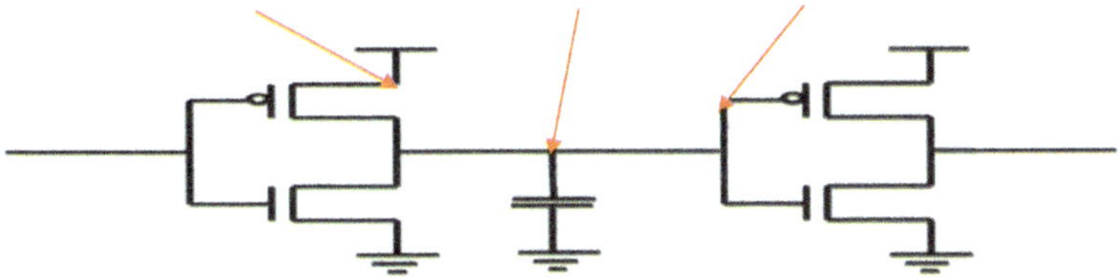

Fig. 5 Major capacitance contributor in switching power

4 Results and Discussion

4.1 Power Results

The power results shown in Table 1 shows that out of the total power consumed by the block, the maximum power is the dynamic power which is expected. The power consumption due to leakage is zero, and the smaller portion of power is consumed by short circuit of 4%. The leakage current contribution can be understood from the fact that the blocks uses the SVT cells considerably high in compared to the LVT cells. The area to focus from these results are the switching power, and more efforts are required to minimize these power consumption.

The cap results in Table 2 suggest that more than 50% of the cap comes from the interconnect wires. The devices are getting smaller, and the more numbers of transistors are being packed by keeping the area almost constant. The wires used to connect these cells have not been scaled by the same amount. These interconnect nets therefore contribute maximum amount of caps and are major contributor of power losses.

The clock contribution in the dynamic power consumed is maximum because clock has the AF of one. The clock distribution network design plays an important role in determining the overall power when a particular projects power budget is defined. Many times several techniques are implemented so that the cap values seen by the clock driver are less because that will directly impact the power consumption. Clock gating is one of the techniques used to reduce the power consumed by the clock network. As seen in Table 3, the maximum cap contribution comes from the interconnect cap on the nets for clock distribution and also the maximum power consumed.

Table 1 Power consumption distribution

	Before optimization (%)	After optimization (%)
Total power	100	100
Dynamic power	96.1	95.2
Short circuit power	3.5	4.2
Leakage power	0.4	0

Table 2 Capacitance distribution

	Before optimization (%)	After optimization (%)
Total capacitance	100	100
Gate capacitance	27.3	25.1
Diffusion capacitance	8.7	12.2
Interconnect capacitance	53.1	51.2
External capacitance	9.2	9.2
Input gate capacitance	0.5	0.6
Output diffusion capacitance	0.1	0.3
Output total capacitance	1	1
Virtual repeater capacitance	0	0

Table 3 Clock power consumption

Output power	0%
Clock power	100%
IC power	54%
Non-IC power	19%
Non-IC sequential power	23%
Leakage power	0%
Non-ICUKN power	4%

Table 4 shows the number of cells present and the total Z (width) of those cells.

For keeping the power consumed within the blocks, the SVT cells are preferred over the LVT cells because of the less leakage current in SVT cells. After the cells are converted in SVT cells from LVT cells, the impact will be on timing as the delay in those paths will increase and it may be possible that earlier some path which was not violating will start violating the timing on them and the margin can go negative. So, after the power of blocks are adjusted within the budget, the timing is re-evaluated and compared with earlier runs. The paths where margin became negative are fixed by following different methods. Table 5 shows that the SVT is 66.76%, and other cells is 33.24%.

Table 4 Total transistor and their width

Total transistor	33,018
Total Z	3347.198

Table 5 Multi-threshold cells

	LVT percentage	HVT percentage
Before optimization	45.3	54.7
After optimization	33.2	66.76

4.2 *Timing Results*

Figure 6 shows the histogram representation of the setup (max) and hold margin (min) on the Fub. As seen in the reports, the number of paths having positive max margin are greater in number in archive area as compared to the latest run as the archive area has the final convergence of timing done. The timing convergence involved all the setup fixes like buffer resizing, clock push, cap reduction by driver spillting, critical nets being pushed to higher metal layers and the buffer removal. The min histogram shown in Fig. 6 that archive area has less number of positive paths than the current area, it may be because of max fixing of the other Fubs and within the Fub. The archive area has almost zero negative min path as expected.

The accumulated max margin and min margin shown in Fig. 7 shows that the accumulated margin on max for archive is less as compared to current and previous

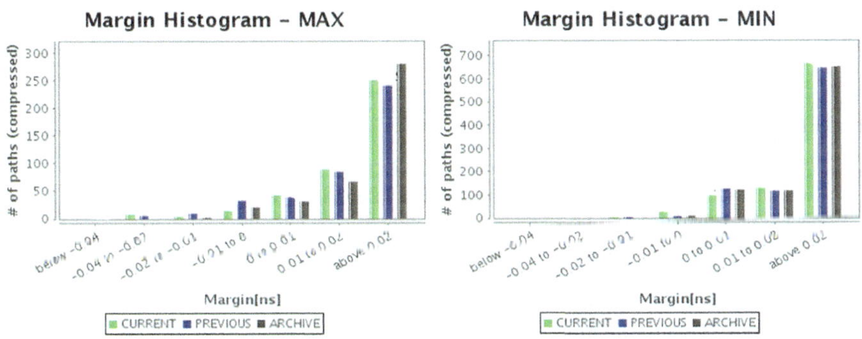

Fig. 6 Setup and hold margin histogram

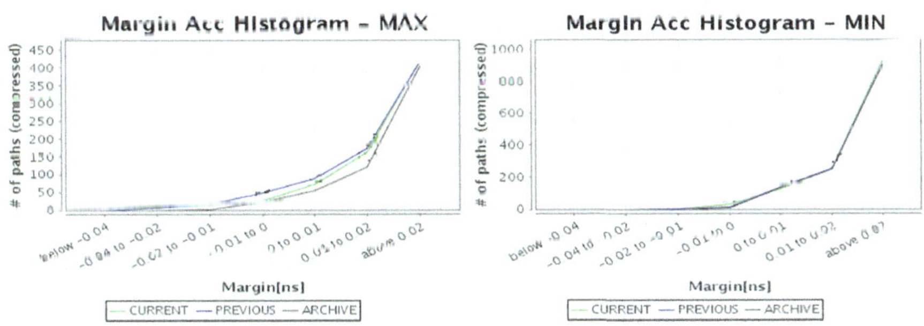

Fig. 7 Accumulated setup and hold margin histogram

runs because the negative paths were fixed in design or from section level. The accumulated negative margin for min is zero for archive and some paths are negative in the current run, these paths may be internal or external. The external negative paths are fixed based on the section recommendation and may not always be possible to get fixed from the Fub side. The reason may be because these negative paths will have very little scope to be fixed in the design or its fix may create the new paths in max margin. If these cases are there the section may ask the neighboring Fubs to improve margin on these paths and after new timing release, these path will be fixed.

5 Conclusion

The performance parameters timing and power were properly converged. The power reduction in dynamic from 96.1 to 95.2% as well as leakage power from 0.4 to 0% was achieved using the power reduction techniques such as use of SVT cells, power gating, and voltage reduction. The short circuit power increased by 0.7% after optimization as device delay increased. The timing was converged and all the possible paths from functional unit block were converged. The positive paths greater than 0.2 ps were increased from 252 to 281, and the negative paths were 0 after the max fixes. The negative internal paths were reduced from 32 to 0 after the min fixes.

References

1. Kulkarni, R., Kulkarni, S.Y.: Power analysis and comparison of clock gated techniques implemented on a 16-bit ALU. In: International Conference on Circuits, Communication, Control and Computing, 21–22 Nov 2014
2. Kumar, H., Kumar, A., Deolia, V.K.: Enabling concurrent clock and power gating in 32 bit ROM. In: 9th in International Conference on Computing, Communication and Networking Technologies (ICCCNT) 10–12 July 2018
3. Dwivedi, S.N., Seth, A.K., Grover, A.: Design and benchmark of single bit and multi bit sequential elements in 65 nm for low standby power consumption. In: 24th International Symposium on VLSI Design and Test (VDAT) 23–25 July 2020
4. Sheu, B.J., et al., BSIM: Berkeley short-channel IGFET model for MOS transistors. IEEE J. Solid State Circ. **SC-22**(4), 558–566 (1987)
5. Roy, K., et al., Leakage current mechanisms and leakage reduction techniques in deep-submicrometer CMOS circuits. Proc. IEEE, **91**(2), 305–327 (2003)
6. Chandrakasan, A.P., Mehra, R., Potkonjak, M., Rabaey, J., Brodersen, R.W.: Optimizing power using transformations. In: IEEE Trans. On CAD, pp. 13–32 (1995)
7. Lanuzza, M., Corsonello, P., Perri, S.: Low-power level shifter for multi-supply voltage designs. IEEE Trans. Circ. Syst. II Exp. Briefs **59**(12), 922–926 (2012)
8. Zhao, W., Alvarez, A.B., Ha, Y.: A 65-nm 25.1-ns 30.7-fJ robust subthreshold level shifter with wide conversion range. IEEE Trans. Circ. Syst. II: Exp. Briefs **62**(7), 671–675 (2015)
9. Hosseini, S.R., Saberi, M., Lotfi, R.: A low-power subthreshold to above-threshold voltage level shifter. IEEE Trans. Circ. Syst. II Exp. Briefs **61**(10), 753–757 (2014)

10. Wen, L., Cheng, X., Tian, S., Wen, H., Zeng, X.: Subthreshold level shifter with self-controlled current limiter by detecting output error. IEEE Trans. Circ. Syst. II Exp. Briefs **63**(4), 346–350 (2016)
11. Le, V.L., Kim, T.T.-H.: An area and energy-efficient ultra-low-voltage level shifter with pass transistor and reduced-swing output buffer in 65-nm CMOS. IEEE Trans. Circ. Syst. II Exp. Briefs **65**(5), 607–611 (2018)
12. Roy, K., Prasad, S.C.: Low-Power CMOS VLSI Circuit Design. Wiley, New York, pp. 214–219 (2000). (Ch. 5)
13. Mead, C.: Scaling of MOS technology to submicrometer feature sizes. Analog Integr. Circ. Sig. Process. **6**, 9–25 (1994)
14. Dennard, R. et al.: Design of ion-implanted MOSFET's with very small physical dimensions. IEEE J. Solid-State Circ. **SC-9**, 256 (1974)
15. Kumar, D., Tiwari, N.: VLSI designs for low power applications. Int. J. Eng. Sci. Adv. Res. **1**(1), 71–75 (2015)
16. Rao, R.M., Burns, J.L., Brown, R.B.: Circuit techniqus for gate and subthreshold leakage minimization in future CMOS technologies. In: Proceedings of ISLPED, pp. 70–73 (2002)
17. Khan, Z.A., Burney, S.A., Naseem, J., Rizwan, K.: Optimization of power consumption in VLSI circuit. IJCSI Int. J. Comput. Sci. Issues **8**(2) (2011)

State-Space Approach of Automatic Generation Control of Two-Area Multi-source Power Systems

P. Vidya Sagar and M. S. Krishna Rayalu

Abstract Tie-lines connect every multi-area power system network in the country. Automatic generation control (AGC) is being utilized to reduce tie-line power and frequency variations while maintaining nominal frequency. The focus of this paper is on a two-area power system with multiple sources in each zone. Hydro, thermal, gas, wind, and solar energy are all examined in each location. *Due to the increasing demand for wind and solar power plants, they are considered as sustainable green energy. But the operation of wind and solar plants allied with stability problem the disturbance is injected. The PI and PID control strategy based on optimal AGC regulators like linear quadratic regulators (LQR) are designed. The system's dynamic performance can be evaluated for a 1% step load disturbance in one or more control areas.*

Keywords Linear quadratic regulator (LQR) · Proportional-integral (PI) · Proportional-integral-derivative (PID) controllers · Frequency and tie-line variations · Two-area power system

1 Introduction to AGC

Every power system attempts to provide consumers with reliable electricity by curbing frequency and tie-line power errors and maintaining a frequency within a specified range. But due to increased demand, there is a mismatch between generation and load demand [1]. Automatic generation control (AGC) can be used to regulate those mismatches. When the system load fluctuates constantly, the job of AGC is to automatically adjust the generation to bring frequency back to a predefined

P. Vidya Sagar (✉) · M. S. K. Rayalu
Department of E.E.E, Velagapudi Ramakrishna Siddartha Engineering College, Kanuru, Andhra Pradesh 520007, India

M. S. K. Rayalu
e-mail: mskr@vrsiddhartha.ac.in

K. N. Das et al. (eds.), *Proceedings of the International Conference on Computational Intelligence and Sustainable Technologies*, Algorithms for Intelligent Systems,
https://doi.org/10.1007/978-981-16-6893-7_41

range and to divide the generation among different generating stations in order to make it economical.

To get better performance and to improve performance of the control loops, PI and PID controllers are used, and they are tuned by using modern artificial intelligence techniques like particle swarm optimizations. To overcome the drawbacks of PI and also PID controller, by using state-space modelling, optimal regulators like LQR [2] are designed.

1.1 Multi-source, Multi-area Power System with Automatic Generation Control

Figure 1 depicts the classic multi-area power system, which includes multiple sources in each area. The above-depicted three-area system uses thermal, hydro, gas, and renewable energy sources such as wind and solar for generation. The regulation of generator output power among different control areas in response to frequency deviations and tie-line power errors is defined as AGC. In order to balance the load and stabilize the system, the nominal frequency must be kept within the prescribed limits. By using feedback of the area control error (ACE) to decrease errors to zero, the frequency as well as tie-line anomalies may be adjusted within the limitations [3].

$$\mathrm{ACE}_i = \Delta P_{\mathrm{tie},i} + B_i \Delta F_i \tag{1}$$

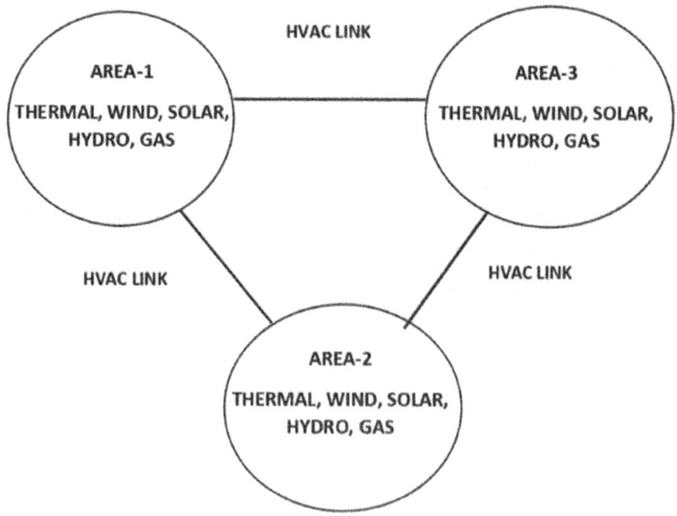

Fig. 1 Multi-area multi-source power system conventional block diagram

Here, there is more than one source in each area generating the power. For ith area, $H_{ti}, H_{hi}, H_{wi}, H_{gi}$, and H_{si} are constituting the generation by conventional plants like gas, hydro, thermal, diesel, etc. and renewable energy sources such as solar and wind. The sharing factor's value is determined by load demand and scheduled economic load dispatch. $P_{Gti}, P_{Ghi}, P_{Ggi}, P_{Gwi}$, and P_{Gsi} are generating power by the thermal, gas, hydro, wind, and solar plants for the ith area. And the total generation for ith area [2–10] given as

$$P_{Gi} = P_{Gti} + P_{Ghi} + P_{Gwi} + P_{Ggi} + P_{Gsi} \tag{2}$$

where

$$\begin{aligned} P_{Gti} &= H_{ti} P_{Gi}, P_{Ghi} = H_{hi} P_{Gi}, P_{Gwi} \\ &= H_{wi} P_{Gi}, P_{Gsi} = H_{si} P_{Gi}, P_{Ggi} = H_{gi} P_{Gi} H_{ti} \\ &\quad + H_{hi} + H_{wi} + H_{si} + H_{gi} = 1 \end{aligned}$$

For small perturb in the load change in the ith area, Eq. (2) can be written as

$$\Delta P_{Gi} = \Delta P_{Gti} + \Delta P_{Ghi} + \Delta P_{Gwi} + \Delta P_{Ggi} + \Delta P_{Gsi} \tag{3}$$

2 AGC of 2-Area Diverse Source Power System Using PID Controller

The PID controller illustrated in Fig. 2 is part of a two-area multiple source power system, and the data for it is included in the appendix. To balance generation and load demand, the tie line and frequency variations must be reduced to zero. Typically, the PID controller is employed as backup. A PID can be tuned manually or automatically using any of today's artificial intelligence tools. Particle swarm optimization (PSO) approach has been used to alter the values of PID controller automatically [5, 10]. The PID controllers help in controlling the system without having complete information of system characteristics or the transfer functions, and it is easy to implement and feasible.

$$\mathrm{Gc}(s) = K_{p_i}\left[1 + \frac{1}{T_i} + k_{di} s\right] \tag{4}$$

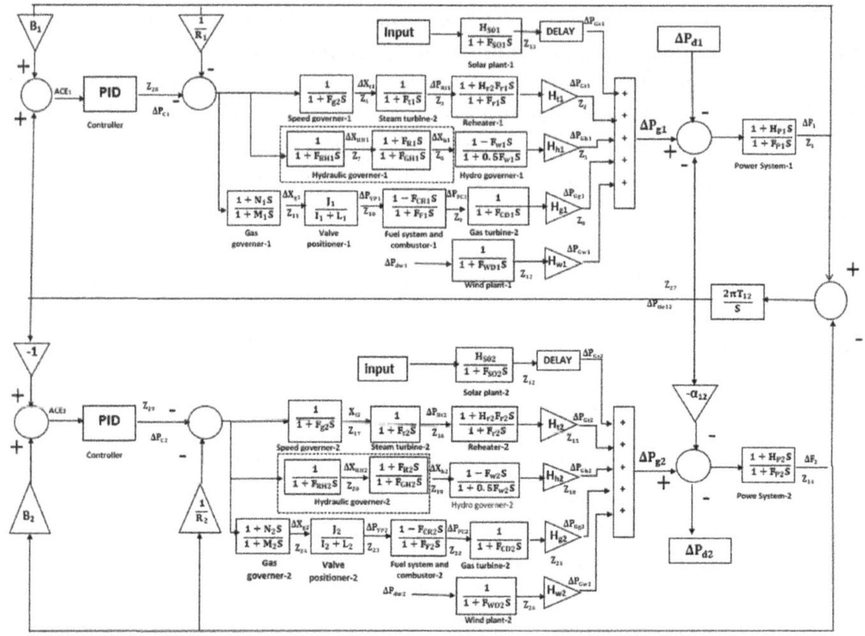

Fig. 2 Multiple area AGC with PID controller integration of multiple source power system

3 Representation of a Two-Area Integrated Multiple Source Power System in State-Space

State variables: Table 1 illustrates the state variables of a two-area multiple source system [3, 10].

Control inputs

$$u_1 = P_{c1} \text{ for area} - 1 \text{ and } u_2 = P_{c2}$$

Table 1 State variables

Area-1	$O_1 = \Delta F_1$, $O_2 = \Delta P_{Gt1}$, $O_3 = \Delta P_{Rt1}$, $O_4 = \Delta X_{t1}$, $O_5 = \Delta P_{Gh1}$, $O_6 = \Delta X_{h1}$, $O_7 = \Delta X_{RH1}$, $O_8 = \Delta P_{Gg1}$, $O_9 = \Delta P_{FC1}$, $O_{10} = \Delta P_{VP1}$, $O_{11} = \Delta P_{GW1}$, $O_{13} = \Delta P_{GS1}$, $O_{28} = \int ACE_1 \mathrm{d}t$
Area-2	$O_{14} = \Delta F_2$, $O_{15} = \Delta P_{Gt2}$, $O_{16} = \Delta P_{Rt2}$, $O_{17} = \Delta X_{t2}$, $O_{18} = \Delta P_{Gh2}$, $O_{19} = \Delta X_{h2}$, $O_{20} = \Delta X_{RH2}$, $O_{21} = \Delta P_{Gg2}$, $O_{22} = \Delta P_{FC2}$, $O_{23} = \Delta P_{VP2}$, $O_{24} = \Delta X_{g2}$, $O_{25} = \Delta P_{GW2}$, $O_{26} = \Delta P_{GS2}$, $O_{29} = \int ACE_1 \mathrm{d}t$, $O_{27} = \Delta P_{tiei,j}$

Disturbance inputs

$$d_1 = \Delta P_{d1}, \Delta P_{dw1}, \Delta P_{ds1} \text{ for area} - 1$$
$$d_2 = \Delta P_{d2}, \Delta P_{dw2}, \Delta P_{ds2} \text{ for area} - 2$$

State equations: The state equations below are all based on [3, 10].

1. $$\begin{aligned}\dot{z}_1 = &-\frac{1}{F_{P_1}} z_1 + \frac{H_{P_1}}{F_{P_1}} z_2 + \frac{H_{P_1}}{F_{P_1}} z_5 + \frac{H_{P_1}}{T_{P_1}} z_8 \\ &+ \frac{H_{P_1}}{F_{P_1}} z_{12} + \frac{H_{P_1}}{F_{P_1}} z_{13} - \frac{H_{P_1}}{F_{P_1}} z_{27} - \frac{H_{P_1}}{F_{P_1}} \Delta P_{d1}\end{aligned}$$
2. $\dot{z}_2 = -\frac{1}{F_{r_1}} z_2 + \left[\frac{H_{t1}}{F_{r1}} - \frac{H_{r_1} H_{t_1}}{F_{t_1}}\right] z_3 + \frac{H_{r_1} H_{t_1}}{F_{t_1}} z_4$
3. $\dot{z}_3 = -\frac{1}{F_{t1}} z_3 + \frac{1}{F_{t1}} z_4$
4. $\dot{z}_4 = \frac{1}{F_{g1} R_1} z_1 - \frac{1}{F_{g1}} z_4 - \frac{1}{F_{g1}} \Delta P_{C1}$
5. $$\begin{aligned}\dot{z}_5 = &\frac{2H_{h1} F_{R1}}{F_{GH1} F_{RH1} R_1} z_1 - \frac{2}{F_{W1}} z_5 + \left[\frac{2F_{h1}}{F_{W1}} + \frac{2H_{h_1}}{F_{GH_1}}\right] z_6 \\ &+ \left[\frac{2H_{h1} F_{R1}}{F_{GH1} F_{RH1}} - \frac{2H_{h_1}}{F_{GH_1}}\right] z_7 + \frac{2H_{h1} F_{R1}}{F_{GH1} F_{RH1}} \Delta P_{C1}\end{aligned}$$
6. $$\begin{aligned}\dot{z}_6 = &-\frac{F_{R1}}{F_{GH1} F_{RH1} R_1} z_1 - \frac{1}{F_{GH_1}} z_6 \\ &+ \left[\frac{1}{F_{GH1}} - \frac{F_{R_1}}{F_{GH_1} F_{RH1}}\right] z_7 - \frac{F_{R_1}}{F_{GH_1} F_{RH1}} \Delta P_{C1}\end{aligned}$$
7. $\dot{z}_7 = -\frac{1}{F_{RH1} R_1} z_1 - \frac{1}{F_{RH1}} z_7 - \frac{1}{F_{RH1}} \Delta P_{C1}$
8. $\dot{z}_8 = -\frac{1}{F_{CD1}} z_8 + \frac{H_{g1}}{F_{CD1}} z_9 - \frac{H_{g1} F_{CR1}}{F_{CD1} F_{F1}} z_{10}$
9. $\dot{z}_9 = -\frac{1}{F_{F1}} z_9 + \left[\frac{1}{F_{F1}} + \frac{F_{CR1}}{(F_{F1})^2}\right] z_{10}$
10. $\dot{z}_{10} = -\frac{N_1}{L_1 R_1 M_1} z_1 - \frac{1}{L_1} z_{10} - \frac{1}{L_1} z_{11} - \frac{N_1}{L_1 M_1} \Delta P_{C1}$
11. $\dot{z}_{11} = \left[\frac{N_1}{R_1 M_1^2} - \frac{1}{R_1 M_1}\right] z_1 - \frac{1}{M_1} z_{11} - \left[\frac{1}{M_1} - \frac{N_1}{M_1^2}\right] \Delta P_{C1}$
12. $\dot{z}_{12} = -\frac{1}{F_{WD1}} z_{12}$
13. $\dot{z}_{13} = -\frac{1}{F_{st1}} z_{13}$
14. $$\begin{aligned}\dot{z}_{14} = &-\frac{1}{F_{P_2}} z_{14} + \frac{H_{P_2}}{F_{P_2}} z_{15} + \frac{H_{P_2}}{F_{P_2}} z_{18} + \frac{H_{P_2}}{F_{P_2}} z_{21} \\ &+ \frac{H_{P_2}}{F_{P_2}} z_{25} + \frac{H_{P_2}}{F_{P_2}} z_{26} - \frac{H_{P_2}}{F_{P_2}} \alpha_{12} z_{27} - \frac{H_{P_2}}{F_{P_2}} \Delta P_{d1}\end{aligned}$$
15. $\dot{z}_{15} = -\frac{1}{F_{r_2}} z_{15} + \left[\frac{H_{t2}}{F_{r2}} - \frac{H_{r_2} H_{t_1}}{F_{t_2}}\right] z_{16} + \frac{H_{r_2} H_{t_2}}{F_{t_2}} z_{17}$
16. $\dot{z}_{16} = -\frac{1}{F_{t2}} z_{16} + \frac{1}{F_{t2}} z_{17}$
17. $\dot{z}_{17} = \frac{1}{F_{g2} R_2} z_{14} - \frac{1}{F_{g2}} z_{17} - \frac{1}{F_{g2}} \Delta P_{C2}$

18. $$\dot{z}_{18} = \frac{2H_{h2}F_{R2}}{F_{GH2}F_{RH2}R_2}z_{14} - \frac{2}{F_{W2}}z_{18} + \left[\frac{2H_{h2}}{F_{W2}} + \frac{2H_{h_2}}{F_{GH_2}}\right]z_{19} + \left[\frac{2H_{h2}F_{R2}}{F_{GH2}F_{RH2}} - \frac{2H_{h_2}}{F_{GH_2}}\right]z_{20} + \frac{2H_{h2}F_{R2}}{F_{GH2}F_{RH2}}\Delta P_{C2}$$

19. $$\dot{z}_{19} = -\frac{F_{R2}}{F_{GH2}F_{RH2}R_2}z_{14} - \frac{1}{F_{GH2}}z_{19} + \left[\frac{1}{F_{GH2}} - \frac{F_{R_2}}{F_{GH2}F_{RH2}}\right]z_{20} - \frac{F_{R_1}}{F_{GH2}F_{RH2}}\Delta P_{C2}$$

20. $\dot{z}_{20} = -\frac{1}{F_{RH2}R_2}z_{14} - \frac{1}{F_{RH2}}z_{20} - \frac{1}{F_{RH2}}\Delta P_{C2}$
21. $\dot{z}_{21} = -\frac{1}{F_{CD2}}z_{21} + \frac{H_{g1}}{F_{CD2}}z_{22} - \frac{H_{g2}F_{CR2}}{F_{CD2}F_{F2}}z_{23}$
22. $\dot{z}_{22} = -\frac{1}{F_{F2}}z_{22} + \left[\frac{1}{F_{F2}} + \frac{F_{CR2}}{(F_{F2})^2}\right]z_{23}$
23. $\dot{z}_{23} = -\frac{N_2}{L_2R_2M_2}z_{14} - \frac{1}{L_2}z_{23} - \frac{1}{L_2}z_{24} - \frac{N_2}{L_2M_2}\Delta P_{C2}$
24. $\dot{z}_{24} = \left[\frac{N_2}{R_2M_2^2} - \frac{1}{R_2M_2}\right]z_{14} - \frac{1}{M_2}z_{24} - \left[\frac{1}{M_2} - \frac{N_2}{M_2^2}\right]\Delta P_{C2}$
25. $\dot{z}_{25} = \frac{-1}{F_{WD2}}z_{25}$
26. $\dot{z}_{26} = \frac{-1}{F_{st2}}z_{26}$
27. $\dot{z}_{27} = 2\pi T_{12}z_1 - 2\pi T_{12}z_{14}$
28. $\dot{z}_{28} = B_1z_1 + z_{27}$
29. $\dot{z}_{29} = B_2z_{14} + \alpha_{12}z_{27}$

By using all the above state equations, the matrix can be represented in the form of,

$$\dot{z} = Az + Bu$$

The dimensions of matrix $A = (29 \times 29)$.
The dimensions of matrix $B = (29 \times 2)$.
Non-zero elements of Matrix A: order of (29×29).

$a_{1,1} = \frac{-1}{F_{P1}}$	$a_{1,2} = \frac{H_{P_1}}{F_{P_1}}$	$a_{1,5} = \frac{K_{P_1}}{T_{P_1}}$
$a_{1,8} = \frac{H_{P_1}}{F_{P_1}}$	$a_{1,12} = \frac{H_{P_1}}{F_{P_1}}$	$a_{1,13} = \frac{H_{P_1}}{F_{P_1}}$
$a_{1,27} = -\frac{H_{P_1}}{F_{P_1}}$	$a_{2,2} = -\frac{1}{F_{r_1}}$	$a_{2,3} = \left[\frac{H_{t1}}{F_{r1}} - \frac{H_{r_1}H_{t_1}}{F_{t_1}}\right]$
$a_{2,4} = \frac{H_{r_1}H_{t_1}}{F_{t_1}}$	$a_{3,3} = -\frac{1}{F_{t1}}$	$a_{3,4} = \frac{1}{F_{t1}}$
$a_{4,1} = \frac{1}{F_{g1}R_1}$	$a_{4,4} = -\frac{1}{F_{g1}}$	$a_{5,1} = \frac{2H_{h1}F_{R1}}{H_{GH1}F_{RH1}R_1}$
$a_{5,5} = -\frac{2}{F_{W1}}$	$a_{5,6} = \left[\frac{2H_{h1}}{F_{W1}} + \frac{2H_{h_1}}{F_{GH_1}}\right]$	$a_{5,7} = \left[\frac{2H_{h1}F_{R1}}{F_{GH1}F_{RH1}} - \frac{2H_{h_1}}{F_{GH_1}}\right]$
$a_{6,1} = -\frac{F_{R1}}{F_{GH1}F_{RH1}R_1}$	$a_{6,6} = -\frac{1}{F_{GH_1}}$	$a_{6,7} = \left[\frac{1}{F_{GH1}} - \frac{F_{R_1}}{F_{GH_1}F_{RH1}}\right]$
$a_{7,1} = -\frac{1}{F_{RH1}R_1}$	$a_{7,7} = -\frac{1}{F_{RH1}}$	$a_{8,8} = -\frac{1}{F_{CD1}}$
$a_{8,9} = \frac{H_{g1}}{F_{CD1}}$	$a_{8,10} = -\frac{H_{g1}F_{CR1}}{F_{CD1}F_{F1}}$	$a_{9,9} = -\frac{1}{F_{F1}}$

(continued)

(continued)

$a_{9,10} = \left[\frac{1}{F_{F1}} + \frac{F_{CR1}}{(F_{F1})^2}\right]$	$a_{10,1} = -\frac{N_1}{L_1 R_1 M_1}$	$a_{10,10} = -\frac{1}{L_1}$
$a_{10,11} = -\frac{1}{L_1}$	$a_{11,1} = \left[\frac{N_1}{R_1 M_1^2} - \frac{1}{R_1 M_1}\right]$	$a_{11,11} = -\frac{1}{M_1}$
$a_{12,12} = -\frac{1}{F_{WD1}}$	$a_{13,13} = -\frac{1}{F_{st1}}$	$a_{14,14} = -\frac{1}{F_{P2}}$
$a_{14,15} = \frac{H_{P_2}}{F_{P_2}}$	$a_{14,18} = \frac{H_{P_2}}{F_{P_2}}$	$a_{14,21} = \frac{H_{P_2}}{F_{P_2}}$
$a_{14,25} = \frac{H_{P_2}}{F_{P_2}}$	$a_{14,26} = \frac{H_{P_2}}{F_{P_2}}$	$a_{1,27} = -\frac{H_{P_2}}{F_{P_2}}\alpha 12$
$a_{15,15} = -\frac{1}{F_{r_2}}$	$a_{15,16} = \left[\frac{H_{t2}}{F_{r2}} - \frac{H_{r_2} H_{t_2}}{F_{t_2}}\right]$	$a_{15,17} = \frac{H_{r_2} H_{t_2}}{F_{t_2}}$
$a_{16,16} = -\frac{1}{F_{t2}}$	$a_{16,17} = \frac{1}{F_{t2}}$	$a_{17,14} = \frac{1}{F_{g2} R_2}$
$a_{17,17} = -\frac{1}{F_{g2}}$	$a_{18,14} = \frac{2H_{h2} F_{R2}}{F_{GH2} F_{RH2} R_2}$	$a_{18,18} = -\frac{2}{F_{W2}}$
$a_{2,4} = \frac{K_{r_1} K_{t_1}}{T_{t_1}}$	$a_{3,3} = -\frac{1}{T_{t1}}$	$a_{3,4} = \frac{1}{T_{t1}}$
$a_{18,19} = \left[\frac{2H_{h2}}{F_{W2}} + \frac{2H_{h_2}}{F_{GH_2}}\right]$	$a_{18,20} = \left[\frac{2H_{h2} F_{R2}}{F_{GH2} F_{RH2}} - \frac{2H_{h_2}}{F_{GH_2}}\right]$	$a_{19,14} = -\frac{F_{R2}}{F_{GH2} F_{RH2} R_2}$
$a_{19,19} = -\frac{1}{F_{GH_2}}$	$a_{19,20} = \left[\frac{1}{F_{GH2}} - \frac{F_{R_2}}{F_{GH_2} F_{RH2}}\right]$	$a_{20,14} = -\frac{1}{F_{RH2} R_2}$
$a_{20,20} = -\frac{1}{F_{RH2}}$	$a_{21,21} = -\frac{1}{F_{CD2}}$	$a_{21,22} = \frac{H_{g2}}{F_{CD2}}$
$a_{21,23} = -\frac{H_{g2} F_{CR2}}{F_{CD2} F_{F2}}$	$a_{22,22} = -\frac{1}{F_{F2}}$	$a_{22,23} = \left[\frac{1}{F_{F2}} + \frac{F_{CR2}}{(F_{F2})^2}\right]$
$a_{23,14} = -\frac{N_2}{L_2 R_2 M_2}$	$a_{23,23} = -\frac{1}{L_2}$	$a_{23,24} = -\frac{1}{L_2}$
$a_{24,14} = \left[\frac{N_2}{R_2 M_2^2} - \frac{1}{R_2 M_2}\right]$	$a_{24,24} = -\frac{1}{M_2}$	$a_{25,25} = -\frac{1}{F_{WD2}}$
$a_{26,26} = \frac{-1}{F_{st2}}$	$a_{27,1} = 2\pi T_{12}$	$a_{27,14} = -2\pi T_{12}$
$a_{28,1} = B_1$	$a_{28,27} = 1$	$a_{29,14} = B_2$
$a_{29,27} = \alpha_{12}$		

Non-zero elements of Matrix B: order of (29×2).

$b_{4,1} = -\frac{1}{F_{g1}}$	$b_{5,1} = \frac{2H_{h1} F_{R1}}{H_{GH1} F_{RH1}}$	$b_{6,1} = -\frac{F_{R_1}}{F_{GH_1} F_{RH1}}$
$b_{7,1} = -\frac{1}{F_{RH1}}$	$b_{10,1} = -\frac{1}{F_{g1}}$	$b_{11,1} = -\left[\frac{1}{M_1} - \frac{N_1}{M_1^2}\right]$
$b_{16,2} = -\frac{1}{F_{g2}}$	$b_{17,2} = \frac{2H_{h2} F_{R2}}{F_{GH2} F_{RH2}}$	$b_{18,2} = -\frac{F_{R_2}}{F_{GH_2} F_{RH2}}$
$b_{19,2} = -\frac{1}{F_{RH2}}$	$b_{22,2} = -\frac{1}{F_{g2}}$	$b_{23,2} = -\left[\frac{1}{M_2} \quad \frac{N_2}{M_2^2}\right]$

$$C = [10000000000000000000000000000$$

$$000000000000001000000000000000]$$

$$D = [0\,0; 0\,0]$$

4 Design of Optimal Quadratic Regulator

The following is the performance index (J) for the supplied by:

$$J = \frac{1}{2}\int_0^{\infty} (z^T Qz + u^T \mathrm{Ru})\mathrm{d}t. \tag{5}$$

where Q is a real symmetric matrix that is positive semidefinite, and R is a positive definite and symmetric matrix [2]. And the significance of the incorrect amount of energy used is determined by Q and R.

Optimum control for complete state feedback, according to optimal control law, was given as,

$$u = -Kz.$$

By solving P, we get the optimum gain matrix from the matrix Riccati equation [2, 11].

$$A^T P + PA - PBR^{-1}B^T P + Q \tag{6}$$

$$K = \text{Optimal gain matrix} = R^{-1}B^T P. \tag{7}$$

5 Results

The appendix contains the various parameter data for the multiple area multiple source power system, and the simulation for the PID controller is depicted in Fig. 2. An AI method called the PSO approach has been used to adjust the PID controller automatically. Figures 3, 4, and 5 exhibit comparative waveforms of frequency and tie-line deviations using PI, PID along with LQR. Matrices Q and R were designed using the trial-and-error approach for performing the LQR technique. Except for diagonal elements, the elements of the matrix Q are all zeros.

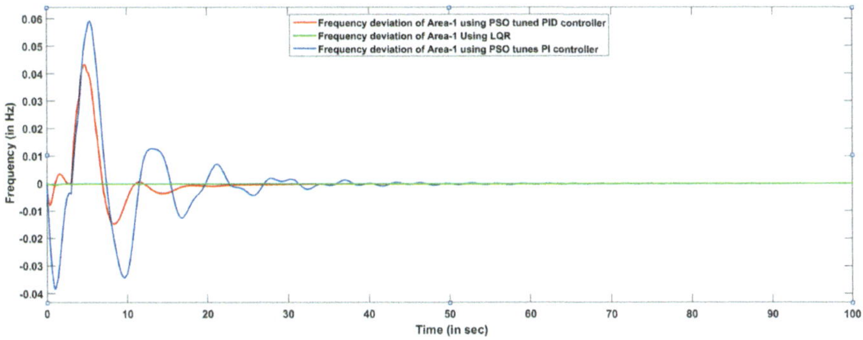

Fig. 3 Area-1 frequency deviations employing PI, PID, and LQR

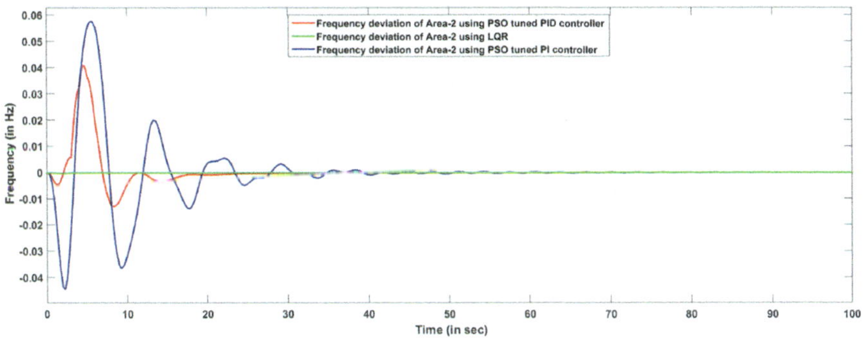

Fig. 4 Area-2 frequency deviations employing PI, PID, and LQR

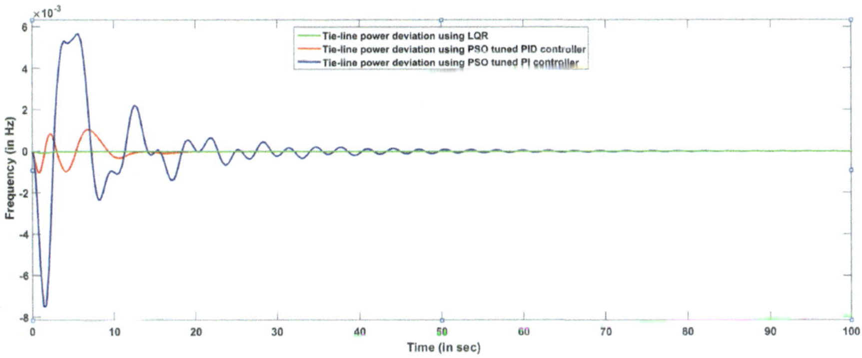

Fig. 5 Tie-line power deviations employing PI, PID, and LQR

The obtained gain values are, for 1% load change in area 1,

$$K_{p1} = 1.0,\ K_{i1} = 0.99,\ K_{d1} = 0.83,\ K_{p2} = 0.93,\ K_{i2} = 1.0,\ K_{d2} = 0.59$$

The non-zero diagonal elements of matrix Q are,

$q_{1,1} = 0.1$, $q_{2,2} = 0.1$, $q_{3,3} = 0.1$, $q_{4,4} = 0.1$, $q_{5,5} = 0.1$, $q_{6,6} = 0.1$, $q_{7,7} = 0.1$, $q_{8,8} = 0.1$, $q_{9,9} = 0.1$, $q_{10,10} = 0.1$, $q_{11,11} = 0.1$, $q_{12,12} = 10$, $q_{13,13} = 1$, $q_{14,14} = 1$, $q_{15,15} = 1$, $q_{16,16} = 1$, $q_{17,17} = 1$, $q_{18,18} = 1$, $q_{19,19} = 1$, $q_{20,20} = 1$, $q_{21,21} = 1$, $q_{22,22} = 1$, $q_{23,23} = 1$, $q_{24,24} = 1$, $q_{25,25} = 1$, $q_{26,26} = 1$, $q_{27,27} = 10$, $q_{28,28} = 10$, $q_{29,29} = 1$

$$R = [0.0001\ 0;\ 0\ 0.0001]$$

The settling time of the PI controller is 70 s, the PID is 15 s, and the LQR is 0.0010 s, based on the aforementioned waveforms.

6 Conclusion

In this paper, a power system with multiple sources within every area is evaluated. Every region has traditional hydro, thermal, and gas sources, as well as renewable sources such as wind and solar plants for generation. The frequency and power in Hz are taken into account while modelling the power system. The frequency and tie-line variations for the load injection in area-1 were observed through incorporating various controllers such as PI and PID. Better performance was achieved by using state-space modelling an optimal regulator like LQR is designed. And the results were compared. The PID controller showed better settling time than the PI controller, and LQR showed the better results than the PID controller. For further research, different controllers can be designed to get better performance for multi-area system.

Appendix

Data [3] for the simulation diagram showed in Fig. 2:

$$P_{dj} = 2000\,\text{MW},\ H_j = 5.0\,\text{MW/MVA},\ \text{Fr}_e = 50\,\text{Hz},$$

$$
\begin{aligned}
B_j &= 0.425(\text{p.u})\text{MW/Hz}, T_{i,j} = 0.0433, \\
D_j &= 8.33 \times 10^{-3}(\text{p.u})\text{MW/Hz}, \alpha_{ij} = -1,\ H_{Pj} = 120, \\
F_{Pj} &= 20, H_{tj} = 0.60, H_{Pj} = 0.20, H_{gj} = 0.10, \\
H_{wj} &= 0.10, H_{soj} = 1.8\,\text{s}, F_{gj} = 0.08\,\text{s}, F_{gj} = 0.30\,\text{s}, \\
F_{rj} &= 10.0\,\text{s}, H_{rj} = 0.3\,\text{s}, F_{RHj} = 41.6\,\text{s}, F_{Rj} = 5\,\text{s}, \\
F_{GHj} &= 0.513\,\text{s}, F_{Wj} = 1.0\,\text{s}, F_{WDj} = 5.0\,\text{s}, F_{soj} = 1.8\,\text{s}, \\
N_j &= 0.60\,\text{s}, M_i = 1.0\,\text{s}, L_j = 0.05\,\text{s}, I_j = 1.0\,\text{s}, \\
F_{Fj} &= 0.23\,\text{s}, F_{CRj} = 0.01\,\text{s}, T_{CDj} = 0.20\,\text{s}, \\
R_J &= 2.4\,\text{Hz/MW(p.u)}, J_j = 1
\end{aligned}
$$

References

1. Kothari, P., Nagrath, I.J.: Modern Power System Analysis. TMH (1980)
2. Nayak, D.B., Rayalu, M.K.: Automatic generation control of two area thermal power system with PI controllers using state space approach. SSRG Int. J. Electr. Electron. Eng. **7**(6), 17–21 (2020)
3. Hakimuddin, N., Nasiruddin, I., Bhatti, T.S., Arya, Y.: Optimal automatic generation control with hydro, thermal, gas, and wind power plants in 2-area interconnected power system. Electr. Power Compon. Syst. (2020)
4. Nizamuddin, I., Bhatti, T.S.: AGC of two area power system interconnected by AC/DC links with diverse sources in each area. Int. J. Electr. Power Energy Syst. **55**, 297–304 (2014)
5. Charitha Reddy, Y.V.L., Krishnarayalu, M.S.: AGC of multi source multi area deregulated power system using PID controller. J. Electr. Eng. (JEE). (Article 18.1.40). www.jee.ro
6. Arya, Y., Kumar, N.: Optimal control strategy-based AGC of electrical power systems. A comparative performance analysis. Optim. Control Appl. Methods **38**(6), 982–992 (2017)
7. Talukdar, B.K., et al.: Design of state space model and optimal controller for automatic generation control of two area thermal power system. Int. J. Adv. Res. Innov. Ideas Edu. (IJARIIE). **4**(3) 2018
8. Arya, Y.: AGC performance enrichment of multi-source hydrothermal gas power systems using new optimized FOFPID controller and redox flow batteries. Energy **127**, 704–715 (2017)
9. Kothari, M.L., et al.: Application of optimal control strategy to automatic generation control of a hydrothermal system. In: IEE Proceedings, vol. 135, Pt. D, no. 4 (1998)
10. Ibraheem, et al.: AGC of two area interconnected power system with diverse sources in each area. J. Electr. Eng. (JEE) (2013)
11. Ogata, K.: Modern Control Engineering 5/e. PHI (2010)

A Neural Network Model to Estimate Parameters of DBSCAN for Flood Image Segmentation

Rohitash Verma and Satyasai Jagannath Nanda

Abstract Density-based spatial clustering applications with noise (DBSCAN) is a very popular technique proposed by Ester et al. in 1996. It does not require apriori information about the number of clusters present in the dataset, and the number of clusters automatically gets evolved based on the density of points present in the dataset. This automatic clustering along with detection of undesirable samples (noise points) present in the dataset is determined by DBSCAN with two parameters: minimum number of points and radius (ϵ). Generally, researchers set these parameters individually for any of real-life application. In this manuscript, an attempt is made to use a neural network to estimate these two parameters of DBSCAN for segmentation of specific category of flood images. The idea behind this is similar kind of flood images (where river and ocean occupy more blue regions due to water) may posses certain common property with respect to pixels representing water density. Twenty flood images from various parts of the world have been collected form Landsat8 satellite data. Though this sample size of twenty is small, but primary study carried out here reveals that the estimated parameters with neural network many times do match with the true parameters of DBSCAN for flood image segmentation.

Keywords DBSCAN clustering · Parameter estimation · Neural network · Flood images segmentation

1 Introduction

After 25 years of its existence, the DBSCAN algorithm [1] proposed by Ester et al. is still a preferred choice for the data science researchers (supported by more

R. Verma · S. J. Nanda (✉)
Department of Electronics and Communication Engineering, Malaviya National Institute of Technology, Jaipur, Rajasthan 302017, India
e-mail: sjnanda.ece@mnit.ac.in

R. Verma
e-mail: 2018pec5174@mnit.ac.in

K. N. Das et al. (eds.), *Proceedings of the International Conference on Computational Intelligence and Sustainable Technologies*, Algorithms for Intelligent Systems,
https://doi.org/10.1007/978-981-16-6893-7_42

than 21,000 citations in the Google scholar and increasing over the years). The algorithm has ability to detect arbitrary shape clusters and also capable in identifying the undesirable samples (noise points) present in any real-life dataset. With the progress of time, several researchers have reported many variants of DBSCAN algorithms which include : incremental DBSCAN by Ester et al. [2] for incremental addition and deletion of clusters when there is a large database which gets updated periodically, ST-DBSCAN by Birant and Kut [3] for clustering spatial-temporal datasets, VDBSCAN by Liu et al. [4] which deals with varied density of samples present in the dataset, HDBSCAN by Campello et al. [5] for hierarchical estimation of the different level of density present in the dataset, computationally efficient DBSCAN by Nanda and Panda [6] to reduce the computational complexity of DBSCAN by introducing new small clusters merging conditions. A variable ϵ in DBSCAN is introduced by Vijay and Nanda [7] to vary the radius depend upon the density of the clusters present in the dataset. These variants and the original algorithm have been extensively applied on several real-life applications [8], and observing the popularity of the group by Ester et al. in 2017 [9] makes a summary and highlights the significance with recently popular article,"DBSCAN revisited, revisited: why and how you should (still) use DBSCAN."

Color image segmentation is a problem of partitioning pixels into different groups (clusters) based on their color similarity and spatial relation. The DBSCAN algorithm is initially used for color image segmentation in 2003 by Ye et al. [10]. Shen et al. [11] employed the DBSCAN clustering for real-time superpixel image segmentation with 50 frames per second. Loke et al. [12] recently introduced a parallelized DBSCAN algorithm for accelerated superpixel image segmentation. The DBSCAN-based image segmentation has been employed in solving several real-life application includes : smoke detection [13], parking-slot detection [14], LiDAR images segmentation [15], brain MR images analysis [16], carotid artery ultrasound images analysis [17].

One of the challenging tasks in DBSCAN-based color image segmentation is to set the parameter of the algorithm which is : minimum number of points and radius (ϵ). These two parameters are set by the programmer on hit and trail basis which suit best for particular type of image. In this paper, an attempt is made to estimate these two DBSCAN parameters for segmentation of a specific category of flood images using a neural network. The concept is that flood images (in which lakes, rivers, oceans, other water containing regions occupy more blue zones) may share some basic characteristics in terms of pixels signifying water density.

The paper organization is as follows. Section 2 describes the application of DBSCAN clustering algorithm for flood image segmentation. The proposed neural network-based model to estimate the parameters of DBSCAN clustering algorithm is presented in Sect. 3. The obtained results and discussions are highlighted in Sect. 4. The concluding remarks are given in Sect. 5.

2 DBSCAN Clustering Algorithm for Flood Image Segmentation

The flood images segmentation which is a type of color image segmentation [10] is dealt here as a clustering problem. The pixels of flood image are segregated into several clusters with DBSCAN algorithm using the procedure narrated in Algorithm 1.

Algorithm 1: DBSCAN clustering for flood image segmentation

```
Input: Flood image for analysis, minimum points associated with cluster,
cluster radius (ϵ)
Result: Desired segmented flood image, cluster accuracy
Set minimum points associated with cluster, cluster radius (ϵ);
while ϵ>0 do
    if pixel p in flood image is visited then
        move to next pixel p;
    else
        mark pixel p as visited;
        if number of pixels within ϵ<minpts then
            mark those pixel as noise;
        else
            make new cluster;
        end
    end
    connect all cluster;
end
```

3 Proposed Neural Network Model to Estimate Parameters of DBSCAN-based Segmentation

The block diagram of proposed neural network model to estimate parameters of DBSCAN algorithm is shown in Fig. 1. Step-by-step procedure for modeling is nar rated below:

Step 1 : Input flood image : Twenty flood images have been collected from NASA Landset8 satellite image [18] used for analysis. These images taken are included in Table 1.

Step 2: True parameter estimation with DBSCAN clustering; The DBSCAN algorithm is allowed to perform segmentation on each flood images. The two parameters of DBSCAN algorithm: Minimum points and radius (ϵ) are set manually and corresponding clustering accuracy is noted. Both the parameters are manually adjusted in such manner that more than 90% accuracy is recorded for each flood image. At this point, the set value of minimum points and radius (ϵ) represents the true value of DBSCAN algorithm for each flood image.

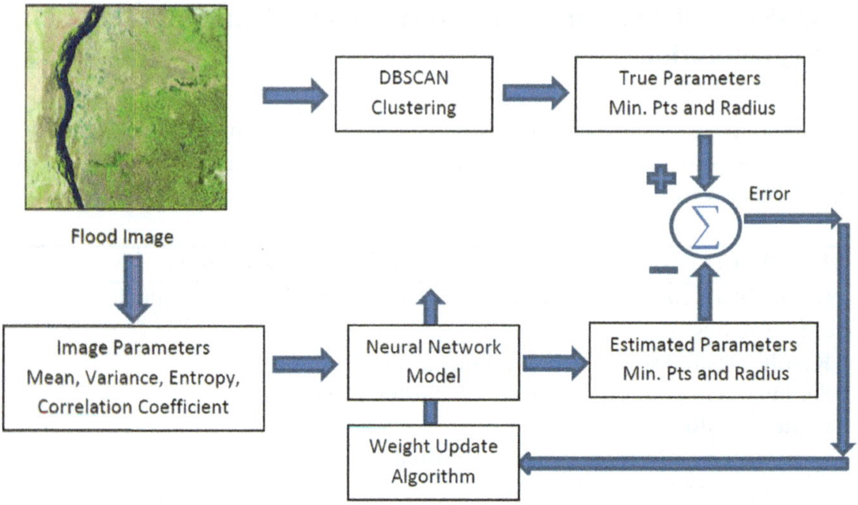

Fig. 1 Block diagram of proposed neural network model used for parameter estimation of DBSCAN algorithm for flood image segmentation

Step 3: Computing statistical parameters of flood image: The statistical parameter like mean, variance, entropy, and correlation coefficient are calculated for each input flood image.

Step 4: Neural network modeling: A two-layer neural network is constructed using MATLAB neural network toolbox [19], and the detail structure is shown in Fig. 2. The transfer function used in Layer 1 (hidden layer) is Tangent Sigmoid (TANSIG), and in Layer 2 (output layer), it is Pure Linear (PURELIN). The number of neurons set in hidden and output layers is one thousand and two, respectively. The input to the neural network is the four statistical parameters obtained for each flood image. The output is estimated value of minimum points and radius (ϵ). These two values are subtracted from the true parameter values of DBSCAN obtained in Step 2, and error values are calculated. The error values are used in backpropagation-based learning algorithm to update the weights of neural network model.

Step 5: Iteration and stopping criteria: The process of Steps 1 to 4 is carried out for a number of iteration. In the first iteration, all the twenty flood images are applied and the weight of neural network is trained. In the next iteration, the same images are applied again and the weight of neural network is further allowed to fine-tune. The process continues till the error values are significantly reduced.

Step 6: Testing of neural network model: After the training is completed, testing is performed using the same twenty flood images. The obtained estimated values of minimum points and radius (ϵ) are recorded and reported in Table 1.

Table 1 True parameter settings of DBSCAN and estimated parameters obtained after training the neural network model

Sl.	Flood image	True parameters (DBSCAN)		DBSCAN	Est. parameters (NN)	
		Min.pts	Radius	Accuracy (%)	Min.pts	Radius
1	Sudan	4	0.66	97.27	3	0.672
2	Morganza	4	0.5	96.48	3	0.5813
3	South dakota	4	0.6	95	3	0.5813
4	Missisipi River	4	0.8	98.44	4	0.789
5	Pakistan	4	0.7	91.80	3	0.692
6	Dakota	4	0.55	95.31	4	0.532
7	Southeast India	3	0.8	99.21	3	0.793
8	Colombia	3	0.46	100	3	0.489
9	United kingdom	3	0.765	100	3	0.774
10	South Asia	3	0.796	99.60	2	0.803
11	Kerla	4	0.4	92.02	3	0.382
12	Burma	4	0.7	93.75	3	0.712
13	Bihar	4	0.5	99.61	4	0.495
14	Carolina	3	0.46	100	3	0.463
15	St. John River	4	0.62	97.24	3	0.631
16	Arkansas River	4	0.556	98.05	4	0.545
17	Missouri Basin	3	0.45	98.05	3	0.433
18	Eastern Nebraska	3	0.45	98.83	3	0.436
19	Danube River	3	1.2	90.23	3	1.134
20	Indian Ocean	4	0.69	93.75	3	0.631

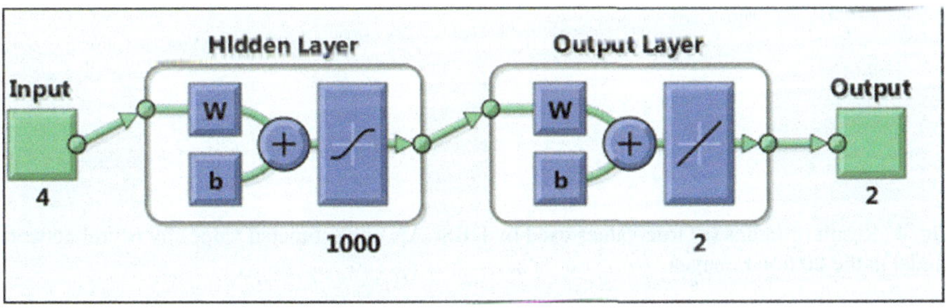

Fig. 2 Detail structure of two-layer neural network model used with MATLAB NN toolbox for parameter estimation

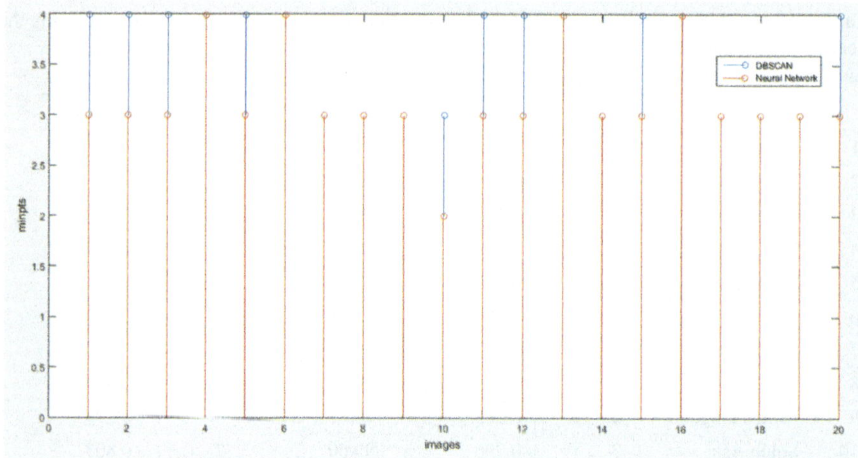

Fig. 3 Result of minimum number of points true values used by DBSCAN and estimated values by neural network model in the 20 flood images

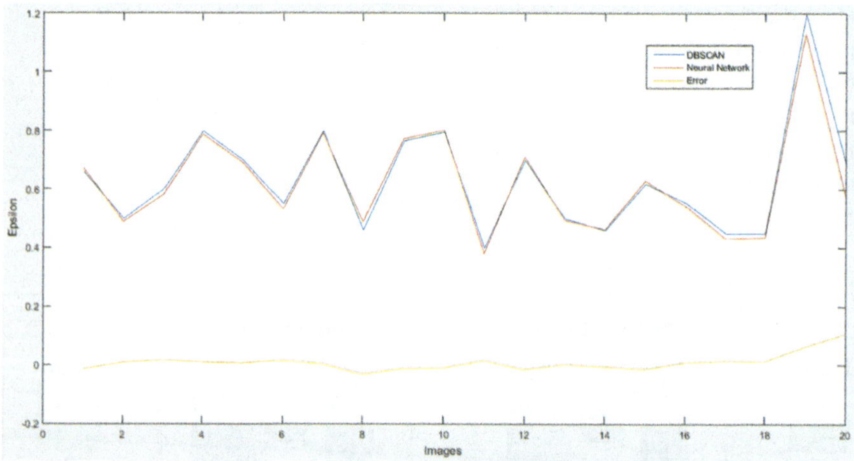

Fig. 4 Result of radius (ϵ) true values used by DBSCAN and estimated values by neural network model in the 20 flood images

4 Result and Discussions

The simulation of the proposed model is carried out in MATLAB environment on an DELL laptop with Intel core i7 processor, 3.5 GHZ clocked CPU with 16GB of inbuilt RAM in Windows 10, 64-bit environment. The twenty flood images are applied for analysis, and the true parameters with accuracy obtained using Step 2 are reported in Table 1. The estimated parameters after testing of proposed neural

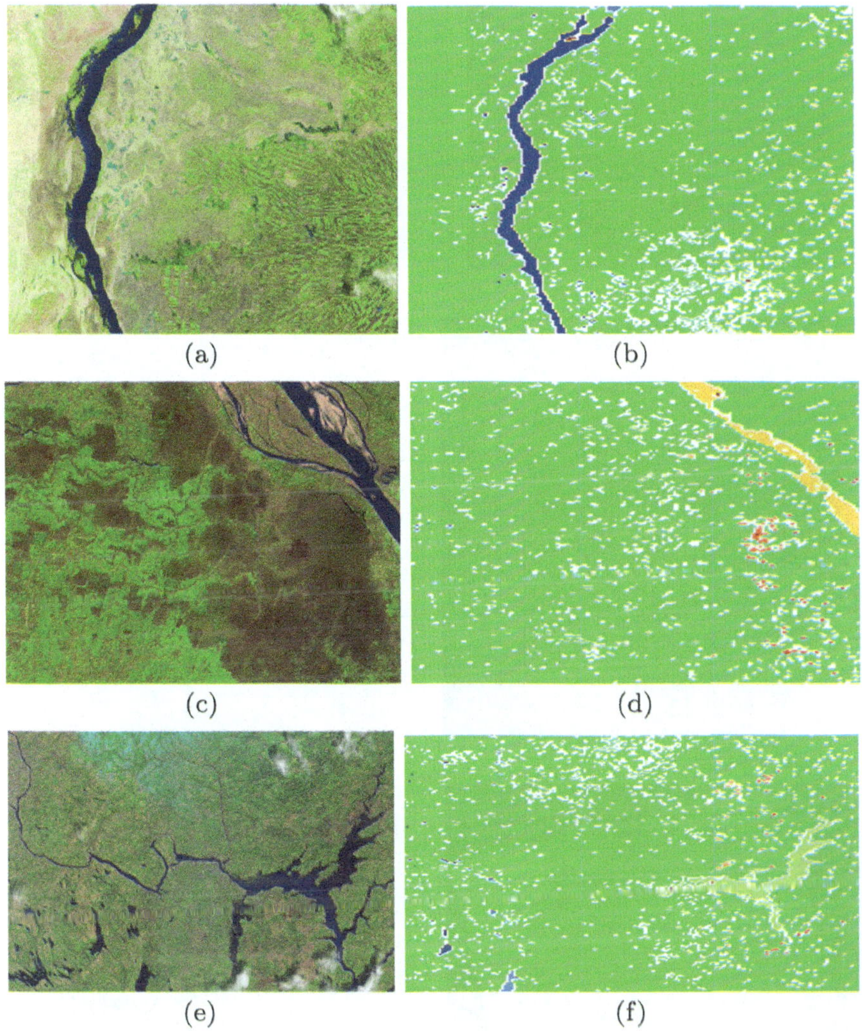

Fig. 5 True satellite captured image and segmented image with DBSCAN clustering for floods occurring in rivers: Image 1: Sudan **a** True, **b** Segmented; Image 13: Bihar **c** True, **d** Segmented; Image 15: St. John River **e** True, **f** Segmented

network model are also included in Table 1. It is observed that for minimum number of points out of twenty observations, in eleven cases, the obtained prediction is correct in Table 1. It is also supported by resemblance of overlapping blue and red point curves in Fig. 3. In rest nine cases, there is single unit difference in minimum number of points. The other parameter radius (ϵ) value obtained with the neural network model closely matches with the true DBSCAN value reflected in Table 1. This is also supported by Fig. 4 in which there is close matching between blue (DBSCAN) and

Fig. 6 True satellite captured image and segmented image with DBSCAN clustering for floods occurring near oceans: Image 7 : South East India **a** True, **b** Segmented; Image 9 : United Kingdom **c** True, **d** Segmented; Image 20 : Indian Ocean **e** True, **f** Segmented

red (neural network) curves. In this figure, it is also observed that the difference in error reflected by deep yellow line rattles around almost zero value. This justifies the close prediction capability of the proposed neural network model.

Three examples of satellite captured true images for floods occurring in rivers at Sudan (Image 1), Bihar (Image 13), and St. John river (Image 15) are shown in Figs. 5a, c, and e, respectively. Corresponding obtained segmented images with DBSCAN clustering techniques are included in Figs. 5b, d, and f. It can be observed that the identified riverbanks have been effectively highlighted with different color combination with respect to a green background.

Similarly, three case studies of satellite images obtained for floods occurring near oceans are included in Fig. 6. These cases' true images are shown in Fig. 6a for Southeast India, Fig. 6c for United Kingdom, Fig. 6e for Indian Ocean. The resulted segmented images with DBSCAN clustering techniques are presented in Figs. 6b, d, and f. The ocean banks with flood waters are printed out with other color combination compared to the green background.

5 Conclusions

In this paper, popular density-based algorithm DBSCAN is employed to segment the twenty flood images collected form NASA Landsat8. These images' statistical parameters like mean, variance, entropy, and correlation coefficient are computed. Considering these parameters as input, a neural network-based model is trained to estimate two parameters of DBSCAN clustering algorithm (minimum number of points and radius). Simulation study reveals in eleven cases that the obtained prediction is correct for minimum number of points. In case of radius estimation, there is close matching of response between DBSCAN and neural network output. This signifies that the proposed neural network-based model is effective for estimating the parameters. Here, it is a primary study, and thus, the sample size for analysis is kept very small only for twenty images. In upcoming days, the researchers are going to do more rigorous analysis by increasing large sample size and use coherent images for accurate estimation of parameters of a specific category of images.

References

1. Ester, M., Kriegel, H.-P., Sander, J., Xiaowei, X., et al.: A density-based algorithm for discovering clusters in large spatial databases with noise. Kdd **96**, 226–231 (1996)
2. Ester, M., Wittmann, R.: Incremental generalization for mining in a data warehousing environment. In: International Conference on Extending Database Technology, pp. 135–149. Springer (1998)
3. Birant, D., Kut, A.: St-dbscan: an algorithm for clustering spatial-temporal data. Data Knowl. Eng. **60**(1), 208–221 (2007)

4. Liu, P., Zhou, D., Wu, N.: Vdbscan: varied density based spatial clustering of applications with noise. In: 2007 International Conference on Service Systems and Service Management, pp. 1–4. IEEE (2007)
5. Campello, R.J.G.B., Moulavi, D., Zimek, A., Sander, J.: Hierarchical density estimates for data clustering, visualization, and outlier detection. ACM Trans. Knowl. Disc. Data (TKDD) **10**(1), 1–51 (2015)
6. Nanda, S.J., Panda, G.: Design of computationally efficient density-based clustering algorithms. Data Knowl. Eng. **95**, 23–38 (2015)
7. Vijay, R.K., Nanda, S.J.: A variable ϵ-dbscan algorithm for declustering earthquake catalogs. In: Soft Computing for Problem Solving, pp. 639–651. Springer (2019)
8. Vijay, R.K., Nanda, J.: Seismicity analysis using space-time density peak clustering method. Patt. Anal. Appl. **24**(1), 181–201 (2021)
9. Schubert, E., Sander, J., Ester, M., Kriegel, H.P., Xu, X.: Dbscan revisited, revisited: why and how you should (still) use dbscan. ACM Trans. Database Syst. (TODS) **42**(3), 1–21 (2017)
10. Ye, Q., Gao, W., Zeng, W.: Color image segmentation using density-based clustering. In: 2003 International Conference on Multimedia and Expo. ICME'03. Proceedings (Cat. No. 03TH8698), vol. 2, pp. II–401. IEEE (2003)
11. Shen, J., Hao, X., Liang, Z., Yu., Liu, W.W., Shao, L.: Real-time superpixel segmentation by dbscan clustering algorithm. IEEE Trans. Image Proc. **25**(12), 5933–5942 (2016)
12. Loke, S.C., MacDonald, B.A., Parsons, M., Wünsche, B.C.: Accelerated superpixel image segmentation with a parallelized dbscan algorithm. J. Real-Time Image Proc. 1–16
13. Sheng, D., Deng, J., Xiang, J.: Automatic smoke detection based on slic-dbscan enhanced convolutional neural network. IEEE Access **9**, 63933–63942 (2021)
14. Lee, S., Hyeon, D., Park, G., Baek, I., Kim, S.-W., Seo, S.-W.: Directional-dbscan: parking-slot detection using a clustering method in around-view monitoring system. In: 2016 IEEE Intelligent Vehicles Symposium (IV), pp. 349–354. IEEE (2016)
15. Wang, C., Ji, M., Wang, J., Wen, W., Li, T., Sun, Y.: An improved dbscan method for lidar data segmentation with automatic eps estimation. Sensors **19**(1), 172 (2019)
16. Baselice, F., Coppolino, L., D'Antonio, S., Ferraioli, G., Sgaglione, L.: A dbscan based approach for jointly segment and classify brain mr images. In: 37th Annual International Conference of the IEEE Engineering in Medicine and Biology Society (EMBC), pp. 2993–2996. IEEE (2015)
17. Latha, S., Samiappan, D., Muthu, P., Kumar, R.: Fully automated integrated segmentation of carotid artery ultrasound images using dbscan and affinity propagation. J. Med. Biolog. Eng. **41**(2), 260–271 (2021)
18. Nasa landset 8 images. https://landsat.visibleearth.nasa.gov/. Accessed: 30 May 2021
19. Matlab neural network toolbox. https://ww2.mathworks.cn/help/deeplearning/ref/nntool.html?requestedDomain=cn. Accessed: 30 May 2021

Quality Assessment of Public Transport: A Review

Saikat Deb and Debasish Das

Abstract Urban growth and population expansion have resulted in a rapid increase of motor vehicles, especially private vehicles across the city premises. To control problems like pollution and congestion etc., it is vital to reduce the number of private vehicles. The promotion of the use of public transit can help to achieve this. But, in almost all the cities of India the quality of the public transport is not up to the expectation levels of the users which leads to the use of private vehicles. Many researchers argued that public transport can potentially reduce the use of private vehicles if it is well maintained. Improvement of the service quality of public transport is therefore important. But the challenging step in the improvement process is to assess the quality and identification of parameters to be considered. Therefore, in this paper, an attempt has been made to analyze the previous literatures associated with quality assessment of the public transport is discussed and some conclusion is drawn.

Keywords Public transport · Subjective measures · Objective measures · Quality assessment · Level of Service

1 Introduction

Urban development along with populace extension in the ongoing years over the world is bringing about an expanded interest for city transportation. This increased demand in city or urban transportation is always not meet by public transportation, which caused an expanded utilization of the private vehicles across the urban area resulting congestion. The marginal quality of the public transport has additionally quickened the development of private vehicles [1]. It is therefore, important to monitor and measure the service quality of the existing public transport system. The

S. Deb (✉)
Civil Engineering Programme, Assam Down Town University, Guwahati, Assam, India

D. Das
Department of Civil Engineering, JIS College of Engineering, Kalyani, West Bengal, India

K. N. Das et al. (eds.), *Proceedings of the International Conference on Computational Intelligence and Sustainable Technologies*, Algorithms for Intelligent Systems,
https://doi.org/10.1007/978-981-16-6893-7_43

estimation of the performance of public transport is an imperative worry for designating assets among various transit classes. Performance measurement is essential to measure the performance of the transit service, to identify the management problems regarding costs of the service, and to improve the service as per the community expectations [2]. Performance measurement should be carried out over time and space [3]. Measurement of the transit performance can be a qualitative or quantitative approach depending on the estimation of a particular characteristic of a transit service [4]. There are numbers of methods which are used to define and assess service quality. Every measure related to the estimation of transit performance has its own set of indicators. In this manner, it is important to identify most suitable indicators to define transit performance and different methods associated with the estimation of performance of these indicators. The measurement of the performance of a transit depends on perception of the researchers [4]. The methods for measuring the transit performance can broadly be divided in two parts—objective measures and subjective measures. The previous research works in subjective and objective analysis and their drawbacks are discussed in this chapter in consecutive subsections.

2 Objective Measures of Transit Quality

The quality of the service can be measured by a series of straight forward quantified calculations that can be used to measure the ability of the transit infrastructure to provide consistency that meets the needs of riders [4]. These estimates are quantitative estimates communicated as a range of values, which does not give any details as to how "great" or "terrible" particular service is and must therefore be matched to a standard or past value [5]. This particular type of estimation is called objective service quality. Examples of the objective indicators are a level of service (LOS) indicators. These are widely recognized as standard indicators from the point of view of the transit authority. [5]. Some of the previous literatures describing the objective methods are discussed below.

2.1 Review of Previous Literatures

The concept of LOS was first coined in the Highway Capacity Manual (HCM) in 1965 [6]. The early concept of the LOS had a very limited concept of users' perception rather it stretched on the importance of the drivers' behavior in association with the better LOS [6]. The later versions of the HCM incorporated the idea of users' perceptions in determining the LOS. In the initial versions of the HCM, LOS was being used mainly to define the conditions of the highways. However, the concept is used to measure the quality of terminals in the airports [7], public transportation [4], pedestrian sidewalks [8] etc.

National Committee on Urban Transportation (NCUT), Chicago was the first to propose some general standards for loadings, headway, routing, stop frequency, speed and regularity of the service etc. as a measure of service quality of the transit facility in the year 1958 [9].

San Antonio Transit system also developed their own criteria to measure the transit service quality. In addition to NCUT guidelines, they used residential density as the basis for extending the service.

Botzow [10] first defined LOS for the public transit on the basis of some indicators on comfort and journey time. He identified 6 LOS scale ranging from LOS A to F defining best to worst service levels. The overall service level in this case was estimated by weightage average method. Time related variables like speed and delay were given 40% weightage whereas comfort related variables like passenger density, vehicle acceleration, jerk, temperature, noise level and ventilation was given 60% weightage. This attempt of the Botzow [10] was similar to the definition of LOS as prescribed in the Transit Cooperative Research Program (TCRP) report [4].

Taylor and Brogan [9] developed guidelines for a fixed route and fixed scheduled bus service which includes service area density, LOS for the service area and some system characteristics like load factor, headways, bus stop spacing etc. They suggested that if any of the indicators attain the worst service level, the entire transit system is assigned as the worst performing system. They argued that if any one of the indicators performs poorly then the entire trip of the customers become intolerable.

The Transportation Research Board (TRB) also specified some guidelines for the assessment of LOS for public transport [4]. Six performance indicators were mentioned in this report. Based on the opinion of some experts, quality of these indicators was characterized by six categories of LOS, ranging from LOS A to LOS F [11, 12].

Bertini and El-Geneidy [13] recommended quality estimation technique for the bus service to help the bus operators measure the performance of bus service and improve the performance accordingly. They suggested some indicators related to accessibility, mobility and economic development.

MoUD [14], GOI likewise started some performance estimation procedure for the public transportation based on some indicators. Service levels of these indicators were categorized under four LOS classes in particular LOS 1 to LOS 4 characterizing best to worst performance level.

Badami and Haider [1] also demonstrated some quantitative measures built on bus transit provision, utilization 7 of 5 the service, performance of the service in terms of financial angle and bus and labor productivity and fuel efficiency for the urban bus transit operation for different Indian cities. They found that operating costs and traffic revenues per passenger-kilometer in the small and medium cities are higher than the metropolitan cities. They also found that bus service faces loss due to increase in operating costs and declining productivity. The losses occurred because of the declining ridership.

2.2 *Drawback of Objective Measures and Importance of Users' Perceptions*

Objective measures are directly associated with different features of transportation system like frequency, service hours, load factor, the transit network coverage etc. The performance of these indicators is settled based on past performance without allowing the customer expectations [15]. Each of these quantitative measures are based on a variety of expert judgments or some earlier standard values [5]. As a result, they are unreliable in showing the overall performance of the transit service. Many researchers [16–19] concluded that the customer's interpretation was the most important for estimating efficiency. Berry et al. [20] found that "users can only judge the quality of service". Quality of service is strongly linked to passenger satisfaction, which reflects user perception [21]. In this connection, the LOS standards must be re-established using users' perception.

3 Subjective Measures of Service Quality

3.1 *User Perception as a Measure of Service Quality*

The estimation of transit quality from the view point of the users can be termed as subjective measures of the transit quality. Subjective measures are carried out using customer satisfaction surveys to help transit regulators identifying factors seemed to be important for the customers [22–24]. A customer satisfaction survey could be used to identify future operational efficiency enhancement initiatives, to measure the extent that past initiatives have succeeded and over time to monitor changes in service quality [4]. Passengers express their judgments through the customer satisfaction survey regarding some attributes which describe the services on a predetermined assessment scale [5]. In order to quantify the responses, users are asked to give scores according to their experience. The quantification of the passengers' responses gives a better understanding of their level of satisfaction. These sets of quantified customer satisfaction data are analyzed with statistical tools to know priorities and some factors which affect the satisfaction of the users. Passengers' perception could be used as a relative measure of the performance of the urban transport. Friman [26] analyzed the effects of quality improvements and perceived service characteristics in Swedish transit services. Passengers were asked to evaluate the transit service based on their perception on nine-point scale. The study indicated that the satisfaction of the customers was influenced by the quality improvements, only to a limited extent. Based on the study, it was suggested that the users of the bus service should be properly informed about quality improvement measures. When the users would be better informed about the service improvements, their expectations become more realistic. Moreover, it has been found that, the demand of the public transport is also affected by the user's perception. For example, Paulley et al. [28] studied effect of fares, quality of service, income and car ownership on the demand for public transport in Great Britain. They found that all these variables affected the demand. Income and car ownership growth were found to be fundamental in assessing demand for public transport. With increasing income, car ownership was found to be increased which would reduce the demand for public transport. In some of the studies it has been found that, improvement in frequency, accessibility of the transport system is required to increase the user base. For example, Tyrinopoulos and Antoniou [30] used factor analysis to find out the factors affecting passenger satisfaction on the transit service of Athens. They further used ordered logit model to find out the importance of different attributes of the transit service. They found that the primary objective of the policymakers in Athens should be to enhance transit service frequency, accessibility and a well-ordered transport system. Ratanavaraha and Jomnokwao [37] used CFA to model users' expectations of the drivers of sightseeing buses in Thailand to improve the traveler safety. The results show that the driver's age, experience, driving licence, driving skills, training and cigarette/drinking habits affect the users' demands in rural areas. Joewono et al. [38] used SEM to find out the policies required to improve the public transport system of three major towns of Indonesia as per users' perceptions. They indicated that the users' preferences were affected by service importance, negative experience and dissatisfaction.

3.2 *Effect of Users' Perception on Loyalty of Using Public Transport*

Lin et al. [29] estimated correlation between behavioral intention and quality of the urban buses on Taiwan based on loss aversion using SEM. The results indicated that repurchase intention was affected by the loss of the service quality more than the gain in service quality. The authors postulated that a key element in the loyalty of old passengers was the value off the service offered. They suggested that the authority should concentrate on enhancing current operation and services of the urban buses to check the declining ridership. In order to detect the factors affecting 'passengers' repurchase intentions on urban busses in Taipei metropolitan area in Taiwan Jen and Hu used a confirmatory factor analysis and path analysis [25]. 26 bus service-related attributes were used as indicators reflecting various dimensions of bus service which were categorized under four factors. They found that the factors—interaction with passengers, tangible service equipment, the convenience of service and operating management support had affected the customer. Furthermore, this study indicated that passengers repurchase intention in city bus services were influenced by perceived value and the attractiveness of alternatives. More specifically, if the perceived value of the service was good then passenger desired to repurchase the service which would ultimately increase the bus ridership. Wen et al. [27] used SEM to estimate the passenger loyalty towards the intercity bus service of Taiwan. They found that service quality, service value, satisfaction, sacrifice, attractiveness of competitors, switching costs and trust are some important factors which affected users' loyalty towards public transport services. Satisfaction was found to be a most significant parameter which influenced passenger loyalty. They found that cleanliness and noise levels of the vehicle significantly affected users' perception.

3.3 Effect of Users' Perception on Global Satisfaction Index

Many researchers used different statistical tool like, SEM, ordered logit and probit model, regression analysis to estimate the effect of users' perception on global satisfaction index. These types of analysis help to estimate the weightage of each parameter. In order to comprehensively evaluate the basic features of the perceived service offered to users in Belgrade Filipovic et al. [31] used the importance and satisfaction surveys. The results of the study exhibited that reliability and comfort were the two important factors for the users. Moreover, they found that satisfactions of the users are largely effected by the environment conditions. During the summer period, the satisfaction of the users was found to be lesser than the other periods. Dell' olio et al. [32] studied the variation of the passengers' perception of quality with the information available to them for the bus service. They used ordered logit model to analyze the qualitative variables collected from both the on-board buses and at bus stops. Rojo et al. [33] used ordered logit and probit model to model the variation in the perception of overall service quality between male and female passengers. The results indicated that female passengers made more frequent and shorter trips than the male passengers. The most relevant parameters for their model were found to be a relationship between price and quality, the bus condition and the service frequency. The role of price equity in transit service was examined by Eriksson and Bjørnskau [34]. They found that fairness in the price of transit increases Sweden's acceptability of a lower fair price for transit. Diana [22] used correlation and correspondence analysis to measure the satisfaction of transit and private vehicle users for local transit services in Italy. No correlation was found between overall satisfaction rates and the frequency of usage. The results of the study also indicated that users of the smaller towns had higher satisfaction with the transit service than the users of the metropolitan cities. Rojo et al. [16] estimated interurban bus transit quality from the users' viewpoint for buses of Spain. They found that except elderly people, users' do not consider the price of the ticket as an important factor. The results also indicated that the bus stop locations and adequate linking of the bus stops with other public transport services were relevant for bus users. de Oña et al. [35] used SEM for estimating the perceived service quality of a bus transit in Spain. They found that the overall satisfaction was affected by three latent variables. The research also discovered that the passengers can evaluate the overall service quality more effectively after they would be imitated with the service attributes. Kaplan et al. [36] investigated the effect of travel mode fairness, price and spatial equality in transit provision on the perceived habitual frequency of use, willingness to pay and transit service quality. The questionnaire data collected through revealed preference survey was analyzed by SEM. Youths participated in the questionnaire survey in Copenhagen and Lisbon. The findings revealed that improved equity with respect to transit improved the perceived efficiency and willingness to pay for transit. Nwachukwu [18] investigated the effect of passengers' attitude on the satisfaction of the bus transit service in Abuja, Nigeria. The questionnaire survey data were evaluated with factor analysis and regression analysis. He found that the comfort, accessibility, bus stop facilities and adequacy of the capacity of bus transit affected the users' satisfaction. Among the four factors, comfort was found to be the most influential one.

It is clear from the above discussions that the subjective analysis of the transit service quality is effectively used by various researchers. They use the subjective data analysis methods to explain why and how various characteristics related to the transit service are important to the users. The subjective analysis also used to identify the underlying or latent factors affecting the users' perception, relationship between the qualitative attributes and the factors and importance of the latent factors to the users.

3.4 Importance of Users' Expectations in Estimating Subjective Service Quality

In view of the subjective measures of service quality in previous literature, the perception of user is important, but expected service quality is also significant in order to improve service delivery. The quality anticipated and experienced are different because the quality predicted is not a function of the passengers' everyday encounters, but rather reflects the preferences of public transport users. As per Parasuraman et al. [39], the satisfaction of the users occurs when perceived service quality match with the desired or expected service quality. Transport planners and operator will also seek to make the perceived quality as close to the desired quality, so that the service is more desirable to the customers. But very few researchers [40, 41] have studied the effect of perception and expectation together. Sam et al. [40] studied the effect of perception and expectation on the satisfaction level of the users on the transit service of Kumasi, Ghana. They used SERVQUAL [17] method to analyze the questionnaire data which contains multiple questions related to service quality of the transit service. The study found that there are significant differences between the perception and expectation of the users which led to the dissatisfaction of the users. Among the different factors, reliability and responsiveness of the transit service were found to be the most important parameters which affect users' dissatisfaction. Deb and Ahmed [41] found that both the perception and expectation of the users have significant effect on the overall satisfaction of the bus service. They estimated dissatisfaction of the users based on the differences between perception and expectation and eventually estimated the service quality of the city bus service on the basis of users' dissatisfaction. Therefore, while estimating subjective service quality, it is required to consider both the perception and expectation of the users.

3.5 Variation of the Perception with Respect to Demographic and Socio-Economic Characteristics

For all the users, the customer experience may not be the same. It differs from one individual [11, 32, 42–45] to another, based on socioeconomic variables [42, 43]. Customers' opinion about the service performance is influenced by their desired and acceptable performance levels [45]. The performance of the service for the same service type may not always be identical, but may be diverse in nature. The acceptance of variability in service performance depends on users' levels of tolerance, which rely on their demographics and degrees of expectation [45]. These users' expectations rely also on their knowledge, attitudes, expertise and specific requirements [46]. The differences or the heterogeneity in the perception of the users for the same service depends on their socioeconomic characteristics [11, 43], their attitude towards the service [11], their demographic characteristics and their trip habits [43, 47]. Ahmed and Vaidya [48] investigated the influence on the willingness to pay for time savings in travel by age, sex and travel purpose. The variance in willingness to pay for various trips was also documented by Phanikumar and Maitra [49]. The disparities in the perceptions of captive riders about willingness to pay for the bus service in Kolkata, India were examined by Maitra et al. [50]. Hu et al. [47] observed a number of differences in the users' understanding of the public transport accessibility, trustworthiness, security and comfort for individual travel purposes. Deb and Ahmed [51] examined how consumers' impression of bus service affects their socio-economic and demographic variables. They discovered that users' assessment of the bus service affected by their age, sex, earnings, work status and automobile ownership. In their survey, the most essential factors in the male respondents were comfort, whilst females were more concerned with the security and trustworthiness of the service. Therefore, while conducting subjective analysis, it is required to study the effect of socioeconomicvcharacteristics on users overall satisfaction of the service so as to make the service more appealing for different users groups.

3.6 Drawback of Subjective Study

As the users' perception changes with their mode and socioeconomic conditions, therefore, users' perception data cannot be considered as 100% reliable [35]. Therefore, both the subjective and objective measures should be considered while estimating the service quality [5]. Very few studies have however taken into account both subjective and objective measures for calculating the transit serviceequality. For example, Tyrinopoulos and Aifadopoulou [52] developed a methodology to estimate the service quality of public transport based on both the objective and subjective measures. This study provided a general overview of the methodology for estimating the service quality and performance of public transport developed by the Institute for Hellenic Transportation. A customer satisfaction measure was incorporated into the compound indicators measure. Factor analysis and multinomial logistic regression were suggested as an analysis tool to investigate the effect of performance on customer satisfaction.

A system for assessing the service quality of rail services has been established by Nathanail [53]. This methodology was based on 22 indicators categorized under six indicators. The data for this study was collected from the railway operators and by conducting a questionnaire survey. The overall performance of the system was based on the combined indicators comprised of both the subjective and objective measures.

A methodology for assessing the transit services quality has been developed by Eboli and Mazzulla [5], which combines both the subjective and the objective measures. The service levels of the subjective measures were calculated by averaging of the satisfaction rates expressed by the users. The objective measures were calculated on the basis of some predefined standard values. Later on, these two types of measures were combined to get the service quality of the service based on both the subjective and objective measures.

Das and Pandit [12] estimated the service levels of the bus transit system of Kolkata city on the basis of nine indicators. These indicators were waiting time, Crowding level, seat availability, number of mode transfer, on-time performance, delay in total journey time, daily service hours, and boarding and alighting time. The performance of these indicators was estimated on the basis of users' perceptions rather than the expert opinions.

Deb and Ahmed [54] estimated the service quality of the city bus service based on subjective and objective measures. The indicators for the subjective measures were estimated based on users' perception and expectation while the indicators for the objective measures were adopted from MoUD handbook [14].

Therefore, it can be said that, while evaluating transit performance, the researchers should consider both the subjective and objective measures.

4 Conclusion

Estimating the subjective level of service depends on the understanding of the customer. Through questionnaire surveys these perception data are collected. Previous researchers identified various statistical methods for analyzing the data on perception. Methods of measuring subjective service quality can be categorized in two different ways. The first category includes techniques of statistical analysis like factor analysis, structural equation modeling etc. to identify different latent factors affecting users' perceptions. The second category includes some methods to estimate the coefficients of different factors on global satisfaction index. Different techniques like regression analysis, structural equation models, logit models and probit models are used to correlate global service quality to the attributes or factors. Logit and probit models are also used to estimate the effect of users' perception on their mode preferences. The strategy for selection of any particular technique depends on the property of the data and the study objectives. Moreover, the selection of a set of qualitative attributes for the questionnaire survey should also represent users' needs and expectations. Users' needs and expectations depend on their socioeconomic and demographic characteristics. Therefore, selection of any particular attribute depends on the requirements of the users for that particular study area and it varies with changes in demographic and socioeconomic characteristics of the users. For these reasons, before conducting the questionnaire survey, it is required to perform some preliminary surveys to know the user's requirements from the service and accordingly the qualitative attributes of the questionnaire can be finalized.

On the other hand, estimation of the objective measures of the service quality depends on the operational performance of the transit service. To measure the operational performance, it is required to estimate the performance of some quantitative indicators which relate to the operational performance of the transit service. For example, speed, total number of vehicles, average waiting time, average crowding level, transit coverage etc. relate to the operational performance of the service. The performances of these factors are estimated by comparing the present performance levels with past performance levels. On the basis of these comparisons, benchmark is set for future reference.

The assessment of the efficiency of transit service may be measured by both subjective and objective measures even though it offers quality metrics from a different point of view. The combination of these two forms of measurement of service quality may thus provide a useful and accurate transit output measuring instrument.

References

1. Badami, M.G., Haider, M.: An analysis of public bus transit performance in Indian cities. Transp. Res. Part A: Policy Pract. **41**, 961–981 (2007). https://doi.org/10.1016/j.tra.2007.06.002
2. TRB, The Role of Performance-Based Measures in Allotting Funding for Transit Operations, TCRP Synth, National Academy Press, Washington, DC, USA, 1994.
3. De Borger, B., Kerstens, K., Costa, A.: Public transit performance: what does one learn from frontier studies? Transp. Rev. **22**, 1–38 (2002)
4. TRB, Transit capacity and quality of service manual: TCRP report 100, 2nd ed. Transportation Research Board, Washington, D.C (2003). http://www.trb.org/Main/Blurbs/153590.aspx.
5. Eboli, L., Mazzulla, G.: A methodology for evaluating transit service quality based on subjective and objective measures from the passenger ' s point of view. Transp. Policy **18**, 172–181 (2011). https://doi.org/10.1016/j.tranpol.2010.07.007
6. Roess, R.P., Vandehey, M.A., Kittelson, W.: Level of Service. Transp. Res. Record J. Trans. Res. Board. **2173**, 20–27 (2010). https://doi.org/10.3141/2173-03
7. Correia, A.R., Wirasinghe, S.C., de Barros, A.G.: Overall level of service measures for airport passenger terminals. Transp. Res. Part A: Policy Pract. **42**, 330–346 (2008). https://doi.org/10.1016/j.tra.2007.10.009
8. Tan, D., Wang, W., Lu, J., Bian, Y.: Research on Methods of Assessing Pedestrian Level of Service for Sidewalk. J. Transp. Syst. Eng. Inf. Technol. **7**, 74–79 (2007). https://doi.org/10.1016/S1570-6672(07)60041-5

9. Taylor W., Brogan, J.: Level-Of-Service concepts in urban public transportation, Lansing, MI: Michigan Transportation Research Program, Highway Safety Research Institute, the University of Michigan. (1978). http://deepblue.lib.umich.edu/%0Abitstream/2027.42/591/2/41443.0001.001.pdf.
10. Botzow, H.: Level-of-service concept for evaluating public transport. Transp. Res. Record. **519**, 73–84 (1974). http://tram.mcgill.ca/Research/Publications/TPM_Portland.pdf
11. Cirillo, C., Eboli, L., Mazzulla, G.: On the Asymmetric user perception of transit service quality. Int. J. Sustain. Transp. **5**, 37–41 (2011). https://doi.org/10.1080/15568318.2010.494231
12. Das, S., Pandit, D.: Determination of level-of-service scale values for quantitative bus transit service attributes based on user perception. Transportmetrica A Transp. Sci. **9935**, 1–21 (2015). https://doi.org/10.1080/23249935.2014.910563
13. Bertini, R.L., El-geneidy, A.M.: Using archived data to generate transit performance measures. Transp. Res. Record J. Transp. Res. Board. **2003**, 109–119 (1841)
14. MoUD, Service level benchmarks for urban transport at a glance, New Delhi, India, 2009. moud.gov.in/upload/uploadfiles/files/VoulmeI_Methodologyreport_final03
15. Eboli, L., Mazzulla, G.: Structural equation modelling for analysing passengers ' perceptions about railway services, in: social and behavioral sciences, University of Calabria, Rende, Italy, pp. 96–106 (2012). https://doi.org/10.1016/j.sbspro.2012.09.729
16. Rojo, M., dell'Olio, L., Gonzalo-Orden, H., Ibeas, Á.: Interurban bus service quality from the users' viewpoint. Transp. Plan. Technol. **36**, 599–616 (2013). https://doi.org/10.1080/03081060.2013.845432
17. Parasuraman, A., Zeithaml, V.A., Berry, L.L.: SERVQUAL: a multiple-item scale for measuring consumer perceptions of service quality. J. Retail. **64**, 12–40 (1988). https://doi.org/10.1016/S0148-2963(99)00084-3
18. Nwachukwu, A.A.: Assessment of passenger satisfaction with intra-city public bus transport services in Abuja, Nigeria. J. Public Transp. **17**, 99–119 (2014). https://doi.org/10.5038/2375-0901.17.1.5
19. Sezhian, V.M., Muralidharan, C., Nambirajan, T., Deshmukh, S.: Attribute-based perceptual mapping using discriminant analysis in a public sector passenger bus transport company: a case study. J. Adv. Transp. **48**, 32–47 (2014). https://doi.org/10.1002/atr.200
20. Berry, L.L., Zeithaml, V.A., Parasuraman, A.: Five imperatives for improving service quality. Sloan Manage. Rev. **31**, 29–38 (1990)
21. Wang, I.-M., Shieh, C.-J.: The relationship between service quality and customer satisfaction: the example of CJCU library. J. Inf. Optim. Sci. **27**, 193–209 (2006). https://doi.org/10.1080/02522667.2006.10699686
22. Diana, M.: Measuring the satisfaction of multimodal travelers for local transit services in different urban contexts. Transp. Res. Part A Policy Pract. **46**, 1–11 (2012). https://doi.org/10.1016/j.tra.2011.09.018
23. Iseki, H., Smart, M.J.: How do people perceive service attributes at transit facilities? Examination of perceptions of transit service by transit user demographics and trip characteristics. Transp. Res. Record J. Transp. Res. Board. **2274**, 164–174 (2012). https://doi.org/10.3141/2274-18
24. de Oña, J., de Oña, R.: Quality of service in public transport based on customer satisfaction surveys : a review and assessment of methodological approaches. Transp. Sci. **49**, 605–622 (2014). https://doi.org/10.1016/j.tranpol.2013.07.001
25. Jen, W., Hu, K.: Application of perceived value model to identify factors affecting passengers' repurchase intentions on city bus: a case of the Taipei metropolitan area. Transportation **30**, 307–327 (2003). https://doi.org/10.1023/A:1023983627092
26. Friman, M.: Implementing quality improvements in public transport. J. Public Transp. **7**, 49–65 (2004). https://doi.org/10.5038/2375-0901.7.4.3
27. Wen, C.H., Lan, L.W., Cheng, H.L.: Structural equation modeling to determine passenger loyalty toward intercity bus services. Transp. Res. Record J. Transp. Res. Board. **1927**, 249–255 (2005). https://doi.org/10.3141/1927-28

28. Paulley, N., Balcombe, R., Mackett, R., Titheridge, H., Preston, J., Wardman, M., Shires, J., White, P.: The demand for public transport: the effects of fares, quality of service, income and car ownership. Transp. Policy **13**, 295–306 (2006). https://doi.org/10.1016/j.tranpol.2005.12.004
29. Lin, J.H., Lee, T.R., Jen, W.: Assessing asymmetric response effect of behavioral intention to service quality in an integrated psychological decision-making process model of intercity bus passengers: a case of Taiwan. Transportation **35**, 129–144 (2008). https://doi.org/10.1007/s11116-007-9139-3
30. Tyrinopoulos, Y., Antoniou, C.: Public transit user satisfaction : variability and policy implications. Transp. Policy **15**, 260–272 (2008). https://doi.org/10.1016/j.tranpol.2008.06.002
31. Filipović, S., Tica, S., Živanović, P., Milovanović, B.: Comparative analysis of the basic features of the expected and perceived quality of mass passenger public transport service in Belgrade. Transport **24**, 265–273 (2009). https://doi.org/10.3846/1648-4142.2009.24.265-273
32. dell'Olio, L., Ibeas, A., Cecín, P.: Modelling user perception of bus transit quality. Transp. Policy **17**, 388–397 (2010). https://doi.org/10.1016/j.tranpol.2010.04.006
33. Rojo, M., Gonzalo-Orden, H., dell'Olio, L., Ibeas, A.: Modelling gender perception of quality in interurban bus services. In: Proceedings of the Institution of Civil Engineers - Transport. Vol. 164, pp. 43–53 (2011). https://doi.org/10.1680/tran.9.00031
34. Eriksson, L., Bjørnskau, T.: Acceptability of traffic safety measures with personal privacy implications. Transp. Res. Part F: Psychol. Behav. **15**, 333–347 (2012). https://doi.org/10.1016/j.trf.2012.02.006
35. de Oña, J., de Oña, R., Eboli, L., Mazzulla, G.: Perceived service quality in bus transit service: a structural equation approach. Transp. Policy **29**, 219–226 (2013). https://doi.org/10.1016/j.tranpol.2013.07.001
36. Kaplan, S., de e Silva, J.A., Di Ciommo, F.: The relationship between young peoples transit use and their perceptions of equity concepts in transit service provision. Trans. Policy. **36**, 79–87 (2014). https://doi.org/10.1016/j.tranpol.2014.08.004
37. Ratanavaraha, V., Jomnonkwao, S.: Model of users' expectations of drivers of sightseeing buses: confirmatory factor analysis. Transp. Policy **36**, 253–262 (2014). https://doi.org/10.1016/j.tranpol.2014.09.004
38. Joewono, T.B., Tarigan, A.K.M., Susilo, Y.O.: Road-based public transportation in urban areas of Indonesia: What policies do users expect to improve the service quality? Transp. Policy **49**, 114–124 (2016). https://doi.org/10.1016/j.tranpol.2016.04.009
39. Parasuraman, A., Zeithaml, V.A., Berry, L.L.: A conceptual model of service quality and its implications for future research. J. Mark. **49**, 41–50 (1985)
40. Sam, E.F., Hamidu, O., Daniels, S.: SERVQUAL analysis of public bus transport services in Kumasi metropolis, Ghana: core user perspectives. Case Stud Transp. Policy **6**, 25–31 (2017). https://doi.org/10.1016/J.CSTP.2017.12.004
41. Deb, S., Ali Ahmed, M.: Determining the service quality of the city bus service based on users' perceptions and expectations, Travel Behav. Soc. **12,** 1–10 (2018). doi:https://doi.org/10.1016/j.tbs.2018.02.008.
42. Bhat, C.R., Guo, J.Y., Sen, S., Weston, L.: Measuring access to public transportation services: review of customer-oriented transit performance measures and methods of transit submarket identification. CTR Res. Report 0-5178-1. **7**, (2005). http://www.ce.utexas.edu/prof/bhat/REPORTS/5178_R1_rev_Sept05.pdf.
43. Dantas, A, Antunes, R, Yamashita, Y, Lamar, M.: Marketing in public transportation: neural networks approach. In: 9th World Conference in Transportation Research: D1–4137, University of Canterbur, Seol, South Korea, 2001. https://ir.canterbury.ac.nz/handle/10092/604.
44. Das, S., Pandit, D.: Transport reviews : a transnational importance of user perception in evaluating level of service for bus transit for a developing country like India : a review. Transp. Rev. A Trans. Transdisciplinary J. **33**, 37–41 (2013). https://doi.org/10.1080/01441647.2013.789571
45. Zeithaml, V.A., Berry, L.L., Parasuraman, A.: The Nature and determinants of customer expectations of service. J. Acad. Mark. Sci. **21**, 1–12 (1993). https://doi.org/10.1177/0092070393211001

46. Davidow, W.H., Uttal, B.: Service companies: focus or falter, Harvard Bus. Rev. **67**, (1989). https://hbr.org/1989/07/service-companies-focus-or-falter
47. Hu, X., Zhao, L., Wang, W.: Impact of perceptions of bus service performance on mode choice preference. Adv. Mech. Eng. **7**, 1–11 (2015). https://doi.org/10.1177/1687814015573826
48. Ahmed, F., Vaidya, K.G.: The valuation of travel time savings in least developed countries: theoretical and empirical challenges and results from a field study. In: World Conference of Transportation Research, Greene, Istanbul, Turkey, (2004)
49. Phanikumar, C.V., Maitra, B.: Modeling willingness-to-pay values for rural bus attributes under different trip purposes. J. Transp. Res. Forum. **45**, 31–44 (2006) http://journals.library.oregonstate.edu/index.php/trforum/article/viewFile/911/806.
50. Maitra, B., Dandapat, S., Chintakayala, P.: Differences between the perceptions of captive and choice riders toward Bus service attributes and the need for segmentation of bus services in urban India. J. Urban Planning Dev. **141**, 1–8 (2014). https://doi.org/10.1061/(ASCE)UP.1943-5444.0000205
51. Deb, S., Ahmed, M.A.: Variability of the perception about safety, comfort, accessibility and reliability of city bus service across different users ' groups. Int. Civil Eng. Technol. **9**, 188–204 (2018)
52. Tyrinopoulos, Y., Aifadopoulou, G.: A complete methodology for the quality control of passenger services in the public transport business. **38**, 1–16 (2008)
53. Nathanail, E.: Measuring the quality of service for passengers on the hellenic railways. Transp. Res. Part A Policy Pract. **42**, 48–66 (2008). https://doi.org/10.1016/j.tra.2007.06.006
54. Deb, S., Ahmed, M.A.: Quality assessment of city bus service based on subjective and objective service quality dimensions: case study in Guwahati India. Benchmarking Int J **26**(2), 567–589 (2019). https://doi.org/10.1108/BIJ-11-2017-0309

Blockchain Technology Used in the Mid-Day Meal Scheme Program Supply Chain Management

Pooja Singhal, Parth Goel, Jatin Sachdeva, Nitin Teotia, and Mohak Gogia

Abstract This study aims to apply blockchain technology in the supply chain through a mid-day meal scheme run in India to provide better nutritional value to the children in primary and upper primary schools. This technology has gained a lot of attention from researchers, engineers, and practitioners in different sectors around the world. Current mid-day meal scheme in India has lot of unchecked processes which can lead to low quality and unhealthy meals for the students. There is a lot of corruption too in this supply chain of mid-day meal, and it is difficult to keep track at which point the discrepancy might have occurred. Having previous applications of the blockchain technology in the supply chain, the exploration of this technology in mid-day meal schemes has a lot of potential. What does it do? This digital technology can make the system immutable, fully tolerant, transparent, and fully traceable. It should keep in check the movement of the meals over the supply chain and the data for which will abstain from tampering. The leveraging of this technology should redefine the supply chain operations giving better performance, distributed governance, and process automation. The trust and performance issues with the previous intermediaries will be resolved. The smart contracts, Web portals, and blockchain frameworks in the supply chain would make the transactions smooth, efficient, and trustworthy. The future application for this can include integrating IoT to help keep the quality of the meal under check and AI for better analysis and automation.

Keywords Blockchain · Smart contract · Supply chain · Transparency · Caterer · Transporter · School · Database

1 Introduction

1.1 About the Current System and Blockchain Technology

The present system is unreliable, non-auditable, opaque, and untraceable. The need for blockchain arose, as dishonesty and trust issues among individuals have become

P. Singhal · P. Goel (✉) · J. Sachdeva · N. Teotia · M. Gogia
Inderprastha Engineering College, Ghaziabad, India

K. N. Das et al. (eds.), *Proceedings of the International Conference on Computational Intelligence and Sustainable Technologies*, Algorithms for Intelligent Systems,
https://doi.org/10.1007/978-981-16-6893-7_44

rampant. Due to this, children are getting lower quality and insufficient amounts of food.

Lack of proper supervision over the supply chain further heightens and numerous fake enrolments are done to embezzle money. Blockchain brings transparency and security to the system as it will store the information from every agent in the supply chain, and hence, they can be held accountable in future. The centralization problem faced by the system can be overcome with the use of a distributed system like blockchain. Blockchain also provides us with high fault tolerance [1]. The security challenges faced in storing the information are dealt with by the hashing algorithms.

The proposed system focuses on tackling the challenges faced by the current mid-day meal system. This project aims to build a mid-day meal system using blockchain technology which would ensure a secure, traceable, and immutable distribution. The efficiency of the system would also be increased, and transparency would be maintained.

The proposed system has three modules: blockchain technology, the Web application, and the feedback module. The concept of blockchain in the existing system allows us to keep track of the meals over the decentralized network where the transactions are stored. The transactions over the blockchain are stored to keep the movement of food in check. The distributed technology should reduce risk, security, and increase transparency, availability, scalability, openness, and speed in the supply chain and improve trust and integrity among the agents on different levels of the supply chain. The verification of the data can be done through blockchain so that the system is not tampered with. A Web application is used to access the whole system and provide a feedback module to keep the quality of the meal in check.

The following properties of blockchain could help it overcome the issues faced by the current mid-day meal system:

i. **Decentralized framework**: This approach does away with the centralized system, thereby improving the performance of the system [2].
ii. **Security**: Transactions are secured in the blockchain system. Since blockchain protects the data against any tampering, it is one of the best approaches to securing information [2].
iii. **Identification**: Every block in the blockchain is uniquely identified. Therefore, blockchain is a trusted technology that provides uniquely identified data stored in a public ledger [2].
iv. **No Paperwork**: The blockchain technology integrated with the mid-day meal system will hold all the information securely over the blockchain network, and it can be viewed anytime from anywhere.
v. **Efficiency**: The transactions over the blockchain are faster than the current manual system. Multiple transactions are processed within a few minutes by the use of this technology.

1.2 Problem Definition

The present system under which the mid-day meal scheme has been implemented has a lot of issues in it, namely lack of trust and transparency, difficult traceability, time inefficiency, and subpar quality of food being served.

2 Literature Review

When we look at the supply chain management systems, we can see that there are three basic flows—the material flow, the information flow, the goods flow. Material flow is the one that focuses on moving the products or the goods from one point to the other. The information flow is the one that has the communication process that happens from the manufacturing point to up until the consumption point. The third flow, the money flow, includes the amount of each product or goods that are manufactured [3].

Although the idea of blockchains first appeared in [4], it was Satoshi Nakamoto in 2008 who developed the first blockchains in use, by adding blocks in a chain [5] (Table 1).

Blockchain databases have numerous advantages over conventional centralized database systems. These advantages include a decentralized trust model, high-security standards, high public access, and transferable identities. As it is evident from the above table, privacy varies anywhere from low to high among different blockchain technologies, hence designers have greater flexibility to choose suitable blockchain technology that may be custom-built for its intended purpose. This cannot be achieved with the conventional centralized database system as privacy cannot be configured based on the use case scenario.

In the blockchain system, the solution for privacy is a private blockchain network; the solution for integrity is the use of a hashed-chain structure for verification over the blockchain, and the solution for the availability is storing the transaction data (hashes) in the tamper-proof smart contracts [6] (Figs. 1, 2, and 3).

Since the explosion of blockchain technology in supply chain management, this technology has been leveraged for financial, legal paper, and agriculture supply

Table 1 Types of blockchains (difference between various kinds of blockchain)

Types of blockchain	Efficiency	Decentralized	Accord Growth	Immovableness	Reading	Determining
Private	Good	No	Yes	Can be	Can be publicly	Only one industry
Public	Worse	Yes	No	No	Yes	All miners
Consortium	Good	Sometimes	Yes	Can be	Can be publicly	IoT devices

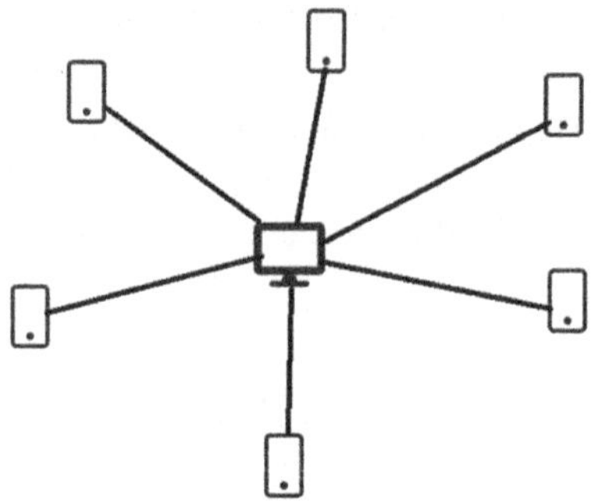
Fig. 1 Centralized blockchain

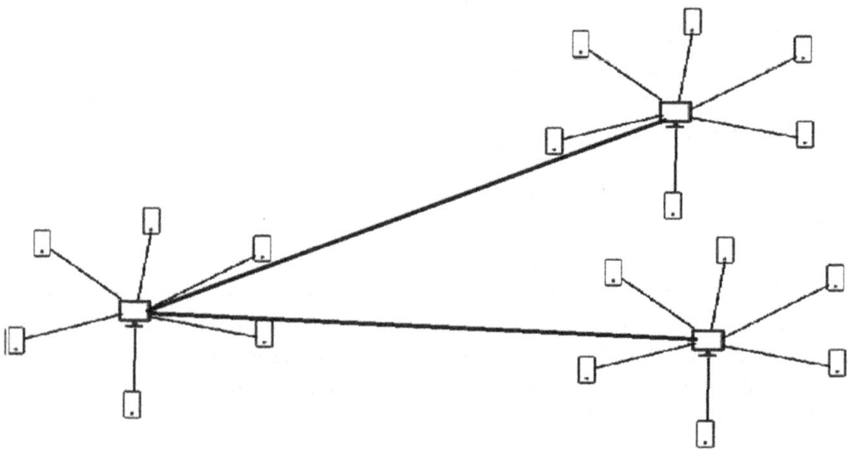
Fig. 2 Decentralized blockchain

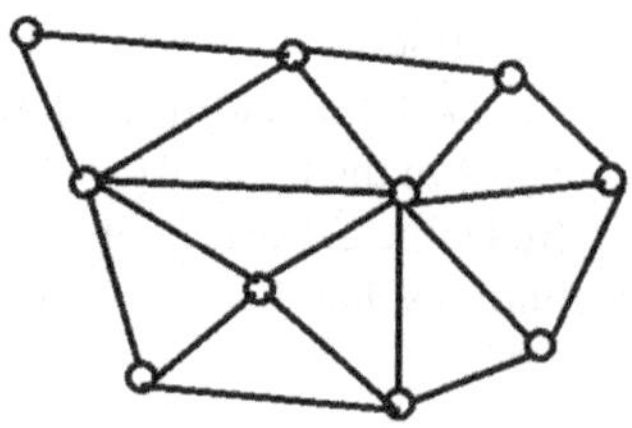
Fig. 3 Distributed blockchain

chains. The qualities of the blockchain providing tamper-proof immutable record storage, enabling potential trust among the participants, being reliable, efficient, secure, and transparent have made many industries dive deep into it [7]. There has been an adoption of the IoT devices with the blockchain in the supply chain management which has attracted a lot of attention.

The distributed blockchain technology would reduce risk, increase security, and transparency, availability, scalability, and speed in the supply chain. It should also improve trust and integrity among the agents on different levels of the supply chain [8]. The environment will be decentralized, and no central authority would be able to tamper with the data.

The use of a Web portal should allow the entering of the data to a blockchain distributed ledger through the use of smartphones that can track all the data and make it accessible in a matter of seconds.

3 Motivation and Objective

3.1 Existing System

The current mid-day meal scheme program faces a lot of issues due to inefficient implementation. These issues include subpar quality of food, rampant corruption in supply and distribution, tampering with official paperwork, and lack of transparency in the network to name a few.

The present system does not have the following characteristics which need to be addressed:

3.1.1 Enhanced Security

The supply chain data in a mid-day meal scheme is highly sensitive and crucial, and the use of blockchain technology prevents any fraud risks by creating a record that is encrypted end-to-end and can't be altered. The characteristic of blockchain also allows keeping the personal data anonymous and preventing access by the use of permissions. Since the information is not centralized, it makes it difficult for hackers to view the data.

3.1.2 Greater Transparency

Blockchain technology has a distributed ledger that makes available identical data or transactions over multiple locations. This prevents organizations from keeping a separate database. The same information is accessible to all the permissioned participants of the network which provides full transparency of the data. All of these transactions are immutable and are time and date-stamped. Members can view the transaction history and eliminate any fraud.

3.1.3 Instant Traceability

Blockchain allows for detailed documentation at different stages of the process. In fields that are concerned with environmental or human rights, this characteristic of blockchain technology can come in very handy. Data stored over the blockchain can be analyzed to track and resolve weaknesses within the supply chain. Since the data cannot be tampered with, it can also be presented as proof in situations involving counterfeiting or fraud.

3.1.4 Increased Efficiency and Speed

Conventional paper-based systems have time-consuming processes which are prone to human error, and they are further slowed down with third-party intervention. This problem can be dealt with very easily with the use of blockchain in place of paper-based systems. The blockchain architecture allows for faster, more efficient transactions by streamlining the process. Process documentation and transaction details can be stored on the blockchain allowing for a much faster settlement of claims. This would eliminate the need for paper and also facilitate faster clearance.

3.1.5 Automation

Blockchain technology offers the feature of "smart contracts" which help automate the transactions while increasing the efficiency and speed of the system even further. The only requirement in an automated system would be to meet the pre-specified conditions, and the next step in the transaction or process would be automatically triggered. This would do away with human intervention, hence increasing the speed of the system and also eliminate the need for verification of terms and conditions in the contract by a third party. A similar implementation can be seen in the case of insurance, wherein upon providing the requisite details and proof, insurance claims for a customer are automatically settled.

3.2 *Proposed System*

The proposed system follows a single-window approach, a Web portal. The proposed is auditable, efficient, and traceable. Here is a summary of the working of the proposed system:

1. The supervisor log in to the portal and add a new school, caterer, transporter, meal code, or assign the meal. These details will be stored in a MySql database, wherein each of them have its own table.

2. After the supervisor has assigned the meal, the caterer log in using the caterer ID and view the meals assigned. The caterer adds the prepared meal details which is also stored in our database. After the meal has been prepared, the caterer clicks on add and dispatch meal which means that the meal has been dispatched to a transporter.
3. The transporter checks the assigned meals and accept shipments entering meal ID, school ID, and transporter ID. The hash code is calculated and pushed over the blockchain with the information available from the transporter, and the transaction hash is calculated.
4. The school checks the assigned meals and accept meals—entering meal ID, caterer ID, school ID, transporter ID, meal code, and the number of students present. The hash code is calculated and pushed over the blockchain with the information available from the school, and the transaction hash is calculated. The school should add the students to the database who can add their feedback for the quality analysis using—school ID and student ID.
5. To submit the feedback, we may go to "Submit the Feedback" on the login portal and log in using the student ID, school ID, and password. The hash code is calculated from the information available with the feedback and is pushed over the blockchain, and the transaction hash is also calculated.
6. We can verify meal details using the meal ID, which is the most important so that we can verify that data are not being tampered with. Enter the meal ID which will display us the meal details after verifying the blockchain hash with the hash stored in the database.

The proposed system uses the following major components to make a secure and transparent system:

SHA256 algorithm—This cryptographic algorithm works with information broken down into pieces of 512 bits (or 64 bytes in other words). It produces its cryptographic "mixing" and then issues a 256 bit hash code. The algorithm includes a relatively simple round, which is repeated 64 times. SHA 256 is used to calculate the hash of the information of the agents in the supply chain, thereby securing each transaction.

The transaction details of the system are stored in a distributed database. The blockchain database is immutable and acts as a secure digital ledger.

4 Methodology

At first, we studied the current mid-day meal system.

For the project, we build a Web application portal so that we can bring the all the actors in our system on one platform.

With each actor, there are some attached variables which when hashed using SHA 256 gives us the data that have been used to prove the immutability feature of the blockchain.

Whenever the meal is passed onto from one actor to the next, a hash is calculated and transacted over the blockchain. We also generate a hash and store it in the blockchain at the student feedback level which helps us determine the quality of the meal.

Using these hashes, we can prove the immutability at every point in the supply chain as if any value gets tampered with then we can prove this discrepancy. And by combining all this, we can make sure that no middle man and no administration can tamper with the data of our supply chain.

5 Result and Discussion

5.1 Prototype

The summary of the prototype for the proposed working model is given below:

5.1.1 Login Portal

Here is the main login screen of our prototype, here, various entities like supervisor (administrator authority), school admin, transporter admin, and caterer admin can log in to view their profiles and meals assigned to them and other options as seen in the screenshot (Fig. 4).

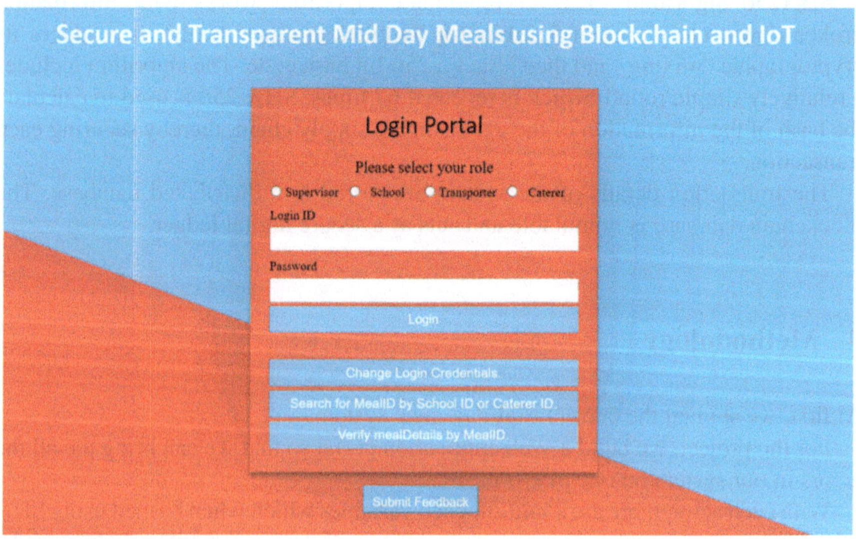

Fig. 4 Login portal

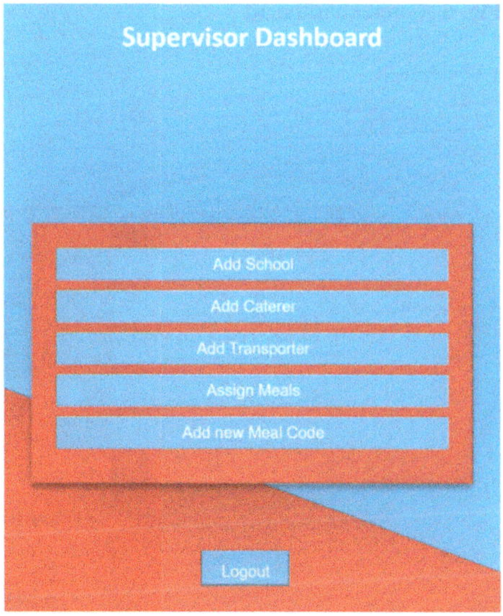

Fig. 5 Supervisor dashboard

5.1.2 Supervisor Dashboard

The supervisor dashboard has various features as you can see in the screenshot. He/she can add a school, caterer, transporter, and new meals to the database and also assign meals to various entities of this prototype (Fig. 5).

5.1.3 Other Entities' Dashboards

Other entities have dashboards like these where they can accept meals assigned to them. After the caterer has dispatched the meal, the transporter and school will log in to their dashboards to accept the meals. Also, at every transaction point, an IoT metric score (1–10) is added manually in the current prototype which would be automatically calculated in future updates (future work) via an IoT device to give a cumulative quality score of the meal based on temperature, weight, etc. (Fig. 6).

5.1.4 Student Feedback

Here, a student can log in to submit feedback on a meal, and when enough feedback is submitted by the students, a cumulative feedback hash is calculated and posted on a smart contract through solidity code (Fig. 7).

Fig. 6 Caterer dashboard

5.1.5 Hashing of a Transaction

After every transaction, the hash of the transaction is calculated at the receiver side of the meal. In the above screenshot, we have taken the case of a transporter. We compute a buffer by combining the details of the meal at that point with the details of the meal receiving entity and then create a hash of those details using SHA256. We then set this data as the transaction hash using the smart contract function from the solidity code (Fig. 8).

5.1.6 Solidity Code

In the solidity code, we have a string variable for storing the hash value of the transaction and two functions getData() and setData() for retrieving the hash from the blockchain and for setting the hash, respectively. Along with that, there is an

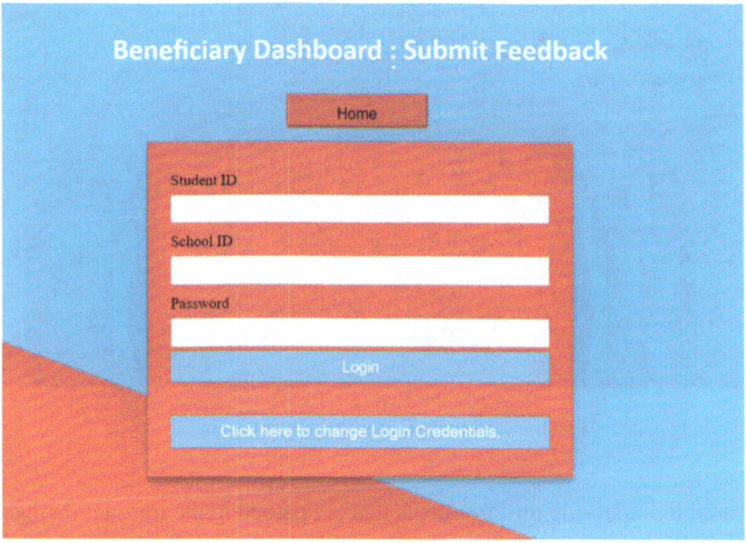

Fig. 7 Student feedback portal

Fig. 8 Hashing of the transaction by transporter

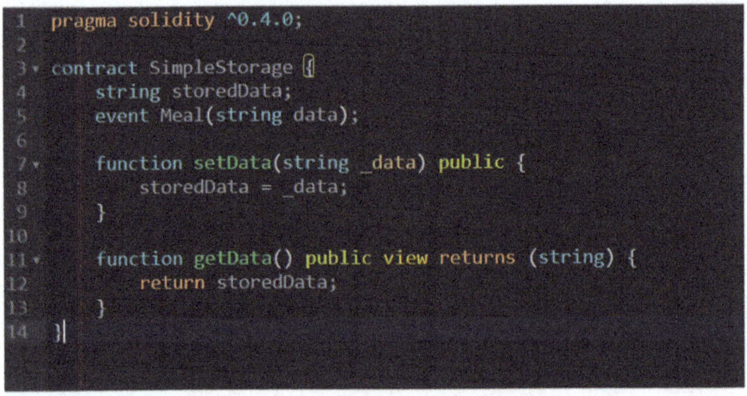

Fig. 9 Solidity code to send transaction hash over to the blockchain

event variable for storing the arguments that are passed in the transaction logs of the block (Fig. 9).

5.1.7 Transaction History

The above screenshot shows the details about the transaction taken from the Ropsten test network where we tested our blockchain app. It shows various details like transaction hash, block no, timestamp of creation, and gas price. of the first transaction that is generated when the transporter receives the shipment from the caterer. Likewise, the screenshots given below show the information about the second and third transaction of the school accepting the shipment and students' feedback being stored on the blockchain, respectively (Figs. 10, 11, and 12).

5.1.8 Transaction Verification

Here, an auditor can enter meal ID to audit and verify meals to check if any tampering is done in the data stored in the database (Fig. 13).

Here is the output of the verification (Fig. 14).

Here, we again create hashes of different entities using data stored in the database and match them with the stored hash in Etherscan. If they are different, then we can say the data have been tampered with. This ensures secure and immutable storage of the data.

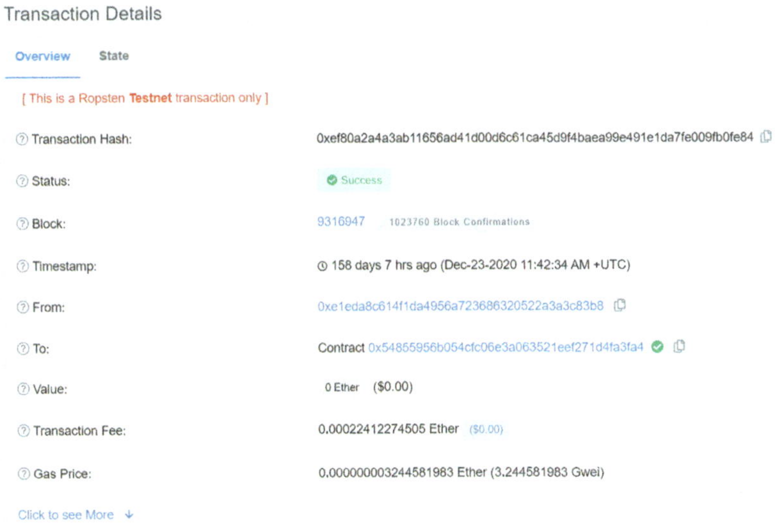

Fig. 10 Transaction details for transporter hash

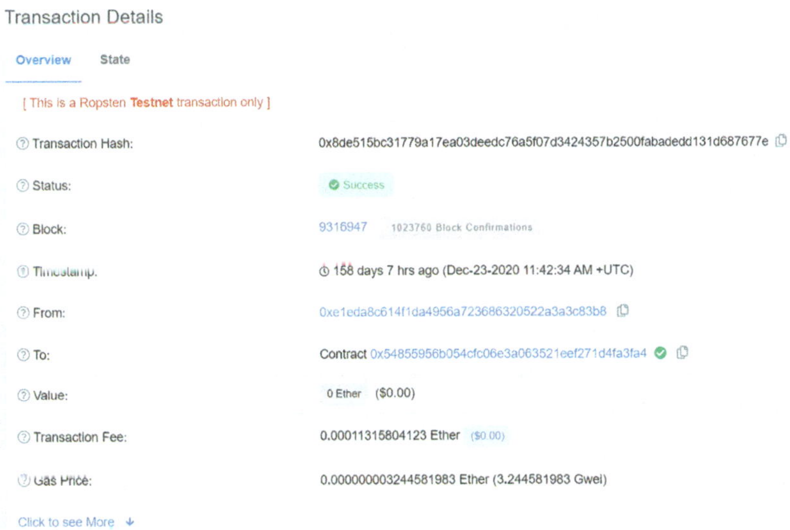

Fig. 11 Transaction details for school hash

6 Conclusion

Here, we have discussed the challenges faced by the mid-day scheme system. The upcoming disruptive technologies, namely blockchain have been implemented and explored in this project, and potential advantages have been highlighted. Blockchain

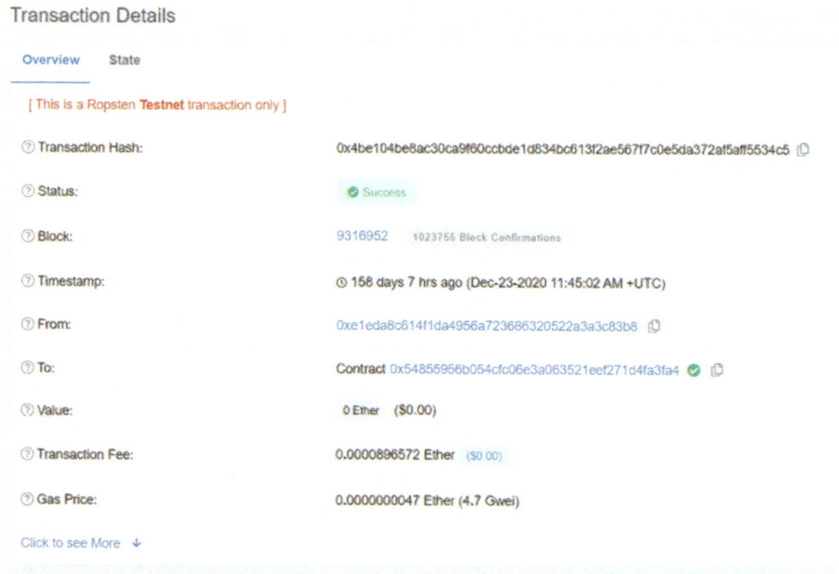

Fig. 12 Transaction details for feedback hash

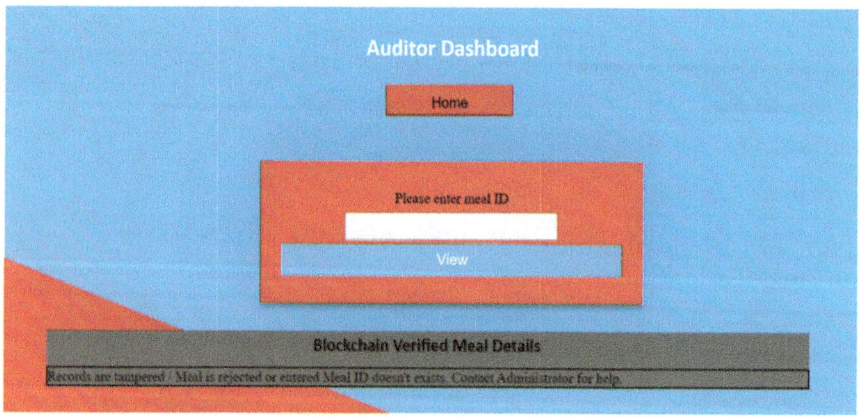

Fig. 13 Auditor dashboard to verify the transaction

in the supply chain is under implementation by various organizations in different projects. The proposed system would reduce the time and cost of the current system, but more importantly, it should provide for a secure and transparent network. The transparency of the system is a crucial requirement and also allows to keep the quality of products in check. There will be a rise in the efficiency of the current mid-day meal distribution system, and any infiltration of the system would be almost impossible.

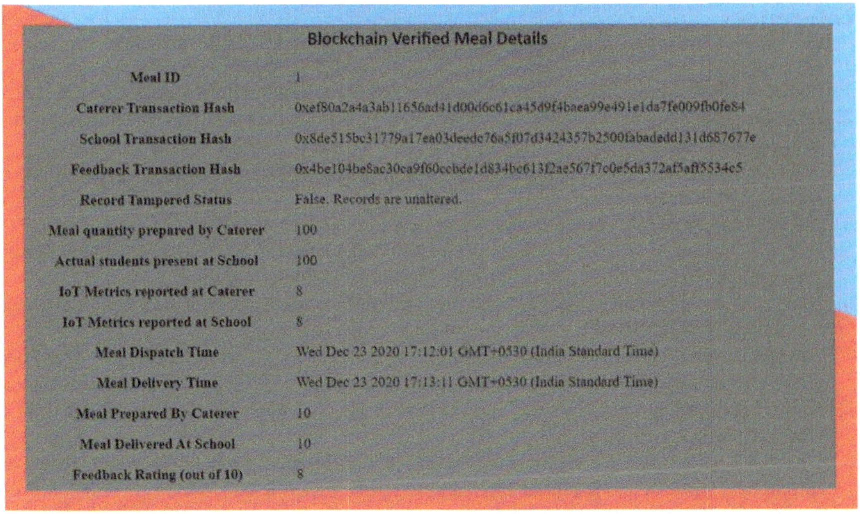

Fig. 14 Verification details of the transaction over the blockchain displaying immutability

We have looked at the advantages that the implementation of blockchain in the mid-day meal system could bring. Use of the new system reduces fraud and involvement of middlemen in the supply chain. It also helps overcome any adulterations in the quality of food and corruption present in the supply chain [9]. Monitoring of meals can be done with the development of an efficient system that would provide for security and transparency [9]. The new system also eliminates the need for extra inspection, workforce, or paperwork as the data are stored in a secured ledger of the blockchain.

7 Future Work

Supply chain management is adopting various other technologies like IoT and machine learning separately. Our further aim is to leverage these disruptive technologies to build a more secure network for all types of supply chains. This implementation should help us work against any fraud risks and can also be fully automated with the use of AI. Such a system could be more scalable, effective, and efficient. The quality of food in the current system is compromised, and hundreds of children are hospitalized each year due to food poisoning and other ill effects. AI can allow the prevention of such accidents in future with the use of smart cameras which would use object detection, face recognition, and classification using ML models [10]. IoT could help integrate hardware architecture and sensors to enable real-time tracking and security in the delivery system.

References

1. Badhe, S.D.: A supply chain management overview : today and tomorrow. **9519**, 66–69 (2013)
2. Alam, T.: Blockchain and its role in the Internet of Things (IoT). (2019). https://doi.org/10.31219/osf.io/cmza5
3. Aliyu, S., Tom, A., Taiye, M., Barakat, M.: The role of blockchain technology applications in supply chain management. 2456–1983 (2018)
4. Nakamoto, S.: Bitcoin: A peer-to-peer electronic cash system. (2008)
5. Haber, S., Stornetta, W.S.: How to time-stamp a digital document. In: Conference on the Theory and Application of Cryptography. Springer, Berlin, Heidelberg (1990)
6. Xiong, et al.: Blockchain technology for agriculture: applications and rationale J front. Blockchain **3**, A7 (2020)
7. Caro, M.P., Ali, M.S., Vecchio, M., Giaffreda, R.: Blockchain-based traceability in Agri-Food supply chain management: a practical implementation. In: 2018 IoT Vertical and Topical Summit on Agriculture -Tuscany (IOT Tuscany), pp. 1–4.https://doi.org/10.1109/IOT-TUSCANY.2018.8373021 (2018)
8. Chang, S.E., Chen, Y.: When blockchain meets supply chain: a systematic literature review on current development and potential applications. IEEE Access **8**, 62478–62494 (2020). https://doi.org/10.1109/ACCESS.2020.2983601
9. Bakare, S., Shinde, S., Hubballi, R., Hebbale, G., Joshi, V.: A Blockchain-based framework for Agriculture subsidy disbursement. (2021)
10. Giridharan, S., Manikandan, S., Kumari, M.: Mid-Day meal analytics using machine learning. J. Phy. Conf. Series. **1717**, 012035 (2021)https://doi.org/10.1088/1742-6596/1717/1/012035

Economic Load Dispatch: A Holistic Review on Modern Bio-inspired Optimization Techniques

Akriti Agrawal, Priyanka Paliwal, and Tripta Thakur

Abstract In power system analysis, economic load dispatch (ELD) is a basic optimization challenge. Over the years, the ELD problems have been resolved with a large range of optimization strategies. Many researchers have reviewed AI techniques in their past literatures. However, newly discovered bio-inspired techniques are yet to be explored particularly in power system. This paper discusses some of the recently developed bio-inspired algorithms and their application in solving combined economic emission dispatch problem. From the comparative analysis provided in this paper, it can be inferred that these bio-inspired optimization techniques prove to be most competitive and successful in solving CEED problems.

Keywords Combined economic emission dispatch · Wind integrated thermal power system · Bio-inspired optimization techniques · Salp swarm optimization (SSO) · Exchange market algorithm (EMA) · Whale optimization algorithm (WOA)

1 Introduction

The present competitive electricity markets need careful planning and management processes to prevent congestion and ensure that the grid runs effectively, safely, and cheaply. *Economic load dispatch (ELD)* attempts to maximize the power generated from various generating units and plan the power performance of all the generators so that load demand is met at minimal costs and system losses, while all system constraints are met. Earlier, ELD problem formulation was done considering only thermal units. But, due to the rising environmental concerns and global desire to achieve sustainable growth, inclusion of green energy resources into the current power grids is emphasized. Also, countries are focusing on reducing the *greenhouse gases (GHGs)* emissions into the atmosphere particularly from thermal power plants. Hence, optimization is applied not only for minimizing the fuel cost but also for the

A. Agrawal (✉) · P. Paliwal · T. Thakur
Department of Electrical Engineering, Maulana Azad National Institute of Technology, Bhopal, Madhya Pradesh 462003, India

K. N. Das et al. (eds.), *Proceedings of the International Conference on Computational Intelligence and Sustainable Technologies*, Algorithms for Intelligent Systems,
https://doi.org/10.1007/978-981-16-6893-7_45

minimization of emission cost by incorporating non-conventional sources of energy along with the traditional thermal and diesel generators. This complex multi-objective optimization problem is called as *combined economic emission dispatch (CEED).*

Many researchers have used numerous optimization methods for solving CEED problem on various RE systems such as wind-solar-hydro-thermal [1], wind-thermal [2], solar-thermal [3], and wind-solar-PV-diesel [4]. The choice of optimization techniques and various combinations of non-conventional resources and conventional sources provide a substantial scope for research. These techniques can be categorized into three distinct techniques: (1) classical analytical methods, (2) artificial intelligence-based techniques, and (3) hybrid methods.

Mathematical-based analytical techniques, such as lambda iteration (LI) [5], linear programming (LP) [6], and quadratic programming (QP) [7], had been applied to solve the CEED problem. Analytical methods have some advantages over AI methods such as ease of implementation, non-iterative in nature, and absence of any algorithm-specific parameters. However, they lack robustness especially when applied in solving non-smooth and complex optimization problems. They also have a tendency to get trapped in local optimum leading to pre-mature convergence, large computation time, and less accuracy.

Artificial intelligence-based methods have played a crucial role in resolving the problems faced in classic CEED problem. Due to their potential to produce numerous optimal solutions in single iteration, AI-based computing approaches are appealing for solving multi-objective (MO) problems. These methods include *evolutionary techniques* such as genetic algorithm (GA) [8] and differential evolution (DE) [9], *population-based techniques* such as particle swarm optimization (PSO) [10], glowworm optimization (GWO) [11], and elephant herding optimization (EHO) [12], *human intelligence-based techniques* such as teaching-learning-based optimization (TLBO) [13] and sine-cosine algorithm (SCA) [14], *ecology-based optimization* such as water cycle algorithm (WCA) [15] and flower pollination algorithm (FPA) [16], *physical science-based techniques* such as gravitational search algorithm (GSA) [17] and black hole (BH) [18]. These modern approaches give harsh competition to conventional numerical techniques, which are limited to convex and smooth objective functions and also suffer from sensitivity toward initial solution.

However, AI techniques have their own shortcomings like early convergence and reliance on various algorithmic factors. These problems are dealt with by developing better variants of the original algorithms and eliminating the particular limitations. Also, to incorporate the best features of various techniques and to get better functionality, two or more techniques are integrated to form various hybrid methods. *Hybrid techniques* are found to be more efficient and provide better convergence characteristics for complex CEED problems. Some of the hybrid methods are PSO-GA [19], BBO-DE [20], PSO-ACO [21], etc.

This paper presents the state-of-the-art review of three of the recently developed BI techniques and the revolutionary concepts and philosophies behind them. The organization of this paper is as follows: CEED problem formulation and also the different types of constraints involved in the system are presented in Sect. 2. Section 3 discusses the modeling of wind integrated thermal power system. Three of the recently developed bio-inspired optimization algorithms are given in Sect. 4. Section 5 contains the key conclusions of the study and provides guidance for the future studies.

2 CEED Problem Formulation

CEED problem aims to find optimum power dispatch for a hybrid renewable energy system, by minimizing the two differing targets, fuel expense and unhealthy emissions simultaneously and also maximize the profits. At the same time, all stability constraints and bounds of the generation must be satisfied [1]. In this paper, objective function is formulated considering wind-thermal power plant.

Cost function is mathematically formulated as

$$F_{\cos t} = \sum_{n=1}^{N} F_n(P_n) + E_n(P_n) \tag{1}$$

And can be further defined as

$$F_n(P_n) = a_n + b_n P_n + c_n P_n^2 \tag{2}$$

$$E_n(P_n) = \alpha_n + \beta_n P_n + \gamma_n P_n^2 \tag{3}$$

Here, F_n (P_n) is the cost function of nth generator; E_n (P_n) is the emission cost function of nth generator; a_n, b_n, and c_n are cost coefficients, and α_n, β_n, and γ_n are emission coefficients of the nth unit; P_n is the real output of nth generating unit in MW, N is the total number of generators connected.

However, practically, cost function is not a smooth quadratic function. It involves a sinusoidal term due to the effect of valve switching, and many other constraints are also involved. Some constraints associated with the power dispatch are given below:

2.1 Generation System Constraints

These are also called as power balance constraints and fall under equality constraints. This is essentially a balance equation which determines that total generation must be equal to the load demand and transmission loss over a given interval "t," i.e.,

$$\sum_{i=1}^{n} P_{i,t} = P_{D,t} + P_{L,t} \tag{4}$$

Here, $P_{D,t}$ and $P_{L,t}$ are power demand and power losses, respectively, in MW. $P_{L,t}$ is obtained using B-coefficients, given by

$$P_{L,t} = \sum_{i=1}^{N} \sum_{j=1}^{N} P_i B_{ij} P_j \tag{5}$$

Here, B_{ij} is the loss coefficient [2].

Along with this, other generation system constraints are spinning reserve requirement, thermal unit constraints, must-run unit constraints, fuel constraints, etc.

2.2 *Generation Limit Constraints*

These are operational constraints called as capacity constraints. It define the upper and lower limit of active and reactive power generation [4].

$$P_{g,i}^{\min} \le P_{g,i} \le P_{g,i}^{\max} \tag{6}$$

$$Q_{g,i}^{\min} \le Q_{g,i} \le Q_{g,i}^{\max} \tag{7}$$

2.3 *Ramp Rate Limits*

Ramp rate is the time in which the output is adjusted from level to level while maintaining the thermal cap. It is determined for two conditions:

i. When power output increases:

$$P_i - P_i^{t-1} \le UR_i \tag{8}$$

ii. When power output reduces:

$$P_i^{t-1} - P_i \le DR_i \tag{9}$$

P_i^{t-1} is the previous power output of unit i in MW, URi is upper ramp limit, and DRi is down ramp limit of ith generating unit in MW/h [1]. Generator limit constraints become following when combined with ramp rate limits:

$$\max(P_{i\,\min} -- P_i^0 - DR_i -- DR_i) \le P_i \le \min(P_{i\,\max}, P_i^0 + UR_i) \tag{10}$$

2.4 Valve-point Loading Effects

Practically, the steam flow through the turbines is regulated by sudden opening and closing of different valve groups. This changes the value of fuel cost function. The rippling effect of valve groups is embedded in the cost function by using a sinusoidal term [1]. The non-smooth cost function is given as

$$F_n(P_n) = a_n + b_n P_n + c_n P_n^2 + |d_n * \sin(e_i * (P_i^{\min} - P_i))| \tag{11}$$

where *dn* and *en* are valve-point loading effect coefficients. $P_i^{\min}$ is the minimum power output of the ith generator [2]. Other constraints include prohibited operating zones (POZ), transmission limits, and system stability constraints [1–4].

3 Modeling of Wind Energy System

Wind is one of the cheapest and cleanest sources of energy. But the intermittent nature of wind brings many challenges to forecast the power output from the system and its integration with the conventional power plants. So, the first step in wind-thermal integration is to predict the power generated from the wind system and its mathematical modeling. Wind's speed at the moment determines the power generated from the wind power plant, and also certain turbine parameters such as shape and scale factor affect the power output. The probability distribution function for wind speed is drawn using Weibull distribution [2]. It is given as

$$\mathrm{pdf}(v) = \frac{\beta}{\alpha}\left(\frac{v}{\alpha}\right)^{\beta-1} \exp\left[-\left(\frac{v}{\alpha}\right)^{\beta}\right] \text{for } 0 < v < \infty \tag{12}$$

where α and β are parameters for scale and shape of PDF. Many authors use a linear function to calculate the wind power output on the basis of wind speed and turbine ratings.

$$p_w(v) = \begin{cases} 0, & \text{for } v \langle v_{\text{in}} \text{ and } v \rangle v_{\text{out}} \\ p_{wr}\left(\frac{v - v_{in}}{v_r - v_{in}}\right), & \text{for } v_{\text{in}} \le v \le v_r \\ p_{wr}, & \text{for } v_r < v \le v_{\text{out}} \end{cases} \tag{13}$$

where *pwr* is the rated power of the turbine. *vin*, v_{out}, and v_r are the cut-in speed, cut-out speed, and rated speed of the wind turbine, respectively. The cost of wind power is calculated by including additional charges for incorrect estimation of wind power output [2].

4 Bio-inspired Techniques

In recent decades, bio-inspired (BI) technologies have undergone great success, exceptional development, and numerous applications, especially in power system analysis. The majority of these methods for optimization have their inspiration in the natural phenomena which can be grouped on the basis of natural evolution, population or swarm intelligence-based, ecology-based, human intelligence-based, and physical or earth science-based. The taxonomy of BI techniques is depicted in Fig. 1.

4.1 Salp Swarm Optimization (SSO)

Salp swarm algorithm is derived from the navigation and conduct of the salps, found in deep oceans. They form a swarm known as salp-chain for their coordinated and self- organized movement in search of food [22]. The salp-chain is divided into two groups: (1) leaders and (2) followers. The leader forms the center of the chain. It is responsible for the exploration and exploitation of the search space for the food sources (optimal solutions). Eventually, the follower salps shift toward their leader to a better position. In each iteration, the leader salp changes its position depending on the location of food sources. In this way, salps reach at the global best position without falling into local optimum.

Vishal et al. [23] have implemented SSO in multi-area economic dispatch (MAED) problem. The authors had tested the algorithm on four systems considering different dimensions and constraints such as VPL effects, RRL, and POZ. The results are compared for 16 and 40 thermal units as well as combination of wind-thermal units applied to two-area and four-area system. SSA performs better than other algorithms in terms of computational cost and efficiency. It attains smoother and faster convergence while satisfying system complexity and constraints. SSA has better search

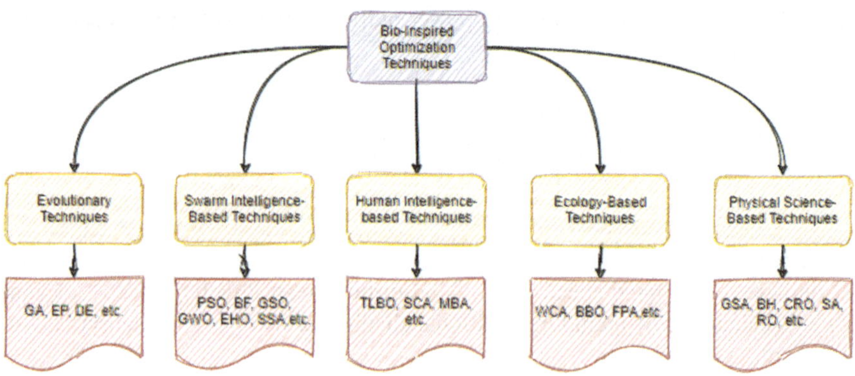

Fig. 1 Taxonomy of bio-inspired techniques

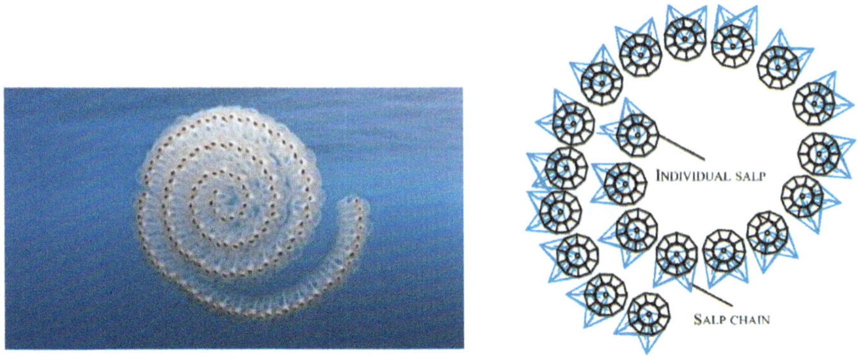

Fig. 2 Salp-chain [24]

capability and also provides better quality of solutions. A salp-chain is shown in Fig. 2.

However, the performance of SSA largely depends on some control parameters such as population size. On increasing the size and convexity of the objective function, after a certain limit, the cost achieved almost remains the same but the computational time increases. These limitations can be eliminated by incorporating suitable modifications in the control parameters, thus providing substantial scope for future research [22–24].

4.2 *Whale Optimization Algorithm (WOA)*

Mirjalili and Lewis [25] proposed WOA, which imitates the social and foraging behavior of humpback whales. The authors mathematically modeled the bubble-net hunting, a rare conduct shown by humpback whales (shown in Fig. 3) and applied it for solving general optimization problems. The search process consists of three

Fig. 3 Bubble-net hunting by humpback Whales [26]

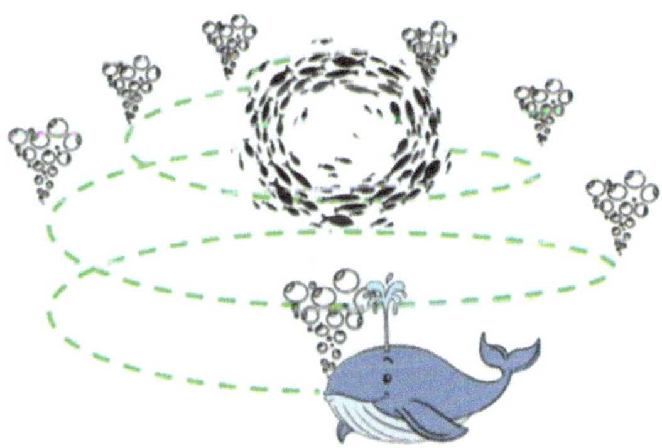

phases: encircling of prey, two ways for exploitation of search space, and then exploration (search for prey). Humpback whales can identify and encircle the prey's position, then swim around the prey in a diminishing circle or simultaneously in a spiral course until they capture it [25, 26].

Later, Chandan et al. [26] embedded chaos behavior in WOA to improve the performance of the original method and proposed a modified algorithm called as chaotic whale optimization algorithm (CWOA). The authors used a set of ten different chaos maps and integrated it with WOA to improve its global search ability and speed of convergence. The proposed methodology is then implemented to solve CEED problem with integrated wind energy system having various constraints. The obtained results on comparison with other classical techniques, such as hybrid PSO, TLBO, GSO, and original WOA, prove superiority of CWOA [25, 26].

4.3 Exchange Market Algorithm (EMA)

EMA is a human intelligence-based optimization method formulated by Ghorbani and Babaei [27]. This algorithm is derived from the stock exchange trading practices used by shareholders in the stock market to earn maximum profit. EMA optimizes the objective function considering two different market phases: (1) balanced and non-oscillatory phase and (2) unbalanced or oscillatory phase. In both the phases, shareholders are divided into three distinct groups: primary or elite (best ones), middle, and end shareholders. In varying market environments, shareholders exchange various shares in the virtual capital market. The financial exchange is constantly subjected to changes due to the country's political and economic policies. The behavior of the economy is easier to forecast under balanced or non- oscillated conditions, and therefore shareholders maximize their earnings without any risks. However, under oscillatory phase, market prices cannot be forecasted and it involves taking certain risks in trading. In both the phases of optimization, primary or elite shareholders remain unaffected and can prosper even in poor market conditions as they are the experienced ones. The profit making ability decreases down the groups and hence, second and third group are sensitive toward unstable market conditions. They try to remain close to the first group shareholders and rely on their skills. The most prominent feature of EMA is that it undergoes two exploitation and exploration stages at each iteration, thereby avoiding getting trapped in local optimum [27–29].

Hagh et al. [28] applied EMA to solve CEED problem with integrated wind power system and tested it on IEEE standard 40-units system with two wind turbines. Results show that EMA provides faster and superior convergence quality than other techniques such as PSO, BBO, and FA. Rajan et al. [29] applied EMA to solve multi-objective dispatch problem on IEEE systems. The results on comparison with recent developed techniques prove the better convergence characteristics and better performance of EMA. Due to limited research and applications performed in power system problems, EMA provides opportunities to explore further. Table1 depicts the comparison between some of the recent optimization techniques [30, 31].

Table 1 Comparison between recent bio-inspired techniques

S. No.	Techniques	Constraints considered	Remarks	References
1	Glow-worm optimization	✓ Transmission losses ✓ Generator's limits constraints ✓ POZs	The authors proposed a multi agent GWO and applied it on a large-scale hydropower system. The technique emerges out to be stronger than other evolutionary algorithms in terms of smooth convergence. It has stronger search capability and also proves robust for complex objective functions	Xiaoyu et al. [11]
2	Elephant herding optimization	✓ Transmission losses ✓ Valve-point loading ✓ Generator's limits constraints ✓ POZs	EHO is capable of handling the system constraints and non-linearity. It converges smoothly and provides optimum solution. It gives tough competition to classical algorithms such as PSO, DE, and GA	Li et al. [12]
3	Water cycle algorithm	✓ Transmission losses ✓ Generator's limits constraints ✓ Ramp rate limits ✓ POZs ✓ Valve-point loading	The simulation results prove the computational efficacy and robustness of WCA. Also, WCA provides competitive solutions and steady convergence when used for large-scale power systems involving various system constraints	Elhameed et al. [15]

(continued)

Table 1 (continued)

S. No.	Techniques	Constraints considered	Remarks	References
4	Flower pollination algorithm	✓ Transmission losses ✓ Valve-point loading ✓ Generator's limits constraints	FPA outperforms the popular hybrid algorithms in terms of efficiency, computation time, and achieves global minimum cost for various dimensions of system	Abdelaziz et al.[16]
5	Salp swarm optimization	✓ Transmission losses ✓ Generator's limits constraints ✓ Ramp rate limits ✓ POZs ✓ Valve-point loading	SSA is applied to power dispatch of multi-area system. It proves to be efficient for larger and complex systems also. It provides smooth and stable cost convergence as compared to other recent techniques	Chaudhary et al. [23]
6	Whale optimization algorithm	✓ Transmission losses ✓ Generator's limits constraints ✓ Ramp rate limits ✓ POZs ✓ Valve-point loading	WOA is highly robust technique and its performance remains unaffected from the uncertainties caused due to system constraints and RES intermittency. CWOA improves the efficacy of WOA in fuel cost minimization. The algorithm performs superior than other modern techniques	Paul et al. [26]
7	Exchange market algorithm	✓ Transmission losses ✓ Valve-point loading ✓ Generator's limits constraints	EMA gives the minimum fuel cost for both single and multi-objective function. The effect of changing the control setting parameters on EMA performance is also shown in the results. EMA provides competitive and economical solutions for all the test systems	Rajan et al. [29]

5 Conclusion

This paper presents an exhaustive study of recently proposed bio-inspired optimization techniques to tackle CEED challenge. A brief classification of the optimization techniques is shown and some of them are discussed in this paper. The present competitive electricity markets need careful planning and management processes to prevent congestion and ensure that the grid runs effectively, safely, and cheaply. With the integration of RES into the system, the complexity increases and optimization becomes more challenging. Owing to the discontinuous nature of RES such as wind energy, and solar power, ELD becomes probabilistic in nature. As a consequence, more advances are made to formulate and optimize the load dispatch problem with various computing techniques.

No single technique can be declared as the best, since every technique has its own benefits and limitations. The performance of these optimization techniques largely relies on the complexity and dimensions of the system. An algorithm may prove to be most efficient and provide best results for a certain system but fail to converge on changing the system constraints. Hence, researchers are provided with ample scope to develop new algorithms or propose better versions of the existing ones by modifying the control parameters.

References

1. Biswas, P.P., Suganthan, P.N., Qu, B.Y., Amaratunga, G.A.J.: Multiobjective economic-environmental power dispatch with stochastic wind-solar-small hydro power. Energy **150**, 1039–1057 (2018)
2. Pathania, A.K., Mehta, S., Rza, C.: Economic load dispatch of wind thermal integrated system using dragonfly algorithm. In: 2016 th India International Conference on Power Electronics (IICPE), Patiala, India, 2016, pp. 1–6. https://doi.org/10.1109/IICPE.2016.8079422
3. Khan, N.A., Awan, A.B., Mahmood, A., Razzaq, S., Zafar, A., Sidhu, G.A.S.: Combined emission economic dispatch of power system including solar photo voltaic generation. Energy Convers. Manage. **92**, 82–91 (2015)
4. Ashok, S.: Optimised model for community-based hybrid energy system. Renewable Energy, Issue **7**(32), 1155–1164 (2007)
5. Zhan, J.P., Wu, Q.H., Guo, C.X., Zhou, X.X.: Fast λ -Iteration method for economic dispatch with prohibited operating zones. IEEE Trans. Power Syst. **29**(2), 990–991 (2014). https://doi.org/10.1109/TPWRS.2013.2287995
6. Lazzerini, B., Pistolesi, F.: A linear programming-driven MCDM approach for multi-objective economic dispatch in smart grids. In: 2015 SAI Intelligent Systems Conference (IntelliSys), London, 2015, pp. 475 484https://doi.org/10.1109/IntelliSys.2015.7361183
7. Subathra, M.S.P., Selvan, S.E., Victoire, T.A.A., Christinal, A.H., Amato, U.: A hybrid with cross-entropy method and sequential quadratic programming to solve economic load dispatch problem. IEEE Syst. J. **9**(3), 1031–1044 (2015). https://doi.org/10.1109/JSYST.2013.2297471
8. Sahay, K.B., Sonkar, A., Kumar, A.: Economic load dispatch using genetic algorithm optimization technique. In: 2018 International Conference and Utility Exhibition on Green Energy for Sustainable Development (ICUE), Phuket, Thailand, 2018, pp. 1–5.https://doi.org/10.23919/ICUE-GESD.2018.8635729

9. Jebaraj, L., Venkatesan, C., Soubache, I., Rajan, C.C.A.: Application of differential evolution algorithm in static and dynamic economic or emission dispatch problem: a review. Renew. Sustain. Energy Rev. **77**, 1206–1220 (2017)
10. Selvakumar, A.I., Thanushkodi, K.: A new particle swarm optimization solution to nonconvex economic dispatch problems. IEEE Trans. Power Syst. **22**(1), 42–51 (2007). https://doi.org/10.1109/TPWRS.2006.889132.s
11. Wang, X., Yang, K.: Economic load dispatch of renewable energy-based power systems with high penetration of large-scale hydropower station based on multi-agent glowworm swarm optimization. Energy Strategy Rev. **26**, 100425 (2019). ISSN 2211–467X
12. Li, W., Wang, G.G., Alavi, A.H.: Learning-based elephant herding optimization algorithm for solving numerical optimization problems. Knowl-Based Syst. **195**, 105675 (2020). ISSN 0950–7051
13. Ghanizadeh, R., Hojber kalali, S.M., Farshi, H.: Teaching–learning-based optimization for economic load dispatch. In: 2019 5th Conference on Knowledge Based Engineering and Innovation (KBEI), Tehran, Iran, pp. 851–856 (2019). https://doi.org/10.1109/KBEI.2019.8734963
14. Mirjalili, S., SCA: a sine cosine algorithm for solving optimization problems. Knowl-Based Syst. **96**, 120–133 (2016). ISSN 0950–7051
15. Elhameed, M.A., El-Fergany, A.A.: Water cycle algorithm-based economic dispatcher for sequential and simultaneous objectives including practical constraints. Appl. Soft Comput. **58**, 145–154 (2017). ISSN 1568-4946
16. Abdelaziz, A.Y., Ali, E.S., Abd Elazim, S.M.: Implementation of flower pollination algorithm for solving economic load dispatch and combined economic emission dispatch problems in power systems. Energy **101**, 506–518 (2016). ISSN 0360–5442
17. Rashedi, E., Nezamabadi-Pour, H., Saryazdi, S.: GSA: a gravitational search algorithm. Inf. Sci. **179**(13), 2232–2248 (2009). ISSN 0020–0255
18. Hatamlou, A.: Black hole: a new heuristic optimization approach for data clustering. Inf. Sci. **222**, 175–184 (2013). ISSN 0020–0255
19. Soni, S.: Using hybrid PSO for solving EELD problem. In: 2020 International Conference on Power Electronics & IoT Applications in Renewable Energy and its Control (PARC), Mathura, Uttar Pradesh, India, pp. 153–158 (2020). https://doi.org/10.1109/PARC49193.2020.236578.
20. Lohokare, M.R., Panigrahi, B.K., Pattnaik, S.S., Devi, S., Mohapatra, A.:Neighborhood search-driven accelerated biogeography-based optimization for optimal load dispatch. In: IEEE Transactions on Systems, Man, and Cybernetics, Part C (Applications and Reviews), vol. 42, no. 5, pp. 641–652 (2012).https://doi.org/10.1109/TSMCC.2012.2190401
21. Santra, D., Mukherjee, A., Sarker, K., Chatterjee, D.: Hybrid PSO-ACO algorithm to solve economic load dispatch problem with transmission loss for small scale power system. In: 2016 International Conference on Intelligent Control Power and Instrumentation (ICICPI), Kolkata, 2016, pp. 226–230. https://doi.org/10.1109/ICICPI.2016.7859707
22. Mirjalili, S., Gandomi, A.H., Mirjalili, S.Z., Saremi, S., Faris, H., Mirjalili, S.H.: Salp Swarm algorithm: a bio-inspired optimizer for engineering design problems. Adv. Eng. Softw. **114**, 163–191 (2017). ISSN 0965-9978
23. Chaudhary, V., Dubey, H.M., Pandit, M., Bansal, J.C.: Multi-area economic dispatch with stochastic wind power using salp swarm algorithm. Array **8**, 100044 (2020). ISSN 2590-0056
24. El Sehiemy, R.A., Selim, F., Bentouati, B., Abido, M.A.: A novel multi-objective hybrid particle swarm and salp optimization algorithm for technical-economical-environmental operation in power systems, Energy **193**, 116817 (2020). ISSN 0360-5442
25. Mirjalili, S., Lewis, A.: The whale optimization algorithm. Adv. Eng. Softw. **95**, 51–67 (2016). ISSN 0965-9978
26. Paul, C., Roy, P.K., Mukherjee, V.: Chaotic whale optimization algorithm for optimal solution of combined heat and power economic dispatch problem incorporating wind, Renew. Energy Focus. **35**, 56–71 (2020). ISSN 1755-0084
27. Ghorbani, N., Babaei, E.: Exchange market algorithm for economic load dispatch. Int. J. Electr. Power Energy Syst. **75**, 19–27 (2016). ISSN 0142-0615

28. Hagh, M.T., Kalajahi, S.M.S., Ghorbani, N.: Solution to economic emission dispatch problem including wind farms using exchange market algorithm method. Appl. Soft Comput. **88**, 106044 (2020). ISSN 1568-4946
29. Rajan, A., Malakar, T.: Optimum economic and emission dispatch using exchange market algorithm. Int. J. Electri. Power Energy Syst. **82**, 545–560 (2016). ISSN 0142-0615
30. Islam, M.M., Nagrial, M., Rizk, J., Hellany, A.: Review of application of optimization techniques in smart grids. In: 2018 2nd International Conference On Electrical Engineering (EECon). https://doi.org/10.1109/eecon.2018.8541016
31. Dubey, H.M., Pandit, M., Panigrahi B.K.: An overview and comparative analysis of recent bio-inspired optimization techniques for wind integrated multi-objective power dispatch. Swarm Evol. Comput. **38**, 12–34 (2018). ISSN 2210-6502

Enhanced Multigradient Dilution Preparation

Meenakshi Sanyal and Somenath Chakraborty

Abstract In our paper, the new algorithm enhanced multigradient dilution preparation (EMDP) is discussed. This new algorithm is reported with a laboratory on-chip or digital microfluidic biochip to operate multiple operation on a tiny chip. We can use digital microfluidic biochip to operate multiple operation on a tiny chip. Samples are very costly which are used in any biochemical laboratory protocols. For the case of fast and high throughput application, it is essential to minimize the cost of operations and the time of operations and that is why one of the most challenging and important phase is sample preparation. In our proposed algorithm, we have hide to reduce sample droplets and waste droplets and for this purpose, waste recycling is used, when different series of multigradient targets concentration factors (CFS) are generated. We have compared our proposed algorithm with recent dilution techniques such as MTC, REMIA, and WARA. For the storage of intermediate droplets which, and generated during this process, on-chip storage space 0(n) is needed.

Keywords Digital microfluidic biochip · Drug discovery · Sample preparation · Electrowetting

1 Introduction

Digital microfluidic (DMF) is biochip or "Lab On Chip (LOC)" is one of the most emerging fields of research in VLSI technology that is very large-scale integration technology. In "More than Moore" technical trends, for the case of diabetes, cardiovascular diseases, DMF biochips help to reduce the healthcare cost. Compare to all biochemical laboratory protocols, DMF biochips are faster.

M. Sanyal
Saroj Mohan Institute of Technology, Hooghly, India

S. Chakraborty (✉)
The University of Southern Mississippi, Hattiesburg, USA
e-mail: Somenath.Chakraborty@usm.edu

K. N. Das et al. (eds.), *Proceedings of the International Conference on Computational Intelligence and Sustainable Technologies*, Algorithms for Intelligent Systems,
https://doi.org/10.1007/978-981-16-6893-7_46

Current flows in the biochip for routing the microfluidic droplets in biochip while in general information processing current does flow in the integrated circuits.

In our proposed algorithm, we have prepared multimode target dilution with minimum number of waste droplets and sample droplets. Sample droplets are very much costly and that is why, in our proposed algorithm, we have tried to minimize it for any number of dilution gradients.

Digital microfluidic biochip is also essential in many real-life application, and drug discovery is one of them. With a certain concentration, we can charge the antibiotic, so that we can kill the bacteria. But if we use the antibiotics excessively, them some biological disorder can be occurred in human body. So for avoiding the effect of toxicity caused by various kinds of biochemical, it is essential to apply the popper amount of drug. So for generating the exact concentration of antibiotics, we have to apply the laboratory process for generating the multigradient dilution preparation. But this process takes a lot of time as well as to generate multigradient dilution, more reagent is needed.

In our proposed algorithm, for preparing the multigradient dilution, we have discussed the linear, harmonic, logarithmic, and parabolic series. In the DMF biochip, our main objective is to reduce the usage of the sample droplets, because in most of the cases, they are very costly, and our another objective is to optimize the time and which is possible only with biochips. A generic algorithm, Mitra ET AL proposed, to prepare the multigradient dilution for digital biochip, a generic algorithm. To reduce the usage of sample droplets, waste droplets and number of mix-split steps, all the efforts are given.

The rest of the paper is organized by the following process. In Sect. 2, the basic idea of DMF biochip is discussed.

The architecture of the DMF Biochip cells, the mixing, and splitting of the droplets are discussed in Sect. 3; we have discussed the review work. The proposed algorithm is discussed in Sect. 4. In Sect. 5, we have discussed the result and comparisons. At the end in Sect. 6, the conclusion has been drawn.

2 Literature Review

Recent years, digital microfluidic supported biochips are widely used in many different kinds of fields like clinical diagnostics, biology manufactory, environmental monitor, analytical chemistry, and military [1–3]. Digital microfluidic-based biochips have received considerable attention as a promising platform for lab-on-a-chip [4] realizations. Such composite systems manipulate fluids at a nanolitre or at picolitre volume scale [5]. By this procedure, the digital microfluidic-based biochips simplify cumbersome laboratory procedures.

In various platforms [6–24], these systems can be used. For real-time bio-molecular detection, and recognition [25], for massively parallel DNA analysis, and for faster and cheaper clinical diagnosis, digital microfluidic-based biochips are widely used. These systems can be used for immediate point-of-care health services, for the management of bio-terrorism threats, and for real-time environmental toxicity monitoring [4]. In the chip design, the placement digital microfluidic-based biochips is one of the key points. The physical potion of each biochemical analysis operation can be found with the smallest biochip area and the shortest completion, through the digital microfluidic-based biochips [26–28].

3 Review Work

In this century, various kinds of sample preparation have been established. waste recycling algorithm (WARA), reactant minimization algorithm (REMIA), and multi-target concentration (MTC) algorithm are most valuable among them. The first algorithm is REMIA. For the preparation of sample, it reduces the usage of the reactant. The skewed mixing tree generation method is the main idea of REMIA. For "More and Moore" sample minimization, according to this method, the source droplet with much more higher concentration will be kept closer to the root. To explain the ex-potential dilution phase, REMIA uses exponential dilution trees. In this method, the concentration value of the droplet is denoted by each node of exponential dilution trees. In this method, from the source droplet to the target droplet concentration one directed edge is pointed. And the concentration value of the root node is 1. Since two resultant droplets are generated for exponential dilution, so for this case the out-degree is 2. In this method, one buffer droplet is needed for each operation.

The exponential dilution operation is represented by each branch note. Therefore, the number of the dilution operation is implied by the total number of branch node. In WARA, it focused mainly on waste recycling for waste as well as sample minimization. The interpolated dilution process by divided by WARA into three phases:

1. Mixing tree generations
2. Droplet sharing
3. Droplet replacement.

Nowadays, a proposed algorithm is for multi-target concentration. To ensure the highest optimization in number of mix-split steps, number of waste droplets, and the number of sample droplets, a dilution graph-based approach has been used.

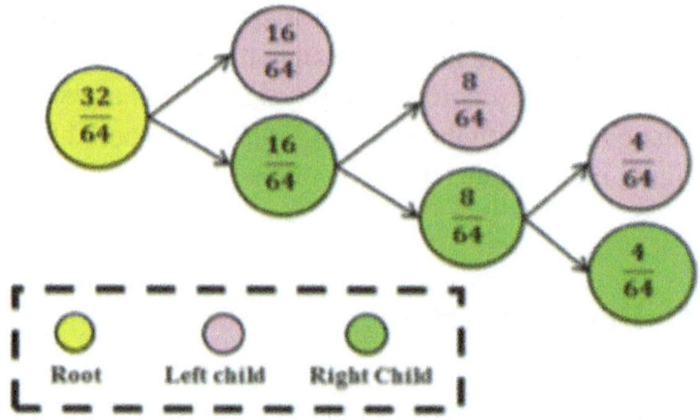

Fig. 1 Droplet creation [20]

4 Proposed Work

4.1 Algorithm

In our proposed algorithm, EMDP here sample droplets (100%) dilute buffer (0%) and recycles the waste droplets for generating the required CF. Here in our proposed method, we divide the dilution process into three steps.

(1) Generation of serial dilution tree.
(2) Creation of droplets with required CF.
(3) Generating of the target series.

4.2 Generating Serial Dilution Tree

We construct serial dilution tree (SDT) by the process serial dilution 1:1 mixing model is followed by the serial dilation tree (SDT). The nodes of the SDT represent the concentration factor of the droplets. The root of the SDT is the source droplets. Firstly, the SDT involves one root droplet and one buffer droplet, and it produces two identical droplets. Among them one droplet for the next dilution and the another droplet for storage. We have mentioned the method in Figs. 1, 2, 3, and 4.

4.3 Droplet Creation

Here, in use proposed algorithm, we will create the required CF by diluting the immediate higher concentration droplet with buffer droplet or any other immediate droplet.

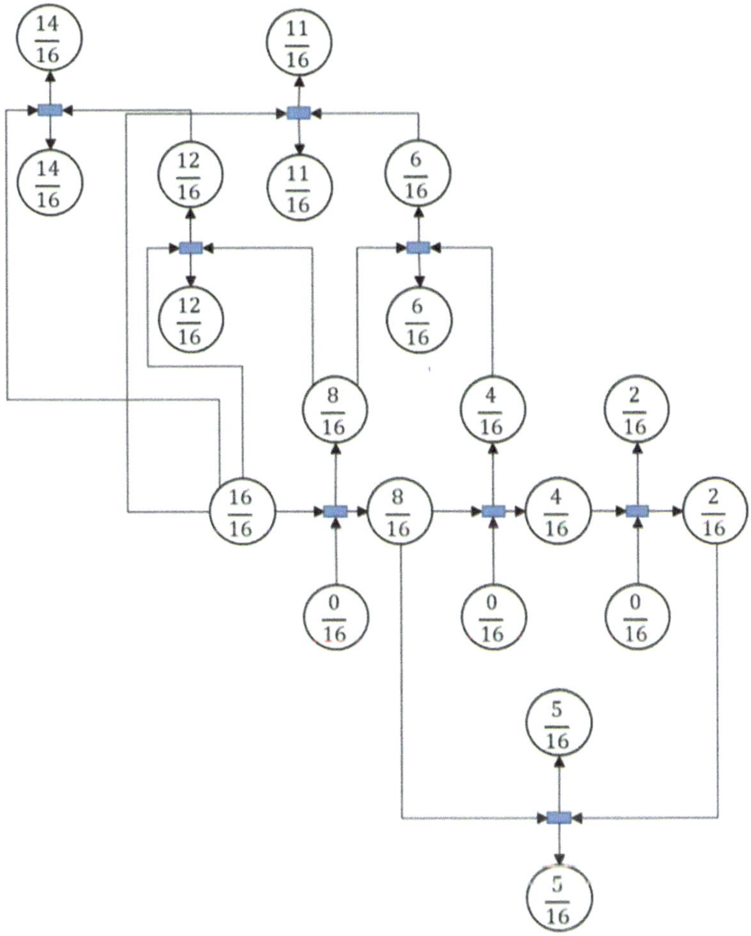

Fig. 2 Dilution Tree 1

4.4 EMDP (Enhanced Multigradient Dilution Preparation):

Algorithm 1 *Serial Dilution Tree (root, target)*

Input: Source droplet and target droplet concentration, sufficient buffer droplet
Output: A set of CFs

1. If root concentration = sample concentration
2. Number of sample = number of sample + 1
3. End if
4. If (root/2) > = target
5. Create two child note using mix and split(root,0)
6. Insert left child to storage

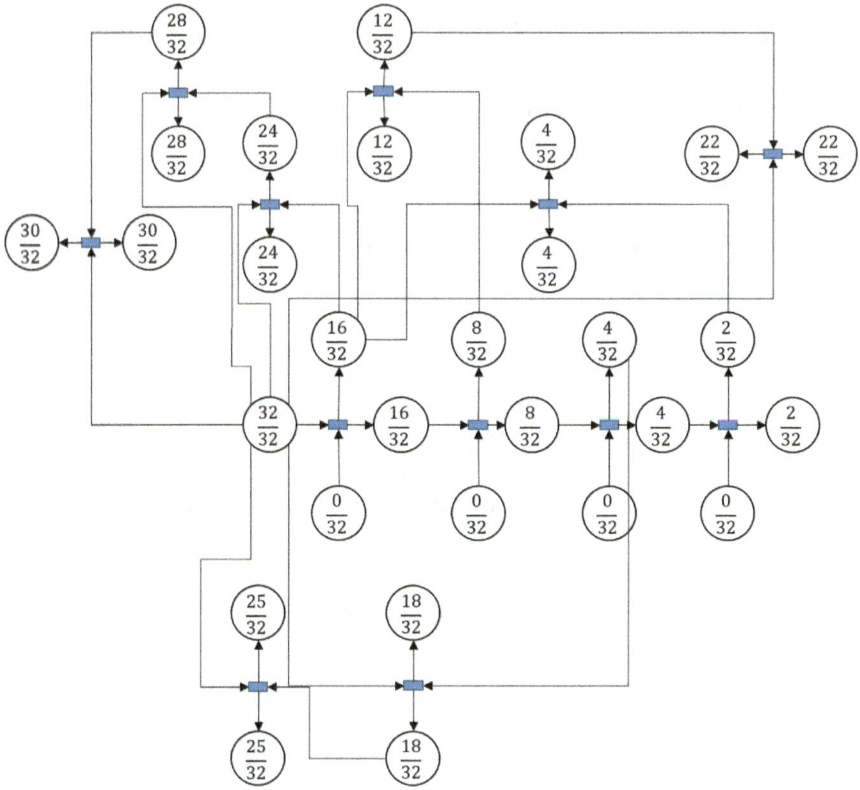

Fig. 3 Dilution Tree 2

7. Number of buffer = number of buffer + 1
8. End while
9. Insert source into storage; circular queue

Algorithm 2 *Create Droplet*

1. Let least higher CF value be stand into value and initially
   Value = 0.
2. While value = 0.
3. Search immediate higher concentration in the storage for required CF.
   Value = resultant droplet CF
4. If search is unsuccessful.
5. SDT (Sample concentration, required CF)
6. Else if value > (2 * CF of required droplet) + 1/2^n.
7. Generate series (resultant droplet CF, required CF).
8. Value 0.
9. End if.

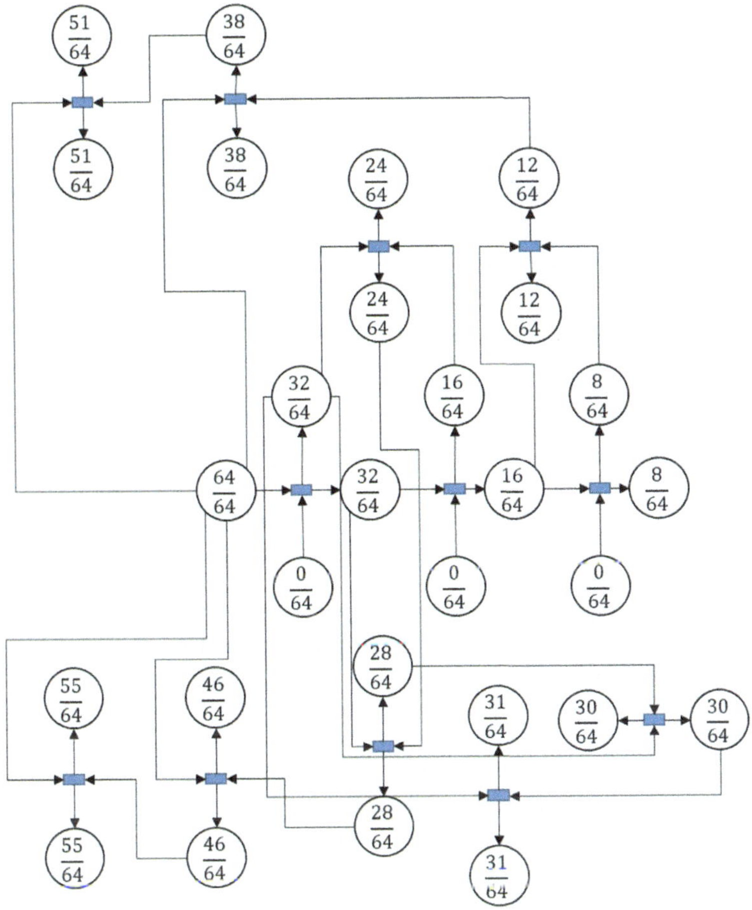

Fig. 4 Dilution Tree 3

Algorithm 3 CF Search.

1. do until all CFs are generated.
2. if CF value = sample concentration.
3. number of sample = number of sample + 1.
4. end if
5. Search the CF value
6. if search is unsuccessful, continue with next CF
7. else
8. create droplet
9. end if

4.5 Dilution Trees

$TS_1 = \left\{\frac{5}{16}, \frac{11}{16}, \frac{14}{16}, \frac{16}{16}, \frac{14}{16}, \frac{11}{16}, \frac{5}{16}\right\}$ Sample: 5, Buffer: 4, Waste: 2, Steps: 8

Explanation of Dilution Tree 1:

An example is shown in the above figure, where the required CF values are as follows:

$$\left\{\frac{5}{16}, \frac{11}{16}, \frac{14}{16}, \frac{16}{16}, \frac{14}{16}, \frac{11}{16}, \frac{5}{16}\right\}$$

Initially, the number of sample droplet = 0, the number of buffer droplet = 1.

1. In the required CFs, the first required droplet concentration 5/16 is not equal to the sample concentration, i.e., 16/16 (100%). Initially, storage is empty.
   So, {16/16, 0/16} mixed split operation is executed until an intermediate higher CF droplet is found. 8/16, 4/16, 4/16 are stored in the storage for further use. The series can be generated with two sample droplet and two buffer droplets in two steps. Now the immediate higher droplet of 5/16 is 8/16
2. According to the algorithm 2*5/16 – 8/16 = 2/16 is another CF, which is required to create 5/16 and it is not present in the storage.
   So, it needs to create with two sample procedure. The immediate higher CF of 2/16 is 4/16 which is available in the storage which is equal to the twice of 2/16. So, a buffer droplet will be used to dilute the immediate buffer droplet and produce 2/16.
3. {2/16, 8/16} are mix split to produce 5/16 using one step. Now the extra unused droplets are stored within the storage. Hence, 2/16 is in the storage at present.
4. Now the next CF value, 11/16, need to be generated. {8/16, 4/16} are mix split to produce 6/16 in one step. Now the extra one unused droplet 6/16 is stored in the storage.
   Now, another droplet 6/16 is there. We mix split {16/16, 6/16} to produce 11/16 in one step.
5. In this way, the remaining droplets will be generated.

Finally, the total dilution process is complete in eight steps with five sample droplets and four buffer droplets.

$$TS_2 = \left\{\frac{9}{32}, \frac{18}{32}, \frac{25}{32}, \frac{30}{32}, \frac{32}{32}, \frac{28}{32}, \frac{22}{32}\right\}$$

Sample: 8, Buffer: 5, Waste: 4, Steps: 12.

$$TS_3 = \left\{\frac{31}{64}, \frac{55}{64}, \frac{64}{64}, \frac{51}{64}\right\}$$

Sample: 6, Buffer: 4, Waste: 8, Steps: 12.

5 Result and Discussion

Here in Table 1, we present the computed result of EMDP algorithm. In this section, we have described the results of multigradient dilution preparation with EMDP algorithm. There we have reported, the series of dilutions TS_1 Ts_{10} and use have shown the results which is Table 1. Here we have found that, for the test series TS_1, it requires the numbers of sample droplets are 5, number of buffer droplets are 4, and the number of waste droplets are 2. Similarly for TS_2 TS_{10}. We have shown the results in Tables 1 and 2.

Table 1 Test series for multigradient dilution preparation with EDMP algorithm

	# Waste (W)						
Test Series	RTWM	REMIA	WARA	RSM	MTC	SWDM	EMDP
TS1	14	6	6	8	8	3	2
TS2	9	2	2	11	12	5	4
TS3	14	4	4	10	10	6	8
	#Sample (S)						
Test Series	RTWM	REMIA	WARA	RSM	MTC	SWDM	EMDP
TS1	15	8	8	7	8	6	5
TS2	17	11	11	9	11	7	8
TS3	17	6	6	7	9	6	6
	#Buffer(B)						
Test Series	RTWM	REMIA	WARA	RSM	MTC	SWDM	EMDP
TS1	9	5	3	3	6	4	4
TS2	10	6	5	5	8	3	5
TS3	8	3	3	4	3	4	4

Table 2 Comparative study of different algorithms RTWM, REMIA, WARA, RSM, and MTC [20]

Test series	CF_s	S	B	W	Steps
TS_1	$\left\{\frac{5}{16}, \frac{11}{16}, \frac{14}{16}, \frac{16}{16}, \frac{14}{16}, \frac{11}{16}, \frac{5}{16}\right\}$	5	4	2	8
TS_2	$\left\{\frac{9}{32}, \frac{18}{32}, \frac{25}{32}, \frac{30}{32}, \frac{32}{32}, \frac{28}{32}, \frac{22}{32}\right\}$	8	5	4	12
TS_3	$\left\{\frac{31}{64}, \frac{55}{64}, \frac{64}{64}, \frac{51}{64}\right\}$	6	4	8	12

6 Conclusion

In our proposed algorithm EMDP, it can be used for sample and waste aware biochemical on chip protocol. In this method, the prime requirement is the recycling of waste droplets from storage for successful multigradient dilution preparation. The major contribution of our proposed algorithm is to minimize the sample droplets as well as the waste droplets to generate different multigradient sample preparation. We have shown the result for different test series Ts_1-Ts_{10}. In comparison with the others, the costly sample requirement is much less and also the lesser amount of waste it produces and that is why the cost of the sample droplets is minimized. Here we have compared the result between multigradient dilution preparation algorithm and the proposed EMDP. The comparative study shows that overall nearby 50% minimization in the use of sample droplets and waste droplets are achieved using the proposed EMDP. Hence, it can be concluded that the proposed EMDP algorithm and may be used for losing cost sample preparation for on-chip bio-chemical process.

References

1. Hull, H.: Smallpox and bioterrorism. Public-health Responses. Lab. Clin. Med.**142**, 4 221–228 (2003)
2. Zeng, X.F., Yue, R.F., Wu, J.G., et al.: Creating liquid droplets by electro wetting-based actuation for digital microfluidics-biochips. Sci China Tech Sci. **36**(1), 105–112 (2006)
3. Yang, J.S., Zuo, C.C., Lian, J., et al.: Droplet scheduling algorithm for digital microfluidics-based biochips. J. Jilin Univ. (Eng. Technol. Edn) **37**(6), 1380–1385 (2007)
4. Chakrabarty, K., Xu, T.: Digital microfluidic biochips: design and optimization. CRC Press, Boca Raton, FL, USA (2010)
5. Herold, K.E., Rasooly, A.: Lab-on-a-Chip technology (Vol. 1):fabrication and microfluidics. Norfolk, U.K.: Caister Academic Press (2009)
6. Yuh, P.H., Yang, C.L., Chang, Y.W.: Placement of digital microfluidic biochips using the t-tree formulation. In: 43rd Design Automation Conference. San Francisco, pp. 4–28 (2006)
7. Yuh, P.H., Yang, C.L., Chang, Y.W.: Placement of defect-tolerant digital microfluidic biochips using the t-tree formulation. ACM J. Emerg. Technol. Comput. Syst. **33**, 13–45 (2007)
8. Roy, S., Bhattacharya, B.B., Chakrabarty, K.: Optimization of dilution and mixing of biochemical samples using digital microfluidic biochips. IEEE Trans Comput.-Aided Des. Integr. Circuits Syst. **29**(11), 1696–1708 (2010)
9. Griffith, E.J., Akella, S., Goldberg, M.K.: Performance characterization of a reconfigurable planar-array digital microfluidic system. IEEE Trans Comput.-Aided Des. Integr. Circuits Syst. **25**(2), 345–357 (2006)
10. Thies, W., Urbanski, J.P., Thorsen, T., Amarasinghe, S.: Abstraction layers for scalable microfluidic biocomputing. Nat. Comput. **7**(2), 255–275 (2008)
11. Roy, S., Bhattacharya, B.B., Chakrabarty, K.: Waste-aware dilution and mixing of biochemical samples with digital microfluidic biochips. In: Proceeding of IEEE/ACM DATE Conference, Grenoble, France, pp. 1059–1064 (2011)
12. Chiang, T.-W., Liu, C.-H., Huang, J.-D.: Graph-based optimal reactant minimization for sample preparation on digital microfluidic biochips. In: Proceeding of International Symposium VLSI-DAT, Hsinchu, Taiwan, pp. 1–4 (2013)

13. Hsieh, Y.-L., Ho, T.-Y., Chakrabarty, K.: On-chip biochemical sample preparation using digital microfluidics. In: Proceeding of International Conferences BioCAS, San Diego, CA, USA, pp. 301–304 (2011)
14. Chakraborty, S., Murali, B.: A Novel Medical Prognosis System for Breast Cancer. In: Mandal J.K., Buyya R., De D. (eds) Proceedings of International Conference on Advanced Computing Applications. Advances in Intelligent Systems and Computing, vol 1406. Springer, Singapore (2022). https://doi.org/10.1007/978-981-16-5207-3_34
15. Huang, J.-D., Liu, C.-H., Chiang, T.-W.: Reactant minimization during sample preparation on digital microfluidic biochips using skewed mixing trees. In: Proceedings of IEEE/ACM ICCAD, San Jose, CA, USA, pp. 377–384 (2012)
16. Bhattacharjee, S., Banerjee, A., Bhattacharya, B.B.: Sample preparation with multiple dilutions on digital microfluidic biochips. IET Comput. Digit. Tech. **8**(1), 49–58 (2014)
17. Huang, J.-D., Liu, C.-H., Lin, H.-S.: Reactant and waste minimization in multitarget sample preparation on digital microfluidic biochips. IEEE Trans Comput.-Aided Design Integr. Circuits Syst. **32**(10), 1484–1494 (2013)
18. Chakraborty, S., Zhang, C.:Survival prediction model of renal transplantation using deep neural network. In: 2020 IEEE 1st International Conference for Convergence in Engineering (ICCE), pp. 180–183 (2020). https://doi.org/10.1109/ICCE50343.2020.9290695
19. Mitra, D., Roy, S., Bhattacharjee, S., Chakrabarty, K., Bhattacharya, B.B.: On-Chip sample preparation for multiple targets using digital microfluidics. IEEE Trans. Comput. Aided Des. Integr. Circuits Syst. **33**(8), 1131–1144 (2014). https://doi.org/10.1109/TCAD.2014.2323200
20. Podder, D.., Das, K.:Harmonic dilution gradient preparation for digital microfluidic biochip. In: 2015 Annual IEEE India Conference (INDICON), pp. 1-6 (2015).https://doi.org/10.1109/INDICON.2015.7443726
21. Chakraborty, S., Murali, B.: Investigate the correlation of breast cancer dataset using different clustering technique. ArXiv abs/2109.01538 (2021)
22. Chakraborty, S.: Category identification technique by a semantic feature generation algorithm. In: Deep Learning for Internet of Things Infrastructure, pp. 129–144. CRC Press
23. Chakraborty, S., Saha, B., Bandyopadhyay, S.K.: Dynamic channel allocation in IEEE 802.11 networks. Int. J. Comput. Appl. **975**, 8887 (2016)
24. Chakraborty, S., Bandyopadhyay, S.K.: Scene text detection using modified histogram of oriented gradient approach. IJAR **2**(7), 795–798 (2016)
25. Srinivasan, V., Pamula, V.K., Fair, R.B.: Droplet-based microfluidic lab-on-a-chip for glucose detection. Anal. Chimica Acta **507**(1), 145–150 (2004)
26. Srinivasan, V., Pamula, V.K., Fair, R.B.: An integrated digital microfluidic lab-on-a-chip for clinical diagnostics on human physiological fluids. Lab Chip. **4**(4), 310–315 (2004)
27. Su, F., Chakrabarty, K.: Module placement for fault-tolerant microfluidics-based biochips. ACM Trans. Des. Autom. Electron Syst. **11**(3), 682–710 (2006)
28. Su, F., Chakrabarty, K.: Unified high-level synthesis and module placement for defect-tolerant microfluidic biochips. In: 42nd Design Automation Conference. California, vol. 49, pp. 825–830 (2005)

Breakdown Voltage Improvement in AlGaN/GaN HEMT by Introducing a Field Plate

Pichingla Kharei, Achinta Baidya, and Niladri Pratap Maity

Abstract Exceptional features for high-frequency, power, low noise, and adverse environment applications have been demonstrated by gallium nitride (GaN) based on high electron mobility transistors (HEMTs). GaN devices are capable of high-power performance owing to their high breakdown voltages (Vbr). The use of a field plate (FP) increases Vbr which overall improves the device performance. Different GaN HEMT structures and their characteristics are studied with the help of TCAD simulator. The study represents the improvement of breakdown voltage characteristics using a field plate. The comparison between different positions of field plates is also carried out. The acquired Vbr are 260 V, 259 V, 991 V, and 977 V for without-FP, drain-FP, source-FP, and gate-FP, respectively.

Keywords HEMT · Field plate (FP) · Breakdown voltage (Vbr)

1 Introduction

A major component of any electronic power system is the power switching device, the qualities of which generally dictate what frequencies and power levels an electronic power system work at. For a long time, silicon (Si) was the chosen material for power devices due to its ease of processing, quality, and abundance of information on its material properties [1]. However, Si devices, identified by the intrinsic properties of the material, surpass their operating limits. Given this, it is important to evaluate new materials for the production of power semiconductors. For devices in power switching applications, particular attention is given to material properties which affect the performance directly. GaN and associated alloys, which are nitride-based wide bandgap semiconductors, have attracted considerable attention as a basis for the production of high-frequency, high-speed, high-voltage, and high-power electronic devices [2]. GaN has been recognized by researchers worldwide as the chosen large band gap semiconductor material, due to its high saturation velocity, high mobility,

P. Kharei (✉) · A. Baidya · N. P. Maity
Department of Electronics and Communication Engineering, Mizoram University, Aizawl, Mizoram 796004, India

K. N. Das et al. (eds.), *Proceedings of the International Conference on Computational Intelligence and Sustainable Technologies*, Algorithms for Intelligent Systems,
https://doi.org/10.1007/978-981-16-6893-7_47

high thermal conductivity, and high electric field [3]. The wide band gap has the capacity of sustaining a high electric field before breakdown. The 2DEG is formed in the channel by introducing fixed positive charge at AlGaN/GaN interface [4]. Owing to the high saturation velocity and the presence of 2DEG, the high output current can also be realized [5]. To improve power devices' Vbr, various techniques such as guard rings, junction termination extension [6], and FPs [7–9] are employed. Among these various techniques, FP has the least fabrication complexity [10, 11]. There are also different field plate structures, such as single-FP [9], dual-FP [12, 13], inner [14], slant [15], multi-finger [16], multistep [17], and air-bridge [18] FPs. Challenges faced by the current researchers include improving the Vbr, enlarging the cut-off frequency (f_T), and reducing the device's ON-resistance (R_{ON}). Increasing the Vbr often results in a trade-off between the f_T and R_{ON}. As a result, a good physical design for a device is required to improve the Vbr while minimizing R_{ON} and increasing f_T [19, 20]. With the TCAD software [21], the influence of the FPs on the Vbr is demonstrated in this paper.

2 Device Structure

The proposed structure includes an n-doped AlGaN barrier layer with a thickness of 20 nm and a GaN channel of 50 nm. The SiN passivation layer thickness of 1200 nm, deposited over the barrier layer. The length of the gate is 0.5 μm. Space between gate and drain (Lgd) is 5.1 μm and that of source and gate (Lsg) is 0.4 μm. The work function of the Schottky gate contact is 5.2 eV. FP length is varied from 1 to 2.6 μm for different AlGaN/GaN HEMT structures to study the Vbr. For simulations, ATLAS TCAD device simulator is used. The AlGaN and GaN channel doping concentrations are 1×10^{16} cm^{-3} and 1×10^{15} cm^{-3}, respectively, and the AlGaN barrier has a mole fraction of 0.29. In this experiment, various models such as Lombardi (CVT) model, Shockley–Read–Hall recombination (srh) model, field mobility model (fldmob), and Selberherr impact ionization models (selb) are used. Different structures of AlGaN/GaN HEMTs and their cross-sectional view are given in Figs. 1 and 2, respectively; in Table 1, physical parameters used in this experiment are listed. SiN is used as the passivation layer due to the fact that it provides 38% more Vbr than SiO_2 [22].

3 Result and Discussion

Different architectures of AlGaN/GaN HEMT are considered for the current investigation, and the details of the structures are mentioned in Table 2. These structures transfer characteristics are shown in Fig. 3, at $V_{DS} = 0.05$ V and at gate voltage (V_G) sweeping to 0 V from -4 V.

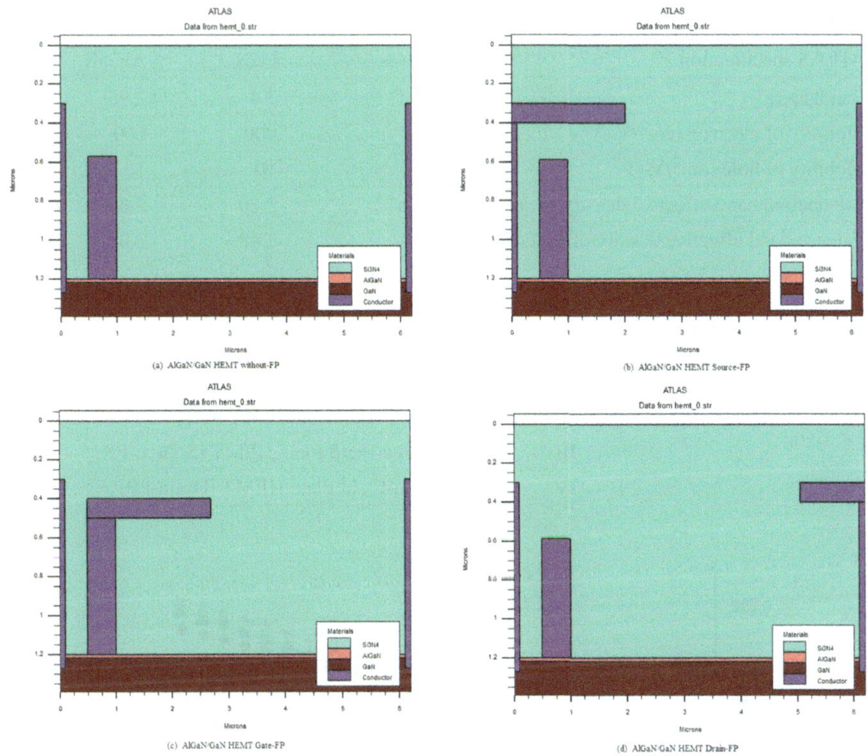

Fig. 1 Different AlGaN/GaN HEMT structures with and without-FP

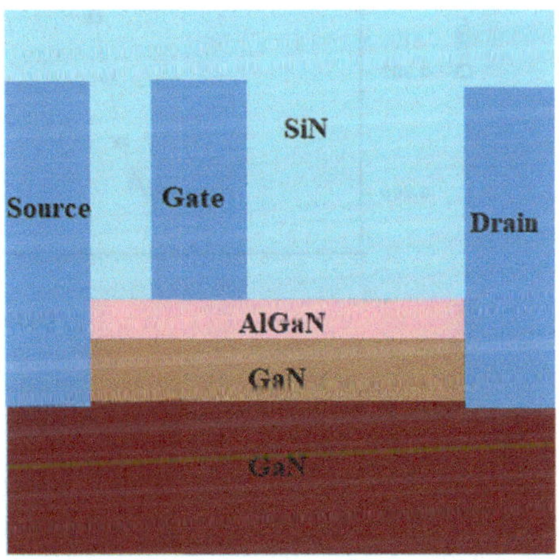

Fig. 2 View of a cross section of a standard AlGaN/GaN HEMT without-FP

Table 1 AlGaN/GaN HEMT physical specification

ATLAS specification	GaN	AlGaN
Bandgap (eV)	3.4	3.96
Mobility of electron (cm^2/V-s)	900	600
Mobility of hole (cm^2/V-s)	10	10
Conduction band effective density of states (10^{18} cm^{-3})	2.20	2.68
Valence band effective density of states (10^{19} cm^{-3})	2.48	2.03

Table 2 Various FP architectures of AlGaN/GaN HEMT

Cases	Architectures
I	Without-Field Plate HEMT (Without-FP)
II	Gate-Field Plate HEMT (Gate-FP)
III	Source-Field Plate HEMT (Source-FP)
IV	Drain-Field Plate HEMT (Drain-FP)

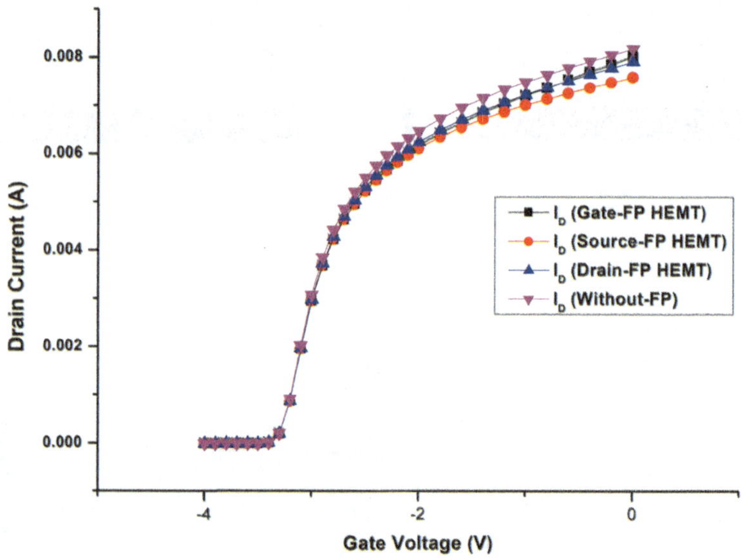

Fig. 3 AlGaN/GaN HEMT transfer characteristics for different structures

Vbr for each AlGaN/GaN HEMT structure is measured with varying lengths of the FP. As for the first case, i.e., AlGaN/GaN without-FP HEMT structure, the Vbr is limited to Vbr = 260 V as given in Fig. 4. For case II, the highest Vbr = 977 V is attained at $l = 1.7$ μm. Vbr at various FP lengths ranging from 1.1 μm to 2.6 μm are given in Figs. 5 and 6 for AlGaN/GaN HEMT Gate-FP and source-FP structures.

For AlGaN/GaN HEMT source-FP structure, Vbr = 991 V is the highest value and it is obtained at FP length of $l = 1.9$ μm. It is also the highest Vbr obtained in this

Fig. 4 Vbr for without-FP structure

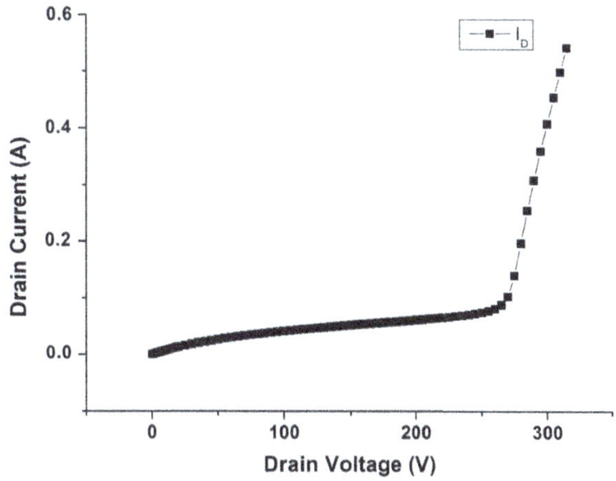

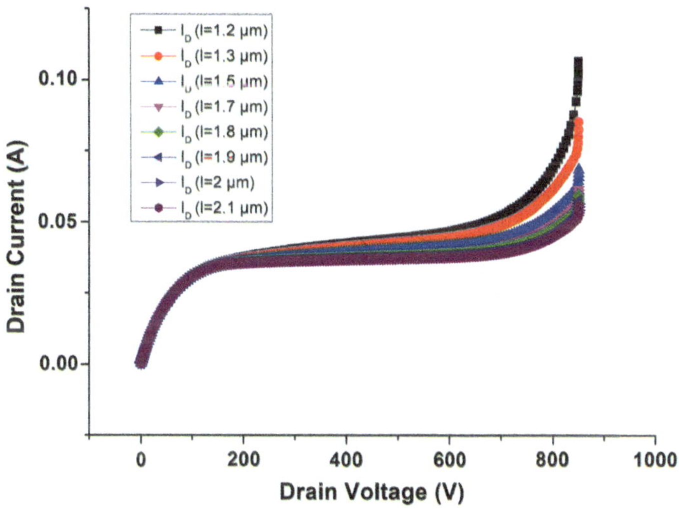

Fig. 5 Vbr for Gate-FP structure at various FP lengths

comparative analysis. For AlGaN/GaN drain-FP HEMT structure, no improvement can be seen in the Vbr as compared to the standard structure with no FP. There is in fact, no significant change in the Vbr as we vary the length of the FP. Different properties of these structures are measured and are given in Table 3. It is observed that R_{ON} also varies with the change in FP length. It increases as we increase the FP length. R_{ON} is highest in AlGAN/GaN HEMT Source-FP structure with a value of 6.58 Ω. Leakage current is found to be higher in standard structure without-FP and in drain-FP structure with 0.54 A and 0.50 A, respectively. It was shown that reduction in the drain leakage current leads to improvement in Vbr [23, 24] (Figs. 7, 8, and 9).

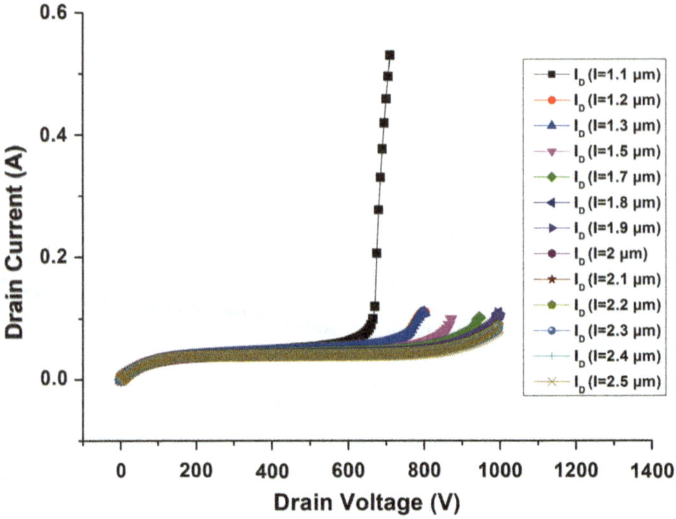

Fig. 6 Vbr for Source-FP structure at various FP lengths

Table 3 Comparison of parameters of various AlGaN/GaN HEMT structures

AlGaN/GaN HEMT structure	R_{ON} (Ohms)	Cgs_max (F/μm)	Cgd_max (F/μm)	gm_max 1/Ohms	f_T (Hz)	fmax (Hz)	Vbr (V)	Ids_Leakage (A)
Without-FP	6.11	2.21×10^{-12}	7.27×10^{-13}	0.010	0.55×10^9	1123	260	0.54
Gate-FP HEMT	6.23	2.34×10^{-12}	8.44×10^{-13}	0.007	0.39×10^9	828	977	0.12
Source-FP HEMT	6.58	2.48×10^{-12}	7.00×10^{-13}	0.007	0.39×10^9	866	991	0.10
Drain-FP HEMT	5.32	2.24×10^{-12}	7.35×10^{-13}	0.008	0.43×10^9	884	259	0.50

In AlGaN/GaN HEMTs, intrinsic capacitance is essential and is utilized to define the f_T. For several V_G, from −5 V to 5 V, the gate-to-drain capacitance (Cgd) and gate-to-source capacitance (Cgs) are studied and the data is examined. For this purpose, AC analysis of various AlGaN/GaN HEMT architectures given in Table 2 was done. Due to electrons accumulation in the channel, the capacitance for the AlGaN/GaN HEMT structure with FPs increases. The maximum value of transconductance (gm) as well as the highest fmax is obtained in the structure without-FP (Fig. 11). The degradation in fmax as compared to the conventional one is 0.73, 0.77, 0.78 in AlGaN/GaN HEMT with Gate-FP, Source-FP, and Drain-FP, respectively (Figs. 10 and 12).

Fig. 7 Vbr of drain-FP structure for different FP lengths

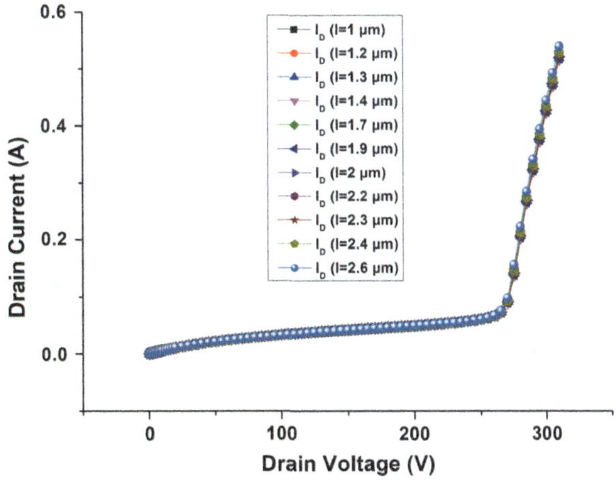

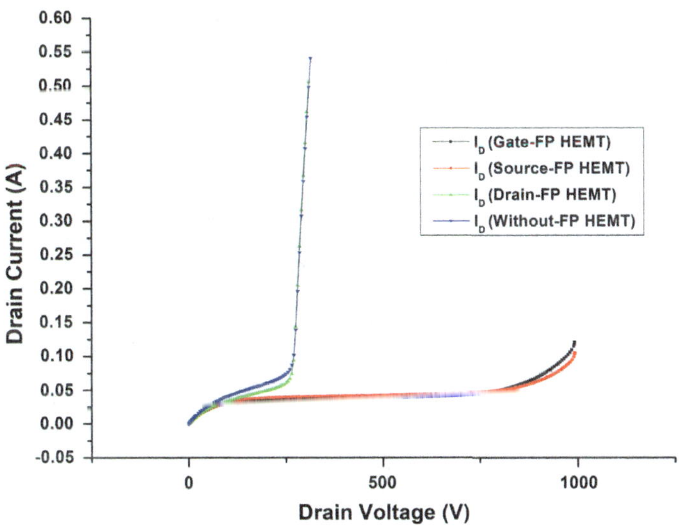

Fig. 8 Breakdown characteristics for different positions of field plate

4 Conclusion

The breakdown voltage of AlGaN/GaN HEMT with SiN passivation and FP structures are investigated. According to the analysis, the device without a FP has a smaller breakdown voltage as compared to the device with a FP. HEMT structures with different FP positions are investigated. The comparative analysis shows that the structure with gate-FP and source-FP AlGaN/GaN structures provides higher breakdown voltage compared to the structure without-FP. Even the filed plate positioned at drain did not perform well with respect to the breakdown voltage. Moreover, SiN as

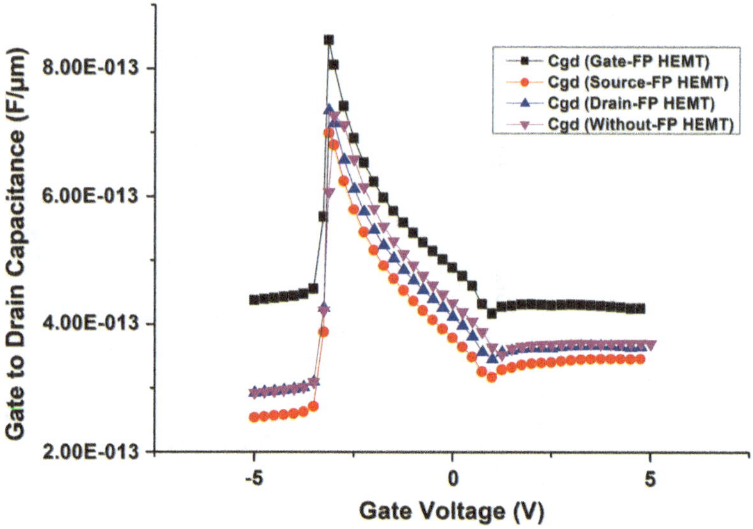

Fig. 9 Gate to drain capacitance of with and without-FP AlGaN/GaN HEMT

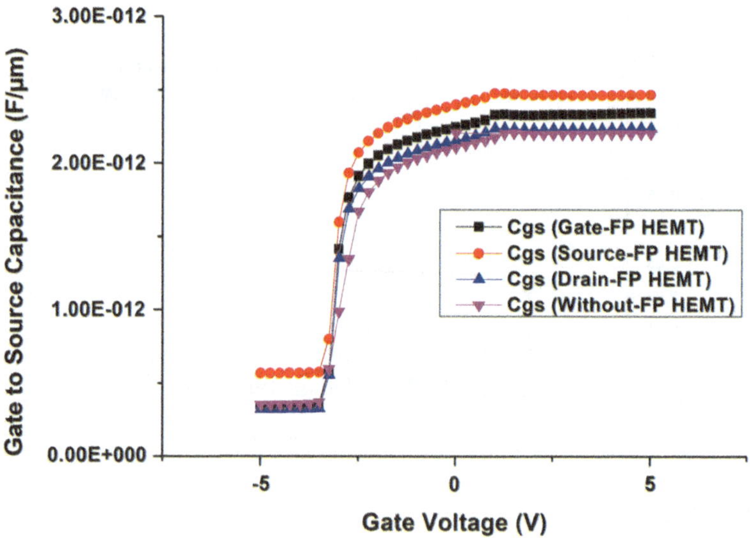

Fig. 10 Gate to source capacitance of with and without-FP AlGaN/GaN HEMT

the passivation layer increases breakdown voltage as compared to SiO_2. Further analysis of other analog characteristics like transconductance, capacitance, and maximum frequency shows that inclusion of field plate in the structure has negative effect on these properties. So, the field plate and its position should be optimized to get suitable performance of the HEMT.

Fig. 11 Transconductance of with and without-FP AlGaN/GaN HEMT

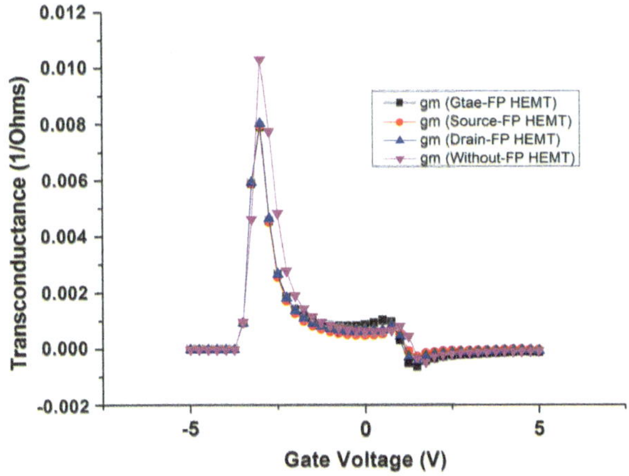

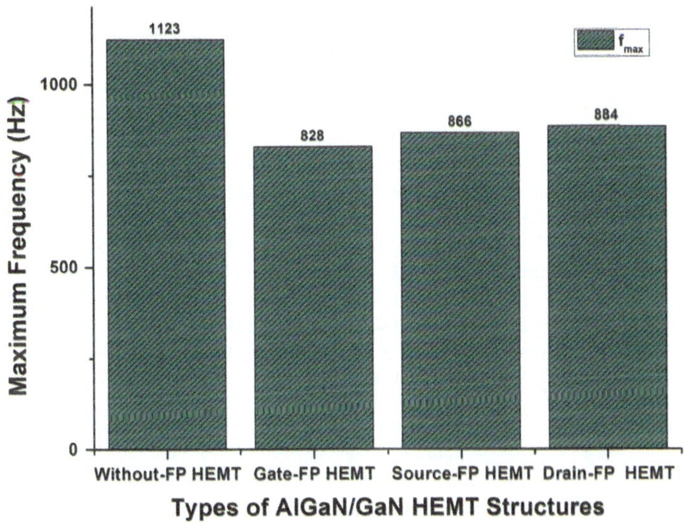

Fig. 12 Comparison of maximum frequency, fmax for different structures

References

1. Flack, T., Pushpakaran, B., Bayne, S.: GaN technology for power electronic applications: a review. J. Electron. Mater. **45** (2016).https://doi.org/10.1007/s11664-016-4435-3
2. Sahoo, D.K., Lal, R.K., Kim, H., Tilak, V., Eastman, L.F.: High-Field effects in silicon nitride passivated GaN MODFETs. IEEE Trans. Electron Devices **50**(5), 1163–1170 (2003)
3. Mishra, U.K., Parikh, P., Wu, Y.-F.: AlGaN/GaN HEMTs-an overview of device operation and applications. In: Proceedings of the IEEE, vol. 90, no. 6, 1022–1031 (2002). https://doi.org/10.1109/JPROC.2002.1021567.
4. Chander, S., Gupta, S., Ajay, Gupta, M.: Enhancement of breakdown voltage in AlGaN/GaN HEMT using passivation technique for microwave application. Superlattices Microstruct. **120**,

217–222, ISSN 0749–6036. https://doi.org/10.1016/j.spmi.2018.05.039
5. Cheng, W., Lei, S., Li, W., Zhao, F., Chan, M., Yu, H.: Improving the drive current of AlGaN/GaN HEMT using external strain engineering. In: Electron Devices Technology and Manufacturing Conference (EDTM), Singapore, pp. 374–376 (2019). https://doi.org/10.1109/EDTM.2019.8731108
6. Chiu, H.-C., Yang, C.-W., Wang, H.-C., Huang, F.-H., Kao, H.-L., Chien, F.-T.: Characteristics of AlGaN/GaN HEMTs with various field-plate and gate-to-drain extensions. IEEE Trans. Electron Devices **60**(11), 3877–3882 (2013)
7. Kumta, A.S., Jinghua, X.: Field-plate-terminated 4H-SiC Schottky diodes using Al-based high-k dielectrics. IEEE Trans. Electron Devices **56**(12), 2925–2934 (2009)
8. Saxena, V., Su, J.N., Steckl, A.J.: High-voltage Ni- and Pt-SiC Schottky diodes utilizing metal field plate termination. IEEE Trans. Electron Devices **46**(3), 456–464 (1999)
9. Ando, Y., Okamoto, Y., Miyamoto, H., Nakayama, T., Inoue, T., Kuzuhara, M.: 10-W/mm AlGaN-GaN HFET with a field modulating plate. IEEE Electron Device Lett. **24**(5), 289–291 (2003). https://doi.org/10.1109/LED.2003.812532
10. Chiu, H.-C., Cheng, C.-S., Lin, S.-W., Wei, C.-C.: A high-linearity single-pole-double-throw pseudomorphic HEMT switch based on tunable field-plate voltage technology. IEEE Trans. Electron Devices **56**(4), 541–545 (2009)
11. Bandi´c, Z.Z., Bridger, P.M., Piquette, E.C., McGill, T.C., Vaudo, R.P., Phanse, V.M., Redwing, J.M.: High voltage (450 V) GaN Schottky rectifiers. Appl. Phys. Lett. **74**(91), 1266–1268 (1999)
12. Ando, Y., et al.: Novel AlGaN/GaN dual-field-plate FET with high gain, increased linearity and stability. In: IEEE International Electron Devices Meeting. IEDM Technical Digest., Washington, DC, pp. 576–579 (2005). https://doi.org/10.1109/IEDM.2005.1609412
13. Hu, Q., Zeng, F., Cheng, W.-C., Zhou, G., Wang, Q., Yu, H.: Reducing dynamic on-resistance of p-GaN gate HEMTs using dual field plate configurations. In: IEEE International Symposium on the Physical and Failure Analysis of Integrated Circuits (IPFA), pp. 1–4 (2020).https://doi.org/10.1109/IPFA49335.2020.9260581
14. Kwak, H.-T., Chang, S.-B., Kim, H.-J., Jang, K.-W., Yoon, H.-S., Lee, S.-H., Lim, J.W., Kim, H.-S.: Operational improvement of AlGaN/GaN high electron mobility transistor by an inner field-plate structure. Appl. Sci. **8**, 974 (2018). https://doi.org/10.3390/app8060974
15. Kobayashi, K., Hatakeyama, S., Yoshida, T., Piedra, D., Palacios, T., Otsuji, T., Suemitsu, T.: Current collapse suppression in AlGaN/GaN HEMTs by means of slant field plates fabricated by multi-layer SiCN. Solid-State Electron. 101 (2014). https://doi.org/10.1016/j.sse.2014.06.022.azA
16. Rawal, D.S., Sharma, S., Kapoor, S., Liashram, R., Chaubey, R., Vinayak, S., Amit, M.: Design and fabrication of multi-finger field plate for enhancement of AlGaN/GaN HEMT breakdown voltage. Defence Sci. J. 68. 290–294 (2018). https://doi.org/10.14429/dsj.68.12134
17. Feiler, W., Falck, E., Gerlach, W.: Multistep field plates for high voltage planar p-n junctions. IEEE Trans. Electron Devices **39**(6), 1514–1520 (1992)
18. Xie, G., et al.: Breakdown voltage enhancement for power AlGaN/GaN HEMTs with air-bridge field plate. In: IEEE International Conference of Electron Devices and Solid-State Circuits, pp. 1–2 (2011).https://doi.org/10.1109/EDSSC.2011.6117720
19. Uren, M.J., Karboyan, S., Chatterjee, I., Pooth, A., Moens, P., Banerjee, A., Kuball, M.: Leaky Dielectric" model for the suppression of dynamic R_{ON} in carbon-doped AlGaN/GaN HEMTs. IEEE Trans. Electron Devices **64**(17), 2826–2834 (2017)
20. Jin, D., del Alamo, J.A.: Methodology for the study of dynamic ON-resistance in high-voltage GaN field-effect transistors. IEEE Trans. Electron Devices **60**(10), 3190–3196 (2013)
21. Silvaco Atlas TCAD tool version 3.10.18 R.
22. Karmalkar, S., Mishra, U.K.: Very High voltage AlGaN/GaN high electron mobility transistors using a field plate deposited on a stepped insulator. Solid State Electron. **45**(9), 1645–1652 (2001)
23. Hasegawa, H., Inagaki, T., Ootomo, S., Hashizume, T.: Mechanisms of current collapse and gate leakage currents in AlGaN/GaN heterostructure field effect transistors. J. Vac. Sci. Technol. **21**(4), 1844–1855 (2003)

24. Parvesh, G., Kaur, R., Pandey, S., Haldar, S., Gupta, M., Gupta, R.S.: Modeling and analysis of fully strained and partially relaxed lattice mismatched AlGaN/GaN HEMT for high temperature applications. Superlattices Microstruct. **44**, 781–793 (2008)

Engineering Optimization Using an Advanced Hybrid Algorithm

Raghav Prasad Parouha

Abstract Among numerous metaheuristic algorithms, differential evolution (DE) and particle swarm optimization (PSO) are found to be efficient and powerful optimization techniques. Similarly, it has been observed that their hybrid algorithms provide a reliable estimate to global optimum. Therefore, based on multi-swarm approach, an advanced hybrid algorithm (*h*aDEPSO) is introduced to solve engineering optimization problems, instead of naïve way. Proposed aDE and aPSO (an advanced DE and PSO) are incorporated in *h*aDEPSO. Also, the population of one is combined with the other in a predefined way in *h*aDEPSO, to create balance between global and local search capability. Introduced hybrid *h*aDEPSO and its integrating component aDE and aPSO have been applied over two engineering optimization problems. Results show supremacy of introduced algorithms comparing to so some modern algorithms. Finally, on basis of performance, the introduced *h*aDEPSO is endorsed in solving engineering optimization problems.

Keywords Hybrid algorithm · Differential evolution · Particle swarm optimization · Engineering optimization

1 Introduction

These days, almost all design optimization problems in engineering are becoming difficult and complicated as a result of involvement of mixed (i.e., continuous and discrete) variables in complex constraints. In general, such situations are multifaceted constrained problems, so they cannot be settled using traditional methods effectively. Currently, to control the shortcomings in conventional optimization methods, an important group of optimization methods called metaheuristics (MAs) have been established. As per the mechanical changes, the MAs are classified into four types—(i) swarm intelligence algorithms (SIAs): These are motivated by activities of social insects or animals like PSO [1], artificial bee colony algorithm (ABC) [2], cuckoo

R. P. Parouha (✉)
Indira Gandhi National Tribal University, Madhya Pradesh, Amarkantak 484886, India

K. N. Das et al. (eds.), *Proceedings of the International Conference on Computational Intelligence and Sustainable Technologies*, Algorithms for Intelligent Systems,
https://doi.org/10.1007/978-981-16-6893-7_48

search (CS) [3], Krill Herd (KH) [4], gray wolf optimizer (GWO) [5], dragonfly algorithm (DA) [6]; (ii) evolutionary algorithms (EAs): These are motivated after biology like DE [7] and genetic algorithm (GA) [8]; (iii) physics-based algorithms (PBAs): These are motivated from the rules which are governing a natural phenomenon such as harmony search (HS) [9], gravitational search algorithm (GSA) [10], and water cycle algorithm (WCA) [11]; (iv) human behavior-based algorithms (HBAs): These are inspired from the human being like teaching-learning-based optimization (TLBO) [12], search and rescue optimization (SAR) [13].

In various MAs, DE and PSO have been broadly worked in difficult optimization problems. This DE has notable performance so that it is turned out to be an effective optimizer in the research area dealing with real-world problems. Though, some concerns are there with it, like local search ability and convergence rate probability. Nowadays, removing its flaws, many efficient DEs have been introduced in the collected works [14–20]. As well, PSO is now famous in solving numerous complex optimization problems because it has effective search ability along with simplicity. Still, it can simply get stuck in a local optimum result area. To get over such concerns, various alterations of PSO are introduced in the literature [21–27]. Also, the idea of hybrid is very important research directions in increasing the effectiveness of single algorithm. So, in improving the efficiency of PSO and DE, several hybrid techniques have been introduced in the collected works [28–33]. Yet, to overcome their individual flaws, the techniques of hybrid are currently preferred over their individual effort.

After extensive vigorous review of literature on various alternates of PSO and DE with their hybrids, succeeding results are observed and inspired from them. (i) PSO and DE have harmonizing properties, so their hybrids are very popular currently. As per our knowledge, discovering methods to combine PSO and DE is an open research area today. (ii) Mutation and crossover tactic with associate control factors of DE are used in order to get the global best solution and that is favorable to improve convergence performance. Hence, best suitable strategies and the parameter values associated in DE are considered as an important research study. (iii) Performance by PSO significantly relies on associated parameters of it. For example, acceleration coefficients and inertia weight which direct particles to balancing diversity and get optimum results. Therefore, scholars have made an effort to adjust control factor of PSO in order to gain optimum results.

Major contribution: Encouraged by observations discussed above and literature survey, an advanced hybrid algorithm *h*aDEPSO with the following suggested component is for solving engineering optimization problems. (i) aDE: It consists of novel operators with its linked parameter, and (ii) aPSO: It implicates unique gradually changing (increasing and/or decreasing) factors.

This paper is arranged as: Sect. 2 briefly explains basics of PSO and DE. Proposed algorithms are described in Sect. 3. In Sect. 4, application of proposed algorithms is presented. Section 5 concludes this study.

2 Brief on DE and PSO

2.1 Differential Evolution (DE)

After doing initialization, DE is conducted in three vital operations given below.

Mutation: At the iteration t, for every $x_{i,j}^t$ (target vector), a $v_{i,j}^t$ (mutant vector) is produced as below.

$$v_{i,j}^t = x_{r_1}^t + F\left(x_{r_2}^t - x_{r_3}^t\right) \quad (1)$$

where $r_1, r_2, r_3 \in \{1, 2, \ldots \text{np}\}$ are arbitrarily chosen integers with $r_1 \neq r_2 \neq, r_3 \neq i$ and F signifies the scaling vector.

Crossover: Here, a $u_{i,j}^t$ (trial vector) is produced by merging $x_{i,j}^t$ and $v_{i,j}^t$ as below.

$$u_{i,j}^t = \begin{cases} v_{i,j}^t; \text{ if rand} \leq C_r \\ x_{i,j}^t; \text{ otherwise} \end{cases} \quad (2)$$

where rand and C_r (crossover rate) $\in$ random digit between 0 and 1, $i \in [1, \text{np}]$, and $j \in [1, D]$.

Selection: It is stated as

$$x_{i,j}^{t+1} = \begin{cases} u_{i,j}^t; \text{ if } f\left(u_{i,j}^t\right) \leq f\left(x_{i,j}^t\right) \\ x_{i,j}^t; \text{ otherwise} \end{cases} \quad (3)$$

Here, mutation, crossover, and selection operators are permitted to offspring repetitively till predetermined stopping condition.

2.2 Particle Swarm Optimization (PSO)

In traditional PSO, a swarm hovers in a *D*-dimensional search area to find inclusive solution. Every *i*th swarm particle has its individual position ($x_i = (x_{i,1}, x_{i,2}, \ldots, x_{i,D})$) and velocity $v_i = (v_{i,1}, v_{i,2}, \ldots, v_{i,D})$. While happening of evolution, every particle tracks its individual best pbest and global best gbest; at each iteration, velocity and position of the *i*th particle are updated as.

$$v_{i,j}^{t+1} = wv_{i,j}^t + c_1 r_1\left(\text{pbest}_{i,j} - x_{i,j}^t\right) + c_2 r_2\left(\text{gbest}_j - x_{i,j}^t\right) \quad (4)$$

$$x_{i,j}^{t+1} = x_{i,j}^t + v_{i,j}^{t+1} \quad (5)$$

where t: iteration index and $v_{i,j}^t$: velocity of ith particle in D-dimension at the iteration t; here, c_1 denotes cognitive acceleration coefficient, c_2 denotes social acceleration coefficient, $r_1, r_2 \in [0, 1]$, and w is the inertia weight.

3 Proposed Methodology

Instead of naïve way, a *h*aDEPSO is introduced for further improvement in solution quality. In this hybrid *h*aDEPSO, whole population is arranged according as the value of fitness function and which is divided in two sub-populations that is pop_1 (best short) and pop_2 (rest part). Because, pop_1 and pop_2 contain top and rest part of the inhabitants which indicates virtuous global and local search competency, respectively. On the separate sub-population (pop_1 and pop_2), it introduced aDE: advanced DE (because of its better local search ability) and also aPSO: advanced PSO (due to its good global search capability) correspondingly. Calculating mutually sub-population, then better solution is obtained in pop_1 (called best) and pop_2 (termed gbest) stored individually. If best < gbest, then pop_2 is fused with pop_1; afterward, this fused population is evaluated by aDE (as it alleviate the possible stagnation). Else, pop_1 is fused with pop_2; later, fused population is evaluated by aPSO (as it recognized to improved movements). Finally, reporting the optimal solution if stoppings condition is encountered, then it stops, otherwise returns to sorting process of population. Continue this whole process until getting desire optimal solution. In Fig. 1, flowchart of the *h*aDEPSO is revealed.

Mainly, *h*aDEPSO works on connecting superior capability of the suggested aDE and aPSO (described as follows).

3.1 aDE: Advanced DE

It consists of modified mutation approach and crossover rate; it also altered selection scheme that is presented which is defined as below.

Mutation:

$$v_{i,j}^t = x_{i,j}^t + \tau \times \text{rand}(0, 1) \times \left(\text{best}_j - x_{i,j}^t\right) \tag{6}$$

where $v_{i,j}^t$: mutant vector, $x_{i,j}^t$: it is target vector, rand(0, 1): random number lying in interval (0,1), best_j: best vector and at last τ: convergence factor (it elects the probing balance for every vector). Moreover, dynamic modifications of τ are specified as: (i) if $\tau = 1$: a vector generated in the range $[x_{i,j}^t, \text{best}_j]$ that may increase the convergence rate of differential evolution (DE); however, note that it may bring the risk of growing likelihood of facing local optima and (ii) if $\tau = \mu(1 - t/t_{\max}) + 1$, where t and $t_{\max}$ are current and total iteration, and μ: positive constant: 1st iteration—$\tau \approx \mu + 1$ (term $t/t_{\max}$ can be ignored as $t = 1$ is much smaller than $t_{\max}$), in max iteration—$\tau =$

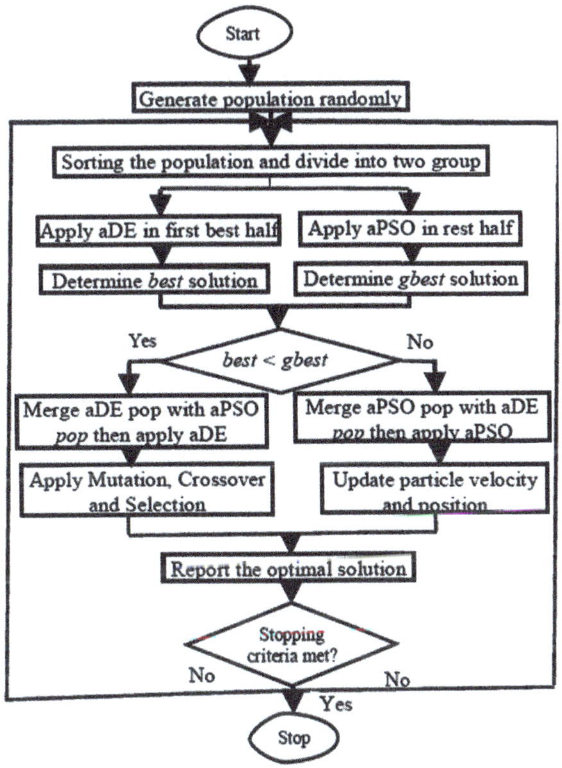

Fig. 1 Flowchart of *h*aDEPSO

1 as $(1 - t/t_{\max}) = 0$. Thus, τ linearly start with $\mu + 1$ to 1 all over the optimization process. That can renovate the convergence and also avert local optima

So, τ is collected of a series of large values that enhance the exploring capability for every vector earlier. Later, it makes surety for exploration and exploitation due to collection of a series of small values.

Crossover:

$$u_{i,j}^{t} = \begin{cases} v_{i,j}^{t};\ \text{if rand}(0,1) \le C_r(\text{crossover rate}) \\ x_{i,j}^{t};\ \text{otherwise} \end{cases} \tag{7}$$

where $C_r = e^{\frac{(t - t_{\max})}{t_{\max}}}$ (it guarantees of separate diversity in initial stage and enhances global search ability in later stage).

Selection:

$$x_{i,j}^{t+1} = \begin{cases} x_{i,j}^{t};\ \text{if } f\left(u_{i,j}^{t}\right) > f\left(x_{i,j}^{t}\right) \text{ and rand}(0,1) < p \\ u_{i,j}^{t};\ \text{otherwise} \end{cases} \tag{8}$$

where $p \in$ rand (0, 1]. In this scheme, collectively innovator vector will get chance to continue and also share its observed information with the others. These encourage searching competences and stabilize essential exploration and exploitation of aDE to produce better quality solutions.

3.2 aPSO: Advanced PSO

Ideally, PSO requires strong exploration ability (particles can move entire search space instead of clustering around the current best solution) and boost exploitation capability (particles can explore in a local region) at primary and later period of the evolution correspondingly. Taking every issue such as benefit, drawbacks, and parameter impacts of the PSO, an aPSO is presented in this work. It depends over new gradually changing (increasing and/or decreasing) parameters ($w, c_1$and c_2) given below.

$$w = w_f + (w_i - w_f)\left(\frac{t}{t_{\max}}\right)^2; c_1 = c_{1f}\left(\frac{c_{1i}}{c_{1f}}\right)^{\left(\frac{t}{t_{\max}}\right)^2}; \text{ and } c_2 = c_{2i}\left(\frac{c_{2f}}{c_{2i}}\right)^{\left(\frac{t}{t_{\max}}\right)^2}$$

where w_i and w_f are initial and final values for w; c_{1i} and c_{1f}: initial and final values of c_1; c_{2i} and c_{2f}: initial and final values of c_2; t and $t_{\max}$: iteration index and highest amount of iteration. Therefore, the velocity and position for the ith particle are updated as

$$\begin{aligned} v_{i,j}^{t+1} &= \left(w_f + (w_i - w_f)\left(\frac{t}{t_{\max}}\right)^2\right)v_{i,j}^t \\ &+ \left(c_{1f}\left(\frac{c_{1i}}{c_{1f}}\right)^{\left(\frac{t}{t_{\max}}\right)^2}\right)r_1\left(p_{\text{best } i,j}^t - x_{ij}^t\right) \\ &+ \left(c_{2i}\left(\frac{c_{2f}}{c_{2i}}\right)^{\left(\frac{t}{t_{\max}}\right)^2}\right)r_2\left(g_{\text{best } j}^t - x_{i,j}^t\right) \end{aligned} \tag{9}$$

$$x_{i,j}^{t+1} = x_{i,j}^t + v_{i,j}^{t+1} \tag{10}$$

4 Application

To examine performance, proposed aDE, aPSO, and *h*aDEPSO are applied to solve two complex engineering optimization problems (EOPs), viz.

(i) Welded beam design (WBD) problem
(ii) Three-bar truss design (TRD) problem.

More details of these problems can be found in [34]. Simulations were done over Intel (R) Core (TM) i7 @ 2.30 GHz, RAM: 6.00 GB, with Windows 10 operating system, also C language. By vast investigation, parameters of the proposed methods are recommended as—$w_i = 0.4$, $c_{1i} = 0.5$ and $c_{2i} = 2.5$ and $w_f = 0.9$, $c_{1f} = 2.5$ and $c_{2f} = 0.5$, and bracket operator penalty [35] ($R = 1e^{03}$) to handle constraint is chosen for this study because it has higher competence. Whole best values in every table are noted with bold letters of the matching methods. To have reasonable judgment, the np (population size) = 100, independent run (30), stopping criteria (240,000 function evaluations maximum) are considered same as comparative algorithms in all cases. The outcomes of proposed methods on two EOPs are equated with PSO [1], ABC [2], CS [3], KH [4], GWO [5], DA [6], GA [8], EO [36], CSDE [37], SCA [38], EPO [39], SHO [40], and GSA-GA [41].

The experiential outcomes of proposed with other methods on particular EOPs are presented in Table 1 (for WBD) and Table 2 (for TRD). From these tables, it is very clear that the proposed methods (aDE, aPSO and *h*aDEPSO) give better and/or equally results on all EOPs. Finally, these aDE, aPSO, and *h*aDEPSO produce less standard for all cases which describe their stability.

The schematic diagram and convergence graphs of proposed and best non-proposed methods are presented in Figs. 2 and 3 on EOPs. These figures clearly depict that proposed methods converge faster than the others. So, projected methods are computationally effective.

Generally, it can be said that performance of the proposed methods (aDE, aPSO, and *h*aDEPSO) is better and/or equally in comparison with others. Though, in proposed three methods, the *h*aDEPSO have superior capability.

5 Conclusion

This work proposes *h*aDEPSO for engineering optimization problems, instead of naïve way. Where, integration of an aDE (advanced DE) and aPSO (advanced PSO) is done in proposed hybrid. As a result—(i) *h*aDEPSO is developed on uniting aDE and aPSO. This is motivated from multi-swarm tactic in which one population is fused with the other according as some predefined way, and it leads to surety of convergence and diversifying solutions. (ii) In improving performance and altering the control factors of DE, aDE is proposed. The novel mutation approach, crossover possibility, and changed selection systems of aDE will assure exploration and exploitation at starting and ending, respectively, and (iii) to escape particles stagnant, an aPSO is proposed which is consisting of new progressively changing (increasing and/or decreasing) parameters. These new parameters can be encourages particles in searching better quality solution of aPSO algorithm.

Table 1 Simulation results for WBD problem

Methods	Best values for variables				Best	Worst	Mean	Std.
	x_1 (H)	x_2 (L)	x_3 (T)	x_4 (b)				
PSO	0.197411	3.315061	10.00000	0.201395	1.820395	3.048231	2.230310	0.324525
ABC	0.205730	3.470489	9.036624	0.205730	1.724852	1.734852	1.741913	0.031
CS	0.2015	3.562	9.0414	0.2057	1.7312065	1.8786560	2.3455793	0.2677989
GWO	0.205678	3.475403	9.036964	0.206229	1.726995	1.727128	1.727564	0.001157
DA	0.194288	3.46681	9.04543	0.205695	1.70808	1.94076	2.52106	0.250234
GA	0.164171	4.032541	10.00000	0.223647	1.873971	2.320125	2.119240	0.034820
EO	0.2057	3.4705	9.03664	0.2057	1.724853	1.736725	1.726482	0.003257
SCA	0.2444	6.2380	8.2886	0.2446	2.3854	6.3996	3.2551	0.9590
EPO	0.205411	3.472341	9.035215	0.201153	1.723589	1.727211	1.725124	0.004325
SHO	0.205563	3.474846	9.035799	0.205811	1.725661	1.726064	1.725828	0.000287
aDE	0.184288	3.26641	8.24133	0.204585	1.72355	1.72874	1.72625	0.000284
aPSO	0.184288	3.26641	8.24133	0.204585	1.72486	1.72698	1.73542	0.000354
*h*aDEPSO	0.184288	3.26641	8.24133	0.204585	**1.69782**	**1.72321**	**1.72421**	**0.000132**

Table 2 Simulation results for TRD problem

Methods	Best values for variables		Best	Worst	Mean	Std.
	x_1 (A_1)	x_2 (A_2)				
GSA-GA	0.788676171219	0.408245358456	263.8958433	263.8958459	263.8958437	5.34e−07
CSDE	0.7886	0.4082	263.148352124271	65,535	263.148352318831	1.44060040776154e−08
PSO	0.7803	0.4330	264.543754826635	2.70864524844245e+05	264.775374387455	1.85577512704186e+00
DE	0.7887	0.4080	263.148352624688	65 535	263.411272728312	1.05476883610644e−01
KH	0.7885	0.4088	263.9	265.0	263.9	1.658e−01
CS	0.7857	0.5945	263.602007628033	5.23525611223402e+12	263.671445662651	1.63380787023616e−02
aDE	0.788526	0.408452	262.6263	263.7845	263.7845	1.0358e−08
aPSO	0.788526	0.408452	263.4263	263.8497	263.8497	1.2458e−08
*h*aDEPSO	0.788526	0.408452	**261.1438**	**262.9796**	**262.5782**	**2.5142e−09**

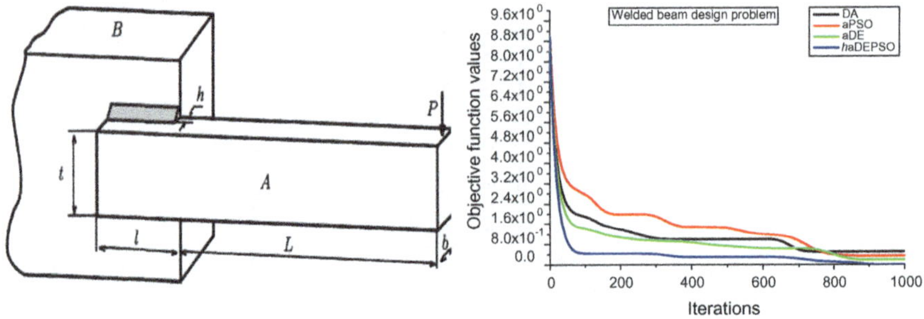

Fig. 2 Schematic diagram and convergence of WBD problem

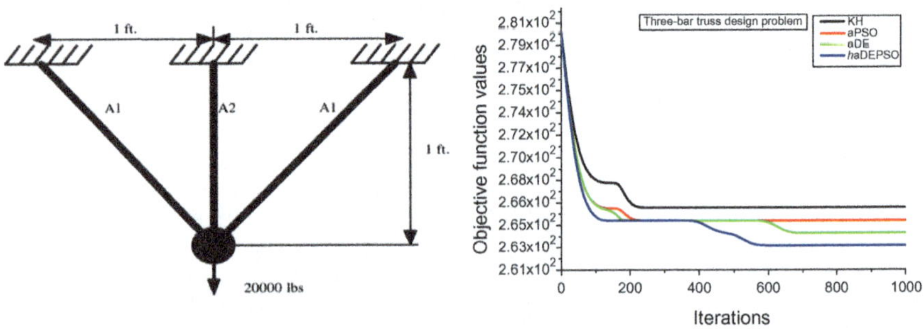

Fig. 3 Schematic diagram and convergence of TRD problem

The efficiency of proposed methods (*h*aDEPSO, aDE, and aPSO) is evaluated on two complex engineering optimization problems. It analyzed numerically, statistically, and graphically the proposed algorithms by comparing with modern algorithms. Results reveal that introduced procedures are more robust with higher effectiveness. Also, *h*aDEPSO outperformed among proposed aDE and aPSO as well as comparative methods, in terms of feasibility, superiority, and solution optimality.

References

1. Kennedy, J., Eberhart, R.C.: Particle swarm optimization. In: Proceeding of IEEE International Conference on Neural Networks, pp. 1942–1948 (1995)
2. Karaboga, D., Basturk, B.: A powerful and efficient algorithm for numerical function optimization: artificial bee colony algorithm. J. Global Optim. **39**(3), 459–471 (2007)
3. Yang, X.S., Deb, S.: Cuckoo search via Lévy flights. In: Proceedings of World Congress on Nature and Biologically Inspired Computing, Coimbatore, India, pp. 210–214 (2009)
4. Gandomi, H., Alavi, A.H.: Krill herd: a new bio-inspired optimization algorithm. Commun. Nonlinear Sci. Numer. Simul. **17**(12), 4831–4845 (2012)
5. Mirjalili, S., Mirjalili, S.M., Lewis, A.: Grey wolf optimizer. Adv. Eng. Softw. **69**, 46–61 (2014)

6. Mirjalili, S.: Dragonfly algorithm: a new meta-heuristic optimization technique for solving single-objective, discrete and multi-objective problems. Neural Comput. Appl. **27**(4), 1053–1073 (2016)
7. Storn, R., Price, K.: Differential evolution—a simple and efficient heuristic for global optimization over continuous spaces. J. Global Optim. **11**, 341–359 (1997)
8. Davis, L.: Handbook of Genetic Algorithms (1991)
9. Geem, Z.W., Kim, J.H., Loganathan, G.V.: A new heuristic optimization algorithm: harmony search. SIMULATION **76**(2), 60–68 (2001)
10. Rashedi, E., Nezamabadi-pour, H., Saryazdi, S.: A gravitational search algorithm. Inf. Sci. **179**(13), 2232–2248 (2009)
11. Eskandar, H., Sadollah, A., Bahreininejad, A., Hamdi, M.: Water cycle algorithm—a novel metaheuristic optimization method for solving constrained engineering optimization problems. Comput. Struct. **110–111**, 151–166 (2012)
12. Rao, R.V., Savsani, V.J., Vakharia, D.P.: Teaching-learning-based optimization: a novel method for constrained mechanical design optimization problems. Comput. Aided Des. **43**(3), 303–315 (2011)
13. Shabani, A., Asgarian, B., Gharebaghi, S.A., Salido, M.A., Giret, A.: A new optimization algorithm based on search and rescue operations. Math. Probl. Eng. **2019**, 1–23 (2019)
14. Yang, X., Li, J., Peng, X.: An improved differential evolution algorithm for learning high-fidelity quantum controls. Sci. Bull. **64**(19), 1402–1408 (2019)
15. Prabha, S., Yadav, R.: Differential evolution with biological-based mutation operator. Eng. Sci. Technol. Int. J. 1–11 (2019)
16. Liu, Z.-G., Ji, X.-H., Yang, Y.: Hierarchical differential evolution algorithm combined with multi-cross operation. Expert Syst. Appl. **130**, 276–292 (2019)
17. Gui, L., Xia, X., Yu, F., Wu, H., Wu, R., Wei, B., He, G.: A multi-role based differential evolution. Swarm Evol. Comput. **50**, 1–15 (2019)
18. Li, S., Gu, Q., Gong, W., Ning, B.: An enhanced adaptive differential evolution algorithm for parameter extraction of photovoltaic models. Energy Convers. Manage. **205**, 1–16 (2020)
19. Hu, L., Hua, W., Lei, W., Xiantian, Z.: A modified Boltzmann annealing differential evolution algorithm for inversion of directional resistivity logging-while-drilling measurements. J. Petrol. Sci. Eng. **180**, 1–10 (2020)
20. Ben, G.N.: An accelerated differential evolution algorithm with new operators for multi-damage detection in plate-like structures. Appl. Math. Model. **80**, 366–383 (2020)
21. Parouha, R.P.: Nonconvex/nonsmooth economic load dispatch using modified time-varying particle swarm optimization. Comput. Intell. 1–28 (2019). https://doi.org/10.1111/coin.12210
22. Hosseini, S.A., Hajipour, A., Tavakoli, H.: Design and optimization of a CMOS power amplifier using innovative fractional-order particle swarm optimization. Appl. Soft Comput. **85**, 1–10 (2019)
23. Kohler, M., Vellasco, M.M.B.R., Tanscheit, R.: PSO+: a new particle swarm optimization algorithm for constrained problems. Appl. Soft Comput. **85**, 1–26 (2019)
24. Khajeh, M.R., Ghasemi, H.G.: Arab, modified particle swarm optimization with novel population initialization. J. Inf. Optim. Sci. **40**(6), 1167–1179 (2019)
25. Ang, K.M., Lim, W.H., Isa, N.A.M., Tiang, S.S., Wong, C.H.: A constrained multi-swarm particle swarm optimization without velocity for constrained optimization problems. Expert Syst. Appl. **140**, 1–23 (2020)
26. Lanlan, K., Ruey, S.C., Wenliang, C., Yeh, C.: Non-inertial opposition-based particle swarm optimization and its theoretical analysis for deep learning applications. Appl. Soft Comput. **88**, 1–10 (2020)
27. Xiong, H., Qiu, B., Liu, J.: An improved multi-swarm particle swarm optimizer for optimizing the electric field distribution of multichannel transcranial magnetic stimulation. Artif. Intell. Med. **104**, 1–14 (2020)
28. Parouha, R.P., Das, K.N.: DPD: an intelligent parallel hybrid algorithm for economic load dispatch problems with various practical constraints. Expert Syst. Appl. **63**, 295–309 (2016)

29. Mao, B., Xie, Z., Wang, Y., Handroos, H., Wu, H.: A hybrid strategy of differential evolution and modified particle swarm optimization for numerical solution of a parallel manipulator. Math. Probl. Eng. 1–9 (2018)
30. Tang, B., Xiang, K., Pang, M.: An integrated particle swarm optimization approach hybridizing a new self-adaptive particle swarm optimization with a modified differential evolution. Neural Comput. Appl. 1–35 (2018)
31. Too, J., Abdullah, A.R., Saad, N.M.: Hybrid binary particle swarm optimization differential evolution-based feature selection for EMG signals classification. Axioms **8**(3), 1–17 (2019)
32. Dash, J., Dam, B., Swain, R.: Design and implementation of sharp edge FIR filters using hybrid differential evolution particle swarm optimization. AEU - Int. J. Electron. Commun. **114**, 153019 (2020). https://doi.org/10.1016/j.aeue.2019.153019
33. Zhao, X., Zhang, Z., Xie, Y., Meng, J.: Economic-environmental dispatch of microgrid based on improved quantum particle swarm optimization. Energy **195**, 117014 (2020). https://doi.org/10.1016/j.energy.2020.117014
34. Liu, H., Cai, Z., Wang, Y.: Hybridizing particle swarm optimization with differential evolution for constrained numerical and engineering optimization. Appl. Soft. Comput. **10**, 629–640 (2010)
35. Deb, K.: Optimization for engineering design: algorithms and examples. Prentice-Hall of India, New Delhi (1995)
36. Faramarzi, A., Heidarinejad, M., Stephens, B., Mirjalili, S.: Equilibrium optimizer: a novel optimization algorithm. Knowl. Based Syst. **191**, 105190 (2020). https://doi.org/10.1016/j.knosys.2019.105190
37. Zhang, Z., Ding, S., Jia, W.: A hybrid optimization algorithm based on cuckoo search and differential evolution for solving constrained engineering problems. Eng. Appl. Artif. Intell. **85**, 254–268 (2019)
38. Mirjalili, S.: SCA: a sine cosine algorithm for solving optimization problems. Knowl. Based Syst. **96**, 120–133 (2016)
39. Dhiman, G., Kumar, V.: Emperor penguin optimizer: a bio-inspired algorithm for engineering problems. Knowl. Based Syst. **159**, 20–50 (2018)
40. Dhiman, G., Kumar, V.: Spotted hyena optimizer: a novel bio-inspired based metaheuristic technique for engineering applications. Adv. Eng. Softw. **114**, 48–70 (2017)
41. Garg, H.: A hybrid GSA-GA algorithm for constrained optimization problems. Inf. Sci. **478**, 499–523 (2019)

Image Enhancement Using Chicken Swarm Optimization

A. Vamsidhar, T. Surya Kavitha, and G. Ramesh Babu

Abstract Image enhancement is a technique used to get a better quality of an image in terms of its clarity, brightness and to provide a better visual impact on the human eye. Data loss is the major factor that diminishes the quality of a captured image. This loss can be recovered by the application of image enhancement techniques like contrast stretching and histogram equalization. In this paper, we propose the application of chicken swarm optimization (CSO) method on different traditional persistent issues and found that CSO is suitable in solving a huge span of optimization including dynamic, stochastic and robust problems for image enhancement. Entropy and peak signal-to-noise ratio (PSNR) are calculated for the CSO enhanced image. The results are compared with the existing histogram equalization (HE) method and found that CSO outperforms.

Keywords Chicken swarm optimization · Histogram equalization · Entropy · Peak signal-to-noise ratio

1 Introduction

Image is a visual representation of an object (object may be place, person or thing). Image can be created, stored or copied in electrical form and expressed in terms of vector graphics or raster graphics. Image can be formed by arranging the pixels in serial order which are represented in binary or gray level format [1]. Digital images are more prone to different quality losses and degradations mainly during the process of image capturing, enhancing, compression and restoration, with an addition of noise sources during their transmission. Eliminating these artifacts is so essential for analysis and processing of images [2]. Image enhancement techniques play a vital role in improving the visual depiction of a digital image. These techniques are having diverse applications in the field of medical images like tumors, etc., video processing, seismic reconnaissance and importantly surveillance systems. A

A. Vamsidhar (✉) · T. Surya Kavitha · G. Ramesh Babu
ECE Department, Raghu Engineering College, Visakhapatnam, India

K. N. Das et al. (eds.), *Proceedings of the International Conference on Computational Intelligence and Sustainable Technologies*, Algorithms for Intelligent Systems,
https://doi.org/10.1007/978-981-16-6893-7_49

Fig. 1 Blur image

large number of these techniques have been developed so far in real time, such as gray scale and histogram-based transformations. But, an application of these image enhancement techniques failed to produce acceptable outcomes when applied on images with low contrast levels [3]. Gray level transformation methods such as logarithmic, power-law, etc., do not provide a balanced contrast and on the other hand histogram-based techniques such as histogram equalization (HE), histogram specification, etc., provide an appearance of pale image affecting its background. Figure 1 shows a blurred image, which does not give exact information of object. Blur of an image occurs due to movement of subject, improper holding of camera, dirty lens, wrong focusing and depth of field, shake due to internal vibrations and focal length of the lens. So, improvement of the image quality relies on understanding the exact information of an object.

There exist different versions of CSO having the features like segmentation, feature extraction and feature reduction, classification, and optimization of the parameters of the input signal. This made CSO a novel approach for the application of images with low contrast levels and is effectively used in this paper for dynamically optimizing the parameters of the image. Hence by the application of CSO, the quality of the image can be improved and makes the objects visible clearly.

This paper comprises of the following sections: Section 2 deals with traditional methods of image enhancement techniques like histogram equalization, and Sect. 3 explains the proposed CSO algorithm and its application for image enhancement. Section 4 depicts the simulation results, followed by conclusion marked as Sect. 5.

2 Existing Image Enhancement Methods

Image enhancement is one of the techniques used to improve the image quality, and belongs to the category of contrast enhancement. Contrast enhancement is the process of enhancing color in an image with a range from 0 to 255, measure by difference in color and brightness of two different objects. There are two methods in contrast enhancement: Global contrast and local contrast enhancement. A well enhanced image is formed by adding global enhanced color image and local enhanced color image on an image.

2.1 Global Contrast

Many color image enhancement methods treat the three-color channels individually and alters the hue of an image to look unnatural. Hue saturation intensity (HSI) is a regular single RGB color model [pixels p with values (r, g, b)] and the hue of p is given as [4]:

$$\mathrm{Hue}(p) = \begin{cases} 0, & \text{if } r = g = b \\ \theta, & \text{if } b \leq g \\ 2\pi - \theta, & \text{if } b > g \end{cases} \tag{1}$$

where

$$\theta = \arccos\left\{ \frac{\frac{1}{2}\{(r-g) + (r-b)\}}{\left\{(r-g)^2 + (r-b)(g-b)\right\}^{\frac{1}{2}}} \right\} \tag{2}$$

Considering a gray image I with number of pixels N and intensity $[0, (L-1)]$, Normalized histogram h_I of the image I is computed by,

$$h_I(S) = \frac{n_s}{N} \tag{3}$$

For the image I, the mapping function T is given by

$$T(i) = \left\lfloor (L-1)\sum_{j=0}^{i} h_I(j) + 0.5 \right\rfloor \tag{4}$$

Figure 2 shows the original image and its global contrast enhanced image. This enhancement method has the capability to increase the brightness of whole image.

2.2 Local Contrast

Local contrast enhancement is to enhance the contrast of the local pixels and preserve the detail information [4]. If $p(i, j)$ is actual image and $b(i, j)$ is a blur image, then mask $m(i, j)$ is measured by subtracting the blur image from original image,

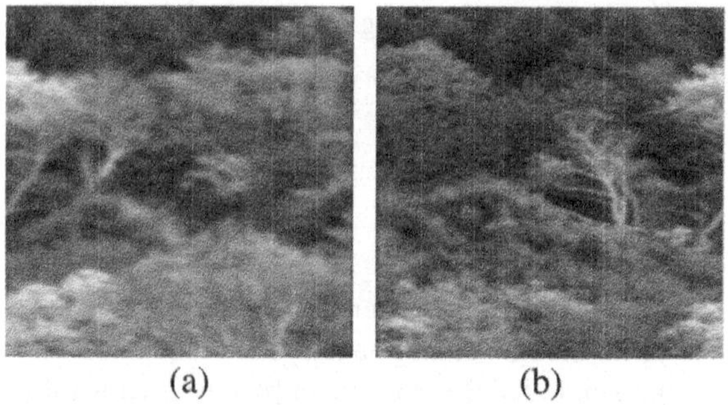

Fig. 2 **a** Original image, **b** global contrast image

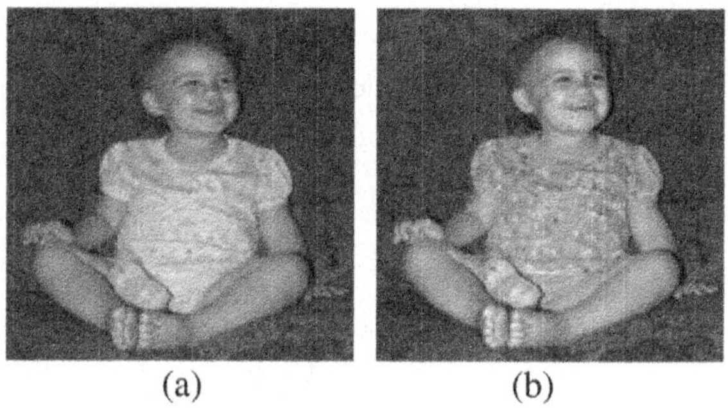

Fig. 3 **a** Original image, **b** local contrast image

$$m(i, j) = p(i, j) - b(i, j) \tag{5}$$

Shaping of image $S(i, j)$ is obtained by considering weighted portion of mask to the original image.

$$S(i, j) = p(i, j) - w^{*}m(i, j) \tag{6}$$

Figure 3 shows the original image and its local contrast enhanced image. This method can enhance many image details by taking different transformations of the same gray level at different places in the original image.

Image enhancement consists of two domains: spatial and frequency domain. In spatial domain, operations directly performed on pixels in an image. There are various techniques used in spatial domain particularly point operation, mask operation and histogram equalization. In point operation, we can perform operation on each and every pixel for changing the brightness or contrast of an image [5]. It can be represented by Eq. 7 as:

$$g(x, y) = T[f(x, y)] \tag{7}$$

where g is processed image, f is the input image, and T is an operation performed on input image. In mask operation, the pixel values are changed by placing the mask over an image. For mask operation generally, a box filter or median filter is used. T operation performs neighborhood of pixels. Histogram equalization technique is used to spread out the gray levels in an image as it reassigns the brightness values of an image, so that the resultant image can be flat. It also provides more visual quality across wide range of image. In frequency domain, frequency commands of an image can be altered by applying the Fourier transform to an image, multiplying by filter function and finally applying an inverse Fourier transform. This domain is broadly classified into two categories namely image smoothing and image sharpening.

2.3 Histogram Equalization

This technique is easy to implement because of the simplicity. Histogram equalization is a widely used spatial domain technique with variety of applications in image processing. It helps in modifying the image intensity to improve its contrast. It extends the image contrast by evenly redistributing the gray level. If there are L number of gray levels in the image, then by using (8), and it can be interpreted as follows [6]:

$$P(r_k) = \frac{n_k}{n} \tag{8}$$

where $k = 0, 1, 2 \ldots, L - 1$, r_k is the kth gray level, total number of pixels is n and n_k is the number of pixels retaining gray level r_k. Although this technique is very easy, its traditional nature contributes to an abnormal over enhancement. The main drawback is that it produces washed-out look in an image and also emphasizes only on global contrast enhancement leading to lack of local information, which can also contribute to image over enhancement.

3 Chicken Swarm Optimization

Chicken swarm optimization (CSO) is a nature aroused algorithm which describes how swarm behaves in the real world. It is a local minimum algorithm which follows hierarchical order, consisting of a group of three different members namely rooster, hens, and chicks. The characteristics of this algorithm are as follows [7]:

(a) ROOSTER: It is head of the group. They search for a food in different places positively. So, it is most powerful in the group.
(b) HEN: It acts as the mother of the chicks and more powerful hens are similar to rooster.
(c) CHICK. It is a weak member in the group, and it follows the hen for searching the food.

From the above-mentioned descriptions, CSO is developed mathematically. The functioning of the chickens is idealized by the following rules [8, 9]:

1. Various groups exist in the chicken swarm with each group comprising of a dominant rooster, along with an adequate number of hens and chicks.
2. Division of the chicken swarm into various groups and finding out the identity of the chickens depend on the fitness values of the chickens. The best fitness valued chickens are selected as roosters, and every rooster will be treated as the group head. Similarly, the least fitness valued chickens are marked as chicks, while all the remaining would be designated as hens. A random selection process will be implemented for hens to choose a group. The mother–child relation between hens and chicks will also be established arbitrarily.
3. In a group, dominance, hierarchical order, and mother–child relation will remain unaltered. The status of the chickens' updates only at every (G) time steps.
4. The chickens follow their group-head (rooster) in search for food, while they may avoid the ones that eat their food. The chicks search for food near their mother (hen). Assuming that chicken steal good food from others, the strong ones always have an advantage. The food searching update formula can be given as [10]:

$$y_{k,l}^{i+1} = y_{k,l}^{i} + \text{rand } n\left(0, \sigma^2\right).y_{k,l}^{i} \tag{9}$$

where $y_{k,l}^{i+1}$ and $y_{k,l}^{i}$ are the position of lth dimension of particle k in $i+1$ and i iterations. rand is a randomly generated value between 0 and 1, and rand $n\left(0, \sigma^2\right)$ is Gaussian distribution function with variance σ^2 and 0 mean, which is represented as:

$$\sigma^2 = \begin{cases} 1, & f_k < f_p \\ \exp\left(\frac{f_p - f_k}{|f_k| + \chi}\right), & f_k \geq f_p \end{cases} \tag{10}$$

where $k, p \epsilon [1, \text{N}]$ and $k \neq p$. f_k and f_p represent the fitness values of rooster particle k and p, respectively. χ indicates a tiny constant counted for ensuring the denominator not equal to zero.
5. Let R_n, H_n, C_n and M_n represent the count of roosters, hens, chicks, and the mother hens, respectively. The best R_n chickens would be designated as roosters, while the worst C_n ones would be treated as chicks, while the remaining are considered as hens.
6. Characterized by their positions $y_{k,l}^{i} (k \in \{1, \ldots, n\}, l \in \{1, \ldots, D\})$ at time step i, all N virtual chickens search for food in a D-dimensional space. In this paper, the best R_n chickens are the ones with R_n minimum fitness values.

The update formula for hens can be given as [11]:

$$y_{k,l}^{i+1} \neq= y_{k,l}^{i} + F1 \times \text{rand}\left(y_{r1,l}^{i} - y_{k,l}^{i}\right) + F2 \times \text{rand}\left(y_{r2,l}^{i} - y_{k,l}^{i}\right) \tag{11}$$

where

$$F1 = \exp\left(\frac{f_k - f_{r1}}{\text{abs}(f_k) + \chi}\right) \tag{12}$$

$$F2 = \exp(f_{r1} - f_k) \tag{13}$$

$r1 \in [1, N]$ indicates the rooster which is ith hen's group mate and $r2 \in [1, N]$ indicates a hen or rooster, which is randomly chosen such that $r1 \neq r2$. The mathematical expression for the chicks following their mother can be given as:

$$y_{k,l}^{i+1} = y_{k,l}^{i} + \text{CM} \times \left(y_{m,l}^{i} - y_{k,l}^{i}\right) \tag{14}$$

where, $y_{m,l}^{i}$ is the position of ith chick's mother and CM is a parameter chosen between 0 and 2, which corresponds to the chick that follow its mother.

3.1 Pseudo-Code

- Input = m, pop, dim, G, rpercent, hpercent, mpercent
  {m = maximal generation; pop = population size; dim = dimension; G = how often the chicken swarm can be updated; rpercent = Roosters population size; hpercent = Hens population size; mpercent = Mother hens population size}
- For i = 1:pop, calculate the fitness value and best position to fit
- For t = 1:m, how chick has to follow their mother for searching the food
- If mod (t,g) = = 1//t = 1 [ans, sort index] = sort p(fit)
- Divide the chicken group into rooster, hen and chicks
- mother lib = rand perm(hnum,mnum) + rnum
  Mother library contain all mother hens. Randomly select mother hen for optimization to maintain strong one corresponding with low fitness value sort index {(1:rnum);(rnum+1+hnum)}
- Rooster are arranged in stored order
  Mate = rand perm (rnum, hnum)
- To assign the one rooster to each group
  Update the hnum values and arrange in sorted order
  Update the cnum values and arrange in sorted order
- If hnum contain best fitness then update as the "rooster"
- If cnum contain best fitness then update as "mother hen"
  End.

Let us consider an image based on population size (pop). We can divide the maximum generations' m with each generation consisting of at least a rooster, many hens, and many chicks. Then individual fitness values are calculated. Based on the values obtained, divide the rooster, hen and chicks so that all the three are arranged in shortest order. Mother lib is the parameter that contains all the mother hen's information to maintain strong one, to low fitness value. To assign the rooster for each group, upgrade the value for G no iteration. After updating the values, if a hen contains more fitness value, then it will be promoted as "rooster" and if a chick contains more fitness value, then can be updated as "Hen."

4 Simulation Results

In this paper, a set of digital images of size 256 × 256 are considered as input samples. The enhanced method of CSO is examined on MATLAB for these digital images containing different noise size, and the parameters of noise are measured to check the efficiency. The results of CSO are dependent on parameters and fine tuning of them provides better results compared to other algorithms. The parameter values in CSO algorithm are as follow: $m = 500$, pop = 100, dim = 256, $G = 10$, rpercent = 0.15, hpercent = 0.7 and mpercent = 0.5.

The acquired digital images are converted to a fixed resolution and stored in data base. For the reduction of storage space and computational complexity during feature extraction, size of the image is lowered and standardized to 256 × 256 pixels. In general, these image formats require huge space compared to other formats, hence supported by commercially available graphic processing algorithms. Five images are considered for simulation, namely: Retina 1, Retina 2, Satellite Image 1, Satellite Image 2 and Moon. Figure 4 shows the original image; its histogram equalization and CSO processed images, respectively. It can be clearly observed from Fig. 4 that histogram equalization method makes an over enhancement of the image making it to appear like a washed-out output. The proposed CSO method is producing a balanced contrast and maintains entropy and PSNR adequate enough better than HE.

Parameters like Entropy and PSNR are measured for all the input samples gathered. Entropy measures the amount of information in a random variable and can be measured by sum of the probability occurrence of events, represented as [12]:

$$H(x) = -\sum p_i \log_2 p_i \tag{15}$$

PSNR is the ratio is used to describe the measurement of quality between compressed and original images. High value of PSNR ensures that the image acquired after compression is of high quality. It is mainly based on mean square error (MSE) and measured as [13],

$$\text{MSE} = \frac{1}{\text{MN}} \sum_{i=0}^{m-1} \sum_{j=0}^{n-1} [I(i, j) - K(i, j)]^2 \tag{16}$$

where I is a monochrome image and K is a noisy approximation image.

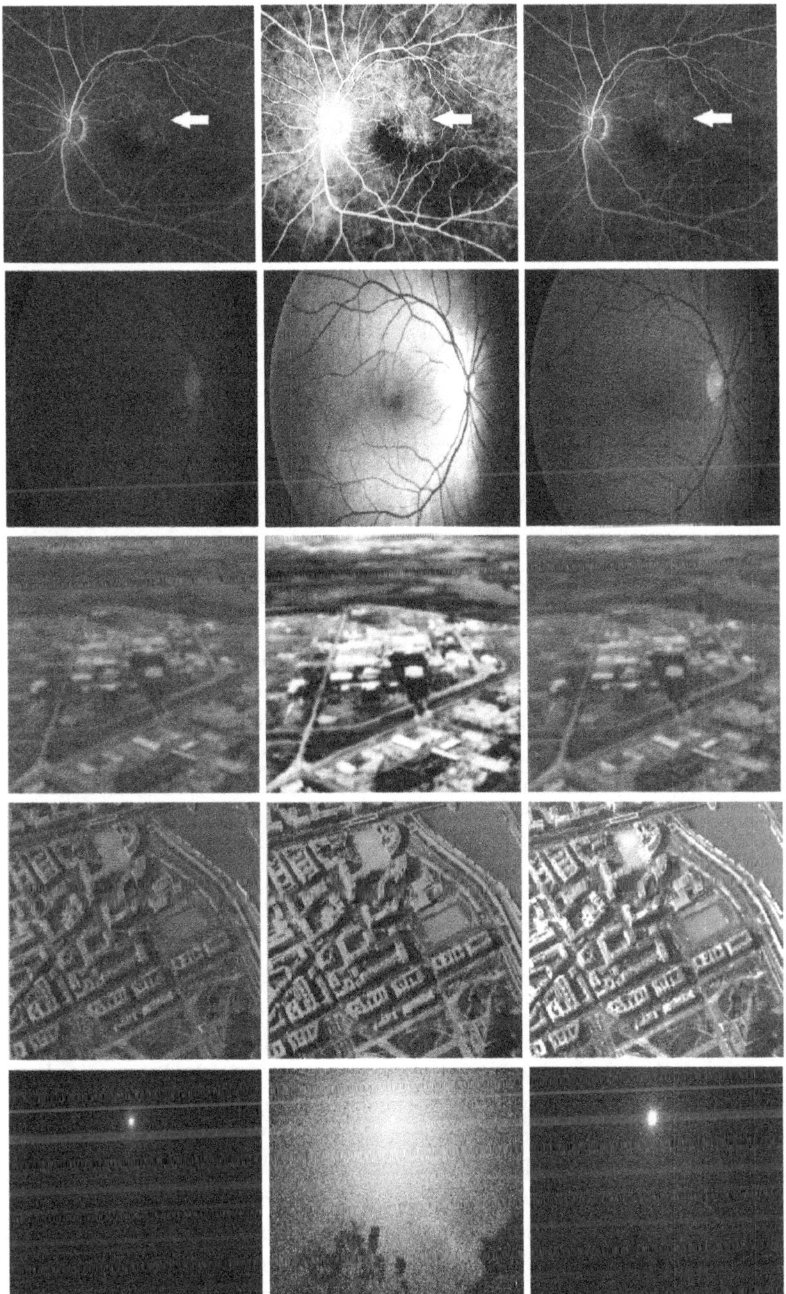

Fig. 4 The original image; its histogram equalization image and CSO processed image (row-wise) of Retina 1, Retina 2, Satellite Image 1, Satellite Image 2 and Moon (Column-wise)

Table 1 Entropy comparison

Image	Original entropy	HE	CSO
Retina 1	6.1778	5.9663	6.4453
Retina 2	5.2114	5.1480	5.9916
Satellite Image 1	6.6151	6.4764	6.8655
Satellite Image 2	6.1049	5.9648	6.1374
Moon	4.5607	4.4786	5.3822

Table 2 PSNR comparison

Image	HE	CSO
Retina 1	24.9584	26.4328
Retina 2	24.0654	24.5701
Satellite Image 1	26.1666	26.7809
Satellite Image 2	24.0986	25.3157
Moon	24.2542	24.5847

$$\text{PSNR} = 20\log_{10}(\max_{1}) - 10\log_{10}(\text{MSE}) \tag{17}$$

where $\max_1$ indicates maximum number of pixels. Tables 1 and 2 provide a comparison in terms of Entropy and PSNR for both the methods.

From these tables, it is clear that the proposed method gives improved results than HE according to the entropy and PSNR values. Entropy of Retina 1 image is 6.1778, whereas histogram equalization results in 5.9663 and CSO provides improvement with the value 6.4453. PSNR value of Retina 1 image after histogram equalization is 24.9584 and CSO results in 26.4328.

5 Conclusion

In this paper, the applications of chicken swarm optimization have been formulated for recovering the digital image. The implemented algorithm was tested on few images which are infected with blur. The experimental results indicate that this approach is a valuable approach, which can significantly support an accurate removal of noise in images with a little computational effort. We can able to see the objects clearly by using chicken swarm optimization algorithm. By using these algorithms, we can improve the two parameters namely Entropy and PSNR. In chicken swarm algorithm, there is less noise adding to an image compared to histogram equalization. It is used in medical field for analyzing the deep nerves inside a body, science field consider a satellite image to see the objects clearly and also used for personal dark images. An extension of this work will focus on developing algorithms in order to obtain the information from a damaged image.

References

1. Gonzalez, R.C., Woods, R.E.: Digital Image Processing, 3rd edn. Prentice Hall (2008)
2. Jaya, V.L., Gopika Kumari, R.: IEM: a new image enhancement metric for contrast and sharpness measurements. Int. J. Comput. Appl. **79**(9), 1–9 (2013)
3. Stark, J.: Adaptive image contrast enhancement using generalizations of histogram equalization. IEEE Trans. Image Process. **9**(5), 889–896 (2000)
4. Chen, S., Suleiman, A.: Scalable global histogram equalization with selective enhancement for photo processing. In: International Conference on Information Technology and Multimedia, Malaysia, pp. 744–752 (2008)
5. Hung, M., Chu, S., Roddick, J.F., Pan, J., Shieh, C.: An effective image enhancement method for electronic portal images. In: International Conference on Computational Collective Intelligence, Technologies and Applications (ICCCI), pp. 174–183 (2010)
6. Singh, R.P., Dixit, M.: Histogram equalization: a strong technique for image enhancement. Int. J. Signal Process Image Process Pattern Recognit **8**(8), 345–352 (2015)
7. Li, Y., Wu, Y., Qu, X.: Chicken swarm-based method for ascent trajectory optimization of hypersonic vehicles. J. Aerosp. Eng. **30**(5) (2017)
8. Meng, X.B., Liu, Y., Gao, X., Zhang, H.: A new bio-inspired algorithm: chicken swarm optimization. In: Proceedings of the International Conference Swarm Intelligence, pp. 86–94 (2014)
9. Deb, S., Gao, X.-Z., Tammi, K., Kalita, K., Mahanta, P.: Recent studies on chicken swarm optimization algorithm: a review. Artif. Intell. (2019)
10. Qu, C.W., Zhao, S., Fu, Y., He, W.: Chicken swarm optimization based on elite opposition-based learning. Math. Problems Eng. **2017** (2017)
11. Yang, X.J., Xu, X.T., Li, R.Y.: Genetic chicken swarm optimization algorithm for solving high dimensional optimization problems. Comput. Eng. Appl. **54**(11), 133–139 (2018)
12. Jain, A.K.: Fundamentals of Digital Image Processing, , 9th ed. Prentice Hall of India (2002)
13. Jayaraman, S., Esakkirajan, S., Veera kumar, T.: Digital Image Processing. Tata McGraw-Hill Education (2011)

AHP-Utility Approach for Mode Choice Analysis of Online Delivery System

Anindita Paul and Mokaddes Ali Ahmed

Abstract The present study optimizes the choice of delivery modes in Agartala city for variable distances and orders. AHP is utilized to evaluate the weightage of the criteria effecting delivery mode choice by formulating pairwise comparison of all the criteria. Time received the weightage of 0.335, followed by environment (0.232), cost (0.191), risk (0.189) and stress (0.053). The utility approach is utilized to find the optimum mode from 1 to 9 km. Bicycle evolved as the optimum mode up to 3 km, whereas bike evolved from 4 to 9 km for single delivery. Further, sensitivity analysis is applied of the results attained by interchanging weightage of two criteria. The concept of AHP-utility and sensitivity analysis has been used to find the optimum mode in case of multiple deliveries (2, 3 and 4 orders). Bicycle evolved optimum mode in some combination of distances, whereas bike alone is optimum for four deliveries.

Keywords Delivery · Mode choice · AHP · Utility · Sensitivity

1 Introduction

Food, medication and other necessities are often required to be delivered within a certain time frame. There has been a problem with maximum delivery staffs that orders are not delivered timely, which acts as a detrimental factor for achieving customer satisfaction. Apparently, the reasons for taking more time of delivery than the expected time (customer) are many, such as poor road condition, congestion due to traffic, vehicle malfunction, improper delivery location, etc. These factors hinder both customer and delivery satisfaction. Thus, if the existing delivery system can be improved then it will satisfy customers, extract more profit, attract more people

A. Paul (✉) · M. A. Ahmed
Department of Civil Engineering, National Institute of Technology Silchar, Silchar, Assam 788010, India

M. A. Ahmed
e-mail: maa@civil.nits.ac.in

K. N. Das et al. (eds.), *Proceedings of the International Conference on Computational Intelligence and Sustainable Technologies*, Algorithms for Intelligent Systems,
https://doi.org/10.1007/978-981-16-6893-7_50

to join, and aid to development of the society as a whole. Few literature related to delivery related transportation studies are discussed herein.

Goyal and Singh [1] determined the relative impact of several factors influencing Indian young consumers' fast food preferences and used multivariate statistical approaches to determine the importance of numerous factors influencing Indian young consumers' choice of fast food establishments. Ma et al. [2] investigated a car driver's data through offline contact from the drivers' psychological perspective to move to dockless bike-sharing. According to the findings, within 2 km distances, the use of private cars was limited. Also, two-thirds of car owners agreed to use dockless bike-sharing. Chandrasekhar et al. [3] determined the impact of online food delivery services such as Swiggy, Foodpanda and Zomato. The findings revealed that consumers prefer uniqueness in terms of price, quality and delivery. Deb and Ahmed [4] evaluated and compared the subjective and objective service quality characteristics for city bus service. Decision-making tools such as analytical hierarchical process (AHP) had been widely used in transportation related problems. Arunyanart et al. [5] applied the AHP approach to find a possible route for transporting export goods from Thailand's northeastern region. Das and Ahmed [6] determined the level of service of on-street parking using AHP based on three criteria: parking characteristic, safety characteristic and design characteristic.

Several delivery service conceptualizations exist, which shows different roles of the customer in the delivery service. The service is not delivered only in an interaction between the customer and the service provider, but can also involve customer interaction with a service provider through intermediaries or technology. This affects the perceptions of the service in many respects and, depending on the interaction, the time and location may vary. To make the delivery system better and run smoothly, it is necessary to optimize the delivery time and delivery mode. The delivery mode and subsequently the time taken for delivery depends on a number of factors such as transportation used, road element (road length, road width, type of median, carriageway, intersection, etc.) path followed for delivery, waiting time for pickup, etc. To evaluate the effect of these factors, data has to be collected, preferably from delivery personnel and customers. In the present study, mode choice of delivery has been addressed by collecting data from delivery personnel through survey and optimized by applying the concept of AHP-utility.

2 Methodology

2.1 Data Collection

Initially, SWOT analysis is carried out to know the strength, weakness, opportunities and threats of delivery at Agartala. Among several opportunities such as delivery mode choice, routing in case of multiple deliveries, introduction of alternative fuel vehicle and bicycle, the delivery mode choice is important aspect which has been

Table 1 Data collected from survey

S. No.	Distance	Bicycle	Bike	Bicycle	Bike	Bicycle	Bike	Bicycle	Bike	Bicycle	Bike
		Time (min)		Risk (1–5)		Cost (Rs.)		Environment (0–5)		Stress (1–5)	
1	1	4	2	1	1	4	4	0	1	1	1
2	2	8	4	1	1	8	8	0	2	1	1
3	3	12	6	2	2	12	12	0	2	2	2
4	4	16	8	3	2	16	18	0	2	3	2
5	5	20	10	4	2	20	24	0	3	3	2
6	6	24	12	4	2	24	30	0	3	4	2
7	7	28	14	5	3	28	36	0	4	4	3
8	8	32	16	5	3	32	42	0	4	5	3
9	9	36	18	5	3	36	48	0	5	5	3

addressed. A questionnaire survey has been conducted from delivery personnel (bike and bicycle user). The main aim is to evaluate the choice of delivery mode for orders at various distances. The attributes considered in the questionnaire survey are delivery time, delivery risk, delivery cost, environment and delivery stress. In total, 40 delivery personnel have participated in the questionnaire survey with equal participation from both category of modes (bike and bicycle). All the data from questionnaire survey and expert from the relevant field are aggregated to their average values and represented in Table 1.

2.2 *Evaluation of Criteria Weight by Analytical Hierarchical Process*

AHP is a multiple criteria decision-making (MCDM) tool that has been used in almost all decision-making applications [7, 8]. In the present study, five criteria are considered in the formulation of AHP model. These are time, risk, cost, environment and stress. The AHP procedure incorporated in the present work has been elaborated in a study reported by Guner [9]. The pairwise comparison matrix and the normalized matrix with priority weight of each criterion is shown in Table 2.

The consistency ratio (CR) is calculated to test the reliability of subjective judgements in AHP. When the CR ≤ 0.1, the consistency of the matrix can be judged as acceptable. The CR of the current model is found to be 0.09, which is <0.1. Thus, the considered pairwise matrix is acceptable, where the weights of the criteria time, risk, cost, environment and stress are found to be 0.335, 0.189, 0.191, 0.232 and 0.053, respectively.

Table 2 Pairwise and normalized pairwise comparison matrices

A. Pairwise comparison matrix

	Time	Risk	Cost	Environment	Stress
Time	1	2	2	2	5
Risk	1/2	1	2	1/2	3
Cost	1/2	1/2	1	2	3
Environment	1/2	2	1/2	1	7
Stress	1/5	1/3	1/3	1/7	1
Sum	2.7	5.83	5.83	5.64	19

B. Normalized pairwise comparison matrix

	Time	Risk	Cost	Environment	Stress	Priority weight
Time	0.370	0.343	0.343	0.354	0.263	0.335
Risk	0.185	0.171	0.343	0.089	0.158	0.189
Cost	0.185	0.086	0.171	0.354	0.158	0.191
Environment	0.185	0.343	0.086	0.177	0.368	0.232
Stress	0.074	0.057	0.057	0.025	0.053	0.053
Total						1

2.3 Selection of Optimum Mode by AHP-Utility

To find the best mode for variable distances, the concept of utility is applied. Initially, the data collected from the questionnaire survey is normalized to make them dimensionless. The normalized matrix of data is shown in Table 3. The normalized values (X_{nor}) for a particular criterion are evaluated by the following equation.

$$X_{\text{nor}}(i) = \frac{X_{\max} - X(i)}{X_{\max} - X_{\min}}, \quad i = 1, 2, \ldots, n, \tag{1}$$

where, $X_{\max}$ and $X_{\min}$ represent the maximum value and minimum value of the criteria; n represents total number of observations.

Thereafter, the AHP weightages obtained for each criterion are utilized to get the AHP-utility (AU) values for each combination of data present in the model by Eq. 2.

$$\begin{aligned}\text{AU}(i) = \frac{1}{k}[&A_{\text{nor}}(i) \times w_A + B_{\text{nor}}(i) \times w_B + C_{\text{nor}}(i)\\ &\times w_C + D_{\text{nor}}(i) \times w_D + E_{\text{nor}}(i) \times w_E]\end{aligned} \tag{2}$$

where A, B, C, D and E represent the criteria time, risk, cost, environment and stress, respectively;w_A, w_B, w_C, w_D and w_D represent the weightage of time, risk, cost, environment and stress, respectively; and k represents number of criteria, which is 5 in the present study.

Table 3 Normalized matrix of data

Mode	Distance (km)	Time	Risk	Cost	Environment	Stress	AHP-utility
Bicycle	1	0.94	1.00	1.00	1.00	1.00	0.98
	2	0.82	1.00	0.91	1.00	1.00	0.92
	3	0.71	0.75	0.82	1.00	0.75	0.81
	4	0.59	0.50	0.73	1.00	0.50	0.69
	5	0.47	0.25	0.64	1.00	0.50	0.58
	6	0.35	0.25	0.55	1.00	0.25	0.51
	7	0.24	0.00	0.45	1.00	0.25	0.41
	8	0.12	0.00	0.36	1.00	0.00	0.34
	9	0.00	0.00	0.27	1.00	0.00	0.28
Bike	1	1.00	1.00	1.00	0.80	1.00	0.95
	2	0.94	1.00	0.91	0.60	1.00	0.87
	3	0.88	0.75	0.82	0.60	0.75	0.77
	4	0.82	0.75	0.68	0.60	0.75	0.73
	5	0.76	0.75	0.55	0.40	0.75	0.63
	6	0.71	0.75	0.41	0.40	0.75	0.59
	7	0.65	0.50	0.27	0.20	0.50	0.44
	8	0.59	0.50	0.14	0.20	0.50	0.39
	9	0.53	0.50	0.00	0.00	0.50	0.30
Weight		0.335	0.189	0.191	0.232	0.053	–

3 Results and Discussion

Initially, AHP-utility has been applied for optimum mode choice of a delivery at distance ranging from 1 to 9 km. The results achieved from the AHP-utility model is plotted in Fig. 1. It is observed that bicycle evolved as the best alterative from 1 to

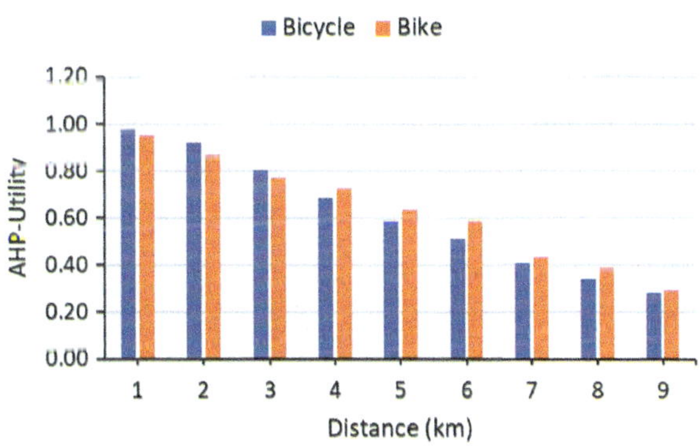

Fig. 1 AHP-utility of modes at various distances

3 km. But bike evolved as the best alternative from 4 to 9 km. To further apprehend the results achieved from AHP-utility, the concept of sensitivity analysis is implemented.

Higher value of utility yields the optimum mode for a particular distance. But it has been observed that as the criterions have their individual weightage (evaluated by AHP), the selection depends upon these weight values. So, any change in the criteria weight may result in change in optimum selection. To overcome this situation of inconsistency, the selection must be validated, which would make it effective and robust. A sensitivity analysis can be utilized to verify the robustness of the model. The idea of sensitivity analysis is to understand the effect on the model with the variation of control variables. In other words, sensitivity analysis helps in determining the significance of criteria and their impact on mode choice [10]. In this study, sensitivity analysis exploited interchange of two criteria weights with each other, while others are kept constant. This provides total 11 combinations for 5 criteria, considering the existing condition. The AHP-utility value of all the modes after conducting sensitivity analysis is shown in Fig. 2. Further, the distance wise result of optimum mode is shown in Fig. 3. In all the distances where bicycle is the optimum mode (Fig. 1), the result has deviated once in 11 conditions after conducting sensitivity analysis. So, out of possible 33 cases, 30 times bicycle evolved as the optimum mode, which reflects 90.91% accuracy. Similarly, for bike as optimum mode, the accuracy is 80.30%. Thus, sensitivity analysis validates the solution of the present problem and verifies that bicycle is the optimum mode up to 3 km, whereas bike is the optimum mode from 4 to 9 km.

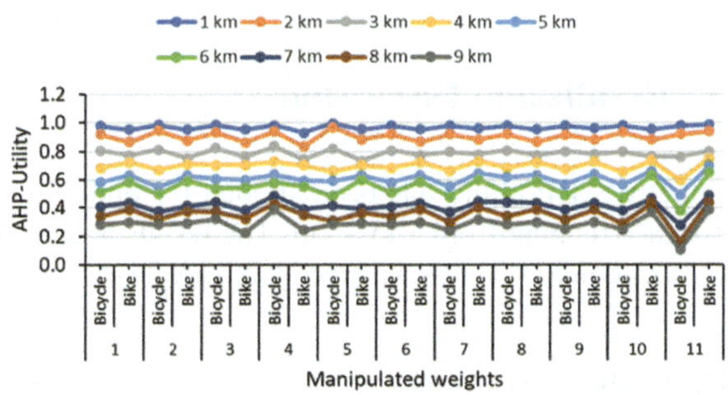

Fig. 2 Ranking of modes after sensitivity analysis

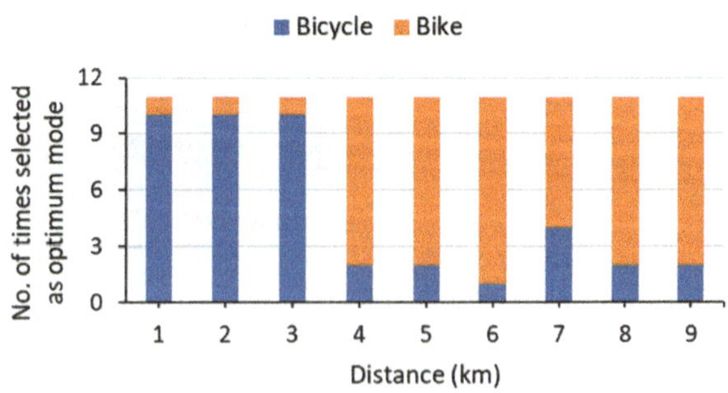

Fig. 3 Result of sensitivity analysis

In the next stage, mode choice for multiple deliveries is addressed. In the present study, two, three and four deliveries are considered. So, the mode has to be chosen between bicycle and bike for delivering the multiple orders at different distances. In this analysis, all possible combination of distances is considered for a total aggregated distance of ≤6 km from the origin to the last destination.

Initially, two deliveries are considered, and then the concept of AHP-utility is applied to compare the performance of both the modes. The result of the AHP-utility values of modes for two deliveries at various distances is plotted in Fig. 4. Bicycle evolved as the optimum mode at all combination of distances, where initial delivery is at 1, 2 and 3 km, and the subsequent delivery is at 1, 2 and 3 km. Sensitivity analysis has been conducted by interchanging weightage of two criterions at a time and the ranking achieved by the modes is plotted in Fig. 5. Further, the result of sensitivity analysis has been plotted in Fig. 6. From both the plots (Figs. 5 and 6), it is observed that the result obtained by the initial assignment of weightage by AHP is reflected in maximum cases after conducting sensitivity analysis. In case of bicycle as optimum mode (Fig. 4), after interchanging weightage of two criterions, only 17 out of 99 times, bike evolved as the optimum mode (Fig. 6), which reflects 82.3% accuracy. Similarly, for bike as optimum mode (Fig. 4), only 16 out of 66 times, bicycle evolved as the optimum mode, which reflects 75.76% accuracy.

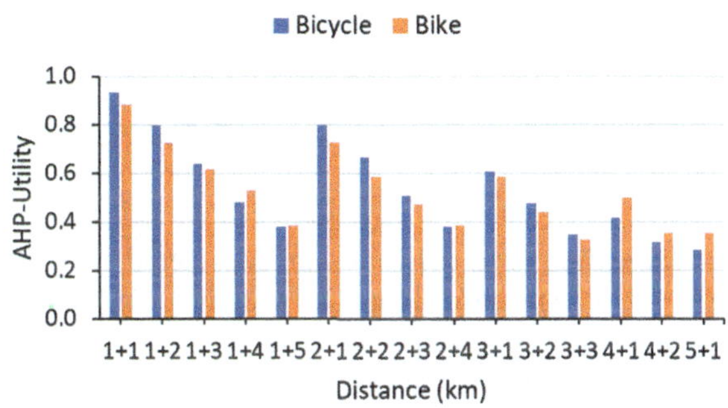

Fig. 4 AHP-utility of modes for two deliveries at various distances

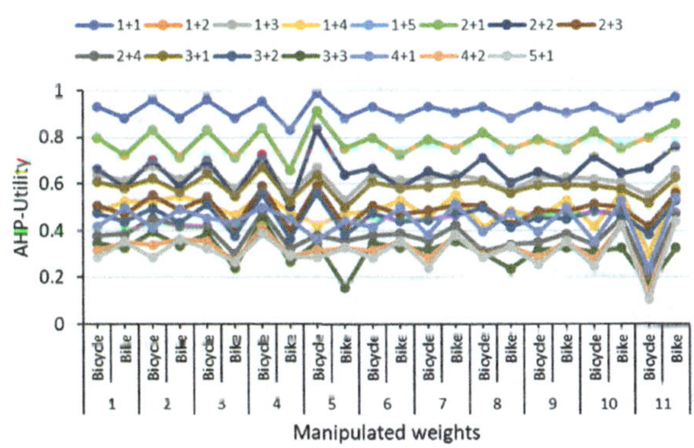

Fig. 5 Ranking of modes for two deliveries after sensitivity analysis

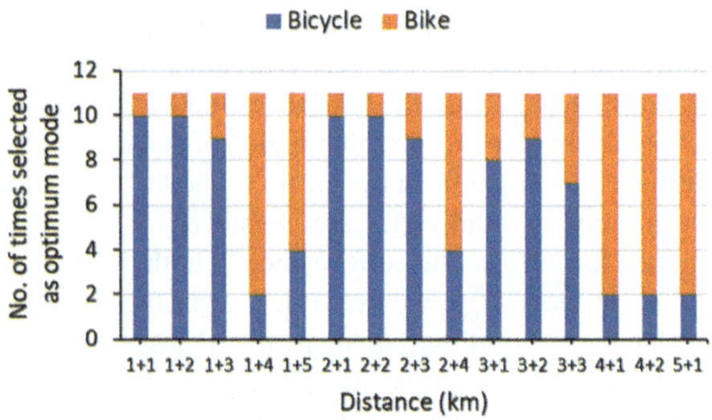

Fig. 6 Result of sensitivity analysis (two deliveries)

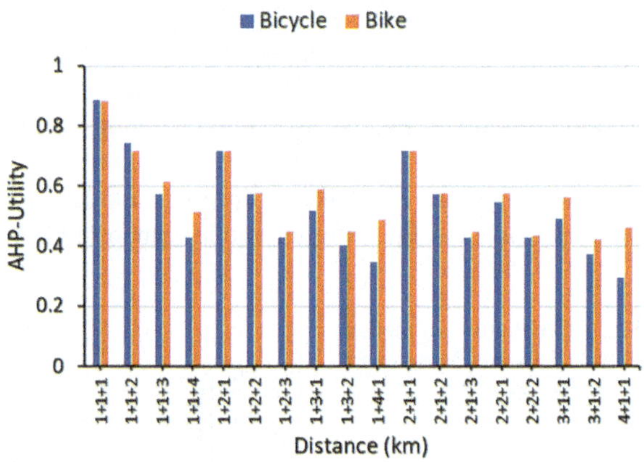

Fig. 7 AHP-utility of modes for three deliveries at various distances

Thereafter, three deliveries are considered, and similar to the previous cases, the concept of AHP-utility is applied to compare the performance of both the modes. The result of the AHP-utility values of modes for three deliveries at various combination of distances is plotted in Fig. 7.

Bicycle evolved as the optimum mode when the distance of the third delivery is 1 and 2 km, whereas both the first and second deliveries are the shortest (1 km). Further, bicycle also possessed higher utility at 2 km of first delivery, followed by shortest second and third deliveries (1 km), and 1 km of first delivery followed by 2 km of second delivery and 1 km of third delivery. With subsequent increase in the distances, the utility of bike is higher as compared to bicycle. From sensitivity analysis, (Figs. 8 and 9), it is observed that the result obtained by the initial assignment of weightage by AHP is reflected in maximum cases. In case of bicycle as optimum mode (Fig. 7), only 15 out of 44 times, bike evolved as the optimum mode, which reflects 65.91% accuracy. Similarly, for bike as optimum mode (Fig. 7), only 49 out of 154 times, bicycle evolved as the optimum mode reflecting 68.18% accuracy.

Lastly, four deliveries have been considered and AHP-utility is applied to evaluate the optimum mode. The result of the AHP-utility values of modes for four deliveries at various combination of distances is plotted in Fig. 10. In case of four deliveries,

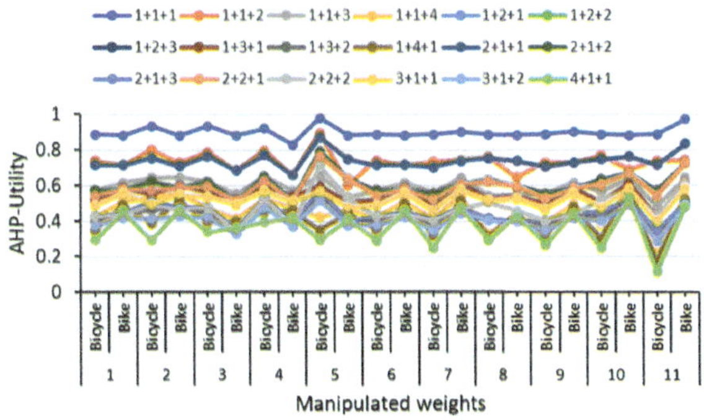

Fig. 8 Ranking of modes for three deliveries after sensitivity analysis

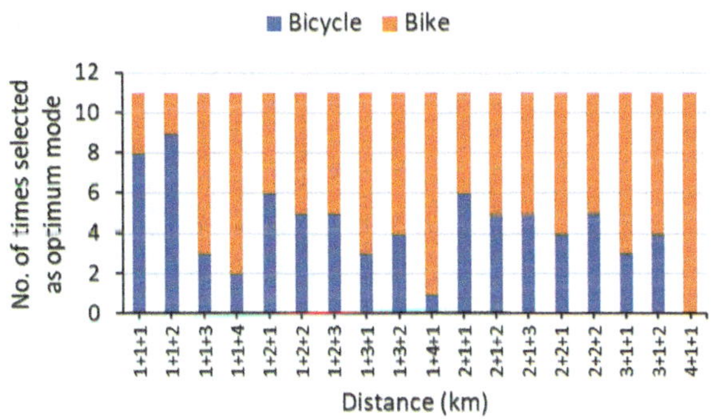

Fig. 9 Result of sensitivity analysis (three deliveries)

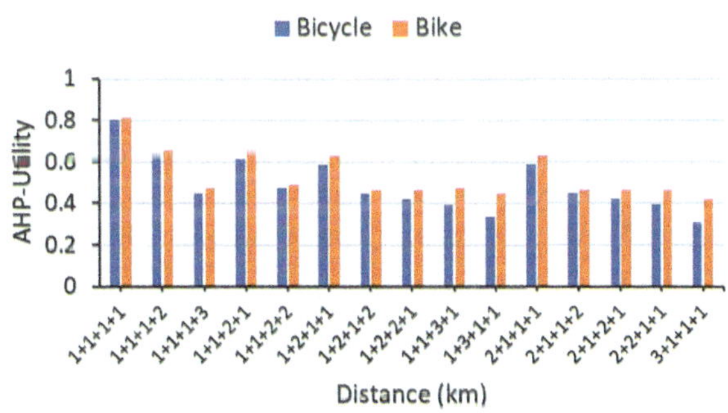

Fig. 10 AHP-utility of modes for four deliveries at various distances

bike evolved as the best mode in each combination of the distances considered in the present study. The ranking of the modes at various combination of distances after sensitivity analysis is shown in Fig. 11. Figure 12 shows the result of sensitivity analysis conducted for four deliveries. It is observed that 63 times bicycle evolved as optimum mode out of 165 possible times, which suggests an accuracy of 61.82%.

Overall, it can be inferred that although bicycle evolved as optimum mode up to 3 km in single delivery system, and it has the potential to deliver multiple orders (two and three) with higher efficiency as compared to bike. But there exist some constraints regarding the combination of distances, where bicycle evolved as the better alternative

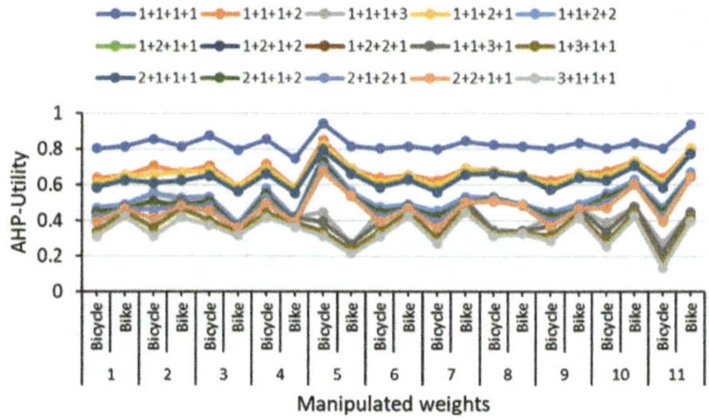

Fig. 11 Ranking of modes for four deliveries after sensitivity analysis

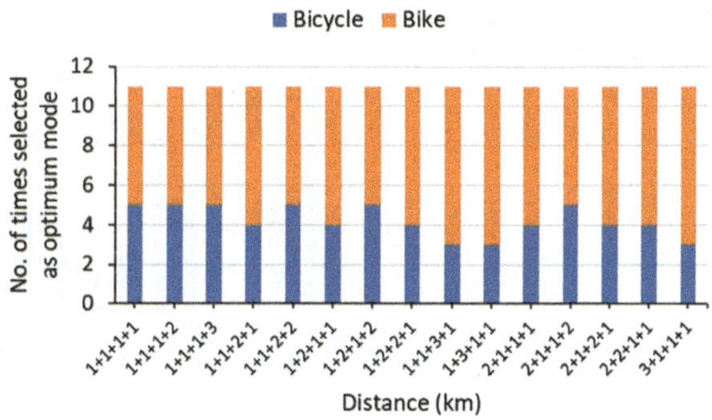

Fig. 12 Result of sensitivity analysis (four deliveries)

as compared to bike. Using bicycle in these combinations of distances would be profitable to the delivery personnel without hampering the satisfaction of customers and the overall delivery system as a whole. Bicycle is also environment friendly and incurs less maintenance, which aids it to the cause of delivery as compared to bike. Further, the economically backward people can earn a living through delivery of orders by their bicycles, if a policy of augmenting the use of bicycle is implemented in future.

4 Conclusions

The present study discussed the degree of usefulness of two modes, i.e., bicycle and bike in online delivery at Agartala city. From AHP, maximum weightage of 0.335 has been allocated to time, followed by 0.232 to environment, 0.191 to cost, 0.189 to risk and 0.053 to stress, respectively. The AHP model's CR is 0.06, indicating that it is acceptable. For trips ranging from 1 to 9 km, the utility technique is used to determine the best mode between bicycle and bike. Up to 3 km, the bicycle evolved as the ideal mode, but from 4 to 9 km, the bicycle evolved as the ideal mode. Sensitivity analysis is used to ensure that the obtained results are reliable. The AHP-utility model's estimate of the optimal mode choice is backed up by the results of the sensitivity analysis. In the event of two and three deliveries, the bicycle emerged as the best form of transportation, although the bike alone is best for four deliveries. According to the AHP-utility findings of multiple deliveries, the bicycle evolved as the ideal mode in combinations of shorter distances for three of the deliveries. However, when longer distances are involved, biking has emerged as the best means of transportation.

References

1. Goyal, A., Singh, N.P.: Consumer perception about fast food in India: an exploratory study. Br. Food J. **109**, 182–195 (2007). https://doi.org/10.1108/00070700710725536
2. Ma, X., Cao, R., Wang, J.: Effects of psychological factors on modal shift from car to dockless bike sharing: a case study of Nanjing, China. Int. J. Environ. Res. Public Health. **16**, 3420 (2019). https://doi.org/10.3390/ijerph16183420
3. Chandrasekhar, N., Gupta, S., Nanda, N.: Food delivery services and customer preference: a comparative analysis. J. Foodserv. Bus. Res. **22**, 375–386 (2019). https://doi.org/10.1080/15378020.2019.1626208
4. Deb, S., Ahmed, M.A.: Quality assessment of city bus service based on subjective and objective service quality dimensions. Benchmarking An Int. J. **26**, 567–589 (2019). https://doi.org/10.1108/BIJ-11-2017-0309
5. Arunyanart, S., Ohmori, S., Yoshmoto, K.: Selection of export route option in GMS region: recommendation for current situation. J. Japan Ind. Manag. Assoc. **67**, 193–201 (2016). https://doi.org/10.11221/jima.67.193
6. Das, D., Ahmed, M.A.: Level of service for on-street parking. KSCE J. Civ. Eng. **22**, 330–340 (2018). https://doi.org/10.1007/s12205-017-1538-1
7. Vaidya, O.S., Kumar, S.: Analytic hierarchy process: an overview of applications. Eur. J. Oper. Res. **169**, 1–29 (2006). https://doi.org/10.1016/j.ejor.2004.04.028
8. Saaty, T.L.: Analytic heirarchy process. In: Wiley StatsRef: Statistics Reference Online, pp. 1–11. Wiley, Chichester, UK (2014). https://doi.org/10.1002/9781118445112.stat05310.
9. Güner, S.: Measuring the quality of public transportation systems and ranking the bus transit routes using multi-criteria decision making techniques. Case Stud. Transp. Policy. **6**, 214–224 (2018). https://doi.org/10.1016/j.cstp.2018.05.005
10. Das, D., Mokaddes, P., Ahmed, A., Sen, S.: Controlling on-street parking demand using sensitivity analysis: a case study at Kolkata. J. Transp. Syst. **1**, 1–13 (2016)

Medical Electronics Device to Provide Non-Invasive Therapy to Treat Deep Vein Thrombosis Using BLE and Embedded Systems

Manasa Murali, A. B. Sreelakshmi, K. S. Swathi, K. Divyatha Shetty, and K. Chethana

Abstract A blood clot in a vein deep beneath the skin causes deep vein thrombosis (DVT). It is often treated by boosting blood flow in the veins of the affected limb. Therefore, it is necessary to focus on a device which is small and has efficient circuitry to make it portable and implementable. In this study, an automated test equipment (ATE) is built to do system testing of deep vein thrombosis (DVT) healthcare equipment. The ATE confirms the functioning of the device and collects data from it, using Microcontroller-CYBLE which is a low-power Bluetooth component in the device. The test begins with the installation of a mobile application that uses the Bluetooth Low Energy (BLE) protocol to communicate with the ATE. The testing system then opens the device under test (DUT). A servo motor is used by ATE to start the gadget. It then uses BLE to send different pressure values to the device and reads the pressure sensor readings. The readings are analyzed, and test results are generated which are sent to the mobile app. The test was deemed successful based on the medically permitted pressure range and inflation time. As a result, our device dramatically reduces the amount of time and effort required for testing.

Keywords Automated test equipment (ATE) · Deep vein thrombosis (DVT) · Bluetooth low energy (BLE) · Device under test (DUT)

1 Introduction

Deep vein thrombosis (DVT) is the most prevalent form of venous thrombosis, and if left unchecked, it can lead to pulmonary embolism (PE) and fatality [1]. When a blood clot (thrombus) forms in one or more of your body's deep veins, typically in your legs, DVT develops. A thrombus can develop spontaneously or as a consequence of clinical situations such as surgery, trauma, or prolonged bed rest [2]. DVT can be diagnosed by various methods such as imaging techniques which include ultrasonography or venography, D-dimer test, and magnetic resonance imaging (MRI).

M. Murali (✉) · A. B. Sreelakshmi · K. S. Swathi · K. Divyatha Shetty · K. Chethana
Department of Electronics and Communication, Jyothy Institute of Technology, VTU, Bangalore, India

K. N. Das et al. (eds.), *Proceedings of the International Conference on Computational Intelligence and Sustainable Technologies*, Algorithms for Intelligent Systems,
https://doi.org/10.1007/978-981-16-6893-7_51

Fig. 1 Device is in-built in the sole of the shoe

One of the effective treatments in general patients could be prophylaxis with low-dose anticoagulants such as heparin or warfarin followed by Vitamin-K antagonists [3]. However, using anticoagulants in neurosurgical patients could be problematic as it leads to serious bleeding [4, 5]. Other invasive treatments include administration of thrombolytic directly into the clot or insertion of the filter into vena cava in the abdominal region [5]. Invasive treatments will always be accompanied by side effects. Hence, the best way of treating DVT would be using non-invasive methods such as the usage of compression stockings that reduce the chance of blood pooling and forming a clot [6–8]. The major drawbacks with the present non-invasive therapies to treat DVT are: The patient must devote time exclusively for the treatment and passive exercises are provided only to the ankle and foot. In this paper, we have discussed a device that follows bio-mimicking to provide continuous compressions to the injured limb [9]. The device is compact and portable making it compliant for DVT patients traveling for a longer duration and not able to take compression therapy in the clinical environment. A mobile application that communicates with the device provides the ease to set the inflation and deflation pressure values when the device is placed in the sole of the patients' shoe as shown in Fig. 1.

2 Instrumentation

Portable medical devices have transformed people's ability to track and assess their health and well-being [9, 10]. These devices can provide continuous and non-invasive monitoring of health parameters and real-time updates to healthcare providers [11].

In order to make the device compact, a multi-layered PCB is used. They have high capacity and better speeds due to their innate electrical properties. The higher density of assembly and reduction in the number of connectors used facilitate simple construction and reduce the weight of the device. We have made use of 0402 smd type package components to reduce the form factor as much as possible.

The hardware components (Figs. 2 and 3) used here are:

1. Microcontroller-CYBLE: Bluetooth Low Energy (BLE) can communicate over short distances of 100 m in open areas.
2. ADP5062: Digital input/output USB interfaces.
3. MMA8653FC: A 3-axis, low-power 10-bit digital accelerometer.
4. MP3V5050GC6U: Pressure sensor.
5. ADP151: Linear voltage regulator.
6. DRV8839: Dual half H-bridge driver IC which drives maximum of 1.8 A current.

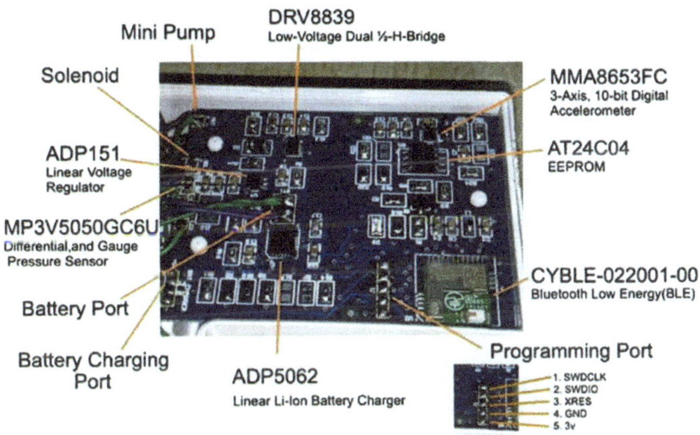

Fig. 2 Circuitry

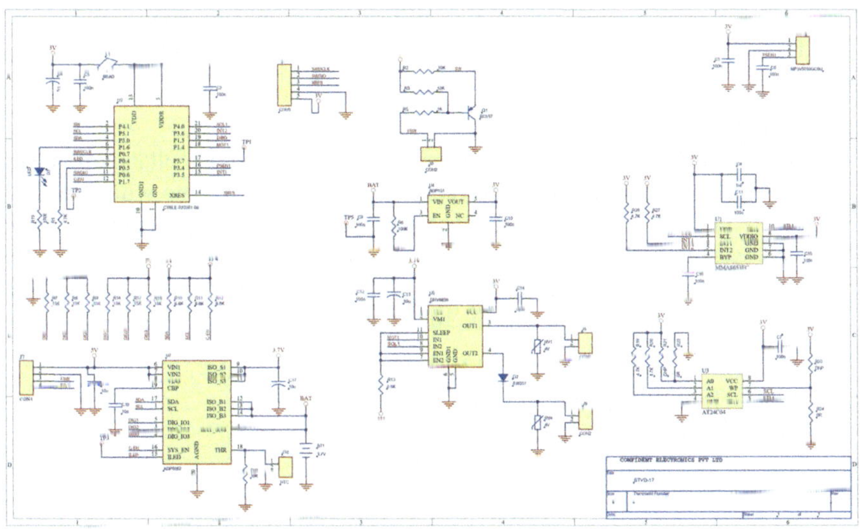

Fig. 3 Schematic

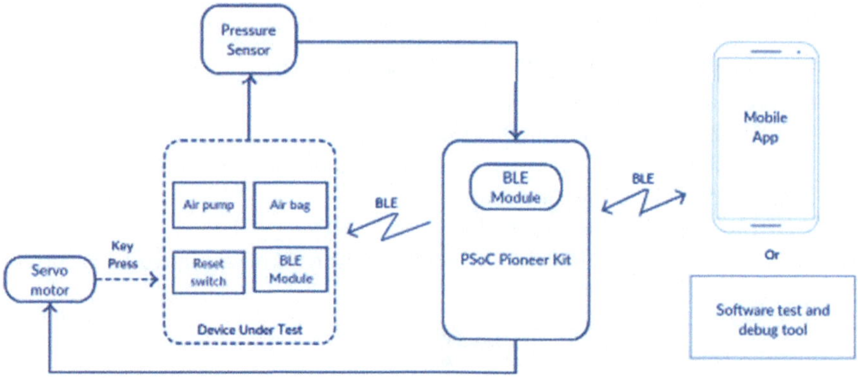

Fig. 4 Schematic diagram

7. AT24C04: EEPROM data storage.

Schematic diagram of the device is as shown in Fig. 4:
It consists of the following:

1. A PSoC 4 pioneer
2. A BLE module
3. Pressure sensor
4. A Mobile app called CySmart.

The PSoC (programmable system-on-chip) Creator is a software that assists us in running the application in the mobile app. The PSoC produces two files: APK and HEX files. The APK file is to run the mobile app, whereas the HEX file is for the device. PSoC can be easily configured for Bluetooth communication as well [12].

BLE operates over the 2.4 GHz ISM band. It is a low-power wireless technology that enables devices to communicate with one another [13, 14].

Mobile Application

CySmart is a Bluetooth Low Energy application developed in collaboration with Cypress Semiconductor and Infineon Technologies. Any BLE chip will work with CySmart. In general, it is an app that uses Bluetooth technology to connect to an embedded system. This app is available for both Android and iOS devices. The CySmart Android app (Fig. 5) can be used with any BLE products including BLE development kits from Cypress like the following ones:

1. CY8CKIT-042-BLE-A PSoC® 4 BLE Pioneer Kit
2. CY8CKIT-062-BLE PSoC® 6 BLE Pioneer Kit
3. WICED BT Kits.

Fig. 5 CYSMART app

3 Device Communication with Mobile App

Steps to establish a communication between the device and the mobile app are as follows:

1. Firstly, the device (STVD_B) is powered up and is connected with the mobile app via Bluetooth as shown in Fig. 6.
2. Then there is a feature called Generic Attribute Profile (GATT) that specifies how two Bluetooth Low Energy devices exchange data using concepts like services and characteristics (Fig. 7).
3. Services are used to divide data into logic entities that contain specific data chunks termed as characteristics.
4. Characteristics consist of 15 ports to control and communicate with the device.
5. Currently, we have used only five ports in our device, namely,

Port f1: This port is used to set the inflation pressure of the airbag.
Port f6: This port is used to turn the device ON and OFF using 1 and 0 as its controls, respectively.
Port fb: This port is used to set the hold time of the airbag.
Port fd: This port is used to describe the battery condition.
Port ff: This port informs the user about the number of cycles the device has undergone.

The entire steps are summarized as (Fig. 8):

- The smartphone application is used to start the test. The mobile app must be configured as server, and the device must be configured as client.
- As soon as the device is turned on, the inflation pressure value, inflation time (in seconds) and number of cycles can be set using the mobile app. The solenoid

Fig. 6 STV_D device

valve is open, allowing the motor to start pumping air into the airbag, thereby inflating it.

- Once the airbag pressure level becomes equal to the pre-set pressure value, the solenoid valve closes. This prevents the escape of air from the airbag thereby retaining pressure for some time, which is referred to as hold time.
- Post which the solenoid valve opens allowing escape of air from the airbag causing deflation, and the motor again starts pumping air into the airbag and the cycle continues.

The inflation and deflation pressure values have to be set taking the following points into consideration:

1. Location of the clot
2. Depth at which the clot is present
3. Blood pressure of the patient.

It is not always necessary for the patient to have his mobile handy as once the inflation pressure value is set, the device can be turned on and off even with a manual key press. This ensures uninterrupted compression therapy for the patient.

Fig. 7 GATT

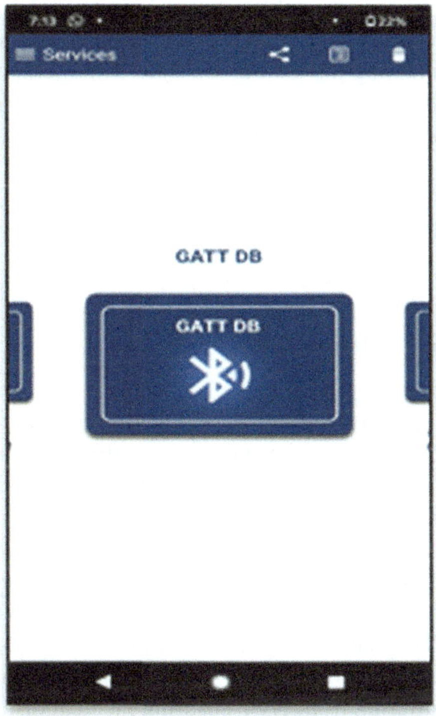

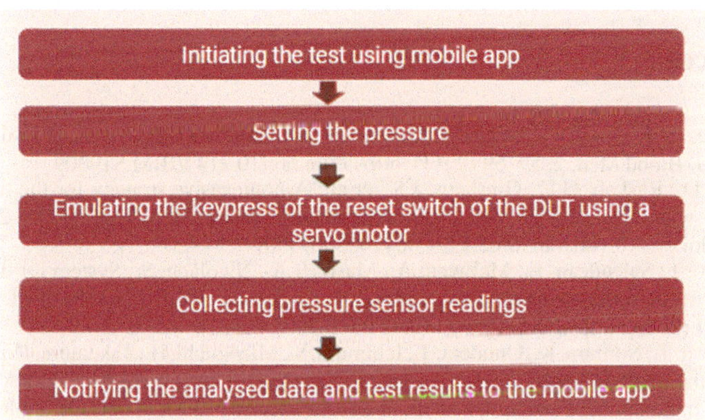

Fig. 8 Steps followed when the device communicates with the mobile app

4 Result

Hospitals now routinely use intermittent pneumatic compression devices to prevent DVT. The differential pressure at the limits produces discomfort to the patient (between the ankle and the knee) since the stocking only applies pressure to a certain section of the leg. The device demonstrated in this study improves blood flow by imitating the natural muscular contractions in airbag placed on the wounded limb. The user can control the airbag inflation pressure via a smartphone app using BLE protocol. After that, the airbag cycles through inflation and deflation.

The device provided best results for a wide range of pressure values (5–200 mmHg) during clinical trials. The GUI of the Cysmart app ensured seamless therapy trials with strict medical distancing.

5 Conclusion

The purpose of this research is to make the lives of DVT patients easier by reducing the number of hospital visits and also facilitating comfortable long-distance travels. The method employed in this study includes the collection of pressure values and their application to improve production efficiency. There is no need for human intervention because the entire treatment method is automated. Furthermore, the complete automation process reduces the treatment costs as adjustable airbag is used unlike in thromboembolic stockings which needs to be replaced with variation in DVT levels. The results showed that our proposed technique was both efficient and effective. If enhanced further, our method could help in post-invasive treatment care of people suffering from DVT ensuring medical distancing.

References

1. Kesieme, E., Kesieme, C., Jebbin, N., Irekpita, E., Dongo, A.: Deep vein thrombosis: a clinical review. J. Blood Med. **2**, 59–69 (2011). https://doi.org/10.2147/JBM.S19009
2. Hull, R.D., Raskob, G.E., Ginsberg, J.S., et al.: A noninvasive strategy for the treatment of patients with suspected pulmonary embolism. Arch Intern Med. **154**(3), 289–297 (1994). https://doi.org/10.1001/archinte.1994.00420030093009
3. Guerrero, J., Salcudean, E., McEwen, A., Masri, B.A., Nicolaou, S.: System for deep venous thrombosis detection using objective compression measures. IEEE Trans. Biomed. Eng. **53**(5), 845–854 (2006). https://doi.org/10.1109/TBME.2005.863878
4. Yonezawa, T., Nomura, K., Onodera, T., Ichimura, S., Mizoguchi, H., Takemura, H.: Evaluation of venous return in lower limb by passive ankle exercise performed by PHARAD. In: 37th Annual International Conference of the IEEE Engineering in Medicine and Biology Society (EMBC), pp. 3582–3585 (2015). https://doi.org/10.1109/EMBC.2015.7319167
5. Effectiveness of intermittent pneumatic compression in reduction of risk of deep vein thrombosis in patients who have had a stroke (CLOTS 3): a multicentre randomised controlled trial. Lancet **382**(9891), 516–524 (2013). ISSN 0140-6736. https://doi.org/10.1016/S0140-6736(13)61050-8. https://www.sciencedirect.com/science/article/pii/S0140673613610508
6. Chang, T., et al.: Design of a wearable deep vein thrombosis prevention device using thin McKibben muscles. In: 18th International Conference on Mechatronics—Mechatronika (ME), pp. 1–6 (2018)
7. Vinay, C.K., Vazhiyal, V., Rao, M.: Design of a non-invasive pulse rate controlled deep vein thrombosis prophylaxis lower limb device. In: 41st Annual International Conference of the

IEEE Engineering in Medicine and Biology Society (EMBC), pp. 5407–5410 (2019). https://doi.org/10.1109/EMBC.2019.8856441
8. Low, F., Yeow, R.C.H., Yap, H.K., Lim, J.H.: Study on the use of soft ankle-foot exoskeleton for alternative mechanical prophylaxis of deep vein thrombosis. IEEE Int. Conf. Rehabil. Robot. (ICORR) **2015**, 589–593 (2015). https://doi.org/10.1109/ICORR.2015.7281264
9. Zhao, S., Liu, R., Guan, D.: Development of an intelligent digital monitoring and biofeedback system for intermittent pneumatic compression therapy device. In: IEEE 8th International Conference on Fluid Power and Mechatronics (FPM), pp. 709–713 (2019). https://doi.org/10.1109/FPM45753.2019.9035788
10. Dhatri, M.P., Shivram, R.: Development of a functional testing system for test automation and statistical analysis of the behavior of health care device used to treat deep vein thrombosis. In: 3rd IEEE International Conference on Recent Trends in Electronics, Information and Communication Technology (RTEICT), pp. 1715–1719 (2018). https://doi.org/10.1109/RTEICT42901.2018.9012553
11. Caprini, J.A.: Mechanical methods for thrombosis prophylaxis. Clin. Appl. Thromb. Hemost. 668–673 (2010). https://doi.org/10.1177/1076029609348645
12. Hrgetic, M., Krois, I., Cifrek, M.: Accuracy analysis of dissolved oxygen measurement system realized with cypress PSoC configurable mixed signal array. In: Proceedings of the IEEE International Symposium on Industrial Electronics, ISIE, vol. 3, pp. 1105–1110 (2005). https://doi.org/10.1109/ISIE.2005.1529078
13. Sairam, K.V.S.S.S.S., Gunasekaran, N., Redd, S.R.: Bluetooth in wireless communication. IEEE Commun. Mag. **40**(6), 90–96 (2002). https://doi.org/10.1109/MCOM.2002.1007414
14. Gomez, C., Oller, J., Paradells, J.: Overview and evaluation of bluetooth low energy: an emerging low-power wireless technology. Sensors **12**(9), 11734–11753 (2012). https://doi.org/10.3390/s120911734

Applying Efforts for Behavior-Based Authentication for Mobile Cloud Security

Vivek Kumar and Sangram Ray

Abstract Mobile devices have become smart for their ability to access different and complex applications. Now, these devices can also sense human behavior and differentiate among the two subjects' habits. The security concerns become more critical when the unauthorized user accessing the network services, using a paired network device (i.e., smart phone or laptop). Therefore, an authentication system is required that continuously observe the activities of the end device user, using a continuous authentication system. The proposed authentication model involves the machine learning-based techniques to analyze the application usages patterns to discover the misleading user behavior. In addition, the correlation coefficient is used for feature ranking. In this work, a data mining model is developed to provide a behavior-based authentication system for mobile devices. The proposed data model includes the PSO for feature selection or dimensionality reduction of the behavioral dataset. Additionally for learning and classification of behaviors, C4.5 or J48 decision tree is used. The experimentations are conducted on a keystroke dataset, and the classification accuracy achieved is 90.01%.

Keywords Mobile cloud · Authentication · Behavior classification · Machine learning · Decision trees

1 Introduction

Cloud computing is a gift for human beings. Using the ability of the cloud, humans also scale their ability of computation, business, and other prospects. When cloud abilities are merged with mobility it opens new ways of computation. In this work, mobile cloud computing is the key area of investigation. More specifically, we investigate the security of mobile devices in terms of authentication using mobile cloud computing. Basically, the new generation smart mobile devices are used for a number of tasks such as accessing the internet services, banking services, and many more.

V. Kumar (✉) · S. Ray
Department of Computer Science and Engineering, National Institute of Technology Sikkim, Ravangla, Sikkim 737139, India

K. N. Das et al. (eds.), *Proceedings of the International Conference on Computational Intelligence and Sustainable Technologies*, Algorithms for Intelligent Systems,
https://doi.org/10.1007/978-981-16-6893-7_52

Due to these, significantly essential and private data reside inside these devices. Therefore, the loss of mobile devices can harm someone socially as well as financially. In this context, a new security technique is required by which device can automatically detect which one is accessing services through mobile devices. That sounds good but not an easy task. That becomes feasible through the behavior-based authentication techniques [1].

In behavior-based authentication techniques, devices sense the user activities and behavior of utilizing applications and devices [2]. If the behavior differs from the previous behavioral attribute, the device stop working; otherwise, it works normally [3]. In this context, the defining user behavioral attributes are the key challenges. Additionally, matching and/or evaluation of the previous behavioral attributes with the current attributes are complicated tasks. The available techniques try to match both kinds of attributes but in common practice, a similar behavior representation is not a feasible thing; therefore, some deviations are possible when the same person using the device.

In this work, we are working with a machine learning model that captures the user behavioral attributes [4]. Additionally, on the basis of past behavior variations update and preserve the attribute values. After behavioral observations, the machine learning algorithm learns over the preserved attributes. Finally, when the test set applied to the learned model it tries to replay the behavior that is similar to the last user or not.

The rest of the paper is organized as follows. In Sect. 2, the different research efforts are reported that works for similar contributions to work. The proposed work is presented in Sect. 3. The performance analysis of the work is given in Sect. 4, and the discussion on the comparative results to show the advantages of the proposed scheme is given in Sect. 5. Finally, Sect. 6 concludes the paper.

2 Related Works

This section provides different research efforts for the smart mobile authentication system. Smart mobile devices are not only used for communication it can able to access different applications. Due to this, essential and private data reside in mobile devices. Therefore, the security of mobile devices is essential for securing the device users. In this context, strong passwords [5] are required for validating the end-user who usages the device. Most of the passwords can be memorized enabled and due to this, authentication techniques are failed in different tests. In this context, behavioral passwords are a unique method by which the target user can be distinguishable according to the device interactions [6]. There are different techniques and methods are available that are recommending the behavioral authentication and among them biometric [7], keystroke [8], touch gesture [9], and behavior profiling [10] based techniques are top-rated techniques [11].

In this work, the aim is to investigate the keystroke based behavioral authentication. During literature collection, the keystroke based authentication requires the three key components—data collection, data processing, and behavior classification [12]. The authors in [13] show how experimentally and observationally we can collect the data from mobile devices for authentication purposes. But in some articles, [15] "Keystroke Dynamics—Benchmark Data Set" [14] is used for design and investigation.

On the other hand for analysis and classification of such kinds of data, various machine learning-based models exist [16, 17]. In addition, some of them recommends optimizing the dimensions and selection of essential features from "Keystroke Dynamics—Benchmark Data Set" [18, 19]. In most of the behavioral authentication techniques, data mining and machine learning models are used for the classification tasks [20, 21]. Therefore in this work, we propose to work with the two popular classifiers for finding accurate patterns namely C4.5 or J48 [22] and Bayesian classifier [23].

3 Proposed Work

This section provides details about the proposed behavior-based authentication system. The basic assumptions, the methodology of system functions, and the proposed algorithm are explained in the below subsections.

3.1 Problem Statement

The proposed machine learning-based authentication system is motivated by the research article [15]. In that scheme, author described a Keystroke Dynamics Based User Authentication which is efficient, convenient and working effectively. Moreover, the authors only consider the Keystroke Dynamics with feature selection technique. But in order to enhance the model for incrementing the authentication strength, we introduce two more factors with the existing selected features namely—(i) touch gesture and (ii) typing speed based on user behavioral attribute. These two parameters are used to identify and recognize the user behavior of utilizing the device.

3.2 System Overview

The behavior-based authentication needs the behavioral characteristics or behavioral features which are classified/recognized through machine learning techniques. Therefore, first we need to organize the list of attributes that represents the characteristics of the mobile operational behavior of a target user. There are a number of behavioral signs such as biometric, touch gesture, keystroke dynamics, and behavioral profiling. Among them, the Keystroke Dynamics based authentication is the most common and popular technique for a strong authentication system. In our work, the Keystroke Dynamics based authentication system is the key area of work. In this context, the "Keystroke Dynamics—Benchmark Data Set" [14] is used for experimentation and process design. The dataset contains the 56 subjects (users) behavior with their multiple session data observations. In addition to that it contains 20,400 total instances and 34 attributes for learning with the data mining models. Therefore, the dimension of the dataset is significantly higher and needs significant time and memory resources for accurate classification of data. Hence, the proposed work includes the features selection technique for minimization of dimensions of the dataset. In addition to that after dimensionality reduction, the classification algorithm is implemented for classifying the behavioral patterns of the end-user. This subsection describes the basic overview of the required data model for the classification of user behavior according to the utilization of device habits. In the next subsection, the detailed modeling of the proposed system is explained.

3.3 Methodology

This section explains the given diagram in terms of processing involved in finding the required outcome. The proposed system and its working flow are demonstrated in Fig. 1.

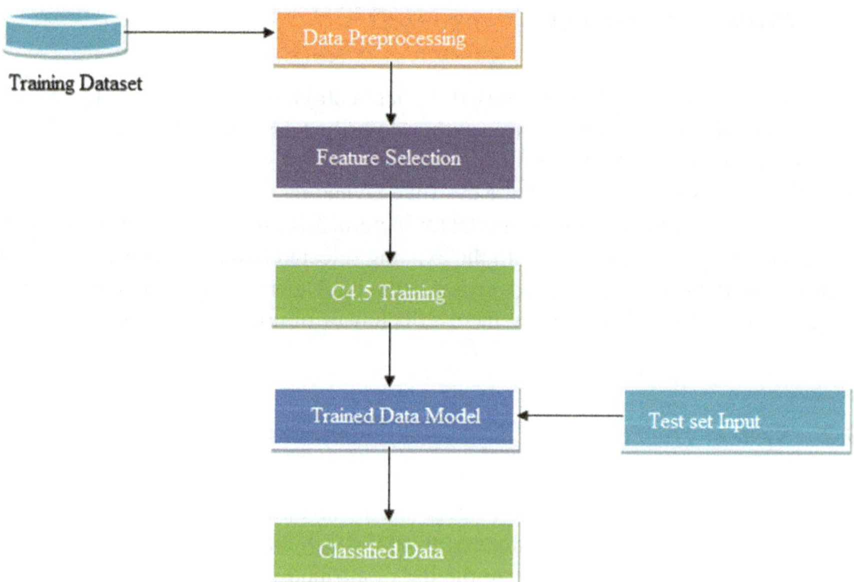

Fig. 1 Proposed classification model

Training dataset: The data mining and machine learning techniques support both kinds of data processing techniques, i.e., supervised and unsupervised. The supervised learning techniques are much accurate as compared to unsupervised learning techniques. But the supervised learning techniques need some predefined data patterns for performing training. The training data contains attributes and the corresponding class labels. Each data set instance is processed for learning in further process steps.

Data preprocessing: The preprocessing techniques are utilized with most of the data mining and machine learning algorithms. The main aim of this process is to optimize the input data in such a manner by which the algorithm can accept and train with the data. Therefore, different kinds of processes are involved in this technique. In addition, it is also motivated to reduce the amount of noise or unwanted data instances and attributes which are not having significance for the classifier's learning process. In this approach, only the missing attribute and null value based data instances are removed from the dataset. The process is used for processing the initial input data as described in Table 1.

Feature selection process: The higher dimension of data is complex in nature. Additionally, learning with a large amount of data attributes has become more critical. That impacts the performance of the learning algorithm. In addition to that, it consumes additional time and memory resources for processing all the attributes of the training dataset. Therefore in this work, the PSO (particle swarm optimization) algorithm is implemented over the preprocessed data for optimizing the amount of data set attributes. Using the PSO algorithm, those attributes are identified that are more significant than the other attributes. Additionally, the less valuable attributes are removed from the dataset. The remaining dataset attributes are further used for the training of the classifier.

Table 1 Data pre-processing

Input: *input dataset D*
Output: *preprocessed data P*

Process:

1. $[row, col] = readInputData(D)$
2. $for(i = 1; i \leq row; i++)$
    a. $for(j = 1; j < col; j++)$
        i. $if(D_{i,j} == null)$
            1. Remove D_i
        ii. Else
            1. $P_{i,j} = D_{i,j}$
        iii. End if
    b. End for
3. End for
4. Return P

C4.5 training: The experimental dataset contains a significant amount of classes or subjects. Therefore, we need an efficient and accurate classifier to classify data in multiple classes. In this context, the decision tree algorithm is found more appropriate for working with multiple classes. The C4.5 decision tree algorithm is an extension of the traditional ID3 algorithm. It splits the data attributes and instances using the entropy and information gain values. After the partition of data, the higher information gain based values are mounted first indecision. Finally, the generated tree is pruned for minimizing the number of rules and tree height.

Trained data model: The C4.5 decision tree processes the data attributes and the corresponding values for preparing the tree data structure. This tree structure is termed here as a trained data model that contains all the training set attributes and classes in the form of a tree.

Testing set: The test dataset is prepared separately for conducting experiments and obtaining the performance of the trained classifier. 30% of randomly selected data instances are collected as test dataset. Each dataset instance is applied to the decision tree model to know or predict the subject names available in tree leaf node.

Classified data: That is the final outcome of the proposed system in terms of class label prediction. The test dataset instances are used to invoke the decision tree branches. Additionally, at the end of each tree branch, the subject name is labeled as the classification label. That is predicted as the outcome of classification. Using the classified attributes, the performance of the classifier is computed.

3.4 Proposed Algorithm

This subsection describes the algorithm steps which are followed for processing the data to classify them to get the current behavior of end user. Table 2 contains the required method's summary in terms of algorithm steps.

Table 2 Proposed algorithm

Input: *training dataset D, test dataset* T_p Output: *class labels*
Process: 1. $R_n = readDataset(D)$ 2. $P_n = preProcessData(R_n)$ 3. $D_m = PSO.OptimizeDataset(P_n)$ 4. $T_{model} = C45.MakeTree(D_m)$ 5. $for(i = 1; i \leq p; i++)$ a. $C_i = T_{model}.Classify(T_i)$ 6. $end\ for$ 7. Return C

The proposed algorithm as demonstrated in Table 2 takes two different inputs—first is the training dataset, and second is the testing dataset. The training dataset is first preprocessed and the outcome of processing is stored in a variable. After preprocessing, the data is used with PSO (particle swarm optimization) algorithm for optimizing the size of the dataset. In this process, only the valuable attributes are preserved and the remaining attributes are removed. Therefore, the size of data is changed. The M dimensional data is used with the classifier C4.5 for constructing a decision tree. The decision tree is denoted as the trained model. This trained model is used for testing of the system. Therefore, the test dataset is applied in this model for traversing of the tree. The tree contains the nodes and edges with the attribute values. Additionally, in the leaf node, the decisions are attached in form of subject class labels as the prediction values.

4 Performance Analysis

This section provides the analysis of different parameters that demonstrate the performance of proposed classification technique. The different parameters are described below.

4.1 Accuracy and Error Rate

Accuracy is the measurement of a classifier's performance in terms of how correctly that classifies or recognizes the given samples. That can evaluate using the following formula:

$$\text{Accuracy\%} = \frac{\text{correctly classified samples}}{\text{total samples to classify}} \times 100$$

Similarly, the error rate is the measurement of the classifier's performance in terms of the misclassification rate of the algorithm. In other words, the algorithm incorrectly identifies the patterns which are termed as error rate. The error rate of an algorithm can be computed using the following formula:

$$\text{Error rate\%} = \frac{\text{misclassified samples}}{\text{total samples to classify}} \times 100$$

Or

$$\text{Error Rate\%} = 100 - \text{Accuracy}$$

The accuracy of the proposed system is compared with two other techniques to justify the performance of the proposed approach. Figure 2 and Table 3 show the accuracy and error rate of all three approaches for classifying the behavioral patterns. Both parameters (i.e., accuracy and error rate) of classification are measured here in terms of percentage (%). To represent the performance of the system using the line graph, *X-axis* contains the number of dataset instances, and the *Y-axis* shows the obtained accuracy and error rate in terms of a percentage value. The proposed C4.5 and PSO based approaches show the improved results as compared to both the traditional classification approaches. In Table 3 and Fig. 2, we can see the proposed technique shows fewer error rate and higher accuracy with respect to the traditional approaches. Therefore, we can conclude that the proposed technique is superior than other two implemented classical models.

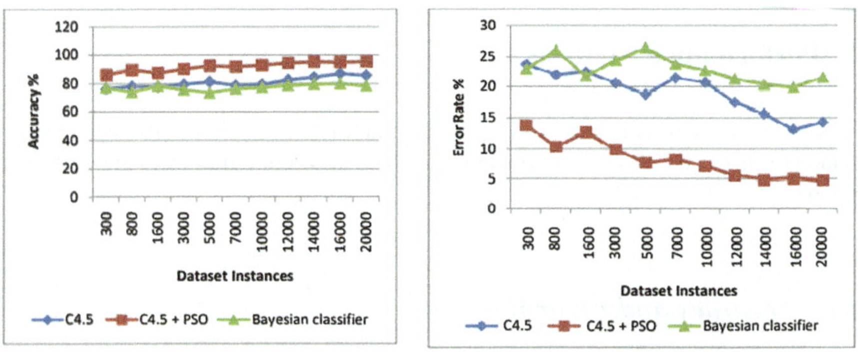

(A) Accuracy (%) (B) Error Rate (%)

Fig. 2 Accuracy and error rate

Table 3 Accuracy and error rate

Dataset Instances	C4.5 (%)		C4.5 + PSO		Bayesian classifier	
	Accuracy	Error rate	Accuracy	Error rate	Accuracy	Error rate
300	76.3	23.7	86.2	13.8	77.1	22.9
800	78.1	21.9	89.6	10.4	73.9	26.1
1600	77.6	22.4	87.3	12.7	78.3	21.7
3000	79.4	20.6	90.1	9.9	75.6	24.4
5000	81.2	18.8	92.3	7.7	73.5	26.5
7000	78.6	21.4	91.7	8.3	76.2	23.8
10,000	79.3	20.7	92.9	7.1	77.3	22.7
12,000	82.5	17.5	94.4	5.6	78.7	21.3
14,000	84.3	15.7	95.1	4.9	79.6	20.4
16,000	86.9	13.1	94.9	5.1	80.09	19.91
20,000	85.7	14.3	95.3	4.7	78.5	21.5

4.2 Memory Usage and Time Consumed

Every process needs some amount of main memory resource for the successful execution of given task. This amount of main memory unit is known as memory usages by the target algorithm. In JAVA applications, the process's memory usages can be computed using the following formula:

$$\text{Memory Usage} = \text{Total memory} - \text{free memory}$$

The amount of time required for authenticating a target user by classifying the user's current behavior is termed as the time complexity of the proposed data mining algorithm. The time complexity of the learning algorithm is given using the following formula:

$$\text{Time Required} = \text{Algorithm End time} - \text{Start time}$$

The memory utilization by a process in terms of KB (kilobytes) is computed as the memory usages or space complexity of the algorithm. For a demonstration of comparative performance among the implemented techniques in terms of time and memory utilization by the algorithms, the Fig. 3 and Table 4 have been prepared. The *X-axis* of the diagram contains the training dataset instances, and the *Y-axis* shows the used memory and time consumed with the algorithms. According to the obtained performance, the proposed technique consumes less memory as compared to the traditional approaches. The less memory consumption is achieved due to reduction of data set dimension and less values of data evaluation. The time measurement is performed in terms of milliseconds (MS). The diagram shows the proposed PSO + C4.5 model reduces time consumption with respect to the other two implemented

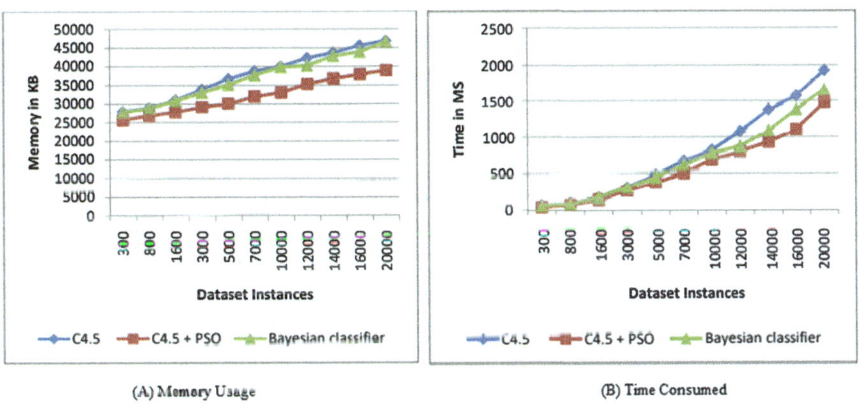

Fig. 3 Memory usage and time consumption

Table 4 Memory usage and time consumption

Dataset instances	C4.5		C4.5 + PSO		Bayesian classifier	
	Memory	Time	Memory	Time	Memory	Time
300	27,948	67	25,737	41	27,810	59
800	28,940	92	26,810	73	28,847	84
1600	31,029	173	27,773	136	30,947	169
3000	33,928	317	29,101	263	33,084	308
5000	36,718	489	30,039	378	35,188	451
7000	38,818	681	32,019	511	37,739	617
10,000	40,039	819	33,013	692	39,910	779
12,000	42,261	1079	35,280	798	40,294	871
14,000	43,728	1382	36,791	928	42,846	1092
16,000	45,621	1568	37,885	1109	43,949	1386
20,000	46,799	1927	38,964	1482	46,625	1652

algorithms. Therefore, we can conclude that the proposed scheme is more acceptable as compared to two other methods.

5 Result and Discussion

The experiments performed with the implemented system demonstrate that the proposed work is acceptable for authentication purpose. In this context, the comparative analysis of mean outcomes is also performed. Figure 4a shows the comparative average accuracy of the proposed system with other two algorithms. According to the results, the proposed model improves the accuracy on average 10–15% with respect to the other two traditional algorithms. Similarly, the error rate of implemented techniques is demonstrated in Fig. 4b. The results show the clear difference among all three algorithms. The proposed algorithm minimizes the 10–15% misclassification rate as compared to the traditional approaches.

The results in terms of average memory consumption are demonstrated in Table 5 and Fig. 4c where the memory is computed in kilobytes (KB). The results demonstrate that the proposed technique preserves the memory approximately 4000 KB with respect to the other implemented algorithms. The comparative time consumption is also given in Fig. 4d and Table 5 which concludes that the proposed approach preserves approximately 50–80 MS with respect to the other two algorithms.

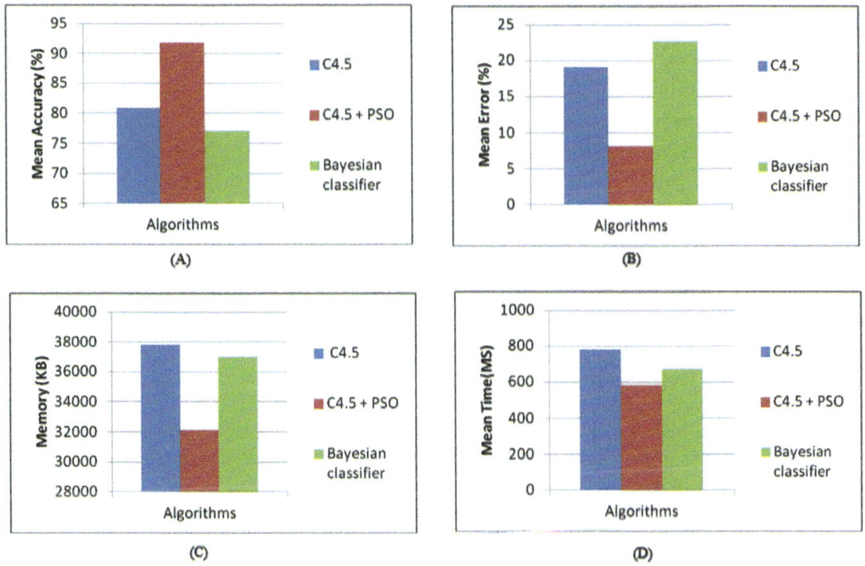

Fig. 4 Mean performance **a** accuracy, **b** error rate, **c** memory usage, **d** time consumed

Table 5 Mean performance

S. No.	Parameter	C4.5	C4.5 + PSO	Bayesian classifier
1	Accuracy (%)	80.9	91.8	77.16
2	Error rate (%)	19.1	8.2	22.84
3	Memory (KB)	37,802.63	32,128.36	37,021.72
4	Time (MS)	781.27	582.81	678.90

6 Conclusion

The proposed work aims to provide authentication for smart mobile devices using the data mining technique. In this context, a new data mining model is proposed in this paper for design and implementation. Basically, the proposed model is based on the concept of mobile cloud computing where using the mobile devices user behavioral attributes are collected and preserved. The data collection is performed in multiple user sessions to consider the difference in behavior. Additionally by a collection of different user behavioral attributes, the training dataset is prepared. In the next phase, the data is learned with the learning model. After learning, the test data is produced for providing final prediction about the user behavior. In this context, a keystroke behavior-based dataset is used for experimentation and system design. The key issue with this dataset is the dimension of data which is significantly higher and can affect the performance of the classification algorithm. Therefore, first the dataset is optimized with the PSO algorithm for selecting the essential attributes from the dataset and next using the selected attributes the C4.5 decision tree takes training and helps to classify the test dataset. According to the classification outcomes with the increasing number of training dataset instances, it is found that the proposed technique improves the accuracy of the system as compared to similar two algorithms, i.e., Bayesian classifier and traditional C4.5 algorithm. In addition to that the proposed scheme preserves the time and memory resource significantly. In future, the performance of the proposed scheme will be enhanced using the ensemble learning technique.

References

1. Kayacik, H.G., Just, M., Baillie, L., Aspinall, D., Micallef, N.: Data driven authentication: on the effectiveness of user behaviour modelling with mobile device sensors. In: Proceedings of the Third Workshop on Mobile Security Technologies (MoST) (2014)
2. Wang, X., Yu, T., Mengshoel, O., Tague, P.: Towards continuous and passive authentication across mobile devices: an empirical study. In: WiSec '17 , Boston, MA, USA, © (2017). ACM 978-1-4503-5084-6/17/07
3. Li, F., Clarke, N., Papadaki, M., Dowland, P.: Behaviour profiling for transparent authentication for mobile devices (2011)
4. Maia, M., Almeida, J., Almeida, V.: Identifying user behavior in online social networks. In: Proceedings of the 1st Workshop on Social Network Systems, pp. 1–6 (2008)
5. Florencio, D., Herley, C.: A large-scale study of web password habits. In: Proceedings of the 16th International Conference on World Wide Web, pp. 657–666 (2007)
6. Lee, W.-H., Liu, X., Shen, Y., Jin, H., Lee, R.B.: Secure pick up: implicit authentication when you start using the smartphone. In: Proceedings of the 22nd ACM on Symposium on Access Control Models and Technologies, pp. 67–78 (2017)
7. Trewin, S., Swart, C., Koved, L., Martino, J., Singh, K., Ben-David, S.: Biometric authentication on a mobile device: a study of user effort, error and task disruption. In: Proceedings of the 28th Annual Computer Security Applications Conference, pp. 159–168 (2012)
8. Kambourakis, G., Damopoulos, D., Papamartzivanos, D., Pavlidakis, E.: Introducing touch stroke: keystroke-based authentication system for smartphones. Secur. Commun. Networks **9**(6), 542–554 (2016)
9. Sae-Bae, N., Ahmed, K., Isbister, K., Memon, N.: Biometric-rich gestures: a novel approach to authentication on multi-touch devices. In: Proceedings of the SIGCHI Conference on Human Factors in Computing Systems, pp. 977–986 (2012)
10. Gonzalez-Manzano, L., De Fuentes, J.M., Ribagorda, A.: Leveraging user-related internet of things for continuous authentication: a survey. ACM Comput. Surv. (CSUR) **52**(3): 1–38 (2019)
11. Alotaibi, S., Furnell, S., Clarke, N.: Transparent authentication systems for mobile device security: a review. In: 10th International Conference for Internet Technology and Secured Transactions (ICITST), pp. 406–413. IEEE (2015)
12. Henze, N., Rukzio, E., Boll, S.: Observational and experimental investigation of typing behaviour using virtual keyboards on mobile devices. CHI'12, Austin, Texas, USA (2012). Copyright 2012 ACM 978-1-4503-1015-4/12/05
13. Niels, H., Rukzio, E., Boll, S.:Observational and experimental investigation of typing behavior using virtual keyboards for mobile devices. In: Proceedings of the SIGCHI Conference on Human Factors in Computing Systems, pp. 2659–2668 (2012)
14. Antal, M., Nemes, L.: The mobikey keystroke dynamics password database: Benchmark results. In: Computer Science On-line Conference, pp. 35–46. Springer, Cham (2016)
15. Krishnamoorthy, S., Rueda, L., Saad, S., Elmiligi, H.: Identification of user behavioral biometrics for authentication using keystroke dynamics and machine learning. In: Proceedings of the

2018 2nd International Conference on Biometric Engineering and Applications (pp. 50–57) (2018, May)
16. Mahfouz, A., Mahmoud, T.M., Eldin, A.S.: Bimodal behavioral authentication framework based on decision fusion. In: 2017 8th International Conference on Information and Communication Systems (ICICS), pp. 368–373. IEEE (2017, April)
17. Wu, C., Ding, W., Liu, R., Wang, J., Wang, A.C., Wang, J., Li, S., Zi, Y., Wang, Z.L.: Keystroke dynamics enabled authentication and identification using triboelectric nanogenerator array. Mater. Today **21**(3), 216–222 (2018)
18. Mondal, S., Bours, P.: Person identification by keystroke dynamics using pairwise user coupling. IEEE Trans. Inf. Forensics Secur. **12**(6), 1319–1329 (2017)
19. Alzubaidi, A., Kalita, J.: Authentication of smartphone users using behavioral biometrics. IEEE Commun. Surv. Tutorials **18**(3), 1998–2026 (2016)
20. Khan, H., Atwater, A., Hengartner, U.: Itus: an implicit authentication framework for android. In: Proceedings of the 20th Annual International Conference on Mobile computing and networking, pp. 507–518 (2014, September)
21. Rajashekar, D., Zincir-Heywood, A.N., Heywood, M.I.: Smart phone user behavior characterization based on autoencoders and self organizing maps. In: 2016 IEEE 16th International Conference on Data Mining Workshops (ICDMW), pp. 319–326. IEEE (2016, December)
22. Hssina, B., Merbouha, A., Ezzikouri, H., Erritali, M.: A comparative study of decision tree ID3 and C4.5. Int. J. Adv. Comput. Sci. Appl. **4**(2), 13–19 (2014)
23. Sucar, L.E.: Probabilistic graphical models. In: Advances in Computer Vision and Pattern Recognition. Springer, London (2015). 10(978), 1

A Comprehensive Study on SCADA Based Intake, Water Treatment Plant and Intermediate Pumping Station of Existing Water Supply System

C. Rohmingtluanga, S. Datta, and N. Sinha

Abstract In this paper, study of existing SCADA based Water Supply System (WSS), i.e., Greater Aizawl Water Supply Schemes phase II (GAWSS P-II) in Aizawl, India is done so as to analyze the water parameters at each pumping stations viz. Intake, Water Treatment Plant (WTP), Intermediate Pumping Station (IPS) and Reservoir. Field equipment such as pH analyzer, Chlorine analyzer, Turbidity sensor, electromagnetic flow meter, Open channel flow meter, Level meter, Energy meter, PLC with HMI, optical fiber system and GSM modem are installed at different locations. The central SCADA station is located at Dihmunzawl (Aizawl, India) collect and monitor all the field data from the above equipment and analyze (i) raw and treated water quality, (ii) raw and treated water quantity at each pumping stations, (iii) pump optimum operation (iv) triggering of backwash scheduling and (v) Energy utilization of each pump sets. Apart from this, alarm system is also implemented so as to inform the duty personnel when timing for a particular operation or failure occurs at any time. The SCADA system monitors the WSS continuously through online and remote using Personal Computer, and the study shows the whole process of SCADA based WSS in Aizawl and how the entire process can be improved in future to get optimum utilization of it.

Keywords GAWSS phase II · SCADA system · Intake · WTP · IPS · Sensors · Pump scheduling · Alarm system

1 Introduction

SCADA system is normally seen in Military, industry and various plants for monitoring and control of the whole system. The field equipment installed at different locations collect data and send to the Programmable Logic Controller (PLC) for

C. Rohmingtluanga (✉) · S. Datta
Department of Electrical Engineering, Mizoram University, Aizawl, Mizoram 796004, India

N. Sinha
Department of Electrical Engineering, NIT Silchar, Silchar, Assam 788010, India

K. N. Das et al. (eds.), *Proceedings of the International Conference on Computational Intelligence and Sustainable Technologies*, Algorithms for Intelligent Systems,
https://doi.org/10.1007/978-981-16-6893-7_53

further computation and decision making as per requirement of the plant. Similarly, the SCADA based Water Transmission System collects data from the field equipment at various locations through PLC and further processed to improve water quality and quantity of Raw water and Treated water [1, 2]. The GAWSS P-II commissioned in 2007 consists of Intake, WTP, IPS and Reservoir for pumping clean water to the Aizawl City. Later on, Supervisory Control and Data Acquisition (SCADA) system was installed in 2020 for general monitoring and control of the Water Supply System (WSS).

Generally, the Public Water Supply system (WSS) consists of Intake, Water Treatment Plant (WTP), Intermediate Pumping Station (IPS) and Reservoir for providing and pumping of clean and drinking water to the people. The intake structure located near the raw water source collected and pump raw water to the Water Treatment Plant (WTP) for further filtration and disinfection. The Intermediate Pumping Station (IPS) placed between the Water Treatment plant and Reservoir act as a booster delivers clear water to the main Reservoir, IPS play a very important role in improving pumping efficiency as well [3, 4].

However, energy consumption of most WSS is huge and effects the financial efficacy of the treatment process. Similarly, losses in the transmission and distribution system have huge financial impact as well, this needs to be improved while reducing transmission and distribution losses in the system. Generally, duties related to optimization, quality monitoring and minimization of losses are under water supply department/companies. Most of the old/existing WTP does not have a SCADA system for monitoring and control of the water supply and distribution networks [5–7].

In this paper, SCADA system application in Intake, Water Treatment Plant, Intermediate Pumping Station and Reservoir for GAWSS phase II is studied by mainly focusing on monitoring of raw and treated water quality and water quantity at each station. Additional features such as pump scheduling schemes and alarm triggering system for filter bed are also studied. It can be seen that SCADA system improves the performance of the existing WTP and contribute a very important role in the plant process.

2 SCADA System Architecture for GAWSS Phase II

The SCADA system architecture of GAWSS phase II shown in Fig. 1 consists of four stations each having PLC, HMI (Human Machine Interface), RTU (Remote terminal unit), analyzer panel, sensors and UPS battery backup with isolation transformer. The central SCADA system located at Dihmunzawl serves as the master station, each station is interlinked with each other using optical fiber and GSM/GPRS device as shown in Fig. 1. Electromagnetic flow meter, open channel flow meter and pressure transmitter are installed at most of the pipelines and channels to monitor the quantity, pressure and flow of water. These equipments serve three purposes: (i) to monitor the flow rate, (ii) health of the pipe lines and (iii) water losses between outlet of the pump

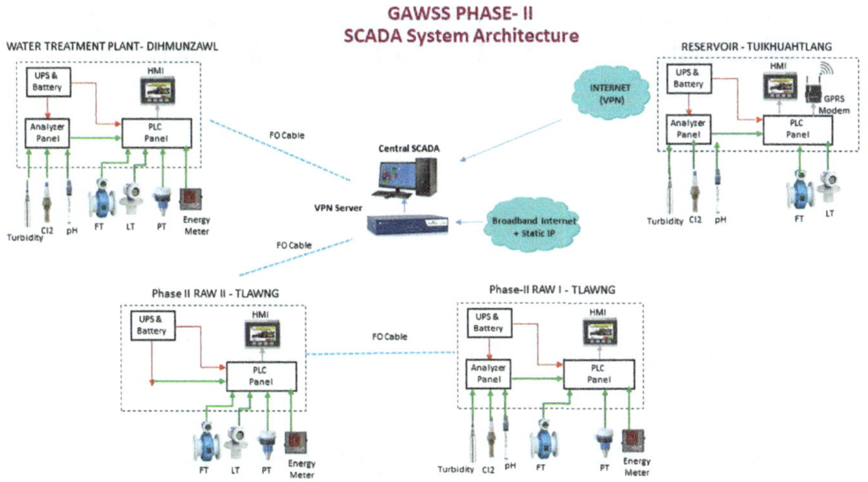

Fig. 1 SCADA system architecture for GAWSS phase II

house and delivery at the reservoir. Analyzer panel consisting of pH, turbidity and chlorine sensors are also installed at intake, WTP and reservoirs so as to monitor the quality of raw water and treated water before and after WTP. Energy meter is also installed for monitoring optimal pump scheduling and power consumption. Level detector was also installed to monitor the water quantity at each reservoir, each filter bed and backwash tank.

3 SCADA System Application in GAWSS Phase II

The GAWSS phase-II consists of two Intake system, WTP, and Reservoir. The SCADA screen for each pumping station is explained in the following sections. Monitoring of the overall real-time pumping status, monitoring of water quantity and quality at each pumping stations is done through the SCADA screen.

3.1 Intake

The intake of phase II consists of two pumping stages, namely (i) phase II raw I and (ii) Phase II raw II. The raw water collected from phase II raw I get settled for a while in pre-settling tank so as to remove sludge present in the raw water and improve the water quality; the raw water further flows to the phase II raw II by using Gravity method. From this, the raw water gets further pumped to the main WTP at Dihmunzawl using electrical pump motors.

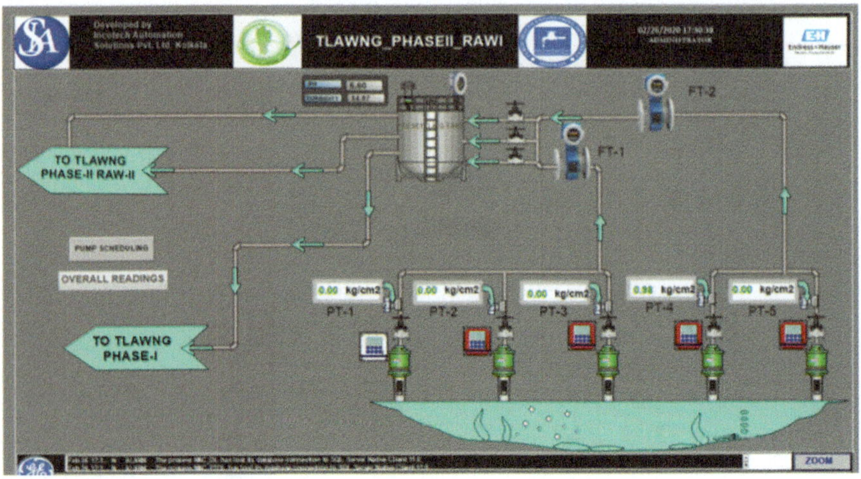

Fig. 2 GAWSS phase II raw I

3.2 *Phase II Raw I*

The phase II raw I consists of five submersible electric driven pumps located inside Jack well as shown in Fig. 2. Raw water from Tlawng River were collected and stored in the pre-settling tank. Energy meter, electromagnetic flow meter, pH analyzer, turbidity analyzer and level meter were installed for measuring the pump efficiency and water parameters. Optical fiber is used as a communication media for transferring data to the main SCADA through PLC and HMI.

3.3 *Phase II Raw II*

GAWSS phase II raw II consists of raw water sump and three VT pumps as shown in Fig. 3. Energy meter, level meter, flow meter and PLC with HMI were installed for monitoring pump data and water parameters. Pump scheduling and overall readings can be seen on the SCADA screen. Optical fiber is used as a communication media for data transfer to the main SCADA center.

3.4 *Water Treatment Plant*

The main WTP of GAWSS phase II located at Dihmunzawl receives water from phase II raw II pumping station as shown in Fig. 4. The raw water further processed

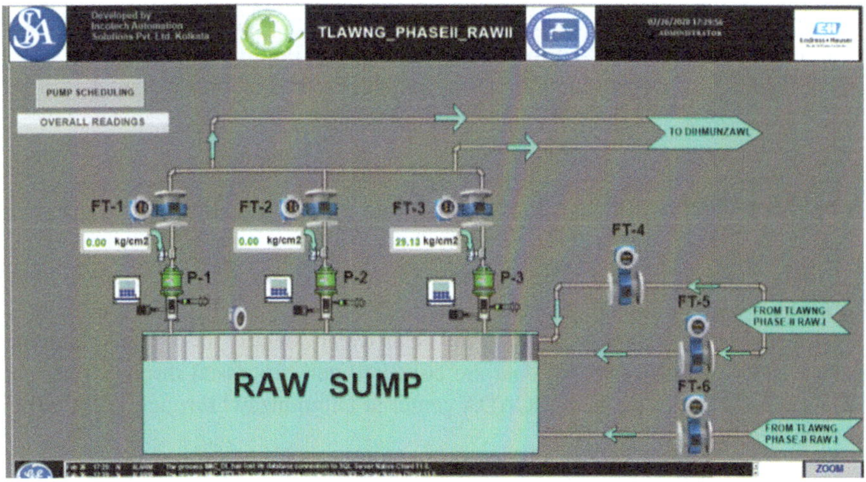

Fig. 3 GAWSS phase II raw II

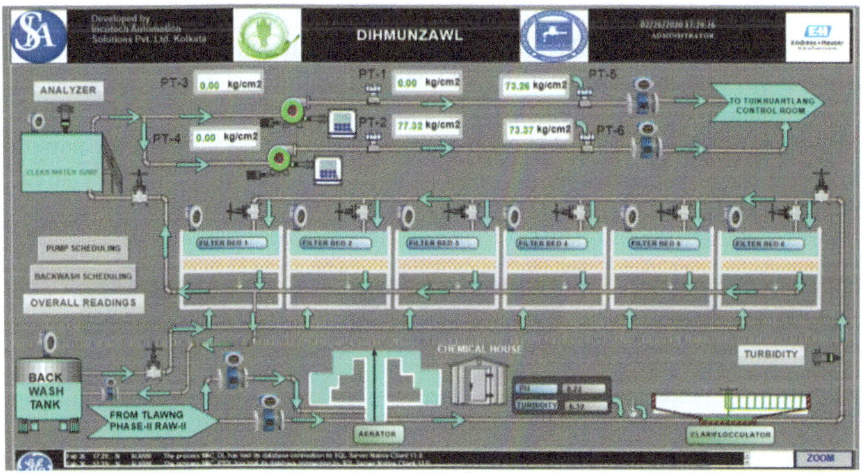

Fig. 4 GAWSS phase II WTP

through aerator, chemical house, clariflocculator, filter house, and further pumped to the main reservoir at Tuikhuahtlang.

The raw water parameter such as turbidity and pH are monitored just after aerator. Turbidity analyzer located after clariflocculator measures the quality of the raw water entering the filter bed. Level meter and open channel flow meter are installed at each filter bed so as to check the health of the filter bed. Energy meter, pressure meter and electromagnetic flow meter are installed for monitoring pump parameters and flow rate of water from the pump house. Analyzer panel is also used to monitor pH, turbidity and chlorine content of the treated water. Here, optical fiber is used as a communication media through PLC and HMI. Pump scheduling, backwash scheduling and overall readings can be seen and operated on this screen.

3.5 *Reservoir*

The main reservoir located at Tuikhuahtlang consists of two main reservoirs coming from phase I and phase II. Electromagnetic flow meter, level meter and analyzer panel were installed so as to measure the clear water quantity, flow rate at inlet and water quality at Reservoir. All data are collected and processed through PLC and further transferred to the main SCADA center at Dihmunzawl through GSM/GPRS as shown in Fig. 5.

4 Optimization of Pump Scheduling

The SCADA system of GAWSS phase II is designed to monitor water quality and quantity through online and notify faults through alarm systems. Apart from this, improvement in the overall efficiency of the WTP is done by implementing pump scheduling block in the SCADA system. This block gives priority of the pump to be operated for each pumping schedule so as to improve pump efficiency at each pumping period.

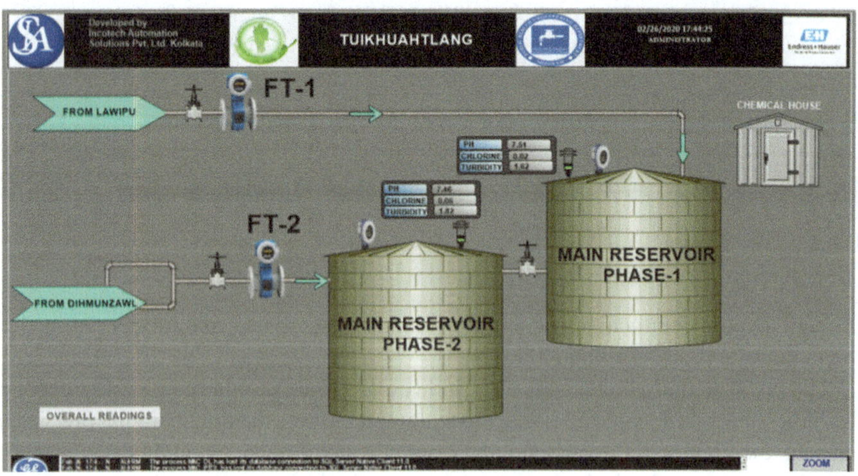

Fig. 5 GAWSS phase I and II main reservoir

4.1 Pump Scheduling

Pump scheduling plays a very important role in the WSS for improving plant efficiency and to reduce operational cost. In order to implement this important unit in SCADA system, priority method is used so as to prioritize the most efficient pump for pumping the water, which will be followed by the next lesser efficient unit and so on. The main goal of pump scheduling is to consume less energy per pumping water flow quantity while improving efficiency [8–10].

In order to understand the concept of pump scheduling, optimization cases for a particular Dihmunzawl area are shown in Table 1. Where, four cases are considered for selecting the priority of the pump to be operated first. Firstly, for Case-I, out of four pump sets, pump 1 and pump 2 are electrical driven pump and pump 3 and 4 are diesel engine driven pump; therefore, electrical driven pumps are considered higher priority due to less operational cost and higher efficiency. For Case-II, pump 1 and 2 are almost similar and manufactured from the same company, both the pump uses soft starter and are considered as same priority. For Case-III, the electrical energy consumption versus outflow rate is calculated from the data collected from Energy meter and outlet flow meter during normal operating condition, as pump 2 obtained lesser ratio rate. Therefore, pump 2 is considered as higher priority compare to pump 1. Case-IV prioritizes the age of the pump to be used for water pumping, as new equipment comes with higher efficiency, and the pump with lesser age is considered as higher priority compare to the old pump. Therefore, considering all cases pump no. 2 is most prioritized for scheduling and, followed by pump no. 1, as the pipeline allows two pumps to operate at the same time, pump no. 2 will be operated first and after obtaining rated speed, pump no. 1 will be started again so as to utilize maximum pump efficiency. The same procedure is followed in all pumping stations so as to obtain optimum pump scheduling.

5 Alarm Triggering System

The alarm system plays a very important role in SCADA system as it continuously monitors the plant process for a particular given task or any unwanted accident or faults happen at any instant of time. The alarm set for a particular task is compared

Table 1 Optimization cases for pump scheduling

Priority	Case-I	Case-II	Case-III	Case-IV (years)
Pump 2	Electrical	Soft starter	11.9	15
Pump 1	Electrical	Soft starter	13.8	15
Pump 3	Diesel	DOL	–	15
Pump 4	Diesel	DOL	–	15

Table 2 Alarm settings for level transmitter

S. No.	Description	Low–Low LL (m)	Low L (m)	High H (m)	High–High HH (m)	Reference (m)
1	Filter bed 1 LT	0.5	1.0	3.0	3.05	3.35
2	Filter bed 2 LT	0.5	1.0	3.0	3.05	3.35
3	Filter bed 3 LT	0.5	1.0	3.0	3.05	3.35
4	Filter bed 4 LT	0.5	1.0	3.0	3.05	3.35
5	Filter bed 5 LT	0.5	1.0	3.0	3.05	3.35
6	Filter bed 6 LT	0.5	1.0	3.0	3.05	3.35
7	Backwash LT	0.5	1.0	3.0	3.05	3.35
8	Clear sump LT	0.5	1.0	3.0	3.05	3.35

from the field equipment data or sensors so as to trigger the alarm if the sensor values are above or below the set point values [11–13]. The alarm system set point adopted for filter house at WTP is discussed for understanding the logic operations.

5.1 *Filter House Alarm System*

Table 2 gives the alarm setting values for Filter bed, Backwash sump and Clear sump at WTP, the threshold values are set for Level Transmitter values of the water level, Here, Low (L) set as 1.0 m indicates the Low set point value for first warning, and Low–Low (L–L) indicates the possible lowest limit value for the system, if the water level crosses below this point, the SCADA system will stop the plant and operate the alarm loop by blinking on the screen till the particular alarm is acknowledged or reset. Similarly, set point for High (H) set as 3.0 m and High–High (H–H) set as 3.35 m is applied for the upper water level values. If the data from the level transmitter crosses above the set point values, alarm will be initiated similar to the low set point values as discussed above.

5.2 *Ladder Logic for Alarm System in Filter Bed*

The ladder logic for alarm system of filter bed is shown in Fig. 6. The GE Proficy machine edition software is used to implement alarm system in the SCADA system. The left and right rails of the ladder logic are called as power lines and the horizontal lines where the component blocks are placed are called as rungs, the PLC processes the code from left to right and top to bottom [14, 15]. Initially, the program will start from the first rung and then compare the real time water level in the filter bed with the given set point level through Greater Than REAL block, if the water level is greater than the set point value, high water level alarm will be triggered by blinking the

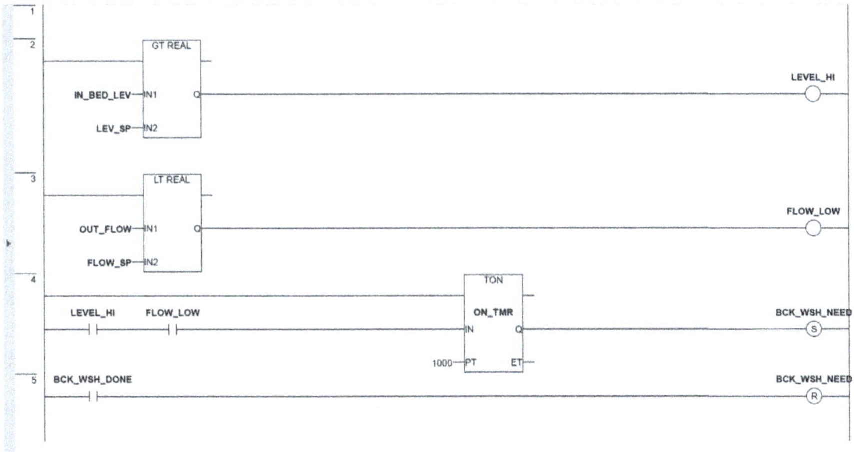

Fig. 6 Ladder logic for filter bed alarm system

water level indicator. Secondly, the program will trace the second rung and compare the water outflow rate of a particular filter bed with the given set point through Less Than REAL block, if the outflow rate is less than the set point value, the low-level flow alarm will be displayed. Similarly, the program will trace the third rung and monitor both conditions of the output of the first rung and second rung, and set the timer for backwash alarm by counting up to 1000 times, if the condition is true for both the cases it will blink the backwash alarm continuously till 1000 count or till it is acknowledged or reset whichever is earlier, if the duty personnel acknowledged the backwash and reset the set point conditions by pressing backwash done as shown in fifth rung, the alarm will stop blinking and halted.

6 Conclusion

SCADA system in existing GAWSS phase II improves the water supply system in many ways due to its accuracy and flexibility. Data collected from PLCs can be formulated to bring optimal pump scheduling and many other plant process improvements, the optimal pump scheduling plays a very important role in improving plant efficiency. Apart from this, alarm system implemented in the SCADA continuously monitor the plant as per set point values, and can be used to prevent pipe leakage and improve operation of the filtration processes, etc.

The main disadvantage of SCADA system is requirement of skilled engineers and field staffs for attending technical issues of the field equipment which may arise during the plant process. Moreover, SCADA system can be improved by improving internet connectivity and providing skilled and efficient duty personnel for assessing data and handling of operation and maintenance. Further, online mobile application can also be incorporated in the system so as to have easy and remote access from any locations, this will provide platform such as smart solutions, and quick response for any problems in the WSS process.

References

1. Tshshla, K.S., Haman, Y., Mahfouz, A.M.A.: State estimation in water distribution network: a review. In: Proceedings of the International Conference on Industrial Informatics, pp. 1247–1252. IEEE, Emden Germany (2017)
2. Sakarya, A.B.A., Mays, L.W.: Optimal operation of water distribution pumps considering water quality. J. Water Resour Plan Manage **126**, 210–220 (2000)
3. Lagu, S.S., Deshmukh, S.B.: Raspberry Pi for automation of water treatment plant. In: International Conference on Advance Computing, Communications and Informatics, pp 1999–2003 (2014)
4. Zaev, E., Babunski, D., Tuneski, A.: SCADA system for real-time measuring and evaluation of river water quality. In: Proceedings of the Mediterranean Conference on Embedded Computing, pp. 12–16, 83–86. Bar IEEE, Montenegro (2016)
5. Gray, A.D.L., Pisica, I., Taylor, A.G., Whitehurst, L.A.: Standardised modular approach for site SCADA applications within a water utility. IEEE Access 17177–17187 (2017)
6. Els, L.A., Pelzer, R., Schutte, A.J.: Load management on a municipal water treatment plant. In: Proceedings of the International Conference on the Industrial and Commercial Use of Energy, pp. 1–4. IEEE, Cape Town South Africa, (2015)
7. Central Public Health and Environmental Engineering Organization (CPHEEO) manual on water supply and treatment, India, (1999) [Online]. Available: http://cpheeo.gov.in/cms/manual-on-water-supply-and-treatment.php
8. Firoozshahi, A., Mengyang, L.: Water treatment plant intelligent monitoring in large gas refinery. In: Proceedings of the International Conference on Computational Technologies in Electrical and Electronics Engineering, pp. 785–789. Listvyanka Russia (2010)
9. Tan, W., Sun, Y., Azad, I.I., Chen, T.: Design of univariate alarm systems via rank order filters. Control. Eng. Pract. **59**, 55–63 (2017)
10. Lakhoua, M.N.: SCADA applications in thermal power plants. Int. J. Phys. Sci. 1175–1182 (2010)
11. Goel, P., Datta, A., Mannan, M.S.: Industrial alarm systems: challenges and opportunities. J. Loss Prev. Process Ind. **50**, 23–36 (2017)
12. Vitkovsky, J.P., Simpson, A.R., Lambert, M.F.: Leakage detection and calibration using transients and genetic algorithms. J. Water Resour. Plann. Manage. 262–265 (2000)
13. Hotta, Y., Sato, M., Kimura, H., Matsuoka, S., Boku, T., Takahashi, D.: Profile-based optimization of power performance by using dynamic voltage scaling on a PC cluster. In: Proceedings of the International Parallel and Distributed Processing Symposium, pp. 298–298. IEEE, Rhodes Island Greece, (2006)
14. da Silva, M.J., Pereira, C.E., Götz, M.: A dynamic approach for industrial alarm systems. In: International Conference on Computer, Information and Telecommunication Systems (CITS), pp. 1–5 (2016)
15. Hajji, M., Fares, A., Glover, F., Driss, O.: Water pump scheduling system using scatter search, Tabu search and neural networks the case of Bouregreg water system in Morocco. Providence RI 822–832 (2010)

Multimodal Summarization: A Concise Review

Hira Javed, M. M. Sufyan Beg, and Nadeem Akhtar

Abstract Rapid increase in multimodal data from online sources necessitates the need for the development of methods and techniques that use diverse modes along with text to generate summaries. Meeting recordings, CCTV footages, and sports coverages are vast information resources that incorporate modes such as speech, text, and images. Multimodal data contains rich information which could help in the improving the quality of summaries. Recent developments in the field of neural networks have furthered the research in this area. This paper is an attempt to contribute to the growing body of research in multimodal summarization by drawing attention to state-of-the-art techniques, datasets, and evaluation strategies. Besides covering the recent work in the field, the paper also highlights the major issues and challenges.

Keywords Multimodal · Summarization · Review · Neural networks

1 Introduction

Summarization aims at extracting the principal and representative content from source document. It deals with reductive transformation through extraction and generation. Text summarization aims at helping the machine to understand a text written in natural language. With the recent development of tools, techniques, and algorithms, automatic summarization has shown great advancement. Content from audio, video, images along with the text gives more insight while extracting features. Visual representations further help people to better understand, memorize, and relate

H. Javed (✉) · M. M. Sufyan Beg · N. Akhtar
Department of Computer Engineering, Zakir Husain College of Engineering and Technology, Aligarh Muslim University, Aligarh, India
e-mail: hira.javed@zhcet.ac.in

M. M. Sufyan Beg
e-mail: mmsbeg@myamu.ac.in

N. Akhtar
e-mail: nadeemakhtar@zhcet.ac.in

K. N. Das et al. (eds.), *Proceedings of the International Conference on Computational Intelligence and Sustainable Technologies*, Algorithms for Intelligent Systems,
https://doi.org/10.1007/978-981-16-6893-7_54

to events. Getting critical information from images is faster for users. Grasping of events is better done by image representations, while the detailed understanding of event is fulfilled by text representations [1]. The output can be represented in the form of image, text, and video shots like a teaser of a film or it could also be a combination of two or more modes. This work presents a review of the current work in the field of multimodal summarization along with a brief account on the techniques, applications, tools, datasets, evaluation strategies, challenges, and issues.

2 Related Work

2.1 Image Captioning/Image Annotation

Much work has been done in recent years in the field of image captioning. The process uses two modes, i.e., image and text to generate the captions. Two approaches for image captioning are retrieval based and generation based. In the first approach, captions are generated by exploring images similar to the given image, while in generation-based approach a learned model is used to generate captions. Generation-based models are flexible, popular, and able to generate novel captions from naive images.

2.2 Video Captioning

Automatic video captioning task generally takes place in two steps; in the first step, video clips are sent as continuous image frames and visual recognition task is performed for identifying objects, performer, and action. In the second step, the extracted features are put in a vector, and captions are generated using deep learning techniques [2]. Video captioning task can also be modeled to image captioning by extracting key frames from the video.

2.3 Multimodal Sentiment Analysis

Some recent works have utilized sentiments in videos for carrying out summarization. In [3], key players and key events in a match have been identified using the news articles and speech transcriptions derived from the audio. Sentiment analysis of online videos has been conducted using perplexed Bayes classification in [4]. A review of the sentiment analysis for multimodal data has been given in [5].

2.4 *Cross-Modal Retrieval*

In cross-modal retrieval whenever a query is given in one mode, its counterpart is obtained in other mode. Image text matching is an example of cross-modal retrieval and plays a key role in semantics. The authors in [6] have utilized the stacked multimodal attention network for the retrieval of cross-modal similarity score.

3 Multimodal Summarization Applications

3.1 *Sports*

Summarization of sports videos has gained popularity in recent years. Cinematic features have been exploited by the authors in [7] to determine the play breaks in sports such as football, tennis, soccer, and basketball. In [8], the authors have extracted highlights from a golf match by focusing on excitement features. Adversarial networks have been used for the enhancement of extracted features in e-sports videos.

3.2 *Meetings*

Meeting recordings is one of the earliest domains that were explored for summarization. To identify the arguments and meeting discussions, high audio amplitude has been combined with sound localization output of meeting recorder system in [9]. High motion is identified when someone joins meeting or participant gives presentation. Large luminance change in a small window has been identified and is taken as a measure of significant change. Actions in meeting like presentations, general discussions, note taking, etc., and interactions have been modeled using hidden Markov model-based approaches after extracting audio–visual features in [10].

3.3 *News*

News descriptions are elaborative as compared to other descriptions of images as they contain details of events which is usually represented in textual form. The news text is summarized using the encoder–decoder framework in [11] based on the query image. The authors in [12] have performed summarization on news dataset by using joint representation for visual features. For the audio, speech transcriptions have been taken into consideration. The problem has been transformed to an optimization problem, wherein the summary is obtained by tuning the submodular functions.

3.4 Movie Summarization

Long length movies can be made shorter by extracting relevant video skims as is the case of trailer of film. In [13], movie summarization has been done based on audio, text, and visual features. Gabor filters have been extracted for visual analysis, and aural saliency has been found by waveform modulations. Teager Kaiser Energy operator has been used for extracting features from audio, and part of speech tagging has been used for text. Evaluation has been done using KNN.

3.5 Others

Summarization can be also be done on the basis of query given by user. Authors in [14] have proposed a multimodal architecture for the adaptation of multimedia document on the basis of the preference and conditions given by the user. In [15], work has been done on visual question answering using self-attention mechanism of neural architecture. Multimodal transformer architecture has been used for jointly modeling all the modes.

4 Approaches

4.1 Algebraic Modeling and Evolutionary Computation Techniques

The problem of summarization can be remodeled to find solutions through diverse computational and modeling techniques. In [16], the problem of trimming shots to produce video skims (short version of original video stream) has been modeled as matrix column subset selection problem. In [17], text and visual content is brought in the same semantic space and modeled as the problem of non-negative factorization of two matrices of text and images.

4.2 Deep Learning

Advancement in deep learning techniques has escalated the research in multimodal summarization. The models used today are generative in nature and usually incorporate encoder–decoder architecture using convolution neural network and recurrent neural network. In case of images and videos as input, in order to get a textual output, the key idea is to encode features and generate meaningful sentences. Generative adversarial network is a popular model used for image captioning, nowadays. The authors in [18] have generated summary and headlines from news articles by using hybrid pointer generator network. Coverage mechanism has been used to generate abstractive summaries.

4.3 Probabilistic Models

Probabilistic graphical models are conventional models aimed at maximizing the joint distribution over modalities, when used for multimodal summarization. In [19], the authors have used multimodal LDA to discover subtopics from microblogs by exploiting the correlations among the different media types. Given the image, the likelihood of the sentence has been maximized [20]. In deep representations, the commonly used probabilistic graphical models include deep belief networks and Boltzmann machines [21].

4.4 Graph Based

Graph-based algorithms have been popular for quite a long time and are commonly used in clustering and visualization. The authors in [22] have used graph-based clustering algorithm to identify dense graphs. These graphs correspond to events in social media streams. For every event, diverse images are ranked by their importance, and top ranked images are retrieved for representation.

4.5 Feature Extraction

A number of features are considered while dealing with the different modes. High motion as a measure of significant change can be identified when someone joins meeting or a participant gives presentation [9]. Face++ tool [23] can be used to extract features like, eye left center, eye right center, etc., from the face. Opensmile [24] tool has been used to extract features from the audio in [4].

4.6 Classical Models

Classical modeling includes the usage of language rules or templates for obtaining annotations for images. The descriptions can also be borrowed from other visual contents. Generation of language models for the purpose of adaptation from visual to text has also been incorporated.

4.7 Multimodal Representations

Multimodal representation can be categorized into joint modeling, coordinated representation, and encoder–decoder representation. In joint modeling, the modalities are brought together in a common subspace, while in case of coordinated representation, modalities are kept independent and the correlations between them are identified. Conditional distribution among modalities are focused in case of encoder–decoder representation [25].

5 Challenges

5.1 Noise in Data

In multimodal summarization, each mode poses distinct difficulties like quality of audio, video, noise in images, etc., which requires solution. At certain instances, accents do not match acoustic models. Speeches from imperfect speakers and background noises also cause problems. Speeches containing out of vocabulary words pose a great challenge in summarization. Apart from this, handling large datasets is also challenging due to noise.

5.2 Joint Modeling

Building joint representations for the visual and language domains is a challenging task. Currently little work has been done on the creation of theories and models that jointly deal with multimodal data. In multimodal summarization, the basic idea is to bridge the semantic gap between different modalities and to build a system that can improve the quality of summarization through automatic speech recognition, natural language processing, and computer vision techniques. Joint processing of visual and language data is difficult.

5.3 Dealing with Asynchronous Data

Asynchronous data which do not contain captions and subtitles for the images and videos, respectively, is difficult to model.

5.4 *Ranking Modality*

If the audio, video, and visual skims are to be represented in time order, importance of each modality must be known. Ranking the importance of the modalities is a challenging task. A number of factors including the domain and themes of the videos contribute to this.

5.5 *Limited Datasets*

Adequate amount of data is not available to train for the recognition tasks in meeting recordings, sport videos, news, etc. Standard datasets are required for evaluating methods of summarization. Construction of image and video datasets with reference summaries on a large scale is challenging. Devising datasets and algorithms that can work well on different domains are required [5].

5.6 *Limited Features*

Diverse features from different modes need to be explored for good quality summaries. Object point detectors, histogram of gradients, and scale-invariant feature transform (SIFT) are commonly used for identifying visual content in videos. Combination of different features, algorithms and techniques are needed to find a good summary.

5.7 *Developing Domain Independent Approach for Summarization*

Generation of effective domain independent techniques for summarization without compromising for the coherence and cohesion in texts has been a major challenge.

5.8 *Comprehension of Visual Data*

A number of activities are complex and lengthy in videos. Automatic extraction of information in those cases is a challenging task.

Table 1 Multimodal datasets

Dataset	Description	Dataset link
Multimodal summarization dataset v1.0 [41]	Documents containing image, text, audio, and videos of News topics	–
UIUC Pascal dataset [26]	1000 images with five descriptions per image	https://vision.cs.uiuc.edu/pascal-sentences/
Flickr 8 K [27]	8000 images of humans and animals with five sentences per image image	https://www.kaggle.com/adityajn105/flickr8k
Flickr 30 K image dataset [28]	31,873 images consisting of everyday activities and events with 158,915 captions	https://www.kaggle.com/hsankesara/flickr-image-dataset
Microsoft COCO(MS-COCO) [29]	complex images of regular scenes and objects	https://cocodataset.org/#home
SBU dataset [30]	Captioned photo graph dataset with 1 million image URLS and captions	http://www.cs.virginia.edu/~vicente/sbucaptions/
Microsoft research video description corpus (MSVD) [31]	2000 short video clips from Youtube	https://vsubhashini.github.io/s2vt.html
Saarbrucken corpus [32]	Corpus of textually annotated cooking scenes (TACoS) Video	http://www.coli.uni-saarland.de/projects/smile/page.php?id = tacos

6 Datasets

Several datasets offer a combination of multiple modes. Few such publicly available datasets are given in Table 1.

7 Evaluation Methodologies

There are a number of techniques that are used for evaluation of summaries. One such method is word error rate of audio transcripts (WER) method. In this method, audio transcriptions are evaluated against human transcriptions of audio. word error rate is obtained from Levenshtein distance that works at word level instead of phenome level [42]. In manual evaluation method, volunteers are given the task to evaluate the summary and provide scores. Scores are based on different parameters like informativeness, readability, etc. Information retrieval test is also done on summaries to measure the accuracy. Retention of key information is checked by running information retrieval test on the retrieved summaries [33]. BLEAU [34], METEOR [35], ROUGE [36], SPICE [37], CIDEr [38], and WMD [39] are some of the most commonly used metrics that are used for evaluation tasks like matching the reference sentences with the generated ones, accuracy of description of images, dissimilarity between text documents, etc.

8 Conclusion and Future Research Directions

Multimodal summarization is in its infancy and has huge scope in terms of the development of techniques, advanced tools, datasets, and evaluation strategies. Tasks like presentation of summaries, long video summarization, generation of coherent sentences and large paragraphs, selection of video segment lengths in skims, and ranking the importance of modality need more exploration. One of the major challenges is to bridge the gap between low level features from audio, images, video, and the natural language. Moreover, deep reinforcement learning techniques must be developed for translation among different modes. Visual question answering systems and visual reasoning systems need to be constructed [40]. Since separate frameworks with loose connections are currently available, a unified framework must be designed for the conversion of images to sentences or vice-versa. Furthermore, in generative methods of summarization, construction of bigger semantic units is required.

References

1. Zhu, J., Li, H., Liu, T., Zhou, Y., Zhang, J., Zong, C.: MSMO: multimodal summarization with multimodal output. In: Proceedings of the 2018 Conference on Empirical Methods in Natural Language Processing, pp. 4154–4164 (2018)
2. Amirian, S., Rasheed, K., Taha, T.R., Arabnia, H.R.: Automatic image and video caption generation with deep learning: a concise review and algorithmic overlap. IEEE Access **8**, 218386–218400 (2020)
3. Tjondronegoro, D., Tao, X., Sasongko, J., Lau, C.H.: Multi-modal summarization of key events and top players in sports tournament videos. In: 2011 IEEE Workshop on Applications of Computer Vision (WACV), pp. 471–478. IEEE (Jan 2011)
4. Gupta, P., Mehrotra, T., Bansal, A., Kumari, B.: Multimodal sentiment analysis and context determination: Using perplexed Bayes classification. In: 2017 23rd International Conference on Automation and Computing (ICAC), pp. 1–6. IEEE (Sept 2017)
5. Ullah, M.A., Islam, M.M., Azman, N.B., Zaki, Z.M.: An overview of multimodal sentiment analysis research: Opportunities and difficulties. In: 2017 IEEE International Conference on Imaging, Vision & Pattern Recognition (icIVPR), pp. 1–6. IEEE (Feb 2017)
6. Ji, Z., Wang, H., Han, J., Pang, Y.: SMAN: Stacked multimodal attention network for cross-modal image-text retrieval. IEEE Trans. Cybern. (2020)
7. Ekin, A., Tekalp, M.: Generic play-break event detection for summarization and hierarchical sports video analysis. In: 2003 International Conference on Multimedia and Expo. ICME'03. Proceedings (Cat. No. 03TH8698), vol. 1, pp. I–169. IEEE (July 2003)
8. Merler, M., Joshi, D., Nguyen, Q.B., Hammer, S., Kent, J., Smith, J.R., Feris, R.S.: Automatic curation of golf highlights using multimodal excitement features. In: 2017 IEEE Conference on Computer Vision and Pattern Recognition Workshops (CVPRW), pp. 57–65. IEEE (July 2017)

9. Erol, B., Lee, D.S., Hull, J.: Multimodal summarization of meeting recordings. In: 2003 International Conference on Multimedia and Expo. ICME'03. Proceedings (Cat. No. 03TH8698), vol. 3, pp. III–25. IEEE (July 2003)
10. McCowan, I., Bengio, S., Gatica-Perez, D., Lathoud, G., Monay, F., Moore, D., Bour lard, H.: Modeling human interaction in meetings. In 2003 IEEE International Conference on Acoustics, Speech, and Signal Processing, 2003. Proceedings (ICASSP'03), vol. 4, pp. IV–748. IEEE (Apr 2003)
11. Chen, J., Zhuge, H.: News image captioning based on text summarization using image as query. In: 2019 15th International Conference on Semantics, Knowledge and Grids (SKG), pp. 123–126. IEEE (Sept 2019)
12. Li, H., Zhu, J., Ma, C., Zhang, J., Zong, C.: Read, watch, listen, and summarize: multi-modal summarization for asynchronous text, image, audio and video. IEEE Trans. Knowl. Data Eng. **31**(5), 996–1009 (2018)
13. Evangelopoulos, G., Zlatintsi, A., Potamianos, A., Maragos, P., Rapantzikos, K., Skoumas, G., Avrithis, Y.: Multimodal saliency and fusion for movie summarization based on aural, visual, and textual attention. IEEE Trans. Multimedia **15**(7), 1553–1568 (2013)
14. Khallouki, H., Bahaj, M.: Multimodal context based-adaptive architecture for multimedia documents. In: 2015 11th International Conference on Innovations in Information Technology (IIT), pp. 140–145. IEEE (Nov 2015)
15. Hu, R., Singh, A., Darrell, T., Rohrbach, M.: Iterative answer prediction with pointer-augmented multimodal transformers for textvqa. In: Proceedings of the IEEE/CVF Conference on Computer Vision and Pattern Recognition, pp. 9992–10002 (2020)
16. Mademlis, I., Tefas, A., Nikolaidis, N., Pitas, I.: Multimodal stereoscopic movie summarization conforming to narrative characteristics. IEEE Trans. Image Process. **25**(12), 5828–5840 (2016)
17. Camargo, J.E., González, F.A.: Multimodal image collection summarization using non-negative matrix factorization. In 2011 6th Colombian Computing Congress (CCC), pp. 1–6. IEEE (May 2011)
18. Agrawal, A., Saraswat, S., Javed, H.: A pointer Generator network model to automatic text summarization and headline generation. Int. J. Eng. Adv. Technol. **8**(5S3), 447–451 (2019)
19. Bian, J., Yang, Y., Chua, T.S.: Multimedia summarization for trending topics in microblogs. In: Proceedings of the 22nd ACM international Conference on information & knowledge management, pp. 1807–1812 (Oct 2013)
20. Vinyals, O., Toshev, A., Bengio, S., Erhan, D.: Show and tell: Lessons learned from the 2015 mscoco image captioning challenge. IEEE Trans. Pattern Anal. Mach. Intell. **39**(4), 652–663 (2016)
21. Salakhutdinov R., Hinton G.: Deep Boltzmann machines. In: Proceedings 29th International Conference Artificial Intelligence Statistics, pp. 448–455 (2009)
22. Schinas, M., Papadopoulos, S., Petkos, G., Kompatsiaris, Y., Mitkas, P. A.: Multimodal graph-based event detection and summarization in social media streams. In Proceedings of the 23rd ACM International Conference on Multimedia, pp. 189–192 (Oct 2015)
23. https://www.faceplusplus.com/. Last Accessed 9 Mar 2021
24. https://www.audeering.com/opensmile/. Last Accessed 9 Mar 2021
25. Guo, W., Wang, J., Wang, S.: Deep multimodal representation learning: A survey. IEEE Access **7**, 63373–63394 (2019)
26. Farhadi et al. A.: Every picture tells a story: Generating sentences from images. In: Proceedings European Conference Computer Vision, pp. 15–29 (2010)
27. Kumari, K.A., Mouneeshwari C., Udhaya R.B., Jasmitha R.: Automated image captioning for Flickr8K Dataset. In: International Conference on Artificial Intelligence, Smart Grid and Smart City Applications, pp. 679–687. Springer, Cham (2019)
28. Young, P., Lai, A., Hodosh, M., Hockenmaier, J.: From image descriptions to visual denotations: New similarity metrics for semantic inference over event de-scriptions. Trans. Assoc. Comput. Linguistics **2**, 67–78 (2014)
29. Lin et al. T.-Y.: Microsoft COCO: Common objects in context. In: Proceeding European Conference Computer Vision, pp. 740–755 (2014)

30. Ordonez V., Kulkarni G., Berg T.L.: Im2text: Describing images using 1 million captioned photographs. In: Proceedings International Conference Advance Neural Information Processing System, pp. 1143–1151 (2011)
31. Chen D.L., Dolan W.B.: Collecting highly parallel data for paraphrase evaluation. In: Proceeding 49th Annual Meeting Association Computational Linguistics, Human Language Technologies, vol. 1, pp. 190–200 (2011)
32. Regneri, M., Rohrbach, M., Wetzel, D., Thater, S., Schiele, B., Pinkal, M.: Grounding action descriptions in videos. Trans. Assoc. Comput. Linguist. **1**, 25–36 (2013)
33. Valenza, R., Robinson, T., Hickey, M., Tucker, R.: Summarisation of spoken audio through information extraction. In: ESCA Tutorial and Research Work-shop (ETRW) on Accessing Information in Spoken Audio (1999)
34. Papineni K., Roukos S., Ward T., Zhu W.J.: BLEU: a method for automatic evaluation of machine translation. In: Proceedings of the 40th Annual Meeting on Association for Computational Linguistics, pp. 311–318 (2002)
35. Denkowski M., Lavie A.: Meteor universal: Language-specific translation evaluation for any target language. In: 9th Workshop on Statistical Machine Translation, pp. 376–380 (2014)
36. Lin, C.Y.: Rouge: A package for automatic evaluation of summaries. In: Text Summarization Branches Out, pp. 74–81 (July 2004)
37. Anderson, P., Fernando, B., Johnson, M., Gould, S.: Spice: Semantic propositional image caption evaluation. In: European Conference on Computer Vision, pp. 382–398. Springer, Cham (Oct 2016)
38. Vedantam R., Zitnick C.L., Parikh D.: Cider: Consensus based image description evaluation. In: CVPR, pp. 4566–4575 (2015)
39. Kusner M., Sun Y., Kolkin N., Weinberger K.: From word embeddings to document distances. In: Proceedings International Conference Machine Learning, pp. 957–966 (2015)
40. Li, S., Tao, Z., Li, K., Fu, Y.: Visual to text: Survey of image and video captioning. IEEE Trans. Emerg. Top. Comput. Intell. **3**(4), 297–312 (2019)
41. Li, H., Zhu, J., Ma, C., Zhang, J., Zong, C.: Multi-modal summarization for asynchronous collection of text, image, audio and video. In Proceedings of the 2017 Conference on Empirical Methods in Natural Language Processing, pp. 1092–1102 (Sept 2017)
42. https://en.wikipedia.org/wiki/Worderror rate. Last Accessed on 9 Mar 2021

Power Quality Issues with Penetration of Wind Energy and the Role of LDC: A Mini Review

Prakhar Srivastava, Mohit Bajaj, Youcef Belkhier, Naveen Kumar Sharma, Subhashree Choudhury, and B. Srikanth Goud

Abstract Renewable energy specifically winds and solar energies are playing an imperative role being connected to the utility at LV, MV or higher voltage levels. But the incorporation of wind and solar energy to the distribution grid poses power quality issues due to their stochastic nature and owing to utilization of power electronic devices. The major power quality problems are as follows: (1) Variable voltage because of stochastic nature of wind. (2) The presence of harmonic and reactive power in the grid as a result of nonlinear loads. According to equality (E) and inequality (I) constraints, the grid can be divided into five states, namely: normal, alert, emergency, extremis or restorative states. The load dispatch center (LDC) tries to be in normal state. The paper presents the widespread literature review regarding power quality affairs because of wind energy integration and the role of the LDC. Three methods to alleviate the power quality problems owing to integration of wind energy are discussed in this paper.

Keywords LDC · Power quality · Grid integration · Voltage control · Frequency control · STATCOM · DVR · UPQC

P. Srivastava
Stony Brook University, New York 11790, USA

M. Bajaj (✉)
National Institute of Technology, Delhi 110040, India
e-mail: mohitbajaj@nitdelhi.ac.in

Y. Belkhier
Faculté de Technologie, Université de Bejaia, 06000 Bejaia, Algeria

N. K. Sharma
I. K. Gujral Punjab Technical University, Jalandhar 144603, India

S. Choudhury
Siksha 'O' Anusandhan (Deemed To Be University), Bhubaneswar 751030, India

B. Srikanth Goud
Koneru Lakshmaiah Education Foundation, Guntur 522502, India

K. N. Das et al. (eds.), *Proceedings of the International Conference on Computational Intelligence and Sustainable Technologies*, Algorithms for Intelligent Systems,
https://doi.org/10.1007/978-981-16-6893-7_55

1 Introduction

The integrated amount of wind energy is snowballing in many countries. In 2020, it is predicted that 93 GW of wind generating capacity will be built across the world (Fig. 1). Adjusting distribution grid to function smoothly with renewable energy sources (RESs) is highly problematical. At wind energy flank, production is non-dispatchable and discontinuous with great variations because of fluctuating nature of wind, while at the grid flank, there are turbulences, for instance, voltage fluctuations resulting from short circuit faults and frequency deviations owing to load and generation alteration [1, 2]. It would cause issues at the point of common coupling (PCC) and undesirable operating circumstances.

Power quality has become a great factor to disturb the complete stability and reliability of the future smart grid [3, 4].

The paper has the following sections: Sect. 2 covers the role of LDC. In sect. 3 the causes of poor power quality owing to interconnection of wind energy to power grid are discussed. Section 4 discusses the mitigation methods for enhancing power quality. Future research directions are given in conclusion.

2 The Role of Load Dispatch Center

If a wind energy system is linked to the grid, it must adhere to certain grid codes [5]. For the satisfactory grid operation, wind energy system at the distribution network must follow the rules defined by IEC-61400-21. The electricity grid can be divided into five states, namely: normal, alert, emergency, extremis and restorative Fig. 2. The Grid can be in any one of these five states depending upon the fulfilment or otherwise of the Equality (E) and Inequality (I) constraints. A power system is expected to operate in NORMAL state in which equality and inequality constraints are satisfied. If E and I are in brim, system is ALERT, and preventive actions are to be taken by LDC to bring it in NORMAL. If some violation or fault occurs in ALERT state, the system enters in EMERGENCY state with the violation of inequality constraints, and necessary emergency control actions are to be taken by the LDC [6]. EMERGENCY state could lead to the EXTREMIS state in which the grid has totally collapsed with blackout or islanded operation. In the RESTORATIVE state, LDC follows the restoration procedure in which black state power is delivered to the needy stations, skeleton grid is formed, the generation and load are resumed in small steps, and the system is brought back to NORMAL state. This state can take hours to days to complete results and discussions (Figs. 1 and 2).

A. *Wind generator in the NORMAL state*

Wind power is directly proportional to the cube of the wind speed. Meanwhile the wind speed is variable, the generation also varies with it. The wind turbine generator is stopped by applying brakes if the wind speed exceeds the furling speed, and the

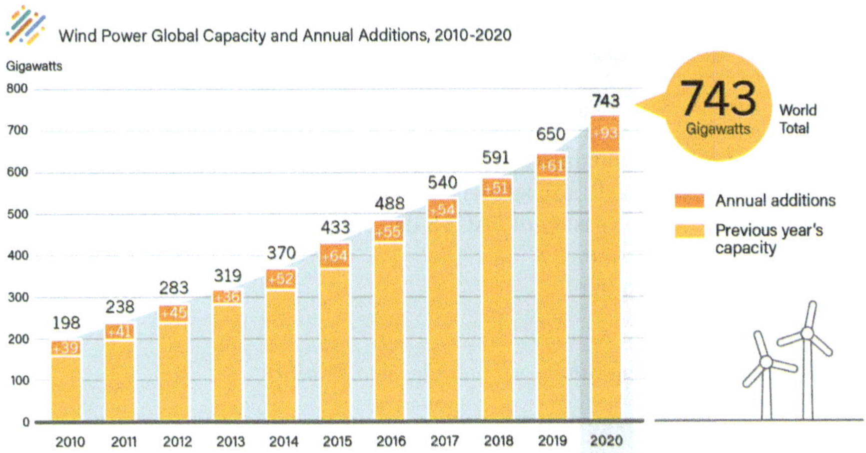

Fig. 1 Annual additions to wind power capacity in the world, 2010–2020 [7]

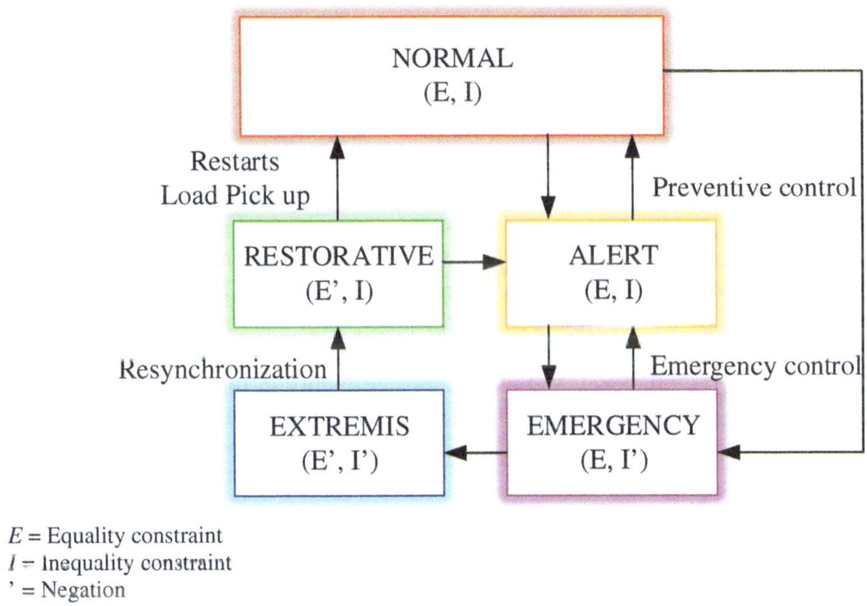

Fig. 2 Diagram of a power system's state transition

generation is zero [8]. This may result in the variation in frequency either up or down. In such cases, load frequency control (LFC) comes into play, and bulk generators of the grid adjust their generation through their speed governors and correct the variation to zero. LDC will decide now whether the system is in normal state or alert state and will initiate suitable preventive control if needed [9]. If the variation is large which the governors cannot handle, the emergency or extremis state is chosen by the computer-controlled operator of an area [10].

B. *The maintenance of voltage profile in the grid*

The voltage profile is maintained by the action of Automatic Voltage Regulator (AVR) in the generators, On Load Tap Changers (OLTC) of transformers and the application of shunt capacitors or provision of FACTS devices like DVR, STATCOM, etc.

E = Equality constraint
I = Inequality constraint
' = Negation

C. *Impact of wind generators on dynamics and stability*

If the voltage drops by more than 20% during a fault, the wind turbine may disconnect from the grid. But with greater grid penetration of wind energy conversion system (WECS), grid operators want the wind turbine generator to stay connected even during the faults and continue to support the power system. Such capability of WPP is called fault-ride-through capability [8, 11].

D. *Wind generators in Restorative state*

In then restorative state, the generators in the grid are expected to supply active and reactive power support [8]. In the case of a grid outage, the wind generator will turn off and will not be able to participate in the initial restoration phase. However, the generating sites with battery storage facility could contribute to the restoration process.

Our traditional power generating system is designed around massive, centrally regulated power plants. Renewable energy, on the other hand, is distributed, self-regulated and sporadic. Consequently, adjusting power distribution grid to function steadfastly with RESs can be extremely convoluted. Smart grids are important for improving energy efficiency and electric power quality in future, especially given the likely large expenditure over the next several years in maintaining and expanding the power infrastructure to set up renewable energy generation [12].

3 Power Quality Issues During Wind Energy Integration

As the penetration level of wind energy increases day by day, power quality concerns have become a major worry for the wind energy system. The increasing use of sensitive power electronics devices is enforced to transform the conventional power system ideas.

The product of voltage and the reactive component of current is reactive power. For active, reactive and perceived power, there is a power triangle. However, for loads with nonlinear characteristics, this definition of reactive power is insufficient. As a result, nonlinear loads need modifying the power factor definition. There are two portions to the power factor. The cosine of phase difference between fundamental supply current and voltage is the displacement power factor. Another is the distortion power factor, often known as the harmonic factor, which is defined as the ratio of RMS harmonic content to RMS value of fundamental component represented as a percentage of the fundamental.

The main power quality issues for wind energy integration refer to

1. Poor power factor due to reactive power.
2. Voltage fluctuation instability due to voltage sag, swell, voltage dip and flickering.
3. Harmonics due to frequency converter, thyristor controller, capacitor, etc.

70% power quality problems are due to voltage sag. The IEC-61400 standard (International Electro-Technical Commission) provides various criteria and measurements as a guiding concept [13]. The power quality problems are generated from both source side and grid side. Because the power generated by a wind turbine is directly related to the cube of wind speed, electricity generation at the wind energy source is intermittent. Consequently, it results in fluctuation in voltage and frequency. When the degree of penetration of wind energy is low, the power quality problems are at sensitive power electronic devices which can be solved at device specifically [14]. But when the degree of penetration is high, the power quality problems are not only limited to device, and more grid level solutions are required. The reliability of the power grid is related to the power quality issues that have been reported. A new definition for voltage sag trip was published in IEEE standard 1547a in 2014 [15]. According to this definition, if the voltage sag is within the range of default to maximum, the distribution system will not permit the trip.

A. *Impact of integration of wind energy on power factor*

The definition of power factor can be defined by power triangle. But this definition is insufficient for the system having nonlinear loads. So, the modified definition of power factor is required. The revised power factor is divided into two parts. One is displacement factor, and another is distortion factor [16]. The increasing application of power electronic devices and nonlinear loads has raised the reactive power in the system, which is responsible for poor power factor [17, 18].

B. *Impact of integration of wind energy on voltage variation*

Voltage varies due to the variation in wind speed in wind farm [19, 20]. The reason for voltage fluctuation on the grid is variation in load. It is described as a shift in the voltage's RMS value over a brief period of time. Voltage fluctuations on the grid are primarily due to variation in loads and power produced. The X/R ratio for transmission line is significantly greater than one [17]. Hence, for LV system, the transfer of active power has a crucial impact on the voltage profile, while reactive power is main factor on HV systems [21].

Voltage fluctuations cause unpleasant variations in the luminance of lamps, which are referred to as flicker. In fixed speed turbines, these are created by the wake of the tower.

IEC 61400-21 is recommended for the wind farms; the 10-min average of voltage fluctuation should be inside ± 5% of its nominal value.

The IEC 61400-4-15 stipulates a flicker meter which can be deployed to quantify flicker unswervingly. The voltage quality of a wind turbine may be evaluated in terms of the following criteria under normal operating conditions [15]:

- Steady-state voltage under continuous production of power
- Voltage fluctuations (i). Flicker during operation (ii). Flicker owing to switching.

C. *Impact of integration of wind energy on harmonics*

The integer multiple of the fundamental frequency is called a harmonic. Harmonics are mostly produced by inverters used in variable-speed turbines [22]. According to the IEC-61400-36 guideline, harmonic voltage and current should be kept below the boundary [23]. The presence of harmonics in the system lowers the power quality at the consumer's end. Harmonics increase power loss and cause communication system failure [18, 24].

4 Mitigation Methods of Poor PQ

The major causes of poor power quality owing to penetration of wind energy are because of fluctuation in voltage and harmonics in current and voltage. The reason is the increasing application of sensitive power electronics devices. It is observed that 70% of PQ problems are voltage sag. Active power filter (APF) technology mitigates many PQ issues [25]. Following are the methods to mitigate PQ problems:

A. DVR
B. STATCOM
C. UPQC.

A. *Dynamic Voltage Restorer (DVR)*

A DVR is a series-connected power electronic device that injects a series of three phase voltages into an electrical power distribution system to correct for voltage sags [26, 27] (Fig. 3).

Usually, DVR uses GTO switches in a PWM inverter. DVRs may be a decent option for power quality improvement. But they are not applied usually in a system that has reactive power shortages remaining for a long time and in the system that is exposed to the likelihood of voltage fall suddenly [28].

B. *Static Compensator (STATCOM)*

STATCOM is a shunt-connected FACTS device used to reduce load current harmonics and compensate for reactive power [29]. It shows better performance at low voltage. STATCOM is based on voltage source converter [30, 31]. It uses

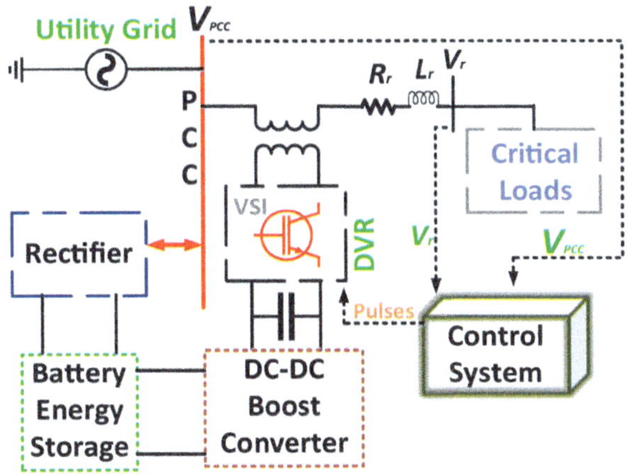

Fig. 3 Schematic diagram of DVR

IGBT, GTO, etc. STATCOM integrated with energy storage system (ESS) significantly enhances the power system operation and control. The greatest approach for increasing the power that can be injected into a poor system network is to integrate a battery energy storage system (BESS) [32] (Fig. 4).

C. *Unified Power Quality Conditioner (UPQC)*

The APF family includes UPQC. Both series and shunt APFs are combined in UPQC. Shunt APF (STATCOM) is the most dependable way for dealing with current-related problems, whereas series APF (DVR) is the greatest approach for dealing with voltage-related issues. For the mitigation of PQ problems which mainly occur due to variable voltage and harmonic in current, both STATCOM and DVR are required to be installed. But this is not a cost-effective solution.

UPQC is the only device that can address both variable voltage and current harmonics issues [33]. UPQC replaces the two devices, namely DVR and STATCOM. UPQC is installed in distribution system. UPQC uses two voltage source inverters (VSIs), coupled to a common DC link. Figure 5 shows a single line illustration of UPQC [34]. The capacitor bank voltage is maintained by using series and shunt filters and hence the reactive power of the DG. Fig. 5 represents the block diagram of UPQC. Here, a transformer is used which converts 690 V 4.16 kV (IEEE 34 bus system voltage).

The selection of controller plays a very significant role in refining the performance of UPQC. PI controller requires a linear mathematical model of a system. It becomes very tough to achieve a mathematical model under the variable parameters and nonlinear loads. This problem can be solved by using fuzzy logic controller as it does not require mathematical modeling.

There are basically two control techniques for UPQC. One is Frequency domain method and other is Time domain method. Frequency domain method is not as successful as it takes large computation time and delay for fast Fourier transform calculation. Time domain technique uses instantaneous derivative of either voltage

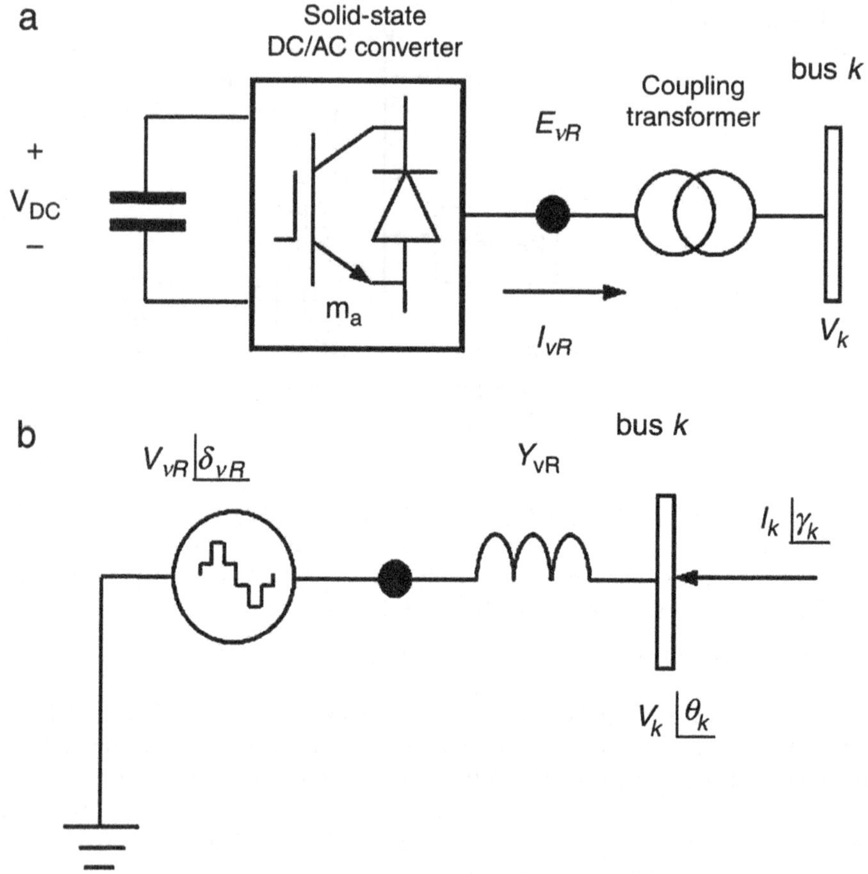

Journal of Applied Research and Technology. 2017;15:36–44

Fig. 4 Simplified diagram of STATCOM with its equivalent circuit

or current signals. There are two popular time domain techniques for controlling UPQC [35].

a. *Three-phase pq theory*
b. *Three-phase dq theory*

Three-phase pq theory or active and reactive power theory has some limitations, but now, it is modified to pqr theory. The dq theory of synchronous reference frame theory is also modified so that the advantages of both theories can be taken out for single-phase UPQC. It is beneficial to use UPQC as it solves both voltage-related and current-related power quality problems. Numerous scholars have used dissimilar nomenclature for UPQC. Paper [36] has given the list of 12 types of UPQC.

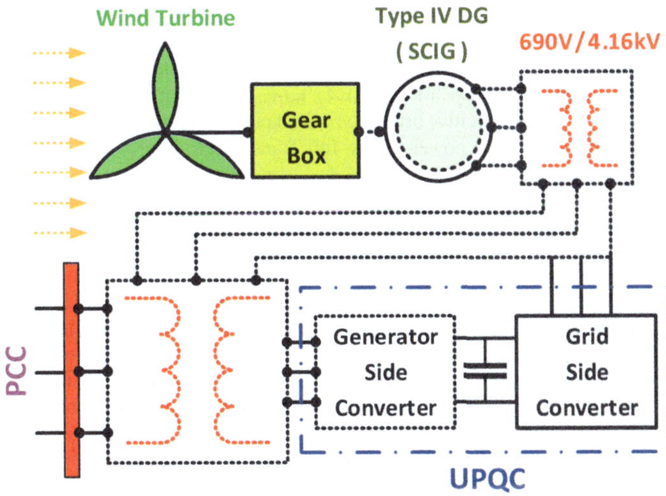

Fig. 5 UPQC-connected grid system

5 Conclusion

A widespread review on the power quality issues as a result of penetration of wind and the role played by the LDC during such conditions is reported in this paper. The stochastic nature of wind creates PQ problems which may be local or widespread in nature. The major quality issues are discussed in the paper. The problems can be solved by FACTS devices. UPQC is the best solution for majority of the power quality problems. This review about PQ issues will be useful for the researchers who are working on enhancing the power quality caused by incorporation of wind energy systems.

References

1. Mandiтereza, P.T., Bansal, R.: Renewable distributed generation: The hidden challenges—A review from the protection perspective. Renew. Sustain. Energy Rev. **58**, 1457–1465 (2016). https://doi.org/10.1016/j.rser.2015.12.276
2. Naderi, Y., Hosseini, S.H., Ghassem Zadeh, S., Mohammadi-Ivatloo, B., Vasquez, J.C., Guerrero, J.M.: An overview of power quality enhancement techniques applied to distributed generation in electrical distribution networks. Renew. Sustain. Energy Rev. **93**, 201–214 (2018). https://doi.org/10.1016/j.rser.2018.05.013
3. Bajaj, M., Singh, A.K.: Grid integrated renewable DG systems: A review of power quality challenges and state-of-the-art mitigation techniques. Int. J. Energy Res. **44**, 26–69 (2020). https://doi.org/10.1002/er.4847
4. Bajaj, M., Aggarwal, S., Singh, A.K.: Power quality concerns with integration of RESs into the smart power grid and associated mitigation techniques. In: 2020 IEEE 9th Power India

International Conference, IEEE, pp. 1–6 (2020). https://doi.org/10.1109/PIICON49524.2020.9113008
5. Kumar, P., Singh, A.K.: Grid codes: Goals and challenges. In: Renewable Energy Integration Green Energy Technologies, pp. 17–39 (2014). https://doi.org/10.1007/978-981-4585-27-9_2
6. Firouzi, M.: A modified capacitive bridge-type fault current limiter (CBFCL) for LVRT performance enhancement of wind power plants. Int. Trans Electr. Energy Syst. **28**, e2505 (2018). https://doi.org/10.1002/etep.2505
7. REN21: Renewables 2021-Global Status Report Technical Report (2021)
8. Firouzi, M., Gharehpetian, G.B.: LVRT performance enhancement of DFIG-based wind farms by capacitive bridge-type fault current limiter. IEEE Trans. Sustain. Energy **9**, 1118–1125 (2018). https://doi.org/10.1109/TSTE.2017.2771321
9. Lezhniuk, P., Teptya, V., Komar, V., Rubanenko, O.: Principle of least action in models and algorithms of optimisation states power system. In: Model Control and Information Technologies, pp. 173–6 (2019). https://doi.org/10.31713/MCIT.2019.38
10. Kumar, R., Diwania, S., Singh, R., Ashfaq, H., Khetrapal, P., Singh, S.: An intelligent hybrid wind–PV farm as a static compensator for overall stability and control of multimachine power system. ISA Trans. (2021). https://doi.org/10.1016/j.isatra.2021.05.014
11. Wang, S., Chen, N., Yu, D., Foley, A., Zhu, L., Li, K., et al.: Flexible fault ride through strategy for wind farm clusters in power systems with high wind power penetration. Energy Convers. Manag. **93**, 239–248 (2015). https://doi.org/10.1016/j.enconman.2015.01.022
12. Gandoman, F.H., Ahmadi, A., Sharaf, A.M., Siano, P., Pou, J., Hredzak, B., et al.: Review of FACTS technologies and applications for power quality in smart grids with renewable energy systems. Renew. Sustain. Energy Rev. **82**, 502–514 (2018). https://doi.org/10.1016/j.rser.2017.09.062
13. Emanuel, H., Schellschmidt, M., Wachtel, S., Adloff, S.: Power quality measurements of wind energy converters with full-scale converter according to IEC 61400-21. In: 2009 10th International Conference Electrical Power Quality and Utilsation, IEEE, pp. 1–7 (2009). https://doi.org/10.1109/EPQU.2009.5318814
14. Rolim, L.G.B., Ortiz, A., Aredes, M., Pregitzer, R., Pinto, J.G., Afonso, J.L.: Custom power interfaces for renewable energy sources. IEEE Int. Symp. Ind. Electron. **2003**, 2673–2678 (2007). https://doi.org/10.1109/ISIE.2007.4375030
15. 1547a-2020—IEEE Standard for Interconnection and Interoperability of Distributed Energy Resources with Associated Electric Power Systems Interfaces--Amendment 1: To Provide More Flexibility for Adoption of Abnormal Operating Performance Category III (2020). https://doi.org/10.1109/IEEESTD.2020.9069495.
16. Bajaj, M., Singh, A.K.: A global power quality index for assessment in distributed energy systems connected to a harmonically polluted network. Energy Sources, Part A Recover. Util. Environ. Eff. 1–27 (2021) https://doi.org/10.1080/15567036.2021.1929577
17. Bajaj, M., Kumar S.A.: Hosting capacity enhancement of renewable-based distributed generation in harmonically polluted distribution systems using passive harmonic filtering. Sustain. Energy Technol. Assessment **44**, 101030 (2021). https://doi.org/10.1016/j.seta.2021.101030
18. Lakshmi, G.S., Rubanenko, O., Divya, G., Lavanya, V.: Distribution energy generation using renewable energy sources. 2020 IEEE India Council International Subsections Conference, IEEE, pp. 108–13 (2020). https://doi.org/10.1109/INDISCON50162.2020.00033
19. Bajaj, M., Singh, A.K.: Power quality challenges associated with distributed generation planning: A simulation-based demonstration. In: 2019 International Conference Electrical Electronic Computer Engineering, Aligarh, IEEE, pp. 1–6 (2019). https://doi.org/10.1109/UPCON47278.2019.8980032
20. Bajaj, M., Singh, A.K.: A single index for voltage quality ranking in the distribution power networks using multiple criteria decision-making. Int. Trans. Electr. Energy Syst. (2021). https://doi.org/10.1002/2050-7038.12965
21. Gundebommu, S.L., Hunko, I., Rubanenko, O., Kuchanskyy, V.: Assessment of the power quality in electric networks with wind power plants. In: 2020 IEEE 7th International Conference Energy Smart System, IEEE, pp. 190–4 (2020). https://doi.org/10.1109/ESS50319.2020.9160231

22. Bajaj, M., Singh, A.K., Alowaidi, M., Sharma, N.K., Sharma, S.K., Mishra, S.: Power quality assessment of distorted distribution networks incorporating renewable distributed generation systems based on the analytic hierarchy process. IEEE Access **8**, 145713–145737 (2020). https://doi.org/10.1109/ACCESS.2020.3014288
23. Liang, S., Hu, Q., Lee, W.J.: A survey of harmonic emissions of a commercially operated wind farm. IEEE Trans. Ind. Appl. **48**, 1115–1123 (2012). https://doi.org/10.1109/TIA.2012.2190702
24. Bajaj, M., Singh, A.K.: An analytic hierarchy process-based novel approach for benchmarking the power quality performance of grid-integrated renewable energy systems. Electr. Eng. **102**, 1153–1173 (2020). https://doi.org/10.1007/s00202-020-00938-3
25. Bajaj, M., Rautela, S., Sharma, A.: A comparative analysis of control techniques of SAPF under source side disturbance. In: 2016 International Conference Circuit, Power Computer Technology, IEEE, pp. 1–7 (2016). https://doi.org/10.1109/ICCPCT.2016.7530196
26. Bajaj, M.: Design and simulation of hybrid DG system fed single-phase dynamic voltage restorer for smart grid application. Smart Sci. 1–15 (2020). https://doi.org/10.1080/23080477.2020.1748928
27. Bajaj, M., kumar, S.A.: Designing of a solar energy based single phase dynamic voltage restorer using fuzzy logic controlled novel boost inverter. In: 2020 IEEE 9th Power India International Conference, IEEE, pp. 1–6 (2020). https://doi.org/10.1109/PIICON49524.2020.9112965
28. Agalar, S., Kaplan, Y.A.: Power quality improvement using STS and DVR in wind energy system. Renew. Energy **118**, 1031–1040 (2018). https://doi.org/10.1016/j.renene.2017.01.013
29. Jain, A.K., Joshi, K., Behal, A., Mohan, N.: Voltage regulation with STATCOMs: Modeling, control and results. IEEE Trans. Power Deliv. **21**, 726–735 (2006). https://doi.org/10.1109/TPWRD.2005.855489
30. Bajaj, M., Flah, A., Alowaidi, M., Sharma, N.K., Mishra, S., Sharma, S.K.: A Lyapunov-function based controller for 3-phase Shunt active power filter and performance assessment considering different system scenarios. IEEE Access 1–1 (2021). https://doi.org/10.1109/ACCESS.2021.3075274
31. Bajaj, M., Rana, A.S.: Harmonics and reactive power compensation of three phase induction motor drive by photovoltaic-based DSTATCOM. Smart Sci. **6** (2018). https://doi.org/10.1080/23080477.2018.1505114
32. Li, W., Joós, G., Abbey, C.: Wind power impact on system frequency deviation and an ESS based power filtering algorithm solution. In: 2006 IEEE PES Power System Conference Exposition PSCE 2006—Proceedings vol. 2, pp. 2077–84 (2006). https://doi.org/10.1109/PSCE.2006.296263
33. Prakash Mahela, O., Gafoor, S.A.: Topological aspects of power quality improvement techniques: A comprehensive overview. Renew. Sustain. Energy Rev. **58**, 1129–1142 (2016). https://doi.org/10.1016/j.rser.2015.12.251
34. Goud, B.S., Rao, B.L.: Power quality enhancement in grid-connected PV/Wind/Battery using UPQC: Atom search optimization. J. Electr. Eng. Technol. **16**, 821–835 (2021). https://doi.org/10.1007/s42835-020-00644-x
35. Goud, B.S., Rao, B.L., Reddy, B.N., Rajesh, N., Anjan, B., Reddy, C.R.: Optimization techniques in PV-wind based distribution generation- a brief review. In: 2020 IEEE 17th India Council International Conference, IEEE, pp. 1–6 (2020). https://doi.org/10.1109/INDICON49873.2020.9342518
36. Khadkikar, V.: Enhancing electric power quality using UPQC: A comprehensive overview. IEEE Trans. Power Electron. **27**, 2284–2297 (2012). https://doi.org/10.1109/TPEL.2011.2172001

Strategies on Smart Energy Management Policies in India

Sujit Kumar, Anuj Banshwar, Naveen Sharma, Bharat Bhushan Sharma, and Mohit Pathak

Abstract About 6% of the world's essential vitality is being ingested by India. Fast urbanization and fast population growth have drastically impacted the capacity to deplete essential resources and emit carbon dioxide. Specifically, cities must accomplish a broad and interlinked set of activities that stem from the interconnectedness of cities (construction atmosphere, flexibility, water, and new management and community facilities). Smart energy management (SEM) incorporation is grave for the continuation of less-carbon metropolitan expansion. Furthermore, energy footmark management is one of the utmost difficult challenges cities face, and as municipalities develop and evolve into smart cities, SEM will be a needed. This article describes the many applications of SEM in the fields of building, conveyance, communal utilities, water and waste, regulations, and the current obstacles and possibilities that need to be overcome in order to implement new SEM policy rules in India. Energy efficiency and emissions reduction go hand in hand with regards to urban planning.

Keywords Smart energy managing · Flexibility · Aquatic energy connection

S. Kumar (✉)
Department of Electrical and Electronics Engineering, Jain (Deemed-To-Be University), Bengaluru, Karnataka, India
e-mail: k.sujith@jainuniversity.ac.in

A. Banshwar
Government Polytechnic Tada Gulabrai, Madhotanda Road, Puranpur (Pilibhit), U.P, India

N. Sharma
IKG Punjab Technical University, Main Campus, Jalandhar, Punjab, India

B. B. Sharma
School of Automation, Banasthali Vidyapith, Vanasthali, Rajasthan, India

M. Pathak
Noida Institute of Engineering and Technology, Greater Noida, India

K. N. Das et al. (eds.), *Proceedings of the International Conference on Computational Intelligence and Sustainable Technologies*, Algorithms for Intelligent Systems,
https://doi.org/10.1007/978-981-16-6893-7_56

1 Introduction

Urban India is on the rise. One of the primary factors for the rising demand for energy is urban expansion. Eighteen % of the world's populace and six % of the world's main energy usage are represented by the country [1]. In India, the proportion of the people who live in cities is predictable to expand from 30 to 58% by 2050, with consequences for energy usage and carbon productions [2]. Urban planners must also contend with the job of balancing and lowering energy request, energy usage, and corresponding carbon productions. Sustainable urban transit, access to renewable and dependable energy, green and strong arrangement, and excess supervision are critical in this environment because of their impact on urban issues such as smart energy management (SEM) [3]. Moreover, different businesses, like housing, conveyance, water, garbage, and community facilities, would benefit from incorporating SEM values in order to decrease carbon productions and meet the maintainable growth goals [4].

Combined SEM also uses all available resources as efficiently as possible, and reduces the need of using unsustainable energy sources. In order to achieve this, TERI and UNSW held Smart Energy Managing for Maintainable Cities (SEMMC) seminars between Australia and India to improve cooperation on SEM for Maintainable Cities. The experts and practitioners who work in local institutions such as think tanks, policy organizations, academic institutions, and research institutes, as well as industry, all took part in the training sessions.

Participants believed that panel discussions organized by industry experts may help them think through the problem [5]. Smart cities will rely more on local resources and should be more independent in order to cut down on their energy impact. The article in this issue discusses smart cities, which use energy efficiency as an important part of the endeavor. Smart city is a smart, environmentally friendly, livable, and vibrant city that uses resources to foster its citizens' well-being. Bringing energy technology programs to the mainstream comprises the introduction of programs to facilitate energy policies, initiatives, and structural transformation, awareness-raising, training, and capacity development [6].

2 Indian Government Policy

2.1 Conservation Policies

The Ministry of Power (MoP) passed the Energy Conservation (EC) Act, as revised in 2010, to reduce energy intensity (energy use per unit of economic output). Office of Energy Efficiency (OEE) was created in 2002 as the federal legislative body responsible for the execution of the EC Act. Energy-intensive sectors are subject to the provisions of the Energy Preservation Construction Code (EPCC), machine specifications and appliance markings, and energy consumption regulations under the ECBC [7]. An estimated 50% reduction in energy usage is expected by 2030 as a result of implementing the ECBC [8].

It is permitted under the act for the government and OEE to help the public use less energy in all areas. Also, federal agencies that have been granted legislation must ensure that the law is upheld and use energy efficiency in their day-to-day operations. With this act, the OEE has implemented a variety of energy redeemable enterprises, including as improving the efficiency of household lights, improving profitable building efficiency, developing standards for various products, and standardizing home appliance labeling [8].

2.2 National Policies

The National Energy Efficiency Mission (NMEE) was established under the National Weather Variation Act (NWVA). Energy Efficacy Facilities Limited (EEFL) is founded as a corporate body with the assignment of market guidance in supplying energy efficiency services [7].

The NMEEE's mission aims to satisfy the country's energy demands and to implement energy efficacy agendas. NMEEE estimates that throughout the whole implementation phase, it will reduce annual fuel consumption by 23 million tons and minimize carbon emissions by 95.55 million tons. The efficiency, achievement, and trade project's first phase, which ran from 2012 to 2015, resulted in 31 million metric tons of carbon dioxide emissions being curbed, or around 1.93% of India's overall emissions [7].

2.3 Sustainable Habitat for National Mission

The National Sustainable Habitat Mission (NSHM) in the NWVA strives to include weather variation edition and extenuation into town preparation. Solid waste, increasing household energy efficiency, and using public transportation are some of the ways to achieve sustainability. Energy usage will be reduced by 50% by 2030 according to the code [8]. The design of energy efficiency for multi-story residential structures is important to OEE, and this aim is implemented by setting rules for such types of buildings in hot and dry areas [8]. The Green Building Code has been made available as part of a handbook for the central works department (CWD) to help the various state and territory governments to include the relevant requirements in their building by-laws, and a draft building by-laws set has also been issued. A wide variety of actions beneath the NSHM were planned for the Atal Retrieval and Urban Development Assignment (ARUDA) [8].

2.4 Solar Nationwide Assignment

The Solar Nationwide Assignment (SNM), created under the NWVA, aims to cover all of the grid and off-grid solar applications with a coordinated plan. In order to enable the nation to get more exposure to solar power, the "Ministry of New and Renewable Energy (MNRE)" is aiming to develop an appropriate environs for solar technology diffusion. The SNM's original purpose was to help install and connect 22 GW of grid-connected solar electricity. The first objective had been set at 20 GW in 2014, and that number was raised to 100 GW by 2022 [9].

2.5 Smart Mission Cities

The Smart Mission Cities, which was established by the Ministry of Casing and Urban Activities (MoCUA) in 2015, pursue to upsurge the competence of urban government and infrastructure, as well as progress the excellence of lifetime in urban parts. To realize sustainable urban expansion while pursuing energy efficiency and lower costs, the mission uses Internet of things (IoT) technology. Additionally, the city's new smart city plans will incorporate two important aspects: an area-based development (ABD) project that focuses on putting together an integrated traffic management system (ITMS), automated city-level garbage collection, e-government, and rooftop solar panels in government buildings [10].

2.6 National Policy of Energy

National policy of energy (NPE) focuses on increasing admittance to reasonable, dependable energy for everybody, decreasing the use of fossil fuels, boosting the generation of low-carbon energy, and creating an environment for sustainable economic development. To reach national energy objectives, this policy endeavors to implement collaboration amongst multiple ministries to govern energy. The purpose of the NEP is to provide electricity to the masses, giving customers more energy options, and helping India become ready for future energy demands. The economy will be energy ready by 2040, according to recent predictions [11].

Incorporation of demand-driven energy management strategies in sustainable cities has been recommended for conveyance, building, and domestic (cookery) sectors, beneath the NPE. Policies like encouraging the use of public transportation, shifting away from driving, promoting the use of electric and hybrid cars, and implementing information and communications technology (ICT) solutions are being considered in the transport industry [11]. A combination of various elements in the buildings sector has been proposed for making buildings more energy efficient, including ECBC implementation, urban planning that encourages compact, efficient development, the use of high-performance lighting technologies and appliances, and shifting away from traditional cooking fuels to modern ones [11].

2.7 *National Draft Cooling Action Plan*

Over the next decade, cities in India will demand 2.2–3 times (2019 baseline) more cooling. As a result, there will be a greater capacity for generating electricity, demanding the peak load, and having a significant impact on carbon footprint. This is why the Ministry of Atmosphere, Forestry, and Weather Alteration (MoAFWA) has created a national draft strategy for alleviating the drought, which includes draught action at several levels (NCAP) [8]. The NCAP's mission is to motivate safer and smarter cooling technique National Cooling Action Plan use across all of Canada by the year 2020. It encompasses many development scenarios and methodologies, including cooling development, transporting, refrigeration, cold chain, air conditioning/refrigeration systems, and industry sectors [12].

3 Application of SEM in Various Fields

A wide array of businesses exist in cities, including construction, transportation, water, garbage, and public services. Delhi focuses on discovering hurdles and possibilities to adoption of SEM by cities in India so as to have an inclusive thoughtful of the possibilities of an integrated SEM. The talks focused on current industry rules and future strategies for different sectors. In this part, we explore SEM solutions and tactics in the various fields, as well as barriers to the use of SEMs in Indian cities [13].

3.1 *Building Sustainable Energy Management*

It is a primary purpose of buildings that deal with diverse energy services to avoid the detrimental environmental impact on buildings through the life cycle and to cut energy expenses [9]. In order to best utilize available energy and demand, they should be able to balance comfort for the inhabitants and assist foster energy independence in the home [6]. SEM solutions applied to structures reduction into three clusters depending on their applicability: (i) building energy-efficiency explanations, which provide energy effectual building control of energy schemes; (ii) demand response solutions, which include incorporating passive solar architecture and sustainable materials; and (iii) low-cost solutions that lower the total cost of building systems. Since [6], the combination of these technologies allows for improved comfort, usefulness, and adaptability. Efforts to optimize operations and manage may reduce building energy consumption by 20–30% without making changes to the system or hardware structure (ibid.). Space heating, aeration, and air conditioning systems have now focused on mutable speed coolers, home-based heat control systems, and intelligent fuzzy ease controllers to provide personalized comfort in buildings. Typical smart lighting systems include illumination controls, structures like appliance control gears, daylight addition through construction info modeling software, tenancy instruments, photometric character fittings, and LED bulbs [10].

3.2 Wasteland Resource Depot

The methodology is integrated in order to guarantee the energy and aquatic segments are not being achieved as separate entities. A complete approach to managing this valuable resource is referred to as a water SEM strategy, and methods to manage supply, treatment, and distribution are referred to as water SEM infrastructure (Environmental Leader, 2018) [11]. This process is used to determine the amount of energy needed for water consumption, supply, and distribution: whether it is utilized for water supply or distribution. Examples of smart water technologies include: Automatic instruments (meters and sensors) that capture and distribute information in real time and work under the supervision of "supervisory control and data acquisition (SCADA)" systems that develop evidence and distantly run and enhance schemes [11].

3.3 Effective and Less-Carbon Agility

The goal of the two primary vehicle fuel efficiency programs is to expand the market for fuel-efficient, low-carbon road transport by driving more sales, via various initiatives including the automobile fuel efficacy platform and the national electric flexibility assignment plan (NEFAP) 2020. India plans to implement these guidelines on a gradual basis for both passenger and freight transportation on the country's roads [12]. The use of fuel efficiency regulations to achieve a reduction in the overall fuel consumption of automobiles is an essential legislative instrument under the Automotive Fuel Efficiency Program (AFEP). Passenger cars equipped with light-duty fuel economy criteria were the first set of such criteria to be introduced by the Ministry of Road Transport and Highways (MoRTH) in 2017. The fuel consumption and efficiency objectives used to calculate these criteria which are established on the commercial average fuel budget (CAFB) criteria and use the CAFB standards as benchmarks. Various European nations have subsidized the purchase of electric vehicles and set aside funds for the construction of tram lines in order to promote the use of taxi aggregators and/or ride-sharing services such as Uber and Ola [12]. Cities will need to increase their support for inventive information and communications technology (ICT) methods and strategies that deal with compelling data and energy use. A transport strategy should not be formed without considering all aspects of it. Consideration should be given to contributions from all areas (including policy) [12].

3.4 Sustainable, Waste-Free Practices

About 43% of India's urban regions are separating their garbage at the source according to 2018 statistics from the Swachh Bharat Mission (SBM). Door-to-door collection was covered by 54% in 2017, rising to 80% in that year. Separate trash is transferred to a processing plant in locations like as Panaji, Indore, Mysore, and Muzaffarpur [12]. Wet waste compost is produced while inert garbage is sent to the landfill. In Alappuzha, Kerala, the predominant mentality is one of segregation and re-use of garbage, resulting in the city being one of the cleanest in India [13].

A total of 80% of residences have biogas plants and pipe composting systems. Clean Home Clean City (2013) set up an aerobic composting factory for the disposal of animal corpses in Alappuzha, Kerala, India [14]. The main difficulty confronting the trash business in India is the need to boost staffing at the level of collection. Separation of trash is a critical concern, and this problem continues an uphill struggle. Waste separation, especially at the home level, is unfortunately still lacking. Economic and user-friendly technical solutions must be used by the industry. The table on the next page summarizes the most important SEM solutions in a number of different sectors [12] (Table 1).

4 Conclusion

Three findings should be mentioned in the research, as stated in the results. Following the performing of the Energy Preservation Act, 2001, several legislative and regulatory measures have been put in place to assist with the SEM in India. Jawaharlal Nehru national solar mission (JNNSM) and the smart metropolises assignment have played a crucial part in revolutionizing India's energy management environment, bringing the country to the forefront of the energy industry's development [15]. To begin, the numerous stakeholders engaged in the Australia–India Information Altercation Plant in India and a full prose analysis allow distinct sectors to be highlighted in the complex world of search engine marketing. Additional issues are further compounded by the fact that certain previously strategy, executive, socio atmosphere, skill, budget, and law concerns are already in place. As a result of these issues, resolution will also assist to speed the proliferation of SEM projects and performs in India. Third, there are several activities aimed at addressing the issues caused by SEM. Governing governments should begin by following these rules, which are not intended to be comprehensive or prescriptive. A helpful resource for policymakers

Table 1 Smart energy management solutions in various fields

Buildings	Water	Transport	Waste	Public services
• Connections to smart grid • Improving air conditioning system • Improve lightning systems and controls • Improve heat recovery and storage • Hybrid ventilation systems • High efficiency generators • Solar active solutions	• Real-time data acquisition and monitoring • Smart metering • Leak detection • Efficient pumping system • İntegrated operating system • Power control and security systems integration • Water treatment and reuse facilities	• EV charging infrastructure and supervision services • Shift vehicle technology—EVs and HEVs • İntelligent traffic management systems • Tolling and congestion charging • İntegrated mobility, multimodal and shared transport • İmprove transportation infrastructure	• Waste to energy conversion • Intelligent monitoring of general waste • Smart dustbins • Dynamic management and routing • Smart recycling facilities • Biological and advanced thermal treatment • İndustrial heat recovery • Solar integration	• Smart sensors • Public safety • Video surveillance • Digital city services • Public street lighting management • Smart district heating and cooling • Automated distributed control • Renewable and distributed energy generators • Electrical and thermal energy storage

is provided in this report, especially for those who are tasked with creating SEM policies and finally advocating emission reduction objectives.

References

1. IEA. India Energy Outlook; International Energy Agency (OECD/IEA): Paris, France (2015)
2. UN DESA. United Nations, Department of Economic and Social Affairs, Population Division (2018)
3. Danish, M.S.S., Sabory, N.R., Ershad, A.M., Danish, S.M.S., Yona, A., Senjyu, T.: Sustainable architecture and urban planning trough exploitation of renewable energy. Int. J. Sustain. Green Energy **6**, 1–7 (2017)
4. Cajot, S., Peter, M., Bahu, J.M., Koch, A., Maréchal, F.: Energy planning in the urban context: Challenges and perspectives. Energy Procedia **78**, 3366–3371 (2015)
5. Energy Lens. The What, Why, and How of Energy Management (2018)
6. Calvillo, C.F., Sánchez-Miralles, A., Villar, J.: Energy management and planning in smart cities. Renew. Sustain. Energy Rev. **55**, 273–287 (2016)
7. Rattani, V.: Coping with climate change: An analysis of India's National Action plan on climate change. In:Mathur, T. (ed.) Centre for Science and Environment (CSE). New Delhi, India (2018)
8. AEEE. The Need for a National Cooling Action Plan (2018)
9. Mosannenzadeh, F., Bisello, A., Vaccaro, R., D'Alonzo, V., Hunter, G.W., Vettorato, D.: Smart energy city development: A story told by urban planners. Cities **64**, 54–65 (2017)
10. Danish, M.S.S., Senjyu, T., Ibrahimi, A.M., Ahmadi, M.: Abdul Motin Howlader A managed framework for energy-efficient building. J. Build. Eng. **21**, 120–128 (2019)

11. Development Asia. Sustainable Water Management for Smart Cities (2017)
12. Sambyal, S.S., Agarwal, R.: Is Swachh Bharat mission ensuring waste segregation systems? (2018)
13. EMBARQ Network. Four Cities' Solutions to Sustainable Garbage Processing (2019)
14. Sambyal, S.S.: Waste Smart Cities (2016)
15. Yenneti, K.: The grid-connected solar energy in India: Structures and challenges. Energy Strategy Rev. **11**, 41–51 (2016)

Stand-Alone Bidirectional Encoder Representations from Transformer-Based Fake News Detection Model

Eduri Raja, Badal Soni, Shivangi Mishra, and Ashish Dwivedi

Abstract Fake news is the kind of news that is made deliberately to deceive the readers. It is a sort of purposeful publicity which is distributed as veritable information. Fake news is spread out through traditional news media and social media. Fake news has been an issue for quite a while. With the introduction of social media, the spread of fake news is increased, and it got hard to separate between true news and fake news. The spread of fake news involves worry as it manipulates public opinions. The widespread of fake news can adversely affect people and society in general. The problem of fake news has increased significantly in recent years. The scope and impact of social media have shifted dramatically. On the one hand, it is low cost and simple accessibility, as well as the ability to quickly exchange information attract people to read news from it. On the other hand, it empowers widespread fake news, which is only false data to deceive people. A stand-alone bidirectional encoder representations from transformers model is used in this paper for fake news detection. We try to pose the problem as a text classification problem and build a deep learning model for achieving the objective. We used the BERT model to develop fake news or real news classification solution for our solution in this model. We achieved more than 95% accuracy on the test set and a remarkable by the area under the curve (AUC) by a stand-alone BERT model.

Keywords News detection · BERT · Deep learning · Machine learning · Social media · Transformer · NLP

E. Raja (✉) · B. Soni
National Institue of Technology, Silchar, India
e-mail: eduri_rs@cse.nits.ac.in

B. Soni
e-mail: badal@cse.nits.ac.in

S. Mishra · A. Dwivedi
United Institute of Technology, Allahabad, India

K. N. Das et al. (eds.), *Proceedings of the International Conference on Computational Intelligence and Sustainable Technologies*, Algorithms for Intelligent Systems,
https://doi.org/10.1007/978-981-16-6893-7_57

1 Introduction

Nowadays, information is freely accessible to everyone. For spreading false news, social media is used rapidly. The Internet provides an enormous measure of data; however, data's believability relies on many factors: the authenticity of the news, the author's credibility, etc. Every day in online and print media, information is published, yet it is difficult to tell whether it is real or fake. It requires an in-depth analysis of the data, including sources of the data, etc. The fabricated data is a purposeful attempt to damage or favor particular organizations, individuals, reputations, or be entirely beneficial financially or politically. Fake news [1] means fabricated data, which deceives people. Recently, in Indian elections campaigns, countless created posts and fake news articles and manipulated photographs circulated on social media.

The spread of false news has resulted in a loss of trust in the news media and political world. Fake news has an impact on people's choice about who to vote for in elections. With the help of social media bots, fake news spread quickly. A social bot is a social media account that has been designed to post information and interact with humans and other bots. These bots influenced significantly online election discussions, according to researchers. Fake news obstructs serious media coverage and makes reporting important news events more difficult for journalists. It is a challenging task to combat fake news. It is no easy task to determine whether a news story is fake by examining the veracity of each piece of information because the reality of the facts occurs on a continuum and is strongly reliant on the nuances of human language, which are tough to comprehend in true/false dichotomies. Sloppily written information with grammatical errors may indicate that the article was not authored by a journalist and is therefore likely inaccurate. Fake news can be published or broadcast by an unknown media organization or publication. These elements, however, do not provide certainty; thus, definitions and types of fake news must be properly understood and classified.

2 Related Work

2.1 Text Classification and Fake News Detection

Text classification [2] is the sub-task of natural language processing [3]. It has evolved from rule-based to static-based techniques, with machine learning [4] and deep learning [5] being used in most studies nowadays. Text classification includes the problem of detecting fake news as one of its subtasks. Fake news detection is the task of categorizing news, whether it is fake or real. Fake news will be news that is intentionally and undoubtedly false, misleading viewers and readers. This definition has two important components: "intention" and "credibility." First, fake news disseminates false information. Second, fake news is made with the misleading objective to deceive the public. Fake news is often divided into three categories. The first category is actual fake news, which the article's author creates. The second type of false news is satire; it is created solely to entertain rather than deceive its readers. Consequently, purposefully deceptive and deceptive fake news is not quite the same as clear parody or satire. "Poorly reported news that serves an agenda" is the third category. This type of data contains some true information, but it is not accurate or well-planned, and it is primarily used for political purposes. For text classification, we can use machine learning and deep learning algorithms also. Some deep learning algorithms are CNN [6], RNN [7], LSTM, etc., but the BERT [8] model is better than these models. Text-dependent speaker verification is given in [21], and multiple forgery detection in images has given in [22].

2.2 Pretrained Model and BERT

BERT abbreviated as bidirectional encoder representations from transformers [9]. It is a deep learning framework developed by Google. BERT is used in natural language processing (NLP). In BERT, B stands for bidirectional, this means that it learns information from both sides, i.e., right and left sides of a word or token in natural language processing. It is more efficient to understand the context. Next, ER stands for encoder representations, and this means that a program or algorithm is used to learn a representation from a set of data to an encoder. In BERT's [10], model is having a large set of data corpus, and the data set is enormous, drawing from Wikipedia [11] (2500 million words) and Google's book corpus (800 million words). In the pretraining [12] phase, a huge number of words are used for training the model. It means that BERT has developed for understanding how language works, and it is a handy tool in NLP. Next, T stands for transformer [13], and it means that the BERT [14] framework is based on the transformer architecture. There are two main steps involved: Create a language model by pretraining it on a vast text data set. The second one is fine-tuning. Fine-tuning is used for models to accumulate without the need for excessive computing power.

3 Methodology

It is a challenging task to combat fake news. It is no easy task to determine whether a news story is fake by examining the veracity of each piece of information because the reality of the facts occurs on a continuum and is strongly reliant on the nuances of human language, which are tough to comprehend in true/false dichotomies. Sloppily written information with grammatical errors may indicate that the article was not authored by a journalist and is therefore likely inaccurate. Fake news can be published or broadcast by an unknown media organization or publication. These elements, however, do not provide certainty; thus, definitions and types of fake news must be properly understood and classified.

In this paper, we use BERT model for fake news detection. It is a multilayer bidirectional transformer encoder based on Vaswani et al. [15] original version, which was released in the tensor2tensor library [16]. Because the usage of transformers has become widespread, we will forgo a thorough background study of the model architecture and refer readers to Vaswani et al. [15] as well as key resources such as "The Annotated Transformer." The original [17] is substantially identical to our implementation. L is for the number of layers (i.e., Transformer blocks), H stands for the concealed size, and A stands for the number of self-attention heads [18]. The two model sizes we usually report on are $\text{BERT}_{\text{BASE}}$ ($L = 12, H = 768, A = 12$, total parameters = 110 M) and $\text{BERT}_{\text{LARGE}}$ ($L = 24, H = 1024, A = 16$, total parameters = 340 M).

3.1 Dataset Descriptions

There are some datasets for detecting fake news that has been collected. In our model, the data are json files stored on the Web. Therefore, we have to download it and convert it into a panda's data frame. For our model, the dataset contains four files. Those are training—set of real news, testing—set of real news, Training—set of fake news, and testing—set of fake news. The **real news data** contains 800 rows and 3 columns for real news in the training set. For simplicity, we will only use the "text" column for modeling. If this model does not perform well, we can use multiple features like URLs. However, we achieved good performance by using text data alone only. The **fake news data** also contains 800 rows. So we can see that the number of real news and fake news are identical in our dataset. It will not be an imbalanced classification problem.

3.2 Data Preprocessing

We labeled our data as 0 (negative) for actual news and 1 (positive) for fake news during data preprocessing. We label fake news as positive because the primary purpose of the modeling is to detect fake news. We then remove non-alphanumeric characters as well as convert them to all lower case from the text. We use a train set to perform exploratory analysis. First, we want to look at the word count for each news and see the difference between real and fake news. Next, we like to see the most common words in real/fake news to discover some patterns. Word cloud is a popular way to visualize it.

Top*K* Word Proportion: We have to see the Top*K* word proportion of the real/fake news. In other words, we like to see how many of the words used in the news are from the top 10 common words, top 100, and so on. The reason to do so is that we suppose fake news is machine generated, and it uses many high-frequency words comparing to real news. Splitting Data to Train/Validation: We should split our training set into training and validation sets in an 8:2 ratio. This way, we can use the validation to tune our model and finally predict the hold-out test set.

3.3 *Long Document Preprocessing*

The BERT algorithm can only accept sentence length up to 512 words; we need to preprocess our data (long news) to feed the algorithm. Segment each of the text into multiple subtexts of no longer than 150 words. The subtexts will have some overlapping; specifically, the last 30 words for the first subtext will be the first 30 words of the second subtext. A lengthy document is split into a list of multiple subtexts. Next, we have to take our original data into a larger dataset where each row contains a subtext and its corresponding label and index. After applying the long document preprocessing, the training set now has 5621 subtexts extracted from 1280 texts. The validation set now has 1305 subtexts extracted from 320 texts. The testing set now has 1568 subtexts extracted from 400 texts.

4 Experimental Result Discussions and Analysis

4.1 *Building BERT Model*

The batch size in the training procedure is set to 16, and the maximum input sequence length is 150. The total number of train epochs is 5, with each epoch including 352 steps. Learning rate of optimizer (lr) is set to 1e-5. In this BERT classification model, the dropout layer is used to prevent over fitting, and the dense layer is used to predict probability. First, we like to load the pretrained weight of BERT and fine-tune it. Next, we preprocess our original text into input features BERT [19] can read. The process is tokenizing and converting our original text into token ids that the algorithm can read. The words are tokenized based on the vocabulary dictionary it pretrained on (about 30,000 words), and unknown words are broken down into smaller words in the dictionary. Maximum sequence length is also specified so we can pad all sequences into the same length. Note: The final sequence length would be more significant than specified since the BERT tokenizer [20] will break unknown words into multiple small known words.

Table 1 describes that the total number of parameters is 108,891,650, and the trainable parameters are 108,891,650, and the non-trainable parameters are zero.

Table 1 BERT classification model

Layer (Type)	Output shape	Param #
input_ids (InputLayer)	(None, 319)	0
bert (BertModelLayer)	(None, 319, 768)	108890112
lambda (Lambda)	(None, 768)	0
dropout (Dropout)	(None, 768)	0
dense (Dense)	(None, 2)	1538

4.2 Model Training

When we train our model, validation sets are used to find the optimal training epochs. The model will stop training when validation accuracy does not improve for a number of epochs. Here, ***epoch*** means it is one complete pass through training data. Then, the model will restore weights to its best validation performance. The maximum number of training epochs in our model is set to 5, and the patience (number of non-improving epochs before the model stops) is set to 1. In model training, first epoch training loss is 0.5591, training accuracy is 0.7315, validation loss is 0.3903, and validation accuracy is 0.9188. In model training, second epoch training loss is 0.4074, training accuracy is 0.9018, validation loss is 0.3747, and validation accuracy is 0.9349. In model training, third epoch training loss is 0.3818, training accuracy is 0.9295, validation loss is 0.3578, and validation accuracy is 0.9540. In model training, fourth epoch training loss is 0.3611, training accuracy is 0.9516, validation loss is 0.3638, and validation accuracy is 0.9487. In the fourth epoch, validation accuracy is not improved. That is the reason we stopped the training.

4.3 Evaluation Observations

After the model is done training, we evaluate the training set, validation set, and test set. The metric used is accuracy. The training loss is 0.3516, and the training accuracy is 0.9596 in this model evaluation on the training set. The validation set the training loss is 0.3578, and the accuracy is 0.9540. The test set loss is 0.3658, and the accuracy is 0.9439.

We can see that we have about approximately 96.0% accuracy for the training set and about 95.40% accuracy for our validation set. It seems to be good. However, there are two more things we should do. The prediction is on the augmented test set, but we care more about the prediction on the original text. Therefore, for each original text, we should average the probability of each subtext and obtain a final prediction for that piece of news. Other metrics such as the area under the curve (AUC) to get a more thorough evaluation of our model. Figure 1 shows the characteristics for original test set, and Table 2 shows the metrics for the original test set.

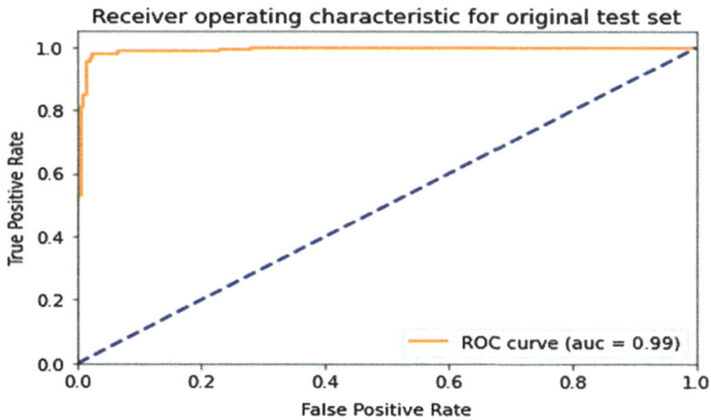

Fig. 1 Receiver operating characteristic (ROC) curve

Table 2 Experimental results

Accuracy	AUC (Area under the ROC curve)	Precision	Recall	F1 score
0.9725	0.9928	0.9655	0.98	0.9727

Predicting on a new text—"NEWS": The goal of the building model is to predict new coming instances. Below model is predicting the result of the input document.

Table 3 describes the total number of parameters is 108,891,650, and the number of trainable parameters is 108,891,650, and the number of non-trainable parameters is zero.

With an accuracy of 98.90%, our proposed model (stand-alone BERT) produced more accurate results. If we run an example text from the original test set, it will predict if the news is fake or real, as well as classify the predicted news label as fake or real.

Table 3 New trained model

Layer (Type)	Output shape	Param #
input_ids (Input Layer)	[(None, 319)]	0
bert (Bert Model Layer)	(None, 319, 768)	108890112
lambda_5 (Lambda)	(None, 768)	0
dropout_5 (Dropout)	(None, 768)	0
dense_5 (Dense)	(None, 2)	1538

5 Conclusions and Future Scope

A stand-alone BERT model for fake news detection is given in this paper. We try to pose the problem as a text classification problem and build a deep learning model for achieving the objective. We used the BERT model to develop fake news or real news classification solution for our solution in this model. We achieved more than 95% accuracy on the test set and a remarkable by the area under the curve (AUC) by a stand-alone BERT model. The performance of the proposed model (stand-alone BERT) for detecting fake news in this study is satisfactory. This proposed model is based on a pretrained word embedding model that uses a bidirectional transformer encoder (BERT). The findings of the BERT classification show that stand-alone BERT produces more accurate results, with an accuracy of 98.90%. With real-world fake news datasets, stand-alone BERT outperforms current state-of-the-art models in terms of accuracy. In future work, we will design a fake news detection model using advanced transformer models. Because using transformer models, we can get accurate results than machine learning and deep learning methods.

References

1. Conroy, N.J., Rubin, V.L., Chen, Y.: Automatic deception detection: methods for finding fake news. In: Proceedings of the 78th ASIS&T Annual Meeting: Information Science with Impact: Research in and for the Community, pp. 82. American Society for Information Science (2015)
2. Wang, B., Liu, K., Zhao, J.: Inner attention based recurrent neural networks for answer selection. In: Proceedings of the 54th Annual Meeting of the Association for Computational Linguistics (ACL), vol. 1, pp. 1288–1297 (2016)
3. Dinu, L.P., Iuga, I.: The best feature of the set. In: Gelbukh, A. (ed.) Computational Linguistics and Intelligent Text Processing. CICLing (2012)
4. Tewari, A.: Fake news detection using machine learning. Int. J. Res. Appl. Sci. Eng. Technol. **9**, 1476–1479 (2021). https://doi.org/10.22214/ijraset.2021.33900
5. Kong, S.H., Tan, L.M., Gan, K.H., Samsudin, N.H.: Fake news detection using deep learning. In: 2020 IEEE 10th Symposium on Computer Applications & Industrial Electronics (ISCAIE), pp. 102–107 (2020). https://doi.org/10.1109/ISCAIE47305.2020.9108841
6. Liu, Y., Wu, Y.F.B.: Early detection of fake news on social media through propagation path classification with recurrent and convolutional networks. In: Thirty-Second AAAI Conference on Artificial Intelligence (2018)
7. https://medium.com/engineering-zemoso/text-classification-bert-vs-dnn-b226497c9de7
8. Moshkin V., Konstantinov A., Yarushkina N.: Application of the BERT Language Model for Sentiment Analysis of Social Network Posts. In: Kuznetsov S.O., Panov A.I., Yakovlev K.S. (eds.) Artificial Intelligence. RCAI (2020)
9. Devlin, J., Chang, M.-W., Lee, K., Toutanova, K.: BERT: Pre-training of deep bidirectional transformers for language understanding. arXiv preprint arXiv: 1810.04805 (2019)
10. Horev, R.: BERT explained: State of the art language model for NLP. https://towardsdatascience.com/bert-explained-state-of-the-art-language-model-for-nlp-f8b21a9b6270. Accessed 21 Apr 2020
11. https://blog.floydhub.com/knowledge-distillation/

12. Devlin, J., Chang, M.W., Lee, K., Toutanova, K.: BERT: pre-training of deep bidirectional transformers for language understanding. In: Proceedings of the 2019 Conference of the North American Chapter of the Association for Computational Linguistics: Human Language Technologies (NAACL-HLT) (2019)
13. https://huggingface.co/transformers/model_doc/bert.html
14. http://jalammar.github.io/illustrated-bert/
15. Vaswani, A., et al.: Attention is all you need. In: Advances in Neural Information Processing Systems (NIPS), pp. 5998–6008 (2017)
16. https://github.com/tensorflow/tensor2tensor
17. http://nlp.seas.harvard.edu/2018/04/03/attention.html
18. https://lilianweng.github.io/lil-log/2020/04/07/the-transformer-family.html
19. Jwa, H., Oh, D., Park, K., Kang, J., Lim, H.: exBAKE: Automatic fake news detection model based on bidirectional encoder representations from transformers (BERT). Appl. Sci. **9**, 4062 (2019). https://doi.org/10.3390/app9194062
20. https://albertauyeung.github.io/2020/06/19/bert-tokenization.html
21. Soni, B., Debnath, S., Das, P.K.: Text-dependent speaker verification using classical LBG, adaptive LBG and FCM vector quantization. Int. J. Speech Technol. **19**, 525–536 (2016). https://doi.org/10.1007/s10772-016-9346-4
22. Soni, B., Das, P.K., Thounaojam, D.M.: MultiCMFD: Fast and efficient system for multiple copy-move forgeries detection in image. In Proceedings of the 2018 International Conference on Image and Graphics Processing (ICIGP 2018), pp. 53–58. Association for Computing Machinery, New York, NY, USA (2018). https://doi.org/10.1145/3191442.3191465

A Tactical Analysis of SSPT Topology under Entanglement with Smart Grid Renewal Energy System

Shivam Sharma, Gagandeep Kaur, and Naveen K. Sharma

Abstract Power conversion devices are becoming popular for domestic as well Industrial Applications. Smart controlling ability and high-power output make conversion units very friendly and cost-effective at the consumer end. All power conversion modules create power quality issues both at the consumer and utility end. Solid-State Power Transformer answers every power quality issue created by non-linear loads and also provides a good integration facility with renewable energy resources. This paper provides different topologies of SSPT with their integration with renewable sources and power quality issues generated by individual modules. Later in the discussion, a proposed model is addressed which provides a good entanglement of the renewable energy system with domestic appliances in both on-grid and island mode of operation. This paper will also provide a list of application areas where SSPT is utilized.

Keywords Solid-State Semi-conducting Switches (SSSD) · Duty Modulation Power Supply (DMPS) · Conventional Line Frequency Transformer (CLFT) · Solid-State Power Transformer (SSPT)

1 Introduction

Use of conventional line frequency transformer in high-power application area is deeply embedded in the main frame system. Apart from limitation still huge block of research and manufacturing is diverted toward the conventional transformer. Mostly core design and material are major part of research in medium frequency transformer. From last couple of decades, research community view point is shifted toward the utilization of SSSD based converter units which can work as CLFT in almost all

S. Sharma (✉) · G. Kaur · N. K. Sharma
Department of Electrical Engineering, I.K. Gujral Punjab Technical University, Main Campus, Kapurthala, India

G. Kaur
e-mail: gaganee@ptu.ac.in

K. N. Das et al. (eds.), *Proceedings of the International Conference on Computational Intelligence and Sustainable Technologies*, Algorithms for Intelligent Systems,
https://doi.org/10.1007/978-981-16-6893-7_58

applicable area where CLFT is presently installed. Our focus will be toward the design constraints of all SSPT topologies including design constraint, application area and modules designed both commercially and institutional research.

Presently most of the consumer-based appliances are either completely or partially SSSD dependent, which results in large bandwidth of distortion inserted in power grid parameters [1, 2]. Not only the power quality parameters are affected by the used of non-linear loads but also affects the security and stability of On-going grid system [3, 4]. Apart from creating distortion in wave-structure of grid parameters, in comparison to linear loads non-linear loads are consumer friendly and highly efficient. Therefore, changing the nature of consumer end loads from non-linear to linear is out of focus now. Replacing the CLFT with SSPT results in elimination of all the power quality issues and provides additional benefit of integration with renewable based energy sources [5].

Any entanglement of renewable sources with smart grid requires stable controlling system, assessable energy storage unit, highly efficient SSSD [6, 7]. Future of smart grid is not possible without the utilization of SSPT for integration with renewable sources which required to deal with features such as fault isolation, power quality control, on-line harmonic suppression and other function [8]. Conventional power transformers provide galvanic isolation enclosed with heavy solid metal core, which itself comes with many issues such as heavy weight, large volume, high initial cost and less efficient power controlling apparatus. With the elimination of conventional line transformer in the up-coming smart grids and energy Internet system, accessibility of power and other quality parameter will become easy and less expensive. Apart from benefits of SSPT integrating with renewable system and smart grids, because of SSSD used in the process working efficiency of SSPT is still less in comparison to the CLFT [9, 10]. It has been realized that with the replacement of conventional line frequency power transformer with SSPT results in increase of running efficiency in fault or no-fault condition will be more than 60% [10]. Approach of this paper will be to realize the efficiency model of different SSPT designs and their cost-saving model to understand the behavior of SSPT when entangled with renewable-smart grid system.

From last two decades, hot topic for research area in the field of grid renewable entanglement is solid-state power transformer. Most of the important feature of SSPT is to make sure that power quality parameters may not get affected by the variation in load nature and rating. In many commercial areas, SSPT is either used as unified power flow device or buffer between grid and load [11]. In [12, 13], work is pushed forward on the analysis of high-frequency galvanic transformer used in SSPT for the availability of galvanic isolation and turn ratio to increase the magnitude of load end voltage. Many renewable sources like wind-based energy system required buffer device which can smoothen the profile of voltage in AC-genre from rough and random values, this problem is analyzed and resolved by the use of SSPT [14]. For many traction applications SSPT is used and implemented till date either for extracting power from rolling stocks or transferring power to drive [15]. In many power backing systems, SSPT is used either as UPSC, UPS, or power management system to eliminate the slightest possibility of power quality issues [16]. Any device having AC in input section and AC in output section is a power transformer, with used of SSSD as basic blocks then device termed as SSPT.

In this paper detailed analysis of conversion units for different renewable energy sources are discussed in Sect. 2 along with issues faced by every conversion unit. In Sect. 3, different types of SSPT modules are discussed along with their application area and issues while integrating between utility and load. Power quality issues related to synchronization of different renewable sources with on-going electrical System is discussed in Sect. 4. Finally, conclusion is made for every module of SSPT on the bases of power quality issues and application areas.

2 Types of Conversion Units Used for Different Renewable Sources

The conversion units installed in renewable energy plant plays very important in the extraction of electrical energy from the renewable sources. Apart from solar-electric generating stations, all other renewable sources require mechanical system for the production of electrical energy. Main ingredient in mechanical system for the production of electrical energy is kinetic energy. Wind power plant, hydro power plant, geothermal power plant and bio-fuel power plant uses synchronous generator for the production of electrical power. The output of synchronous generator is always AC which can be synchronized with on-going grid power by proper selection of phase, voltage and frequency.

Wind power plant is made of large numbers of wind mills which gathers the kinetic energy of wind and convert it to the electrical power by using Doubly-fed Induction Generator (DFIG). Direct connection of DFIG with the mains supply results in the introduction of harmonics or disturbances in the main supply which finally decrease the power quality of on-going electrical power. Similarly, solar photovoltaic output follows the pattern of solar irradiance. If the intensity of sunlight is low then the output DC power from photovoltaic cells will be low. For the synchronization or connection of solar power plant with on-going grid supply, output of all the photovoltaic cells should follow the wave nature of grid supply with almost identical power parameters (voltage, frequency and phase). Figure 1 shows the basic block diagram of solar-electric energy system with entire power conditioning bocks. In solar power plant, we do require solid-state converter units which changes the nature of output power from solar photovoltaic (waved-DC) to AC power with proper value of power parameters for complete synchronization with on-going power grid.

For the conversion or conditioning of power parameters, we utilize four types of converting units. For better power conditioning either we use one of them or we use all of them form the proper conversion and conditioning:

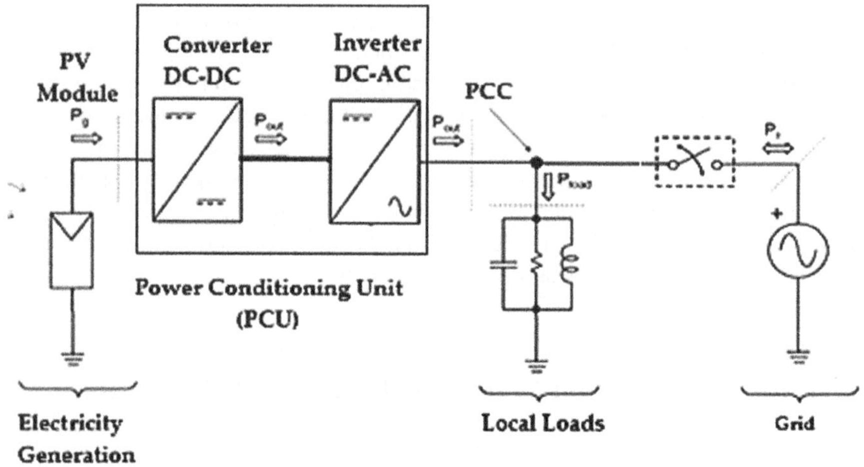

Fig. 1 Block diagram of solar power plant

2.1 AC-DC Converter Unit

This unit converts any periodic or alternating supply in to constant supply with single direction of current flow. This type of device mostly used where DC supply is mandatory like computers, LCD, etc. For highly efficient transmission of long-distance power, we use AC-DC units. Figure 2 shows the basic schematic of AC-DC converter system where sinusoidal supply is converter is DC supply with the use of solid-state devices (diode) and low-pass filter.

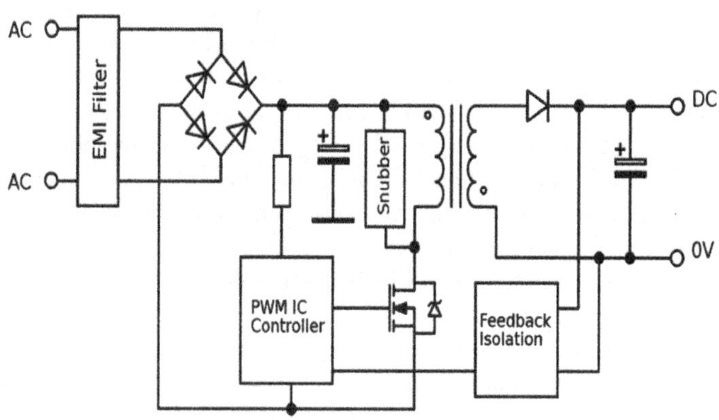

Fig. 2 Schematic of full-wave diode-bridge rectifier with parallel R–C shunt filter

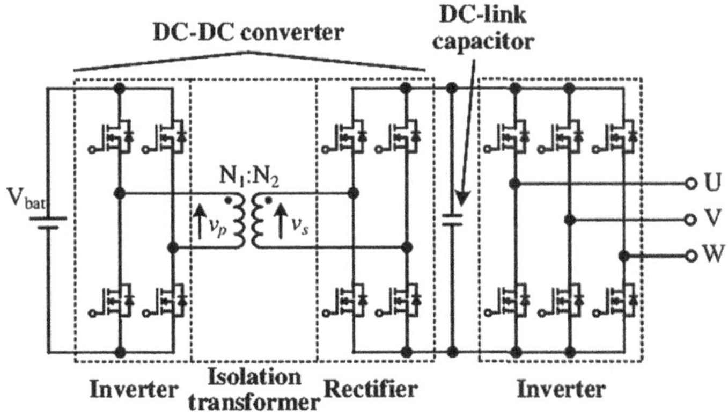

Fig. 3 Block schematic of DC-AC converter module

2.2 DC-DC Converter Module

Just like AC-DC converter, DC-DC converter works on single genre supply (DC supply) and only used for conditioning of parameters. In advanced solar power plants, DC-DC converter is used to change the waved-DC supply to constant DC supply for further conversion. As we discussed in introduction part of this chapter, the output power of solar photovoltaic follow the pattern of solar irradiance. This device maintains the constant voltage irrespective of sunlight intensity.

2.3 DC-AC Converter Module

DC-AC converter module is also known as inverter unit which basically changes the nature of input power and convert it to alternating power. In solar photovoltaic system after DC-DC converter module, we DC-AC module for the synchronization of solar based electric power with on-going grid power. Figure 3 shows the basic schematic of DC-AC converter module with least description of actual design.

2.4 AC-AC Converter Module

This module is also known to be AC-DC-AC converter module as it involves all the three modules of solid-state converters. This typical type of module is mostly used in the application where output parameters should be either same as input or different from input parameters. AC-AC converter module is used in wind power stations where it is installed with every single wind mill. The output of DFIG is provided to this module where conversion and conditioning of power take place.

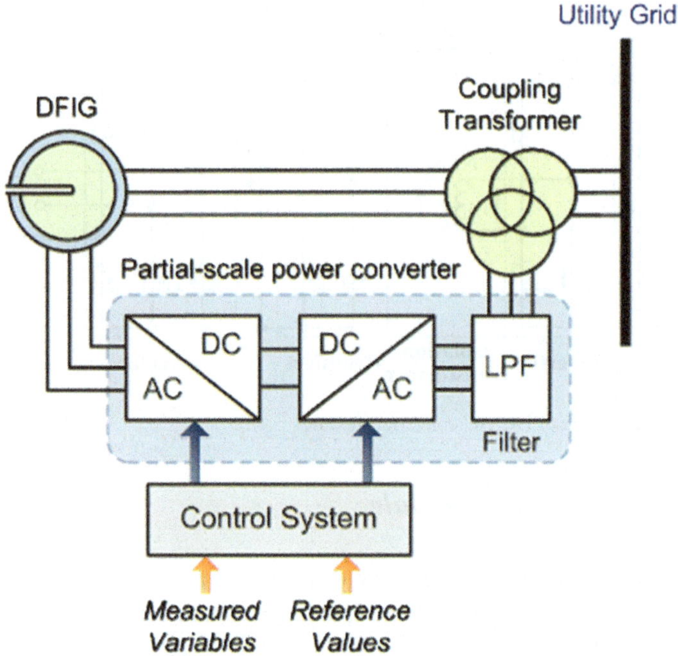

Fig. 4 Block diagram schematic of AC-AC converter module coupled with DFIG and on-going grid power including power transformer

Output of this module provides wide range of changeable power parameters for proper synchronization with on-going grid power. Generalized topology for AC-AC converter unit under renewable application area is shown in Fig. 4.

For the synchronization of renewable based electric power with on-going grid power, there are some necessary conditions:

- Voltage of both the sources should be identical
- Frequency of both the sources should be identical
- Phase-sequence of both the sources should be identical

With the use of DC-AC converter unit in solar power plant, we can achieve all the necessary conditions for the synchronization with on-going grid supply. Although achieving the above conditions exactly the way they meant to is completely identical. Difference between the desired and actual output from the inverter unit leads to the introduction of disturbances in the on-going grid supply.

3 Topologies Used for the Design of SSPT

This section will focus on the different topologies over which SSPT is designed for different application areas. Before dealing with recent advancements in the field of SSPT, thorough visit is necessary to understand basic building structure of solid-state power transformer. In some of the topologies, high frequency/moderate frequency transformer are utilized either for only isolation purpose or providing even turn ratio more than unity for enhanced voltage profile. Before understanding the behavior of every topology, effect of different transformer core must be acknowledged for every frequency domain. Table 1 shows the core material and frequency range including every citation provided in the SSPT

3.1 Single-Stage AC Buck Based SSPT

An American institute design first SSPT with no conversion of state and have no isolation between source and load. This results in limited application area with only variation in RMS values of voltage and current. On the other hand, this topology of SSPT delivers huge number of harmonics in load and source side which leads to less efficient functioning. Diagram describing the AC buck based SSPT topology is shown in Fig. 5a [12].

3.2 DMPS/Flyback Based SSPT

This topology of SSPT is almost similar to AC buck topology, the only difference is the no. of SSSD used and value of harmonics inserted both at load and source side. In this method, duty-cycle based modulation is provided to change the waveshape of voltage provided to source side transformer. Because of electromagnetic induction at the load side winding of transformer, it creates almost sinusoidal voltage waveshape at load side which leads to less indulgence of harmonics at load side parameters.

Table 1 Foray of literature related to material used in core of SSPT

Core Material Type	Frequency range						Quotation number
	a	*b*	*c*	*d*	*e*	*f*	
Silicon ferrite	1	1					[17, 18]
Amorphous	1	1					[19–22]
Nano-crystalline		1	1	1	1	1	[23–27]
Ferrite		1	1	1	1	1	[28–32]

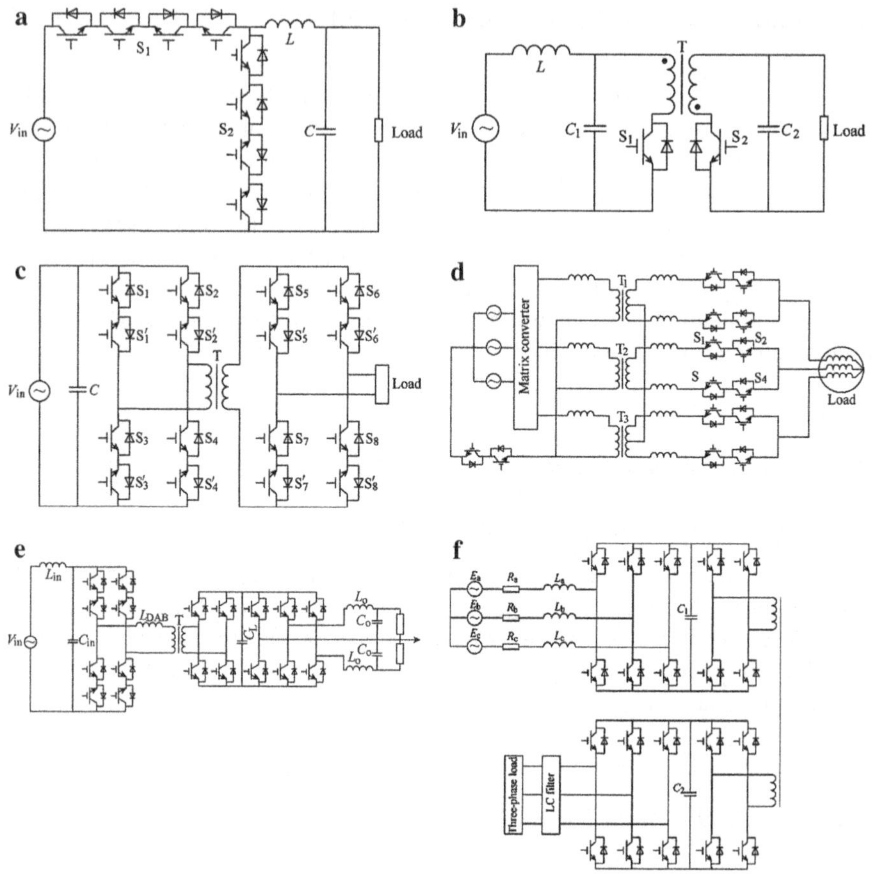

Fig. 5 **a** Single-Stage AC-AC buck based SSPT, **b** DMPS/Flyback based SSPT, **c** AC-AC with AC-link topology used in SSPT, **d** AC-AC with AC-link topology used in SSPT, **e** Polyphase Single-Stage Matrix based SSPT, **f** Two-Stage DC-Link Based Topology used for SSPT

Figure 5b shows the basic schematic of duty modulation power supply based SSPT [33].

3.3 AC-AC with AC-Link Based SSPT

In both previous topologies of SSPT, indulgence of harmonics both at load and source side is common which makes them less efficient in overall working state. If we eliminate the losses taken forward by SSSD at both conduction and switching state, we can manage to make an ideal model of SSSD with efficiency more than CLFT in changing load nature conditions. This step can only be accomplished by providing more than one changing stage in the SSSD module. AC-link based SSPC have either high frequency AC-link or medium frequency AC-link between two conversion stages. Both these stages are cyclo-conversion state which is designed to make sure that wave nature of both source and load current should not be changed. The only limitation of this topology, it cannot provide same efficiency results in different load natures. Figure 5c shows the basic schematic of AC-link SSPT [13].

3.4 Matrix Based SSPT

As name present, matrix based SSPT work with polyphase supply system and involves phase shifting of AC-parameters across load. Just like AC-link SSPT, matrix based SSPT also have only single conversion unit which is used to convert modular frequency signal to high frequency signal for reducing size of six winding transformer. Main benefit of using matric topology is usage at the time of unbalanced load. Figure 5d shows the basic schematic of matric SSPT topology along with phase shift section [34].

3.5 DC-Link Based SSPT

One of the advanced and reliable topologies of SSPT is DC-link topology. In this module, three-stage conversion is taken place with DC-link as controlling and stable stage. In this process, changing nature of load will not affect the waveshape of input current and voltage which makes this module very reliable and highly efficient in comparison to conventional line frequency transformer for changing load nature. Figure 5e shows the basic schematic of two-stage DC-link based topology used for SSPT [35].

3.6 Dual DC-Link Based SSPT

In order to solve the problem of linking SSPT with renewable and smart grid system, topology used in the SSPT must have reasonable working efficiency and high probability of eliminating all harmonica from the source and loads side beside of what load type is connected and to isolate the source-load system in the state of fault with minimal fault current values. Reason behind the name suggested is the involvement of two different voltage level DC-links which are isolated from each other by the mean of high frequency AC-link shown in the Fig. 5f [33].

4 Power Quality Issues Related to Synchronization of Different Renewable Sources with On-Going Electrical System

Electric power quality or power quality involves voltage, frequency and waveform. With the utilization of non-linear load, power quality also relates to the waveform of source current also. Good power quality can be defined as a steady supply voltage that stays within the prescribed range, steady AC-frequency close to the rated value and smooth voltage curve waveform (resembles a sine wave). In general, it is useful to consider power quality as the compatibility between what comes out of an electric outlet and the load that is plugged into it [39]. Table 2 shows the power quality issues provided by every SSPT module with their complexity, cost and conversion stages.

For the elimination of most dominating power quality issues provide by SSPT, designing of entire SSPT module for different application rea is necessary. Choosing single structure of dual DC-link SSPT module may eliminate most dominating power quality issues but it is still difficult to integrate renewable based energy system in-between SSPT module for domestic as well commercial purpose.

Table 2 Design approach of SSPT with power quality issues

SSPT Module	Stages	Complexity	Cost	Power Quality issues
Single-Stage AC buck based SSPT	AC-AC	Less	€	Source and load current wave-structure, even harmonics, notches, short surges, neutral voltage fluctuation
DMPS/Flyback based SSPS	AC-AC	Less	€	Source and load current wave-structure, even harmonics, notches, short surges
AC-AC with AC-link based SSPT	AC-AC	Moderate	€€	Variation in source-load current waveshape in different load conditions, source side notch
Matrix based SSPT	AC-AC	High	€€	Voltage fluctuation, notches, source current wave-structure, neutral voltage
DC-link based SSPT	AC-AC-DC-AC	High	€€€	Source current wave-structure, load current wave-structure in different load nature, notches
Dual DC-link based SSPT	AC-DC-AC-DC-AC	Very High	€€€	Notches, short surges

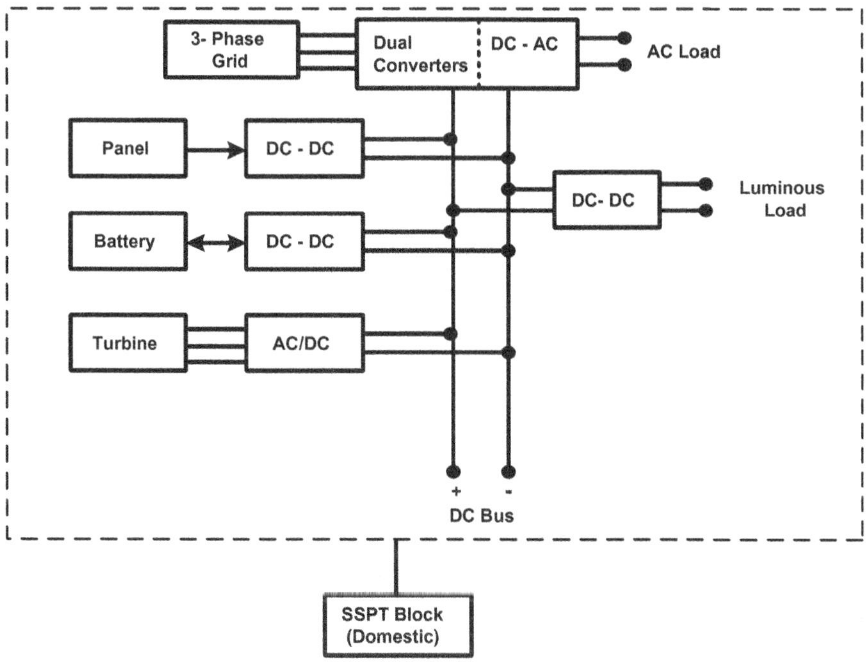

Fig. 6 Proposed model for domestic usage including all possible renewable energy resources

Figure 6 demonstrate the block representation of SSPT module in domestic application including 2 renewable based energy sources and one uninterrupted storage unit. This module is best suited for house hold usage and its main purpose will be to eliminate all power quality issues created by domestic appliances on main grid system. This module will also provide on-grid supply of power from renewable source to grid system for elimination of short coming load shading. All the conversion blocks are chosen with most sophisticated working mechanism to reduce in-house losses and increase overall module efficiency.

5 Conclusion

Fundamental motive behind designing of SSPT is to provide buffer between load and source, nature of load can be either of same genre as source or different. In this process most basic requirement is to reduce the switching loss and harmonic entanglement between load/source and SSPT module. Proposed model discussed in Sect. 4 will eliminate all the issues related to power quality in both load and source side. With this model any type of load can be connected with source and no indulgence of harmonica and other power quality issues. For better working and reduced cost/complexity, all the luminous load provided in domestic area are gathered

Table 3 Implementation areas of designed SSPT

Application Area	Quotation number
Railway traction	[21, 36]
Rolling stocks	[37]
EVs	[38]
Battery charger	[38, 39]
DC electric power system	[40]
Battery energy storage system	[41]
Grid connected renewable energy system	[42]
High-frequency power distribution system	[43]
Adjustable speed drive system	[17]
Airborne passenger system	[44]
Ship propulsion system	[45]
Wind energy extraction system	[46]
Solar energy extraction system	[47]

at one place with only DC supply provided. In this manner additional conversion units came with every luminous appliance will be eliminated and will reduce burden of eliminating harmonics from SSPT module. Table 3 shows the utilization area of SSPT module in every sector of industrial, commercial and domestic applications.

References

1. Ramos-Carranza, H.A., Medina-Ríos, A., Segundo, J., et al.: Suppression of parallel resonance and mitigation of harmonic distortion through shunt active power compensation. Int. J. Electr. Power Energy Syst. **75**, 152–161 (2016)
2. Farokhnia, N., Fathi, S.H., Yousefpoor, N., et al.: Minimisation of total harmonic distortion in a cascaded multilevel inverter by regulating voltages of dc sources. IET Power Electron. **5**(1), 106–114 (2012)
3. Li, Z., Tao, Y., Abu-Siada, A., et al.: A new vibration testing platform for electronic current transformers. IEEE Trans. Instrum. Meas. **68**(3), 704–712 (2019)
4. Doweck, Y., Amar, A., Cohen, I.: Joint model order selection and parameter estimation of chirps with harmonic components. IEEE Trans. Signal Process. **63**(7), 1765–1778 (2015)
5. Dannier, A., Rizzo, R.: An overview of power electronic transformer: Control strategies and topologies. In: International Symposium on Power Electronics Power Electronics, Electrical Drives, Automation and Motion, pp. 1552–1557 (2012). https://doi.org/10.1109/SPEEDAM.2012.6264573
6. She, X., Yu, X., Wang, F., Huang, A.Q.: Design and demonstration of a 3.6kVc120V/10kVA solid-state transformer for smart grid application. IEEE Trans. Power Electron. **29**(8), 3982–3996 (2014)
7. Yu, X., She, X., Zhou, X., Huang, A.Q.: Power management for DC microgrid enabled by solid-state transformer. IEEE Trans. Smart Grid **5**(2), 954–965 (2014)
8. Wang, D., Tian, J., Mao, C., Lu, J., Duan, Y., Qiu, J., Cai, H.: A 10kV/400V 500kVA electronic power transformer. IEEE Trans. Ind. Electron. **99**, 1–1 (2016)

9. Ronan, E.R., Sudhoff, S.D., Glover, S.F., Galloway, D.L.: A power electronic-based distribution transformer. IEEE Power Eng. Rev. **22**(3), 61–61 (2007)
10. Wang, D., Mao, C., Lu, J.: Modelling of electronic power transformer and its application to power system. IET Gener. Transm. Distrib. let **1**(6), 887–895 (2007)
11. Gu, C., Zheng, Z., Xu, L., Wang, K., Li, Y.: Modeling and control of a multiport power electronic transformer (PET) for electric traction applications. IEEE Trans. Power Electron. **31**(2), 915–927 (2015)
12. Ronan, E.R., Sudhoff, S.D., Glover, S.F., Galloway, D.L.: A power electronic-based distribution transformer. IEEE Trans. Power Deliv. **17**(2), 537–543 (2002)
13. She, X., Huang, A.Q., Burgos, R.: Review of solid-state transformer technologies and their application in power distribution systems. IEEE J. Emerg. Sel. Top. Power Electron. **1**(3), 186–198 (2013)
14. Hosseini, S.H., Sabahi, M., Goharrizi, A.Y.: Multi-function zero-voltage and zero-current switching phase shift modulation converter using a cycloconverter with bi-directional switches. IET Power Electron. **1**(2), 275–286 (2008)
15. Sabahi, M., Hosseini, S.H., Sharifian, M.B.B., Goharrizi, A.Y., Gharehpetian, G.B.: A three-phase dimmable lighting system using a bidirectional power electronic transformer. IEEE Trans. Power Electron. **24**(3), 830–837 (2009)
16. Kang, M., Enjeti, P.N., Pitel, I.J.: Analysis and design of electronic transformers for electric power distribution system. IEEE Trans. Power Electron. **14**(6), 1133–1141 (2002)
17. Sandoval, J.J., Krishnamoorthy, H., Enjeti, P., Choi, S.: High power density adjustable speed drive topology with medium frequency transformer isolation. In: 2015 IEEE Energy Conversion Congress and Exposition (ECCE), pp. 6986–6992 (2015). https://doi.org/10.1109/ECCE.2015.7310638
18. Villar, I., Viscarret, U., Etxeberria-Otadui, I., Rufer, A.: Transient thermal model of a medium frequency power transformer. In: IEEE-34th Annual Conference of Industrial Electronics (IECON), pp. 1033–1038 (2008). https://doi.org/10.1109/IECON.2008.4758096
19. Villar, I., Viscarret, U., Etxeberria-Otadui, I., Rufer, A.: Global loss evaluation methods for nonsinusoidally fed medium-frequency power transformers. IEEE Trans. Industr. Electron. **56**(10), 4132–4140 (2009). https://doi.org/10.1109/TIE.2009.2021174
20. Hunziker, C., Schulz, N.: Potential of solid-state transformers for grid optimization in existing low-voltage grid environments. Electric Power Syst. Res. **146**, 124–131 (2017). https://doi.org/10.1016/j.epsr.2017.01.024
21. Villar, I., Mir, L., Etxeberria-Otadui, I., Colmenero, J., Agirre, X., Nieva, T.: Optimal design and experimental validation of a Medium Frequency 400kVA power transformer for railway traction applications. In: 2012 IEEE Energy Conversion Congress and Exposition (ECCE), pp. 684–690. https://doi.org/10.1109/ECCE.2012.6342754
22. Krishnamoorthy, H.S., Garg, P., Enjeti, P.N.: A new medium-voltage energy storage converter topology with medium-frequency transformer isolation. In: 2012 IEEE Energy Conversion Congress and Exposition (ECCE), pp. 3471- 3478. https://doi.org/10.1109/ECCE.2012.6342320
23. Kauder, T., Hameyer, K.: Iron loss comparison of standard SiFe and nanocrystalline materials for power transformers in a dual active bridge converter. In: 18th European Conference on Power Electronics and Applications (EPE'16 ECCE Europe) (2016). https://doi.org/10.1109/EPE.2016.7695371
24. Drofenik, U.: A 150kW medium frequency transformer optimized for maximum power density. In: 2012 7th International Conference on Integrated Power Electronics Systems (CIPS), pp. 1–6 (2012)
25. Farnesi, S., Marchesoni, M., Vaccaro, L.: Advances in locomotive power electronic systems directly fed through AC lines. In: 2016 International Symposium on Power Electronics, Electrical Drives, Automation and Motion (SPEEDAM), pp. 657–664. https://doi.org/10.1109/SPEEDAM.2016.7525932
26. Watson, A.J., Wheeler, P.W., Clare, J.C.: Field programmable gate array based control of dual active bridge DC/DC converter for the UNIFLEX-PM Project. IEEE-14th European Conference on Power Electronics and Applications (EPE), pp. 1–9 (2011)

27. Nikolov, G.T., Valchev, V. C.: Nanocrystalline magnetic materials versus ferrites in power electronics. In: The 6th International Conference on Mining Science & Technology Procedia Earth and Planetary Science, vol. 1, pp. 1357–1361 (2009). https://doi.org/10.1016/j.proeps.2009.09.209
28. Moonem, M.A., Pechacek, C.L., Hernandez, R., Krishnaswami, H.: Analysis of a multilevel dual active bridge (ML-DAB) DC-DC converter using symmetric modulation. Electronics **4**(2), 239–260 (2015). https://doi.org/10.3390/electronics4020239
29. Nakahara, M., Wada, K.: Analysis of hysteresis and eddy current losses for a medium-frequency transformer in anisolated DC-DC converter. In: IEEE International Power Electronics Conference (IPEC -Hiroshima 2014 - ECCE ASIA), pp. 2511–2516 (2014). https://doi.org/10.1109/IPEC.2014.6869942
30. Garcia-Bediaga, A., Villar, I., Etxeberria-Otadui, I., Barrade, P., Rufer, A.: Comparison of multi-phase SAB converters vs. multi-phase SRC converters. In: 2013 IEEE Energy Conversion Congress and Exposition, pp. 5448–5455. https://doi.org/10.1109/ECCE.2013.6647440
31. Muhlethaler, J., Biela, J., Kolar, J.W., Ecklebe, A.: Improved core-loss calculation for magnetic components employed in power electronic systems. IEEE Trans. Power Electron. **27**(2), 964–973 (2012). https://doi.org/10.1109/TPEL.2011.2162252
32. Soltanzadeh, K., Tavakoli, A., Arbab, P.B.: Investigation of DC-DC resonant converter constructed nanostructure magnetic core transformer for power distribution system. In: 2012 Proceedings of 17th Conference on Electrical Power Distribution (2012)
33. Manjrekar, M.D., Kieferndorf, R., Venkataramanan, G.: Power electronic transformers for utility applications. In: Conference Record of the Industry Applications Conference, vol.4, pp. 2496–2502 (2000)
34. Wheeler, P.W., Clare, J.C., Empringham, L., Bland, M.: Matrix converters. IEEE Ind. Appl. Mag. **10**(1), 59–65 (2002)
35. Falcones, S., Mao, X., Ayyanar, R.: Topology comparison for solid state transformer implementation. In: Power and Energy Society General Meeting, pp. 1–8 (2010)
36. Drofenik, U., Canales, F.: European trends and technologies in traction. In: IEEE-International Power Electronics Conference (IPEC), pp. 1043–1049 (2014). https://doi.org/10.1109/IPEC.2014.6869715
37. Parchomiuk, M.: Single phase voltage converter for traction applications—simulation and investigation. PRZEGLĄD ELEKTROTECHNICZNY (Electr. Rev.) **88**(1), 81–85 (2012)
38. Rosekeit, M., van der Broeck, C., De Doncker, R.W.: Dynamic control of a dual active bridge for bidirectional AC charging. In: 2015 IEEE International Conference on Industrial Technology (ICIT), pp. 2085- 2091. https://doi.org/10.1109/ICIT.2015.7125404
39. Villar, I., Iruretagoyena, U., Rujas, A., Garcia-Bediaga, A., Perez de Arenaza, I.: Design and implementation of a SiC based contactless battery charger for electric vehicles. In: 2015 IEEE Energy Conversion Congress and Exposition (ECCE). https://doi.org/10.1109/ECCE.2015.7309841
40. Ortiz, G., Biela, J., Kolar, J.W.: Optimized design of medium frequency transformers with high isolation requirements. In: IECON 2010—36th Annual Conference on IEEE Industrial Electronics Society, pp. 631–638. https://doi.org/10.1109/IECON.2010.5675240
41. Gao, N., Chen, Q., Li, R., Cai, X.: Multi-level cascaded medium frequency isolated power conversion system based on dual active bridge. In: 2014 International Power Electronics and Application Conference and Exposition. https://doi.org/10.1109/PEAC.2014.7038004
42. Islam, M.R., Lei, G., Guo, Y., Zhu, J.: Optimal design of high-frequency magnetic links for power converters used in grid-connected renewable energy systems. IEEE Trans. Magn. **50**(11) (Nov 2014). https://doi.org/10.1109/TMAG.2014.2329939
43. Iyer, K. V., Robbins, W. P., Mohan, N.: Winding design of a high power medium frequency transformer. In: 2014 International Symposium on Power Electronics, Electrical Drives, Automation and Motion, pp. 665–669. https://doi.org/10.1109/SPEEDAM.2014.6871956
44. Roboam, X., Sareni, B., Andrade, A.D.: More electricity in the air: Toward optimized electrical networks embedded in more-electrical aircraft. IEEE Ind. Electron. Mag. **6**(4), 6–17 (2012). https://doi.org/10.1109/MIE.2012.2221355

45. Gunther, W., Malbasic, Z.: Nanocrystalline cores for common mode current suppression in electrical ship propulsion systems—a case study. In: PCIM Europe 2018, International Exhibition and Conference for Power Electronics, Intelligent Motion, Renewable Energy and Energy Management, pp. 1146–1150 (2018)
46. Shen, W.: Design of high-density transformers for high-frequency high-power converters control and design of a high voltage solid state transformer and its ıntegration with renewable energy. (Doctoral dissertation, Virginia Polytechnic Institute and State University, 2006). 3232303, pp. 1–15 (2006)
47. Krishnamoorthy, H., Daniel, M., Jorge Ramos-Ruiz, J., Prasad Enjeti, P., Liming Liu, L., Aeloiza, E.: Isolated AC–DC converter using medium frequency transformer for off-shore wind turbine DC collection grid. IEEE Trans. Ind. Electron. **64**(11), 8939–8947 (Nov 2017). https://doi.org/10.1109/TIE.2017.2652405

Smart Student Participation Assessment Using Fuzzy Logic

Sudheer Kumar Nagothu, Pudota Bindu Sri, and Rajyalakshmi Koppolu

Abstract In the light of the ongoing COVID-19 epidemic, a lockdown has been instituted and academic calendars have been suspended for all educational systems. This obstacle has necessitated the development of novel teaching approaches, among which online classes play a significant role. As a result, teachers must employ a variety of strategies to encourage and motivate students to attend and participate actively in class. Thus, we must analyse students' participation in online classes and, through the use of certain factors, encourage them to connect with lecturers. We measure student engagement in this work by examining several input factors such as total length, poll votes, and chats. Student participation is classified into four levels based on input parameters: much active, active, satisfactory, and poor. With the proposed fuzzy model accuracy of 97% is achieved. The benefit of this proposed approach is that students will participate more actively in class and absorb more knowledge.

Keywords Student assessment · Student participation · Fuzzy inference system · MATLAB

1 Introduction

Nowadays, all educational institutions assign internal grades to students based on their academic attendance. As a result of this, pupils attend lessons without actively participating. The difficulty with evaluating student participation is that it cannot be dependent on the student's physical presence; even if attendance and participation are related, they may be distinct. Students' participation in online classes with the instructor can take numerous forms, including engagement, thought exchange, explanation of doubts, and discussion. Pupils' attendance has become mandatory, and as a result, students are engaging passively in classrooms without interaction with professors. If a student's participation is insufficient daily, the professors will

S. K. Nagothu (✉) · P. B. Sri · R. Koppolu
RVR&JC College of Engineering, Chowdavram, Guntur, Andhra Pradesh 522019, India
e-mail: nsudheerkumar@rvrjc.ac.in

K. N. Das et al. (eds.), *Proceedings of the International Conference on Computational Intelligence and Sustainable Technologies*, Algorithms for Intelligent Systems,
https://doi.org/10.1007/978-981-16-6893-7_59

warn him and he will be marked absent. Thus, participation assessment is necessary to ensure that students participate actively in class. This method benefits both students and teachers. Students will take ownership of their learning and increase their knowledge in the classroom by conversing and raising questions, and teachers will be able to address students' uncertainties.

Several earlier studies have used a GPS sensor to determine the presence on-site of students [1, 2]. The student's physical position is transmitted to the server, which compares it to the student's predetermined location [3, 4]. The person's presence will be recorded if the student is present in a predefined area at the allotted time [5]. Only attendance is recorded in this case, not the student's participation [6]. In this current research paper, we are evaluating the student's engagement using a variety of characteristics [7, 8]. Measuring participation is critical to ensuring that students participate actively in class to learn knowledge. A combination of fuzzy logic and IoT systems are used in various applications such as agriculture, smart homes, [9, 10].

2 Materials and Methods

In this research article, to evaluate participation of students, fuzzy logic inference is used, which is nothing more than a simulation of human decision-making. This option may involve the possibility of an intermediate value between 0 and 1, i.e. yes or no. The concept of fuzziness is introduced here, and membership functions are used to determine the fuzziness of fuzzy sets, i.e. whether the items in the fuzzy set are discrete or continuous. By experience, membership functions are employed to resolve real difficulties [11, 12].

The shape of the membership function is critical when solving a specific problem since it affects the fuzzy inference system. To quantify the input parameters in this research, we use a triangular membership function with each element mapped between 0 and 1. This triangular function can be used to determine the participation of students based on the supplied parameters. The primary advantage of fuzzy logic is that it may be reprogrammed in response to changing circumstances [13, 14]. Figure 1 illustrates a fuzzy model for the current system in MATLAB simulation.

2.1 Fuzzy Logic Inference Implementation

MATLAB simulation is used to implement the currently suggested system. Total time is the input parameter for student participation. The number of polled votes and chats is tallied. Data is collected at various time intervals and an excel sheet with input and output parameters is prepared.

To transform the original raw input into the desired output, a fuzzy inference system is used. In this research, scenarios such as low participation, satisfactory participation, and active participation are considered as outputs. To validate the given

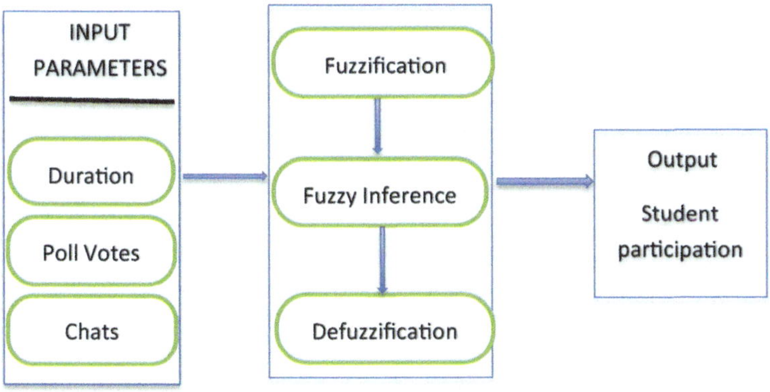

Fig. 1 Fuzzy inference system

inputs and the fuzzy inference model, triangular membership functions are employed. The current research employs the Mamdani engine model to implement the fuzzy inference model. A plot of the membership functions for each input parameter has been provided, including the overall duration, the number of votes polled, and the number of talks. After the input and output membership functions are set, the rules for evaluating student involvement will be constructed automatically. Fuzzy rules are formed through the use of the AND operator on input parameters to measure the output parameter which is student participation.

2.2 *Categorization of Input and Output Parameters*

The participation of students is quantified in this article through the use of input and output characteristics. It necessitates the monitoring of often seen data such as total duration, number of polled votes, and chats. The arguments are easily implemented using the triangle membership function. The estimation will also be quick in this case.

2.2.1 Measuring Total Duration

This parameter indicates the duration of a student's attendance in class. The class duration of 60 min is transformed to a scale of 1–10. If the duration is between 1 and3, participation is low; if the duration is between 4 and 7, participation is satisfactory; and if the duration is between 8 and 10, participation is active. The total duration is provided in tabular format in Table 1. Equations for triangular membership functions are provided. The following Fig. 2 illustrates the graphic representation.

Table 1 Total duration

S.no	Total duration	Category
1	1–3	Low
2	4–6	Satisfactory
3	7–10	Active

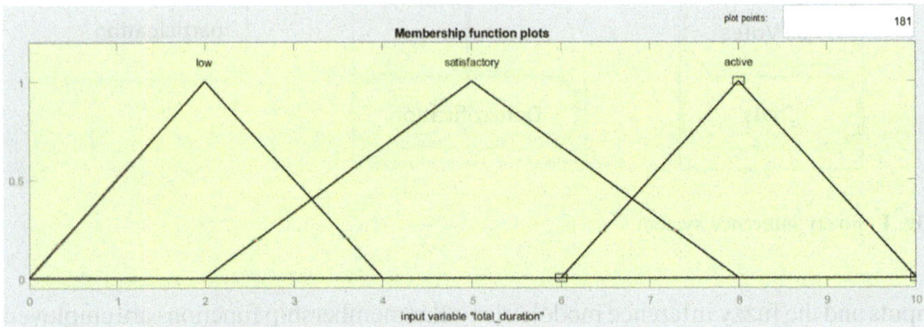

Fig. 2 Total duration membership function

$$\text{Total Duration}_{\text{low}(x)} = \begin{cases} \frac{x}{2}, 0 < x \leq 2 \\ \frac{4-x}{2}, 2 < x < 4 \\ 0, x \geq 4 \end{cases}$$

$$\text{Total Duration}_{\text{Satisfactory}(x)} = \begin{cases} 0, x \leq 2 \\ \frac{x-2}{3}, 2 < x \leq 5 \\ \frac{8-x}{3}, 5 < x < 8 \\ 0, x \geq 8 \end{cases}$$

$$\text{Total Duration}_{\text{high}(x)} = \begin{cases} 0, x \leq 6 \\ \frac{x-6}{2}, 6 < x \leq 8 \\ \frac{10-x}{2}, 8 < x < 10 \\ 0, x \geq 10 \end{cases}$$

2.2.2 Measuring No. of Votes Polled

This statistic measures the number of votes polled by students. The number of votes polled is scaled in the range of 1–10. If a student is not actively participating in polling votes, his or her participation will be poor. The tabular representation of the total number of votes cast is displayed in Table 2, along with triangular membership functions. Figure 3 shows a graphical representation.

Table 2 Number of votes polled

S.no	No. of votes polled	Category
1	0–3	Low
2	4–6	Satisfactory
3	7–10	Active

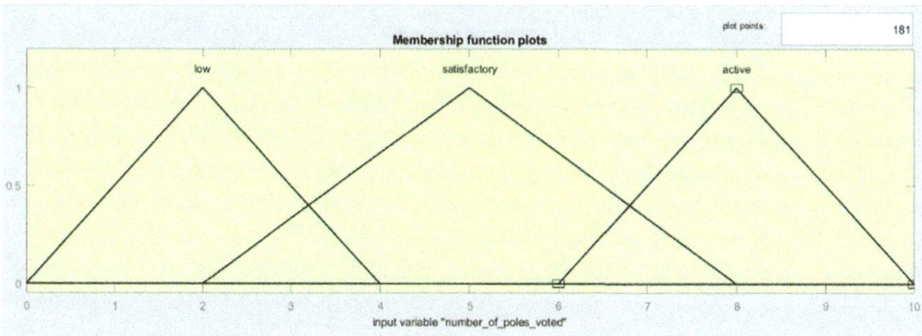

Fig. 3 Number of poles voted membership function

$$\text{Votes polled}_{\text{low}(x)} = \begin{cases} \frac{x}{2}, 0 < x \leq 2 \\ \frac{4-x}{2}, 2 < x < 4 \\ 0, x \geq 4 \end{cases}$$

$$\text{Votes polled}_{\text{Satisfactory}(x)} = \begin{cases} 0, x \leq 2 \\ \frac{x-2}{3}, 2 < x \leq 5 \\ \frac{8-x}{3}, 5 < x < 8 \\ 0, x \geq 8 \end{cases}$$

$$\text{Votes polled}_{\text{high}(x)} = \begin{cases} 0, x \leq 6 \\ \frac{x-6}{2}, 6 < x \leq 8 \\ \frac{10-x}{2}, 8 < x < 10 \\ 0, x \geq 10 \end{cases}$$

2.2.3 Measuring Chats

This parameter counts the number of talks in which students participate actively. The number of chats students involved is scaled in the range of 1–10. The tabular representation of the measured chats is provided in Table 3, together with triangular membership function formulae. Figure 4 illustrates the concept graphically.

Table 3 Number of chats

S.no	Number of chats	Category
1	0–3	Low
2	4–6	Satisfactory
3	7–10	High

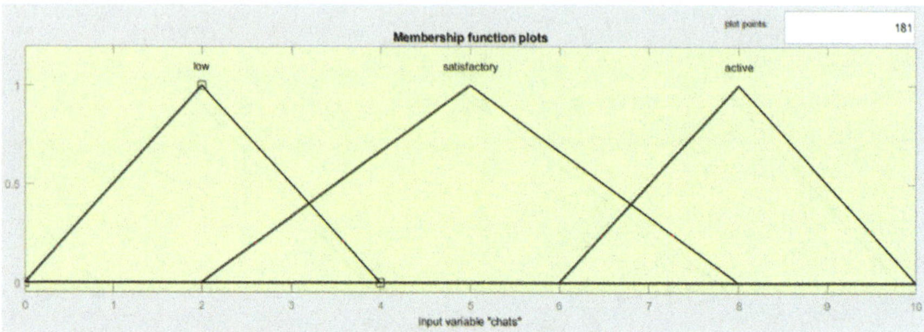

Fig. 4 Number of chats membership function

$$\text{Number of chats}_{\text{low}(x)} = \begin{cases} \frac{x}{2}, 0 < x \leq 2 \\ \frac{4-x}{2}, 2 < x < 4 \\ 0, x \geq 4 \end{cases}$$

$$\text{Number of chats}_{\text{Satisfactory}(x)} = \begin{cases} 0, x \leq 2 \\ \frac{x-2}{3}, 2 < x \leq 5 \\ \frac{8-x}{3}, 5 < x < 8 \\ 0, x \geq 8 \end{cases}$$

$$\text{Number of chats}_{\text{high}(x)} = \begin{cases} 0, x \leq 6 \\ \frac{x-6}{2}, 6 < x \leq 8 \\ \frac{10-x}{2}, 8 < x < 10 \\ 0, x \geq 10 \end{cases}$$

2.3 Output Actions

The suggested system's output parameter is an estimate of student participation. The input parameters are the total duration, the number of polled votes, and the number of talks. When the input values vary, the participation index is calculated using four degrees of activity: poor, satisfactory, active, and highly active. Student Participation Evaluation with categories is given in Table 4. The Student Participation Membership function is shown in Fig. 5.

Table 4 Student participation evaluation with categories

Participation Evaluation values	Range label
0–0.39	Poor
0.4–0.59	Satisfactory
0.6–0.89	active
0.9–1	highly active

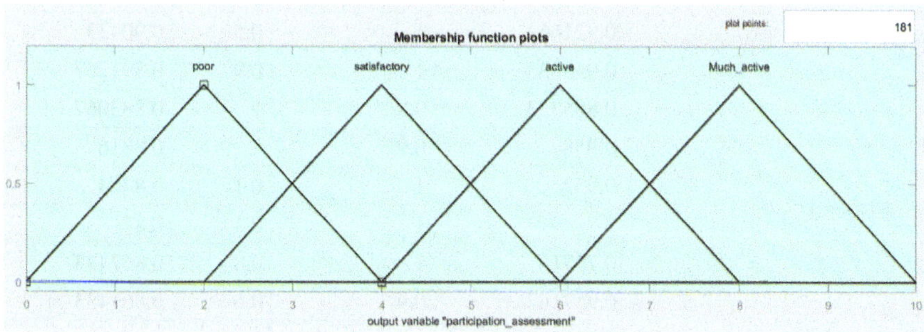

Fig. 5 Student Participation Membership function

$$\text{Student Participation}_{\text{low}(x)} = \begin{cases} \frac{x}{2}, 0 < x \leq 2 \\ \frac{4-x}{2}, 2 < x < 4 \\ 0, x \geq 4 \end{cases}$$

$$\text{Student Participation}_{\text{Satisfactory}(x)} = \begin{cases} 0, x \leq 2 \\ \frac{x-2}{2}, 2 < x \leq 4 \\ \frac{6-x}{2}, 4 < x < 6 \\ 0, x \geq 6 \end{cases}$$

$$\text{Student Participation}_{\text{Active}(x)} = \begin{cases} 0, x \leq 4 \\ \frac{x-4}{2}, 4 < x \leq 6 \\ \frac{8-x}{2}, 6 < x < 8 \\ 0, x \geq 8 \end{cases}$$

$$\text{Student Participation}_{\text{Much Active}(x)} = \begin{cases} 0, x \leq 6 \\ \frac{x-6}{2}, 6 < x \leq 8 \\ \frac{10-x}{2}, 8 < x < 10 \\ 0, x \geq 10 \end{cases}$$

Table 5 contains sample data for the input and output parameters. The following diagram illustrates the fuzzy logic rule set for input parameters and output actions. MATLAB's ruler editor is used to merge fuzzy rules into a fuzzy inference system. The Mamdani inference system is used in this research work. It is used to anticipate

Table 5 Sample input data sets and participation output

Input values			Output value
Total duration student present	Votes polled	Chats	Participation assessment of student
1	1	0.8	0.96
1	1	0.6	0.95
0.9341	1	0.7	0.916833
0.8795	1	0.7	0.9159
0.92166	1	0.5	0.90433
0.958833	1	0.9	0.891767
0.865333	0.8	1	0.883067
0.958	0.9	1	0.8816
0.952	1	0.1	0.8704
1	1	0.4	0.87
0.73716	1	0.8	0.867433
0.957167	0.9	0.6	0.861433
0.9905	1	1	0.85196
1	1	0.6	0.85
1	1	0	0.85
0.912	1	0.8	0.8424
0.985	1	0.9	0.84331
0.94966	1	0.6	0.826933
0.93166	1	1	0.824633
0.82166	0.6	1	0.822533

the presence of four input parameters and a single output parameter. Output is generated using the logic AND operation of input parameters. Table 6 contains fuzzy logic rule sets.

3 Simulation and Results

Figure 6 depicts a simulation of assessing student participation using a fuzzy logic inference method. The total duration, the number of polled votes, and the number of chats are used as input parameters. Students' participation is measured by altering the input parameters. Here, we are utilizing MATLAB and providing input data for the suggested student participation assessment. Data is gathered throughout class time, and the output is seen and compared to the rule base's intended values. Figure 7 illustrates the interface for examining fuzzy rules. Figure 8 illustrates verbal fuzzy rules.

Table 6 Fuzzy logic rule sets

Total duration	No. of votes	Chats	Participation assessment
Low	Low	Satisfactory	Low
Low	Satisfactory	Satisfactory	Satisfactory
Low	Low	Active	Satisfactory
Low	Satisfactory	Active	Active
Satisfactory	Satisfactory	Satisfactory	Satisfactory
Low	Active	Active	Very active
Active	Active	Active	Very active
Active	Satisfactory	Active	Very active
Low	Active	Satisfactory	Satisfactory
Active	Low	Low	Low
Low	Low	Low	Low
Satisfactory	Active	Low	Active
Active	Active	Active	Very active
Low	Active	Low	Satisfactory
Active	Satisfactory	Low	Active
Satisfactory	Active	Satisfactory	Active
Active	Active	Satisfactory	Very active
Active	Low	Active	Active
Satisfactory	Satisfactory	Active	Satisfactory
Active	Low	Satisfactory	Satisfactory

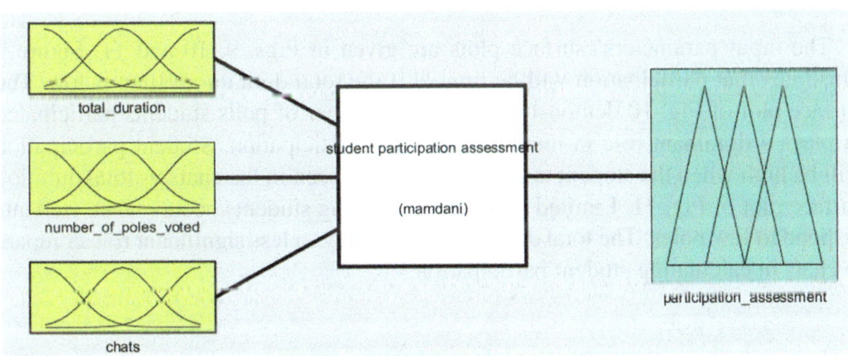

Fig. 6 Student participation assessment using fuzzy logic

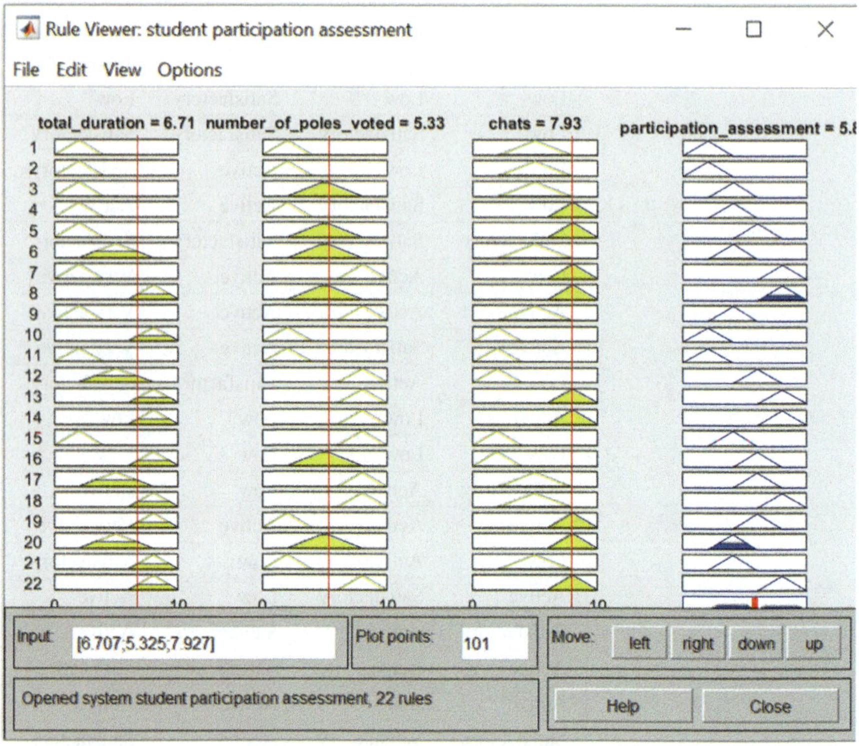

Fig. 7 Fuzzy rule window viewer

The input parameters' surface plots are given in Figs. 9, 10 and 11. Figure 9 illustrates that participation will be limited if the total duration of time is low. The surface plot of Fig. 10 demonstrates that the number of polls students participated in plays a dominant role in measuring student participation. Student participation will be high when the student is active in chats, as seen in the chats vs total duration surface plot of Fig. 11. Limited participation among students occurs when students respond to few poles. The total duration of time plays a less significant role compare to chats in calculating student participation.

4 Conclusion

Using a fuzzy inference approach, this paper examines the clever and intelligent involvement of students in online classes. This fuzzy inference system is used because it is trustworthy and can be reprogrammed to match the needs of various situations. As a consequence of the notification, students will be able to increase their participation in the class as a result of the low student involvement in the class. Students will get

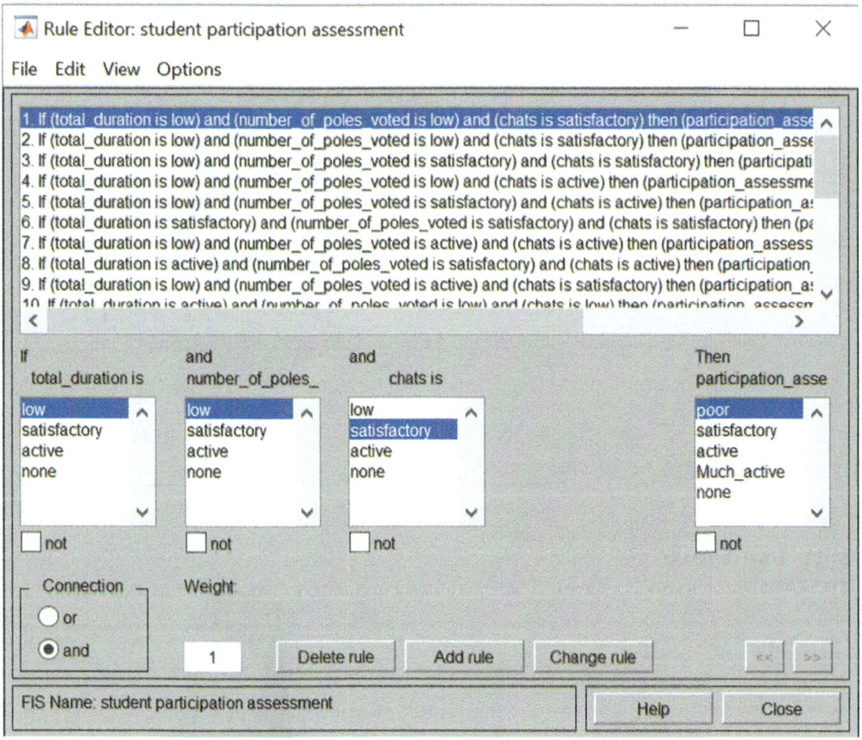

Fig. 8 Verbal fuzzy rules

Fig. 9 Total duration versus votes polled

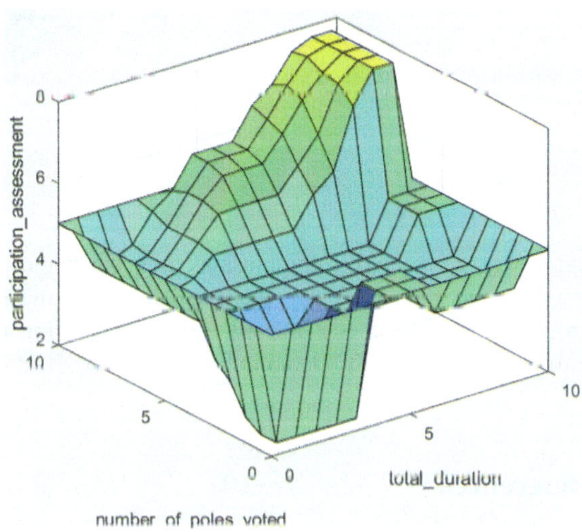

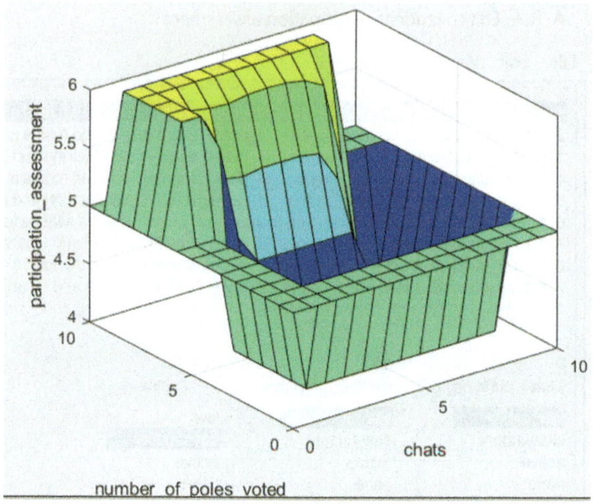

Fig. 10 Chats versus poll votes

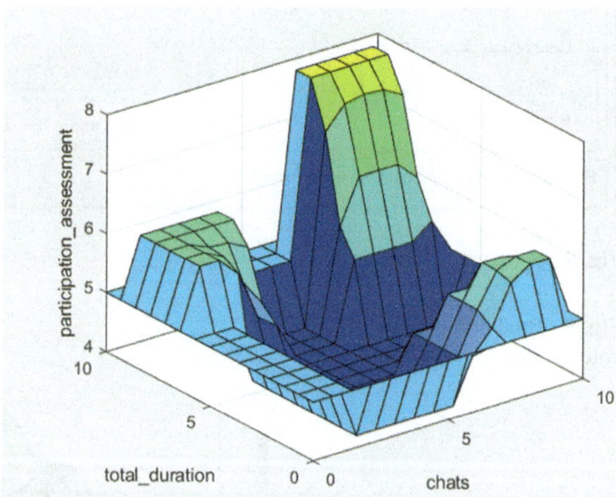

Fig. 11 Total duration versus chats

increasingly involved in the class as their participation is evaluated and grades are issued, rather than remaining passive listeners, as the course advances. In this model, an accuracy of 97% is achieved for the data using the proposed Fuzzy inference system.

References

1. Nagothu, S.K., Kumar, O.P., Anitha, G.: Autonomous monitoring and attendance system using inertial navigation system and GPRS in predefined locations. In: 2014 3rd International Conference on Eco-friendly Computing and Communication Systems, Mangalore, pp. 261–265 (2014). https://doi.org/10.1109/Eco-friendly.2014.60

2. Nagothu, S.K., Anitha, G., Annapantula, S.: Navigation aid for people (joggers and runners) in the unfamiliar urban environment using inertial navigation. In: 2014 Sixth International Conference on Advanced Computing (icoac), Chennai, pp. 216–219 (2014). https://doi.org/10.1109/icoac.2014.7229713
3. Nagothu, S.K., Kumar, O.P., Anitha, G.: GPS aided autonomous monitoring and attendance system. Procedia Comput. Sci. **87**, 99104, ISSN 1877-0509 (2016). https://doi.org/10.1016/j.procs.2016.05.133
4. Nagothu, S.K.: Automated toll collection system using GPS and GPRS. In: 2016 International Conference on Communication and Signal Processing (ICCSP), Melmaruvathur, Tamilnadu, India, pp. 0651–0653 (2016). https://doi.org/10.1109/ICCSP.2016.7754222
5. Nagothu, S.K., Anitha, G.: INS - GPS integrated aid to partially vision impaired people using Doppler sensor. In: 2016 3rd International Conference on Advanced Computing and Communication Systems (ICACCS), Coimbatore, pp. 1–4 (2016). https://doi.org/10.1109/ICACCS.2016.7586386
6. Nagothu, S.K., Anitha, G.: INS-GPS enabled driving aid using Doppler sensor. 2015 International Conference on Smart Sensors and Systems (IC-SSS), Bangalore, pp. 1–4 (2015). https://doi.org/10.1109/SMARTSENS.2015.7873619
7. Nagothu, S.K., Anitha, G.: Low-cost smart watering system in multi-soil and multi-crop environment using GPS and GPRS. In: Proceedings of the First International Conference on Computational Intelligence and Informatics Volume 507 of the series Advances in Intelligent Systems and Computing, pp. 637–643. https://doi.org/10.1007/978-981-10-2471-9_61
8. Nagothu, S.K.: Weather based smart watering system using soil sensor and GSM. In: 2016 World Conference on Futuristic Trends in Research and Innovation for Social Welfare (Startup Conclave), Coimbatore, pp. 1–3 (2016). https://doi.org/10.1109/STARTUP.2016.7583991
9. Nagothu S.K.: Intelligent Control of Aerator and Water Pump in Aquaculture Using Fuzzy Logic. In: Arunachalam V., Sivasankaran K. (eds) Microelectronic Devices, Circuits and Systems. ICMDCS 2021. Communications in Computer and Information Science, vol 1392. Springer, Singapore (2021). https://doi.org/10.1007/978-981-16-5048-2_13
10. Nagothu, S.K.: Smart student assessment system for online classes participation. In: Iyer B., Ghosh D., Balas V.E. (eds.) Applied Information Processing Systems. Advances in Intelligent Systems and Computing, vol 1354. Springer, Singapore (2022). https://doi.org/10.1007/978-981-16-2008-9_51
11. Nagothu S.K., Anitha G.: Automatic Landing Site Detection for UAV Using Supervised Classification. In: Rao P., Rao K., Kubo S. (eds) Proceedings of International Conference on Remote Sensing for Disaster Management. Springer Series in Geomechanics and Geoengineering. Springer, Cham (2019). https://doi.org/10.1007/978-3-319-77276-9_27
12. Nagothu, S.K.: ANFIS based smart wound monitoring system. 2021 978-981-16-5047-5, ICMDCS 2021, CCIS 1392, (499590_1_En, Chapter 15) (2022). https://doi.org/10.1007/978-981-16-5048-2_15
13. Nagothu, S.K.: GPS Aided hassle-free cycling using sensors, 2021 [978-981-16-5047-5, ICMDCS 2021, CCIS 1392, (499590_1_En, Chapter 35)) (2022). https://doi.org/10.1007/978-981-16-5048-2_35
14. Nagothu, S.K.: IoT based sheep guarding system in Indian scenario. In: 3rd International (Virtual) Conference on Recent Trends in Advanced Computing (2022)

Study on Real World Applications of SVM

Vipashi Kansal, Harsiddhi Singhdev, and Bhaskar Pant

Abstract Nowadays, machine learning is the foundational tool for developing and maintaining digital applications. Machine learning tools enabled organizations to identify profitable opportunities and potential risks more quickly. As there are numerous new advancements that aid in improving the accuracy of machine learning technology. Our goal is to identify the domains in which machine learning works well. In this paper, we use a support vector machine model in two real-world scenarios: behavior analysis on social media platforms and object detection. The paper also uses real-time data and does classification and detection with the help of SVM, producing accurate results.

Keywords Machine learning · Support vector machine · Vehicle detection · Mental health · Classification

1 Introduction

Machine learning is a rapidly evolving technology that allows computers to learn from the previous data automatically. It employs a variety of algorithms to create mathematical models and make predictions based on past data or knowledge. It is being utilized for visual identification, speech recognition, email filtering, Facebook auto-tagging, recommendation engines, and many more activities. Basically, it is a subset of AI (Artificial Intelligence) that is primarily concerned with the creation of algorithms that allow a computer to learn from data and previous experiences on its own, similar to how humans learn from their experiences. Machine learning algorithms create a mathematical model with the help of sample historical data, referred to as training data, which aids in making predictions or judgments without being explicitly programmed. For decades, machine learning tools have been used to predict future behavior patterns of feed in data in a range of settings. Nowadays, a huge amount of data is available for creating schemes and maintains sustainability

V. Kansal (✉) · H. Singhdev · B. Pant
Graphic Era Deemed To Be Unviersity, Dehradun, Uttarakhand, India

K. N. Das et al. (eds.), *Proceedings of the International Conference on Computational Intelligence and Sustainable Technologies*, Algorithms for Intelligent Systems,
https://doi.org/10.1007/978-981-16-6893-7_60

therefore, the machine learning tools can be applied to available data, and the results are used for maintaining sustainability. As a result, it is important to bring out sustainable development necessitates, and it is achievable not only by the national policy scheme, but the combined move of different cites, metropolitan and zones. In this research, we used Support Vector Machine (SVM) which is a part of supervised learning algorithm. Supervised learning methods are the most widely used machine-learning methods. The aim of supervised learning methods is to identify a mapping function between input variable (X) and output variable (Y). Supervised learning is a type of machine-learning methods in which machine is trained using "well-labeled" data based upon that predict the output. In this, the learning dataset provided to the machine act as a supervisor instructing the machine on how to accurately predict the results. It uses the same agenda as when a student learns under the guidance of a teacher. SVM classifies the category of data whether it is present to a particular group or not in terms of vector. The SVM algorithm's objective is to find the best line or conclusion dividing line for categorizing n-dimensional space into groups so that new data points can be easily placed in the exact category in future. A hyper plane is a name for the best decision boundary. This research focused to apply the SVM in two different scenarios, i.e., vehicle detection and classification of the mental health related problem and try to find that whether it will work perfectly or not.

1.1 Related Work

In [1], Sivaraman and Trivedi offer a comprehensive overview of recent developments in vehicle detection, tracking, and behavior analysis based on road vision, with a focus on vehicle identification sensors and exemplary vision-based vehicle identification and tracking. In paper [6], the author has used Word2vec and SVM classifier to propose a comment classification model which classifies comments from social media on a rating scale from 1 to 5. This model helps organizations as well as users to get overall satisfaction. The highest accuracy obtained from the removal of stop words and the class elimination method. Another interesting study regarding cameras mounted on vehicles [2], for example, propose "the study of gradient orientations in concentric rectangles for which new descriptor is formed," concerning "a feature space which is very small as compared to already established descriptors, which are unaffordable priced for practical systems or real-world applications. To train the SVM, a new-fangled database is formed which consists the vehicle images".

Huang, on the other hand, offers a deep examination of the background and applies entropy for a motion detection method in [3], and, while it is very nice research, the accuracy gained is quite poor (53.43%). In paper [7], to train the dataset (text data), the authors used semi-supervised learning. They enhance the performance of the classification model by using Support Vector Machine (SVM) on the dataset. Dataset categorizes into two fields, data with no tags called unlabeled data and data with one or more labels called labeled data, the amount of unlabeled data is more as compared to labeled data. By using expert advice, we convert unlabeled data to label data, but there is a lot of unlabeled data so we take a portion from it and label it. It took less time and effort, while maintaining the model's accuracy.

The occlusion handling algorithm in [4] is SVM-based, for training 11 videos were used and for occlusion detection 14 videos were used. Even though this is a narrative methodology, but for SVM classifier it uses picture/images for input space. As a result, its computational complexity is higher than that of other techniques which do not use these images rather than they use complex elements. In [8], the authors applied some models (Naive Bayes, Decision Tree, Neural Networks, Nearest Neighbor, and SVM) for text classification to determine 4 models for feature selection that are the state-of-art. The conclusion of the experiment shows that SVM performs best for text classification. The study discovered information gaps for indicators that determine the nature of sustainability dimensions in the micro territory. A significant amount of data is required, either to feed the machine learning models or to properly characterize each sustainability dimension. As a result, the data extracting segment of this study need more time and assets [5]. The authors used autopsy reports in the form of the dataset to predict the cause of death. They collected autopsy reports from 8 distinct causes of death, ran them through initialization, extraction of attributes, and a choice of feature, and then formed a feature memory with forty three features as per report. At the end, paper concluded that none of the other classifier performs better than the SVM.

2 Support Vector Machine

In these days, machine learning is very popular and used everywhere like in mobile devices, image detection, vision recognition in computers, etc. So we can easily enhance the capabilities of mobile devices and improve the performance by using a machine learning and resulting in a better and more sustainable mobile user experience. To classify various domain-specific challenges, we would employ SVM in different areas like vehicle detection, behavior analysis, and sustainability development.

2.1 SVM in Vehicle Detection

There are some key terms that we have to know to work on vehicle detection like feature extraction, it is the first and most important thing to extract the feature from an image. We used the HOG (Histogram of oriented gradients) feature descriptor to train a classifier (Linear Support Vector Machine (SVM)) and perform the extraction of attributes on an image training set that has been labeled. Then, apply a color transform to the HOG feature vector and binned color features and color histograms should be together. Now perform a sliding-window technique and the classifier which is already trained will be used to find vehicles in images/pictures. To eliminate outliers and track identified vehicles, in a video run pipeline and make a heat map of repeating detection outline by outline, estimate a bounding box for detected vehicles. Even though for this case, SVM classifier works and also it is a finest way to understand the concept behind vehicle detection and tracking, it is also quite slow and computationally expensive, making it difficult to apply this algorithm in real-world scenarios involving pedestrians, bicycles and oncoming vehicles, among other things. The another disadvantage is that to train the SVM, a huge data is needed to ignore bogus positives, as a result, our current algorithm will become slower and if we try to combine multiple windows then it will make computational expensive. So, this will not properly work in other scenarios like pedestrian detection and also it took around 20 min to process this algorithm on vehicle detection (Fig. 1).

2.2 *SVM in Classifier to Identify User Mental Health*

Social media is a popular platform for sharing a user's sentiments. People start conversations on social media sites on a daily basis, and open up and share their opinions from different perspectives. Quite a lot of earlier studies discovered that using SNS (social networking site) we can identify the users who may face mental health problems. In this proposed methodology, the objective is to develop a classification model (using SVM) which predict accurately that users mental disorders if having any (ADHD, Addiction, anxiety, depression). We extracted data from SNS (social networking site), i.e., Reddit, a well-known platform that contains various mental health-related groups (or so-called "subreddits"), such as r/addiction, r/anxiety, and r/adhd. Both methods, namely the semi-supervised method (statistical approach) and expert assessment, we considered only four of the most popular 83 subreddit for mental health related ones.

2.2.1 Data Collection

Data extraction is the first step in the proposed methodology. We use PRAW (python API) to scrape information from our four topmost subreddit, i.e., r/depression, r/Anxiety, r/adhd and r/addiction. User delete some posts related to mental health in the subreddit so the total count of posts bringing to 3300.

2.2.2 Data Pre-Processing Procedure

Data pre-processing is the next step for the collected post. After extracting information, each subreddit was merged with its related post. We used the NLP package (Neattext) implemented in Python to cleaning textual data and text pre-processing. It includes the elimination of void data, hash tags, stop words, white spaces and non-alphabetical characters.

2.2.3 Feature Extraction

Feature extraction is a technique that recognizes relevant features of data. It also reduces the dimensionality of the large datasets without losing any information only important features are taken into consideration. To extract characteristics from initialization information, frequency-inverse document frequency (TF-IDF) is applied. This procedure is divided into two stages. It calculates term frequencies (TF) in the first phase, which is a record of how many times a word appears in a dataset. This activity produces a weight for each term that is corresponding to its count of repetition. The second aspect is the IDF (Inverse Document Frequency), which is polar inverse of term frequencies. Some terms appear more frequently than others, and these terms incorrectly emphasize the document, resulting in lower classification accuracy. As a result, IDF reduces the weight of terms that are used repeatedly. The weights of the words are represented as vectors in the TF-IDF output program listings or program commands in the text are normally set in typewriter font, e.g., CMTT10 or Courier.

2.2.4 Classifications

Support vector machine (SVM) is supervised learning algorithm that classifies the category whether it present to a particular group or not in terms of vectors. Therefore, in previous step, we apply feature extraction technique (TF-IDF) and convert the word into vectors. Now, we apply the SVM classifier for the classification of subreddit. Primarily, it is used for classification problems in machine learning (Fig. 2; Table 1).

3 Conclusion

In this study, we used SVM classifier to classify mental health data and discovered that it works effectively in cases of depression and anxiety with 88% (approx), and in cases of addiction or ADHD with 65% (approx). However, this may not work for large multiclass classification, so we can try another algorithm like a random forest. In another scenario, vehicle detection, we discovered that while the cars were correctly detected, we believe that there are some flaws like, it took higher processing time and provide the minimum accuracy. Furthermore, this may not work perfectly in other scenario involving other objects, such as pedestrians, so we can use other approaches, such as CNN, YOLO, and so on, for better accuracy.

4 Future Scope

Sustainable engineering is the process of modeling or operating systems that allows for efficient use of products and resources at a scale that does not adversely affect the natural environment or future generations' ability to meet their own needs. Machine learning can be used in a variety of fields, including healthcare, agriculture, security, and even everyday life. Machine learning is also important in sustainable computing. Machine learning's development, and even its substantially larger consequences on many markets, necessitate a review of its implications for long-term innovation success.

Fig. 1 Shows that the vehicle is detected

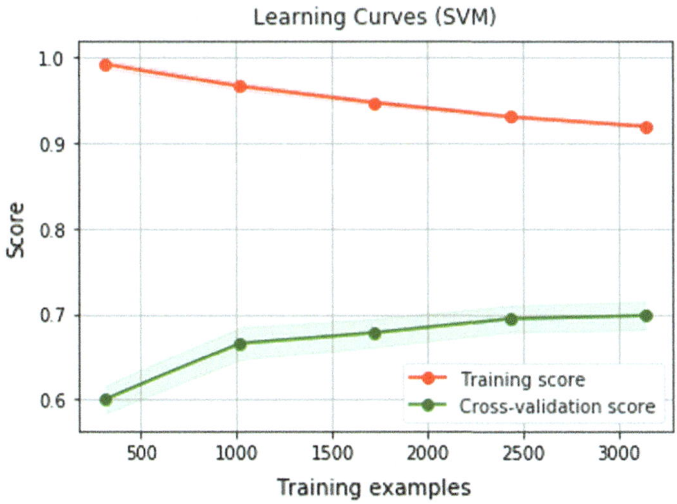

Fig. 2 Shows the SVM model learning a curve on the reddit dataset SVM performance on testing data it gives better performance (87%) as compared to Naive bayes classifier

Table 1 Shows performance metric

Subredit	Performance metric		
	F-measure	Recall	Precision
Anxiety	0.87	0.81	0.84
Addiction	0.70	0.68	0.67
Depression	0.85	0.75	0.70
ADHD	0.67	0.60	0.60

References

1. Sivaraman, et al.: Looking at vehicles on the road: a survey of vision-based vehicle detection, tracking, and behavior analysis. IEEE Trans. Intell. Transp. Syst. **14**, 1773–1795 (2013)
2. Arróspide et al.: Video analysis-based vehicle detection and tracking using an MCMC sampling framework. EURASIP J. Adv. Sig. Process (2012)
3. Huang, S.C.: An advanced motion detection algorithm with video quality analysis for video surveillance systems. IEEE Trans. Circuits Syst. Video Technol. **21**, 1–14 (2011)
4. Shirazi et al.: Vision-Based Vehicle Counting with High Accuracy for Highways with Perspective View, pp. 809–818. Springer, Cham, Switzerland
5. Molina et al.: Using machine learning tools to classify sustainability levels in the development of urban ecosystems. Sustainability (2020)
6. Zhang, Z., et al.: Sentiment classification of Internet restaurant reviews written in Cantonese. Exp. Syst. Appl. **38**(6), 7674–7682 (2011)
7. Goudjil, M., et al.: a novel active learning method using svm for text classification. Int. J. Autom. Comput. (2018)
8. Deng, X., et al.: Feature selection for text classification: a review. Multimed. Tools Appl. (2019)

9. Schueller, S.M., Hunter, J.F., Figueroa, C., Aguilera, A.: Use of digital mental health for marginalized and underserved populations. Curr. Treatment Opt. Psychiatry **6**(3), 243–255 (2019)
10. Twenge, J.M., Joiner, T.E., Rogers, M.L., Martin, G.N.: Increases in depressive symptoms, suicide-related outcomes, and suicide rates among US adolescents after 2010 and links to increased new media screen time. Clin. Psychol. Sci. **6**(1), 3–17 (2018)

Generation of a Trust Model for Secured Data Transfer in IoT

Ananya Roy, Prodipto Das, and Rajib Das

Abstract The potential of the Internet of Things (IoT) is increasing day by day. The Internet of Things has laid its applications on every sector from traditional use to its implementation in household appliances. The perspective of the Internet of Things brings with it various challenges. This paper tried to focus on the various factors that need to be considered for a secured framework for the Internet of Things. A trust model named Trustworthy Mediator Implementation Chip (TMIC) is developed which will act as a shield for all the sensors and mediators which are part of a network. The simulation performed using Cooja shows the successful development of trust among the nodes and the model where data communication takes place through devoted lines resulting in a secured path of data transmission. For the model to be accepted a data safety calculation is performed, the results of which show the positivity of the model. Finally, the comparison of the proposed model with the standard models shows that the model proposed has all the features of a standard model which guarantees its security.

Keywords IoT · Trust · Sensors · TMIC · Cooja

1 Introduction

Internet of Things is defined as a new projecting approach toward the normal internet. The architecture of IoT is such that every tiny item which is linked to the internet has a unique id and can further interconnect with every other which is part of the internet. Each object is embedded with chips that enable them to read their ambiances and bang back to the network or other object. In other words, each object in an IoT network acts both as an information provider or service requester or recommender

A. Roy (✉) · R. Das
Department of Computer Science and Application, Karimganj College, Karimganj, Assam, India

P. Das
Department of Computer Science, Assam University, Silchar, Assam, India

K. N. Das et al. (eds.), *Proceedings of the International Conference on Computational Intelligence and Sustainable Technologies*, Algorithms for Intelligent Systems,
https://doi.org/10.1007/978-981-16-6893-7_61

[1]. In this process, it is generating a massive amount of data that needs to be stored and processed in a secured way.

Some of the widespread applications of IoT viz. smart health, smart environment monitoring, smart education, etc., have taken the graph of IoT to a long height. The important constituent that drives the technology in IoT are sensors, actuators, RFIDs, WSN, etc., through which information is sensed, processed including their security and confidentiality. Along with the accessibility and productivity that IoT has given to users, it has given rise to numerous extortions.

The easy way of attacking IoT is to attack the devices in the network [2]. The devices are dependent upon the battery, and they have a minimum storage capacity. They are not protected against viruses or any malware, so they are always a direct target by the attacker which makes them act as a malicious node and thereby pass important information from the network to others. All this is enough to disrupt the whole communication process as well as its acceptance [3]. Thus, ensuring security in IoT is still a big concern because of their unnatural behavior and the security features available for the traditional networks cannot be applied to IoT.

The applications of IoT have paved the way for a convenient life for every user but it has failed to provide full security. The information gathered from any person may be trickled by a node to a fake node so in total the IoT still lacks security which needs more and more attention. If proper solutions are not developed for the above-mentioned problems then it will hinder the success graph of IoT a lot.

Figure 1 depicts the present style of IoT. In the coming days, IoT devices will not only be restricted to communicate with other local devices or persons with the help of the internet instead they will interact with the objects of the internet directly. A new concept social IoT (SIoT) is emerging which will enable users on social networking sites to connect to the devices and also share their devices on the internet [4].

The overall work of the paper has been organized in the following manner. "Security factors in IoT" section discusses in detail the various security factors in IoT. "Trust of IoT system & Trust implications" section overviews literature review of trust in respect of Internet of Things. "Trust model & Working principle of Trust model" section shows the working principle of the proposed model TMIC. In "Data safety

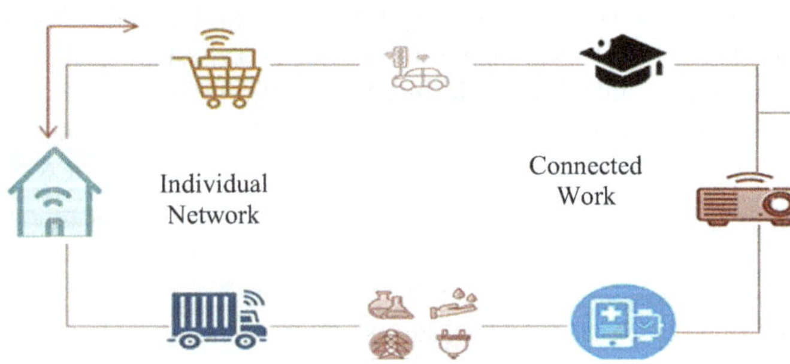

Fig. 1 Present day architecture of IoT

calculation" section, a safety calculation is performed on the model TMIC to guarantee that the model is secured. "Simulation environments & Results" section shows the simulation performed using Cooja simulator along with the results. "Comparisons between Different models" section depicts the comparison among the proposed model and some standard models.

2 Security Factors in IoT

The various challenges that we face in IoT are mainly because of the devices that are part of the network and the security weakness of the network. The nodes or actuators in the IoT network are not at all secured. They can work in any network due to their widespread adaptability to work in any network etc. which makes them prone to attacks. The next is the network itself which in most cases work without following any rules or standards. In any IoT application, security should be ensured from the development to the end of the process in all the devices involved [5]. The safety features to accomplish a protected communication structure for the public, progression in IoT are given below.

The security principles that should be followed to achieve a secure interaction framework for the people, software, processes, and things in an IoT are:

- ***Privacy:*** It is to ensure that the data is accessible only to the permitted handlers. In IoT, data is exchanged between devices of different origins so it is very vital to check the accurateness of the data; that it is not altered during the transmission process.
- ***Validation:*** The objects need to perform some kind of authentication. When an object communicates with another object or person or device for the first time it needs to verify them [6]. The objects should have to be clever enough to perform this authentication for confirming that the objects in the network are all part of it.
- ***Heterogeneous:*** In IoT, the nodes have a different design, different versions, developed by different vendors, support different boundaries, etc. In a network, all these nodes having different functionality and origin are made to work together. As such boundaries must be sketched for them to work in the same or different conditions [7]. The main purpose of IoT is to connect different objects, objects to humans, person to person, etc., and thus, it gears to fix them in a network [8].
- ***Accessibility:*** The spirit of IoT is to connect a large no of devices. Continuing that users must get all the available information from whichever object they need at any point of time. Data is not the only concern in IoT, along with data all the devices, and the facilities must be available.
- ***Procedures:*** To ensure that data will be secured, sheltered, and conveyed in a well-organized way, effective policies should be adopted. To fulfill all such policies a way is needed for assuring that the standard approved in all IoT applications has been followed.

3 Trust of IoT System

Buttyan et al. [9] have well defined it as the capability of one party to forecast the activity of another party in advance. Mayer et al. [10] have defined trust as the inclination of one party to a second party centered on the anticipation that the latter will perform the task which will be useful to the trusting party. According to Corritore et al. [11], trust is an approach of self-assured hope that in any online activity of risk one's weakness will not be misused. Chang et al. [12] have given a clear picture of trust as the belief a proxy has in the trustworthy agent's proficiency to provide an excellent service in any matter of concern in a limited period. According to [13], trust has been defined as generating requirements to meet the prospects and outlook of investors for confidentiality.

In the Internet of Things, trust can be exemplified into different forms such as trust in the nodes, trustworthiness in the services and the most important is trust in the data. So, first of all, if we talk about the device or node trust then–it is a big experiment for us because the nodes come from varied networks, and thus, it is very difficult to establish trust between them. Henceforth, it is necessary to take up methods for a trustful computation. When it comes to the trust of services then we find that different entity shows a different level of trust. By trust of service, we mean the performance or action of any node or person. Thus, the architecture of IoT has to be structured to deal with this issue. Trust of the information means that–in an unstructured network, data may come from some unwanted sources. Also in the network itself, the data collected from the varied sources are themselves suspicious sometimes.

4 Trust Implications

Trust is used in all applications. We cannot define trust technologically but still, it is widely used in various cases like to check uniqueness, grant to access something, etc. It is the belief of one party on another based on the earlier explanations [14, 15]. Trust can be of two types viz: direct and indirect. Direct trust is evaluated based on some defined algorithms and is implemented in any solicitation centered on the annotations of the current and preceding working system. In the case of indirect trust endorsements from other objects are taken into consideration [16]. The trustworthiness of any object or node or system is estimated based on the gradation of the belief or dis-belief between the objects. For ensuring security, it is necessary to define the different factors; be it the various formulas that will enable a fully trust model.

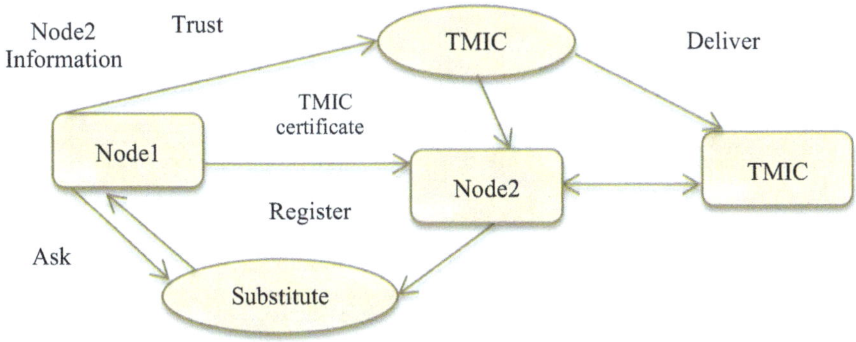

Fig. 2 Mediator protection representation based on TMIC

5 Trust Model

Figure 2 shows the proposed approach. In the mediator IoT system, we will have to implant mediators in the sensor nodes. Mediators can run on the TMIC platforms to perform all the jobs. TMIC helps in shielding the sensor nodes as well as the mediator.

Here, the sensor nodes will record all its details like name, address, safety requirements, etc. TMIC (Trustworthy Mediator Implementation Chip) manufacturer issues the sensor nodes a certificate which contains the various details like TMIC type, the security measures provided by the TMIC, encryption keys. The digital certificates are issued by the manufacturer of TMIC to the sensor nodes and also the sensor nodes have a public key to recertify the certificates issued.

6 Working Principle of Trust Model

The total data flow of the model is shown in Fig. 3. The nodes are created as agents and configured with the network. Some of the nodes are created as servers denoted as TMIC. The clients register their details to the TMIC and exchange keys. If the clients can decrypt the keys then it needs to be confirmed whether the clients are within the range of the server node.

When the clients are within the range then upon connection establishment, it is assumed that the clients have developed trust in the TMIC. Thus, data transfer is started among them using devoted lines. Devoted lines are used to prevent one node from interfering in the communication of another node. On the other hand, if the clients fail to decrypt the keys; they need to register them again to get the process started. TMIC works as a backbone of all the sensor nodes and also issues them certificates generated by the TMIC manufacturer.

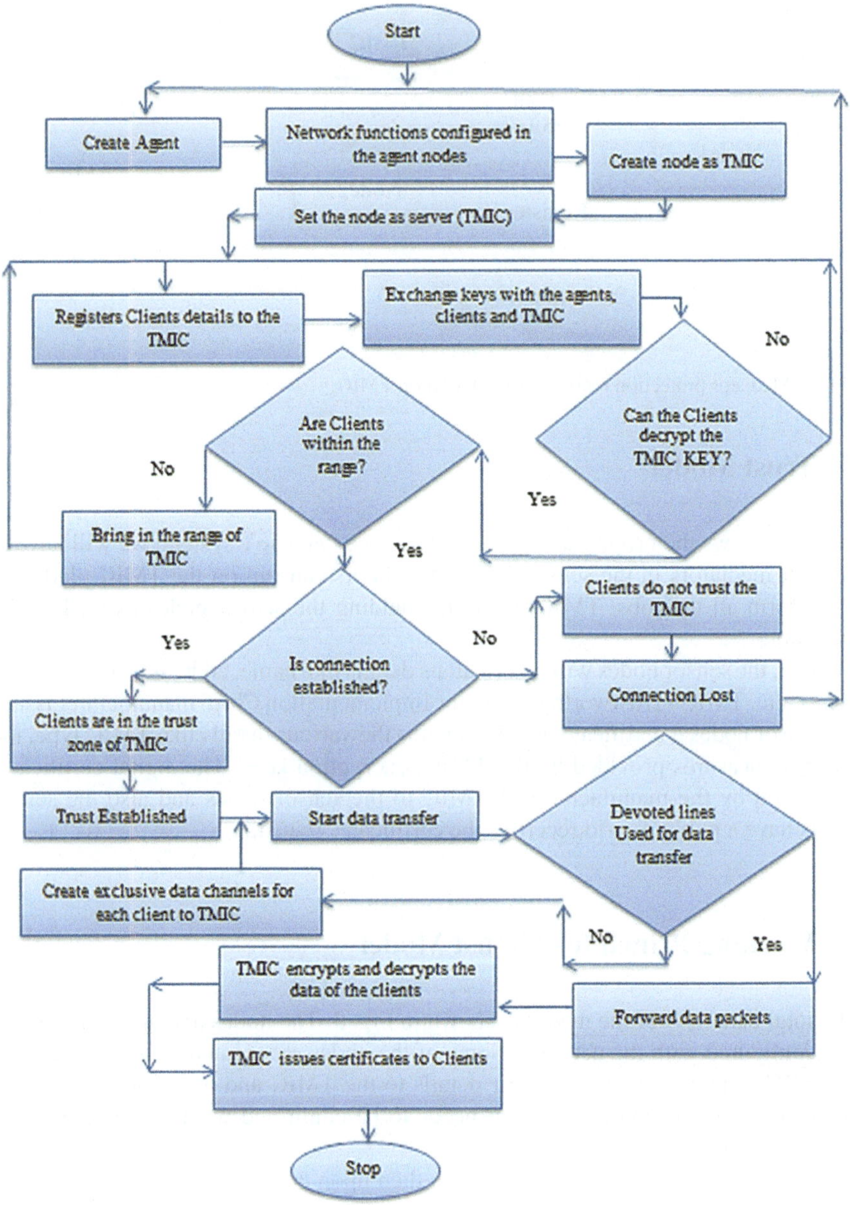

Fig. 3 Data flow of trust model

7 Data Safety Calculation

Based on our model, for it to be accepted among users we have calculated the data concentrating on the performance of the system. This will work for the latest operating system. Here, we have used probability values of 1 (high), 0.5 (medium), and 0.1 (low). And for calculating the effect we have taken 1 (high), 0.4 (medium), and 0.2 (low).

The data safety calculation will have two outputs:

The probability of an attack being successful and the possible impact of an attack.

The formula for calculating data safety is:

$$D_s = 1 - S_t \times S_e \quad (1)$$

where

S_t—Chance of an attack
S_e—The effect of the attack
$S_t \times S_e$—It is the predictable loss.

This calculation is performed for every node or device which is part of the network. By computing the individual scores, the security of the whole system can be predicted.

The aspects that mark the likelihood of an attack are:

The presence of weakness in a system known to attackers
Attention level on devices by the attackers.

In our case, there is no method to check if the TMIC is authenticated, so the base hardware setup has a higher case probability (1) of an attack. The probable effect of such an attack is also high (1). After the installation of TMIC still, there may be chances for attacks from the manufacturer's side. The probable cases of such an attack may be medium (0.5). Conversely, the effect probability is high (1). In the backend, where the data storehouse is present, there is a high level of security, so the probable case of such an attack is low (0.1) and the effect of such an attack will also be low (0.2) as the data is stored locally as well as in the cloud. The local storage system has a probability of attack which is very high (1). The sensors used in the system are also under the threat of attack which is low (0.5) as high-end sensors are used in this model, the probability of such attack will be low (0.5). Lastly, the security of the whole system is calculated by combining all the cases involved in the system. Here, D_s (M) is the security of mediators, D_s (S) is the security of sensors, and D_s (L) is the security of local storage. D_s (M), D_s (S), and D_s (L) are calculated using Eq. 1.

Tables 1, 2, and 3 show the calculation results of mediator, sensor, and local storage, respectively. From the Table 4, we find that the total score varies starting from 0 to 0.648. From the first value, it can be predicted that the system is not prone to attack. The rest values indicate a chance of attack. Due to the trusted environment,

Table 1 Data safety calculation for mediator

S_t	S_e	$D_s(M)$
1	1	0
0.5	0.2	0.9
0.1	0.2	0.9
1	0.4	0.6
0.5	0.4	0.8

Table 2 Data safety calculation for sensor

S_t	S_e	$D_s(S)$
0.5	0.4	0.8
0.1	0.2	0.9
1	0.4	0.6
1	0.2	0.8
0.5	0.2	0.9

Table 3 Data safety calculation for local storage

S_t	S_e	$D_s(L)$
1	1	0
0.5	0.4	0.8
0.5	0.2	0.9
0.1	0.2	0.9
1	0.2	0.8

Table 4 Combined safety calculation

$D_s(M)$	$D_s(S)$	$D_s(L)$	$D_s(M \times S \times L)$
0	0.8	0	0
0.9	0.9	0.8	0.648
0.9	0.6	0.9	0.486
0.6	0.8	0.9	0.432
0.8	0.9	0.8	0.576

it is predicted that decryption of secret keys by the attacker and generation of devoted lines of communication is a difficult task here.

8 Simulation Environments

Here, we have used Cooja which is a Contiki network emulator. With the help of this, we created a network where we can enable the process of data transfer between the sensor nodes. Here, we have used Sky motes. The behavior of the model proposed is inspected with the network created by the motes. With the help of Cooja, wireless sensor networks can be simulated. It ropes certain standards like TI CC2420, Contiki-RPL, etc. Four types of radio medium present in a Cooja simulator are:

- *Unit Disk Graph Medium (UDGM):* Constant Loss—In this range, the motes set up in the range are capable of receiving data packets while those outside cannot receive them.
- *Unit Disk Graph Medium (UDGM):* Distant Loss—An addition of the first medium has led to this medium. Here, data packets are transferred with TX probability of success and data packets are received back with RX probability.
- *Directed Graph Radio Medium (DGRM):* In this medium, the nodes transfer data among them depending upon their edge configuration.
- *Multipath Ray-tracer Medium (MRM):* In this approach, ray-tracing is used to find the strength of the signal between the nodes.

9 Results

The particulars of the node after simulating for 15 min are shown in Table 5. As the clients have developed trust in the TMIC node, we find that no lost data packet in any of the nodes. The maximum time taken for data transfer of packets is approximately less than 2 min. Also, the minimum time requirement is at the highest value of 25 s in the case of the 6th node which has received 13 packets. Node 1 has received the maximum number of packets as it is receiving data from all the nodes.

In the Mote output window (Fig. 4), we can see the data transfer taking place only between the TMIC and those clients who have developed trust.

Here, we have created a network with 20 motes, out of which mote 1 is receiver mote (TMIC), and the rest are client nodes. In the network motes, 2, 3, 4, 6, 9, 10, 11, 12, 13, 14, 15, 16, 17, 18, 20, 21, 22, 23, 24, 25 are in the communication range, whereas motes 5, 7, 8, 11, 19, 25 are in the interference range of TMIC and connection could not be established between them. Thus, trust could not be established among node 1 and nodes 5, 7, 8, 11, 19, 25.

ID:1 represents the TMIC and ID:6, ID:18, ID:9, ID:23, ID:24, ID:14, ID:25, ID:3, ID:20 are the client nodes. The first set of messages is represented as "Hello 1" from the client nodes, followed by "Hello 2," "Hello 3," and "Hello 4." The Mote output shows the actual real time at which the data is sent from the clients and thereafter when is received by the trusted node. Figure 5 shows the difference in the number of packets acknowledged by the nodes after continuing the simulation for 15 min.

Based on the results above, we can conclude that in the proposed model the client nodes are constantly sending data packets to the server node or TMIC. Sensors choose devoted lines to communicate with the server (TMIC). It is evident from the results that those nodes which fail to stay in the safe zone of TMIC are not able to work in the trusted environment. This type of model will be useful in creating a secure working place for IoT applications. The model differs from some of the previous work done

Table 5 Nodes information

Node	Received	Lost	Beacon interval	Min-inter packet time	Max inter-packet time
1	26	0	10 min, 07 s	02 s	1 min, 24 s
2	14	0	10 min, 00 s	05 s	1 min, 35 s
3	11	0	10 min, 22 s	01 s	1 min, 58 s
4	11	0	10 min, 49 s	20 s	1 min, 28 s
6	13	0	10 min, 44 s	25 s	1 min, 46 s
9	12	0	11 min, 15 s	15 s	1 min, 43 s
10	12	0	11 min 30 s	11 s	1 min, 47 s
12	11	0	10 min, 34 s	25 s	1 min, 45 s
13	11	0	10 min, 34 s	07 s	1 min, 41 s
14	12	0	10 min, 24 s	23 s	1 min, 53 s
15	11	0	10 min, 54 s	23 s	1 min, 33 s
16	12	0	10 min, 34 s	12 s	1 min, 44 s
17	12	0	11 min, 41 s	20 s	1 min, 46 s
18	12	0	10 min, 34 s	12 s	1 min, 45 s
20	12	0	11 min, 27 s	19 s	1 min, 38 s
21	12	0	11 min, 23 s	20 s	1 min, 38 s
22	12	0	10 min, 59 s	23 s	1 min, 42 s
23	12	0	11 min, 25 s	14 s	1 min, 45 s
24	12	0	12 min, 00 s	17 s	1 min, 37 s
26	12	0	11 min, 23 s	23 s	1 min, 52 s

```
01:19.503  ID:6   DATA send to 1 'Hello 1'
01:19.699  ID:1   DATA recv 'Hello 1 from the client' from 6
01:21.071  ID:18  DATA send to 1 'Hello 1'
01:21.224  ID:1   DATA recv 'Hello 1 from the client' from 18
01:22.508  ID:9   DATA send to 1 'Hello 1'
01:22.588  ID:1   DATA recv 'Hello 1 from the client' from 9
01:23.526  ID:23  DATA send to 1 'Hello 1'
01:23.630  ID:1   DATA recv 'Hello 1 from the client' from 23
01:23.975  ID:24  DATA send to 1 'Hello 1'
01:24.105  ID:1   DATA recv 'Hello 1 from the client' from 24
01:27.655  ID:14  DATA send to 1 'Hello 1'
01:27.829  ID:1   DATA recv 'Hello 1 from the client' from 14
01:30.618  ID:25  DATA send to 1 'Hello 1'
01:34.886  ID:3   DATA send to 1 'Hello 1'
```

Fig. 4 Mote output

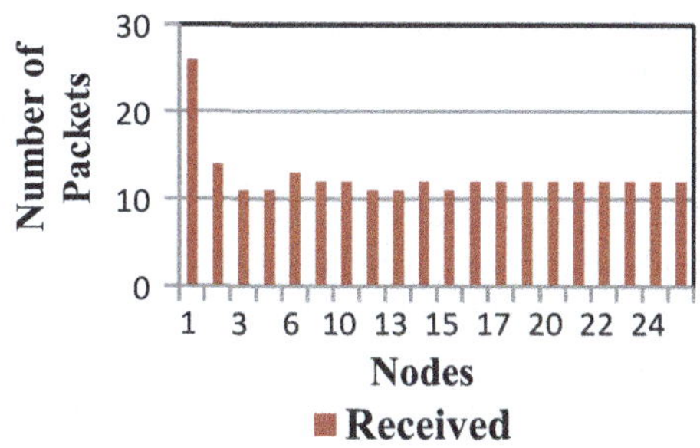

Fig. 5 Data packets received by the nodes

Table 6 Comparison study between proposed and standard models

IoT reference model (Levels)	Proposed model	Dell	Intel	IBM Watson
User verification	Yes	Yes	Yes	Yes
Support for big data	Yes	No	Yes	No
Data accumulation	Yes	Yes	Yes	Yes
Records generalization	No	No	No	No
Security (device) in cloud	Yes	No	Yes	No
Provision to short consumption devices	Yes	Yes	Yes	Yes

which only states that whether a node can trust or distrust another node [17] without actually going in-depth to find out the solutions of disconnection among the client nodes and TMIC which is very much possible in our model.

10 Comparisons Between Different Models

Table 6 shows the various features which the proposed model possesses in comparison with other standard models.

11 Conclusions

In this paper, we have presented the architecture of the trust model for IoT which is TMIC. The senor node records all its details in the TMIC and communicates with it through devoted lines. All the data transfer takes place under the umbrella of the TMIC based on the trust that their information is in safe hands and thus prevents any node with malicious behavior to leak the information. From the simulation, we found that the data transfer occurs only among those nodes which have developed trust in the TMIC. The proposed model will help in building a trusted environment where the sensors upon exchange of keys commit to stay in the safe range zone of

TMIC to work under it. Also, this model will help in securing the IoT environment by securing the data communication between the nodes and thereby help in recovering the various security issue concerns which arise mainly due to the moving nodes of IoT. We also made a security check for the mediator, sensor, and local storage, and the calculated values show less chance of the model being attacked or misutilized and the encryption and decryption of secret keys is a difficult task that guarantees the safety of the model. We compared our model with the standard model like Dell, Intel, etc., which has advanced security features, and found that the proposed architecture supports the properties like user verification, big data support, data accumulation, records generation, etc., which are very much necessary for data safety. Thus, it is concluded that this model can be implemented in all IoT environments. In future, we will work more on the detailed functioning of the TMIC to increase the success rate of the proposed model in the sensor nodes.

References

1. Mohammadi, V., Rahmani, A.M., Darwesh, A.M., Sahafi, A.: Trust-based recommendation systems in Internet of Things: a systematic literature review. In: Human-centric Computing and Information Sciences, vol. 9. Article number: 21 (2019)
2. Xiaohui, X.: Study on security problems and key technologies of the Internet of Things. In: Proceedings of IEEE Fifth International Conference on Computational and Information Sciences (ICCIS), Hubei, China. IEEE. https://doi.org/10.1109/ICCIS.2031.114 (2013)
3. Kanuparthi, A., Karri, R., Addepalli, S.: Hardware and embedded security in the context of Internet of Things. In: Proceedings of the ACM Workshop on Security, Privacy & Dependability for Cyber Vehicles, Berlin, Germany, pp. 61–64. https://doi.org/10.1145/2517968.2517976 (2013)
4. Frustaci, M., Pace, P., Aloi, G., Fortino, G.: Evaluating critical security issues of the IoT world: present and future challenges. IEEE Internet Things J. **5**(4), 2483–2495 (2018). https://doi.org/10.1109/JJIOT.2017.2767291
5. Leo, M., Battisti, F., Carli, M., Neri, A.: A federated architecture approach for Internet of Things security. In: Euro Med Telco Conference (EMTC), pp. 1–5 (2014)
6. Roman, R., Zhou, J., Lopez, J.: On the features and challenges of security and privacy in distributed Internet of Things. Comput. Netw. **57**, 2266–2279 (2013)
7. Zhao, K., Ge, L.: A survey on the Internet of Things security. In: 9th International Conference on Computational Intelligence and Security (CIS), pp. 663–667. https://doi.org/10.1109/CIS.2013.145 (2013)
8. Mahalle, P.N., Anggorojati, B., Prasad, N.R., Prasad, R.: Identity authentication and capability based access control (IACAC) for the Internet of Things. J. Cyber Sec. Mob. **1**, 309–348 (2013)
9. Buttyan, L., Hubaux, J.: Security and Cooperation in Wireless Networks: Thwarting Malicious and Selfish Behavior in the Age of Ubiquitous Computing. Cambridge University Press, New York, NY, USA (2007). Vol. 9780521873710, ISBN (Electronic). 9780511815102 (2016)
10. Mayer, R.C., Davis, J.H., Schoorman, F.D.: An integrative model of organizational trust. Acad. Manag. Rev. **20**(3), 709–734 (1995)
11. Corritore, C.L., Kracher, B., Wiedenbeck, S.: On-line trust: concepts, evolving themes, a model. Int. J. Human-Comput. Stud. **58**(6), 737–758 (2003)
12. Chang, E., Dillon, T., Hussain, F.: Trust and reputation relationships in service-oriented environments. In: Third International Conference on Information Technology and Applications, (ICITA), vol. 1, pp. 4–14 (2005)

13. Hanan, A., Rehman, S.U., Umar, I.: Trust in IoT systems: a vision on the current issues, challenges, and recommended solutions. In: Advances on Smart and Soft Computing, Part of Advances in Intelligent Systems and Computing book series, AISC, vol. 1188 (2020)
14. Suryani, V., Selo W.: A survey on trust in Internet of Things. IEEE. https://doi.org/10.1109/ICITEED.2016.7863238 (2016)
15. Sicari, S., Rizzardi, A., Grieco, L.A., Coen-Porisini, A.: Security. Privacy and trust in Internet of Things: the road ahead. Comput. Netw. (2015)
16. Kounelis, I., Baldini, G., Neisse, R., Steri, G., Tallacchini, M., Pereira, G.: Building trust in the human-Internet of Things relationship. IEEE Technol. Soc. Mag. 1932-4529/14©2014 IEEE. https://doi.org/10.1109/MTS.2014.2364020 (2014)
17. Leister, W., Schulz, T.: Ideas for a trust indicator in the Internet of Things. In: The First International Conference on Smart Systems, Devices and Technologies (2012)

Investigation of Localization Error in Range Free Centroid and DV-Hop Based Localization Algorithms and Optimization Using PSO

Vikas Gupta and Brahmjit Singh

Abstract The essence of localization in wireless sensor networks is the estimation of locations of unknown nodes for proper steering of data to the base station. The DV-Hop and Centroid based range free localization algorithms are widely explored algorithms. But the basic and improved versions of these algorithms still bear a large localization error. First, we simulated the basic forms of these two algorithms, and then, the performance measure the localization error is calculated. Further the swarm based soft computing technique particle swarm optimization (PSO) is applied on these algorithms and the effects of change in communication range, anchor ratio, number of nodes, and network size on localization error is studied. It may be established that the positioning error in both of the algorithms reduces by utilizing the PSO.

Keywords Localization error · Optimization · Centroid based localization algorithm · DV-Hop · etc.

1 Introduction

The wireless sensor networks (WSN) have revealed a notable expansion due to the revolution in the field of sensors and embedded systems. Nowadays, due to tiny dimension, low-priced, and very less power consumption, these sensor nodes have become an appropriate contender for numerous applications in the area of calamity management, agriculture, medicine, and defense, etc. [1]. The key purpose of the localization is to discover the positions of the unidentified nodes those are spread arbitrarily in the region of concern. These nodes communicate with each other and are competent of local processing of the data exchanged. And if the proper location

V. Gupta (✉)
Computer Science and Engineering Department, Chandigarh Engineering College (CGC), Landran, Mohali, India

B. Singh
Electronics and Communication Engineering Department, National Institute of Technology, Kurukshetra, India

K. N. Das et al. (eds.), *Proceedings of the International Conference on Computational Intelligence and Sustainable Technologies*, Algorithms for Intelligent Systems,
https://doi.org/10.1007/978-981-16-6893-7_62

information is not available, the data sent to the base station will be of no use. The estimation of the coordinates of unknown sensor nodes also provides other opportunities like geographic direction-finding, object tracking, etc. These locations finding algorithms may be separated in two broad classes namely the range-free and range based. For location determination, range based class of algorithms need an absolute estimation of angle or distance. Various methods like time of arrival (TOA), time difference of arrival (TDOA) angle of arrival (AOA) and received signal strength indications (RSSI) [2–6] are generally utilized range based position estimation methods. Although these algorithms provide high accuracy in position findings, but the kind of equipment utilized by these methods is costly. And in range free localization techniques, there is no need of any kind of costly hardware, only connectivity information is sufficient to localize the whole network. The basic and PSO based versions of both of these algorithms are evaluated and effects of changing the communication range, anchor ratio, amount of nodes, and the dimension of area of concern wherever the nodes are to be scattered arbitrarily on localization error are investigated. Further the paper contains 9 sections wherein the Sect. 2 gives an insight into the background of the work. Section 3 briefs the formulation of problem. Section 4 discusses the Centroid based localization algorithm in detail. Section 5 explains the DV-Hop algorithm of localization in detail. Section 6 offers the detail of PSO. Section 7 presents the simulation of proposed work. Section 8 describes the results and discussion. Section 9 presents the conclusion, followed by references.

2 Background

Range free localization techniques have covered a big space of literature. Idea of Centroid based localization was projected by Bulusu et al. in [7] which estimate the coordinates of unidentified node by finding the Centroid of reference nodes or anchor nodes. The reference nodes are the nodes those are having GPS like arrangement. Another range-free localization algorithm called as DV-Hop was offered by Niculescu et al. in [8] wherein the unknown coordinates are anticipated using classical distance vector exchange. In 2003, an approximation-point-in-triangulation test (APIT) algorithm was floated by He et al. in [9] where the coordinates of unidentified node is found by estimating the Centroid formed by the beacon triangles. In 2001, Doherty et al. [10] proposed a methodology in which the localization problem is considered as linear or semi-definite program (SDP) whose solution is found using connectivity and angles between the nodes. Another range free localization algorithm the multi-dimension scaling (MDS-MAP) was presented by Shang et al. [11] which employ the connectivity information to estimates the coordinates of the unidentified nodes in WSN. To improve the performance of these range-free localization algorithms, a large number of soft computing techniques such as fuzzy logic systems (FLS), genetic algorithm (GA), artificial neural networks (ANN), and simulated annealing (SA) [12–14], etc., have been reported in past. This motivates us to explore and use particle swarm based optimization algorithm; PSO to optimize the localization error [15–18]. In present work, first both of the algorithms; DV-Hop and Centroid based localization algorithms are simulated using MATLAB then both the algorithms are optimized using particle swarm optimization (PSO).

3 Problem Formulation

Although a big number of research articles have been reported by the research community on range free Centroid and DV-Hop algorithm but the presence of large localization error is the main cause of concern which is addressed in the proposed work. On the other hand, the both of the algorithms in past have been studied separately. Very few works have been reported in literature on the performance comparison of these algorithms which motivates to do a comparison of the algorithms by changing the communication range, anchor ratio, the total number of nodes, and the dimensions of network area. Further a well-known swarm based algorithm PSO is utilized to optimize the performance of both of the algorithms. For both of the algorithms, the performance parameter considered is localization error (LE). The target of the paper is to reduce the localization error as minimum as possible by using soft computing techniques.

4 Centroid Based Localization Algorithm

In 2000, Bulusu et al. in [7] presented Centroid based localization algorithm where connectivity based information is used by the reference nodes to locate the coordinates of unidentified nodes. Centroid is the intersection of reference nodes which is used to localize the unidentified nodes. Figure 1 shows the scenario of estimation of the unknown coordinates using Centroid localization algorithm. It is shown in the figure how the coordinates of unidentified sensor node (*D*) is found using more than one nearby anchor nodes (*A*, *B*, *C*). Using this scenario, the coordinates of the unidentified nodes can be predicted using Eq. 1.

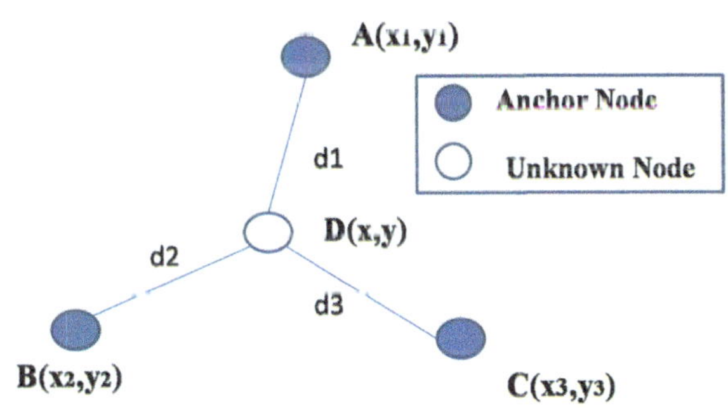

Fig. 1 Estimation of unknown coordinates using Centroid localization algorithm [22]

$$(X_{\text{est}}, Y_{\text{est}}) = \left(\frac{X_{i1} + X_{i2} + \cdots + X_{ik}}{k}, \frac{Y_{i1} + Y_{i2} + \cdots + Y_{ik}}{k}\right) \tag{1}$$

Here, $(X_{\text{est}}, Y_{\text{est}})$ are the estimated values of unidentified node position, and k shows the reference or anchor nodes connected to the node whose coordinates are to be estimated. Although this method is quite simple and straightforward but give large error in location estimation. The LE is given by Eq. 2.

$$\text{LE} = \sqrt{(X_{\text{est}} - X_a)^2 + (Y_{\text{est}} - Y_a)^2} \tag{2}$$

Here, (X_a, Y_a) are the actual coordinates of the unidentified sensor node. In [19], Hay-qing et al. projected a weight based Centroid based algorithm where the coordinates of the unidentified sensor node is premeditated using weights. These weights are directly proportional to the range involved amid transmitter and receiver of sensor node. They estimated the coordinates using Eq. 3.

$$(X_{\text{est}}, Y_{\text{est}}) = \left(\frac{w_1 * X_1 + \cdots + w_n * X_n}{w_1 + \cdots + w_n}, \frac{w_1 * Y_1 + \cdots + w_n * Y_n}{w_1 + \cdots + w_n}\right) \tag{3}$$

Although the weighted Centroid based algorithm decreases the localization error but performance of this algorithm is highly dependent upon optimization of the weights which increases the complexity of algorithm. In 2007, Blumenthal et al. [20] analyzed the localization error on the basis of transmission range. They estimated the unknown coordinates with error of less than 1 m but they did not take into account the environmental clutters, while estimating the unknown positions. In [21], Dong and Xu proposed a weight compensated algorithm which improved the accuracy of the weighted algorithms but still the error present was less. In [23, 24], Gupta et al. studied Centroid based localization algorithm in detail and analyzed the error in location estimation on the basis of various network parameters. Those parameters are utilized in present article to examine the performance of Centroid and DV-Hop based localization algorithms.

5 DV-HOP Localization Algorithm

In DV-Hop [25–28], all the distances are calculated in the form of hops from every unidentified nodes to reference or anchor nodes using classical distance vector exchange method which is the first stage of the algorithm. A table consisting of (X_i, Y_i, h_i) is maintained at each reference node where (X_i, Y_i, h_i) are the coordinates and hop count of reference node i. The reference nodes share this information with their neighbors. Second stage the average step size of one hop is anticipated by finding the distances to other reference nodes. This hop size is utilized as modification by the neighborhood and is found by using Eq. 4.

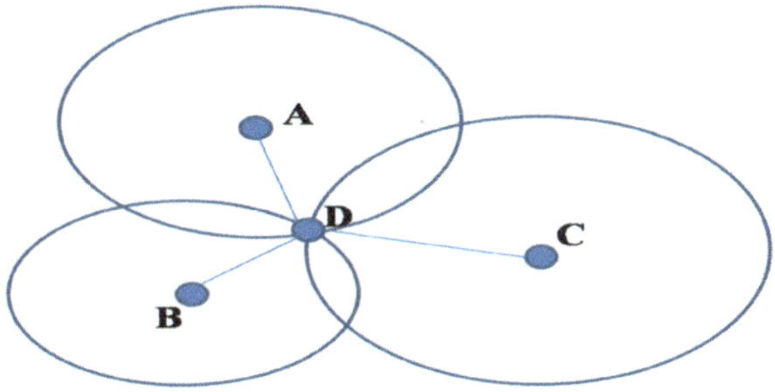

Fig. 2 The trilateration technique [25]

$$\frac{\sum\sqrt{(X_i - X_j)^2 + (Y_i - Y_j)^2}}{\sum h_i}, i \neq j, \text{for all anchors } j \tag{4}$$

Here, (x_i, y_i), (x_j, y_j) represent the coordinates of reference node i and j, and h_i represents the count of hops. Following this correction, unknown node estimate distances to the anchors by utilizing the trilateration technique that is the third stage. Figure 2 shows the estimation of coordinates (x_u, y_u) of unknown node (D) with the help of coordinates of anchors A, B, and C.

These distances from three anchors to unknown node D are d_1, d_2, d_3, and may be calculated using Eq. 5.

$$\left\{\begin{array}{c} (x_1 - x_u)^2 + (y_1 - y_u)^2 = d_1^2 \\ (x_2 - x_u)^2 + (y_2 - y_u)^2 = d_2^2 \\ . \\ . \\ . \\ (x_n - x_u)^2 + (y_n - y_u)^2 = d_n^2 \end{array}\right\} \tag{5}$$

Equation (5) can be expanded as

$$\left\{\begin{array}{l} 2(x_1 - x_n)x_u + 2(y_1 - y_n)y_u = x_1^2 - x_n^2 + y_1^2 - y_n^2 - d_1^2 + d_n^2 \\ 2(x_2 - x_n)x_u + 2(y_2 - y_n)y_u = x_2^2 - x_n^2 + y_2^2 - y_n^2 - d_2^2 + d_n^2 \\ . \\ . \\ . \\ 2(x_{n-1} - x_n)x_u + 2(y_{n-1} - y_n)y_u = x_{n-1}^2 - x_n^2 + y_{n-1}^2 - y_n^2 - d_{n-1}^2 + d_n^2 \end{array}\right\} \tag{6}$$

Equations in 6 are kind of linear equations which are of the form $Ax = B$, where unknown position of D node may be premeditated with the assistance of matrix formulas (Eq. 7)

$$A = 2\begin{bmatrix} x_1 - x_n & y_1 - y_n \\ x_2 - x_n & y_2 - y_n \\ \vdots & \vdots \\ x_{n-1} & y_{n-1} \end{bmatrix}, B = \begin{bmatrix} x_1^2 - x_n^2 + y_1^2 - y_n^2 - d_1^2 + d_n^2 \\ x_2^2 - x_n^2 + y_2^2 - y_n^2 - d_2^2 + d_n^2 \\ \vdots \\ x_{n-1}^2 - x_n^2 + y_1^2 + y_n^2 - d_{n-1}^2 + d_n^2 \end{bmatrix} \quad (7)$$

The unknown coordinates (x_u, y_u) can be found by using matrix formula given in Eq. 8.

$$(x_u, y_u) = \left(A^T A\right)^{-1} A^T B \quad (8)$$

6 Particle Swarm Optimization

Invented in 1995 by Eberhart and Kennedy [29] and explored by Cao et al. [30], Daanoune et al. [31] and Jawad et al. [32]; PSO is a swarm based approach. It is the study of activities of bird flocking where social conduct of birds is studied and replicated in the form of simulation. PSO is very simple to implement as it involves little parameters to be optimized. Originally developed to replicate the activities of birds PSO later proved very helpful to obtain optimized results of many problems. Each candidate solution in PSO is represented by a bird or particle. Initialized with arbitrary populace of the solutions the PSO find the finest result in the exploration space. The individual best among all particles is depicted by personal best (p_{best}). The superior among the whole populace is depicted as global best (g_{best}). Updation in velocity and position in PSO is done with Eqs. 9 and 10.

$$\begin{aligned} v_i(t+1) = {} & w.v_i(t) + c_1.\text{rand}().(Pp_{\text{best}}(t) - x_i(t)) \\ & + c_2.\text{rand}().(Pg_{\text{best}}(t) - x_i(t)) \end{aligned} \quad (9)$$

$$x_i(t+1) = x_i(t) + v_i(t+1) \quad (10)$$

In the above equations, $x_i\ (t+1)$ represents the updation in position and $v_i\ (t+1)$ represents the updation in velocity, w denotes the is weight whose value varies from 0 to 1 in inertia term $w.v_i(t)$. In Eq. 9, second part is recognized as cognition part that shows the private conduct of individual particle. Third part of Eq. 9 is recognized as social part that shows the cooperative conduct of the particles. In both parts c_1 and c_2 represent coefficient of acceleration.

7 Simulation of Proposed Work

Here, first both range-free localization algorithms; Centroid and DV-Hop are simulated and their performance is examined. From the simulation results, the localization error is obtained for both algorithms under different scenarios. Further both the algorithms are explored using swarm intelligence based technique called particle swarm optimization which is widely employed to get optimum solutions of many problems. Both of the algorithms are again simulated using PSO under same scenarios those were used for the basic versions. The performance measure, localization error for both Centroid and DV-Hop based location estimation methods and PSO based Centroid and DV-Hop methods is obtained by changing the values of communication range, amount of nodes, anchor ratio, and the network size. Results obtained after simulation prove that the performance of both of the algorithms improves using PSO. Steps of simulation of both algorithms are given below:

7.1 *Steps of Simulation of Centroid Based Algorithm*

(1) Nodes are distributed (anchor and unknown) arbitrarily in the region of concern.
(2) Using Eq. 11 Euclidean distance (*d*) is calculated between reference and unidentified nodes.

$$d = \sqrt{\left(x_i - x_j\right)^2 + (y_i - y_j)^2} \quad (11)$$

Here, (x_i, y_i) represents the coordinates of reference or anchor nodes, and (x_j, y_j) represents the coordinates unidentified nodes.

(3) The formula (Eq. 1) of Centroid based algorithm is used to estimate unknown coordinates.
(4) Using Eq. 2, the localization error is calculated.
(5) Here for PSO, the localization error is taken as an objective function.
(6) The p_{best} is initialized with random position and Centroid position is assumed as g_{best}.
(7) Updation in position and velocity (Eqs. 9 and 10) provides a new value of objective function.
(8) Objective of using PSO is to reduce the localization error.

7.2 *Steps of Simulation of DV-HOP Algorithm*

(1) Nodes are distributed (anchor and unknown) arbitrarily in the region of concern.
(2) Using Eq. 11, Euclidean distance (*d*) is calculated between reference and unidentified nodes.

(3) Hop matrix is initialized with 0. Shortest path algorithm is utilized to find the number of hops and average hop distance is obtained using Eq. 4.
(4) To determine the positions of unidentified nodes least square scheme is utilized (equations from 5 to 8).
(5) Here, also localization error is assumed as objective function for PSO.
(6) Objective of using PSO is to minimize the location error.

8 Results and Discussion

All the parameters considered for the simulation of Centroid, PSO based Centroid, DV-Hop and PSO based DV-Hop localization algorithms and their numerical values presumed to create simulation background have been put in Table 1. The localization error for both algorithms is calculated using different values of various parameters like communication range, anchor ratio, number of nodes and area of network. On the other hand, optimized values of various parameters of PSO obtained after simulations have been presented in Table 2. The numerical values chosen here are the values for which the PSO gives the best results.

Localization Error versus Communication Range

The plot in Fig. 3 shows the graphical representation of localization error for Centroid, PSO based Centroid, DV-Hop, and PSO based DV-Hop localization algorithms for various values of communication range. From the graph, it can be shown that localization error is large in Centroid based algorithm which reduces by using PSO. And the error is less in DV-Hop algorithm if related with Centroid based algorithm and LE reduces further by using the PSO. In general, it can be shown from the trends of all of the curves that the location estimation error reduces as the communication range expands.

Table 1 Parameters with numerical values presumed to create simulation background

S. No	Parameters	Values
1	Network area	100–500 m^2
2	Number of nodes	25–250
3	Anchor nodes	10–100
4	Communication range	10-100 m

Table 2 Parameters of PSO with their optimized values

S. No	Parameters	Values
1	Swarm size	50
2	No. of iterations	30
3	Weight	0.65
4	Acceleration coefficients	1.35

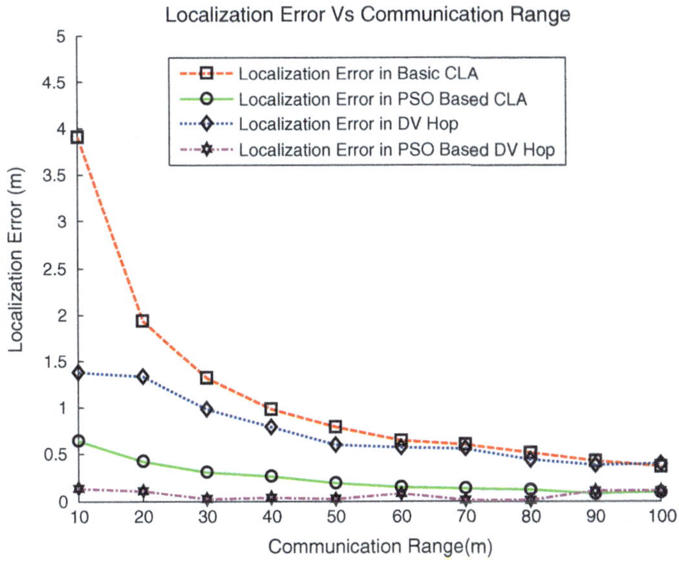

Fig. 3 Localization error versus communication range

Localization Error versus Anchor Ratio

The plot in Fig. 4 shows the graphical representation of localization error for Centroid, PSO based Centroid, DV-Hop, and PSO based DV-Hop localization algorithms for

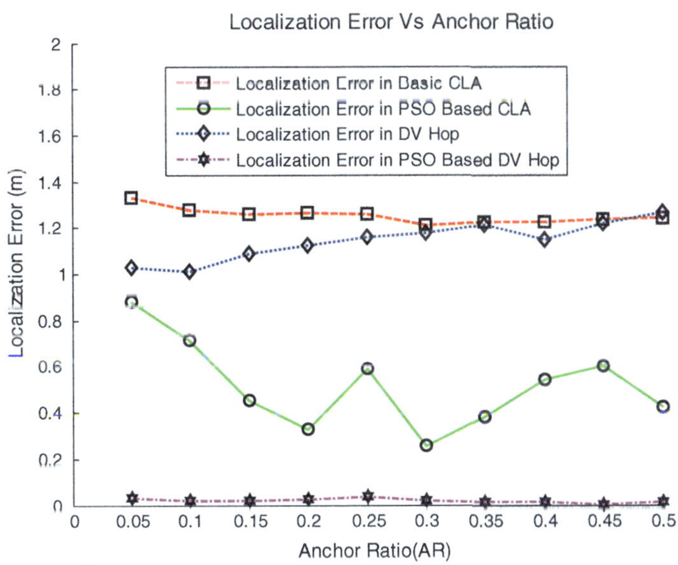

Fig. 4 Localization error versus anchor ratio

various values of anchor ratio (AR). The AR is obtained by dividing the reference or anchor nodes with entire number of nodes. For simulation, communication range is taken as 30 m. Simulation results prove that for all anchor ratios localization error is large in Centroid based algorithm as compared to DV-Hop which further reduces in both of the algorithms if PSO is applied. It is to be noted that as anchor ratio increases the system becomes costly as it will be using more number of anchor nodes having GPS like arrangement. So to determine the locations of unknown nodes that value of AR should be used which gives the least error.

Localization Error versus Number of Nodes

The plot in Fig. 5 shows the graphical representation of localization error for Centroid, PSO based Centroid, DV-Hop, and PSO based DV-Hop localization algorithms for diverse values of whole number of nodes. For finding the LE with respect to number of nodes, the communication range of 30 m and an anchor ratio of 0.25 are considered for simulation. The results prove localization error is large in Centroid based algorithm which reduces by using PSO. And the value of localization error is less in basic DV-Hop algorithm as compared to the Centroid based algorithm which further reduces by using the PSO.

Localization Error versus Network Size

The plot in Fig. 6 shows the graphical representation of localization error for Centroid, PSO based Centroid, DV-Hop, and PSO based DV-Hop localization algorithms for various values of network size. In finding the error for various values of number of

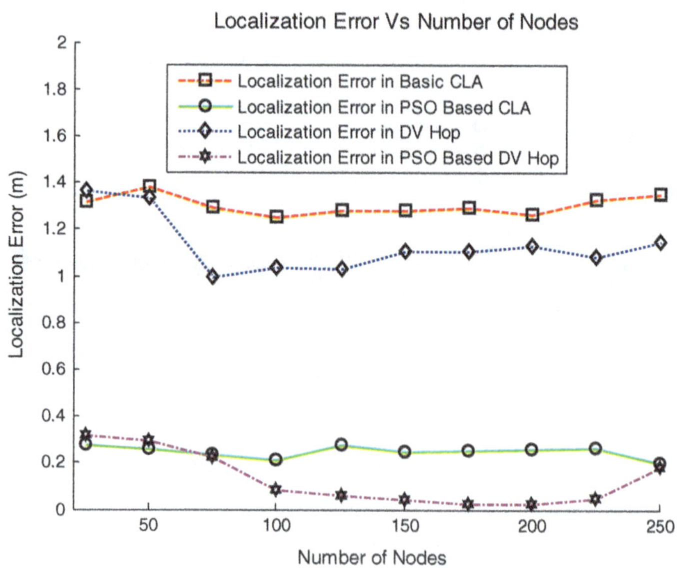

Fig. 5 Localization error versus number of nodes

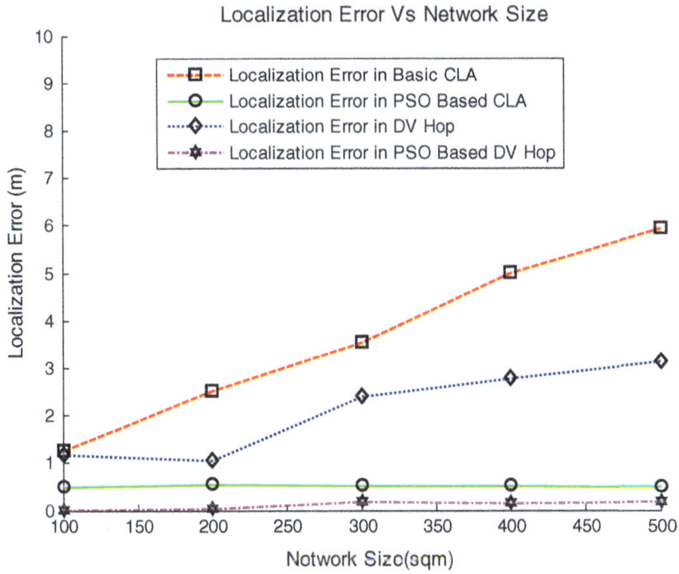

Fig. 6 Localization error versus network size

nodes communication range of 30 m and anchor ratio of 0.25 is considered for simulation. The results demonstrate that LE increases as the size of the network increases in both Centroid and DV-Hop algorithms. However, for PSO based algorithms, the trends are uniform, and localization errors are small.

Average Localization Error

The plot in Fig. 7 shows the graphical representation of average localization error for Centroid, PSO based Centroid, DV Hop, and PSO based DV Hop localization algorithms for various values of communication range, anchor ratio, numbers of nodes and size of network. The graph demonstrates that the average LE is large in Centroid based algorithm compared to DV-Hop. The LE reduces by using PSO. It can also be observed from the plot that the size of network area affects the average localization error more in comparison to other parameters taken.

Standard Deviation of Localization Error

The values of standard deviation of LE for Centroid, PSO based Centroid, DV-Hop, and PSO based DV-Hop localization algorithms for various values of communication range, anchor ratio, number of nodes and size of network is shown in Fig. 8. The bar graph demonstrates standard deviation of error is large in both the Centroid based and DV-Hop algorithms for communication range and size of network area and reduces if these algorithms are used with PSO.

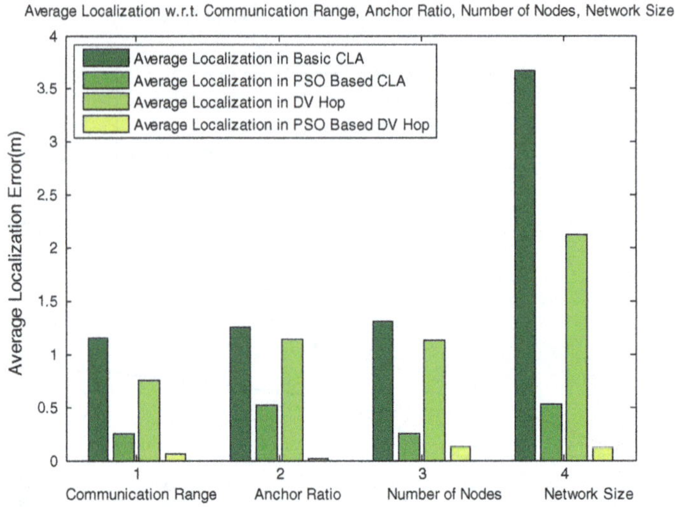

Fig. 7 Plot showing the average localization error

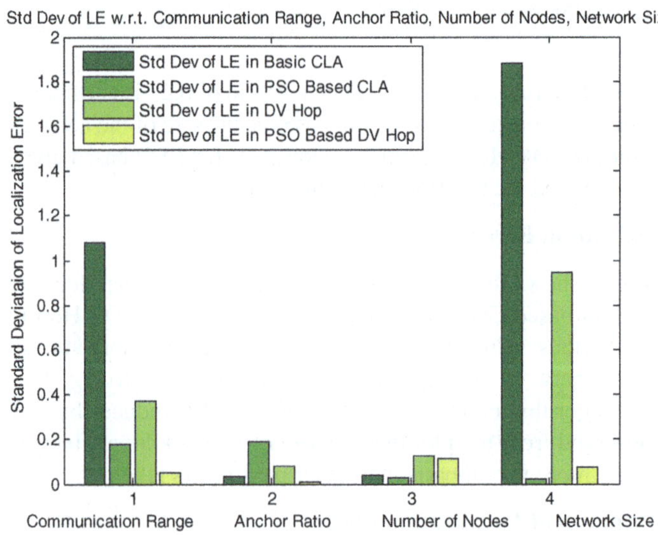

Fig. 8 Plot showing the standard deviation of localization error

9 Conclusion

In the present paper, localization algorithms Centroid, PSO based Centroid, DV-Hop, and PSO based DV-Hop are studied and simulated. The LE is computed for various values of communication range, anchor ratio, number of nodes, and size of network area. One of the important conclusions drawn from simulation results are location estimation error reduces as the communication range expands. PSO reduces the localization error in both of the algorithms for all anchor ratios. For the different number of nodes the basic versions of both the algorithms suffer from large localization error which reduces by using PSO. The localization error increases in both Centroid and DV-Hop algorithms to large values if the network size increases but it reduces and becomes uniform in PSO based Centroid and DV-Hop. In all of the scenarios, it is concluded that although PSO improves the performance of both the algorithms but the DV-Hop algorithm and PSO based DV-Hop performs better as compared to the Centroid and PSO based Centroid algorithm. Some more effort can be applied in this area by employing available machine learning strategies to reduce the error in location estimation.

10 Conflicts of Interest

The authors declare no conflict of interest.

References

1. Akyildiz, I.F., Su, W., Sankarasubramaninam, Y., Cayirci, E.: A survey on sensor networks. IEEE Commun. Mag. **40**(8), 102–114 (2002)
2. Voltz, P.J., Hernandez, D.: Maximum likelihood time of arrival estimation for real-time physical location tracking of 802.1 1 a/g mobile stations in indoor environments ad-hoc positioning system. In: IEEE Conference on Position Location and Navigation Symposium, pp. 585–591 (2004)
3. Kovavisarruch, L., Ho, K.C.: Alternate source and receiver location estimation using TDOA with receiver position uncertainties. In: IEEE International Conference on Acoustics, Speech and Signal Processing, vol. 4, pp. 1065–1068 (2005)
4. Niculescu, D., Nath, B.: Ad hoc positioning system (APS) using AOA. In: Twenty Second Annual Joint Conference of the IEEE Computer and Communication Societies, vol. 3, pp. 1734–1743 (2003)
5. Kumar, P., Reddy, L., Varma, S.: Distance measurement and error estimation scheme for RSSI based localization in wireless sensor networks. In: Fifth IEEE Conference on Wireless Communication and Sensor Networks, Allahabad, pp. 1–4 (2009)
6. Thimmaiah, S., Mahadevan, G.: A radio signal strength based localization error optimization technique for wireless sensor network. Indonesian J. Electr. Eng. Comput. Sci. **11**(3), 839–847 (2018)
7. Bulusu, N., Heidemann, J., Estrin, D.: GPS-less low cost outdoor localization for very small devices. IEEE Pers. Commun. Mag. **7**(5), 28–34 (2000)
8. Niculescu, D., Nath, B.: DV based positioning in ad hoc networks. Telecommun. Syst. **22**, 267–280 (2003)
9. He, T., Huang C., Blum, B., Stankovic, J., Abdelzaher T.: Range free localization schemes for large scale sensor networks. In: MobiCom, ACM Press, pp. 81–95 (2003)
10. Doherty, L., Pister, K.S., Ghaoui, L.E.: Convex position estimation in wireless sensor networks. In: IEEE Conference ICC, Anchorage, pp. 1655–63 (2001)

11. Shang, Y., Ruml, W., Zhang, Y., Fromherz, M.: Localization from connectivity in sensor networks. IEEE Trans. Parallel Distrib. Syst. (2004)
12. Chagas, H.S., Martins, J., Oliviera, L.: Genetic algorithms and simulated annealing optimization methods in wireless sensor networks localization using ANN. In: Fifth IEEE International Midwest Symposium on Circuit and Systems, pp. 928–931 (2012)
13. Rahman, M.S., Park, Y., Kim, K.: Localization of wireless sensor networks using ANN. In: IEEE International Symposium on Communication and Information Technology, pp. 639–642 (2009)
14. Katekaew, W., So-In, C., Rujirakul, K., Waikham, B.: H-FCD: hybrid fuzzy centroid & DV-Hop localization algorithm in wireless sensor networks. In: Fifth IEEE International Conference on Intelligent System, Modeling and Simulation, pp. 551–555 (2014)
15. Gopakumar, A., Jacob, L.: Localization in wireless sensor networks using PSO. In: IET International Conference on Wireless, Mobile and Multimedia Networks, Beijing, pp. 227–230 (2008)
16. Sun, Z., Tao, L., Wang, X., Zhou, V.: Localization algorithm in wireless sensor networks based on multi-objective particle swarm optimization. Int. J. Distrib. Sens. Net. 1–9 (2015)
17. Shunyuan, S., Quan, Y., Baoguo, X.: A node positioning algorithm in wireless sensor networks based on improved particle swarm optimization. Int. J. Future Gener. Commun. Netw. **9**, 179–190 (2016)
18. Kulkarni, R., Venayagamoorthy, G.: Particle swarm optimization in wireless sensor networks: a brief survey. IEEE Trans. Syst. Man Cybern. **41**, 262–267 (2011)
19. Hay-qing, C., Hua, W., Hua-kui, W.: An improved centroid localization algorithm based on weighted average in WSN. In: Third IEEE International Conference on Electronics Computer Technology, pp. 258–262 (2011)
20. Blumenthal, J., Grossmann, R., Golatowski, F., Timmermann, D.: Weighted centroid localization in Zigbee-based sensor networks. In: IEEE International Symposium on Intelligent Signal Processing. Alcala de Henares, pp. 1–6 (2007)
21. Dong, Q., Xu, X.: A novel weighted centroid localization algorithm based on RSSI for an outdoor environment. J. Commun. **9**, 279–285 (2014)
22. Liang, S., Liao, L., Lee, Y.: Localization algorithm based on improved weighted centroid in wireless sensor networks. J. Netw. **9**, 183–186 (2014)
23. Gupta, V., Singh, B.: Centroid based localization utilizing artificial bee colony algorithm. Int. J. Comput. Netw. Appl. **6**(3), 47–54 (2019)
24. Gupta, V., Singh, B.: Study of range free centroid based localization algorithm and its improvement using particle swarm optimization for wireless sensor networks under log normal shadowing. Int. J. Inf. Technol. **12**(3), 975–981 (2020)
25. Cui, H., Wang, Y., Liu, L.: Improvement of DV-Hop localization algorithm. In: Seventh IEEE International Conference on Modeling, Identification and Control, Sousse, pp. 1–4 (2015)
26. Zhipeng, X., Chunwen, L., Huanyu, L.: An improved hop size estimation for DV-Hop localization algorithm in wireless sensor networks. In: Twenty Seventh IEEE International Conference Chinese Control and Decision Control, pp. 1431–1434 (2015)
27. Ji, W., Liu, Z.: An improvement of DV-Hop algorithm in wireless sensor networks. In: IEEE International Conference, Wireless Communication, Networking and Mobile Computing, pp. 1–4 (2006)
28. Guo, W., Wei, J.: Optimization research of the DV Hop localization algorithm. Telkomnika Indonesian J. Electr. Eng. **12**(4), 2735–2742 (2014)
29. Kennedy, J., Eberhart, R.: Particle swarm optimization. In: IEEE International Conference on Neural Networks, Perth, pp. 1942–1948 (1995)
30. Cao, L., Cai, Y., Yue, Y.: Swarm intelligence based performance optimization for mobile wireless sensor networks: survey, challenges and future directions. IEEE Access **7**, 161524–161553 (2019)
31. Daanoune, I., Baghdad, A., Balllouk, A.: A comparative study between ACO based protocols and PSO based protocols in WSN. In: Seventh Mediterranean Congress of Telecommunications, pp. 1–4 (2019)

32. Jawad, H.M., Jawad, A.M., Nordin, R., Gharghan, S.K., Abdullah, N.F., Ismail, M., Alshaeer, M.J.: Accurate empirical path loss model based on particle swarm optimization for wireless sensor networks in smart agriculture. IEEE Sens. J. **20**(1), 552–561 (2020)

Dynamic Network Traffic Management for Secure and Faster Data Transmission

Purna Chandra Sethi and Prafulla Kumar Behera

Abstract The evolution of a large number of network-related emerging technologies leads to the exponential growth in the number of users. Traditional to corporate activities are being done over online medium. Hence, the network administrators' job is becoming tedious. The overall performance of the network depends on various parameters among which network traffic plays a major role. All the network resources are equally shared among all the available users. Malicious activities within the network lead to less availability of network resources for the actual users. Due to this reason, dynamic network traffic monitoring and analysis are becoming essential for identification of the malicious activities in the network. In this paper, we have proposed a dynamic network traffic monitoring and management technique to identify malicious traffic and prioritize the activities in order to assign the bandwidth within the network for network QoS. Finally, the security algorithm is applied to achieve information security.

Keywords Network Traffic Monitoring · Dynamic Bandwidth Allocation · Generalized Frequent Common Gram (GFCG) · SeLeCT · Quality of Service (QoS)

1 Introduction

In the present COVID-19 pandemic situation, smaller to larger businesses are shifted from traditional to online mode over Internet. Organizations progressively depend on Internet for its fundamental tasks. Beside increment in the multifaceted nature of Internet administrations, there is a radical increment of content delivery networks (CDNs) for its versatile Internet utilization. As per Digital 2021 Global Statshot

P. C. Sethi (✉)
Rama Devi Women's University, Bhubaneswar, Odisha 751022, India
e-mail: pcsethi@rdwu.ac.in

P. K. Behera
Utkal University, Vani Vihar, Bhubaneswar, Odisha 751004, India

K. N. Das et al. (eds.), *Proceedings of the International Conference on Computational Intelligence and Sustainable Technologies*, Algorithms for Intelligent Systems,
https://doi.org/10.1007/978-981-16-6893-7_63

Report in April 2021 [1], there are 4.72 billion worldwide Internet users, out of which 332 million users are added in the last 12 months. Each day, more than 900,000 new users are growing. The number of global Internet user growth rate is approximately 7.5%, and on time passing this progress rate is rising exponentially. The average universal Internet user spends nearly 7 h online every day. Accordingly, about 53% of the world populace has Internet connection that is supposed to increase to 55.1% in 2021 which was under 1% in 1995. Out of the above statistics, around 49% of Internet users are from Asia and India, the second-largest country which comes after China with respect to access of Internet by common people. Similarly, if we will consider mobile users, nearly 5.112 billion unique mobile users are there, out of which 4.388 billion have mobile Internet with 3.484 billion active social media users. This exponential growth of Internet user will highly motivate for a dynamic resources management and utilization to improve the average performance rate of system. Network performance depends on many parameters, out of which network traffic and security are vital ones which are focused in our research work.

The traffic engineering (TE) manages the estimation and executives of system traffic in order to structure, upgrade traffic steering process for improving the relative performance. With the increase in number of clients, bottleneck issue of network system resources such as network bandwidth may be raised. Dynamic resource utilization can be considered as one of the solutions to this bottleneck issue. In case of mobile network, remote system design, observation, its organization, etc., are the significant parameters for improving the QoS of system.

Any network system information security is considered as a major parameter of network QoS. In the age of information communication technology (ICT), the person or the organization having deals with more information are more powerful and costly. With the growth in Internet users, the crime over Internet is also increased exponentially. Organizations are spending a major portion of the total cost for providing security to information. In order to provide protection to information from unauthorized user access, various cryptography algorithms are followed by various organizations.

In order to overcome above issues, we have proposed research model that have two major components such as dynamic bandwidth on demand and faster as well as secure information processing. The dynamic bandwidth on demand concept is implemented using a reference model having network traffic estimation and traffic organization as the two basic components. The estimation of network traffic is dependent on the number of available users anticipated with traffic load balancing for improving the overall performance rate of network. We have considered web page rating in order to identify the bandwidth to be assigned to each user, i.e., if a user accesses a higher page rating web page, then relatively higher bandwidth is assigned in contrast to the Gaussian distribution. Apart from this, a pre-calculated guaranteed bandwidth is allocated to every user. The remaining bandwidth is assigned to all the available users based on the rating of web page. When any user application is complete, then the allocated bandwidth is released and assigned to the available users which are participating in the communication process. The traffic management undergoes a three-tire SeLeCT model in order to provide faster and secure searching of information.

The research paper is structured over different sections. Section 2 of this research paper covers a brief insight into the research work done by different researchers for managing network traffic as well as several current cryptography algorithms. Section 3 of this paper focuses on the proposed research work followed by the proposed algorithm explained in Sect. 4. Section 5 enlightens the experimental result and performance rate of the proposed experimental model. Finally, the conclusion and the future scope are specified in Sect. 6.

2 Literature Overview

In context of computer network, the total data transmitting over a network at a particular time is called network traffic. For data transmission, it has to be encapsulated in network packets, which is the prime parameter of network traffic. Network traffic estimation is the process of estimating the amount and the kind of traffic on a specified network. This is particularly significant with regards to effective bandwidth management. This network traffic may happen because of the exponential development of Internet users and restricted accessibility of different Internets users. As per Cisco, the world wide mobile phone traffic increased nearly by ten times in five years starting from 2014 to 2019 (depicted in Fig. 1) [1] and due to the COVID-19 pandemic, the number of mobile phone users increases by many folds in this two years.

According to Cisco, the expected Compound Annual Growth Rate (CAGR), the number of mobile devices will increase by many folds till 2022 and will reach nearly 9.4 billion as depicted in Fig. 2. Due to such large number of Internet users, network traffic increases exponentially and to handle such large number of Internet users, various traffic engineering (TE) techniques are adopted by network traffic engineers and network administrators [1].

Traffic Engineering (TE) manages the investigation of system traffic examination and estimation. Productive directing systems are proposed by network traffic experts

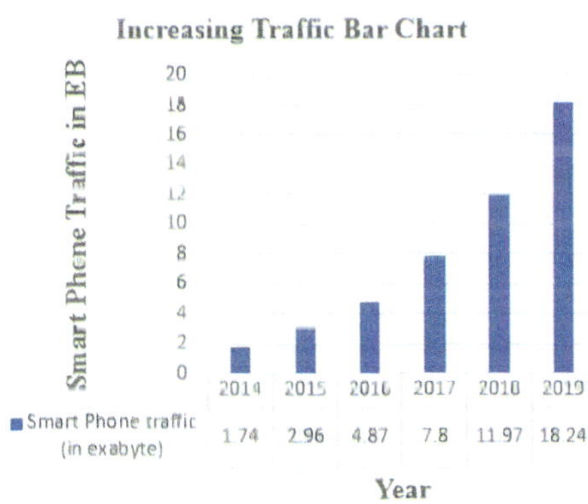

Fig. 1 Smart phone traffic bar chart

Fig. 2 Increasing rate of number of devices (in billions)

to organize the resources; control the traffic in order to improve the QoS. A software-defined network (SDN) is another method used for traffic analysis which works in forwarding and controlling layer. The performance of traffic analysis is done by the administrator in order to forward traffic in to different routes so that traffic congestion will be minimized. Network organization is accomplished for proficient framework availability and execution improvement. We had anticipated a functional system that allows dynamic information handling, traffic analysis and load balancing for burden management to improve the relative QoS of system.

Different dynamic data transfer approaches are studied widely by various researchers [2–4]. The researchers projected a probabilistic method for effective network traffic controlling and streaming [2]. Stream-based investigation becomes useful that decreased the system traffic by separating it into streams generators. Stream-based investigation and observation are made for analyzing traffic streaming consistently. Since stream-based investigation gives poor performance, monitoring technique is integrated in order to examine the traffic and presumed that stream-based monitoring strategy gives efficient traffic management than conventional methodology.

The researcher proposed a technique by measuring the link capacity for traffic management in [3]. Link capacity measurement is utilized to ascertain packet-level estimation and perform traffic sampling for observation. The researcher made a comparison analysis among three packet sampling techniques, namely n-in-N, Bernoulli, and sFlow sampling and resolved that there is adverse effect present due to packet inspected and sampling rate. The precision of framework remains event-independent starting from short time-scales starting from 10 ms to huge dataset utilization around the globe.

In [4], the researcher proposed a dynamic method called UPnP (Universal Plug and Play), in which each node will register first and then continue its networking activities. Due to this protocol, the resource distribution needs not to be done equally, and after network process completion, the resources are released so that it can be used by the current participating nodes.

In [5], the researcher proposed a recommendation technique using web page rating in order to calculate the effective requirement of bandwidth for each user activity.

Before implementation, it analyzes the available traffic by considering the number of current user activities and available bandwidth, and then it calculates the minimum bandwidth called as guaranteed bandwidth. This minimum guaranteed bandwidth is initially calculated and is assigned to individual user that participates in the network. After assigning the calculated amount of minimum guaranteed bandwidth, the excess available bandwidth is estimated and distributed between the available users based on the web page rating which they want to access. When any user activity completes, the assigned bandwidth is released and redistributed among the available users participating in the networking for improving the efficiency of the relative performance of the network.

In [7] is the extension work of [6], in which the researcher proposed a secured packet searching technique following hierarchical pattern matching technique. Apart from [8], the researcher used SeLeCT algorithm that provides faster searching and minimum on-chip memory utilization in order to perform the searching operation. For achieving information security, the SHA-256 algorithm [9] is implemented in the third tier of the proposed model. SHA-256 algorithm is applied because it involves a complex hash function implementation, and it is nearly impossible to regenerate the original information from the encrypted information. Hence, better information security is achieved by the proposed three-tier security model.

In [10], the researchers provided an efficient Internet traffic classification technique for faster data transmission. Security algorithm is integrated for achieving information security. As an extended work, in [11], the researcher proposed a technique in which the network traffic anomalies are identified that will minimize the network traffic for dynamic network traffic management. In [12], the researcher proposed a hybrid group security approach using ECC that not only provides faster processing but also be able to achieve better security towards improvement of network QoS.

3 Proposed Work

The proposed model focuses on two basic parts. The first part deals with improving the quality of service (QoS) by providing a mechanism for efficient utilization of network resources (here, we have considered bandwidth), and the second part deals with faster searching along with security of information. The proposed work flow model can be represented in Fig. 3. Here, we have analyzed the traffic by considering the available user and their operations, especially the web pages they are accessing. According to the web page rating, if a user wants to access a web page with higher rating, then it is assigned with a higher bandwidth, and when the user operation completed, the allocated bandwidth is released and are assigned to the other available users. A minimum guaranteed bandwidth is estimated first and allocated to every available user. Then, the calculated excess bandwidth available undergoes distribution according to the web page rating.

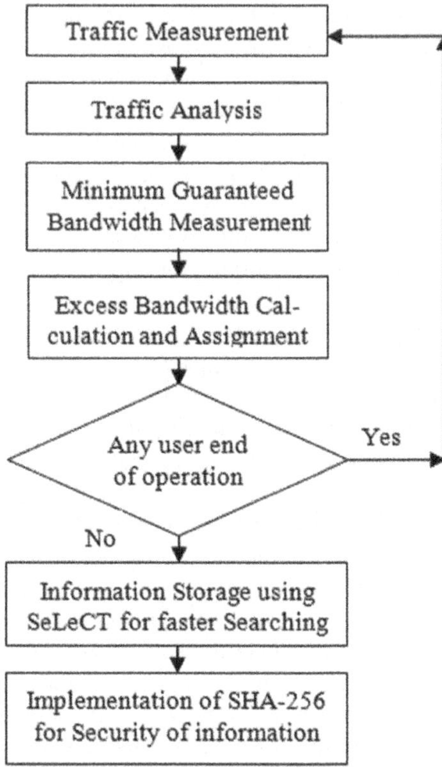

Fig. 3 Flow model of the proposed work

We have considered the MovieLens (http://movielens.org) dataset of UCI repository for our implementation. The dataset is the result of the movies referenced by 138,493 users between 09/01/1995 and 31/03/2015. Each user provided rating between 1 and 5 randomly according to their preference. The total dataset consists of 465,564 tag values with 20,000,263 ratings for the total dataset. Every user had evaluated at least 20 movies. Each user has unique userID without any other data involvement for identification of the users.

The first section of our research work deals with two activities such as: total available bandwidth calculation and dynamic bandwidth allocation. The flowchart for the proposed dynamic bandwidth allocation in order to manage the network traffic is depicted in Fig. 4.

3.1 Dynamic Bandwidth Allocation

Dynamic bandwidth allocation involves calculation of minimum guaranteed bandwidth which is allotted to every user, and then the available excess bandwidth is

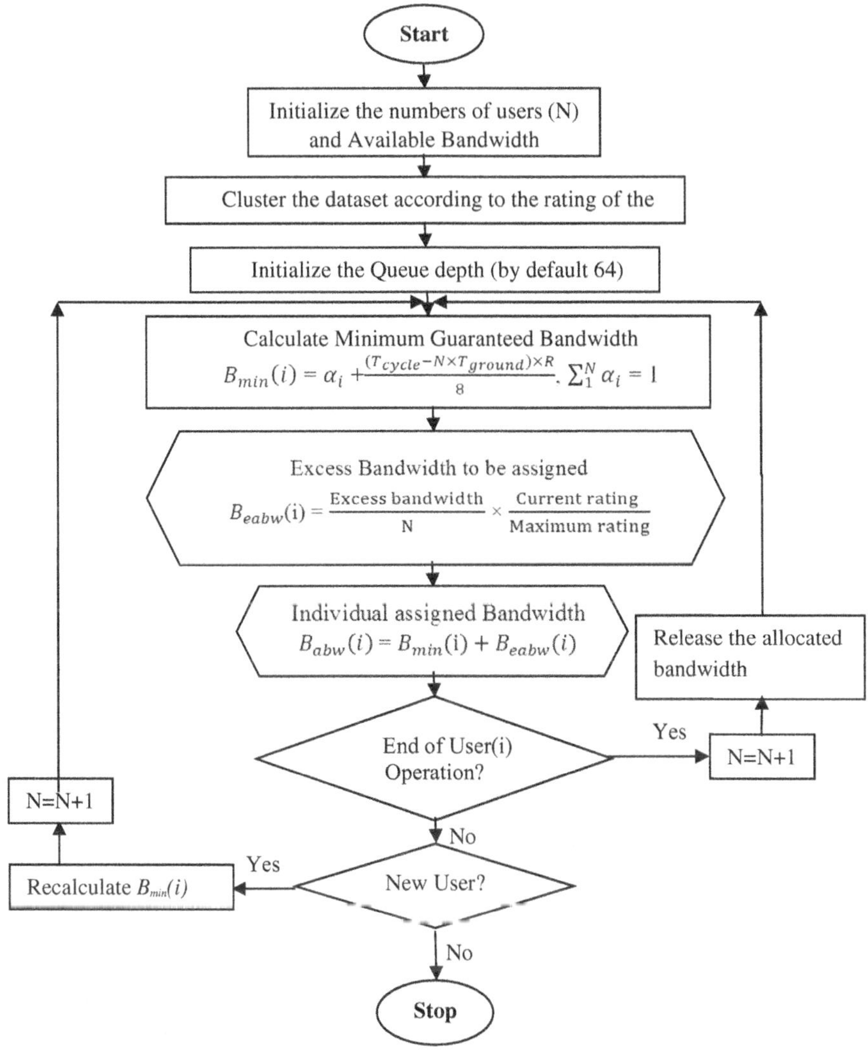

Fig. 4 Flowchart for dynamic bandwidth allocation

distributed among the available users. Since the theoretical and real assigned bandwidth always differs, it is quite a difficult task for traffic engineers to estimate the minimum guaranteed bandwidth. In general, allocated bandwidth is always smaller than theoretically bandwidth. For our research work, the minimum guaranteed bandwidth ($B_{\min}$) is set as constant and is calculated:

$$B_{\min}(i) = \alpha_i + \frac{\left(T_{\text{cycle}} - N \times T_{\text{ground}}\right) \times R}{8}, \sum_{1}^{N} \alpha_i = 1$$

Fig. 5 SHA-256 notations]

⊕	Bitwise XOR
∧	Bitwise AND
∨	Bitwise OR
¬	Bitwise complement
+	Mod 2^{32} addition
R^n	Right shift by n bits
S^n	Right rotation by n bits

where

α_i—The web page rating,
T_{cycle}—Individual maximum transmission cycle,
N—Number of Internet users,
T_{guard}—Guard time (total time between two consecutive accesses),
R—Rate of transmission.

The flowchart for dynamic bandwidth allocation is represented in Fig. 4 depicted below.

3.2 SeLeCT

SeLeCT, is an unsupervised self-learning classifier technique [9]. It is an efficient Internet traffic inspection technique. It follows self-seeding approach to recognize the class of traffic and assigns it to appropriate cluster. It is based on a quantifiable assessment for using for the movement in an automated fashion and achieves nearly 98% precision during networking. It supports adaptive classification in a simple iterative mode with limited complexity. It involves the generalized frequent common gram identification which is used for storage of clustered information. Frequent common grams are the set of bits which occurs frequently in the set of information, and after identification of the frequent common grams, the rest of the elements are stored using that element. The same common gram is used for searching as well as retrieval of information. It can be represented in Fig. 5.

From the GFCGS algorithm, the frequent common gram is identified $\{e, i, r\}$ in which are stored in first layer (H_1), and in the second layer, the other information are stored physically. The third layer is used for providing security to information using SHA-256 algorithm. It is depicted in Fig. 6.

3.3 SHA–256 Implementation

We have used the SHA-256 algorithm [5] for achieving information security in our research work. SHA-256 algorithm consists of the compression function followed

Fig. 6 GFCG identification

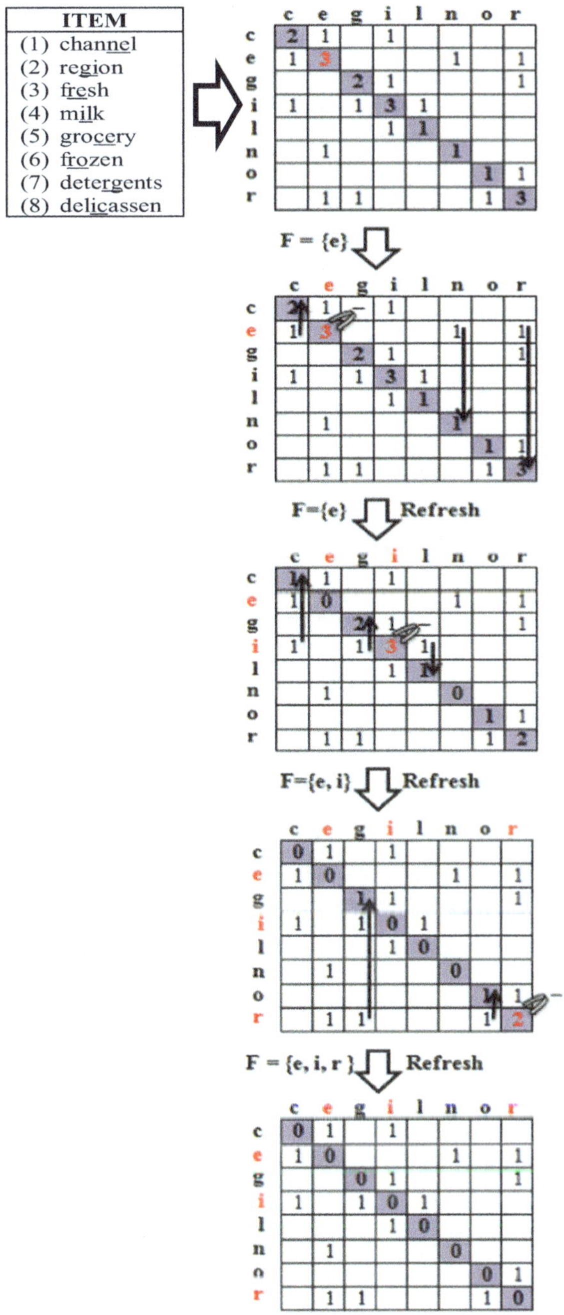

by message schedule as the two components. The compression function works on message block of 512 bits and intermediate hash value of block size 256 bits. The intermediate hash code is encrypted using message block of 256-bit block. It uses various symbols that work on 32-bit words.

The initial hash value $H_i(0)$ is found by considering 32-bit words of the fractional parts of square root of the initial eight prime numbers represented as: "6a09e667, bb67ae85, 3c6ef372, a54ff53a, 510e527f, 9b05688c, 1f83d9ab, and 5be0cd19."

The hash message preparation involves the following steps:

1. If the message length is M then adds the bit "1" at the termination of message followed by k number of "0" bits, where k is the least positive solution for the equation $1 + 1 + k \equiv 448 \bmod 512$. The 64-bit block is appended representing the length of message in binary.

   E.g. If the message is "abc," then length of message in ASCII becomes $8 \times 3 = 24$. So the ASCII equivalent in binary is appended with a "1" bit, then $423(=448-(24+1))$ number of "0" bits, and then the whole length of message is represented using 64 bits in order to make the whole information 512-bit size. The corresponding message is represented as:

$$\underbrace{011000010110001001100011}_{\substack{\text{Binary value of}\\ \text{ASCII equivalent info.}}}\ \underbrace{100\ldots000}_{\substack{423\\ \text{no. of 0 s}}}\ \underbrace{00\ldots11000}_{\substack{\text{64 bits, denoting}\\ \text{length of info.}}}$$

   The length of message must be a multiple of 512 bits.

2. Represent the whole message as multiple of 512-bit blocks in N groups as: $M_j(1)$, $M_j(2),\ldots, M_j(N)$. The sequence of 32-bits message block can be represented as $M_j(i)$, where, $j = 0 \ldots 15$ and $i = 1$ to N.

   The SHA–256 algorithm for N number of padded message block as:

   For $i = 1 : N\{$

   1. Initialize the 8 registers (a to h) by the hash value of $(i-1)$th element

$$a = H_1(i-1)$$
$$b = H_2(i-1)$$
$$\ldots$$
$$\ldots$$
$$H = H_8(i-1)$$

   2. For $j \leftarrow 0$: 63 //SHA-256 Compression function

      {

      Compute $\mathrm{Ch}(e, f, g)$, $\mathrm{Maj}(a, b, c)$, $\sum_0(a)$, $\sum_1(e)$, W_j

      $T_1 = h + \sum_1(e) + \mathrm{Ch}(e, f, g) + K_j + W_j$

      $T_2 = \sum_0(a) + \mathrm{Ma_j}(a, b, c)$

$$h = g$$
$$g = f$$
$$f = e$$
$$e = d + T_1$$
$$d = c$$
$$c = b$$
$$b = a$$
$$a = T_1 + T_2$$

}

3. Find out the hash value of $H(i)$

$$H_1(i) = a + H_1(i-1)$$
$$H_2(i) = b + H_2(i-1)$$
$$\ldots$$
$$\ldots$$
$$H_8(i) = a + H_8(i-1)$$

4. Return $H(N)$ in terms of different calculated hash value of M as ($H_1(N)$, $H_2(N)$,…, $H_8(N)$).

}

SHA-256 algorithm uses six logical functions operates on 32-bit words in order to return 32-bit word block. The functions are defined as (Fig. 7):

$$\mathrm{Ch}(x, y, z) = \left(x^\wedge y\right) \oplus \left(\neg x^\wedge z\right)$$

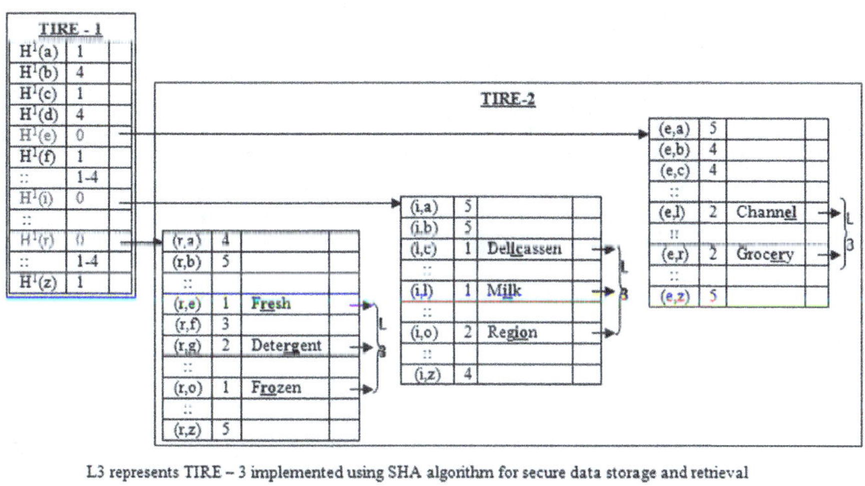

Fig. 7 EHMA implementation using layer-3 for providing information security

$$\text{Maj}(x, y, z) = (x^\wedge y) \oplus (x^\wedge z) \oplus (y^\wedge z)$$
$$\sum_0 (x) = S^2(x) \oplus S^{13}(x) \oplus S^{22}(x)$$
$$\sum_1 (x) = S^6(x) \oplus S^{11}(x) \oplus S^{25}(x)$$
$$\sigma_0(x) = S^7(x) \oplus S^{18}(x) \oplus R^3(x)$$
$$\sigma_1(x) = S^{17}(x) \oplus S^{19}(x) \oplus R^{10}(x)$$

The expanded form of message blocks W_0, W_1, …, W_{63} are calculated using the message scheduler of SHA-256 as:

$W_j = M(i)$ for $j = 0$, 1,……, 15, and
For $j = 16$ to 63
{

$$W_j = \sigma_1(W_{j-2}) + W_{j-7} + \sigma_0(W_{j-15}) + W_{j-16}$$

}

SHA-256 algorithm requires 64 number constant words, K_0, K_1, …, K_{63} considering the fractional parts of the cube roots of first 64 prime numbers for different K_i representations. The hexadecimal equivalents of K_i are used for application in this algorithm. The model for SHA-256 algorithm can be represented as Fig. 8.

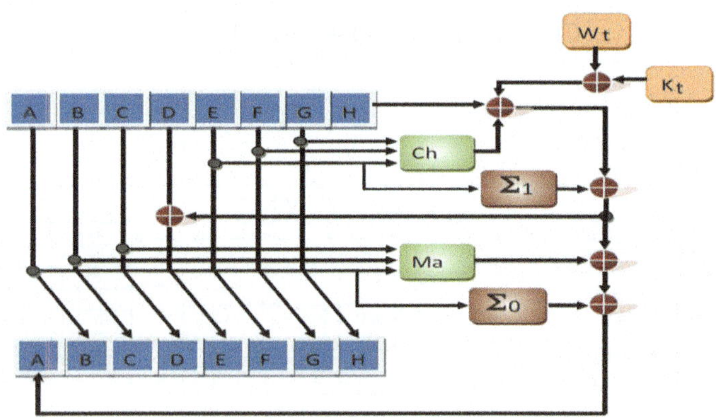

Fig. 8 SHA-256 operation model

4 Proposed Algorithm

The algorithm for dynamic bandwidth allocation for achieving faster and secure data transmission using web pages rating is specified as:

Step 1: Estimate the total number of users (N) along with the available bandwidth.
Step 2: Set the maximum size of cluster and apply Incremental Clustering according to web pages ratings.
Step 3: Set the Queue length to store clusters information.
Step 4: Calculate the Minimum Guaranteed Bandwidth (MGB) and assign to each user application.

$$B_{\min}(i) = \alpha_i + \frac{\left(T_{\text{cycle}} - N \times T_{\text{ground}}\right) \times R}{8}, \sum_{1}^{N} \alpha_i = 1$$

Step 5: Calculate the Excess bandwidth available:
Excess bandwidth ($B_{\text{excess}}(i)$) = Total available bandwidth − $N \times$ MGB.
Step 6: Calculate the Extra bandwidth to be allocated to individual user:

$$B_{\text{eabw}}(i) = \frac{\text{Excess bandwidth}}{N} \times \frac{\text{Current rating}}{\text{Maximum rating}}$$

Step 7: Bandwidth to be allocated to individual user ($B_{\text{abw}}(i)$) = $B_{\min}(i) + B_{\text{eabw}}(i)$.
Step 8: IF (User$_i$ operation End) Then

Release the assigned bandwidth to User i
N− −
Goto Step 5

ELSE IF (New User) Then

Allocate B min to the New User
N++
Goto Step 4

End IF

Step 9: 9.1 Consider the information of each cluster.

9.2 Find the Generalized Frequent Common Gram (GFCG).
9.3 Store the cluster information using the GFCG.
9.4 Apply SHA-256 algorithm on the GFCG to generate group signature for individual cluster
Step 10: Stop

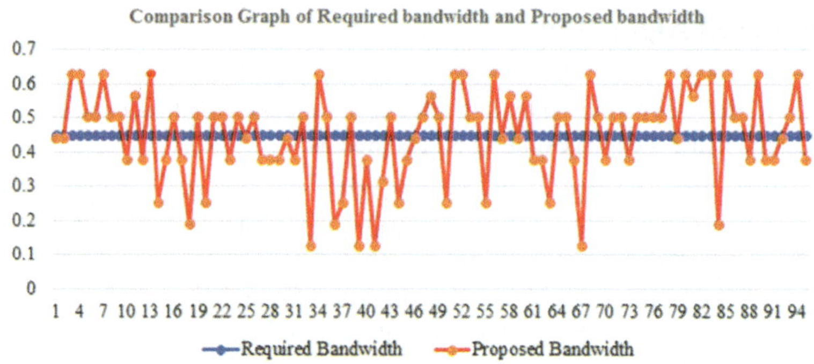

Fig. 9 Comparison graph of required bandwidth and proposed bandwidth

5 Experimental Result and Performance Analysis

The above-proposed algorithm is implemented in Google Colab using Intel core i3 2.20 GHz processor, 8 GB RAM. The first 500 user information (2230 movie) from Movielens dataset is used for implementation of the proposed research work. However, for simplicity of result representation in terms of graph, we have only considered the first instance of each user in a tabular form, and a comparison graph is traced for showing the efficiency of the proposed research work as depicted in Fig. 9.

The performance of the proposed model could be defined on the basis of various parameters. The proposed dynamic bandwidth allocation solves the wastage as well as insufficiency of available bandwidth. The GFGS algorithm provides the representatives (frequent common grams) among the cluster elements for storing as well as faster searching. The SHA-256 cryptographic hash functions produced an encrypted digital data which is nearly impossible to reverse engineer. After verification of user authentication, the information is allowed to be accessed by the valid user. The SHA-256 application on the frequent common grams provided better security to the information in comparison to the traditional approach.

6 Conclusion

Dynamic bandwidth allocation technique solves the even distribution of bandwidth irrespective of information type. Initially, a pre-determined lowest guaranteed bandwidth is considered and allotted to individual user. Then, the additional available bandwidth is determined and is divided among the users according to the web page rating. By the use of clustering technique, the information having similar properties are kept in particular cluster. By the use of SeLeCT algorithm using GFCG approach, the cluster elements can be stored in such a way that searching process becomes relatively faster in comparison to other searching techniques. Finally, SHA-256 implementation makes the system cryptographically strong so that it can be cracked without prior without prior information.

Due to the proposed secured dynamic bandwidth management technique, an intelligent resource management in terms of network bandwidth could be achieved. This leads to the optimized utilization network resources for achieving better bandwidth within the available resource.

References

1. https://datareportal.com/reports/digital-2021-april-global-statshot
2. Jadidi, Z., Muthukkumarasamy, V., Sithirasenan, E., Singh, K.: A probabilistic sampling method for efficient flow-based analysis. J. Commun. Netw. **18**(5), P818-825 (2016)
3. de Oliveira Schmidt, R., Sadre, R., Sperotto, A., van den Berg, H., Pras, A.: Impact of packet sampling on link dimensioning. IEEE Trans. Netw. Serv. Manag. **12**(3), 392–405 (2015)
4. Sethi, P.C.: UPnP and secure group communication technique for zero-configuration environment construction using incremental clustering. Int. J. Eng. Res. Technol. (IJERT) **2**(12), 2095–2101 (2013). ISSN 2278-0181
5. Sethi, P.C., Behera, P.K.: Network traffic management using dynamic bandwidth on demand. Int. J. Comput. Sci. Inform. Sec. (IJCSIS) **15**(6), P369-375 (2017)
6. Sheu, T.-F., Huang, N.-F., Lee, H.-P.: In-depth packet inspection using a hierarchical pattern matching algorithm. IEEE Trans. Depend. Sec. Comput. **7**(2), 175–188 (2010)
7. Sethi, P.C., Behera, P.K.: Secure packet inspection using hierarchical pattern matching implemented using incremental clustering algorithm. In: IEEE International Conference on ICHPCA-2014, 22–24 December 2014
8. Grimaudo, L., Mellia, M., Baralis, E., Keralapura, R.: SeLeCT: self-learning classifier for internet traffic. IEEE Trans. Netw. Serv. Manag. **11**(2), 144–157 (2014)
9. http://www.iwar.org.uk/comsec/index.htm\New hash algorithms (SHA-256, SHA-384, and SHA-512
10. Sethi, P.C., Behera, P.K.: Internet traffic classification for faster and secured network service. Int. J. Comput. Appl. (IJCA) **131**(4), 15–20 (2015)
11. Sethi, P.C., Behera, P.K.: Anomaly detection and optimization using scalable and flexible network data characterization. In: Global Transitions Proceedings, vol. 2(1), pp. 67–72 (2021). https://doi.org/10.1016/j.gltp.2021.01.009
12. Sethi, P.C., Sahu, N., Behera, P.K.: Group security using ECC. Int. J. Inf. Technol. (2021). https://doi.org/10.1007/s41870-021-00613-1

Numerical Modeling of Infiltration Rate

Rajendra Nath Paul, Prashanth Janardhan, and Harish Narayana

Abstract Infiltration is a vital component of the hydrological cycle. Estimation of infiltration directly is very expensive and laborious. Thus, the indirect measure of infiltration is gaining importance. Horton's infiltration equation is one of the famous equations used to measure infiltration rate. However, the prediction of the infiltration rate at a location at a particular time using Horton's model requires the rate of initial infiltration (RII) and the rate of final infiltration rate (RFI). In this study, a soft computing tool viz. artificial neural network (ANN) was used to predict the RII and RFI form soil properties. The results obtained indicate that the ANN model was successfully able to predict the RII and RFI and thus can be employed to predict RII and RFI.

Keywords Infiltration · Double-ring infiltrometer · Artificial Neural Network · Horton's model

1 Introduction

Infiltration has a considerable significance in irrigation system design, dam, culverts design, watershed management, and flood prediction. Downward movement of water through the pores of soil at the ground surface is termed infiltration. It is a natural phenomenon and one of the main components of the hydrological cycle. Infiltration rate depends on so many factors like soil texture, compaction of soil, vegetation cover, the porosity of the soil, moisture content which influences the infiltration rate at any particular site. From the past literature, the infiltration rate is evaluated by experimental tests and numerical models. Champatiray et al. [1] asserted the use of double infiltrometer over single-ring infiltrometer for measuring infiltration rate.

R. N. Paul · P. Janardhan (✉)
Department of Civil Engneering, NIT Silchar, Assam, India
e-mail: prashanth@civil.nits.ac.in

H. Narayana
Department of Civil Engneering, NIT Goa, Goa, India

K. N. Das et al. (eds.), *Proceedings of the International Conference on Computational Intelligence and Sustainable Technologies*, Algorithms for Intelligent Systems,
https://doi.org/10.1007/978-981-16-6893-7_64

Researchers and scientists have developed several physical, empirical, and semi-empirical models to compute the infiltration. Specific empirical models include SCS-CN, Kostiakov [3], modified Kostiakov [2], derived based on data obtained from laboratory and field tests. Horton [4], Overton [5], Grigorjev and Iritz [6] are a few examples of semi-empirical models. Various researchers make several attempts to evaluate the best performing model among those available infiltration models for different field conditions [7–13]. It was observed that various models are site-specific, and no one model was universal. Dagadu and Nimbalkar [14] compared the experimental and numerical infiltration rates and suggested that the Green-Ampt model and Horton's model were the best fitting models to predict infiltration rates for that particular site. Irrespective of the choice of the infiltration model, the determination of model parameters is the first step in using them [15]. These model parameters are dependent on the soil condition and vegetation at the site. These model parameters are determined through field measurements or model calibration, cumbersome in practice and time-consuming.

For the past two decades, soft computing tools are efficiently used in modeling and forecasting [16–20] and can also be used in the determination of model parameters [15]. Many researchers have successfully used soft computing techniques to predict the model parameters and cumulative infiltration rate [15, 21–23].

The above discussion shows that numerical models can be used under specific conditions. The numerical models are also dependent on the laboratory or field experimental data up to a certain extent. The laboratory or field experiments are time-consuming and may include errors. Furthermore, soft computing tools successfully model non-linear hydrological systems [18, 19, 24]. Therefore, this paper aims to use soft computing techniques to predict the parameters of numerical models that can be further used to predict cumulative infiltration rates at any particular site accurately.

2 Methodology

2.1 Study Area

National Institute of Technology, Silchar is one of the premier academic institutes of North-East India situated in the district of Cachar, Assam. The geographical location of the institute is 24.7577° N, 92.7923° E which comes under the southern part of Assam. The annual rainfall is 3875 mm and maximum precipitation occurs during the monsoon season between May to August. The temperature varies between 12 °C in winter and 35 °C in summer. Sub-tropical and humid climate prevails over the district. The soil cover of Cachar varies from alluvial to lateritic in nature. The texture of soil varies from clayey loam to clay. Two types of soil, namely, sandy loam, and clayey loam, are available in Silchar. Eight different locations were selected to study the variation of infiltration rate inside the campus for experimentation.

2.2 Experimental Setup

In the present study, Horton's infiltration model was used to evaluate the infiltration rate. The model parameters were estimated using the experimentally observed field data set. The Horton's equation is given below.

$$m = (m_0 - m_c)e^{-kt} + m_c \tag{1}$$

where m_c is the rate of ultimate infiltration (LT^{-1}); m_0 is the rate of initial infiltration (LT^{-1}), and t is the time (T), and k is the decay factor for soil infiltration.

For measuring the infiltration rates at the site, a double-ring infiltrometer was used. The instrument has two concentric rings of 300 and 450 mm in diameter. The depth of the rings is 300 mm. The soil properties like dry density by core cutter method, moisture content by oven dry method, and soil sample texture by hydrometer apparatus are also calculated and represented in Table 1. The average rate of initial infiltration and final infiltration is shown in Table 2.

Table 1 Physical properties of soil sample of each site

Site no	Sand (%)	Silt (%)	Clay (%)	DD (gm/cc)	MC (%)
1	53	27.5	19.5	1.37	14.20
2	70.91	11.42	17.67	1.64	11.50
3	52	27.63	20.37	1.41	12.96
4	60	17.6	22.4	1.70	13.07
5	54	29	19	1.61	12.57
6	59.48	18.32	22.2	1.45	18.07
7	66	20	14	1.66	13.57
8	72	17	11	1.49	12.07

Table 2 RII and RFI of the study area

Site no	RII (mm/h)	RFI (mm/h)
1	60	8
2	24	8
3	54	4
4	30	2
5	66	14
6	48	6
7	48	3
8	72	14

2.3 Artificial Neural Network (ANN)

Artificial Neural Network (ANN) is a modern type of computing system similar to the biological neural networks of animals [24]. Generally, an ANN is made of an. These types of systems learn from the data that is provided to them. These neural networks learn from the examples and adapt to a particular task without being programmed with task specific rules. Generally, a simple ANN model is made of 3 layers: an input layer, a hidden layer, and an output layer. The layers are interconnected by the weight. Each layer is made of neurons which are fed with external information. The data is further transmitted to the output layer, where the information is analyzed. The output layer gives the predicted information to the outside world [25, 26].

Figure 1 shows the line diagram of an ANN model. Suppose there are 'n' inputs say A_1, A_2, …, A_n. They are connected to neuron 'j' through weights W_{1j}, W_{2j}, …, W_{nj}. The summing junction of an artificial neuron sums up all the weighted inputs. Thus, it works similar to dendrites and the cell body of a biological neuron. Sometimes, the output of the summing junction may be zero, and to avoid this, a bias of fixed value 'b_j' is added to it. The transfer function (or activation function) can be compared to the axon and synapse of a biological neuron. The input to the transfer function 'f' is determined based on the relation,

$$u_j = \sum_{k=1}^{n} A_k W_{kj} + b_j \tag{2}$$

The final output, O_j, can be obtained as,

$$O_j = f(u_j) = f\left(\sum_{k=1}^{n} A_k W_{kj} + b_j\right) \tag{3}$$

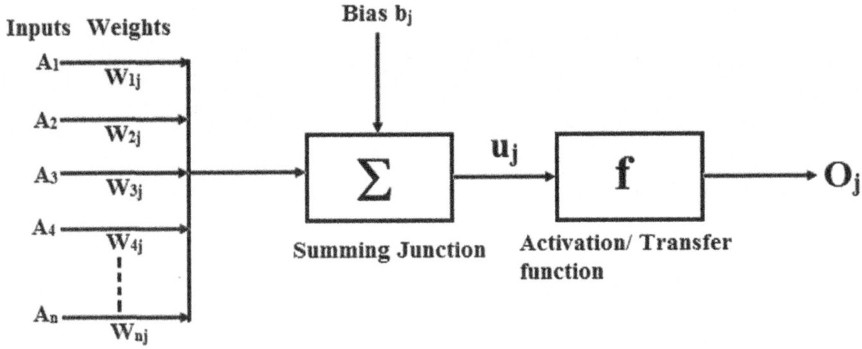

Fig. 1 Line diagram of an ANN model

In any ANN model, the transfer function largely influences the output of the model. There are various transfer functions like linear, tan sigmoid, hard limit, log sigmoid, etc.

2.4 Statistical Analysis

Statistical tools were used to evaluate the goodness of fit of the developed ANN model. In the present study, Root Mean Square Error (RMSE), and Coefficient of Determination (R^2) were used. The equations of the same are given below.

$$R^2 = \frac{\left(\sum_{i=1}^{z}(a_i - \overline{a})(b_i - \overline{b})\right)^2}{\sum_{i=1}^{z}(a_i - \overline{a})^2 \sum_{i=1}^{z}(b_i - \overline{b})^2} \tag{4}$$

$$\text{RMSE} = \sqrt{\frac{\sum_{i=1}^{z}(a_i - b_i)^2}{z}} \tag{5}$$

where, z = number of observations, a = observed value, b = predicted value, $\overline{a}$ = averaged observed values, $\overline{b}$ = averaged predicted values.

3 Results and Discussion

The infiltration test was performed at eight selected sites have different infiltration rates due to variation in soil properties. The rate of initial infiltration (RII) varied from 72 to 24 mm/h, and the moisture content from 18.07 to 11.50%. It was observed that the location with maximum moisture content took minimum time to stabilize the infiltration rates. The parameters of Horton's equation are shown in Table 3. The

Table 3 Horton's equation infiltration parameter

Site No	Horton's parameter (k)
1	0.060
2	0.034
3	0.057
4	0.047
5	0.066
6	0.074
7	0.066
8	0.029

parameters are dependent on the site condition, soil type, moisture content of the soil, compaction of the soil.

3.1 Artificial Neural Network Model

Rate of initial and final infiltration are the important terms in determining the constants of the infiltration models. ANN model was developed to predict these two terms using soil properties like a constituent of soil, its dry density, and moisture content (Table 1). The optimum parameters for the ANN model were obtained through an optimum search approach. The optimum parameters of the model are shown in Table 4. The developed model was trained with 80% data and remaining was used for testing the model. The results obtained were compared with experimental results using statistical tools.

Figure 2a, b shows the comparison graph between experimental and ANN model predicted rate of initial infiltration for both training and testing dataset. It is clear from the graph that the ANN model predicted the RII accurately based on the statistical results. Similar results can be observed for the rate of final infiltration as shown in Figs. 3a, b.

The generalization performance of the ANN model for RII and RFI is shown in Figs. 4a, b and 5a, b, respectively. It is clear that the model could capture the path of measured infiltration rate, which is the target of our prediction.

Table 4 Optimum parameters for the ANN model

Parameter	Value
Hidden layer	1
Hidden neuron	6
Training function	trainlm

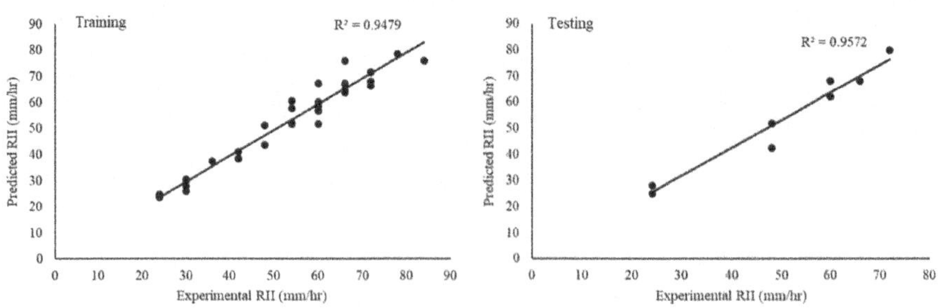

(a) (b)

Fig. 2 **a**, **b** Performance of training and testing dataset for RII

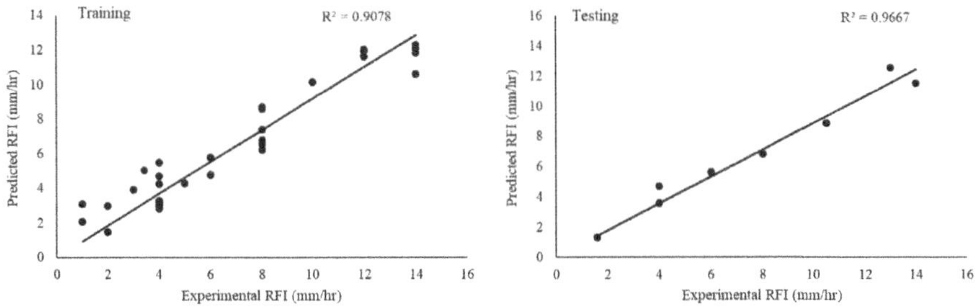

(a) (b)

Fig. 3 **a**, **b** Performance of training and testing dataset for RFI

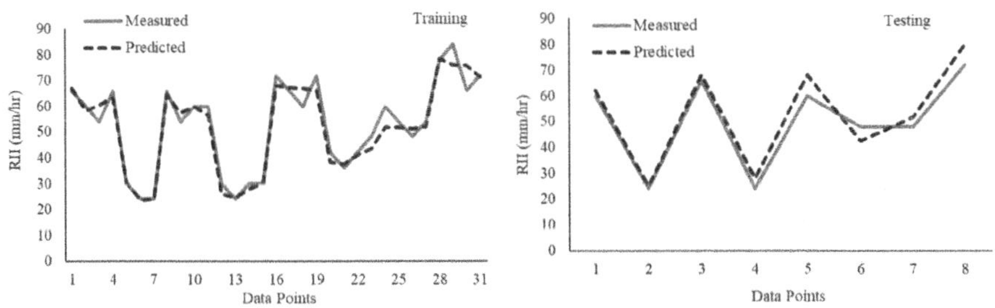

(a) (b)

Fig. 4 **a**, **b** Tracking of measured RII by ANN (training and testing data)

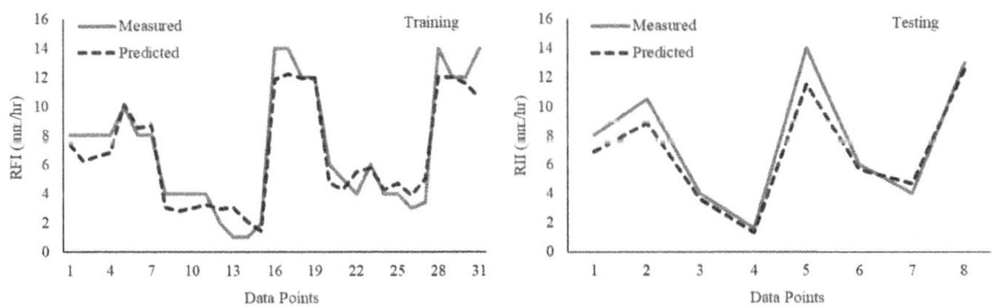

(a) (b)

Fig. 5 **a**, **b** Tracking of measured RFI by ANN (training and testing data)

4 Conclusions

Infiltration rate plays an essential role in the watershed management system and estimation of groundwater recharge. The input for Horton's equation requires RII and RFI, which have to be measured in the field. In the absence of field data, Horton's equation cannot be used.

Hence, in this study, ANN model was developed to predict the RII and RFI using the soil properties. The results clearly show that the ANN model can better predict the infiltration rates. The value of R^2 and RMSE of the ANN model was 95.7% and 4.96 for RII and 96.6% and 1.18 for RFI, respectively. The comparison of measured field data with the data estimated by the ANN method showed that it is closely fitted, and the general nature of both the datasets is also reasonably similar. The minor mismatch in the dataset between the measured and ANN predicted values might be attributed to the non-consideration of macro-pores and soil structure [27, 28]. Thus, the consideration of quantitative information of soil structure may enhance the model performance.

References

1. Champatiray, A., Balmuri, V., Patra, K.C., Sahoo, M.M.: Standard test for determination of infiltration rate of soil using double ring infiltrometer. In: Mishra, G.C. (eds.) Innovative Trends in Applied Physical, Chemical, Mathematical Sciences and Emerging Energy Technology for Sustainable Development, Excellent Publishing House, New Delhi, India, pp. 9–13
2. Smith, R.E.: The infiltration envelope: results from a theoretical infiltrometer. J. Hydrol. **17**, 1–21 (1972)
3. Kostiakov, A.N.: On the dynamics of the coefficient of water percolation in soils and on the necessity of studying it from a dynamic point of view for purposes of amelioration. Trans. Com. Int. Soc. Soil Sci., sixth ed., Moscow, Part A, pp. 17–21 (1932)
4. Horton, R.E.: An approach towards a physical interpretation of infiltration capacity. Soil Sci. Soc. Am. Proc. **5**, 399–417 (1940)
5. Overton, D.E.: Mathematical Refinement of an Infiltration Equation for Watershed Engineering. U.S. Department of Agricultural Service, Washington, DC (1964)
6. Grigorjev, V.Y., Iritz, L.: Dynamic simulation model of vertical infiltration of water in soil. J. Hydrol. Sci. **36**(2), 171–179 (1991)
7. Skaggs, R.W., Huggins, L.E., Monke, E.J., Foster, G.R.: Experimental evaluation of infiltration equation. TASABE **12**(6), 822–828 (1969)
8. Whisler, F.D., Bower, H.: Comparison of methods for calculating vertical drainage and infiltration for soil. J. Hydrol. **10**, 12–19 (1970)
9. Innes, G.: Comparison of infiltration models of disturbed soils using parameter optimization, Unpublished MS thesis, University of Tennessee, Knoxville, p. 83 (1980)
10. Singh, P., Narda, N.K., Singh, A.: Evaluation of Horton and Philip infiltration functions for determining optimum slope of graded check borders. J. Agric. Eng. Res. **29**(1–4), 1–9 (1992)
11. Mishra, S.K., Tyagi, J.V., Singh, V.P.: Comparison of infiltration models. Hydrol. Process **17**, 2629–2652 (2003)
12. Turner, E.R.: Comparison of infiltration equations and their field validation with rainfall simulation. PhD diss. (2006)
13. Sihag, P., Tiwari, N.K., Ranjan, S.: Prediction of unsaturated hydraulic conductivity using adaptive neuro-fuzzy inference system (ANFIS). ISH J. Hydraul. Eng. **25**(2), 132–142 (2017). https://doi.org/10.1080/09715010.2017.1381861
14. Dagadu, J.S., Nimbalkar, P.T.: Infiltration studies of different soils under different soil conditions and comparison of infiltration models with field data. Int. J. Adv. Eng. Technol. **3**(2), 154–157 (2012)
15. Jain, A., Kumar, A.: An evaluation of artificial neural network technique for the determination of infiltration model parameters. Appl. Soft Comput. **6**(3), 272–282 (2006)

16. Bingöl, D., Xiyili, H., Elevli, S., Kılıç, E., Çetintaş, S.: Comparison of multiple regression analysis using dummy variables and a NARX network model: an example of a heavy metal adsorption process. Water Environ. J. **32**(2), 186–196 (2018)
17. Tayfur, G., Aksoy, H., Eris, E.: Prediction of rainfall runoff-induced sediment load from bare land surfaces by generalized regression neural network and empirical model. Water Environ. J. **34**(1), 66–76 (2020)
18. Orzepowski, W., Paruch, A.M., Kowalczyk, T., Pokładek, R., Pulikowski, K.: Assessing the applicability of groundwater monitoring data in the modelling of soil water retention characteristics. Water Environ. J. **33**(2), 192–202 (2019)
19. Poursaeid, M., Mastouri, R., Shabanlou, S., Najarchi, M.: Modelling qualitative and quantitative parameters of groundwater using a new wavelet conjunction heuristic method: wavelet extreme learning machine versus wavelet neural networks. Water Environ. J. **35**(1), 67–83 (2021)
20. Mokhtari, H.A., Bagheri, M., Mirbagheri, S.A., Akbari, A.: Performance evaluation and modelling of an integrated municipal wastewater treatment system using neural networks. Water Environ. J. **34**, 622–634 (2020)
21. Parchami-Araghi, F., Mirlatifi, S.M., Ghorbani Dashtaki, S., Mahdian, M.H.: Point estimation of soil water infiltration process using Artificial Neural Networks for some calcareous soils. J. Hydrol. **481**, 35–47 (2013)
22. Mattar, M.A.A., El Marazky, M.S., Ahmed, K.A.: Modeling sprinkler irrigation infiltration based on a fuzzy-logic approach. Span. J. Agric. Res. **15**(1), 27 (2017)
23. Dewidar, A.Z., Al-Ghobari, H., Alataway, A.: Developing a fuzzy logic model for predicting soil infiltration rate based on soil texture properties. Water SA **45**(3), 400–410 (2019)
24. Cahit, B., Cengiz, D., Atis, H.T., Okan, K.: Predicting the compressive strength of ground granulated blast furnace slag concrete using artificial neural network. Adv. Eng. Soft. **40**(5), 334–340 (2009)
25. Duan, Z.H., Kou, S.C., Poon, C.S.: Using artificial neural networks for predicting the elastic modulus of recycled aggregate concrete. Constr. Build. Mater. **44**, 524–532 (2013)
26. Bondar, D.: Use of a neural network to predict strength and optimum compositions of natural alumina-silica-based geopolymers. J. Mater. Civil Eng. **26**(3), 499–503 (2014)
27. Brady, N.C., Weil, R.R.: The Nature and Properties of Soils, 12th ed. Prentice Hall, Inc., Upper Saddle River, New Jersey, USA
28. Shukla, M., Lal, R., Owens, L., Unkefer, P.: Land use and management impacts on structure and infiltration characteristics of soils in the North Appalachian region of Ohio. Soil Sci. **168**(3), 167–177 (2003)

Hybrid Flower Pollination Algorithm for Optimization Problems

S. Dhivya and R. Arul

Abstract In this article, an evolutionary approach with different objectives based on the improvement of the hybrid GSA (golden section search algorithm) using the flower pollination algorithm (FPA) with fitness distance balance selection (FDS) is proposed. In GSFPAFDS algorithm, the optimization problem is improved in several sub-problems. The pollination of flowers in the population is enhanced in Levy flight with fitness distance to enhance convergence. Therefore, the GSA can be used as a local search in FPA. This simulation is part of the traditional GSA, FPA, GSFPA, and personnel processing to solve the consequences of sequential global optimization problems. A local update strategy is used to increase optimization capacity. Its main objectives are to minimize the overall execution time and the allocation of the balance between local and global optimal zone. It allows global parallelization with the simultaneous performance of the processor to the functions, and determine the ratio of the communication to the functions. The behavior of the GSFPAFDS with few methods such as GSA, FPA, and GSFPA understandability has been showcased to solve high-dimensional technical problems.

Keywords Golden section algorithm (GSA) · Flower pollination algorithm (FPA) · And fitness distance balance

1 Introduction

In real life, several major design issues need to be addressed. The current design problem has some feasible limits and at least one target capacity [1]. The intention is to draw a series of marginal scores to limit the position's target score. The imperative can be an equity requirement or an imbalance constraint. Usually, this area's

S. Dhivya · R. Arul (✉)
School of Electrical Engineering, Vellore Institute of Technology, Chennai 600127, India

S. Dhivya
e-mail: dhivya.s2019@vitstudent.ac.in

K. N. Das et al. (eds.), *Proceedings of the International Conference on Computational Intelligence and Sustainable Technologies*, Algorithms for Intelligent Systems,
https://doi.org/10.1007/978-981-16-6893-7_65

parallelization concerns the simultaneous command assignment to several processors; here increases the correspondence's cost, especially when the correspondence's latency is essential. Therefore, performing the task on some specific process degrades proficiency.

Consequently, the sustainable calculation's primary goal is to achieve harmony between increasing parallelization and reducing pairing costs [2, 3]. The static allocation reservation is a NP-hard problem. To solve this problem in the proposed calculations, heuristic methods are used to obtain an ideal scheme as a whole [4]. The main difficulty with heuristic strategies is getting stuck in a prominent neighborhood. To avoid this difficulty, several metaheuristic strategies are used. These methods exploit the discovery opportunities in search methodologies, leading to a decrease in the chances of finding optimal neighborhood zones and reducing the search space [5]. Swarm intelligence has become an experimental interest for many people. It has recently become an exploratory interest for many exploratory researchers in related fields. Any attempt to develop calculations or gadgets that think critically through the province's social neighbors' general behavior or other organisms' social order depends on the energy efficiency [6]. Business planning focuses on effectively allocating activities to suitable properties.

In sustainable computing, determining the best solution is seen as an NP-complete problem. Each algorithm for planning is based on one or more techniques. Various planning algorithms based on heuristic algorithms have been established for cloud schemes, such as Heterogeneous Earliest Finish Time (HEFT) algorithms and Min–Min, Max–Min [7]. Besides, several metaheuristic activity planning algorithms rely on creating optimal schedules, such as Genetic Algorithm (GA), Firefly Algorithm (FFA), Salp Swarm Algorithm (SSA), Pathfinder Algorithm (PFA), Cheetah based Optimization Algorithm (CBA), Particle Swarm Optimization (PSO), Bacterial Foraging Search Optimization (BFO), and Ant Colony Optimization (ACO) have also been developed [8, 9]. In addition, various resource optimization algorithms are used in cloud computing to optimize resources [10]. In this article, three existing algorithms were analyzed and compared with the proposed hybrid optimization algorithm. The golden section search algorithm (GSA), Flower Pollination Algorithm (FPA), and GSFPA were analyzed in this article.

The paper structure is given as follows.

- In Sect. 1, it describes about briefly optimization planning problems based evolutionay algorithms.
- In Sect. 2, we reported on associated literature work.
- In Sect. 3, we presented the proposed method (GSFPAFDS) and the details of its description.
- In Sect. 4, we analyzed the results and discussions based proposed method.
- In Sect. 5, we have concluded this article along with its future scope.

2 Literature Review

Numerous metaheuristic calculations have been planned and used for task planning in the field of distributed computing. Metaheuristic calculations rely on two strategies to be efficient. The primary "exploitation" is the abuse of the best arrangement from preceding outcomes. The following procedure is "exploration," which examines new regions of the accommodation space. The vast majority of these calculations are recognized and noteworthy, such as the GA, PSO, ACO, and many more algorithms.

Deng et al. [11] have introduced an enhanced ACO dependent on the multi-populace methodology, co-development component, pheromone refreshing procedure, and pheromone dispersion instrument to adjust the combination speed and arrangement variety advances the advancement execution in tackling the enormous scope streamlining issue. The insects in the populace are separated into world-class subterranean insects and ordinary insects to advance the union rate and not fall into the neighborhood's ideal worth. The mobile sales reps issue and the genuine door task issue are chosen to confirm the streamlining execution. Sun et al. [12] have examined a whale enhancement calculation dependent on quadratic addition for high-dimensional worldwide improvement issues. On the one hand, an adjusted investigation measure by presenting another boundary is proposed to proficiently look through the locales and manage the untimely union issue.

On the other hand, the best inquiry specialist's quadratic insertion improves the abuse capacity and the arrangement precision. Besides, the calculation attempts to make a harmony between misuse and investigation. Pelusi et al. [13] have improved moth-fire streamlining calculation with a crossover search stage. The fundamental curiosity of the proposed approach is the meaning of a mixture stage among investigation and abuse. This stage is portrayed by wellness depended on the weight factor for refreshing the moth's positions.

Hajimirzaei et al. [14] carried out an intrusion detection system about a mixture of multilayer perceptron (MLP) and artificial bee colony (ABC) and fuzzy clustering algorithms. The MLP recognizes ordinary and unusually organized traffic packets. Simultaneously, the preparation of the MLP by the ABC calculation is completed by the further development of connection load estimates and settings. Abualigah et al. [15] introduced a hybrid ant lion optimization with tip-based differential expansion to address issues with multi-target activities under distributed processing conditions. In the planned strategy, to which we allude as MALO, the problem's multi-objective nature arises from the need to limit the span instantaneously by increasing resources' consumption. The ant lion's progress calculation has been updated using the differential expansion based on the top peak as a nearby tracking strategy to advance its offensive capability and not get surrounded in the large neighborhood.

The above metaheuristic calculations typically make up a significant arrangement of improvement calculations around the world. Absolutely, heuristic strategy intends to discover by mistake and preliminary constantly. "Meta-heuristics" can be applied to higher-level techniques that have modified and guided heuristic strategies in a way that can lead to agreements and progress beyond what is typically available in the investigation of neighborhood ideals. There are two types of neighborhood and world optimum while addressing improvement issues. Neighborhood optima are the best accommodation found in a region of accommodation spaces, but not quite the best for the entire problem space.

Interestingly, worldwide optima are the best answer for the whole problem space. Tracing the world into the most factual matters is surprisingly problematic, and pleasant and appropriate arrangements are often recognized as a result. Metaheuristic calculations are used to achieve these goals. Simultaneously, ABC, WOA, and PSO's weaknesses can help each other, finally improving the deal's nature. Unique and contrasted analyzer, FPA has a reasonable structure, a more remarkable capability to deal with high measurements and non-straight problems, and has fewer deficiencies to trap in optimal neighbors, in this way to discover a cheaper plan in complex forced design problems. The reproduction outcomes display that the planned GSFPAFDS calculation can acquire supplementary interesting optimal outcomes than other progress calculations on unrestricted standard work. It achieves cheaper design plan results than GS, FPA, and GSFPA on design problems.

3 Proposed Hybrid GSFPAFDS for Optimization Problems

For efficient use of limited resources, a hybrid GSFPAFDS with the following essential functions is recommended. First, a heuristic task interleaving algorithm is developed to analyze the possibility of scheduling tasks for scheduling schemes (particles). It not only separates different types of functions but also covers different reception areas. Meanwhile it offers an efficient way to calculate eligibility that significantly reduces the GSA's computational burden. Second, the GSA and FPA algorithms are combined to determine the optimal scheduling scheme.

The GSA quickly integrates a hybrid algorithm with local FPA optimization to improve efficiency and effectiveness. Third, secluded fitness distance is being used to improve the quality of start-up solutions. Finally, a fitness distance balance selection (FDS) was introduced to visualize group diversity and adjust for inertial weight, the likelihood of pollination, and Lévy flight probability. Therefore, the aggregation rate can be adjusted according to the variety of particles. Besides, research and its skills are balanced. In the GSFPAFDS process, demographic updates are based on segmentation and integration. In the partitioning procedure, we randomly split the total population into two equivalent subcategories, i.e., both subcategories have identical numbers of persons. Once everyone has completed the search, the two subgroups obtained from GSA and FPA are grouped so that people can share position data in the search area. This activity aims to allow everyone to get an excellent explanation in the shortest possible period and memorize valuable data. The basic concept of the GSFPA algorithm is constructed on three aspects: First, the hybrid algorithm and the benefits and drawbacks of the different algorithms can match each other. Second, by merging subsets to separate the best individuals and create new populations, this combination can also share the qualitative characteristics of GSA and FPA with the FDS, thereby preventing individuals from converting prematurely. Thirdly, after completing the GSFPAFDS algorithm, no supplementary limits were added, excluding the basic GSA and FPA algorithm parameters with FDS. The flow diagram of GSFPAFDS is defined in Fig. 1.

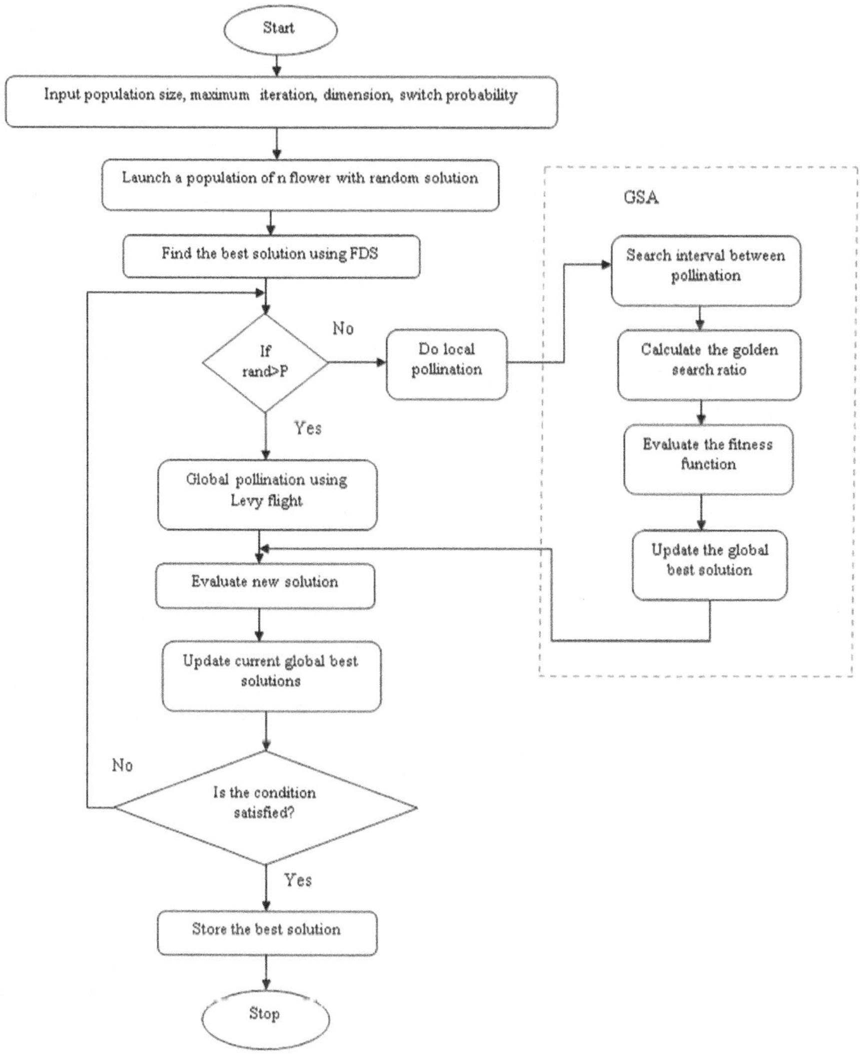

Fig. 1 The flowchart of the proposed method (GSFPAFDS)

3.1 Proposed Hybrid Flower Pollination Algorithm

The superior knowledge of the FPA emerges from the natural fertilization criteria deliberated above. In the FPA, one flower can be compared to a promising arrangement [16]. In the improvement cycle, dispersion through the research site is terminated by antibiotic and cross-fertilization. The pollen transfer through the flight of the Lévy is resolved. FPA calculations provide a quick and competent solution to address complex rationalization issues. Still, standard FPA calculations cannot address improvements. They distinguish the boundaries of an organization's reservation problem in fog calculations. Next, the standard FPA requires discretionary preparation and adopts distinct FPA calculations to solve the assignment planning problem to reassert the levy position and flight speed. The FPA is based on many hypothetical criteria. Biotic, cross-fertilization going about as worldwide hunt through Lévy flight. Abiotic and self-fertilization measures achieve the nearby inquiry. Flower consistency can be included because of the likeness of two blossoms.

Step 1: Initialization: The calculation begins with creating an introductory populace haphazardly, which is assessed to decide the current best arrangement. Here, take the irregular number $R \in [0, 1]$ is created and assuming, the worldwide fertilization and $R < p$ the blossom steadiness (a and c) can happen as following,

$$X_{t+1}^{i} = X_{t}^{i} + \gamma \text{Levy}\left(X_{t}^{i} - g_{\text{best}}\right) \quad (1)$$

At the time, there is an answer t, g_{best} signifies the current best arrangement, γ is the scaling factor, and Levy is a step size drawn from Levy flight.

Step 2: Fitness Function Evaluation: This task planning model assumes that GSA welfare is a cloud worker and this value is determined by the optimal location of FPA improvement and fitness distance selection (FDS). The adjustment increases the competitors' welfare estimates and the applicant's distance to the population's best fit [17]. This ensures that the applicant for the fit is selected with the utmost respect for well-being.

Step 3: Levy flight position:

$$\text{Levy}(s, a) = \frac{\beta \Gamma(\beta) \sin\left(\frac{\pi \beta}{2}\right)}{\pi} \cdot \frac{a}{s_{1+\beta}}, \; |s| \to \infty \quad (2)$$

where $\Gamma(\beta)$ is the standard gamma variable. β is a control boundary for the tail abundancy of the dispersion. The enormous advances ($s \gg s_0 > 0$) which are created by the accompanying nonlinear change as,

$$s = \frac{u'}{|v'|^{\beta - 1}} \quad (3)$$

where u' and v' are two unilateral examples taken from the Gaussian average campaign with mean equal to zero and standard deviation $\sigma_{u'}^2$ and $\sigma_{v'}^2$.

$$u' = \left(0, \sigma_{u'}^2\right), v' = \left(0, \sigma_{v'}^2\right) \tag{4}$$

$$\sigma_{u'} = \left[\frac{\Gamma(1+\beta)}{\beta\Gamma\left(\left(\frac{1+\beta}{2}\right)\right)} * \frac{\sin\left(\frac{\pi\beta}{2}\right)}{2^{\left(\frac{\beta-1}{2}\right)}}\right]^{1/\beta}, \sigma_{v'} = 1 \tag{5}$$

Here, the standard deviation $\sigma_{v'}$ is set to 1. $\sigma_{u'}$ and $\sigma_{v'}$ cannot be selected on its own for the calculation.

Step 4: Update the local pollination using the GSA: The GSA is utilized to refresh the streamlined position esteems for the proposed crossbreed method. GSA calculation is a technique intended for taking care of an advancement issue from a unimodal capacity. GSA is utilized to tackle multi-target advancement issues [18].

To advance the undertaking measure, the embedded assignment needs to follow the conditions underneath,

$$a_{X_{t+1}^i}^{\text{FPA}} + b_{X_{t+1}^i}^{\text{FPA}} = b_{X_{t+1}^i}^{\text{FPA}} + c_{X_{t+1}^i}^{\text{FPA}} \tag{6}$$

$$\frac{a_{X_{t+1}^i}^{\text{FPA}}}{b_{X_{t+1}^i}^{\text{FPA}}} = \frac{c_{X_{t+1}^i}^{\text{FPA}}}{b_{X_{t+1}^i}^{\text{FPA}}} = \frac{b_{X_{t+1}^i}^{\text{FPA}} + c_{X_{t+1}^i}^{\text{FPA}}}{a_{X_{t+1}^i}^{\text{FPA}}} = \phi \tag{7}$$

The ideal proportion φ is inferred by tackling (6) and (7). $\phi = 1.618$. This number is called Golden Ratio and is utilized to decide the new point inside the cloud task planning.

where,

$$X_{t+1}^i = X_t^i + \varepsilon\left(X_j^i - X_k^i\right) \tag{8}$$

Step 5: Store the best solution: Next, it updates the current best and resumes the search iteration until the last iteration.

4 Results and Discussion

In this section, we showed that the simulation results examine the use of the algorithm. This simulation compared the proposed methodology results with three methods: GSA, FPA, and GSFPA. The proposed hybrid GSFPAFDS algorithm works well. The simulation takes place on the MATLAB 2020 (A) platform, implementing the proposed hybrid strategy. Furthermore, the it is presented as the most popular task planning criterion and produces more favorable results than other metaheuristic optimization algorithms. In the simulations presented in this section, the proposed methodology evaluates 100 iterations.

4.1 Test Functions

The Ackley function [19] can be written as,

$$f_1(X) = -20\left[-\frac{1}{5}\sqrt{\frac{1}{\sigma}\sum_{a=1}^{\sigma} X_a^2}\right] - \exp\left[\frac{1}{\sigma}\sum_{i=1}^{\sigma}\cos(2\pi X_a)\right] + 20 + E \quad (9)$$

The simplest of De Jong's functions is the so-called sphere function [19],

$$f_2(X) = \sum_{a=1}^{z} X_a^2, \quad -5.12 \le X_a \le 5.12 \quad (10)$$

Yang's forest-like function [19],

$$f_3(X) = \left(\sum_{a=1}^{\sigma}|X_a|\right)\exp\left[-\sum_{a=1}^{\sigma}\sin\left(X_a^2\right)\right], \quad -2\pi \le X_a \le 2\pi \quad (11)$$

has a global minimum $f^* = 0$ at (0, 0).

For the three test functions above, each function can have a different size; it is a question of which dimensions to be used in simulations. Research suggests that higher-dimensional issues than the standard be more difficult; therefore, we tend to focus on higher-dimensional issues. We can then calculate the difference between the current solutions for the best average solution after 100 independent analyzes. The proposed GSFPAFDS achieved the best result of the four methods and converged the fastest, as given in Table 1.

Table 1 Comparison of the planned method with other algorithms for 3 test functions

Test functions	F_{min}			
	GSA	FPA	GSFPA	Proposed method (GSFPAFDS)
Ackley	0.0021	8.8818×10^{-16}	8.7838×10^{-11}	$\mathbf{2.9310 \times 10^{-14}}$
Sphere	2.4576×10^{-4}	4.2700×10^{-44}	1.4467×10^{-21}	$\mathbf{7.5388 \times 10^{-27}}$
Yang	0.0779	3.5791×10^{-22}	6.5879×10^{-11}	$\mathbf{2.4305 \times 10^{-16}}$

Figures 2, 3 and 4 show convergence chart of objective values for 100 iterations for all 3 test functions. Like the small-scale case, each algorithm's total cost decreases as the number of iterations rises, demonstrating the benefits of all cloud business planning approaches again. Besides, it can be seen that the proposed hybrid algorithm (GSFPAFDS) can accomplish an excellent presentation on the total cost associated with the other three algorithms. This proves the efficiency of the proposed hybrid GSFPAFDS in managing massive business implementations in cloud computing.

Furthermore, as the number of activities increased, the standardized cost of the planned hybrid GSFPAFDS decreased slightly. In contrast, the other three methods increased slightly, indicating that hybrid GSFPAFDS offers optimal solutions available in the occurrence of heavy works or significant activities. All this proves once

Fig. 2 Convergence curve of GSFPAFDS for Ackley

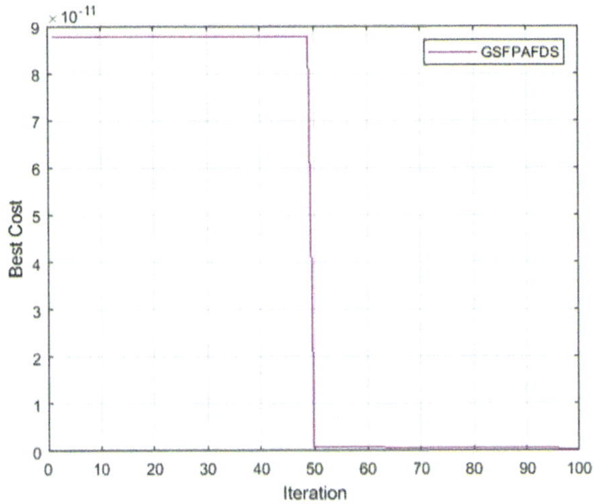

Fig. 3 Convergence curve of GSFPAFDS for Sphere

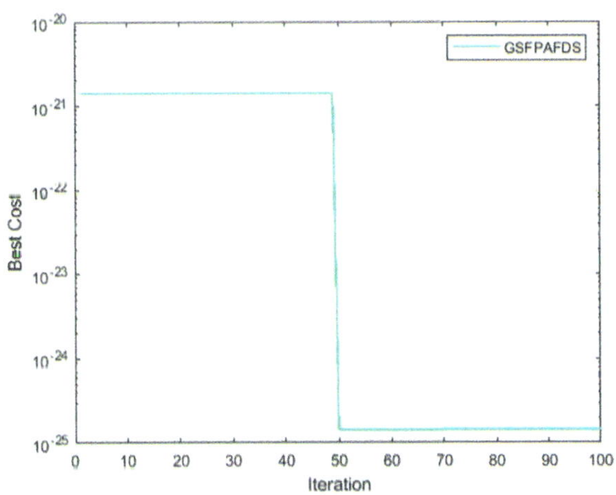

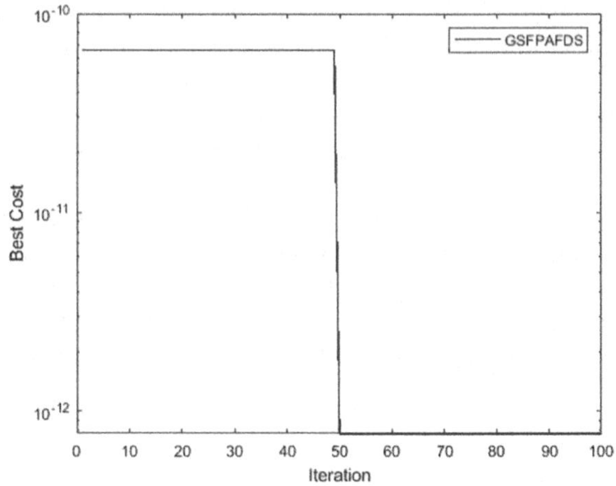

Fig. 4 Convergence curve of GSFPAFDS for Yang

again that the proposed hybrid GSFPAFDS can offer a better solution for cloud computing work than the GS, FPA, and GSFPA algorithms.

5 Conclusion

In the present study, the hybrid GSFPAFDS algorithm's performance was compared to another scalable algorithm for optimization problems with different objectives. The behavior of the hybrid GSFPAFDS algorithm was analyzed under different values of the control parameters. It is known in the field of soft computing for its high efficiency, flexibility, and cost-efficient nature. As a result, it is challenging to change requirements or create the framework for many requirements and processing tasks. This article proposes a new hybrid GSFPAFDS that solves problems avoiding premature false convergence in soft computing approaches. Three test systems were performed on the MATLAB platform. Some algorithms reduce the response time and full time by increasing each algorithm's output in this process. The results showed that GSFPAFDS outperforms other known methods for solving optimization problems.

6 Future Scope

Furthermore, it can be integrated faster for extensive research areas than other approaches, making it ideal for significant planning problems. The simulation results show that the proposed algorithm works better than the specified algorithms and can effectively solve large-scale multidimensional design problems. Given the high efficiency of the proposed algorithm, load balancing and programming the soft computing in a dynamic environment requires a high-speed algorithm to indicate that different algorithms and collective intelligence algorithms can be combined in future.

Acknowledgements The authors are very grateful to the authorities of VIT University, Chennai campus for carrying out this extensive research work in a prosperous manner.

References

1. Engin, O., Güçlü, A.: A new hybrid ant colony optimization algorithm for solving the no-wait flow shop scheduling problems. Appl. Soft Comput. **72**, 166–176 (2018)
2. Liu, B., Li, P., Lin, W., Shu, N., Li, Y., Chang, V.: A new container scheduling algorithm based on multi-objective optimization. Soft Comput. **22**, 7741–7752 (2018)
3. Moon, Y., Yu, H., Gil, J.M., Lim, J.: A slave ants-based ant colony optimization algorithm for task scheduling in cloud computing environments. HCIS **7**(1), 1–10 (2017)
4. Xu, L.Z., Xie, Q.S., Yuan, Q.N., Huang, H.S.: An intelligent optimization algorithm for blocking flow-shop scheduling based on differential evolution. Int. J. Simul. Model. **18**(4), 678–688 (2019)
5. Li, J., Tian, Q., Zhang, G., Wu, W., Xue, D., Li, L., Wang, J., Chen, L.: Task scheduling algorithm based on fireworks algorithm. EURASIP J. Wirel. Commun. Netw. **1**, 1–8 (2018)
6. Nayak, J., Naik, B., Jena, A.K., Barik, R.K., Das, H.: Nature-inspired optimizations in cloud computing: applications and challenges. In: Cloud Computing for Optimization: Foundations, Applications, and Challenges, pp. 1–26. Springer (2018)
7. Guo, W., Liu, T., Dai, F., Xu, P.: An improved whale optimization algorithm for forecasting water resources demand. Appl. Soft Comput. **86**, 105925 (2020)
8. Abdel-Basset, M., Abdle-Fatah, L., Sangaiah, A.K.: An improved Lévy based whale optimization algorithm for bandwidth-efficient virtual machine placement in the cloud computing environment. Clust. Comput. **22**(4), 8319–8334 (2019)
9. Xin-gang, Z., Ji, L., Jin, M., Ying, Z.: An improved quantum particle swarm optimization algorithm for environmental, economic dispatch. Exp. Syst. Appl. **152**, 113370 (2020)
10. Jana, B., Chakraborty, M., Mandal, T.: A task scheduling technique based on particle swarm optimization algorithm in the cloud environment. In: Soft Computing: Theories and Applications, pp. 525–536. Springer (2019)
11. Deng, W., Xu, J., Zhao, H.: An improved ant colony optimization algorithm based on hybrid strategies for scheduling problem. IEEE Access **7**, 20281–20292 (2019)
12. Sun, Y., Yang, T., Liu, Z.: A whale optimization algorithm based on quadratic interpolation for high-dimensional global optimization problems. Appl. Soft Comput. **85**, 105744 (2019)
13. Pelusi, D., Mascella, R., Tallini, L., Nayak, J., Naik, B., Deng, Y.: An improved moth-flame Optimization algorithm with hybrid search phase. Knowl. Based Syst. **191**, 105277 (2020)
14. Hajimirzaei, B., Navimipour, N.J.: Intrusion detection for cloud computing using neural networks and artificial bee colony optimization algorithm. ICT Exp. **5**(1), 56–59 (2019)
15. Abualigah, L., Diabat, A.: A novel hybrid ant lion optimization algorithm for multi-objective task scheduling problems in cloud computing environments. Cluster Comput. 1–19 (2020)
16. Adithiyaa, T., Chandramohan, D., Sathish, T.: Flower pollination algorithm for the optimization of stair casting parameter for the preparation of AMC. Mater. Today Proc. **21**, 882–886 (2020)
17. Kahraman, H.T., Aras, S., Gedikli, E.: Fitness-distance balance (FDB): a new selection method for meta-heuristic search algorithms. Knowl. -Based Syst. **190**, 105169 (2020)

18. Zoubiri, F.Z., Rihani, R., Bentahar, F.: Golden section algorithm to optimize the chemical pretreatment of agro-industrial waste for sugar extraction. Fuel **266**, 117028 (2020)
19. Yang, X.S.: Flower pollination algorithm for global optimization. In: Durand-Lose J., Jonoska, N. (eds.) Unconventional Computation and Natural Computation. UCNC 2012. Lecture Notes in Computer Science, vol. 7445. Springer, Berlin, Heidelberg (2012). https://doi.org/10.1007/978-3-642-32894-7_27

Predicting Autism Using Machine Learning

M. N. Pratik and Md. Azharuddin

Abstract Autism spectrum disorder (ASD) is a brain-based or neurodevelopmental condition that limits social interaction, communication behaviors, and social imagination. A specific range of activities and interests both are unique for every special person and followed repetitively. In this paper, we explain how a 4–11 years old child can be diagnosed with autism or without in a very short time. In many cases, children cannot talk about their problems but we can tell a lot of looking at his face. So, we use pictures as data to determine if someone had autism or not. We first extract the handcrafted features of the images with the help of local binary pattern (LBP) and gray level co-occurrence matrix (GLCM) algorithm and save the handcrafted features in an Excel seat. Then, we use Waikato environment for knowledge analysis (Weka) where different types of machine learning classifier are available.

Keywords Autism · Machine learning · GLCM · LBP · Weka

1 Introduction

Autism spectrum disorder (ASD) problem limits the social interaction (difficulty in making relationships, etc.) and the communication behaviors (difficulty in verbal/non-verbal communication, for instance gestures, body language, etc.), social imagination (difficulty in flexibility of thought, organizing, etc.), and a specific range of activities and interests [1]. Around 75% ASD begin in infancy but its effect carries after childhood and tends to adolescence and adulthood (WHO). Syndromes of ASD are usually identified in infancy up to the age of 3 years of life. Less number of girls are autistic than boys in nature (girls boys ratio 1:4 to 1:5) [2]. It is estimated worldwide that ~0.62% child has ASD (WHO). Till now, experts are trying to find the exact cause of ASD, but they are unable. Research indicates it may be combination of such causes those are genetic, environmental factors, and biological. There are various types of disorders fall under ASD and each of them diagnosed separately, but

M. N. Pratik · Md. Azharuddin (✉)
Department of Computer Science and Engineering, Aliah University, Newtown, Kolkata, West Bengal 700156, India

K. N. Das et al. (eds.), *Proceedings of the International Conference on Computational Intelligence and Sustainable Technologies*, Algorithms for Intelligent Systems,
https://doi.org/10.1007/978-981-16-6893-7_66

currently, all condition grouped together which is known as ASD (1994). In 2014, all disorders grouped together into a single diagnosis DSM-5ASD [2]. 70% of ASD cases are non-genetic [1]. Non-genetic ASD follows the manner of developmental disease includes environmental factors with useful data suggesting a specific contribution by prenatal and newborn exposure [2–4]. Globally, autistic people are mostly subject to stigma, human rights violations, and discrimination. Worldwide supporters and access to service for autistic people are inadequate.

In this paper, we focus on only image. Images are divided into two part such as autistic and non-autistic, both of them divided into two part such as training set and test set using LBP and GLCM. We extract features from images and after that, we apply different machine learning algorithms on the extracted feature for accuracy test.

We organize the paper as follows. In Sect. 2, we discuss related works. In Sect. 3, we discuss the architecture. The proposed work is discussed in Sect. 4. The simulation results are discussed in Sect. 5. Finally, we conclude in Sect. 6.

2 Related Works

Sula et al. [5] proposed a peer-to-peer (P2P) IoT-based system that can recognized autism spectrum disorder (ADS). They implemented a system which identifies various ways for how to improve lifestyle of autistic people. Arthi and Tamilarasi [6] implemented a system which is based on neuro fuzzy for the predicting autism spectrum disorder. The system manually takes physiological data as input. Maite et al. [7] implemented a concept which identifies phonological disorder of an autistic child. They mainly differentiate and characterize the level of keenness of the phonological matter. Each word evaluated as percent score and the outcomes put into a bar diagram. These results can be traded and kept a fast register of specific client movements. However, uncertainty can be caused by the system. The performance can be improved by using Internet of things (IoT). Najwani et al. [8] provided a classifier using multilayer perceptron (MP) and Gaussian mixture model (GLM) for identify autistic children depend on electroencephalography (EEG) signal. This scheme may be used to quickly identifying autistic children for the motive of early intervention. Moreover, there are two groups autistic and normal, signal strength of which has big difference. The system used a senor network for automatic data collection. Dorothée et al. [9] proposed a concept which interacts with human behavior. It adapts and classifies behavior on on-line, according to interaction between human and robot. This approach can classify two modes of interactions, i.e., "strong interaction" and "gentle interaction." The result is showed by a robot dog known as "AIBO." AIBO can recognize which types of interaction were involved. Albinali et al. [10] proposed using three wireless accelerometers put on each carpus of autistic children to detect stereotyped and iterative motor movements to shake hands and the body with 6 autistic children. Westeyn et al. [11] proposed a concept which identifies stereotyped behavior using wireless Bluetooth accelerometer for a child. The functions identified by this system are drumming, hand striking hand flapping, placing, and rocking.

Patenaude et al. [12], for example, utilized both shape appearance and active models into a Bayesian framework because of brain segmentation.

The brain has absorbed both the shape appearance and active models in a Bayesian structure for division. Wang et al. [13] realized using random forest procedure to integral features from different modalities for child brain segment along with probability maps of gray matter (GM) and cerebral white matter (CWM). Liu et al. [14] coined a new paradigm which depends on facial determination, as well as the use of a classification method for diagnosing autism spectrum disorder [15]. Yun Jiao and partner applied four different types of machine learning algorithms to detect ASD. Those are multilayer perceptron, support vector machine (SVM), logistic model trees, and functional trees.

3 Architecture

The architecture for ASD prediction is defined as follows. There are four important modules in classifier for predicting ASD. They are given bellow:

I. Data collection.
II. Data preprocessing
III. Machine learning algorithms
IV. Prediction.

I. **Data collection**:

This is a collecting and measuring process of raw data for target variables on a regular basis. The general data collection process ensures that data are defined and truthful as necessary so that data-based decisions are valid.

This dataset can be used in two ways and both methods are applicable for machine learning algorithms. One standard method is to divide the data into training, test, and validation set. The training set is referred to as train. Train directory holds two subdirectories, i.e., one autistic and another non-autistic. Autistic subdirectory holds 1667 autistic child image in JPG format, 224 × 224 × 3 size. Non-autistic subdirectory holds 1667 autistic child image in JPG format, 224 × 224 × 3 size.

All subdirectories images are labeled in serial numbers (001.jpg, 002.jpg, etc.). "Zero" is provided to save the file order from the stream when using the directory. This creates it fairly easy to relate to a test image prediction with a related image file. Valid directory contains validation images. It similarly separate 50 images of non-autistic children and 50 images of autistic children in the same format as training [16].

II. **Machine learning algorithms**:

Various types of machine learning algorithm are used such as BayesNet, Naïve Bayes multinomial, logistic, SGD, SMO, IBK, KSTAR, and random tree.

III. **Data preprocessing**:

Images are divided into two parts such as autistic and non-autistic, both of them are divided into two parts such as training set and test set. We first extracted the handcrafted features of the images with the help of local binary pattern (LBP) and gray level co-occurrence matrix (GLCM) algorithm and saved the handcrafted features in an Excel seat.

IV. **Prediction**:

Test data for autism can be predicted with classified datasets and positive or negative predictions related to their greater probabilities are obtained as final result.

4 Proposed Work

4.1 GLCM

The gray level co-occurrence matrix (GLCM) is a well-known method for extracting texture-based properties. It extracts the feature according to second-order statistics (SOS). GLCM (Table 1) works on the textural relationship between pixels and describes the frequency of pixel intensity from each part of an image. GLCM

Table 1 GLCM handcrafted features

−48.7523	0.388776	0.122858	2.708531	0.835137
−47.2297	0.419082	0.183752	2.567419	0.822664
−85.8516	0.393061	0.251063	2.267278	0.839966
−46.5527	0.408163	0.148308	2.525418	0.828736
−48.4894	0.98551	0.057699	3.395033	0.666734
−54.681	0.545306	0.173222	2.580199	0.798071
−34.1767	0.545306	0.101387	2.874347	0.783983
−63.2021	0.677245	0.221943	2.562799	0.774000
−76.2112	0.480816	0.172678	2.522507	0.813480
−16.2022	0.572143	0.088927	2.97551	0.769508
−81.2685	0.414082	0.166117	2.345507	0.828939
−86.9227	0.463061	0.198128	2.389384	0.825558
−16.3135	0.719082	0.060385	3.229834	0.722595
−42.9694	0.378367	0.14234	2.53192	0.835883
−135.622	0.488367	0.235419	2.382709	0.825878
−39.8623	0.579286	0.100578	2.896936	0.770315

expresses the properties of an image using the image, converting the image into a matrix (dimension of rows and columns is equal) with numbered gray values of an image. Each value of this matrix depends entirely on the frequency of such pairs of pixels. Each and every pixel value depends on their surrounding pixel pairs. If the intensity values of pixels are wide, then this image creates too big transient matrix. GLCM has acquired a total of twenty-two features on the basis of texture; contrast, self-association, strength, interrelationship, cluster properties, sum entropy, entropy, cluster shade, maximum probability, difference, homogeneity, sum of squares (variation), inverse difference normalization, sum average, difference variability, difference inequality, difference in measuring relationship information, normalizing the opposite difference moment.

4.2 LBP

Local binary pattern (LBP) is an operator used to discuss details about the gray level layout structure of an image. The structural and statistical model of the text structure are calculated mathematically. It produces a numeric value known as LBP code for each pixel in the image (Table 2). It uses 3 × 3 neighboring pixels for LBP labeling. The middle pixel is compared to the values of the adjacent neighboring pixels, if the neighboring pixels are higher than the center pixels, keep 1 and 0 lower.

Table 2 LBP handcrafted features

0.036527	0.000836	0.020454	0.018255	0.016522
0.072037	0.002579	0.010946	0.013810	0.021390
0.045656	0.005633	0.032794	0.018478	0.016979
0.043590	0.013943	0.049811	0.024104	0.020167
0.040740	0.005559	0.023106	0.015250	0.011025
0.074979	0.011536	0.049495	0.035403	0.040404
0.078848	0.002486	0.015433	0.033302	0.033738
0.101414	0.005395	0.020647	0.032502	0.038806
0.062626	0.008483	0.027718	0.028505	0.049364
0.070146	0.004605	0.013864	0.017468	0.012199
0.088970	0.005524	0.020565	0.023707	0.016902
0.088970	0.005524	0.020565	0.023707	0.016902
0.046167	0.006681	0.021617	0.021243	0.031855
0.086370	0.005860	0.030400	0.022368	0.026138
0.065949	0.010159	0.026913	0.025058	0.026650
0.083861	0.007707	0.023407	0.026284	0.029200

The most important features of this algorithm are the quantitative simplicity and the resistance to gray layer change, they are used for real-time application. LBP code value calculates for the pixel as follows:

$$\mathrm{LBP}_{P,R}(k_c) = \sum_{P=0}^{P-1} S(K_P - K_C)2^P, \quad S(x) = \begin{cases} 0, x < 0 \\ 1, x \geq 0 \end{cases}$$

The code value is created by comparing the intensity value K_C of the center pixel with the intensity value K_P ($P = 0, 1, 2, \ldots, P - 1$) of the neighbor pixels P. R is the radius used to determine neighborhoods. The function $S(X)$ is used in the LBP operator.

4.3 LBPGLCM

The local binary pattern gray level co-occurrence matrix (LBPGLCM) is based on combination of LBP and GLCM algorithms. We perform for this operation, firstly, we apply LBP operator on the row image. After analysis, LBP operator produces a numeric value known as LBP code for each pixel in the image (Table 3). Finally, the GLCM applies on the image. GLCM works based on second-order statistics (SOS). After, feature extraction is combined both LBP and GLCM feature in one tabular form. Extracting properties of the GLCM algorithm are each pixel highly depends on

Table 3 LBGLCM handcrafted features

0.090438	0.006703	0.029492	0.006365	0.086274
0.090338	0.006367	0.067303	0.005681	0.066106
0.060719	0.007589	0.075692	0.009064	0.063135
0.079469	0.012589	0.069153	0.010293	0.048345
0.044604	0.007560	0.052965	0.003201	0.054853
0.035800	0.006975	0.049535	0.004819	0.04521
0.029383	0.002278	0.049812	0.002072	0.027898
0.048836	0.004340	0.070672	0.008726	0.052339
0.064595	0.009142	0.068084	0.010980	0.083904
0.061042	0.004453	0.066422	0.011932	0.062325
0.043801	0.006275	0.087942	0.008872	0.040845
0.037552	0.004298	0.065135	0.005336	0.036331
0.053881	0.010097	0.073367	0.010684	0.06823
0.047741	0.011052	0.139667	0.004254	0.045523
0.058835	0.002735	0.057936	0.006870	0.074358
0.032411	0.002813	0.069217	0.004508	0.040294

the neighboring pixel. It does not care about the nearest local patterns in the image. However, using this LBPGLCM method, we can extract the entire special feature (considering all the texture structures).

4.4 *Weka*

Waikato environment for knowledge analysis (Weka) is a free software licensed under the GNU general public license, and the book "data mining: Practical machine learning tools and techniques" developed at the University of Waikato, New Zealand.

Weka supports a number of standard data mining functions, more specifically, data preprocessing, classification, clustering, visualization, regression, and feature selection. All Weka's strategies are predicted with the idea that data are available as a flat file or relationship, where each and every data point is described by a certain number of attributes (generally numerical or nominal attributes, but some other attribute types are also supported). Weka provides access to deep learning with Deeplearning4j.

5 Simulation Results

The proposed algorithm was simulated by using Python on 4.2GHz CPU, Intel CORE i5 processor, 8 GB RAM running on the Microsoft Windows 10 platform. We have used dataset available in [16]. We have implemented various machine learning algorithms to calculate the result in terms of accuracy shown in Fig. 1.

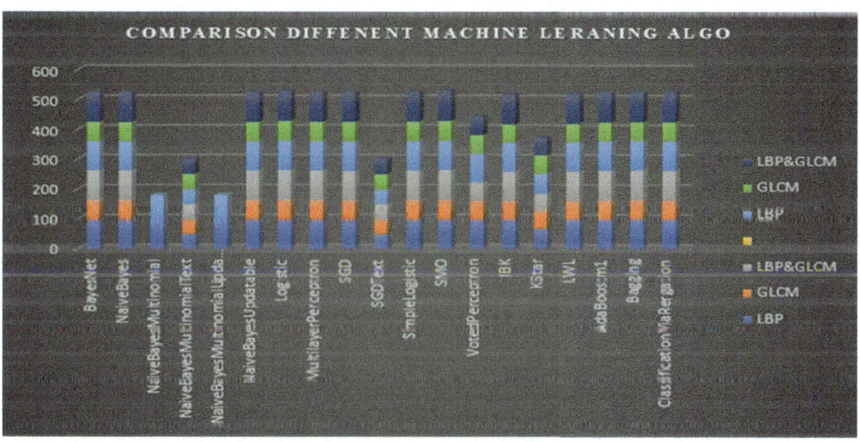

Fig. 1 Comparison of different machine learning algorithm

Table 4 Comparison between deferent machine learning classifier

		5			10	
Classifier	LBP	GLCM	LBP&GLCM	LBP	GLCM	LBP&GLCM
BayesNet	99.9334	63.1702	99.9334	99.9334	63.1036	99.9334
NaiveBayes	99.9334	62.4043	99.9334	99.9334	62.4376	99.9334
NaiveBayesMultinomialText	50.1332	50.1499	50.1499	50.1332	50.1499	50.1499
NaiveBayesUpdatable	99.9334	62.4043	99.9334	99.9334	62.4376	99.9334
Logistic	99.9001	64.8019	99.8002	99.9001	64.6354	99.8335
MultilayerPerceptron	99.8668	63.7363	99.6337	99.8668	63.0703	99.6337
SGD	99.8668	61.9048	99.8668	99.8668	64.369	99.8668
SGDText	50.1332	50.1499	50.1499	50.1332	50.1499	50.1499
SimpleLogistic	99.9334	63.703	99.8335	99.9001	63.703	99.8668
SMO	99.9334	64.5022	99.9334	99.9334	64.8685	99.9334
VotedPerceptron	98.3678	61.9714	61.9714	98.3344	61.7383	61.7383
IBK	99.8668	59.0077	99.2008	99.8668	59.7736	99.3673
KStar	67.5217	59.4073	58.3084	67.9214	60.0067	58.7413
LWL	99.2672	60.706	99.2341	99.2338	61.2721	99.2341
AdaBoostm1	99.9334	62.038	99.9001	99.9334	62.0047	99.9001
Bagging	99.9334	63.37	99.8335	99.8668	62.6041	99.8668
ClassificationViaRergation	99.2338	63.37	99.4006	99.2338	63.8362	99.4339

We use originally four different fold, i.e., 5, 10, 15, and 20, and 42 different machine learning algorithm for experiment. Table 4 discuses about accuracy of different machine learning algorithm, Table 5 about correctly classified autistic or non-autistic cases, and Table 6 about wrongly classified autistic or non-autistic cases.

The system for detecting ASD in children consists of 3014 image data samples and two class. We use three feature extraction algorithm, i.e., (i) GLCM, (ii) LBP, and (iii) LBGLCM.

Total features in different feature extraction algorithm are:

$$\text{(i)}\quad GLCM = j\left(\sum_{i=0}^{n} i\right) \quad, i = 1 \text{ to } n \quad j = 22, n = 1507$$

GLCM AUTISTIC=1507*22=33154
GLCM NON-AUTISTIC=1507*22=33154
TOTAL FEATURE=33154*2=66308

$$\text{(ii)}\quad \text{LBP} = k\left(\sum_{i=0}^{n} i\right), i = 1 \text{ to } n \quad k = 62, n = 1507$$

LBP AUTISTIC=1507*62=93434

Table 5 Correctly classified

		5			10	
Classifier	LBP	GLCM	LBP&GLCM	LBP	GLCM	LBP&GLCM
BayesNet	3011.993	1903.95	3011.993	3011.993	1901.943	3011.9927
NaiveBayes	3011.993	1880.866	3011.993	3011.993	1881.869	3011.9927
NaiveBayesMultinomialText	1511.015	1511.518	1511.518	1511.015	1511.518	1511.518
NaiveBayesUpdatable	3011.993	1880.866	3011.993	3011.993	1881.869	3011.9927
Logistic	3010.989	1953.129	3007.978	3010.989	1948.111	3008.9817
MultilayerPerceptron	3009.985	1921.012	3002.96	3009.985	1900.939	3002.9597
SGD	3009.985	1865.811	3009.985	3009.985	1940.082	3009.9854
SGDText	1511.015	1511.518	1511.518	1511.015	1511.518	1511.518
SimpleLogistic	3011.993	1920.008	3008.982	3010.989	1920.008	3009.9854
SMO	3011.993	1944.096	3011.993	3011.993	1955.137	3011.9927
VotedPerceptron	2964.805	1867.818	1867.818	2963.799	1860.792	1860.7924
IBK	3009.985	1778.492	2989.912	3009.985	1801.576	2994.9304
KStar	2035.104	1790.536	1757.415	2047.151	1808.602	1770.4628
LWL	2991.913	1829.679	2990.916	2990.907	1846.741	2990.9158
AdaBoostm1	3011.993	1869.825	3010.989	3011.993	1868.822	3010.989
Bagging	3011.993	1909.972	3008.982	3009.985	1886.888	3009.9854
ClassificationViaRergation	2990.907	1909.972	2995.934	2990.907	1924.023	2996.9377

Table 6 Wrongly classified

		5			10	
Classifier	LBP	GLCM	LBP&GLCM	LBP	GLCM	LBP&GLCM
BayesNet	2.007324	1110.05	2.007324	2.007324	1112.057	2.007324
NaiveBayes	2.007324	1133.134	2.007324	2.007324	1132.131	2.007324
NaiveBayesMultinomialText	1502.985	1502.482	1502.482	1502.985	1502.482	1502.482
NaiveBayesUpdatable	2.007324	1133.134	2.007324	2.007324	1132.131	2.007324
Logistic	3.010986	1060.871	6.021972	3.010986	1065.889	5.01831
MultilayerPerceptron	4.014648	1092.988	11.040282	4.014648	1113.061	11.04028
SGD	4.014648	1148.189	4.014648	4.014648	1073.918	4.014648
SGDText	1502.985	1502.482	1502.482	1502.985	1502.482	1502.482
SimpleLogistic	2.007324	1093.992	5.01831	3.010986	1093.992	4.014648
SMO	2.007324	1069.904	2.007324	2.007324	1058.863	2.007324
VotedPerceptron	49.19451	1146.182	1146.182	50.20118	1153.208	1153.208
IBK	4.014648	1235.508	24.087888	4.014648	1212.424	19.06958
KStar	978.896	1223.464	1256.5848	966.849	1205.398	1243.537
LWL	22.08659	1184.321	23.084226	23.09327	1167.259	23.08423
AdaBoostm1	2.007324	1144.175	3.010986	2.007324	1145.178	3.010986
Bagging	2.007324	1104.028	5.01831	4.014648	1127.112	4.014648
ClassificationViaRergation	23.09327	1104.028	18.065916	23.09327	1089.977	17.06225

LBP NON-AUTISTIC=1507*62=93434
TOTAL FEATURES=93434*2=18

$$\text{(iii)}\quad \text{LBGLCM} = j\left(\sum_{i=0}^{n} \mathrm{i}\right) + k\left(\sum_{i=0}^{n} i,\right), \quad i = 1 \text{ to } n,$$
$$j = 22, k = 62, n = 1507$$

LBGLCM AUTISTIC=33154+93434=126588
LBGLCM NON-AUTISTIC=33154+93434=126588
TOTAL FEATURE=126588*2=253176

6 Conclusion and Future Research

As research on autism moves toward strength-based approaches, there is a needs to give a voice to the literature of adults and children in the autism spectrum. It allows people in the autism spectrum to disclose their own feelings, thoughts, and experiences and provides invaluable insights into their own experiences. In this study, an attempt was made to diagnose ASD disease in children aged 4–11 years by different machine learning classification method. A dataset of detailed questionnaires was applied in this study. Machine learning is main focus. This paper helps to identify autism in early stage by using machine leaning. In future, deep learning models like CNN can be employed to improve the reliability and detection accuracy of ASD by the proposed system.

References

1. Xiao, Z., Qiu, T., Ke, X., Xiao, X., Xiao, T., Liang, F., Zou, B., Huang, H., Fang, H., Chu, K., Zhang, J., Liu, Y.: Autism spectrum disorder as early neurodevelopmental disorder: evidence from the brain imaging abnormalities in 2–3 years old toddlers. J. Autism. Dev. Disord. **44**, 1633–1640 (2014)
2. American Psychiatric Association: Diagnostic and Statistical Manual of Mental Disorders, 5th edn. American Psychiatric Association, Washington, DC (2013)
3. Multiple recurrent de novo CNVs, including duplications of the 7q11.23 Williams syndrome region, are strongly associated with autism. Neuron **70**, 863–885
4. Grice, D.E., Buxbaum, J.D.: The genetics of autism spectrum disorders. Neuromol. Med. **8**, 451–460 (2006)
5. Randolph-Gips (sula), M.: Autism: a systems biology disease. In: 2011 IEEE First International Conference on Healthcare Informatics, Imaging and Systems Biology, July 2011, pp. 359–366
6. Arthi, K., Tamilarasi, A.: Prediction of autistic disorder using neuro fuzzy system by applying ann technique. Int. J. Dev. Neurosci. **26**(7), 699–704 (2008)
7. Maite, F., Bustos, I., Zapirain, B.G., Zorrilla, A.M.: Computer game to learn and enhance speech problems for children with autism. In: 2011 16th International Conference on Computer Games (CGAMES), July 2011, pp. 209–216

8. Najwani, R., Rahman, A.W.A.: Motor movement for autism spectrum disorder (asd) detection. In: Proceeding of the 3rd International Conference on Information and Communication Technology for the Moslem World (ICT4M) 2010, Dec 2010, pp. E90–E95
9. Dorothée, F., Polani, D., Dautenhahn, K.: On-line behaviour classification and adaptation to human-robot interaction styles. In: Proceedings of the ACM/IEEE International Conference on Human-Robot Interaction. ACM (2007)
10. Albinali, F., Goodwin, M.S., Intille, S.S.: Recognizing stereotypical motor movements in the laboratory and classroom: a case study with children on the autism spectrum. In: International Conference Ubiquitous Computer, pp. 71–80 (2009)
11. Westeyn, T., Vadas, K., Bian, X., Starner, T., Abowd, G.D.: Recognizing mimicked autistic self-stimulatory behaviors using hmms. In: Ninth IEEE International Symposium on Wearable Computers, Proceedings, pp. 164–167 (2005)
12. Saad Talal Alharbi, M.S.H., Monrat, A.A.: A belief rule based expert system to assess autism under uncertainty. In: Lecture Notes in Engineering and Computer Science: Proceedings of the World Congress on Engineering and Computer Science 2015, Dec 2015, pp. 483–490
13. Gong, L., Yang, R., Yan, Q., Sdiun, X.: Prioritization of disease susceptibility genes using LSM/SVD. IEEE Trans. Biomed. Eng. **60** December 2013
14. Liu, W., Li, M., Yi, L.: Identifying children with autism spectrum disorder based on their face processing abnormality: A machine learning framework. Autism Res. **9**(8), 888–898 (2016)
15. Jiao, Y., Chen, R., Ke, X., Chu, K., Lu, Z., Herskovits, E.H.: Predictive models of autism spectrum disorder based on brain regional cortical thickness. Neuroimage **50**(2), 589–599. ISO 690 (2010)
16. https://www.kaggle.com/gpiosenka/autistic-children-data-set-raintestvalidate

Author Index

K. N. Das et al. (eds.), *Proceedings of the International Conference on Computational Intelligence and Sustainable Technologies*, Algorithms for Intelligent Systems,
https://doi.org/10.1007/978-981-16-6893-7

The manufacturer's authorised representative in the EU is Springer Nature Customer Service Centre GmbH, Europaplatz 3, 69115 Heidelberg, Germany. If you have any concerns regarding our products, please contact ProductSafety@springernature.com

Printed and bound by CPI Group (UK) Ltd, Croydon, CR0 4YY
30/09/2026
02225332-0020